T. Fujita T. Kanno S. Kobayashi

The Paraneuron

With 105 Figures

Springer-Verlag
Tokyo Berlin Heidelberg New York London Paris

Tsuneo Fujita
Professor and Chairman
Department of Anatomy
University of Niigata
School of Medicine
Niigata, 951 Japan

Tomio Kanno
Professor and Chairman
Department of Physiology
Faculty of Veterinary Medicine
Hokkaido University
Sapporo, 060 Japan

Shigeru Kobayashi
Professor and Chairman
Department of Anatomy
Yamanashi Medical College
Tamaho, Yamanashi, 409-38 Japan

ISBN 978-4-431-68068-0 ISBN 978-4-431-68066-6 (eBook)
DOI 10.1007/978-4-431-68066-6

Library of Congress Cataloging-in-Publication Data
Fujita, Tsuneo, 1929–
The paraneuron/T. Fujita, T. Kanno, S. Kobayashi.
p. cm. Bibliography: p. Includes index. ISBN-13:978-4-431-68068-0 (U.S.): $170.00 (est.)
1. Paraneurons. I. Kanno, Tomio, 1933–. II. Kobayashi, Shigeru. III. Title.
[DNLM: 1. Neuroendocrinology. 2. Neurosecretory Systems–cytology. 3. Neruosecretory Systems–physiology. WL 102 F9609p]
QP356.4.F85 1988 611'.0188–dc19 DNLM/DLC for Library of Congress

Softcover reprint of the hardcover 1st edition 1988

Typesetting: Asco Trade Typesetting Ltd., Hong Kong

Preface

The regulation of the organism has traditionally been ascribed to two distinct systems—the nervous and the endocrine. Though coordination between the two systems has been acknowledged, researchers and authors have tended to deal with them as comprising separate categories of cells involved in different activities. With this approach, a given regulatory mechanism would be evaluated as to whether it should be accounted for by nervous or endocrine functions.

The past 15 years, however, have witnessed numerous important discoveries and conceptual developments concerning the morphological, physiological, and biochemical relations between the nervous and endocrine systems. Advances in immunocytochemical studies have revealed that there are a wide variety of messenger substances that function in both regulatory systems. As a result, researchers have been stimulated to investigate neuronlike properties of endocrine cells and, conversely, endocrine or secretory features of neurons. It has thus become obvious that the rigidities in the classic criteria of neurotransmitters and hormones may rather impede further advances in these research fields. The activities of neurons are no longer evaluated simply in terms of EPSP, IPSP, and the release of classic transmitters such as acetylcholine, noradrenaline, and GABA. Hormonal actions are no longer analyzed solely with regard to concentrations of classic aminic and peptidic hormones in the systemic blood circulation.

The concept of the paraneuron, which we proposed in 1975, has become one of the theoretical bases for the development of this trend of study. With this concept it has become possible to account for the continuity of the nervous and endocrine systems. It is our hope that this notion will continue to be of service in the elucidation of as yet unknown aspects of regulatory cells and substances.

It is noteworthy that the idea of unifying the nervous and endocrine systems is not new: Starling cited acetylcholine, adrenaline, and bioactive substances in pancreatic islets and gastroenteric mucosa as examples in his famous lecture on the newly proposed concept of "hormone" to the Royal College of Physicians in 1905. In the modern age too, overlapping functions of the nervous and endocrine systems have often been stressed, based on accumulated findings in neuroscience and hormone research. The impact caused by the establishment of the phenomenon of neurosecretion is still fresh in our memories. Nevertheless, such an innovation did not achieve a comprehensive and systematic integration of the research work of neurologists and endocrinologists. The two systems have tended to be investigated in terms of separate ideas by different researchers using different techniques.

Natura non facit saltum, the motto of Charles Lyell in his "Principles of Geology," inspired Charles Darwin, who took the book with him on the Beagle, in formulating his theory of evolution. The same words now encourage us in systematizing in one volume the accumulated data concerning the continuity of previously separate types of regulatory cells and substances. In the first part of this volume, we attempt to review the conceptual and experimental bases for the continuity of neurons and paraneurons. In the second part, we present the structure and function of all paraneurons as well as some neurons with special reference to their common cell-biological features.

We hope that this book will encourage a fruitful exchange of information between the hitherto separate fields of neuroscience and endocrinology. It would be most gratifying if the implications of the concept of the paraneuron presented here stimulated the reader to investigate unanswered problems in regulatory cells and mechanisms.

During the preparation of this book, we were provided with valuable information by many researchers all over the world. Some of them kindly contributed photomicrographs and illustrations. We thank them cordially for their help.

In the departments of each of the three authors, colleagues helped by providing scientific data. In particular, the cooperation of Drs. R. Yui and T. Iwanaga (Niigata), Drs. T. Uchida and Y. Endo (Yamanashi), and Dr. T. Saito (Sapporo) is gratefully acknowledged. Cordial thanks are further due to the competent secretarial work of Mrs Mitoko Honma, Miss Hiroko Nakamura, Miss Megumi Miyoshi, Miss Yuko Kurogi, Miss Masae Ishiguro, and Miss Junko Kubota.

TSUNEO FUJITA
TOMIO KANNO
SHIGERU KOBAYASHI

Table of Contents

I. General Part

II. Special Part

I. General Part

1. Historical Background

1.1 Criteria for Neurons

It was long believed that nerve cells (neurons) were highly differentiated entities with specific structures and functions that could be clearly distinguished from other cells (review: Cajal 1911). However, advances in electron microscopy, cytochemistry, and electrophysiology have gradually revealed that features which had been believed to be neuron-specific could also be shared by nonneuronal cells.

Some early histologists thought that *argyrophil neurofibrils* might be responsible for nervous conduction. After this view lost support, neurofibriles were believed to be specific to neurons, until in the 1950s electron-microscopic studies demonstrated that neurofibrils corresponded to bundles of filamentous and tubular structures. Although they were designated *neuro*filaments and *neuro*tubules, similar structures were found in other, nonneural cells.

The *tigroid bodies of Nissl* were also long regarded as another morphological criterion for neurons. Electron-microscopic studies have, however, revealed that they are nothing more than stacks of the cisterns of rough endoplasmic reticulum and aggregated free ribosomes (review: Palade 1975).

Synapses and similar structures do not constitute a criterion for neurons, as many nonneuronal cells—certain endocrine and sensory cells as well as muscle cells—may form pre- and/or post-synaptic structures.

It is also impossible to define neurons on the basis of their *development*. The neuroectoderm is not such a specific area as early scientists believed. It has been shown that the neural crest in the chicken gives rise not only to neurons, glial cells, and related cells, but also to nonneural, even mesenchymal cells, including osteocytes and chondrocytes of the face and anterior part of the neck (Ichikawa 1933; Johnston 1966; review: Le Douarin 1982).

Conversely, neurons are not always neuroectodermal in origin. The cranial sensory ganglia contain neurons which are derived from the placodal ectoderm. As placodes *per se* are mere thickenings of epithelium and have nothing to do with those neural areas like the neural tube and crest, it is only a quite arbitrary tradition if one regards them as detached territories of the neuroectoderm.

Experimental studies in amphibian, avian, and mammalian embryos have clarified that the trigeminal ganglion forms by mixed contribution of neurons of placodal and neural crest origins (Yntema 1944; Hamburger 1961; Johnston 1966; Noden 1975; Le Douarin 1982; D'Amico-Martel and Noden 1983; Nichols 1986). Furthermore, the ganglia of the facial and statoacoustic nerves consist mainly of neurons of placodal origin, containing only a few neural crest origin neurons, whereas the distal ganglia of the glossopharyngeal and vagal nerves (petrosal and nodose ganglia), exclusively of placodal neurons (Yntema 1937, 1944; Narayanan and Narayanan 1980; D'Amico-Martel and Noden 1983).

Olfactory cells which have been regarded as typical bipolar neurons are long known to be derived from placodal ectoderm. Moreover, recent studies have indicated that in adult nasal epithelium they are constantly replaced by immature cells (Nagahara 1940; Andres 1965; Graziadei 1973; Kratzing 1978; Yamamoto et al. 1976).

Physiologists regarded the generation of *action potential* as a specific ability of neurons and muscle cells. However, there are many neurons which seem to communicate with each other by nonspiking electric activity (review: Hisada et al. 1984b) (p. 69). Moreover, since electrophysiologists discovered the smooth muscle-like membrane potentials in glucose-stimulated B cells of the pancreatic islet (Dean and Matthews 1970a, b; Meissner and Schmelz 1974), this kind of electric activity has been documented in several other endocrine cells, including adrenomedullary cells (Brandt et al. 1976), and adenohypophyseal cells (Taraskevich and Douglas 1978; see Ozawa 1981). Even some nonneuromuscular nonendocrine epithelial cells have been demonstrated to produce action potential-like activities (Kater et al. 1978; review: Anderson 1980).

The final criterion for neurons was their *chemical products*. However, aminic and peptidic substances produced and released by neurons have been successively revealed to be shared by various types of endocrine and sensory cells (see pp. 19, ff.).

More recently, a series of "*neuron-specific*" *proteins*, e.g., neuron-specific enolase (NSE), neurofilament protein (NFP), and brain tubulin, have been identified, but even these substances have been immunocytochemically demonstrated to be shared by certain endocrine (Schmechel and Marangos 1978; Iwanaga et al. 1982; Fujita et al. 1983a) and sensory cells (Fujita et al. 1983a). This problem will be dealt with in greater detail in Chap. 9.

It is thus clear that there are no longer any definitive criteria for identifying neurons.

1.2 Bridges Connecting the Nervous and Endocrine Systems

Homeostasis and reactions to stimuli are regulated by the nervous and endocrine systems which have generally been thought to be clearly separated and distinguishable from each other. For many years, however, researchers have been conscious of two bridges connecting the features of neurons and endocrine cells.

1.2.1 Amines in Chromaffin Cells: APUD Concept

Adrenomedullary cells are endocrine cells which possess features like those of neurons. They produce and release into the blood noradrenaline and adrenaline, typical neurotransmitter substances; moreover, these cells are derived from the neural crest. The occurrence of these amines in adrenomedullary cells can be histochemically demonstrated by the chromaffin reaction. On the other hand, a large part of the endocrine cells in the gastroenteric mucosa are known to be positive in the chromaffin reaction; this nature was ascribed by Erspamer and Asero (1952) to the occurrence in the cells of another neuramine, serotonin (5HT).

The introduction of the formaldehyde-induced fluorescence (FIF) technique (Falck et al. 1962) conspicuously elevated the sensitivity of the detection of intracellular monoamines, and many nonchromaffin neurons and endocrine cells have been found to be FIF-positive. The FIF reaction was improved and gained wider recognition when the cells were observed after the experimental animal was given amine precursors, such as L-dopa; this caused many neurons and endocrine cells to take up the precursors and dacarboxylate them into amines such as dopamine. Pearse (1968, 1969) and Pearse and Polak (1971) categorized cells possessing this "Amine Precursor Uptake and Decarboxylation" ability into the "*APUD series.*" Besides the chromaffin cells, thyroid parafollicular, enteroendocrine, pancreatic islet cells, and some other cells were considered APUD cells.

Thus, the initially narrow bridge represented by adrenomedullary cells was markedly widened by the introduction of modern techniques and views, especially Pearse's APUD concept. However, not all neurons and not all peptide-producing endocrine cells can be demonstrated to possess the APUD ability. As will be discussed later, amines are rather inconstant products of neurons and related cells, and therefore do not seem to constitute specific markers.

1.2.2 Neurosecretion: Peptidergic Neurons

Another bridge connecting the nervous and endocrine systems is found in neurosecretion. Certain hypothalamic neurons were demonstrated to contain endocrinelike granules by Scharrer (1928) in teleosts, and later in a variety of invertebrates and vertebrates by Scharrer and Scharrer (1937, 1940, 1945); this phenomenon was designated neurosecretion. Later, Bargmann (1949) succeeded in staining the neurons extending from the hypothalamus to the neurohypophysis in the dog with chrome-hematoxylin, which was favored at that time for the demonstration of pancreatic B cells, and identified them as vasopressin and oxytocin-containing neurosecretory neurons. Electron-microscopic studies revealed that these neurosecretory neurons contained round, solid granules, called "*elementary granules*" resembling the secretory granules in endocrine cells.

Neurosecretory neurons had long been regarded as an exceptional type of neurons—at least in higher vertebrates. However, it became gradually apparent that nerve terminals contain, intermingled with small synaptic vesicles, a few "*large cored vesicles*" which seemed to be identical in structure to the "elementary granules" in neurosecretory fibers. Moreover, various researchers recorded that in some autonomic nerves, especially frequently in the intestine, these granules occupied the overwhelming part of a nerve terminal. Burnstock (1972) designated this type of nerve "*purinergic*," due to the occurrence of adenosine triphosphate (ATP) which he believed to be the transmitter of the nerves. On the other hand, Baumgarten et al. (1970) considered the polypeptide contents of the large solid granules in gut autonomic nerves to be more important and described such nerves as "*p*" or "*polypeptide type*" nerves. In their study on the pituitary intermediate lobe, Bargmann et al. (1967) noticed that nerves containing elementary granules and presumably synthesizing oxytocin or vasopressin formed synapses on the endocrine cells just as the cholinergic and adrenergic nerves did and called them "*peptidergic* nerves."

Subsequent electron-microscopic studies have indicated that there is perhaps no distinction between the purinergic nerves of Burnstock, the P-type nerves of Baumgarten, and the peptidergic nerves of Bargmann, although it has become clear that many different subtypes could be morphologically distinguished among what are collectively termed "large cored vesicles" or "large solid granules."

In the meantime, advances in immunocytochemistry have revealed the occurrence of a large variety of peptides in the nerves, especially in those containing the P type granules. Physiological studies suggest that these peptides might be transmitters, modulators, or neurohormones. Moreover, many of these neuronal substances have been demonstrated in endocrine cells and are therefore termed "brain-gut peptides." It has now become clear that neurons and endocrine cells may share secretory granules of similar structure and bioactive messenger substances of peptidic nature.

For the reasons described above, the term "peptidergic" proposed by Bargmann has become the most accepted. The term "purinergic" seems inappropriate because recent biochemical studies indicate that ATP and other adenine nucleotides are contained in every secretory granule and vesicle of nervous and endocrine cells (pp. 23, 63).

We can conclude that the initially narrow bridge of neurosecretion connecting the nervous and endocrine elements has been conspicuously widened by the increasing numbers of "peptidergic" neurons and "brain-gut peptides."

1.3 Summary

Nerve cells (neurons) were long believed to be highly specific in structure and function and to be clearly distinguishable from other cells. Advances in electron microscopy, cytochemistry, and electrophysiology, however, have disclosed one after another, that the criteria for neurons are not solely specific for neurons but also apply to certain nonneuronal cells. Ontogenetically, not all neurons are neuroectodermal, while certain nonneuronal cells are neuroectodermal in origin.

The classic studies of chromaffin cells have advanced to modern amine histochemistry, which has revealed the *amine precursor uptake and decarboxylation ability* being shared by a series of neurons and endocrine cells (APUD series, Pearse). The classic studies of *neurosecretion*, on the other hand, have been connected with a growing concept of *peptidergic neurons* and it has been established that a large number of peptidic messengers ("brain-gut peptides") are shared by neurons and endocrine cells, being contained in specific cytoplasmic granules which early neurosecretionists called "elementary granules."

2. What are Paraneurons?

2.1 Origin of the Concept

In the early 1970s, it was felt by many researchers in different countries that new terms and concepts were necessary for the group of cells having common properties with neurons. The APUD series of Pearse (1968, 1969) at that time covered only those cells which showed a positive formaldehyde-induced fluorescence (FIF) reaction after administration of amine precursors such as L-dopa. Thus, the ACTH and MSH cells in the pituitary belonged to the APUD series but other cell types did not (Pearse 1968, 1969); the B cells in the pancreatic islets belonged to the APUD series but A and D cells did not (Pearse 1968). It was in 1974 (Pearse and Polak 1974; Pearse 1975a, b) that some FIF-negative cells were included in the APUD series. The concept of the APUD series, although worthwhile for stressing the correlation between the nervous and endocrine cells as mentioned above, appeared incomplete because the APUD ability, which should be the very base of this concept, was lacking or undetectable in many endocrine cells which, in other respects, would be considered to be related to nerve cells, and some of which occurred intermingled with the FIF-positive APUD cells.

At this time, we were engaged in combined morphological and physiological studies of gut endocrine cells, and our main discovery was that these cells, which project microvillous tufts into the gut lumen, are half-receptive and half-secretory in both structure and function (Fujita and Kobayashi 1973). They recognize chemical information from the food in the gut lumen and respond to it by releasing their hormones—peptides and, in some cells, amines. Here again, the APUD ability is present only in part of the endocrine cells, including enterochromaffin (EC) cells, which had long been known to contain serotonin (5HT). We then realized that it was very necessary to have a new term and concept with which different kinds of cells can be dealt with on the basis of multilateral cell-biological criteria, with regard to their neuron-related features.

In 1975, an "International Symposium on Chromaffin, Enterochromaffin and Related Cells" was held in Gifu, Japan and on this occasion we proposed the term "paraneuron" and defined the criteria of paraneurons (Fujita and Kobaysahi 1975; Fujita 1976). These criteria of paraneurons originally included neuroectodermal origin of the cells, but this idea was later abandoned because it turned out that many paraneurons are derived from nonneural germ layers.

2.2 Criteria for Paraneurons

Paraneurons are cells which generally have not been counted as neurons and yet can be regarded as related to neurons on the basis of their structure, function, and metabolism. The criteria for paraneurons are as follows:

1) A paraneuron is a cell which produces substances identical or related to neurotransmitters or neurohormones.
2) A paraneuron is a cell which possesses neurosecretion-like and/or synaptic vesicle-like granules.
3) A paraneuron is a cell which is *receptosecretory* in function; it releases secretions in response to adequate stimuli acting upon its receptors on the cell membrane.

Although every paraneuron is more or less receptive and simultaneously more or less secretory in nature, some of them may be called "sensory" or "receptor cells" if their receptive function is foremost, while others may be categorized as "endocrine cells" if their secretory nature appears dominant.

The concept of paraneurons inevitably embraces *sensory cells*, which had not been included in the APUD series. Sensory cells had been known to contain synaptic vesicles and/or neurosecretory-like granules, but little attention had been paid to the transmitter substances contained in those granules and vesicles. The view that sensory cells should be included in the family of paraneurons has activated studies on the secretory aspect of these cells.

2.3 The Paraneuron Family

The main members of the paraneuron group are shown in Fig. 2.1.

The gastroenteric endocrine cells (labeled *1* in Fig. 2.1) are the sources of over a dozen gut hormones, including gastrin, secretin, and cholecystokinin-pancreozymin (CCK-PZ) as well as serotonin and other monoamines. The gut endocrine cells are typical basal-granulated cells of the open type: they open onto the gut lumen with a tuft of microvilli, representing the receptor site for the chemical information from the gut contents (Fujita and Kobayashi 1973, 1978). As this type of bipolar chemoreceptive cell can be found already in coelenterates (*Hydra*), the cell in question is regarded as the *prototype of paraneurons* and therefore located at the center of the family in Fig. 2.3.

Basal granulated cells showing similar structure and secreting serotonin are found also in the urogenital epithelia, especially in the *urethra*.

The basal-granulated cells in the bronchopulmonary epithelium (labeled *2* in Fig. 2.1) are variable with regard to their distribution and nervous supply. Some are solitary and others are grouped. Some cells open to the bronchial lumen and others are closed in type and horizontally extend cell processes. Some cells are synaptically connected with afferent and efferent nerve fibers whereas others are free from nerves. A larger mass of basal-granulated cells with some open-type (chemoreceptive) cells and heavy afferent and efferent innervation has been described in the bronchiolar epithelium of some mammals (including human infants) and amphibians under the name of the neuroepithelial body (Lauweryns and Peuskens 1972; Lauweryns et al. 1972; Rogers and Haller 1978). Lauweryns et al. (1977, 1978) demonstrated that the neuroepithelial body cells in the rabbit recognize hypoxia in the inspired air and respond to it by the release of their basal granules.

The gustatory cells of the taste bud (labeled *3* in Fig. 2.1) represent typical chemoreceptor cells connected with afferent nerves.

The olfactory cells of the nasal mucosa (labeled *4* in Fig. 2.1) had been previously regarded as typical bipolar neurons. However, as it has been revealed that the olfactory cell regenerates in the epithelium *in situ* (Graziadei 1973a) and that it differentiates from a basal-granulated cell (Yamamoto et al. 1976), it seems more appropriate to consider this cell as a paraneuron.

The Merkel cells at the base of the epidermis (labeled *5* in Fig. 2.1) are specialized basal-granulated cells synaptically supplied by an afferent nerve. With microvillous knobs extended from the cell body, the cell is believed to receive mechanical stimuli.

The hair cells of the inner ear and lateral line system (labeled *6* in Fig. 2.1) are mechanoreceptive cells supplied with afferent and efferent nerves. They are responsible for static or acoustic sensation by detecting mechanical forces moving their hairs. The lateral line hair cells in lower vertebrates detect the pressure and movement of water on the body surface.

The visual cells of the retina (labeled *7* in Fig. 2.1) possess the photoreceptive outer segment, which is a particular modification of the cilium. Homologous to this cell is the *pinealocyte* (labeled *8* and *9*) which is a melatonin producing endocrine cell no longer photosensitive in mammals (*9*), but in lower vertebrates it represents a transitional form between photoreceptive and endocrine cells (*8*).

The adrenomedullary and other peptide/amine producing *endocrine cells* (labeled *10* in Fig. 2.1) are synaptically supplied with nerves, usually efferent in nature. Many other endocrine cells (labeled *11*) are without direct nervous connection. Parafollicular, parathyroid, and adenohypophyseal cells generally belong to this latter type, whereas the pancreatic islet cell may be richly innervated or free of nerves according to species and genus.

The carotid body chief cell (labeled *12* in Fig. 2.1) is supplied with both afferent and efferent nerves. Ultrastructural studies indicate a reciprocal nature of the synapses (McDonald and Mitchell 1975). The chief cell itself is widely believed to be a chemoreceptor for O_2 and CO_2 tension in the blood, though an alternative function has been proposed, whereby its nerve endings are chemoreceptive and the cell is intercalated in the sensory nerves to adjust the sensitivity of the nerves (McDonald 1976, 1977). In this context, *the SIF cell* (small intensely fluorescent cell) in the sympathetic ganglia can be represented by the same diagram. It is a rounded cell with a few relatively short processes (Coupland 1965). Intercalated between neurons, this cell exerts adrenergic inhibitory modulation in

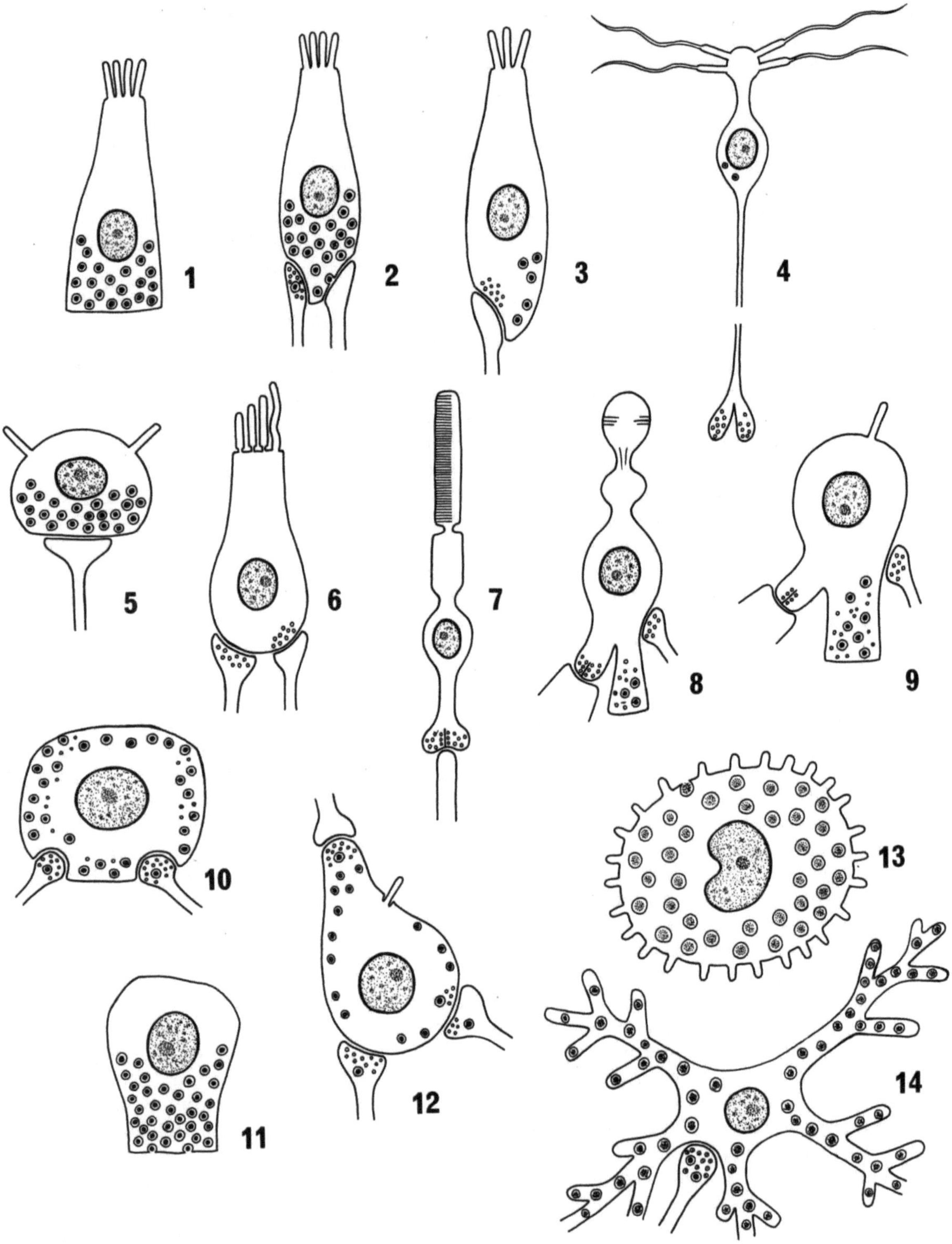

Fig. 2.1. Main representatives of paraneurons. *1* Gut endocrine cell; *2* bronchopulmonary paraneuron; *3* gustatory cell in taste bud; *4* olfactory cell; *5* Merkel cell; *6* hair cell in the inner ear and lateral line organ; *7* visual (rod) cell; *8* pinealocyte in amphibians and reptiles; *9* pinealocyte in birds and mammals; *10* adrenal chromaffin cell; *11* adenohypophysial, parafollicular, parathyroid, or other endocrine cell—pancreatic endocrine cell may be heavily innervated as in *10* or may be as in *11*; *12* carotid body chief cell; *13* mast cell; *14* melanocyte in fish

sympathetic ganglia (see review by Williams et al. 1976). Yamauchi (1977) proposed the term "internuncial paraneurons" for this type of cell.

The mast cell (labeled *13* in Fig. 2.1) may also be regarded as a paraneuron. The cell is filled with granules containing monoamines such as histamine and serotonin (5HT). Further reasons for including this cell of mesenchymal origin in the family of paraneurons will be elaborated in pp. 261, ff.

The melanocyte (labeled *14* in Fig. 2.1) which produces melanin through dopamine may not be qualified as a typical paraneuron because it is not a secretory cell. It does not release its melanin-containing granules. Nevertheless, the granules in the melanocyte in lower vertebrates diverge and converge within the cell according to excitation. This cell in lower vertebrates is elaborately provided with dendritic processes and usually is synaptically connected with an efferent nerve terminal.

2.4 General Structure of Paraneurons

Due to the diverse cell types belonging to the paraneuron family, the cell shapes and morphological features vary conspicuously among them. Some are slender bipolar cells, others are round or polyhedral in shape, and still others extend axonal and/or dendritic processes. In order to provide a visual impression of the appearance of paraneurons, it seems worthwhile to depict briefly the structural features common to these diverse forms. The functional and metabolic features common to paraneurons are omitted here as they are detailed in following chapters.

Since paraneurons are engaged in the production of secretory substances which are partly peptides and proteins, they contain a *protein-producing machinery* in the cytoplasm (Figs. 2.2, 3.2). It consists of cisterns of rough endoplasmic reticulum and the Golgi apparatus. The rough endoplasmic reticulum in paraneurons is generally well developed; occasionally, a large ergastoplasm reminiscent of a Nissl body may occur. The Golgi apparatus is also well developed, often separated in several areas. The apparatus produces small granules packaged in its membrane. These immature granules, while moving toward the cell periphery, grow into mature secretory granules. The secretory granules, ranging from 100 to 500 nm in diameter, are highly variable among paraneurons in fine structure and intracellular population. Many paraneurons possess, in addition, small vesicles resembling the synaptic vesicles in neurons.

Mitochondria, moderate in size and generally elongated in shape, are dispersed in the cell. Lysosomes are also moderate in distribution.

The *cytoskeleton system* is well developed as in neurons. Microtubules morphologically indistinguishable from neurotubules may occur numerously. Intermediate filaments may be richly found, including also elements indistinguishable from neurofilaments.

The *centrioles* in paraneurons are generally paired into a diplosome. One of the centrioles is usually extended, forming a *primary cilium* or *single cilium*. The occurrence of a primary cilium has been demonstrated in a large variety of cells, especially in sensory and endocrine cells categorized as paraneurons; it is difficult to find a cell type that lacks it (review: Wheatley 1982; see also Yamamoto and Kataoka 1986). The cilium of paraneurons is mostly immotile in type, showing a 9+0 (occasionally also a 8+1 or 7+2) pattern of microtubules. For some paraneurons it seems a mere remnant, while for others it represents the *receptive site* of the cell. Some sensory paraneurons possess a highly specialized device as a modified cilium, e.g., the olfactory hair and visual cell outer segment. It is noteworthy that the phylogenetically oldest paraneuron in the ectoderm and endoderm of *Hydra* (see p. 125) extends an apical cilium into the external milieu, apparently to detect chemical and other stimuli. For the reception of stimuli, however, the cell membrane with receptors is essential and the cilium serves only to provide the receptors with an advantageous area and location. The cilium, therefore, is replaced in various paraneurons by *microvilli* and other cell projections which carry receptors on their surface.

The *nucleus* of paraneurons is generally large and contains a more or less conspicuous nucleolus, reflecting the active synthesis of proteinic secretions in numerous ribosomes in the cytoplasm.

2.5 Spectrum of Neurons and Paraneurons

Neurons of different types make up a spectrum with delicate gradations between the elements, and this spectrum is continuous or overlapped with that of paraneurons. Figure 2.3 shows only one example of the arrangement of cells; there may be other possibilities.

SIF cells (labeled *10*) are endocrinelike

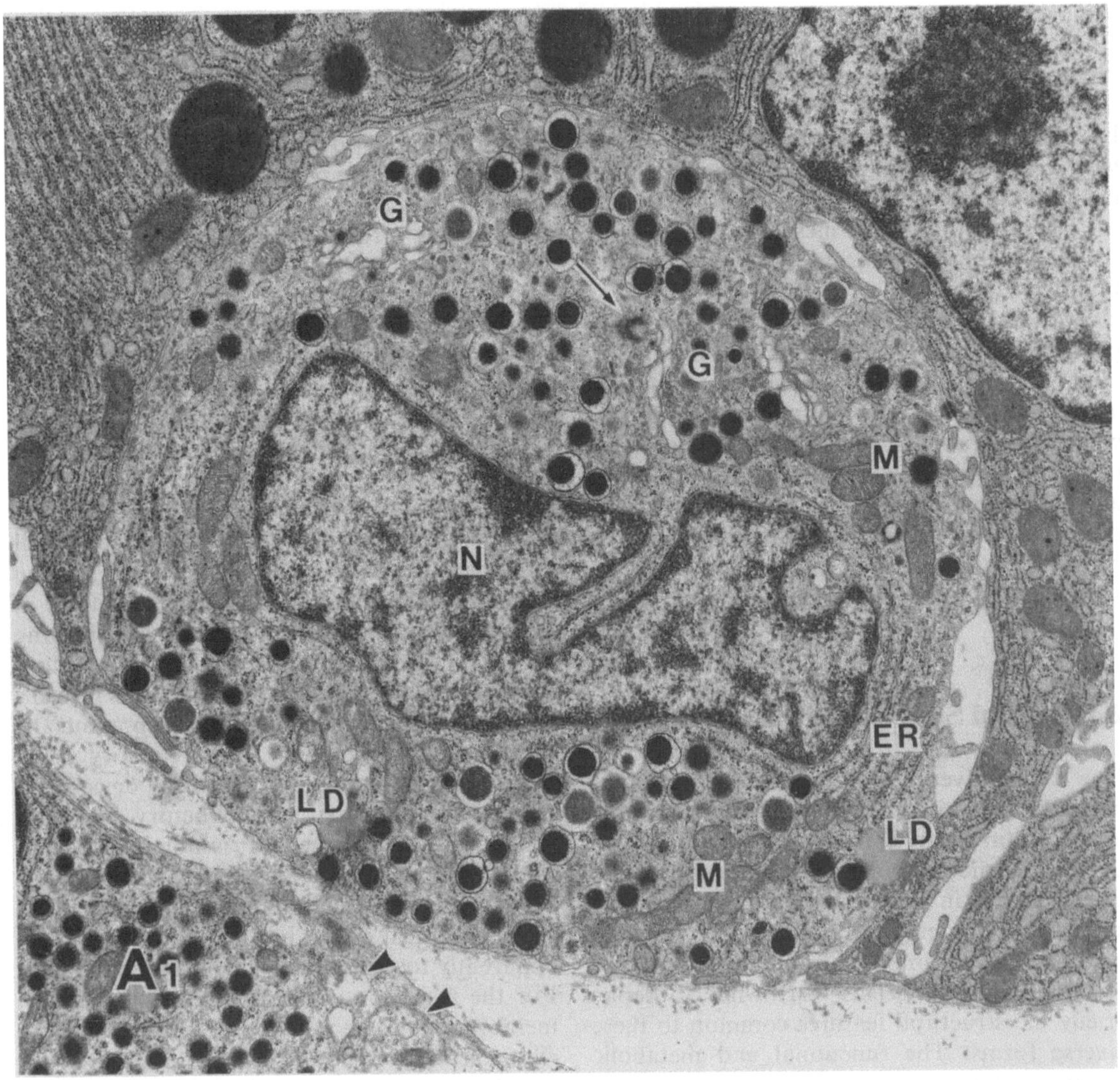

Fig. 2.2. Pancreatic endocrine cell as a typical paraneuron. Electron micrograph showing a glucagon-secreting A cell which occurs outside the pancreatic islet (so-called extrainsular islet cell) in a dog. In the *lower left corner*, there is another A cell (A_1). The latter exists in the pancreatic islet and is partially covered by a thin cytoplasmic process of a Schwann cell together with nerve terminals (*arrowheads*). A nucleus (*N*), mitochondria (*M*), granular endoplasmic reticulum (*ER*), Golgi apparatuses(*G*), a centriole (*arrow*), and lipid droplets (*LD*) are observed. × 12 000

cells intercalated in the sympathetic and parasympathetic nervous system. Together with adrenomedullary (*11*) and carotid body cells, these monoamine-containing paraneurons represent the points of overlap between paraneurons and autonomic neurons. *Olfactory cells* (*8*) possess neuronal and paraneuronal features and may also be located at the point of overlap.

The *neurosecretory neuron* (*6*) possesses paraneuronlike features and may represents one of the points of overlap between neurons and paraneurons.

In his review of one of our publications, Hagadorn (1979), after expressing agreement with our concept of paraneurons, claimed that the neurosecretory neuron had been excluded from the paraneuron family. However, it would seem more acceptable to deal with the neurosecretory cell as a type of peptidergic *neuron* as has been regarded traditionally.

Cerebrospinal fluid-contacting neuron (*7*) also has various paraneuronal features. Although their functions are still largely unknown, Vigh-Teichmann and Vigh (1983), who have investigated these cells most extensively, regard them as

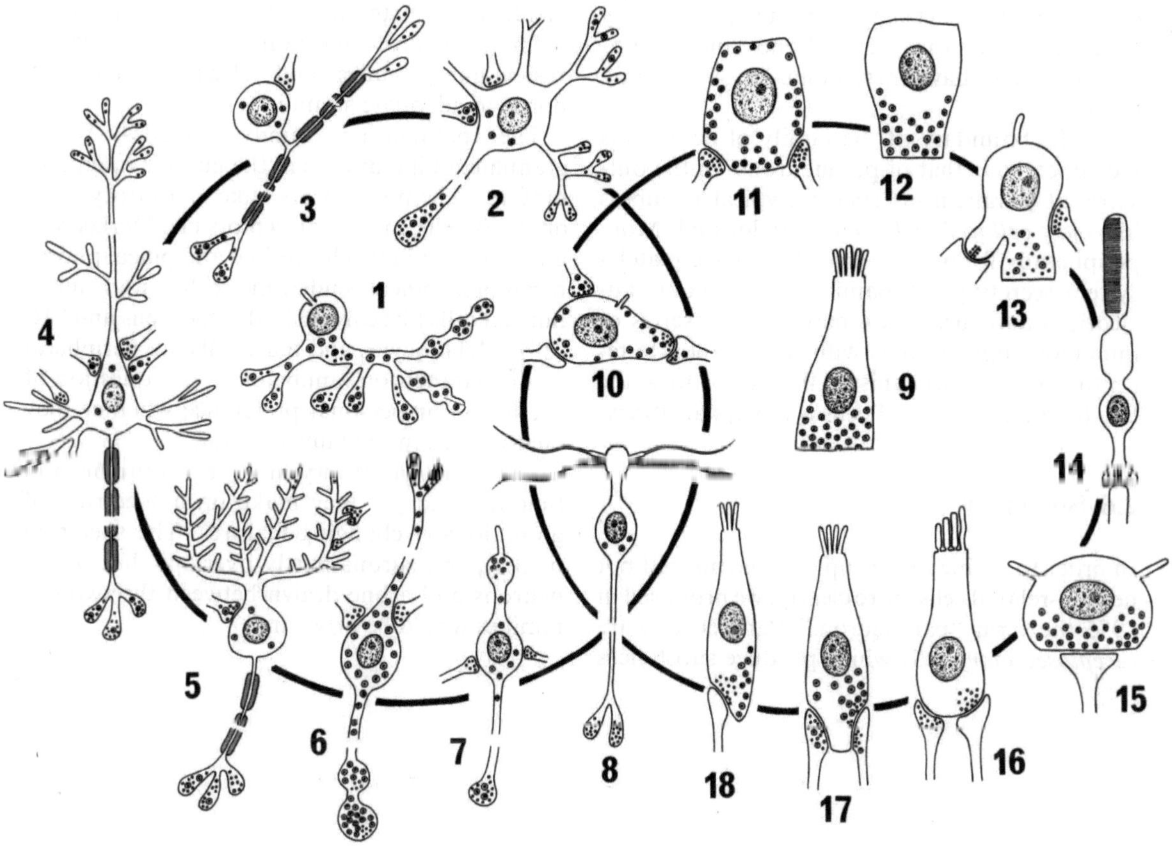

Fig. 2.3. Spectrum of neurons (*left*) and paraneurons (*right*) continuous and overlapping with each other. *1* Enteric intramural neuron, which represents the prototype of neurons—"archineuron"; *2* multipolar autonomic neuron; *3* sensory neuron; *4* pyramidal neuron; *5* Purkinje cell; *6* "neurosecretory" or "peptidergic" neuron; *7* CSF-contacting neuron; *8* olfactory cell; *9* gut endocrine cell, which represents the prototype of paraneurons—"archiparaneuron"; *10* carotid body chief cell and SIF cell; *11* adrenal chromaffin cell; *12* adenohypophysial, parafollicular, or other endocrine cell; *13* pinealocyte; *14* visual cell; *15* Merkel cell; *16* inner ear hair cell; *17* bronchopulmonary paraneuron; *18* gustatory cell

closely related to visual cell in the retina and to pinealocyte.

Intramural neuron in the gut (*1*) is a multipolar cell with beaded and nonbeaded projections filled with different neuropeptides and amines. In addition to this secretorylike feature of the cells, they show electrophysiological characteristics reminiscent of endocrine paraneurons (pp. 44, ff.). For various reasons, above all as this type of neuron occurs already in coelenterates (*Hydra*), this cell is regarded as the *prototype of neurons.*

The gradation in the spectrum of paraneurons becomes more subtle when one takes into account paraneurons with intermediate structures and functions which may be found in ontogenetic and phylogenetic processes of animals. The gap between the *visual cell* and mammalian *pinealocyte* has been filled by the findings in lower vertebrates and their juvenile forms, showing pinealocytes with structural and functional evidence of sensitivity to light and, partly, to color. Conversely, the visual cells in invertebrates are represented by cells with simple microvilli as the receptor sites. Comparative studies of sensory cells, e.g., *ampullary electroreceptors* in primitive fish (Roth and Tscharntke, 1976), indicate that for the same stimulus there are different morphological devices of receptors, such as single cilium and its derivatives, microvilli and their variations, and a simple cytoplasmic swelling.

Moreover, different kinds of stimuli may be perceived by the same receptor cells. The frog, *Xenopus*, that has no tongue, perceives "tastes" of salts, acids and certain amino acids with the hair cells of the lateral line system which is dispersed in patches on the body surface (Yoshioka et al. 1978). This fact suggests that the apparently separate mechanisms of mechano- and chemoreception may have transitional or overlapping forms.

As the boundary of the family of neurons is indistinct, so is that of paraneurons. In the outskirts of paraneurons such untypical members like *mast cell* and *melanocyte* are located. More peripherally follows the megakaryocyte-platelet system secreting monoamines and adenine nucleotides and further we may count a series of immunocompetent cells with their elaborate receptosecretory mechanism. Where to draw the line limiting a group of cells is always arbitrary.

2.6 Summary

In order to define the group of nonneuronal but neuron-related cells more clearly, we proposed in 1975 the term "paraneuron." Paraneurons are *receptosecretory* cells which produce substances identical or related to neuronal secretions, store them in neurosecretion-like and/or synaptic vesicle-like granules, and release them in response to adequate stimuli.

The paraneuron family includes basal-granulated bipolar cells in the gut, air tract, and urethra, chemoreceptors like gustatory and olfactory cells, mechanoreceptors like Merkel cells and inner ear hair cells, retinal and pineal photoreceptors, pineal endocrine cells, and internuncial cells like carotid body chief cells and SIF cells. Melanocytes and mast cells are peripheral to the paraneuron family and it is a question of usefulness or personal preference whether they are included in the family or not.

Paraneurons are very variable in structure and function and yet they make up a spectrum of gradationally changing features. The spectrum of the paraneurons partly overlaps that of the neurons and a line drawn between the two cell families would be arbitrary.

3. Secretions and Secretory Granules

3.1 Secretory Granules

3.1.1 Historical Aspects

Earlier *light-microscopic studies* on classic endocrine organs demonstrated the existence of granular inclusions in the cytoplasm of endocrine cells. In some organs, the granules were recognized to undergo periodical changes in amount and intracellular location due to the secretory activity of the cell. Their histochemical nature was, in a few cases, correlated with the secretory products of the endocrine cell. Thus, it was accepted that these granular inclusions represented the sites of storage of the secretory products which may be collectively called hormones.

Adrenal chromaffin cells were determined by Kohn (1903) to contain granules which stained yellow to brown after treatment with fixatives containing dichromate or chromate. This property of the chromaffin cells was ascribed to adrenaline and related substances in the granules (Ogata and Ogata 1923; see Watzka 1943; Coupland 1965a). The adenohypophyseal cells were divided into three groups—acidophil (α), basophil (β), and chromophobe (γ) cells, based on the reactions of their granules to different stainings, especially to "kresazan" (kresfuchsin plus azan) devised by Romeis (1940). In the thyroid gland, Nonidez (1932) distinguished the parafollicular from the follicular cells. With a silver impregnation technique, he demonstrated fine granules in the parafollicular cells whose hormone was later identified as calcitonin (Matsuzawa 1966; Pearse 1966). Bloom (1931), using the Mallory-azan method, could differentially stain three types of granulated cells in the pancreatic islet. The A cells, now known to be the source of glucagon, possessed coarser granules which stained a brilliant red; the B cells, secreting insulin, contained finer, pale-orange granules; and the D cells were filled with blue-stained granules. It was not until several years ago that the *neuro*peptide somatostatin was localized in the D cells by immunocytochemical methods, and thus the field of study on the brain-gut peptides was opened.

The development of *cell-fractionation* techniques during the 1950s greatly increased our knowledge on the mode of production and storage of secretory substances in endocrine cells. Isolation of a granular fraction from the adrenal medulla by Blaschko and Welsch (1953) and by Hillarp et al. (1953) led to the first successful localization of the hormones, i.e., catecholamines, to the secretory granules. Later studies of a similar nature in other endocrine organs, such as the adenohypophysis (Costoff and McShan 1969), parathyroid (L'Heureux and Melius 1956), and pancreatic islets (Noe et al. 1977), showed that peptidelike hormones are also contained in the specific granule fraction.

In the late 1950s, many investigators studied various endocrine organs by transmission *electron microscopy*, and it became gradually clear that both peptide producing cells like adenohypophyseal, pancreatic islet, and parafollicular cells and amine-producing cells like adrenal chromaffin and enterochromaffin cells share essentially the same ultrastructure characterized by membrane-bound granules derived from the Golgi apparatus. The mode of granule release has been much disputed. Electron microscopists demonstrated images of Ω-shaped granule openings, i.e., exocytosis in different endocrine cells, especially in adenohypophyseal cells (Ichikawa 1959), pancreatic islet cells (Williamson et al. 1961; Lacy et al. 1968), and adrenal chromaffin cells (De Robertis and Vaz Ferreira 1957; Diner 1967).

In the late 1960s, advances in *fluorescence cytochemistry* enabled Pearse and his associates (1969) to demonstrate that monoamines are produced not only by chromaffin cells but also by many other peptide producing endocrine cells, though in lower amounts (see p. 23). Our paraneuron concept (Fujita 1976; Fujita and Kobayashi 1978) extended the view further and established that the amine producing and peptide producing endocrine systems are identical with each other and comprise a family of paraneurons (see p. 19). The endocrine cells in this family form a sharp boundary against steroid-producing endocrine cells; however, they are continuous with and inseparable from neurosecretory and ordinary neurons in their cell-biological features, especially in the nature of their secretory granules. Sensory cells are also included in the paraneuron family, as they possess membrane-bound granules and vesicles and share the same cell-biological features as the peptide/amine-producing endocrine cells.

3.1.2 General Structure

Paraneurons have specific *secretory granules* in their cytoplasm. The granules can usually be stained by appropriate dyes—acidic, basic, or both.

Azan and Masson's trichrome stains as well as different techniques of silver impregnation are the classic methods. Gomori's chrome-hematoxylin (Gomori 1941) and aldehyde-fuchsin (Gomori 1950) provided selective demonstration of the secretory granules of pancreatic B cells as well as those of hypothalamohypophyseal neurosecretory cells, thus suggesting that endocrine and neurosecretory granules may have a similar chemical nature. More recently, lead hematoxylin (Solcia et al. 1969) and the metachromatic reaction after HCl hydration of sections (Solcia et al. 1968) have been available as universal stains of paraneuronal granules. The chemical basis of these staining techniques will be dealt with on p. 26.

Some paraneurons contain numerous secretory granules, whereas others, e.g., parathyroid cells, possess such small numbers that they can only be recognized by electron microscopy. When the cells contain moderate amounts of the granules, these show a more or less clear tendency to accumulate toward blood capillaries. It may generally be judged that the more conspicuous this polarity of the cell, the more active its secretory function. Under the electron microscope, secretory granules of the paraneurons are generally spherical, membrane-bound inclusions, 100–500 nm in diameter, with moderate to high osmiophilic contents.

It should be recollected here that the classic neurosecretory neurons (producing oxytocin and vasopressin) contain essentially the same round, membrane-bound granules of higher electron density, measuring around 300 nm in diameter. This coincidence in granule morphology is one of the bases of the paraneuron concept. Without such a concept, some electron microscopists have favored the term "neurosecretion-like" or even "neurosecretory granules" (Terzakis et al. 1972) for the granules of some endocrine cells, especially dispersed endocrine cells as in the gut and air tract.

In spite of the general features of the paraneuronal granules defined above, their size and shape greatly vary among different cell types. Some paraneurons possess large granules, even exceeding 1 μm, while others have granules of distorted and irregular shapes. The fine structure of the granule contents reflects their physicochemical characteristics, and thus, each kind of paraneurons possesses granules of different appearance.

Electron microscopy has revealed that in addition to the large electron-dense secretory granules, some paraneurons possess *small vesicles* apparently identical in morphology to the synaptic vesicles in neurons (Figs. 2.1, 3.1). These vesicles may have clear contents, but they also may be cored, i.e., possess a small dark core (Kobayashi 1977). The vesicles usually accumulate at a synaptic site or sites of the cell, when such are present. Sensory paraneurons, such as olfactory, gustatory, and visual cells and inner ear hair cells, contain accumulations of the vesicles in addition to a few large cored vesicles, i.e., endocrinelike granules. Pinealocytes, carotid body chief cells, SIF cells, and adrenal chromaffin cells also contain the vesicles as well as the granules.

We (Kobayashi and Uchida 1982) applied the zinc iodide-osmium tetroxide staining method developed by Akert and Sandri (1968) for the electron-microscopic investigation of the fine structure of the enteric endocrine cell. A peculiar network consisting of fine *tubulovesicular structures* was demonstrated in the basal cytoplasm of the cell. Small synaptic-type vesicles similar to those in the carotid body chief cells were also present in the space between the secretory granules. We suggested that these structures represent a device by which the granule membrane returns to the Golgi complex after the exocytotic release of gut hormones (Kobayashi and Uchida 1982). We further examined some paraneurons such as adrenal chromaffin cells, pancreatic islet cells, thyroid parafollicular cells, and adenohypophyseal cells using the same zinc iodide-osmium tetroxide staining method. The tubulovesicular structures identical to those of the enteric endocrine cells were demonstrated in all the paraneurons examined (Kobayashi and Uchida, unpublished data). We suggested that all paraneurons possess the tubulovesicular structures demonstrated by zinc iodide-osmium tetroxide staining and that the "synaptic vesicles" form a part of this membranous system.

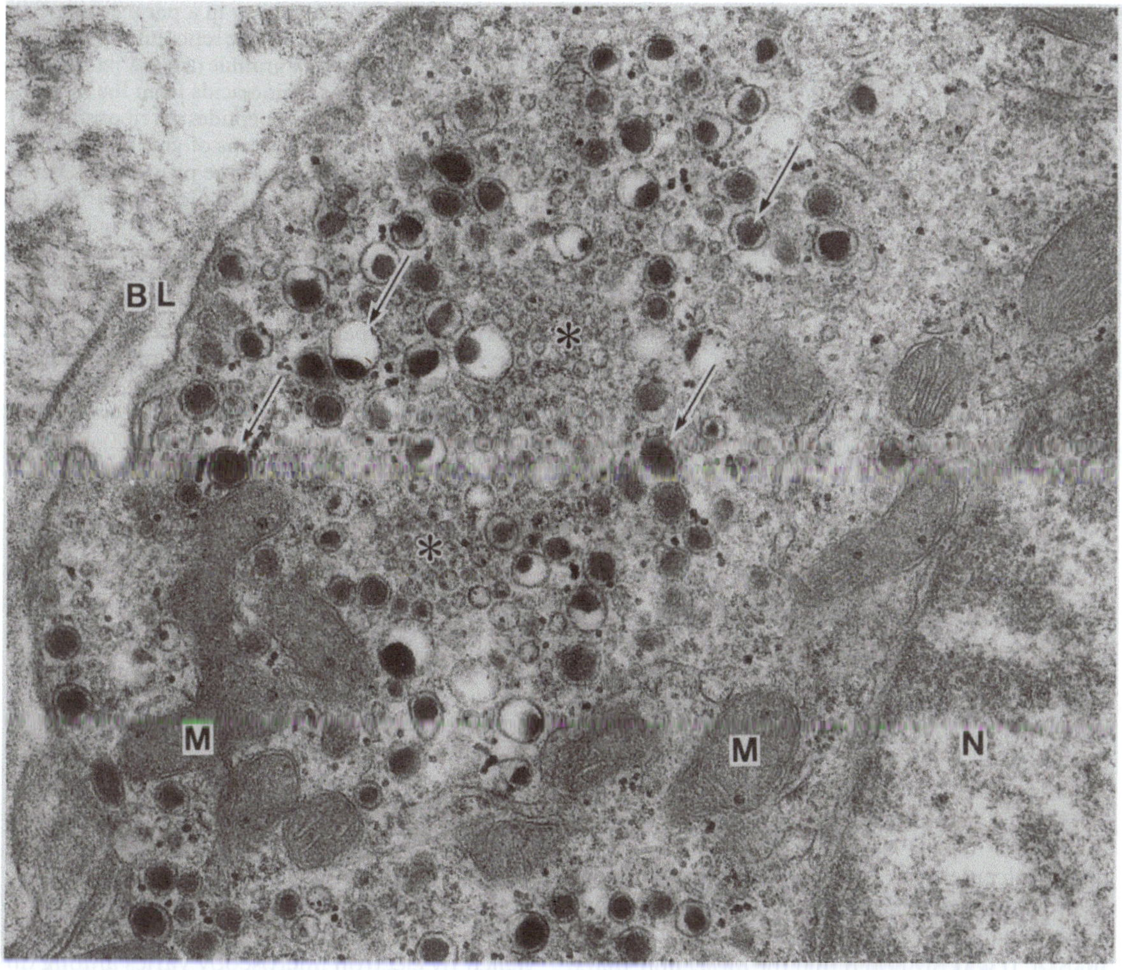

Fig. 3.1. Electron micrograph showing large cored and small synaptic type vesicles in a paraneuron. An SGC cell of the mouse adrenal medulla contains aggregations of small cored vesicles (*asterisks*) together with large secretory type granules (*arrows*). *N* nucleus, *M* mitochondria, *BL* basal lamina. ×40 000

We also observed an identical tubulovesicular system of neurons extending throughout the soma and its processes, including their terminals.

3.1.3 Granule Membrane

The membrane surrounding the secretory granules is called the granule membrane. Under the electron microscope, it appears as an 8- to 10-nm-thick trilaminar sheath consisting of two external electron-dense layers and one clear layer. This is the overall structure common to all biological membranes, such as the plasma membrane and mitochondrial membrane. In high-power electron micrographs, fine transverse bridges across the outer two dark layers and fine globular structures in the middle lucent layer are visible. The surface of the granule membrane facing the granule contents is generally covered by a polysaccharide layer of variable thickness which corresponds to the glycocalyx of the plasma membrane. This layer frequently appears to be continuous with the granule contents and is difficult to discern, especially in those secretory granules situated deep in the cytoplasm. The granule membrane is underlaid with the cytoskeleton on the side of the cytoplasmic matrix.

The granule membrane is characterized by the abundance of lysolecithin, i.e., lecithin from which the unsaturated fatty acid radical has been removed. Winkler (1969) showed that almost all the lysolecithin in the adrenal medulla is localized in the secretory granule membrane. Haydon and Taylor (1963) pointed out that diacylglycerophosphatide moieties of the lysoleci-

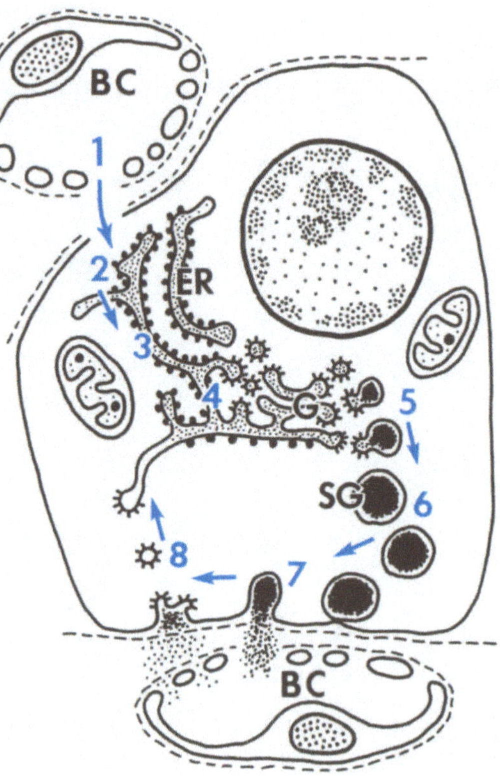

Fig. 3.2. The secretory machinery in a paraneuron, consisting of granular endoplasmic reticulum (*ER*), Golgi apparatus (*G*), and secretory granule (*SG*). A paraneuron takes up material such as amino acids from the bloodstream (*1*), synthesizes prepropeptides on ribosomes (*2*), segregates propeptides into cisterns of granular endoplasmic reticulum (*3*), transports the propeptides, being processed into peptides, toward the Golgi apparatus (*4*), forms young secretory granules on the trans side of the Golgi apparatus (*5*), and costores secretions such as peptides, amines, and adenine nucleotides in the secretory granules in a condensed form (*6*). When an adequate stimulus reaches the paraneuron, it releases the secretions by exocytosis (*7*) followed by the recapture of membrane substances (*8*), which are transported to the Golgi apparatus and reutilized for the formation of a new granule membrane. *BC* blood capillary

thin molecules are approximately rectangular in crosssection, whereas their lyso-compounds (monoacylglycerophosphatides) are wedge-shaped. It was proposed that the molecular shape of lysolecithin is probably related to the specific function of the granule membrane, such as its fusion to the plasma membrane. Lucy (1969) presented a model illustrating how lysolecithin might be involved in the fusion of the secretory granule membrane.

The granule membrane of the adrenal chromaffin cell has been shown to contain proteins, among others chromomembrin B (review: Winkler 1976). The existence of various kinds of ATPase and of cytochrome b-561 is also reported (see Winkler 1976). They may be involved directly or indirectly in the transmembranous transport of substances such as Ca^{2+} and catecholamines.

In the Golgi area many secretory granules are associated with the characteristic coating. Similar coating occurs around the Ω-shaped exocytotic pits on the cell surface (Figs. 3.2, 4.4). This coating is formed mainly by clathrin (Pearse and Bretscher 1981), a 180000-dalton polypeptide which is involved in granule membrane recycling (see p. 54).

3.1.4 Granule Contents

The fine structure of the granule contents as revealed by electron microscopy varies among different paraneurons. In the human pancreatic islet, for example, glucagon-storing A cells, insulin-storing B cells, and somatostatin-storing D cells are easily identified in a single electron micrograph by the appearance of their secretory granules. The granule contents of the A cells consist of two parts—a solid, osmiophilic, round core and a halo of lower electron density. By electronmicroscopic immunocytochemistry, Ravazzola et al. (1979) demonstrated that the granule core contains glucagon, whereas the halo, its precursor glicentin. The B cells frequently have crystalloid cores in their granules. Insulin can be detected in the granule cores by immunocytochemical techniques (Roth et al. 1978). The D-granule contents of the D cells are fine-granular and moderately osmiophilic. Immunocytochemical studies have shown that somatostatin is evenly distributed in these granules (Ravazzola et al. 1983).

In many of the paraneurons, the homogeneous cores of the secretory granules are separated from the granule membrane by a clear halo which mostly appears empty but sometimes contains a moderately dense material. When the halo consists of a clear zone, the possibility cannot be ruled out that the vacant appearance of the space is caused by loss of fluid contents or extraction of less condensed proteins during specimen preparation. Better fixation of the tissue pieces may sometimes preserve visible material in the halo (Mortensen and Morris 1977).

A *crystalloid core* may rarely characterize the secretory granules of paraneurons. Distinct crystalloid cores are known in the pancreatic B cells of a few species, such as man, the dog, bat, and cat. The crystalloid structures in the B-cell granules in rodents are rather indistinct. In other vertebrate species, the B cells possess homogeneous spherical granules similar to those of the pancreatic A and D cells (for literature see p. 159 of Howell and Tyhurst 1982). It is unknown why of the paraneurons only the insulin-producing B cells and only those in certain species form the crystalloid granule core. It is reasonable to believe that the crystalloid cores of the B-cell granules may not be solely ascribed to the nature of insulin, which is easily crystallized in the test tube. The granule core consists not only of insulin, but also of other peptides, such as C-peptide, chromogranins, and converting enzymes. Here again, it is noted that the ultrastructure of the granules of a given paraneuron reflects the physicochemical nature of the entire granule contents, especially their reaction to fixative and staining agents.

3.2 Formation of Secretory Granules

All paraneurons form secretory granules in the *Golgi apparatus* (Fig. 3.2). This cell organelle is usually located in the vicinity of the nucleus and frequently surrounds a pair of centrioles, one of which usually projects a single cilium. The Golgi apparatus is composed of a network of anastomosing canals. Several membrane-bound smooth-surfaced cisterns are flattened into Golgi lamellae which resemble a stack of plates. Each lamella is perforated by quite regular fenestrations, about 70 nm in diameter. Thus, the Golgi stack as a whole forms a three-dimensional tubular reticulum. Edges of the stacked reticulum of the Golgi apparatus are frequently dilated and form Golgi vesicles. The majority of these vesicles are coated by a polyhedral lattice of clathrin. They are termed the coated vesicles.

The Golgi apparatus has a distinct polarity. Structurally and biochemically it has two faces—a *cis*, or forming, face and a *trans*, or maturing, face. The *cis* face is the outer convex aspect of the stacked Golgi lamellae, whereas the *trans* face is the inner concave aspect. The cisterns of granular endoplasmic reticulum associated with the *cis* face of the Golgi apparatus are devoid of ribosomes on the side facing this cell organelle. Histochemically, the Golgi lamellae on the *cis* side reduce osmium tetroxide, whereas the cisterns on the *trans* side contain lysosomal proteins such as acid phosphatase. Novikoff (1964) proposed the term GERL (Golgi apparatus with hydrolase-rich cisterns of smooth endoplasmic reticulum giving rise to lysosomes). The GERL cisterns originate directly from the adjacent granular endoplasmic reticulum and communicate with it via a special tubular system. Coated vesicles swell outward, are pinched off from the edges of the GERL, and form lysosomes.

Among the many secretory products of the cell, peptides, such as prohormones, converting enzymes and chromogranins are synthesized on the ribosomes attached to the endoplasmic reticulum; they are segregated into the lumen of the endoplasmic reticulum and transported to the Golgi region. It is supposed that they enter the outermost cistern of the Golgi lamellae (*cis* face) and move through them toward the inner concave aspect (*trans* face) of the Golgi apparatus.

Recent studies indicate that every paraneuron apparently contains glycosaminoglycans (mucopolysaccharides) in its secretory granules (see p. 28). Although few studies on the production of these substances in general paraneurons have been published, cytochemical and radioautographic studies in cells rich in glycosaminoglycans, such as mast cells and exocrine cells, suggest that these substances are usually synthesized in the Golgi apparatus and added to the proteinic products from the rough endoplasmic reticulum as they pass through the apparatus.

Golgi vesicles originate by pinching off of the swollen edges of the Golgi lamellae. These vesicles, containing secretory proteins and glycosaminoglycans, fuse with each other and grow into so-called young secretory granules. These granules are variable in size and morphology—the youngest is simply a Golgi vesicle. The limiting membrane of the young secretory granules is usually underlaid by a clathrin-coat. Fusion of the young secretory granules and one or more Golgi vesicles frequently takes place. Young granules mostly surrounded by a non-coated membrane may be observed fusing with several coated vesicles.

Condensation of the secretory peptides takes place in the young secretory granules. Some Golgi vesicles carry secretory peptides to growing secretory granules. Thus, the amount of peptides in the secretory granules gradually increases. The condensation of the secretory peptides takes place by a non-energy-consuming process (Berneis et al. 1969).

3.3 Maturation of Secretory Granules

3.3.1 Changes in Size and Electron Density

A single paraneuron contains thousands of secretory granules, each of which is different in size and electron density of its contents. This difference in morphology of the granules reflects the stages of their "maturation," which include the following changes in the granule contents and membrane.

Condensation of the contents. The granule contents coagulate and extraction of water takes place to control osmolarity.

Incorporation of metal ions, adenine nucleotides, and amines. Mature secretory granules not only contain secretory *peptides* and *glycosaminoglycans* which are derived from the Golgi apparatus, but also *ATP*, *other adenine nucleotides*, and *amines*. These substances form a storage complex with other minor components such as Ca^{2+}. Electron-microscopic autoradiographic studies using isotope-labeled amines and amine precursors, such as ^{3}H-dopa, ^{3}H-dopamine, and ^{3}H-5-hydroxytriptophane, have indicated that amines are incorporated into the secretory granules directly from the cytoplasmic matrix without passing through the Golgi apparatus (adrenal chromaffin cells, Coupland and Kobayashi 1976; gut endocrine cells, Kobayashi 1975). ATP is synthesized in the mitochondria, enters the cytoplasmic matrix, and is taken up by the secretory granules (Winkler et al. 1972). Details of the molecular mechanisms of the transport of ATP at the granule membrane remain unknown.

Processing of peptides. There is increasing evidence that messenger peptides are first synthesized as a large-molecule precursor, which is then cut into smaller peptides, including one or more bioactive ones, by proteases during the intracellular transport through the rough endoplasmic reticulum, Golgi apparatus, and secretory granules. This splitting of large-molecule precursors is called "processing." The stage of processing of a certain peptide precursor differs among the secretory granules within a cell, and this may result in differences in the appearance of the contents of individual granules.

Turnover of granule membrane component. The granule membrane is not a stable and rigid structure. The granule contents become coagulated and this causes a reduction in their volume. Conversely, the fusion of many young secretory granules naturally results in an increase in the granule membrane substances. Thus, there is a mechanism by which some membrane components are selectively removed and a certain steady state of the granule membrane is maintained.

3.3.2 Movement Toward the Secretory Pole of the Cell

Young secretory granules formed in the Golgi region become "mature" and move to the periphery of the cell where they release their contents by exocytosis. In paraneurons without long processes, the sites of granule formation and release are close to one another. Nevertheless, passive diffusion would not appear to be adequate as the mechanism of intracellular movement of the secretory granules (Lacy 1970).

In neurons, the site of granule formation is at large distance from the site of exocytotic granule release. "Axonal flows" occur by which newly formed secretory granules are transported toward the tip of the cell processes. It is widely believed that cytoskeletons, such as microtubules and intermediate microfilaments (i.e., actin filaments), could play essential roles in the transport of the secretory granules. Administration of cytochalasin B, which prevents the polymerization of actin into microfilaments (Lin et al. 1980), or colchicine, which disrupts microtubules (Wilson 1975), severely disturb the movement of the secretory granules. Thus, an accumulation of secretory granules in the perikaryon of the neurons takes place if cytochalasin B or colchicine acts on the neurons. These experiments, however, do not indicate an active role of microtubules or microfilaments in providing the driving force for the granule movement because if their arrangement in the narrow cytoplasm is destroyed the movement of the granules might be hindered.

The same mechanism has been suggested for paraneurons as for neurons in that the movement of the secretory granules may be ascribed to functions of the cytoskeleton. Involvement of *microtubules* in granule release in pancreatic B cells has been proposed by Lacy et al. (1968, 1972). They proposed that B cell granules are transported to the secretory pole of the cell by contraction of the microtubules, which, according to the electron-microscopic studies of Lacy et al., extend from the plasma membrane into the cytoplasm and are attached to the membrane sac of the granules. Agents which destroy microtubules, such as colchicine, vinblastine, and deuterium oxide, inhibit glucose-induced insulin release of isolated islets (Lacy et al. 1968; Malaisse et al. 1970).

Movement of the granules has been examined by time-lapse cinematography using monolayer cultures of B cells. The granules tended to occur in rows as if they were lined along the microtubules, and they moved backward and forward along these lines (Somers et al. 1979). Changes have been observed in granule images suggesting exocytotic release (Kanazawa et al. 1976). Colchicine (Kanazawa et al. 1976) and vinblastine (Somers et al. 1979) inhibited the granule movements. Contractile movements (Somers et al. 1979) or projection of cell processes (Kanazawa et al. 1976) were observed in glucose-stimulated B cells but the significance of these changes in the mechanism of granule release is unknown (see also Lacy et al. 1975; Howell and Tyhurst 1982).

3.4 Contents of Secretory Granules

Since Blaschko and Welsch (1953) and Hillarp et al. (1953) succeeded in isolating the granular fraction from the adrenal medulla containing catecholamines, the chromaffin granules of adrenomedullary cells have been extensively studied to determine their chemical contents (Phillips and Apps 1980; Winkler and Westhead 1980; Winkler and Carmichael 1982). Although corresponding investigations have been made much less intensively in other kinds of paraneurons, recent advances in cytochemical and immunocytochemical studies have provided a considerable amount of valuable information on the granule contents.

Today, it is clear that paraneurons do not singly produce amine or peptide hormones but form *numerous categories of substances* (Fujita 1976; Fujita and Kobayashi 1979; Fujita 1983a, b). These include (1) peptides (and their precursors), (2) amines, (3) ATP and other adenine nucleotides, (4) chromogranins and other glycoproteins, (5) protein-processing and amine-converting enzymes, (6) glycosaminoglycans (acid mucopolysaccharides), (7) lipids, and (8) calcium and other ions (Table 3.1). The occurrence of these substances has not yet demon-

Table 3.1. Set of secretory substances produced by neurons and paraneurons

Categories of substances	Intracellular transport	Main substances
Polypeptides/peptides	RER → Golgi apparatus → SG	Fragments of prepropeptides (some of which are bioactive messengers) Acidic carrier proteins (chromogranins, neurophysins, etc.) Enzymes for conversion of granule substances Amine conversion enzymes (DBH, etc.) Proteolytic enzymes for processing preprohormones
Classic messengers	Cytosol → SG	Monoamines (catecholamines, indolamines, and histamine) Acetylcholine Amino acids and their derivatives (GABA, glutamate, glycine, etc.)
Adenine nucleotides	Mitochondria (?) → SG	ATP, ADP, AMP, adenosine, etc.
Glycosaminoglycans	Golgi apparatus → SG	Heparin, etc.
Lipids	Golgi apparatus → SG	
Metal ions	RER(?) → Golgi apparatus → SG	Ca^{2+}, Zn^{2+}, etc.

RER rough endoplasmic reticulum, *SG* secretory granules

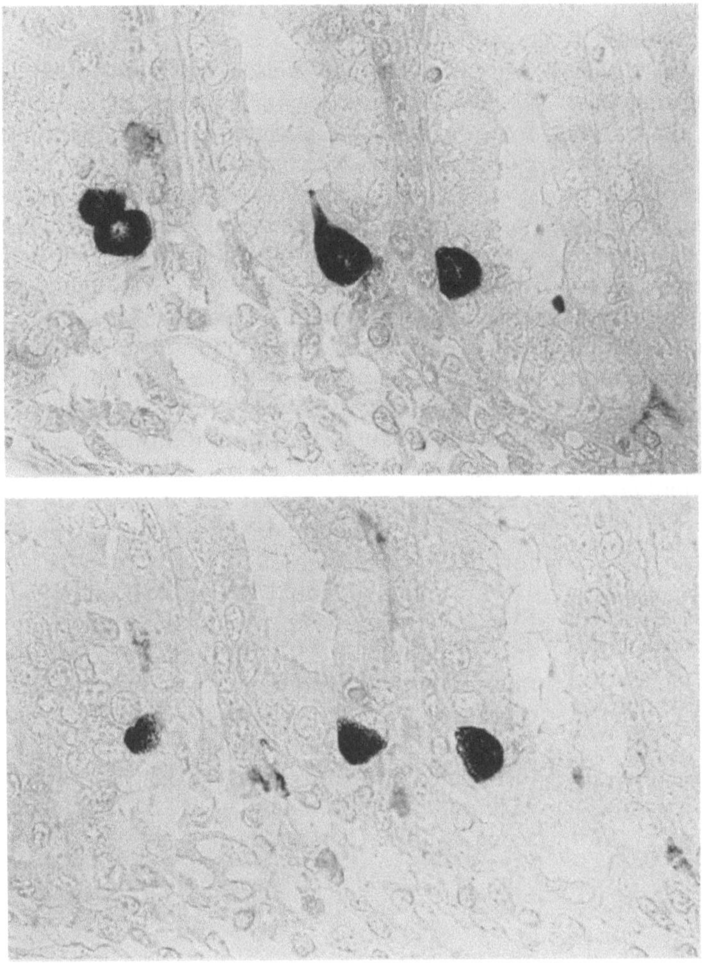

a

b

Fig. 3.3a, b. Coexistence of the peptide enkephalin and the amine serotonin in EC cells in porcine duodenal epithelium. Adjacent serial sections were stained **a** with an antiserum against Met-enkephalin-Arg-Gly-Leu and **b** with an antiserotonin serum. The same cells are immunopositive for both substances. Prepared by T. Iwanaga, Department of Anatomy, Niigata University Medical School

strated for every kind of paraneuron, but it is believed that all paraneurons share, at least in principle, the ability to produce and store a set of these substances. Each of the above listed granule components will be dealt with below.

3.4.1 Peptides and Their Precursors

Bioactive peptides are believed to be the most essential and universal substances secreted by paraneurons. In this context, the typical catecholamine secreting cells in the *adrenal medulla* and *carotid body* should not be exceptions (Fujita and Kobayashi 1973; Fujita 1976). Recent immunocytochemical studies have actually provided evidence that they contain Met- and Leu-enkephalins (Schultzberg et al. 1978a, b; Lundberg et al. 1979a; Varndell et al. 1982) or opioids derived from preproenkephalin A (Kobayashi et al. 1983a–c). Immunocytochemical studies at the electron-microscopic level have demonstrated that enkephalins are located in the chromaffin granules of the cells (Hervonen et al. 1980; Kobayashi et al. 1983b) (Fig. 3.3).

More recent studies have revealed that some other messenger peptides, such as VIP, substance P, neuropeptide Y (NPY), neurotensin, and somatostatin, may be contained in adrenal chromaffin cells (Lundberg et al. 1979a–c, 1982; Linnoila et al. 1980; Varndell et al. 1984; Kondo et al. 1985; Kuramoto et al. 1986). As these peptides generally occur in different cells, we consider that the adrenal medulla comprises endocrine paraneurons containing different peptide messengers gathered together in a mosaic fashion, just as in the adenohypophysis and pancreatic islets. However, immunocytochemical reactions for various neuropeptides in the adrenal chromaffin cells display a curious species variation (Kobayashi et al. 1983a–c). Thus, future studies are needed to substantiate this hypothesis.

Enterochromaffin (EC) cells in the gastroen-

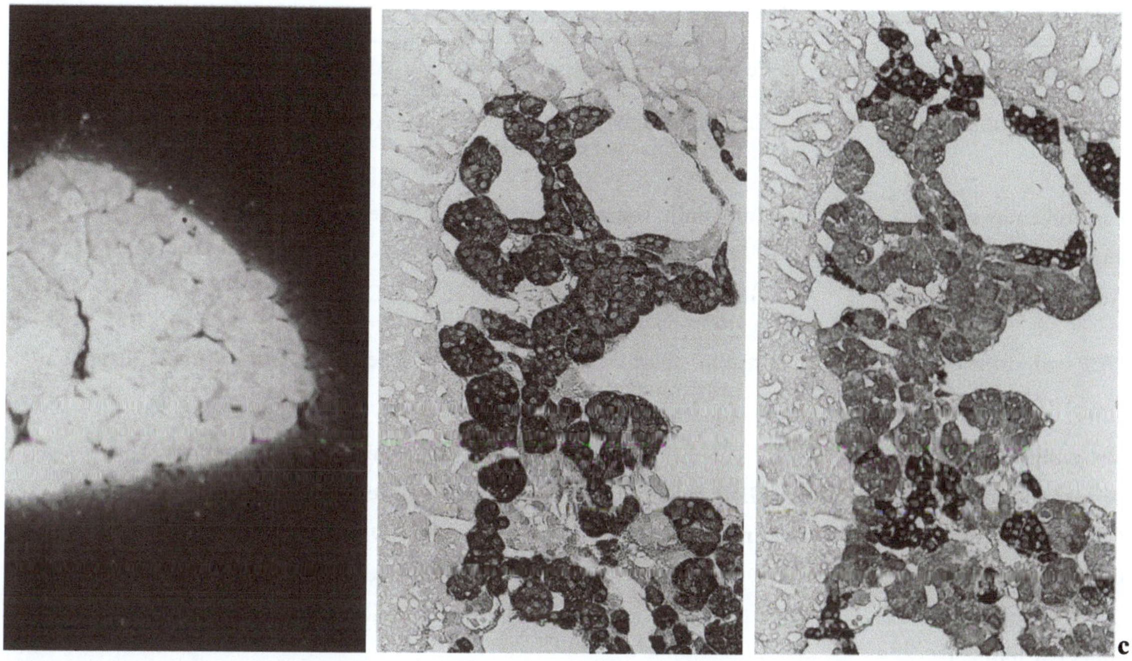

a, b c

Fig. 3.4a–c. Detection of monoamines by the Falck-Hillarp fluorescence cytochemical and immunocytochemical methods. The adrenal medulla of the rat was examined **a** by the Falck-Hillarp fluorescence cytochemical method and by the immunocytochemical peroxidase-antiperoxidase (PAP) method using **b** antiadrenaline and **c** antinoradrenaline sera. Adrenaline-producing cells contain a considerable amount of noradrenaline, which may represent the precursor of adrenaline. *A* cluster of adrenaline-storing or A cells, *BV* blood vessels, *C* adrenal cortex, *NA* noradrenaline-storing or NA cells. Antisera for adrenaline and noradrenaline were kindly supplied by A.A.J. Verhofstad, Department of Anatomy and Embyrology, University of Nijmegen, Netherlands

teric mucosa must, according to our concept, secrete a bioactive peptide besides the indolamine messenger, serotonin (Fujita and Kobayashi 1973; Fujita 1976). Studies into the disputed candidates of such peptides, motilin and substance P, are referred to on p. 167. A more recent result from our research group has indicated that the EC cells, at least in the small intestine of the pig, contain enkephalins derived from preproenkephalin A (Nihei et al. 1983) (Fig. 3.4). Electron-microscopic immunocytochemical investigations have demonstrated that the peptides are located in the chromaffin granules of the cells, together with serotonin (Uchida and Kobayashi, unpublished data).

Although it is stated above that peptides are the essential and universal messengers of paraneurons, knowledge on the possible occurrence of bioactive peptides in sensory paraneurons is very meager (Table 3.2). It has been reported that Merkel cells in the skin contain an enkephalin-like immunoreactivity (Hartschuh et al. 1979), but this result remains to be confirmed. It would be a worthwhile task to identify the substances in the blanks in Table 3.2.

Prepropeptides. Advances in biochemistry have clarified that bioactive peptides correspond to parts of amino acid sequences of larger precursor peptides. In earlier studies, the occurrence of various "big hormones" or "prohormones" (e.g., proinsulin, big gastrin, big big gastrin) was demonstrated, but recently nucleotide sequence analysis of cDNA has given us precise information on the amino acid sequence of the ultimate precursor molecules, which are called "preprohormones." The preprohormones generally have several hundreds of amino acid residues, and the bioactive peptides, mostly consisting of several to a hundred residues, are their fragments produced by processing of the molecules at the sites of the basic amino acids, lysine and arginine (p. 29).

Recent studies have revealed that a prepropeptide may give rise to more than one bioactive peptides by processing. The best examples are the

Table 3.2. Main paraneurons and their peptidic and aminic products

Paraneurons	Peptides	Amines
Adrenomedullary cell	Enkephalins, NPY, substance P, VIP, somatostatin, CGRP, ANP, etc.[a]	Adrenaline, noradrenaline, dopamine, serotonin[b], histamine[c]
Parafollicular cell	Calcitonin, somatostatin[d]	Serotonin, dopamine[e]
Parathyroid cell	PTH	?
Adenohypophyseal cell	GH, ACTH, MSH, TSH, GTH, etc.[e]	(Dopamine)[e]
GEP endocrine cell	Insulin, glicentin/glucagon, gastrin, secretin, CCK, somatostatin, enkephalins, etc.	Serotonin, (dopamine), (histamine)
Urogenital paraneuron	?[f]	Serotonin, dopamine[f]
Bronchopulmonary paraneuron	GRP, calcitonin, (enkephalins?)[g]	Serotonin, (dopamine)[g]
Carotid body chief cell	Enkephalins[h]	Dopamine, noradrenaline, serotonin[h]
Gustatory cell	?	Serotonin, noradrenaline[i]
Olfactory cell	?	Acetylcholine?
Inner ear hair cell	?	?
Merkel cell	Enkephalins, VIP?[j]	?
Visual cell	?	?[k]
Pinealocyte	?[l]	Melatonin

The data are based on studies in humans and other mammals. *Parentheses* indicate an inconsistent or insignificant occurrence

[a] See p. 141
[b] Verhofstad and Johnsson (1983)
[c] Häppölä et al. (1985)
[d] See p. 146
[e] See p. 160
[f] See p. 188
[g] See p. 196
[h] See p. 201
[i] Nagahama and Kurihara (1985); see p. 208
[j] Hartschuh et al. (1983, 1984)
[k] See p. 239
[l] See p. 251

prepropeptides for opioids. A series of identical and related molecules of opioid peptides are processed from a precursor molecule to be stored in the secretory granules. For instance, preproenkephalin A, which occurs in the adrenal medulla, carotid body, and pig EC cells, contains four Met-enkephalin sequences, one Leu-enkephalin, one Met-enkephalin-Arg^6-Phe^7, and one Met-enkephalin-Arg^6-Gly^7-Leu^8 sequence (Noda et al. 1982).

3.4.2 Amines

Historical. Isolation of crystalline *adrenaline* from the adrenal gland by Takamine (1901) was one of the biggest events in the early stage of hormone research, though the term "hormone" was not then used. Fifty years later, the occurrence of noradrenaline in the adrenal medulla was established by Euler and Hamberg (1949), and that of dopamine by Goodall (1951).

Following major advances in biochemistry and histochemistry in the 1950s, it became clear that in the adrenomedullary cells *dopamine* is produced by carboxylation of L-dopa and converted into *noradrenaline* by means of dopamine-β-hydroxylase (DBH); in some cells called A cells, noradrenaline is converted into adrenaline by phenylethanolamine-*N*-methyltransferase (PNMT), while in other cells called NA cells noradrenaline remains as such since they lack the enzyme (review: Kirschner 1975).

After centrifugation of bovine adrenal medulla, Blaschko and Welsch (1953) and Hillarp et al. (1953) detected the concentrated occurrence of *catecholamines in the granule fraction* of adrenomedullary cells and opened a new field of chemical studies of granules (review: Winkler and Smith 1975; Winkler and Carmichael 1982). Cytochemical and radioautographic studies, especially at the electron-microscopic level (Coupland and Kobayashi 1976), also contributed to the demonstration of dynamic events in individual granules (p. 18).

The *chromaffin reaction* of adrenomedullary cells has been known since the work carried out by Henle (1865) (see also Kohn 1903). The cells show a yellow-brownish coloration after fixation of the tissue in a dichromate-containing solution. Since this reaction has been ascribed to the occurrence of adrenaline in the cells (Ogata and Ogata 1923), it has become a histochemical tool for demonstration of those amines presently called catecholamines (Watzka 1943; Coupland 1965a). It was later confirmed that serotonin also produces a chromaffin reaction when Erspamer and Asero (1952) identified this indolamine in enterochromaffin cells in the gastroenteric mucosa. Eränkö (1952, 1955), on the other hand, discovered that adrenal catecholamines are transformed into fluorescent substances by treating the tissue with formaldehyde. This *formaldehyde-induced fluorescent (FIF) reaction* was applied to histochemistry and was especially successful in the Falck-Hillarp technique (Falck et al. 1962). This technique has proved to be far more sensitive than the chromaffin reaction in the detection of catecholamines and indolamines in cells and tissues (Fig. 3.5).

Pearse (1968, 1969) demonstrated that L-dopa administered systemically to an animal is taken up by a group of endocrine cells called the APUD series (p. 4). The L-dopa is then decarboxylated into catecholamines within the cell, which can be detected by the FIF reaction. This method of Pearse made detection of the amine producing and storing ability of cells even more sensitive. Nevertheless, it is now clear that some paraneurons are, like many neurons, free from detectable amounts of biogenic amines.

Recently, some research groups have succeeded in raising *antisera* that react specifically to certain monoamines. Many reports have appeared on the localization of serotonin (Steinbusch et al. 1978; Takeuchi et al. 1982), some on that of noradrenaline (review: Verhofstad et al. 1983), and some on that of histamine (Håkanson et al. 1986). The occurrence of these amines can be examined by electron-microscopic immunocytochemistry, such as the protein A-gold technique, and useful information has been obtained concerning the intragranule distribution and metabolism of these amines (Kobayashi and Uchida 1983).

Distribution of amines. As indicated on p. 6, many but not all paraneurons contain biogenic monoamines (Table 3.2). *Chromaffin cells* in the *adrenal medulla* represent the endocrine sources of noradrenaline and adrenaline. Recently, serotonin (Verhofstad and Jonsson 1983) and histamine (Häppölä et al. 1985) have been found to be contained in some of the cells. *Carotid body chief cells* contain noradrenaline and dopamine (p. 201). *Enterochromaffin (EC) cells* in the gastroenteric, bronchopulmonary, and urogenital mucosa contain serotonin. Histamine was located in *enterochromaffin-like (ECL) cells* only in the murine stomach (Håkanson and Owman 1967); but recent immunocytochemical studies have demonstrated histamine containing endocrine cells in the gastric and partly in the enteric epithelium of various vertebrate species, from fish to mammals (Håkanson et al. 1986). Parafollicular cells in the thyroid contain serotonin and some catecholamines (p. 147). In the GEP endocrine cells, except for EC and ECL cells, and in some adenohypophysial cells small amounts of monoamines may be detected by the FIF reaction, which is enhanced by previous administration of L-dopa to the animal (Pearse 1968, 1969; Carvalheira et al. 1968).

Gustatory cells have been shown to contain serotonin in some mammals (Nada and Hirata 1976; Takeda 1977) and in the frog (Nada and Hirata 1975b, c). Nagahama and Kurihara (1985) have recently suggested that noradrenaline is the most likely candidate for the transmitter involved in the transmission from the taste cells to the sensory nerve terminals in the frog. It is possible that amines are involved in the transmission of other sensory paraneurons, but no evidence supporting this is available.

Pinealocytes, on the other hand, produce melatonin which is a derivative of serotonin. Melanin is a derivative of dopa and is contained in *melanocytes*.

3.4.3 ATP and Other Adenine Nucleotides

Biochemical analysis of isolated granules of chromaffin cells has revealed that they contain a considerable amount of ATP and smaller amounts of ADP and AMP (Hillarp and Thieme 1959; Winkler et al. 1972). It has been reported

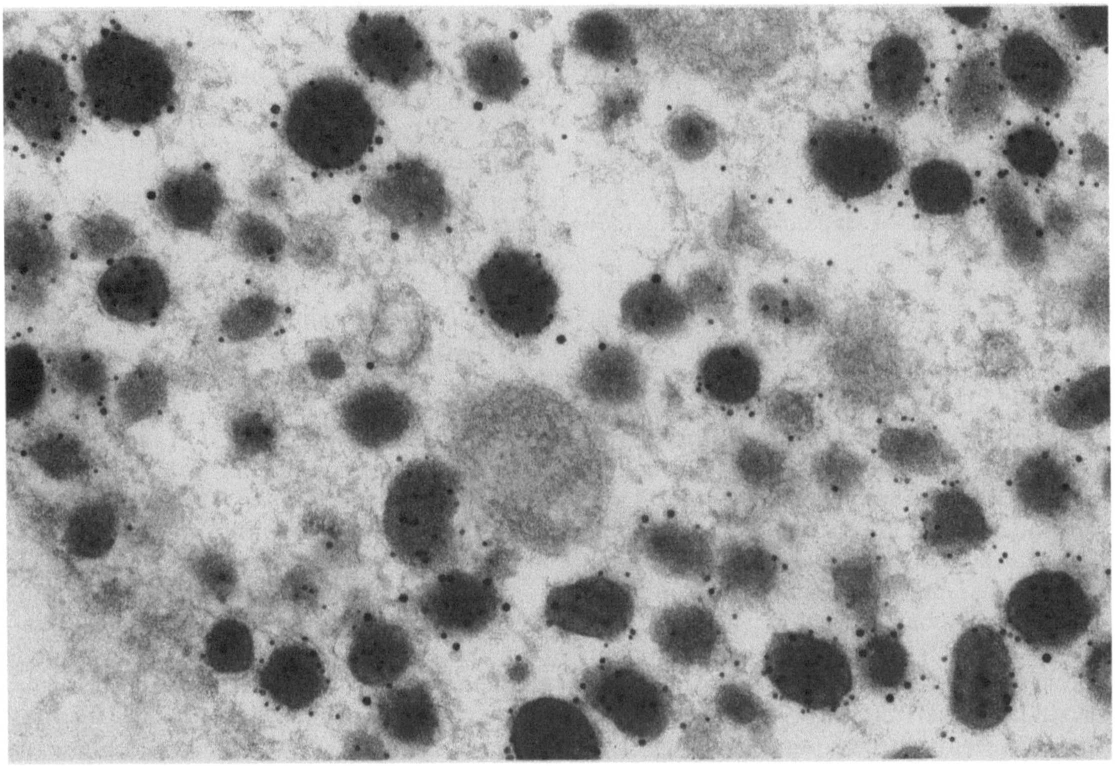

Fig. 3.5. Coexistence of an amine, adrenaline, and a peptide, Met-enkephalin-Arg^6-Gly^7-Leu^8(Met-Enk-8), in secretory granules. An ultrathin section of the canine adrenal medulla was immunostained with the protein A-colloidal gold method. Small gold particles represent adrenaline, whereas large gold particles indicate the localization of Met-Enk-8. Reproduced, with permission, from Kobayashi S, Uchida T (1983) Gunma Symp Endocrinol 20:39–49. ×76 000

that in chromaffin granules, ATP and catecholamines occur at a molecular ratio of 1:4 (Schneider et al., 1967; review: Winkler and Carmichael 1982). Leitner et al. (1975) demonstrated that after experimentally induced degranulation of pancreatic B cells, ATP is released into the blood at a constant ratio to insulin release.

Evidence is accumulating supporting the notion that ATP and other adenine nucleotides are *constant* constitutents of the secretory granules of *every* paraneuron. Böck (1980) demonstrated that a variety of paraneurons fluoresced after staining with *quinacrine*, a fluorescent dye with an affinity for adenine nucleotides. He also showed that paraneuron granules under the electron microscope were positive in the *uranaffin reaction*—a "selective" staining of adenine nucleotides with uranium according to the method of Richards and Da Prada (1977) (Fig. 3.6).

Like Böck (1980), we are skeptical about the specificity of both quinacrine and uranium for adenine nucleotides, although there is circumstantial evidence that the reactions Böck demonstrated in various paraneurons can be ascribed, at least to a large extent, to the occurrence of adenine nucleotides.

A high ATPase activity has been demonstrated in the immediate vicinity of some paraneuronal cells (Rubin 1972; Nakai et al. 1976). This may possibly be involved in the elimination of the granule ATP released from the cell. The possible significance of the granule ATP as a messenger substance of paraneurons will be dealt with in relation to neuronal ATP (p. 63).

3.4.4 Chromogranins and Related Substances

Soluble proteins collectively termed chromogranins are an important constituent of the chromaffin granules of adrenomedullary cells (Blaschko et al. 1967). The major component (about 40%) of the total soluble protein is *chromogranin A* (Schneider et al. 1967; Winkler 1976), which is an acidic polypeptide with a molecular weight of about 75 000 (O'Conner et al. 1984).

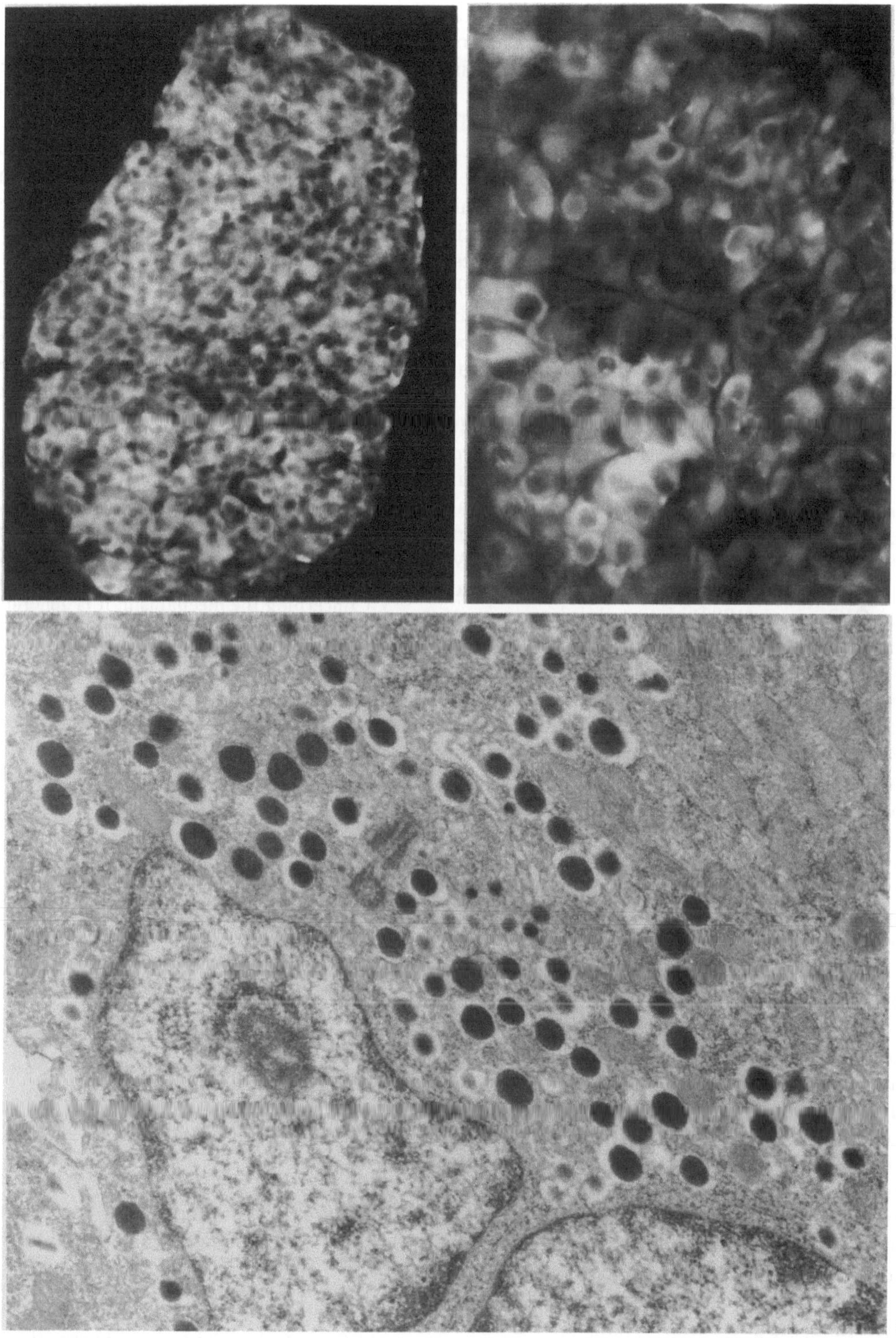

a b c

Fig. 3.6a–c. Histochemical demonstration of ATP in secretory granules of paraneurons. Quinacrine fluorescence of **a** pancreatic islet and **b** adrenomedullary cells in the mouse. The animal was sacrificed 3 days after a single intraperitoneal injection of 200 mg/kg body weight of quinacrine. The tissues were quenched in isopentane cooled with liquid nitrogen. The sectiones (15 µm thick) were freeze-dried and mounted in liquid paraffin. Quinacrine fluorescene is clearly restricted to paraneurons both in the pancreas and adrenal gland. In the latter, the cells showing stronger fluorescence are likely NA cells. Courtesy of Böck P, Institut für Mikromorphologie, Universität Wien. **c** Uranaffin reaction in the secretory granules of pancreatic islet cells. The round granules of the A cell are intensely stained. Prepared by R. Yui, Department of Anatomy, Niigata University Medical School

Chromogranin A and other chromogranins which are also acidic and probably not significantly different in amino acid composition (Winkler and Carmichael 1982), are believed to be the carrier proteins for messenger peptides (and their precursors), amines, and nucleotides in the granules.

Although we have long proposed that similar carrier proteins (or glycoproteins) that are acidic in nature might also be present in the granules of other paraneurons (Fujita and Takaya 1968; Fujita and Kobayashi 1973; Fujita 1976), biochemical analysis of these granules has not been as extensive as that of the chromaffin granules. Recently, it has been demonstrated that the acidic soluble protein contained in the secretory granules of the parathyroid gland, called *secretory protein 1 (SP-1)* or *parathyroid secretory protein (PSP)*, closely resembles chromogranin A in chemical, physicochemical and immunochemical features (Cohn and Elting, 1983). With a sensitive radioimmunoassay of chromogranin A, O'Connor (1983) detected immunoreactive chromogranin A not only in the adrenal medulla, but also in the pituitary, pancreas, intestine, and thyroid. Furthermore, immunocytochemical studies by O'Connor et al. (1983) and Lloyd and Wilson (1983) have detected the chromogranin A-like immunoreactivity in a large variety of paraneuronal elements in the body (Fig. 3.7).

Using an antiserum against gel-eluted chromogranin (75000 daltons) from bovine adrenal chromaffin granules, Nolan et al. (1985) demonstrated occurrence of the chromogranin immunoreactivity not only in a variety of paraneurons but also in many neurons in the central and peripheral nervous systems.

It has been noticed that, e.g., in the adenohypophysis (Rindi et al. 1986; Rundle et al. 1986) and pancreatic islets (Grube et al. 1986; Ehrhart et al. 1986), endocrine paraneurons are clearly positive, some only weakly positive, and others may be completely negative in the immunoreaction for chromogranin A. The difference in reaction depends on the cell types and on the animal species. In agreement with the view of Grube et al. (1986), we conceive of two possibilities to account for the *heterogeneous immunoreactivity* within a paraneuronal organ: One is that the chromogranin molecules, as the antigens, may be different depending on the cell type and species; the other is that even in the negatively immunoreacting granules the antigens may be present but "masked" by one or two substances (amines and/or peptides?) binding with them.

Chromogranin A and other acidic carrier proteins are believed to be responsible for the *metachromatic reaction* of secretory granules, which was first demonstrated in the pancreatic islet cells (Manocchio 1960; Solcia and Sampietro 1965; Fujita and Takaya 1968), then in gut endocrine cells (Solcia and Sampietro 1965; Fujita and Kobayashi 1973) and other paraneurons, including adrenal chromaffin cells (Bussolati et al. 1969; Kobayashi and Watanabe 1971). Other staining techniques such as lead-hematoxylin and aldehyde-fuchsin (which is not valid for all paraneurons), are also presumed to stain the acidic carrier proteins and not the messenger peptides (Fujita et al. 1968; Fujita and Kobayashi 1973).

A useful technique developed by Italian pathologists (Solcia et al. 1968) demonstrates gastro-entero-pancreatic and other paraneurons in an intense and selective metachromatic reaction. In this technique, the paraffin sections are treated in 0.2 *N* HCl at 60°C for a few hours and then stained in toluidine blue, pH 5. Concerning the mechanism, it is generally believed that metachromatic substance(s) with many acidic (mainly carboxyl) groups are contained in the granules in a "masked" form and become unmasked by the hydrolytic action of HCl. Although it has been proposed that the metachromatic substances might be acidic hormones like gastrin (Solcia and Sampietro 1965) or hormone precursors (Bussolati et al. 1969), we believe that they are just the acidic carrier proteins or glycoproteins (Fujita et al. 1968; Fujita and Takaya 1968; Kobayashi and Watanabe 1971; review: Fujita and Kobayashi 1973).

As will be dealt with later (p. 28), glycosaminoglycans or acid mucopolysaccharides have been recently found to be contained in adrenal chromaffin granules and they likely are shared by other paraneurons. Thus, it seems possible that the metachromatic reaction of paraneurons may, at least partly, be ascribed to these substances.

3.4.5 Enzymes

Among the catecholamine-synthesizing and -converting enzymes, *dopamine-β-hydroxylase (DBH)* has been established as a component of the chromaffin granule contents and membrane (review: Skotland and Ljones 1979; Winkler and Carmichael 1982). This localization of DBH was confirmed by immunocytochemistry at the electron-microscopic level (Thomas et al. 1973; Redick et al. 1974; Pickel et al. 1976; Dahlström et al. 1977; Nagatsu et al. 1979; Aunis et al. 1980).

DBH is a glycoprotein (Wallace et al. 1973) with a molecular weight of around 80000 (review: Skotland

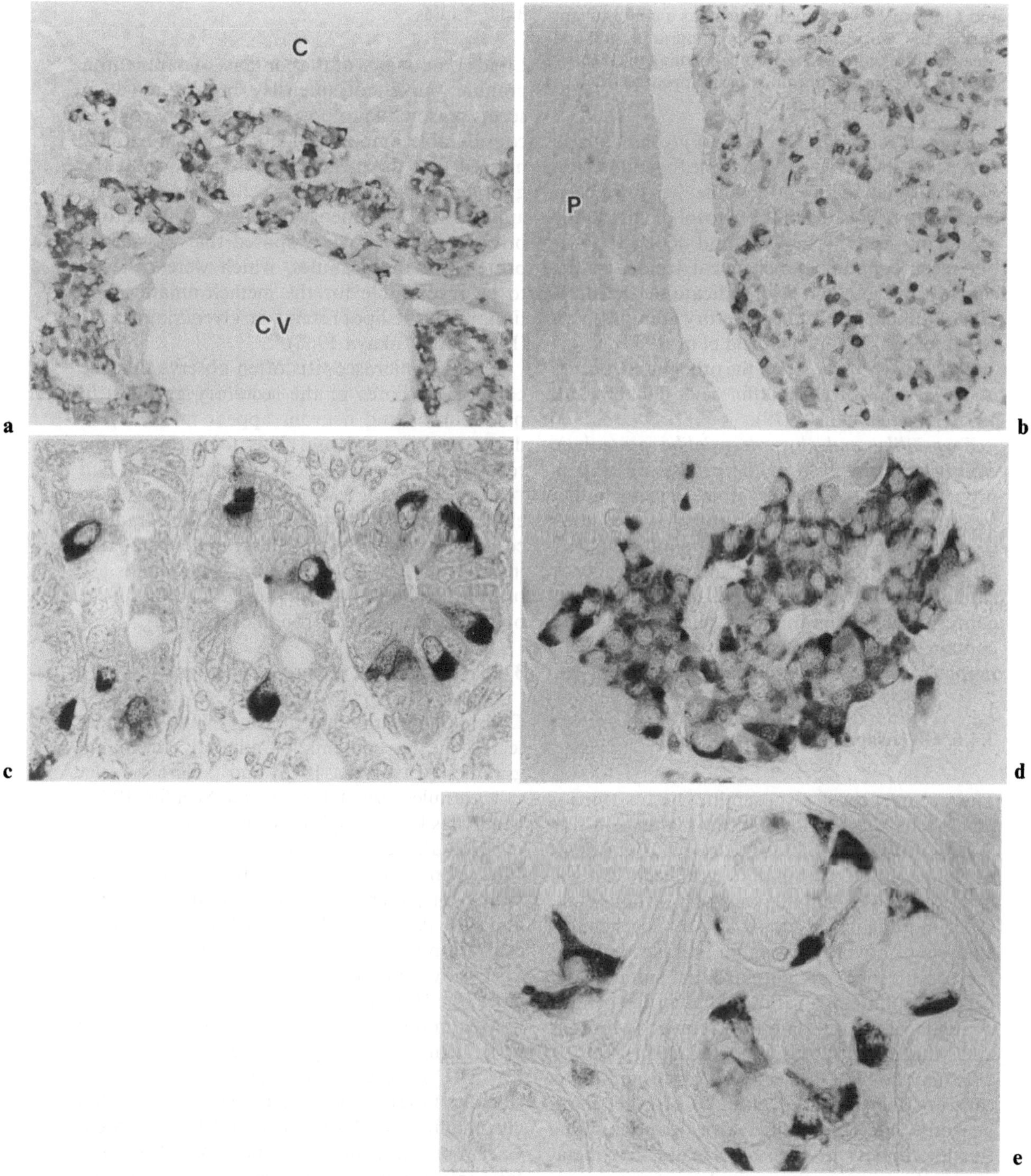

Fig. 3.7a–e. Distribution of chromogranin A in different paraneurons. **a** Human adrenal gland. Chromaffin cells in the medulla surrounding the central vein (*CV*) are immunoreactive, whereas the cortical cells (*C*) are immunonegative. **b** Human fetal pituitary. The anterior lobe (*right*) contains strongly and weakly immunostained cells. The posterior lobe (*P*) is immunonegative, though it is heavily stained with antineurophysin antisera. **c** Human duodenum showing numerous open-type endocrine cells with the immunoreactivity gathered at the cell base. **d** Porcine pancreatic islet consisting of numerous immunopositive and a few negative cells. **e** Antral mucosa of the snake *Elaphe quadrivirgata* (Boie) showing endocrine cells reactive to the antiserum against mammalian chromogranin. The sections were stained with an anti-SP-1/chromogranin A antiserum purchased from Immunonuclear Cooperation, MN, USA. Prepared by T. Iwanaga, Department of Anatomy, Niigata University Medical School

and Ljones 1979). It usually occurs as a contaminant during the purification of chromogranin A, and because of this confused reports concerning the identity of DBH and chromogranin A have appeared (review: Winkler and Carmichael 1982).

Phenylethanolamine-N-methyl-transferase (PNMT), the enzyme responsible for conversion of noradrenaline into adrenaline, has once been claimed to be located in the cytosol of adrenaline producing cells (Nagatsu and Kondo 1974), but recent immunocytochemical stuides at the electron microscopic level indicate the localization of this enzyme in the secretory granules (Van Orden et al. 1977; Brownfield et al. 1985).

Tyrosin hydroxylase, on the other hand, occurs in the cytosol of chromaffin cells (Pickel et al. 1975, Nagatsu et al. 1979).

Trypsinlike and chymotrypsinlike proteolytic enzymes are involved in the processing of prepropeptides into smaller molecules, including messenger substances. This process is presumed to occur partly in the Golgi apparatus, partly also in maturing granules, although evidence supporting this view is meager. It is reasonable to conceive that the granules of paraneurons may contain the enzymes needed for the processing of preprohormones of peptidic messengers.

3.4.6 Glycosaminoglycans

The occurrence of glycosaminoglycans (mucopolysaccharides) in adrenal chromaffin cells was first established by Fillion et al. (1971), who demonstrated that chromaffin granules can be labeled with ^{35}S-sulfate (see also Margolis et al. 1973; Baumgartner et al. 1974). Biochemical analysis of bovine chromaffin granules revealed the occurrence of chondroitin sulfate and heparan sulfate in a ratio of 20:1 in the soluble contents of the chromaffin granules (Margolis and Margolis 1973; Geissler et al. 1977).

It may be hypothesized that glycosaminoglycans are shared by the granules of other paraneurons. No biochemical investigations along this line appear to have been made. The metachromatic reaction recognized in various paraneurons may partly reflect the occurrence of glycosaminoglycans (see above).

The mast cell, which possesses the strongly acidic mucopolysaccharide *heparin* in its granules, may be regarded as a specialized form of paraneuron with regard to its ability to produce glycosaminoglycans (pp. 261, ff.).

3.4.7 Lipids

Lipid components of the contents of paraneuron granules have not intensely been studied. It seems reasonable to postulate that they contain considerable amounts of lipids. We previously noticed that the metachromatic reaction of the granules of pancreatic islet cells completely disappears after treatment of the tissue with acetone prior to fixation. We proposed that the carrier proteins in the granules, which were believed to be responsible for the metachromatic reaction, might be lipoproteins or glycolipoproteins (Fujita and Takaya 1968).

Electron microscopists often observe that the contents or cores of the secretory granules in paraneurons and neurons appear more or less dark or electron-dense. This tendency can be recognized even in sections which have not had "electron staining" with lead or uranium. The electron density in the paraneuron granules could be accounted for by the granules being rich in lipids. It is well known that when lipids are osmificated in the fixation procedure they appear opaque under the electron microscope.

3.4.8 Calcium and Other Cationic Ions

Considerable amounts of Ca^{2+} have been detected in the contents of adrenal chromaffin granules (Ravazzola 1976) and pancreatic islet cell granules (see Klöppel and Schäfer 1976). Mg^{2+} has been found to occur to a lesser extent in chromaffin granules (Winkler and Carmichael 1982). Zn^{2+} has been cytochemically demonstrated in pancreatic B cell granules (Falkmer and Pihl 1968; Yokoh et al. 1969), and its concentration in granule cores has been accounted for by covalent linkage of zinc with insulin molecules (see Howell nad Tyhurst 1982).

Membranes of chromaffin granules have a general impermeability to cations (e.g., Na^+, K^+, Ca^{2+}, H^+) and a greater permeability to anions (Pazoles 1982). Although it has been claimed that chromaffin granules possess an ATP-dependent Ca^{2+} uptake, a thorough analysis of Ca^{2+} uptake into granule ghosts has not provided any evidence for such a mechanism (for references, see Winkler and Carmichael 1982). Over the last few years, several laboratories have examined the physicochemical properties of the chromaffin granule membrane, the generation and maintenance of proton gradients, and the electrochemical gradient for catecholamines. The findings from those laboratories have resulted in the H^+ gradient hypothesis, the basic tenets of which are now generally accepted (for reference see Johnson et al. 1982).

Note. We have concentrated our description on the contents of the granule, and substances forming its membrane ("chromomembrins," phospholipids of membrane, etc.) are not dealt with in this text except for a brief mention on p. 15. This is because the granule contents appear to us to be much more important as they are the cell secretions released by exocytosis; they are compared with neuronal secretions when we analyze the similarity of paraneurons and neurons.

3.5 Contents of Small Vesicles

Little is known about the contents of the small vesicles which occur in some paraneurons, especially of the sensory and internuntial types. By analogy with the synaptic vesicles of neurons with the same morphology, it is suggested that they contain amines or choline and ATP but not peptides and precursors, chromogranins, and peptide-processing enzymes. Besides large-cored granules, the carotid body chief cells contain small vesicles located close to nervous synapses. These vesicles have been demonstrated to contain primary catecholamines such as noradrenaline and dopamine (Kobayashi et al. 1978).

Similar small vesicles are present in such paraneurons as the adrenal chromaffin cells, pinealocytes, SIF cells of the autonomic ganglion, and gustatory cells of the taste bud (Kobayashi 1977). Cytochemical studies in the adrenal small-granule chromaffin (SGC) cells showed that these small vesicles contain primary catecholamines such as noradrenaline and/or dopamine (Kobayashi et al. 1978).

3.6 Biosynthesis of Peptides and Messenger Substances

The contents of the secretory granules introduced in the preceding chapter are synthesized by paraneurons as well as by neurons with their specific intracellular mechanisms. Among the diverse kinds of substances contained in the granules, bioactive peptides and amines, which may serve as hormones, transmitters or, more widely, messenger substances, will be dealt with later as to how they are synthesized, transported, and converted within the cell. This chapter will concentrate on paraneurons as far as possible. However, as the cellular mechanisms involved are identical in both neurons and paraneurons, neurons will also be mentioned. This chapter will also consider the biosynthesis of such small-molecule neurotransmitters as gamma-aminobutyric acid (GABA), glutamic acid, and glycine, as it is a reasonable possibility that these substances may also be contained and released as messenger substances in some paraneurons.

3.6.1 Biosynthesis and Processing of Prepropeptides

The essential details of the intracellular mechanisms of the synthesis and transport of proteinic secretions were established by a series of radioautographic studies on pancreatic exocrine cells by Palade and his associates (Caro and Palade 1964; Jamieson and Palade 1967a, b, 1968a, b; see Palade 1975). It has thus become an accepted view that proteins are synthesized by rough endoplasmic reticulum, transported to the Golgi apparatus, where the secretions are packaged in the membrane sacs of secretory granules. The same mechanisms have been shown to be involved in the production of proteinic hormones in endocrine cells, such as anterior pituitary cells (Farquhar and Rinehart 1954; Racodot et al. 1965; Tixier-Vidal and Picart 1967) and pancreatic islet cells (Howell et al. 1969). It was later demonstrated by Leblond's group that both in exocrine and endocrine cells sugar components are combined to the proteinic secretions as these pass through the Golgi aparatus (Whur et al. 1969; Nakagami et al. 1971).

Recent studies have indicated that a messenger peptide (either a hormone, neurohormone, or neurotransmitter) is usually synthesized, by direct translation of mRNA, as a large-molecule precursor called a *prepropeptide* (or *preprohormone*). The preprohormone contains an NH_2-terminal signal peptide sequence that functions in the transfer of the protein from attached ribosomes into the cisterns of the rough endoplasmic reticulum. The signal peptide is subsequently cleaved by a protease (signal peptidase) in the rough endoplasmic reticulum. The rest of the preprohormone molecule is further cleaved by specific proteases contained in the rough endoplasmic reticulum-Golgi system into smaller molecules, which include one or more messenger peptides. This splitting is called *processing*, and it occurs between the following pairs of *basic amino acid residues*—Arg-Arg, Arg-Lys, Lys-Arg, and Lys-Lys (Fig. 3.8).

At present, the prepropeptides for the follow-

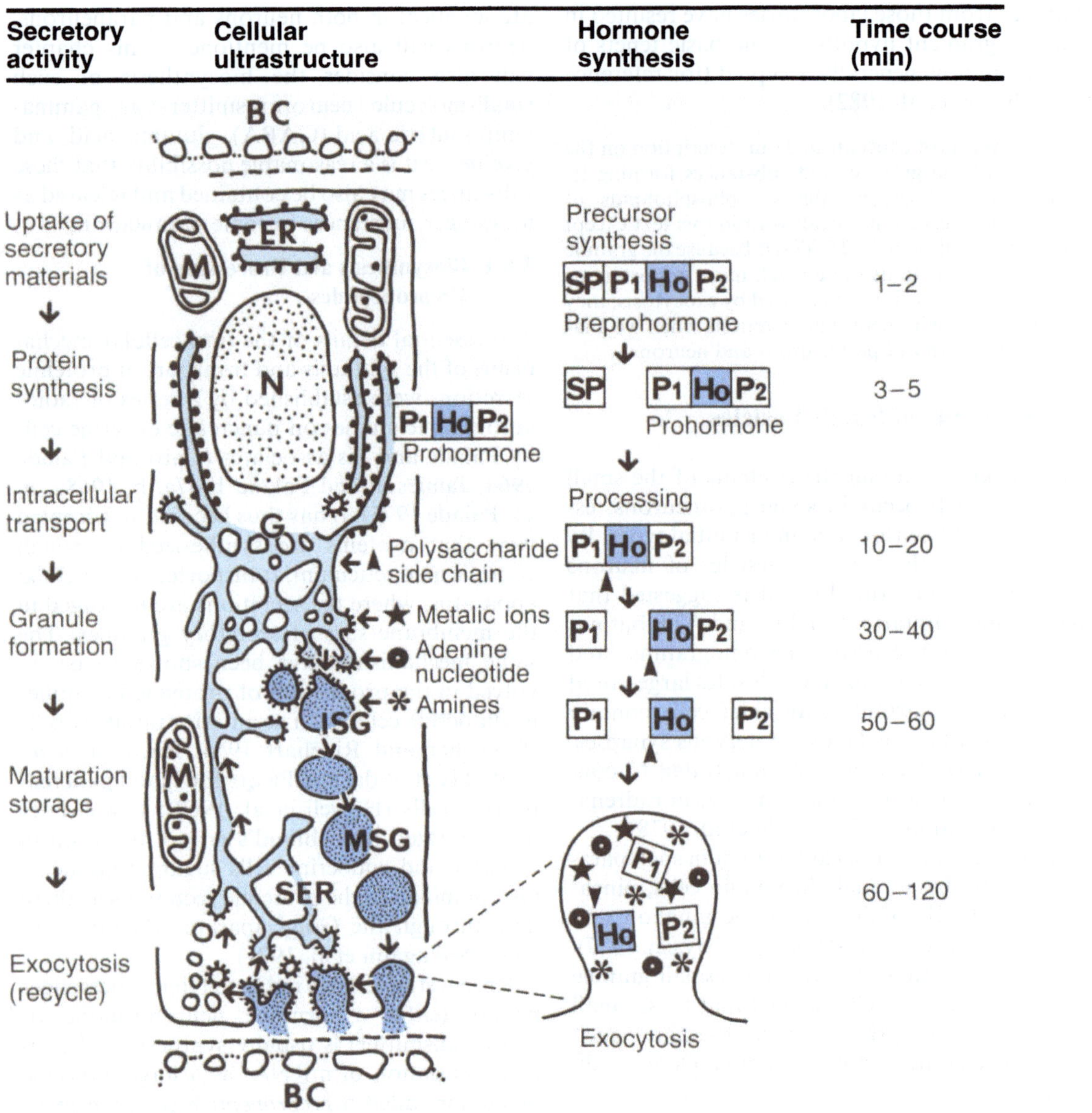

Fig. 3.8. Intracellular mechanism of the production of peptide messengers. Structure and secretory function of a paraneuron are illustrated on the *left*. Time-course of the peptide messenger formation is approximately shown on the *right*. *ER* granular endoplasmic reticulum, *N* nucleus; *G* Golgi apparatus; *ISG* immature secretory granule, *MSG* mature secretory granule, *SER* smooth endoplasmic reticulum; *M* mitochondria, *BC* blood capillary, *SP* signal peptide; P_1, P_2 peptides without messenger actions, *Ho* peptide with messenger action

ing messenger peptides of neurons and paraneurons have been characterized and their processing into small molecules has been analyzed—insulin, parathormone, glucagon/glicentin, somatostatin, VIP/PHI, CCK, gastrin, substance P, enkephalins, etc.

The *biosynthesis of insulin* has been the leading model in the studies of the biosynthesis of peptidic messenger substances in neurons and paraneurons.

Steiner and Oyer (1967) identified an unidentified peak on a gel chromatography profile of radioactive material extracted from incubated slices of human insulinomas. This finding started a line of research that led to the isolation of proinsulin (see Steiner 1977). The conversion from proinsulin to insulin is relatively slow. This, in part, reflects the time required for the transport of the proinsulin newly synthesized in ribosomes on the rough endoplasmic reticulum to the

Golgi region, where proteolytic enzymes appear to be localized and where processing is initiated (Steiner 1977). It then became clear that a precursor larger than proinsulin participates in the biosynthesis of insulin. This form, called preproinsulin (see Steiner 1977), has been less readily detectable because it turns over so rapidly in the cell that it is virtually invisible to the whole-cell-labeling techniques that had been used to study proinsulin biosynthesis and its conversion to insulin. The development of cell-free systems that efficiently translate mRNA in the absence of an appreciable background of endogenous protein synthesis has been of great importance in resolving the problem. One such system that has been widely used is the wheat-germ ribosomal system (see Okamoto et al. 1979).

On the basis of the findings obtained with the cell-free systems, Steiner (1977) summarized the flow of information from the gene, via preproinsulin and proinsulin to insulin. The gene or genes for insulin are located somewhere in the chromosomal DNA. They are transcribed by RNA polymerases in the nucleus, giving rise to the mRNA molecules, which encode the preregion as well as the B-, C-, and A-chain regions of proinsulin. Translation of this mRNA leads initially to the synthesis of the preregion, which presumably aids in the binding of the appropriate ribosomes to the membranes of endoplasmic reticulum of the B cell. After the vectorial discharge of the proinsulin into the cisternal spaces, its subsequent processing and storage occur.

With the availability of pure preproinsulin mRNA, it has become possible to approach the problem of isolating and analyzing the gene(s) for insulin. Several techniques are now available for obtaining large amounts of specific mammalian gene segments by cloning in bacterial hosts, utilizing the techniques of recombinant DNA. These methods hold great potential as a means of preparing sufficient amounts of specific DNA sequences.

Preproteins and proproteins. The intracellularly processed precursors are divided by Docherty and Steiner (1982) into preproteins and proproteins. These are defined by differences in their cleavage kinetics and their intracellular sites of cleavage: A *preprotein* may be rapidly processed in association with the membrane of the rough endoplasmic reticulum, often before the polypeptide is fully translated (i.e., cotranslational cleavage). The *proproteins*, which include all the polypeptide prohormones and some serum proteins (e.g., proalbumin) are post-translationally cleaved after passing the rough endoplasmic reticulum, i.e., in the Golgi apparatus and secretory granules; hence, their proteolytic cleavage occurs more slowly in the cell, in part due to translocation time. The role of the preproteins is to provide a signal sequence at the *N*-terminal of the polypeptide, which is used to translocate the translated product into the cisterns of the rough endoplasmic reticulum. Thus, by this mechanism the segregation process for the routing of secretory polypeptides in Palade's general scheme (Palade 1975) is determined during the translation itself.

The roles of the proproteins are more diverse (Loh and Gainer 1982). Prohormones are usually transferred from the rough endoplasmic reticulum to the *cis* phase of the Golgi apparatus and in this organelle, terminal glycosylation (galactose and fucose incorporation) of the prohormone takes places (Trifaro and Poisner 1982). However, incorporation of some other sugars, for example, mannose, takes place earlier on the *cis* side of the rough endoplasmic reticulum. Moreover, it has been suggested that glycosylation of nascent chains is not only an early event but is also coupled to translation and is dependent on transduction in the initial stage of biosynthesis (Katz et al. 1977; Rothman and Lodish 1977; Toneguzz and Ghosh 1977; Lingappa et al. 1978). The Golgi apparatus also seems to be the site of the initial processing of prohormones or intermediates, although final processing may also take place at the level of the immature secretory granules (Fig. 3.8). Therefore, it seems reasonable to presume that the Golgi apparatus and secretory granules may possess different and specific endopeptidases, capable of the cleavage of final secretory products from their precursors.

Precursors of opioid peptides. Advances in the investigation of the precursors of opioid peptides have revealed the occurrence of three different prepeptides—corticotropin/β-lipotropin precursor, preproenkephalin A, and preproenkephalin B. In these precursors, bioactive residues have been found to be contained repeatedly. This finding is of importance in the notion that the genes for bioactive peptides have often evolved an effective way to enclose repetitive residue sequences in a single molecule of a bioactive peptide.

In an attempt to elucidate the primary structure of the initial prepropeptide, Nakanishi et al. (1979) cloned cDNA encoding bovine cortico-

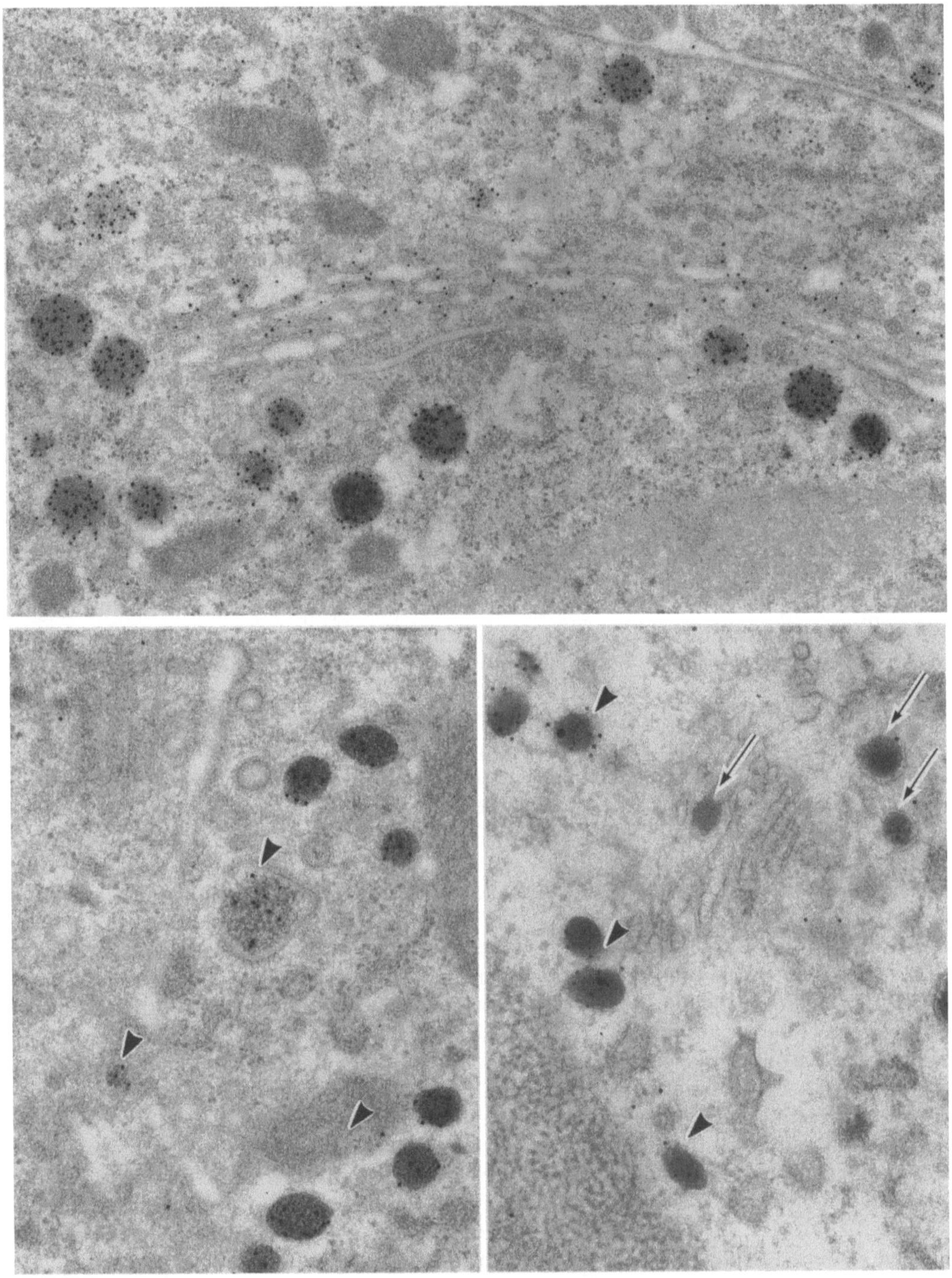

a b c

Fig. 3.9a–c. Occurrence and transport of messenger peptides and amines as demonstrated by the protein A gold method. **a** Localization of insulin immunoreactivity in the Golgi region of a B cell in a rat pancreatic islet. Gold particles representing insulin are seen not only in the secretory granules but also in some Golgi lamellae. Anti-insulin serum was donated by S. Ito, First Department of Medicine, Niigata University Medical School. × 31 000. **b** Localization of immunoreactivity for Met-enkephalin-Arg-Phe in an adrenaline-producing cell in the rat adrenal medulla. The gold particles are seen both in mature and in immature secretory granules (*arrowheads*). Anti-Met-enkephalin-Arg-Phe serum (AP3-311) was donated by H. Imura and K. Nakao. × 41 000. **c** Localization of adrenaline immunoreactivity in an adrenaline-producing cell in the rat adrenal medulla. The gold particles representing the amine are seen only in mature granules (*arrowheads*) at some distance from the Golgi apparatus. Immature granules (*arrows*) are immunonegative. The anti-adrenaline serum (Ap16-2) was donated by A.A.J. Verhofstad. × 52 000. Prepared by T. Uchida, Department of Anatomy, Yamanashi Medical College; reproduced, with permission, from Fujita T et al. (1984) Biomed Res Suppl 5: 1–8

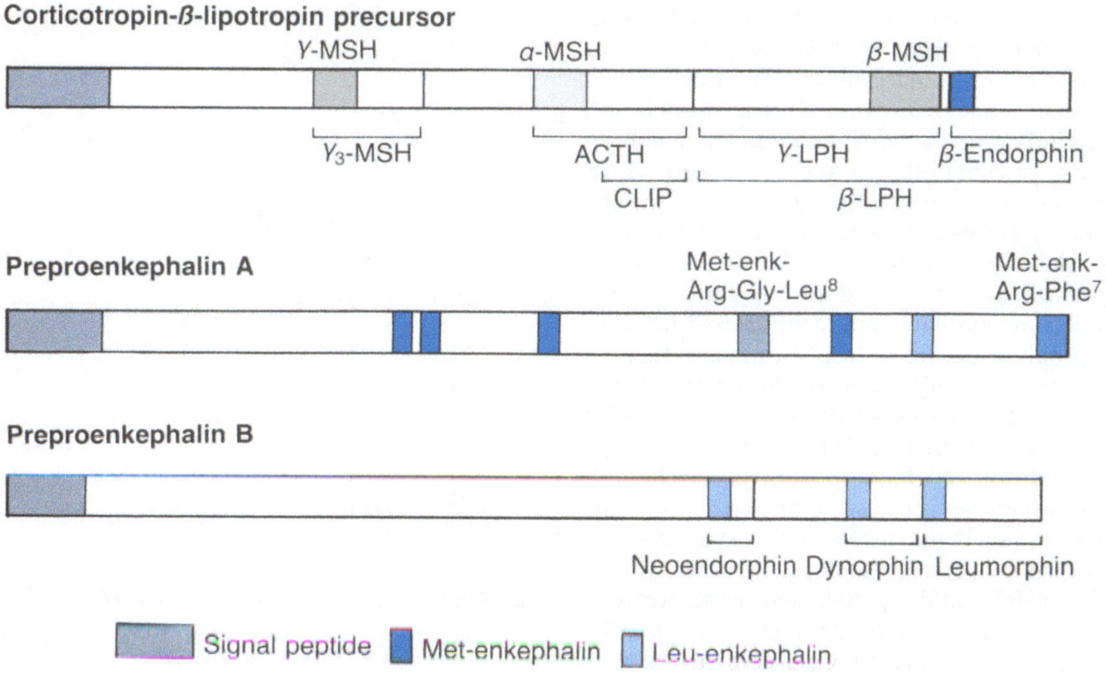

Fig. 3.10. Molecular structure of enkephalins. Based on the reports by S. Numa and his group

tropin/β-lipotropin precursor mRNA. The corresponding amino acid sequence indicates that the prepropeptide consists of repetitive residues, including a γ-melanotropin (γMSH) sequence in its cryptic portion. Pairs of Lys and Arg residues separate the component peptides of the precursor.

Noda et al. (1982) have cloned DNA sequences (cDNA) complementary to bovine mRNA coding for the precursor of adrenomedullary enkephalins, called preproenkephalin A. Nucleotide sequence analysis of the cloned cDNA for preproenkephalin A from the bovine adrenal medulla has indicated that the precursor protein contains four copies of Met-enkephalin and one copy each of Leu-enkephalin, Met-enkephalin-Arg^6-Phe^7, and Met-enkephalin-Arg^6-Gly^7-Leu^8, a previously unknown opioid peptide (Fig. 3.10).

In a similar way, a precursor that contains a β-neoendorphin, dynorphin, and Leu-enkephalin sequence with a carboxyl extension has been detected in the nucleotide sequence of cDNA complementary to the porcine hypothalamic mRNA encoding it (Kakidani et al. 1982). The three bioactive peptides are each bounded by the sequence Lys-Arg. This precursor, referred to as preproenkephalin B, like adrenal preproenkephalin A and the corticotropin/β-lipotropin precursor, contains multiple repetitive units and a cysteine-containing amino-terminal sequence preceded by a signal peptide.

Furutani et al. (1983) have cloned DNA sequences complementary to the ovine hypothalamic mRNA encoding the corticotropin releasing factor (CRF) precursor (referred to as prepro-CRF). The nucleotide sequence of the cloned cDNA has revealed the primary structure of prepro-CRF. Comparison of the amino acid sequence of prepro-CRF with those of the adrenocorticotropic hormone (ACTH)/β lipotropin precursor and the Arg-vasopressin-neurophysin II precursor suggests that these precursor proteins may be evolutionally related.

3.6.2 Biosynthesis of Amines

The principal catecholamines found in the body—noradrenaline, adrenaline, and dopamine—are synthesized by hydroxylation and decarboxylation of the amino acids phenylalanine and tyrosine (Fig. 3.11). The established sequence was first proposed by Blaschko in 1939 and has been confirmed by many observations (for reference see Kirshner 1975). Information on the sequential order of the reactions was obtained from studies using tissue slices, homogenates, and subcellular fractions of adrenal glands. *Tyrosine* is transported into catecholamine-secreting neurons by a concentrating mechanism. In the cytoplasm of the neurons, it is converted to *dopa* and then to *dopamine* by tyrosine hydroxylase and dopa decarboxylase. The decarboxylase, which is also called aromatic L-amino acid decarboxylase, is closely similar but probably not identical to 5-hydroxytryptophan decarboxylase. Dopamine then enters the secretory granules, within which it is converted to *noradrenaline* by dopamine-β-hydroxylase (DBH). DBH is found in particular fractions of cell homogenates, specifically the catecholamine storage granules. The rate-limiting step in the catecholamine synthesis is the conversion of tyrosine to dopa. Tyrosine hydroxylase, which catalyzes this step, is subjected to feedback inhibition by dopamine and noradrenaline, thus providing internal control of the synthetic process.

Synthesis of catecholamines in adrenomedullary paraneurons is presumably identical to that in noradrenergic neurons except that the paraneurons also contain the enzyme phenylethanolamine-*N*-methyltransferase (PNMT). PNMT catalyzes the conversion of noradrenaline to *adrenaline* (Kirshner 1975; Ganong 1985).

In the secretory granules, noradrenaline and adrenaline are bound to ATP and binding proteins called chromogranins. The molecular catecholamine/ATP ratios found in cat and rat granules are around 4, when the adrenal glands of these animals are removed immediately after death (Winkler and Carmichael 1982).

Noradrenaline in nerve endings is partly manufactured there, but partly represents a component that has been released and then taken up again (Axelrod 1971). An active reuptake mechanism is characteristic of catecholaminergic neurons and paraneurons. Circulating catecholamines are also taken up in small amounts by catecholaminergic neurons and paraneurons.

Serotonin is contained in enterochromaffin cells, adrenal chromaffin cells (p. 38), bronchopulmonary and urogenital paraneurons, pinealocytes, mast cells, blood platelets, and in certain neurons in the central and peripheral nervous systems. Serotonin is formed by hydroxylation and decarboxylation of tryptophan, an essential amino acid (Fig. 3.12). Normally, the hydroxylase is not saturated; consequently, increased absorption of tryptophan from the diet can increase the serotonin content in the brain. Much of the serotonin released from serotonin-secreting neurons and paraneurons is taken up again by those cells with an active mechanism. Serotonin is inactivated by monoamine oxidase. In the pineal gland, serotonin is converted to melatonin (see Chap. 25).

Histamine is contained in mast cells in a high concentration (see Chap. 27). Murine enterochromaffinlike (ECL) cells also contain histamine. It has been localized in the anterior pituitary and

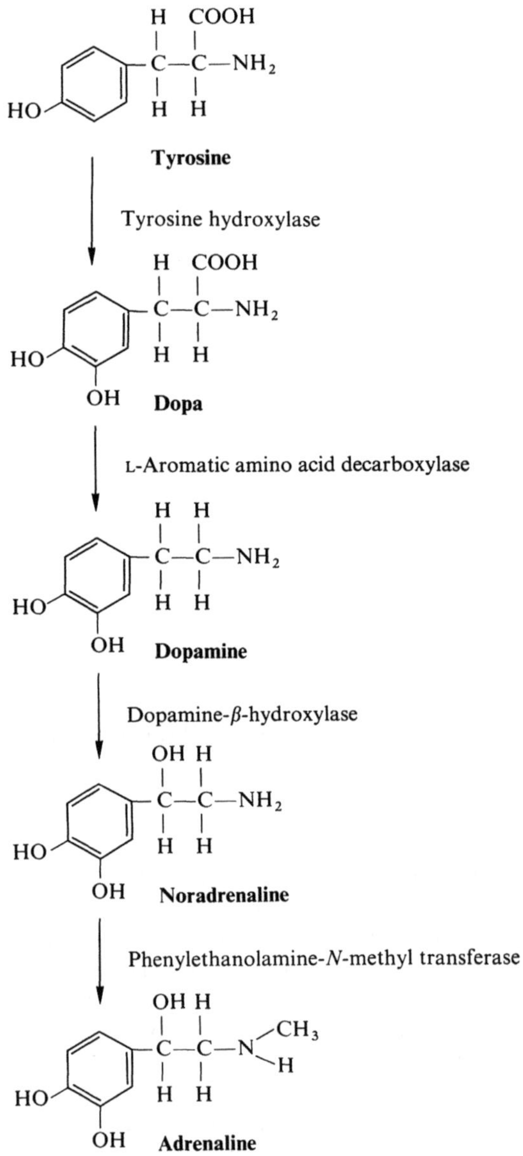

Fig. 3.11. Primary metabolic pathway for the synthesis of catecholamines

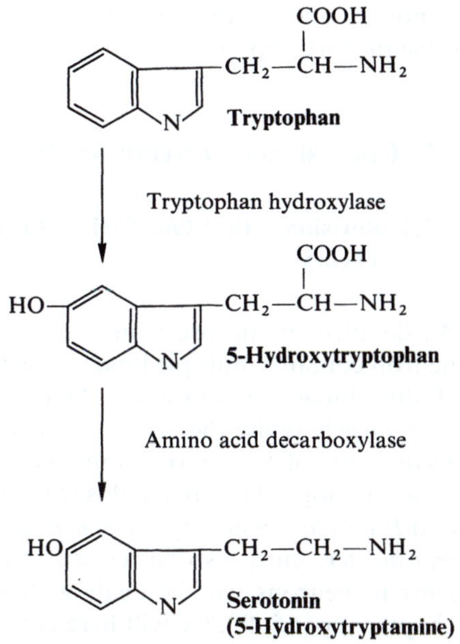

Fig. 3.12. Metabolic pathways for the synthesis of serotonin

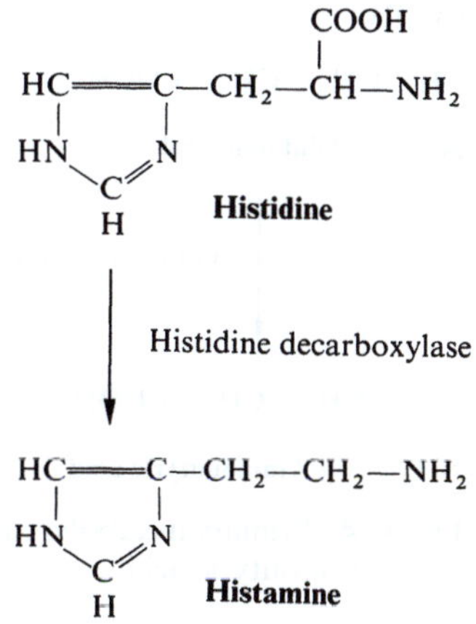

Fig. 3.13. Metabolic pathway for the synthesis of histamine

hypothalamus (Ganong 1985). In other parts of the brain, the histamine content is low, despite the fact that various parts of the brain contain histamine-activated adenylate cyclase. Histamine is formed by decarboxylation of histidine (Fig. 3.13). The enzyme, histidine decarboxylase, that catalyzes this step differs from the L-aromatic amino acid decarboxylases that decarboxylate serotonin and dopa. Histamine is converted to methylhistamine or, alternatively, to imidazoleacetic acid.

Biosynthesis and transport of acetylcholine, gamma-aminobutyric acid, glutamic acid, and glycine are dealt with here, although the occurrence of these neurosubstances in paraneurons has not been confirmed, except for some positive data on acetylcholine.

3.6.3 Biosynthesis of Acetylcholine and Other Classic Messenger Substances

Acetylcholine is the acetyl ester of choline, and is largely enclosed in small clear synaptic vesicles, which are concentrated in the terminals of cholinergic neurons. Biosynthesis of acetylcholine involves the reaction of choline with acetate. There is an active uptake of choline into cholinergic neurons. Choline is also synthesized in neurons (Ganong 1985). The acetate is activated by the combination of acetate groups with reduced coenzyme A. The reaction between active acetate (acetyl-coenzyme A) and choline is catalyzed by the enzyme choline acetyltransferase, which is found in high concentrations in the cytoplasm of cholinergic nerve endings. Acetylcholine per se is not taken up to any appreciable degree, but the choline formed by the action of acetylcholinesterase is actively taken up and recycled.

It seems reasonable to presume that acetylcholine is stored and released by many paraneurons such as the carotid body chief cells (Eyzaguirre et al. 1965), though at present no method is available to verify this hypothesis. The fact that strong acetylcholinesterase activity can be demonstrated in and around different types of paraneurons, such as adrenal chromaffin cells (Mizobe and Livett 1983), would appear to support this possibility.

It is of note that a clonal cell line (PC12), which was established from a transplantable rat adrenomedullary pheochromocytoma, not only possesses (1) many properties characteristic of adrenal chromaffin cells and sympathetic neurons, including synthesis, storage, uptake, and release of catecholamines (Greene and Tischler 1976), (2) but also synthesizes, stores and releases acetylcholine (Greene and Rein 1977). The dual nature of the messenger substance function shown by this type of cell resembles that shown by other paraneuron of normal and neoplastic origin.

Gamma-aminobutyric acid is an inhibitory mediator in the brain and retina and is probably

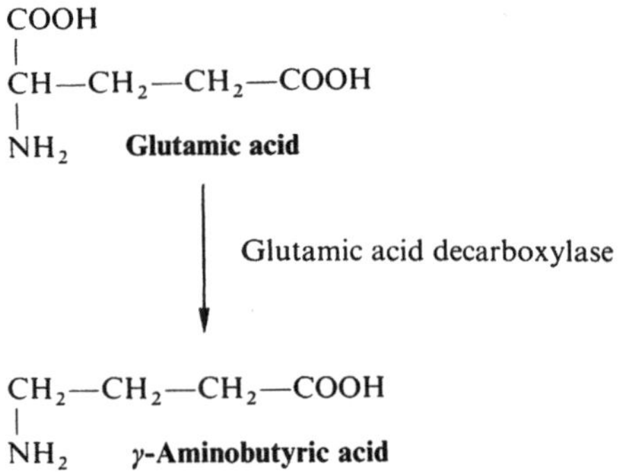

Fig. 3.14. Primary metabolic pathway for the synthesis of γ-aminobutyric acid

the mediator responsible for presynaptic inhibition. GABA is formed by decarboxylation of glutamic acid (Fig. 3.14). The enzyme that catalyzes this reaction is glutamic decarboxylase (GAD), which has been demonstrated by immunocytochemical techniques to be present in nerve endings throughout the central nervous system (Obata 1977). GABA is metabolized primarily by transamination to succinic semialdehyde and then to succinic acid in the citric acid cycle. GABA transaminase (GABA-T) is the enzyme that catalyzes the transamination. Unlike the transamination, the decarboxylation is essentially irreversible, and the GABA content of the brain is reduced in pyridoxine deficiency. Pyridoxine deficiency is associated with signs of neural hyperexcitability and convulsions, although pyridoxine treatment is unfortunately of no value in most clinical cases of idiopathic epilepsy (Ganong 1985).

Glutamic acid (glutamate) has been shown to be the excitatory mediator at myoneural junctions in certain insects and may be the excitatory mediator secreted at the endings of parallel fibers on Purkinje cells in the mammalian cerebellum. Glutamic acid and aspartic acid (aspartate) depolarize many mammalian neurons when these amino acids are applied directly onto neuronal membranes by microelectrophoresis, but they have not been proved to be transmitters at any specific location in mammals (Ganong 1985).

Glycine. Evidence is now accumulating that glycine is the mediator responsible for direct inhibition in the spinal cord. However, one conceptual objection to glycine and other amino acids as transmitters is the fact that these amino acids are probably taken up by passive and active transport and are retained not only in neurons but in most, if not all, living cells. However, this is not at odds with the criteria of messenger substances (Chap. 6).

3.7 Coexistence of Secretory Substances

3.7.1 Revision of the "One Cell—One Hormone" Theory

As detailed in the preceding sections, a paraneuron contains multiple kinds of secretions, including bioactive substances. One must, therefore, largely revise the *"one cell—one hormone" theory*, one of the central principles of modern endocrinology. The same holds true for neurons, and *Dale's law*, whereby *one neuron produces one transmitter*, must also be revised. As the problems in neurons will be dealt with in another chapter (pp. 65, ff), we will here concentrate on those of paraneuron.

The coexistence of granule substances can be classified into two types. The first type of coexistence, which may be termed *heterologous coexistence*, is based on the fact that substances of *different* chemical categories, such as peptides, amines, nucleotides, are stored together in the same granules. The second type, which may be called *homologous coexistence*, indicates that different substances of the *same* chemical categories occur in the same granules.

3.7.2 Heterologous Coexistence of Substances

A paraneuron usually contains different types of substances in its secretory granules: (1) messenger and related peptides, (2) bioactive amines, (3) ATP and other adenine nucleotides, (4) chromogranins, (5) enzymes, (6) acid mucopolysaccharides, (7) lipids, and (8) Ca^{2+} and other ions (Fig. 3.15). These substances are believed to be loosely connected by ionic bonds into a storage complex, representing a rather viscous solution in the granules (Winkler and Carmichael 1982).

Due to the linkage of molecules of different chemical categories, the osmotic pressure of the storage complex formed by them is lower than would be expected from the sum of the analytical concentration of its components (Daniels et al. 1978). If this were not the case, the granules of paraneurons would have to undergo a disadvantageous swelling while they mature by taking up new molecules such as nucleotides and amines.

Amines may be lacking in certain paraneurons, but other substances listed above seem

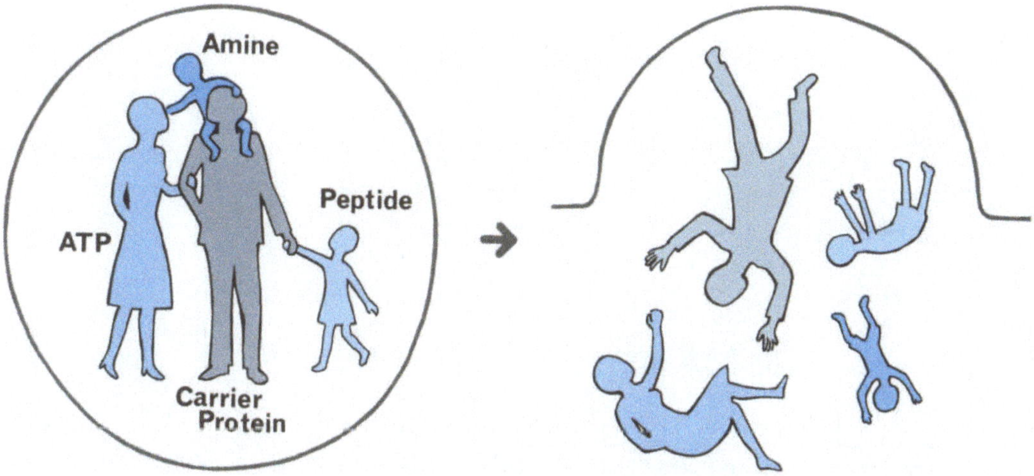

Fig. 3.15. Main secretory substances forming the granule contents in neurons and paraneurons. *Left* Family members, symbolizing the secretory substances, form a storage complex in the secretory granule. *Right* They are separated when released from the cell by exocytosis

to be present in all paraneurons. At any rate, chemical substances of several different categories coexist in each paraneuron, more precisely, in its granules. There is also (heterologous) cosecretion, i.e., costorage and corelease of the secretions.

Table 3.2 shows a few representative paraneurons and the main substances cosecreted by them.

3.7.3 Homologous Coexistence of Substances

Paraneurons and their secretory granules may often contain two or more secretory substances, including messengers or their candidates, of the same chemical category.

Peptides. As already mentioned on p. 33, taking the example of preproenkephalin A processed into several opioid molecules, paraneurons may contain multiple messenger peptides derived from the same prepropeptide. It seems a general rule that peptides from a prepropeptide resemble each other in structure and function, but there are instances in which such a rule might not be valid, as in the case of VIP and PHI. These peptides have considerably different functions and yet are derived from a common precursor, prepro-VIP (Itoh et al. 1983), and coexist in some neurons (Yanaihara et al. 1983).

A type of coexistence of peptides which is more difficult to explain concerns peptides that cannot be ascribed to a common precursor. Many reports on this category of coexistence are available in the literature, but some of them seem to be caused by nonspecific reactions in immunostaining procedures (Grube 1980).

Taking examples from gastroenteropancreatic (GEP) endocrine cells, pancreatic A cells have been claimed to contain, besides glucagon, gastric inhibitory peptide (GIP)- (rat, Smith et al. 1977), cholecystokinin (CCK)- (man and rat, Grube et al. 1978a), and endorphin-like (Grube et al. 1978b) immunoreactivities, whereas pancreatic D cells in humans have been claimed to contain gastrin besides somatostatin (Erlandsen et al. 1976). G cells in the human gastric antrum have been reported to contain, besides gastrin, enkephalin- (Polak et al. 1977) and ACTH-like (Larsson 1981) immunoreactivities. Confirmation of these reports is necessary, but if they are accurate, the functions of these peptides released with their main partner hormones will be important subjects of study.

More widely accepted instances are that pituitary gonadotrophs in man and the rat simultaneously contain follicle-stimulating hormone (FSH) and luteinizing hormone (LH) (Nakane 1970) and that thyroid parafollicular cells in the rat and some other species produce somatostatin and calcitonin in one and the same cell (Van Noorden et al. 1977; Kameda et al. 1982). Recently, we demonstrated that adrenal chromaffin cells are not exceptional in the homological cosecretion of peptides: Cells containing immunoreactivities for enkephalins, NPY, and CGRP partly overlap with each other, indicating that

a 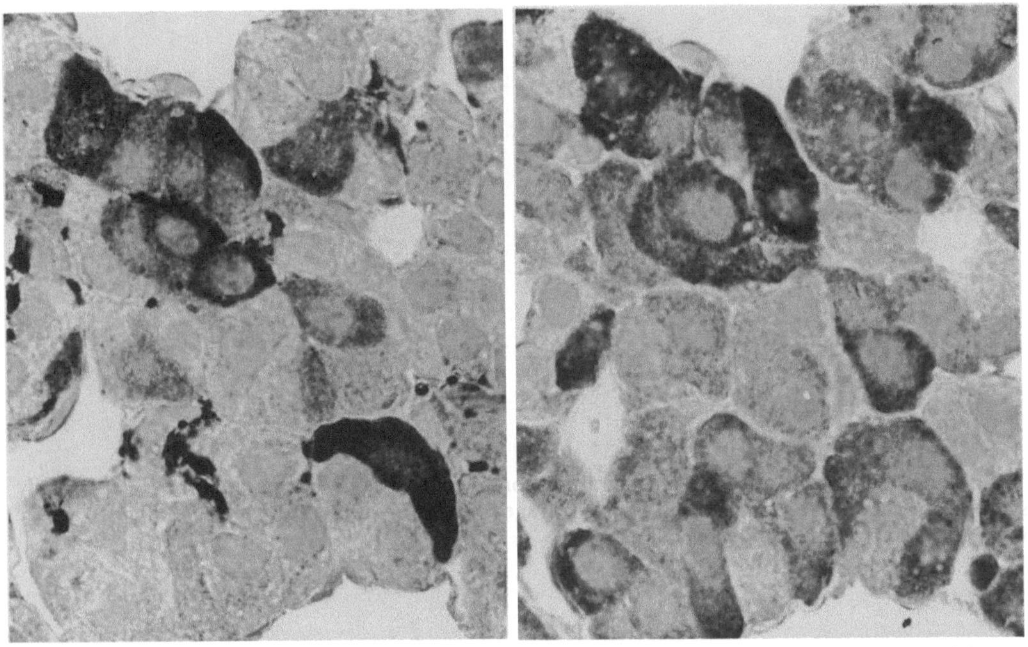 b

Fig. 3.16a, b. Coexistence of different kinds of peptides in the same adrenal chromaffin cells of the rat. Adjacent serial sections were immunostained with an antiserum against **a** NPY and **b** Met-enkephalin-Arg-Gly-Leu. Note that some cells are immunopositive for both peptides, whereas a few others are positive only for one. Prepared by H. Kuramoto, Department of Anatomy, Niigata University Medical School

one cell may produce two or more peptides derived from different precursors (Kuramoto et al. 1986, 1987) (Fig. 3.16).

Amines. It seems to be a rule that *a paraneuron contains more than one monoamine at the same time.* Due to the process of catecholamine conversion, noradrenaline-containing chromaffin cells inevitably possess the precursor, dopamine. Adrenaline-containing cells possess noradrenaline as well as dopamine. In addition to this coexistence of different catecholamines, there are instances of monoamines derived from different precursors. Indolamine, histamine, and catecholamines may coexist in the same paraneuron. The adrenal medulla of rats and mice has been shown to take up exogenous serotonin or its precursor, 5-HTP (Gershon and Ross 1966; Csaba and Baráth 1974). Recent immunocytochemical studies have demonstrated that serotonin (Verhofstad and Jonsson 1983) and histamine (Häppölä et al. 1985) are contained in rat adrenal chromaffin cells. Indolamine-containing enterochromaffin (EC) cells in the gut have been shown to possess a high concentration of dopa decarboxylase (Håkanson et al. 1970). Enterochromaffinlike (ECL) cells in murine animals contain histamine and can produce and store dopamine and serotonin after the natural precursors of these monoamines have been administered (Håkanson and Owman 1967; Aures et al. 1970). Thyroid parafollicular cells in some mammals contain serotonin in addition to dopamine (Falck and Owman 1968; Ericson 1970, 1972).

Adenine nucleotides. As discussed earlier (p. 23), it has been shown in adrenal chromaffin granules that ATP is stored together with smaller amounts of ADP, AMP, and adenosine. It seems reasonable to generalize this finding to other paraneurons.

3.7.4 Significance of Messenger Coexistence

The substances of the homologous and heterologous categories contained in the granule sac are *all released at the same time* from the secretory pole of the cell when the cell is stimulated, and exocytosis occurs. Our knowledge of the functions of the secretions coreleased with the main hormone of paraneurons is very incomplete. In the case of adrenal chromaffin cells, studies have been made almost exclusively on the functions of catecholamines, while little attention has been paid to the possible function of chromogranins and ATP released simultaneously. The recent discovery of the production of enkephalins in the chromaffin cells does not seem to have stimulated the interest of researchers sufficiently in the functions of the opioids released into the circulation together with catecholamines at the time of stress. If we take another example, that of the pancreatic islet cells, previous studies have concentrated on the functions of peptidic secretions, insulin, glucagon, somatostatin, etc. of the cells, and very few investigations have been carried out on the possible functions of other categories of substances, such as amines and ATP released from the cells.

Ca^{2+} and other metal ions have previously been evaluated only on their activity in combining and stabilizing the substances in the granules, and little emphasis has been given to their possible functions, for instance their paracrine effects, as components of paraneuronal secretions.

With this poor knowledge of matters, it seems premature to discuss the significance of the cosecretion of granule substances. However, a broad account of this problem will be given here.

The significance of heterologous cosecretion may be pursued in two directions. The first is the idea that secretions of different chemical categories may work jointly after being released from the cell. Serotonin and other amines are thought to elevate the permeability of blood capillaries and postcapillary venules and, in some cases, move nearby smooth muscles to assist peptide messengers, nucleotides etc. in entering the blood.

Peptides, amines, and nucleotides may act *synergetically* upon identical or related targets.

Dopamine, which is contained in many of the GEP endocrine cells, has been shown to be as potent as secretin in its effect on pancreatic juice flow in the dog (Hashimoto et al. 1971), and it also seems reasonable to postulate similar effects of dopamine upon the secretion of duodenal and other intestinal fluids. The effects of dopamine on the esophageal smooth muscle have also been demonstrated (review: Goldberg et al. 1978). Serotonin is a potent mucus secretagogue and inhibitor of gastric acid secretion (review: Thompson 1971), and its effects stimulating intestinal motility have been shown (Hiatt et al. 1970). ATP, like CCK, cause intestinal muscle contraction (Burnstock 1972).

These are a few examples from the GEP endocrine system in which amines and adenine nucleotides may collaborate or modulate the functions of peptide hormones.

The second possibility for the significance of heterologous cosecretion is a *feedback mechanism* for the release of other substances. Using the ductus deferens of the rat, which is richly supplied with adrenergic nerves which costore noradrenaline and neuropeptide Y (NPY), Ohhashi and Jacobowitz (1983) suggested that the peptide in the intercellular space may suppress, for a considerably long time, the release of noradrenaline from the nerve terminals.

The purpose of the homologous cosecretion seems simpler. As in the case of opioid peptides contained in the same paraneuron or neuron, peptide messengers derived from the same precursor often resemble each other in structure and function. They likely exert their effects in a synergistic fashion. The cases of coexistence of multiple peptides due to occurrence of two or more precursors may be interpreted that the peptides can exert functions effectively or at least without hindrance, being released simultaneously from the same cell by the same stimulus.

Similar functional relations seem to occur among different amines and possibly also among different nucleotides cosecreted from paraneurons.

3.8 Summary

Paraneurons, as peptidergic neurons, are characterized by secretory granules limited by a membrane sac derived from the Golgi apparatus.

As secretory substances, paraneurons produce (1) peptides and their precursors, (2) amines, (3) ATP and other adenine nucleotides, (4) chromogranins (carrier proteins), and some other substances. These substances of different (heterologous) categories are believed to form a storage complex to represent the contents of the secretory granules.

Besides this heterologous coexistence of secretory substances, homologous coexistence of substances is known in paraneurons and in their secretory granules. Messenger peptides derived from either the same or different precursors (prepropeptides) may simultaneously occur in paraneuronal granules. Monoamines of different kinds coexist, and adenine nucleotides of different kinds also coexist in paraneuronal granules.

The secretory substances of paraneurons as well as the phenomena of their coexistence in granules are shared by neurons.

4. Stimulus-Secretion Coupling

4.1 Receptosecretory Nature of Neurons and Paraneurons

The sequence of events in stimulus-secretion coupling in *neurons* can be divided into three main steps—reception of stimuli, conduction of excitation, and extrusion of messenger substances. These three main steps occur in neurons and neurosecretory cells in three functionally different membrane structures: The *receptive* or *input portion* is generally the dendritic zone and the soma (perikaryon), and the *terminal* or *output portion* may have terminal branches or an unbranched process; the *conductile portion* between the input and output sites is generally long (Fig. 4.1)

Neurons may be divided into two general classes on the basis of the type of input—synaptic and sensory neurons (Ralph 1978). The output portion of the presynaptic neuron releases a messenger substance (chemical transmitter) into the cleft of a synapse, and binding of the transmitter with the receptors on the postsynaptic membrane may induce either *depolarization* (excitatory postsynaptic potential, EPSP) or *hyperpolarization* (inhibitory postsynaptic potential, IPSP) in the postsynaptic neuron. When the net electrogenesis of the input is depolarizing, it can initiate *spikes* in the conductile, spike generating portion. These signals, reaching the terminal, output portion of the cell, cause *extrusion of transmitters* which may excite or inhibit the next postjunctional cell (Grundfest 1957). In the sensory neurons, the receptors in the input portion receive various environmental stimuli such as heat, mechanical, electric, and chemical information. Information of these kinds is transduced via the conductile portion to the output portion.

Paraneurons are believed to show essentially identical regional differentiation in the cell membrane, provided that their conductile region is short enough to transmit the net electrogenesis to the output portion without generating spikes (Fig. 4.1) (Kanno 1977, 1980). Paraneurons are also receptive-conductile-secretory cells in nature: they receive adequate stimuli by proper receptors and transduce the stimulus-induced changes in the input region to the output region, from where messenger substances are released by exocytosis.

The entire process between reception of the stimuli and release of the messenger substances (hormones and transmitters) was designated as "*stimulus-secretion coupling*" by Douglas and Rubin (1961). The term "stimulus-secretion coupling" originally expressed a sequence of events set in motion by acetylcholine which results in the release of catecholamines from the adrenal medulla. This term has since been applied to a variety of secretory systems including neurons, peptide/amine producing endocrine cells, proteinic exocrine glands, and mast cells (see Douglas 1968; 1975a, b; Rubin 1970, 1974, 1982). It has become increasingly evident that in the various types of secretion *calcium* is commonly involved as an essential coupler and that the final event of the coupling is extrusion of secretory granules and vesicles by *exocytosis*.

4.2 Receptor and Stimulus Recognition

4.2.1 Receptor Concept

The term *receptor* is adopted in this text to indicate the cellular component which recognizes a specific ligand (e.g., hormone, neurotransmitter, drug, antigen) and sets into motion a sequence of events which culminates in a cellular response characteristic of that ligand. The early works on the receptor defined many of the fundamental aspects of the receptor concept, i.e., that the receptor is a highly specific recognition entity present in or on cells in small numbers and which, in combination with a specific series of compounds, alters some biochemical processes in the cell (probably in a rate-limiting way; Valentine and Hollenberg 1984). Thus, the receptor for a ligand is now regarded as a macromolecule that has the dual function of both recognizing a ligand in a chemically specific manner (*recognition* function) and causing an immediate perturbation of membrane function (*action* function) that in some manner leads to a biological response (Hollenberg 1981, 1982; Valentine and Hollenberg 1984). Direct analysis of the interaction of radiolabeled ligands with the recognition subunits of receptors has advanced in recent years with the use of ^{3}H-labeled ligand probes, which have been used in immunoassays.

4.2.2 Chemical Receptors

Receptors for extrinsic chemical substances. Olfactory and taste receptors are chemoreceptors that respond to chemical substances dissolved in the thin layer of fluids between the external environment and those receptors. Although there is evidence that some taste-inducing molecules bind to specific proteins in the taste buds (Ganong 1985), the mechanism by which the molecules generate receptor potentials in these chemoreceptors is poorly known.

Receptors for intrinsic/extrinsic chemical substances. Sugars, amino acids, and some other nutrients can work as extrinsic signals and, after incorporation into the body, as intrinsic signals. Among the receptors for these dual stimulants, *glucoreceptors* will be dealt with in some detail here, as they have bee investigated more intensively than others. It is an important point that glucoreceptors are *shared by neurons and paraneurons*, supporting the continuity between the two cell groups. Paraneurons that recognize glucose and respond to it promptly include gustatory cells in the taste buds, A and B cells in the pancreas, and L cells (enteroglucagon cells) in the small intestine; neurons showing this feature include glucoreceptive cells in the hypothalamus and neurons terminating in the portal vein. All these cells possess glucoreceptors, although the range of their specificity of signal substances differs according to the type of cell. The glucoreceptors of gustatory cells, for instance, recognize a wide range of sugars and other substances which elicit the sensation of sweetness, whereas those of pancreatic B cells recognize a limited range of substances, including glucose and mannose.

Among various protein fractions obtained from the bovine tongue surface, Dastoli and Price (1966) found one particular "sweet" protein which complexed with sugars and saccharin (see also Beidler 1970). The proteinic nature of the glucoreceptor was supported by the physiological studies of Hiji (1975). This author treated the tongue surface of rats with various proteases and found that pronase E was able to suppress the sensation of sweetness in terms of the electrophysiological response to sugars in the chorda tympani. The same protease put on the human tongue had the same effect.

In analogy to the stereochemical theory for smell recognition (Amoore et al. 1964), (p. 216), it was hypothesized that the glucoreceptors are macromolecules containing a recognition site or sites which stereospecifically fit the molecules of sweet substances. This view was strongly supported by experiments using glucose anomers and pancreatic islet cells.

D glucose occurs in nature and it can be incompletely separated into α- and β-anomers, which show three-dimensionally the reverse site of hydroxyl at position 1. Niki et al. (1974, 1979) examined the effects of these anomers on isolated or pertused pancreatic islets and found that α-anomer is more effective than β-anomer in stimulating insulin release. Subsequently, the same result was obtained more precisely by Grodsky et al. (1974). α-Anomer was found to be more effective in suppressing glucagon release from pancreatic A cells (Grodsky et al. 1975; Matschinsky et al. 1975). Niki et al. (1974, 1979) proposed that these data might evidence the occurrence of stereospecific glucoreceptors in the A and B cells on the cell surface. This "glucoreceptor theory" has since had to conflict with the "metabolism theory," according to which certain products or byproducts of glucose are incorporated and metabolized in the cell to stimulate insulin release or suppress the glucagon release. To substantiate this theory, Malaisse et al. (1976, 1979) demonstrated that α-anomer is preferentially taken up and metabolized by the islet cells. Although it is at present difficult to determine which of these theories is correct (for reviews of this debate: see Niki and Niki, 1980), we support the glucoreceptor theory because it can be better correlated with the findings on the glucoreceptors in other paraneurons and some neurons.

After the above-mentioned reports by Niki and associates were published, it was noticed that previous researchers had recorded that the α-anomer of D glucose causes a stronger sensation of sweetness than the β-anomer, although the results had not been correlated with the nature of the receptor of gustatory cells (Tsuzuki 1947; Pangborn and Gee 1961). Recently, it was re-

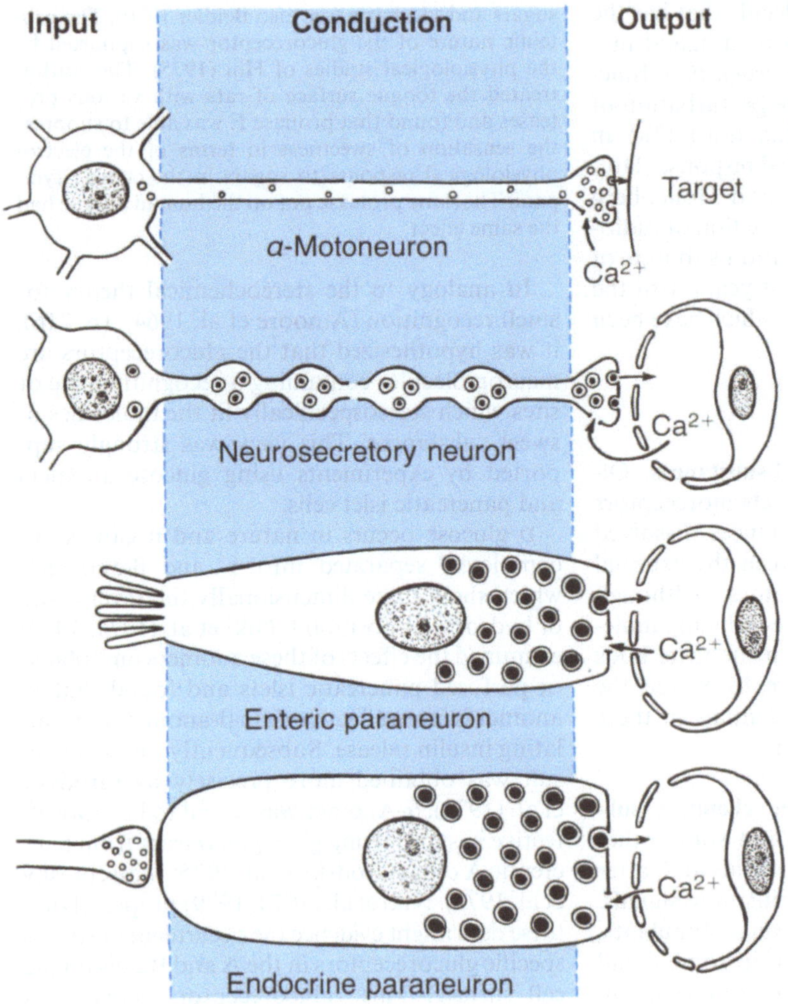

Fig. 4.1. The receptive, conductile, and secretory portions of some representative neurons and paraneurons. Note that the influx of Ca^{2+} occurs on the secretory side of the cell

ported that the release of enteroglucagon is more intensely stimulated by the α-anomer of D glucose than by the β-anomer (Matsuyama et al. 1979). Neurons in the hypothalamic satiety center (ventromedial nucleus) and in the hunger center (lateral area) are known to be sensitive to D glucose (Oomura et al. 1978). Growth hormone-secreting α-cells in the anterior pituitary and somatostatin-producing neurons in the hypothalamus are also sensitive to D glucose. It would be desirable to determine whether an anomeric preference exists in these cells. The neurons in the satiety center were recently observed to be more sensitive to the β-anomer of methyl-D-glucose than to the α-anomer (Oomura et al. 1978). Of importance in this context is the recent study by Niijima et al. (1983), who demonstrated in the guinea pig that suppression of afferent activity of the hepatic vagal nerve by glucose infusion into the portal vein occurred more markedly with β-anomer than with α-anomer. As histological observations indicate the occurrence of free nerve endings in the hepatic portal vessels (Tsai 1958), Niijima et al. (1983) propose that the glucose sensors which may be represented by these nerves make a distinction between α- and β-anomers.

On the basis of the above findings, Niki and Niki (1976, 1980) propose a hypothesis that a series of neurons and paraneurons share glucoreceptors—*neuroparaneuronal glucoreceptor system*—which differentiate between the α- and β-anomers of D glucose.

Maturity-onset diabetics are usually characterized by lowered insulin response to glucose, which can be ascribed to a selective defect of glucose recognition of B cells. Suppression of A-cell glucagon release by glucose is likewise impaired. Enteroglucagon release has been reported to be unusually high in some diabetics (references: Niki and Niki 1980). It was recently reported that the lingual sensitivity to sweetness may be markedly lowered in maturity-onset diabetics suggesting impaired receptors on the gustatory cells (Halter et al. 1975; Lawson et al. 1979). Furthermore, impairment of glucoreceptor cells in the hypothalamus in diabetics has been suggested (Asplund 1976). This may account for the polyphagy and obesity in many adult-onset diabetics. Suppression by glucose of the release of growth hormone has also been demonstrated to be impaired (review: Niki and Niki 1980). It is worthy of note that most of the impairments in the neurons and paraneurons mentioned above can already be found in a prediabetic state (Niki and Niki 1976, 1980). These findings support the view that maturity-onset or type II diabetes mellitus may be due to a generalized impairment of the neuroparaneuronal glucoreceptor system.

Receptors for intrinsic chemical substances. Structure and functions of receptors for chemical substances, e.g., nutrients, hormones and neurotransmitters, in extracellular fluid have been studied with the use of ^{3}H-labeled ligand probes. The molecular structures of receptors for acetylcholine, insulin, and glucagon have been most extensively studied among the various receptors for intrinsic chemical substances.

Detailed information about the molecular structure of recognition and action subunits of receptors has recently been provided from studies on the *acetylcholine receptor* (nicotinic-cholinergic receptor) in the electroplax of the ray, *Torpedo californica*. In detergent solutions, this receptor behaves as a macromolecule (molecular weight about 250000). The entire oligometric structure comprises two recognition subunits and an ion-gating oligomer (action subunit) composed of three distinct but chemically related substituents (Hollenberg 1981). The complete primary structures of the subunit precursors of the *Torpedo* acetylcholine receptor have been elucidated by cloning and sequencing cDNAs for these polypeptides (Noda et al. 1983a). The transmembrane topology of the subunit molecules has also been inferred (Noda et al. 1983a). The molecular structure of the nicotinic-cholinergic receptor in the electroplax unit provides a leading model of recognition and action subunits of various receptors in neurons and paraneurons.

The molecular structure of the *insulin receptor* has been studied in adipose and liver cells. The insulin receptor is a complex protein, which is a tetramer composed of two α- and two β-glycoprotein subunits linked by sulfide bonds. The α-subunit is a protein with a molecular weight of about 135000 and binds with insulin. The β-subunit is a protein kinase with a molecular weight of about 95000 and catalyzes the phosphorylation of tyrosine residues in the β-subunit and other proteins; phosphorylation of these proteins may lead to a change in the number of transporters (Ganong 1985; Kadowaki et al. 1985). The molecular structure of the insulin receptor in the adipose and liver cells may be a prime model of the insulin receptor in the neuron and paraneuron.

Glucagon receptors have been identified in a variety of tissues and partially purified. The molecular weight of the receptor protein is approximately 190000 (Ganong 1985). The receptor for glucagon acts by adenylate cyclase activation and interacts via a guanosine triphosphate (GTP)-binding regulatory subunit to modulate the activity of the catalytic subunit, which catalyzes ATP to cyclic AMP. It is suggested that a receptor region, which may be similar in a variety of cyclase-related hormone receptors, interacts directly with the GTP-binding regulatory subunit. The catalytic subunits of cyclase are shown to have a region for interacting with a specific portion of the regulatory subunit (Hollenberg 1982).

4.2.3 Photomechanical Receptors

The problem of how receptors convert information into action potentials in the sensory nerves has been the subject of intensive study. In the specialized sense organs such as those concerned with vision, hearing, equilibrium, and taste, there are separate receptor paraneurons intercalated between the stimuli and afferent nerves.

Among various receptor paraneurons for photomechanical stimuli, the visual process in the rod cell is best understood as follows (Saibil 1986). The rod cells are the dim-light receptors of the vertebrate retina. Their large size facilitates electric measurements, and they are easy to isolate biochemically. The rod is divided into an inner segment, containing the cell's metabolic machinery, and an outer segment, entirely specialized for the absorption and transduction of light into an electric signal. The photoreceptive disk membranes of the *rod outer segments* of a frog are packed with 10^9 molecules of the visual pigment protein, *rhodopsin*. Excitation of one of these rhodopsin molecules is sufficient to cause the closure of many Na^+ channels in the plasma membrane, leading to the hyperpolarizing response to light. This Na^+ channel is distinct from the excitable Na^+ channels in normal neurons in that it is voltage insensitive, very small, and less cation selective. From the amino acid sequence of rhodopsin, a transmembrane topology of seven hydrophobic α-helices could be predicted, consistent with its physical and chemical properties, with about half the protein mass being distributed in cytoplasmic and intradiskal loop domains.

In most of the cutaneous sense organs, the stimuli are thought to be received by specialized, histologically modified ends of receptor neurons. It appears that a graded electric change in the membrane of the input portion of a sensory paraneuron, or a neuron, is induced by an adequate stimulus to its receptor. How this occurs is unknown except in the following cases (Ralph 1978). In crustacean stretch receptors and visual receptors, stimulation produces a change in permeability of the cell membrane to various ions. Na^+ conductance is increased in the stretch receptors. The transduction process in photoreceptors involves a chemical event is an essential step in the development of the electric event. A change in pressure may cause certain changes in molecular organization, either of the plasma membrane or of hydrated ions on either side of it. For other receptors, the transduction process is even more obscure.

4.3 Various Modes of Stimulus Transformation

4.3.1 Changes in Membrane Potential

In most synaptic neurons and paraneurons studied, the stimulus is transformed to analogous changes of either *depolarization* (EPSP) or *hyperpolarization* (IPSP). The depolarization or hyperpolarization can commonly be detected by intracellular recording of the potentials of neurons or paraneurons. These electric activities are due to *ionic movements across the channels* in the cell membrane. Recent developments in patch clamp techniques enable us to measure single-channel currents carried by various ions (Sakmann and Neher 1984).

The electrophysiological technique of recording *transmembrane potentials* from paraneurons was initially applied to the adrenal chromaffin cell 20 years ago (Kanno et al. 1966; Douglas et al. 1967a, b; Douglas and Kanno 1967; Kanno and Douglas 1967). In this work, the technique of electrophysiology was combined with that of tissue culture. It was thus established that depolarizing responses occur in chromaffin cells following the application of acetylcholine, i.e., the transmitter substance liberated by the secretomotor nerves to the chromaffin cells of the adrenal medulla. The depolarizing effect increased in proportion to logarithmic increase in $[Na^+]_0$ and $[Ca^{2+}]_0$. The effect was partially depressed by amethocaine, a local anesthetic, when the extracellular environment contained both Na^+ and Ca^{2+}, but it was completely blocked by the drug in Na-free environments.

In 1976, the recording of action potentials in *adrenal chromaffin cells* was made simultaneously and independently by Biales et al. (1976) and Brandt et al. (1976). The action potential in the adrenal chromaffin cell is mediated by Na^+ as well as by Ca^{2+} inward currents. Thus, Ca^{2+} entering the cell during the action potential links the acetylcholine stimulation and hormone release. In addition, Ca^{2+} entering the cell through the acetylcholine receptor channels may partly contribute to the exocytosis of chromaffin granules. Acetylcholine receptors are probably localized at the input region (subsynaptic area), which is only a small portion of the cell surface. Ca^{2+} influx through the acetylcholine receptor channels is thus limited to the area close to the synapse, whereas voltage-dependent Ca^{2+} channels may be located over the entire surface of the adrenal chromaffin cell (Kidokoro 1984). Recent studies by patch-clamp technique have demonstrated that the Ca^{2+} entry into the chromaffin cell during stimulation could be through either voltage-dependent Ca^{2+} channels or acetylcholine-receptor channels clustered on the subsynaptic membrane; the former channels, which are more widely distributed on the cell surface, may play a major important role under physiological conditions (Kidokoro 1985). Regional differentiation of receptors (input region) from the output region in the adrenal chromaffin cell is also suggested on the basis of morphological evidence (Coupland and Kobayashi 1976).

The question now is whether *action potentials* are essentially involved in stimulus-secretion coupling in the chromaffin cell. The electric excitability of chromaffin cell was debated some 50 years ago by Cannon and Rosenblueth (1937) and by different groups of French investigators who attempted to answer the question by passing current through denervated glands (see Cannon and Rosenblueth 1937). These earlier experiments yielded conflicting results and the problem has remained unsolved.

Strong evidence for the view that an action potential is involved in stimulus-secretion coupling is as follows. Increases in spike frequency were demonstrated when the external solution contained of acetylcholine, and a good correlation between acetylcholine concentration and hormone release was found in rat adrenal chromaffin cells (Brandt et al. 1976). However, the addition of tetrodotoxin to acetylcholine decreased the acetylcholine-induced adrenaline release to only 70% of control valves. Wakade and Wakade (1982) also demonstrated that a single electric shock to the perfused adrenal gland of the rat resulted in a massive release of adrenaline and noradrenaline, and that the release was directly related to the strength and duration of the applied stimulus over a wide range. However, catecholamine release was unaffected by tetrodotoxin or hexamethonium/atropine but it was abolished by Ca^{2+} lack or 3 mol/l Mn^{2+}. Wakade and Wakade (1982) suggested that the chromaffin cell membrane undergoes nonpropagated electrotonic depolarization on electric stimulation and voltage-dependent Ca^{2+} channels might thereby be opened to initiate catecholamine release.

The *patch-clamp technique* in the "whole cell" recording configuration is a system sensitive enough to measure changes of cell membrane capacitance associated with secretion. A large transient increase in membrane capacitance was linked to the Ca^{2+} current through the membrane as a result of electric stimulation of the chromaffin cell. This was followed by relatively large steps of capacitance decrease. The step-like capacitance increase was considered to reflect individual chromaffin granule exocytosis events, whereas the step in the opposite polarity the formation of an endocytotic vesicle (Neher and Marty 1982).

The electric excitability of chromaffin cells may be increased by a humoral factor, nerve growth factor (NGF). Dichter et al. (1977) showed that NGF inhibited cell division, stimulated extension of neuronlike processes, and increased the electric excitability and acetylcholine sensitivity in a cloned line of rat pheochromocytoma cells (PC12).

The clonal *anterior pituitary cell* (GH_3) is also capable of generating action potentials, which are caused by an increase in membrane permeability to both Na^+ and Ca^{2+} (Ozawa 1981). Continuous recording of the intracellular potential from GH_3 cells revealed that thyrotropine-releasing hormone (TRH), which stimulates the release of prolactin and growth hormone from these cells, enhanced the generation of action potentials following a transient membrane hyperpolarization. The stimulatory effect of TRH on hormone release was inhibited under the conditions in which the Ca^{2+} spike was blocked. These results strongly suggest that TRH may stimulate the release of hormones from the GH_3 cells by increasing Ca entry through the voltage-dependent Ca^{2+} channels of the cell membrane (Ozawa 1981).

Mammotrophs of the pituitary pars distalis (Taraskevich and Douglas 1978) and melanotrophs of the pituitary pars intermedia (Davis and Hadley 1978; Douglas and Taraskevich 1978) also discharge action potentials spontaneously and this discharge is slowed or arrested by dopamine. In both mammotrophs and melanotropes, the electric activity shows a calcium component indicative of Ca^{2+} influx (Taraskevich and Douglas 1978; Douglas and Taraskevich 1980); Ca^{2+} is required for secretory activity (Vale et al. 1977). Action potential frequency decreased when the Ca^{2+} concentration was lowered to 0.1 mmol/l and increased when it was raised from this level to 1 or 2 mmol/l or when Ba^{2+} was introduced. These effects, together with the effects of Ca^{2+}, Ni^{2+}, and Mn^{2+}, are consistent with the possibility that Ca^{2+} participates in the regulation of spike discharge (Douglas and Taraskevich 1982).

Normal endocrine cells isolated from the pituitary of the rat, lizard, and fish and maintained in short-term culture produce action potentials spontaneously in response to depolarizing current pulse of stimulant hypophyseotropic substances (Taraskevich and Douglas 1984). Dopamine, which inhibits basal release from rat melanotrophs and teleostean prolactin cells, inhibited action potential discharge in both these cell types. Serotonin, which stimulates release from lizard pars intermedia cells, stimulated action potentials in these cells. A Ca component that indicates inward Ca^{2+} flux was present in the action potentials in each of the cell types examined, although its contribution varied according to species and gland region. Electric activity may therefore participate in secretory control in adenohypophyseal cells. However, in contrast to neurons where the Na^+ spike is all-important for conveying the signal for transmitter release to the nerve terminal, it may well be that the Na^+ spike has only an adjuvant function in adenohypophyseal cells and that Ca^{2+} fluxes, possibly with small oscillations in membrane potential that seem to be related to spike generation, may be more significant (Taraskevich and Douglas 1984).

The electric activity of *pancreatic B cells* in isolated islets has also been studied with respect to the control of membrane ionic fluxes involved in stimulus-secretion coupling in the cell. (Fig. 4.2) Since the initial observation of Dean and Matthews (1968), evidence has accumulated to support the hypothesis that voltage-dependent electric spikes (Atwater 1980; Dean and Matthews 1970a, b; Matthews and Sakamoto 1975)

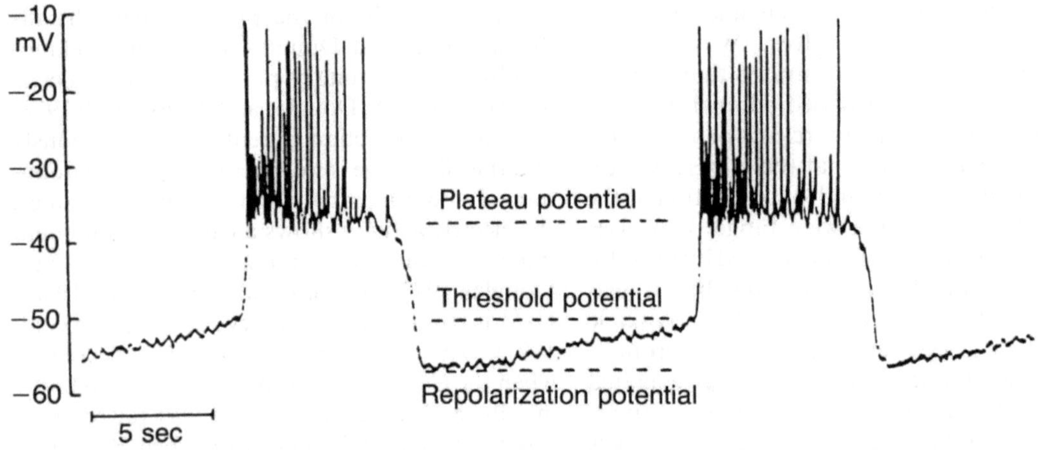

Fig. 4.2. Slow waves of membrane potential induced by 10 mmol/l glucose in a single B cell of a mouse pancreatic islet. Reproduced by courtesy of Henquin and Meissner (1984)

mediate the uptake of extracellular Ca^{2+} into an intracellular pool, which in turn controls the rate of insulin release. The view that the spikes appear to be Ca^{2+} dependent has been supported by the following results: (1) The spike potentials were abolished, at least transiently, by removal of extracellular Ca^{2+} (Atwater 1980; Atwater et al. 1980; Dean and Matthews 1970a, b; Matthews and Sakamoto 1975; Meissner and Schmelz 1974); (2) the spike potentials were accentuated by a high Ca^{2+} environment (Atwater et al. 1980; Dean and Matthews 1970a, b; Meissner and Schmelz 1974; Ribalet and Biegelman 1980, 1981). There appears to be little Na^+ contribution to the spike potentials because they were not diminished by a low Na^+ environment (Dean and Matthews 1970; Ribalet and Beigelman 1982) or by tetrodotoxin (Meissner and Schmelz 1974; Tarvin and Pace 1981). Furthermore, glucose-induced spiking activity correlates closely with glucose-induced Ca^{2+} uptake in islet cells (review: Wollheim and Sharp 1981). Evidence indicates that the voltage-dependent plateau depolarization may also be caused by Ca^{2+} influx (Cook 1984). It is also suggested that this plateau conductance is periodically terminated in turn by a pacemaker current through membrane K^+ channels that are activated by intracellular free Ca^{2+} and by changes in intracellular Ca^{2+}-buffering capacity (Cook 1984); the voltage-dependent entry of extracellular Ca^{2+} is believed to be primarily responsible for the appearance of spikes and the increase in $[Ca^{2+}]_i$ required for the glucose-evoked secretion of insulin, although some channels do exist for the inward transport of Na^+ across the B-cell membrane (Matthews 1985). Evidence for polarization of plasma membrane domains has recently been reported in pancreatic endocrine cells (Lombardi et al. 1985).

4.3.2 Roles of Calcium Ions

The evidence mentioned above clearly indicates the presence of *voltage-dependent* Ca^{2+} *channels* in various paraneurons, including gastroenteric endocrine cells, pancreatic B cells, adrenal chromaffin cells, adenohypophyseal cells, as well as in various neurons. The voltage-dependent Ca^{2+} channels, coupled with the depolarizing effect of excess K^+, provide direct electrophysiological evidence for the common supposition that the Ca^{2+}-dependent release elicited by excess K^+ from various paraneurons and neurons is caused by Ca^{2+} influx through the Ca^{2+} channels that are opened by depolarization (Fig. 4.3).

Direct support for the view that a rise in the $[Ca^{2+}]_i$ suffices to induce exocytosis was obtained by intracellularly injecting Ca^{2+} through micropipettes placed within the presynaptic terminal of squid giant axons (Miledi 1973) and rat peritoneal mast cells (Kanno et al. 1973). Indirect support for this view has been provided by a variety of means. Ca ionophores such as A23187 and X537A have been used to facilitate Ca^{2+} influx in various cells including paraneurons and neurons. Ionophores facilitate the transfer of cations across cell membranes by functioning either as carriers of ions or pore-formers that remain relatively fixed while ions move through them (McLaughlin and Eisenberg 1975; Rubin 1982). In general, the actions of these Ca ionophores mimic those of microinjections of Ca^{2+}

directly into the secretory cells (Rubin 1982).

To make the cell membrane permeable to Ca^{2+}, Baker and Knight (1978, 1981) have developed an electric technique that permits the plasma membrane barrier to be bypassed without impairing exocytosis in the isolated adrenal chromaffin cells by exposing the cells to high-voltage discharges of brief duration. By making the cell permeable, less than 1% of the total cellular catecholamines was released at an ionized Ca concentration of 10 mmol/l or less; but if the ionized Ca concentration was subsequently raised to the micromolar range, up to 30% of the cellular catecholamines appeared in the external medium (Baker and Knight 1981). Permeable cells can also be produced by exposure to low concentrations of detergents (Brooks and Treml 1983; Dunn and Holz 1983; Wilson and Kirshner 1983), which generate quite large holes in the plasma membrane.

Fluorescent dyes have been used for measuring free Ca^{2+} in populations of cells (Tsien 1980, 1981). The mean concentration of resting free Ca^{2+} in bovine adrenal medullary cells monitored with the lipid-soluble Quin 2 acetoxymethyl ester was close to 0.1 mmol/l (Baker and Knight 1981), a value closely similar to that found in a wide variety of cell types. At the peak of the secretory response, free Ca^{2+} was usually in the range of 0.5–1.0 µmol/l (Knight and Kesteven 1983).

The fusion of secretory vesicles in vitro has been regarded as a model of the secretion-release process in various secretory cells in vivo. The incidence of intervesicular fusion of secretory vesicles isolated from pancreatic islets, neurohypophysis, and adrenal medulla was low in media containing less than 0.1 µmol/l Ca^{2+}, the concentration estimated in resting cells. Vesicle fusion increased as the Ca^{2+} concentration was elevated to the level found in stimulated cells (Gratzl et al. 1980).

Since the term "stimulus-secretion coupling" was patterned after the phrase "excitation-contraction coupling" for muscular cells, the possible participation of the actomyosin system in stimulus-secretion coupling has been a focus of studies. A working hypothesis of a molecular mechanism of exocytosis was proposed by Pollard et al. (1982). They assumed that chromaffin granules are fixed by an actin network surrounding the granules in the chromaffin cell in a resting state and that a stimulus-induced increase in $[Ca^{2+}]_0$ might liberate the granules from the actin network and in turn facilitate attachment of the granules to the internal surface of the plasma membrane. This model of Pollard et al. (1982) was based on various biochemical results obtained *in vitro*, but it is not consistent with the finding that Brownian movement of the granules has not been observed in cultured cells under stimulation (Douglas 1974a, b).

The $[Ca^{2+}]_i$ is controlled by inward movement of extracellular Ca^{2+}, cellular sequestration, and outward movement of intracellular Ca^{2+}. It is thought that Ca^{2+} enters cells by three different routes (Rasmussen and Goodman 1977; Wollheim and Sharp 1981): (1) inward diffusion; (2) diffusion via a voltage-dependent (so-called late) Ca^{2+} channel, which depends on the membrane potential and can be specifically inhibited, e.g., by cobalt, verapamil, and methoxyverapamil (D 600); and (3) diffusion via the veratridine-sensitive Na^+ channel, which is blocked by tetrodotoxin. Mechanisms involved in the cellular sequestration of Ca are: (1) Ca bound in the plasma membrane complexed with ATP and phospholipid; (2) Ca buffering by macromolecular components and small molecules of the cytoplasm; (3) Ca uptake into the endoplasmic reticulum, (4) Ca uptake into mitochondria and Ca^{2+} sequestration by secretory organelles (Rubin 1982; Rasmussen and Barrett 1984). There are two mechanisms for outward movement of Ca from cells: (1) the Ca^{2+} pump or Ca^{2+}-activated ATPase; (2) the Na^+-Ca^{2+} exchange mechanism (Wollheim and Sharp 1981). Dynamic events in the regulation of $[Ca^{2+}]_i$ have been discussed in detail in recent reviews and monographs (Rasmussen 1981; Rubin 1982; Rasmussen and Waisman 1982; Anghileri and Tuffet-Anghileri 1982; Campbell 1983; Rasmussen and Barrett 1984).

4.3.3 Roles of Inositol Lipid Cycle

After many years of dispute, the past several years have brought agreement that *receptor-mediated hydrolysis of inositol phospholipids* is a common mechanism for transmembrane signaling when cells respond to external stimuli, such as hormones, neurotransmitters, antigens, light, and some growth factors (Michell 1986). The external stimuli activate an enzyme termed phospholipase C, which breaks down inositol lipids in the plasma membrane; the major lipid is phosphatidylinositol (PI), which represents 5%–10% of the total phospholipids, while two phosphorylated derivatives, phosphatidylinositol-4-phosphate (PIP) and phosphatidylionositol-4,5-bisphosphate (PIP_2), are present as minor components. Metabolically, these three are the most active inositol lipids in the plasma membrane. When suitable agonists bind to the proper receptor sites in the cell membrane, a conformational change occurs, which leads to the activation of a guanosine triphosphate-binding protein (G-protein; Gilman 1984). A number of different G-proteins are present in the cell membrane, and one of these is thought to be responsible for the coupling of information transfer from occupied receptor sites to phospholipase C, which is located on the inside of the cell membrane phospholipid bilayer. Phospholipase C is inactive at resting levels of intracellular free Ca^{2+}, but when activated by the G-protein a very rapid breakdown of PIP_2 occurs with the production of two second messengers with different functions (Fig. 4.3).

One of the two second messengers, inositol-1,4,5-triphosphate (IP_3), is responsible for mobilization of Ca^{2+} from an intracellular, non-mitochondrial, vesicular calcium pool (Berridge and Irvine 1984; Hokin 1985; Williamson et al. 1985). This Ca^{2+}-signaling system is terminated by hydrolysis of IP_3. The other second messenger, 1,2-diacylglycerol, has as its primary signaling role the activation of a Ca^{2+}- and a phospholipid-dependent protein kinase called protein kinase C (Nishizuka 1984). Two processes that are enhanced through protein kinase C activation are Na^+-H^+ exchange across the cell membrane, leading to alkalization of the cell interior, and PI kinase, which replenishes the polyphosphoinositide pool. In many secretory cells, the effects mediated by mobilization of intracellular Ca^{2+} and activation of protein kinase C act synergically to elicit a prolonged functional effect (Takai et al. 1984). However, the mechanism by which this occurs is at present unknown (Williamson 1986).

4.3.4 Roles of Cyclic AMP

In the case of the pancreatic B cell, there is substantial evidence that a rise in $[cAMP]_i$ leads to an enhanced response to a standard change in medium glucose concentration (Hedeskov 1980; Malaisse 1973). Recent studies with dibutyryl cyclic AMP (Henquin and Meissner 1983, 1984a, b) and forskolin (Henquin et al. 1983), an activator of adenylate cyclase, strongly suggest that cyclic AMP also influences the ionic permeabilities of the plasma membrane in pancreatic B cells. However, in B cells, the basic messenger system regulating insulin release is the Ca^{2+}-messenger system; the function of the cyclic AMP messenger is that of determining the set point around which the Ca^{2+}-messenger system operates (Rasmussen and Barrett 1984).

In general, the basic system regulating the stimulus-secretion coupling in neurons and paraneurons is $[Ca^{2+}]_i$, while the function of the cyclic AMP system is only a modulator in the coupling (Fig. 4.3).

4.4 Release of Granule Contents

4.4.1 Different Views on Granule Release Mechanisms

The mechanism as to how different exocrine cells release their secretions has long been a subject of keen interest to light microscopist. Early electron-microscopic studies in the 1950s revealed that distinct morphological changes, more elaborate than those known from light-microscopic studies, occur in secretory cells, including endocrine cells, as their secretions are released from the cell. Kurosumi (1961) was the first to propose a systematic classification of the modes of extrusion of the secretory products of exocrine and endocrine cells at the ultrastructural level. He classified the different modes of secretion release into five types, and this classification is still widely used today with minor modifications.

Extrusion type I: holocrine secretion. The dead cell body itself becomes the secretory product. Various degenerative signs such as pyknosis of the nucleus precede the breakdown of the whole cell. This type of secretion release occurs in a few exocrine glands such as the sebaceous gland in man. However, in neurons and paraneurons, cell death is not an effective mode of extrusion of secretory products.

Extrusion type II: macroapocrine secretion. The secretory products become concentrated at the free end of the glandular cells. The protruding apical poles of these cells are then thrown off into the glandular lumen. This mode of secretion release is represented by the mammary gland. However, concerning neurons and paraneurons no convincing reports of macroapocrine release of secretory products have been presented.

Extrusion type III: microapocrine secretion. Release of the secretory products by pinching off of the expanded tips of cytoplasmic projections, such as microvilli, is designated microapocrine secretion. This mechanism of secretion release is claimed to take place in the choroid plexus, the carpal organ of the pig, and the intrahepatic bile duct. In neurons and paraneurons, microapocrine secretion is unlikely as the release mechanism of the secretory products.

Extrusion type IV: exocytosis. For this type of extrusion, it is essential that the secretory products in a nondiffusible form are isolated by a membrane sac in the secretory granules. The process of exocytosis consists of attachment of the granule membrane to the plasma membrane, fusion of both membranes, extrusion of the contents, and membrane retrieval (see below). This type of secretion release is seen in a large variety of protein-producing exocrine cells (pancreas, salivary gland, etc.), mucus-producing exocrine cells (goblet cells, gastric surface mucous cells, etc.), and peptide/amine-producing endocrine cells, i.e., neurons and paraneurons.

Kurosumi (1961) first termed this type of extrusion of secretory products "opening mechanism." Later, he used the term "eruptocrine secretion." This type of cellular activity was interpreted as a kind of "reversed pinocytosis." This concept is, however, inadequate because the membrane retrieval process cannot be accounted for in this manner. The term "emiocytosis" has also been used by many authors, such as Lacy (1970). However, the most popular term used today to designate this mechanism is exocytosis (DeDuve and Wattiaux 1966).

Extrusion type V: diacrine secretion. This type of secretion release is not associated with any structural changes in the glandular cells detectable by electron microscopy. Substances contained in the cytoplasmic matrix as well as those enveloped in membrane sacs are claimed to be released by the diacrine mechanism. Transfer of the secretory substances thus takes place through the granule membrane and plasma membrane. The secretory products remains either transiently or for some time in the cytoplasmic matrix prior to release.

The mechanism of diacrine secretion occurs at the molecular level. Protein-free lipid bilayers of the plasma membrane are freely permeable to water but impermeable to charged molecules such as metallic ions (Gomperts 1976). There are specific membrane devices such as channel or transport proteins. The extrusion of a charged secretion of a small size may be achieved by these special transmembrane proteins. A previous dispute concerning the diacrine mechanism mainly pertained to substances made up of larger molecules, especially polypeptides. Insulin, gastrin, and some other peptide hormones have been claimed to be released by the diacrine mechanism (Forssmann and Orci 1969). Recent studies, however, have shown electron microscopic evidence for the exocrine release of granules containing these hormones (see p. 173). Also from a theoretical point of view, it would seem most unlikely that secretory products once packed in a sac to be released outside the cytoplasm would escape into the latter. Moreover, some components of the granule contents—such as monoamines—could be harmful to the cell if they leaked into the cytoplasm.

Electron-microscopic observation of glandular tissues frequently reveals a disruption of the plasma membrane. This may often be accompanied by a pale or "empty" appearance of the granules. These images should not be too readily associated with diacrine extrusion of the secretory substances, because membrane disruption and escape of contents in electron micrographs are in many cases due to artifacts during specimen preparation.

A modification of diacrine secretion is the cation exchange mechanism. Uvnäs and Åborg (1980a) demonstrated that the matrix of the chromaffin granules prepared from bovine adrenal glands had properties similar to cation exchanger materials with carboxyls as the cation binding sites. Kinetic analysis of adrenal catecholamine release from the adrenal glands of cats, dogs, and pigs elicited by splanchnic nerve stimulation indicates that catecholamine may be released by a cation exchange between catecholamine$^+$ in the granule matrix and Na^+ in the plasma or possibly K^+ in the cytoplasma. Uvnäs and Åborg (1980a) suggested that cation exchange may be a general principle in the release of biogenic amines, including transmitter amines and other costored charged substances, e.g., polypeptides. However, the experimental basis for this hypothesis is less stable than that for the theory of exocytosis detailed in the following section.

4.4.2 Exocytosis as a Normal and Universal Process

Extrusion of the secretory products of neurons and paraneurons normally takes place by *exocytosis* (Fig. 4.3.) Many electron micrographs showing fusion of the secretory granules with the cell membrane followed by opening of the secretory granules to the extracellular space have been published for a large variety of the cells (Figs. 4.4,4.5). The *stages of the exocytotic chain reaction* of the secretory granule and cell membrane as revealed by electron microscopy may be summarized as follows: (1) Movement of the secretory granules to the periphery of the cell immediately beneath the cell membrane. (2) Fusion of the cytoplasmic half of the granule membrane with the inner half of the plasma membrane, formation of trilamellar or bilamellar adhesion between the two membranes. (3) Disruption of the fused granule and plasma membranes, bringing about a so-called Ω-shaped invagination of the cell surface containing secre-

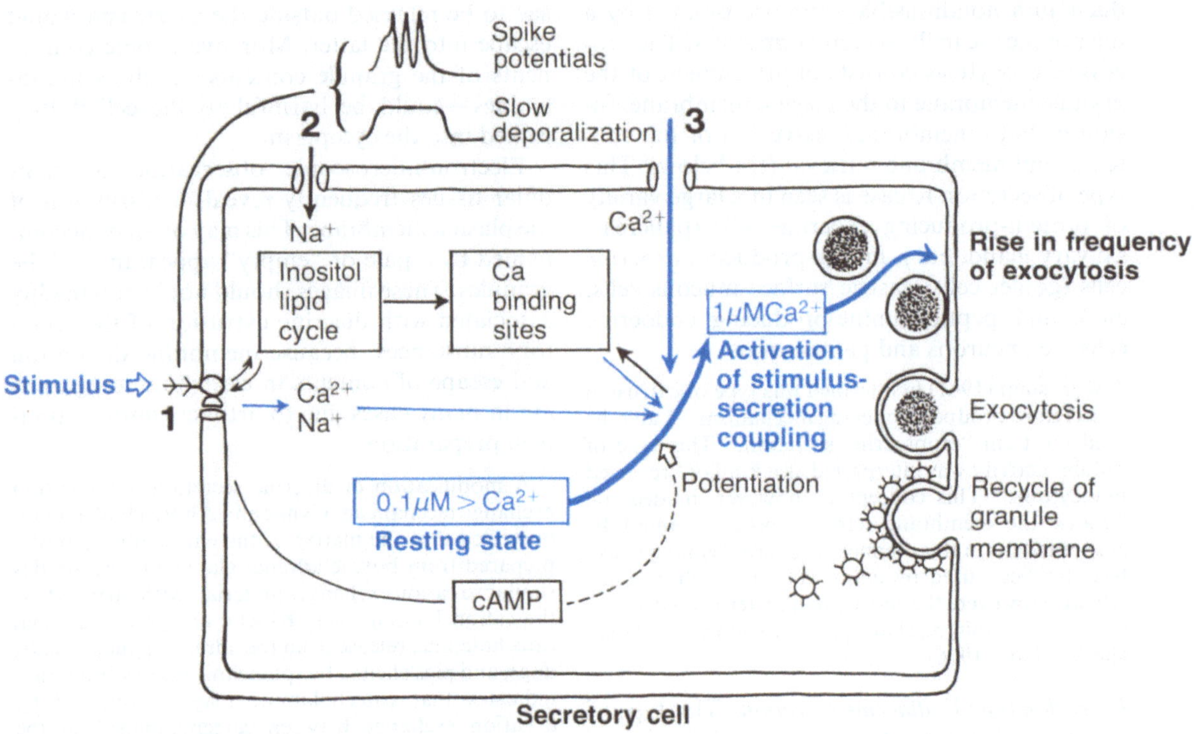

Fig. 4.3. Schematic representation of the main processes of stimulus-secretion coupling of a paraneuron. *1* Complex of receptor and ionophore, *2* tetrodotoxin-sensitive Na^+ channel, *3* voltage-dependent Ca^{2+} channel

tory products. (4) Development of a bristlelike coating on the cytoplasmic side of the invaginated plasma membrane and the incorporation of the granule membrane into the plasma membrane. (5) Small invaginations and pinching off of the coated regions of the plasma membrane; coated pits and vesicles are thus formed as the cell membrane returns to its noninvaginated state (Figs. 4.4, 4.5).

In some kinds of paraneurons when they are intensively stimulated and secretions are massively released, deeper granules open to more superficial ones and granules become connected in tandem. Sometimes, many granule cores are surrounded by a labyrinthine sac, which is caused by fusion of each granule sac (see e.g., Wilbur et al. 1978; Fujita et al. 1983: anterior pituitary cells).

In the tissue specimen examined by electron microscopy, secretory products are more or less coagulated by the fixatives used. It seems a reasonable speculation that different compositions of the granule contents give rise to different appearances of the secretory granules in the electron micrographs. It is observed that the granule contents in the Ω-shaped invaginations on the cell surface are usually looser in texture and lower in electron density than the intracellular granules. This implies that when the secretory products of the secretory granules are extruded into the extracellular space they are gradually dissolved by the tissue fluid and finally vanish.

The concept of the process which is today called exocytosis was first proposed by De Robertis and Vaz Ferreira (1957) in their electron-microscopic study of adrenomedullary cells. The first convicing images of Ω-shaped opening of the secretory granules were presented by Ichikawa (1959) in anterior pituitary cells. Diner (1967) demonstrated the corresponding images in adrenomedullary cells. Later, similar exocytotic images were presented in almost all paraneurons, including pancreatic islet cells (Lacy et al. 1968; Lacy 1970) and gastrointestinal endocrine cells (Fujita and Kobayashi 1971; Osaka et al. 1974, 1975). Many electron microscopists came to notice that exocytotic invaginations were accompanied by a few coated pits and vesicles. In their study on adrenomedullary cells, Nagasawa and Douglas (1972) demonstrated that after injection of collodial thorium dioxide this electron-opaque marker was trapped by the coated pits and vesicles upon reaching the cells from the perivascular space. Similar experiments were also performed in the rat pituitary gland using

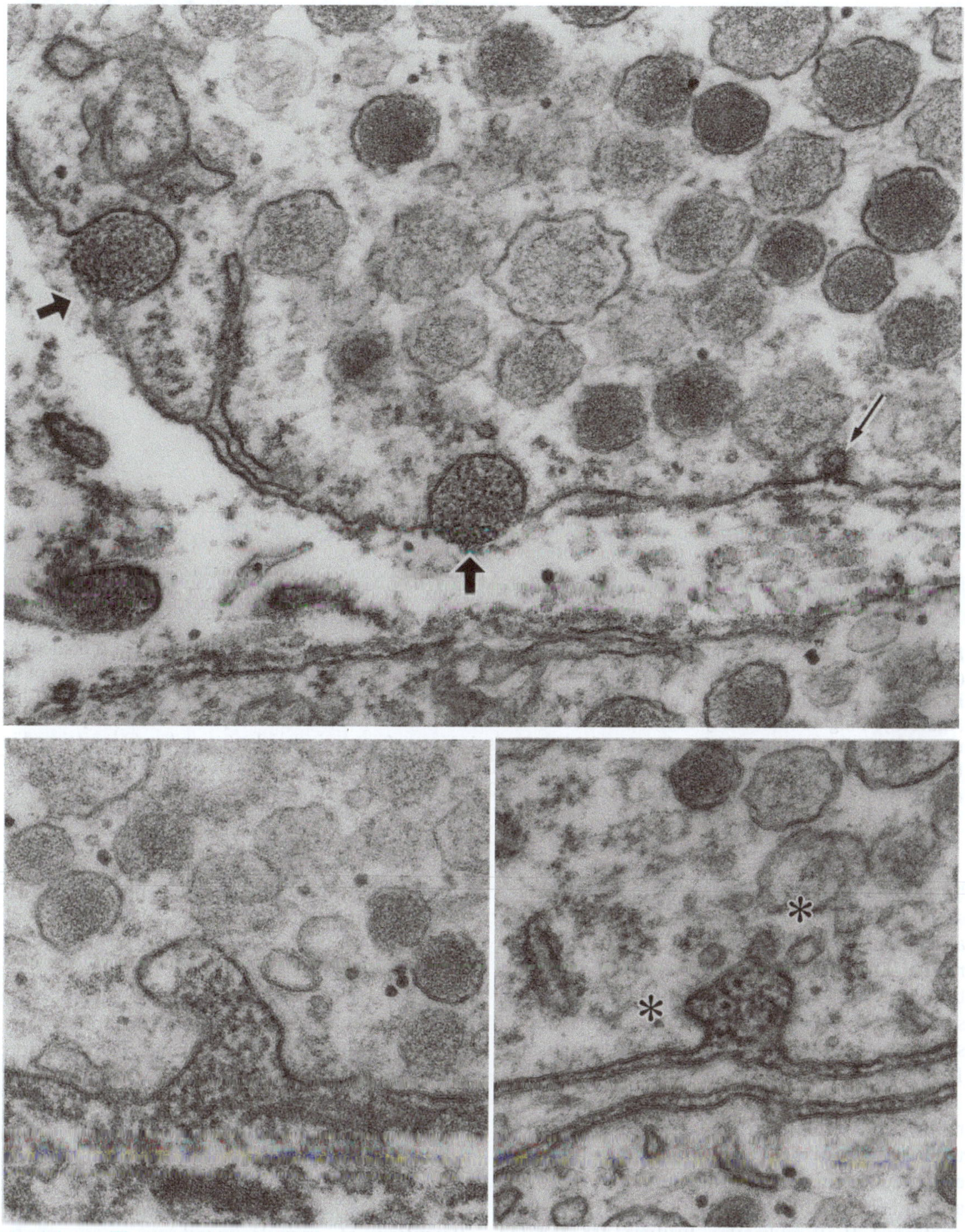

a

b

c

Fig. 4.4a–c. Electron micrographs showing granule release by exocytosis. The various stages of exocytosis were obtained from the rat adrenal medulla. **a** Two Ω-shaped invaginations of the plasma membrane (*thick arrows*) and a coated pit (*thin arrow*) are seen on the surface of the adrenaline-storing cell. Exocytotic granule release seen **b** on the surface of the adrenaline-storing cells facing the tissue and **c** in the intercellular spaces. The cell membrane areas marked by *asterisks* are provided with a clathrin coat. Tannic acid-glutaraldehyde-osmium tetroxide method. Prepared by Y. Endo, Department of Anatomy, Yamanashi Medical College. **a** × 70 000, **b** × 78 000, **c** × 82 000

a 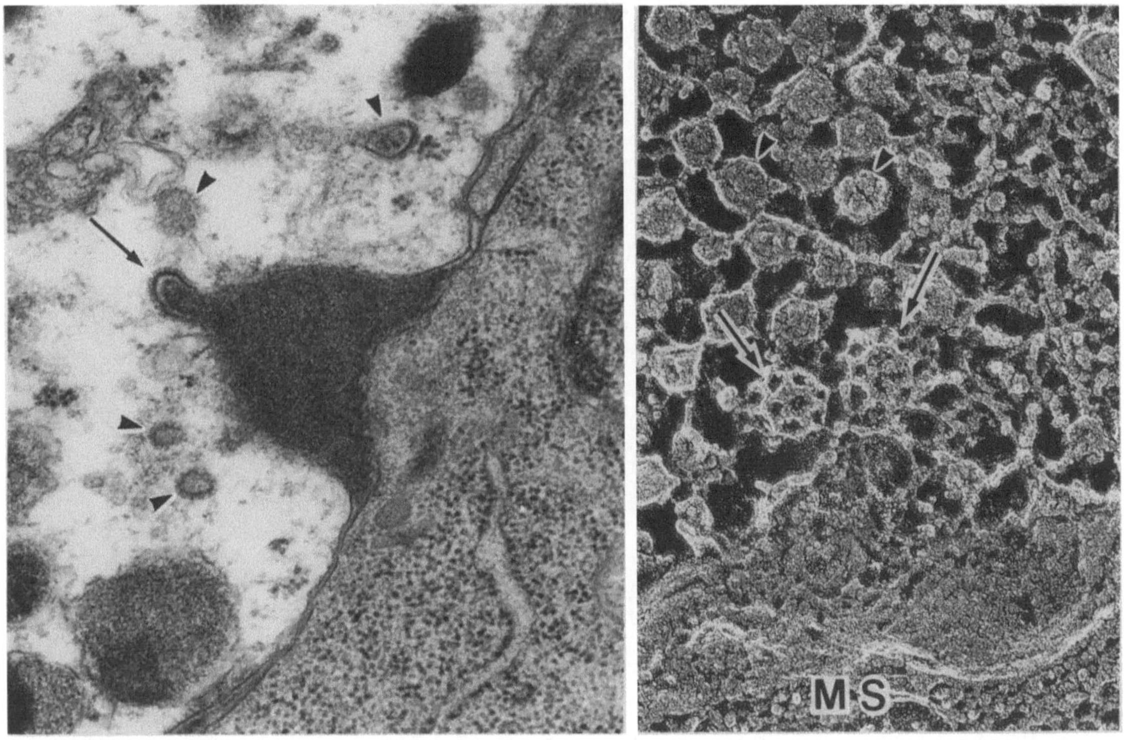 b

Fig. 4.5a, b. Electron micrographs showing the relation between exocytosis and coated vesicles. **a** A coated pit (*arrow*) and several coated vesicles (*arrowheads*) accompanied by an exocytotic figure on the lateral surface of an EC cell of the rabbit duodenum 1 h after cholera toxin infusion. Prepared by M. Osaka, Department of Anatomy, Niigata University Medical School. × 50 000. **b** Coated vesicles (*arrows*) in the presynaptic nerve terminal on the skeletal muscle cell (*MS*). Membrane substance of synaptic vesicles (*arrowheads*) is retrieved in the form of coated vesicles. Freeze-sublimation replica method. Prepared by N. Hirokawa, Department of Anatomy, Tokyo University Medical Faculty. × 80 000

horseradish peroxidase as the marker (Nagasawa et al. 1971). The results obtained were in agreement with the results of previous experiments. Douglas and co-workers proposed that the exocytosis-vesiculation sequence is a universal process of the extrusion of membrane-bound secretory products (Douglas 1974; Nagasawa 1977).

Freeze-fracture studies on neurons and paraneurons have clarified further details of exocytosis and provided additional morphological evidence for this cellular phenomenon. In bovine adrenal chromaffin cells, Aunis et al. (1979) demonstrated that molecular bridges are formed between the granule sac and plasma membrane during a preexocytotic stage. They also showed that intramembranous particles, which probably represent membrane proteins, lack the exocytotic membrane-fusion site. Migration of the plasma membrane-associated particles during exocytosis has also been reported in the pancreatic B cells (Orci et al. 1977), neurosecretory terminals of the pituitary gland (Theodosis et al. 1978), and mast cells (Chi et al. 1976).

4.4.3 Biochemical Evidence for Exocytosis

During exocytosis, the secretion is separated from the cytoplasm and cell membrane. The secretion thus does not contain any cytoplasmic matrix or cell membrane. The following results from biochemical studies of stimulated adrenal chromaffin cells provide evidence for the exocytotic release of the granule contents:

Retention of granule membrane components. Phospholipids and cholesterol, the main lipids of the granule membrane, do not show significant efflux during the stimulation of catecholamine secretion (Trifaró et al. 1967; Schneider et al. 1967). Granule membrane proteins, such as Mg^{2+}-activated nucleotide triphosphatase, are not detectable in the efflux of experimentally stimulated adrenal medulla (Winkler et al. 1970).

Retention of the cytoplasmic constituents. When the secretion of catecholamines in the bovine adrenal medulla is stimulated, there is no detectable release of enzymes such as lactate dehydrogenase (Schneider et al. 1967), phenylethanolamine-*N*-methyltransferase (Kirschner et al. 1967), and tyrosine hydroxylase (Viveros et al. 1968) in the cytosol of the chromaffin cells.

Corelease of catecholamines with other granule contents. The isolated secretory granules of the bovine adrenal chromaffin cells contain catecholamines, ATP and chromogranins in the molar ratio 4:1:0.02. After stimulation of secretory activity, these substances are evident in the perfusates of the adrenal medulla in an essentially identical proportion to that found in isolated secretory granules (Banks and Helle 1965; Kirschner et al. 1966; Schneider et al. 1967). The cosecretion of proteinic secretory products with ATP has also been detected in the pancreatic islets (Leitner et al. 1975).

Bovine adrenal chromaffin cells are the most intensively investigated paraneurons from the point of view of extrusion mechanisms of the secretory products. Results from the biochemical studies mentioned above support the view that the secretory substances are extruded by exocytosis. Biochemical studies of exocytosis in paraneurons other than the adrenal chromaffin cells are few. However, it can be argued that what is true for adrenal chromaffin cells may also apply to other paraneurons.

4.4.4 Implications of Exocytosis

In exocytosis, *selective release* of the secretory granule contents occurs—the secretory products are exclusively extruded. Other modes of extrusion of secretory substances cause nonselective release of cellular constituents; in other words, cytoplasmic and membranous elements are more or less discarded.

In all neurons and paraneurons there is a "secretory machinery," consisting of the endoplasmic reticulum, Golgi apparatus, and secretory granules. Although the secretory machinery is localized within the cell, the secretory substance itself is transferred to the extracellular space. It is reasonable to believe that the secretory substances are always segregated outside the cytoplasmic matrix.

Exocytosis has the merit that the cell can produce secretory substances without directly disturbing the homeostasis of the cytoplasmic matrix. Therefore, exocytosis seems to be the most harmless mechanism for the cell to secrete products. Another advantage of exocytosis is the formation of *storage complexes* of the secretory products. These storage complexes occur in the secretory granules in a loosely bound conglomerate. In adrenal chromaffin cells, catecholamines and ATP form a storage complex with granule proteins such as chromogranins, glycoseaminoglycans, and divalent cations such as Ca^{2+} and Mg^{2+} (see p. 65). Formation of the storage complex may be necessary to lower the osmolarity of the secretory granule interior. Thus, an enormous concentration of catecholamines can be maintained. Exocytosis may be considered as cellular mechanism for extruding large amounts of secretory products in a pelletlike mass. It is then possible in exocytosis that the secretions are *released as quanta*, i.e., one pellet after another, each pellet containing a fairly constant amounts and ratio of secretory substances. The rate of release of the quanta can be controlled by the intensity of stimulation to the cell.

Different lines of evidence point to the synaptic vesicle as the morphological correlate of the electrophysiologically detected quantum. The *vesicle hypothesis* makes specific quantitative and qualitative predictions, most of which have been fulfilled. The kinetics and incidence of exocytosis match the kinetics and incidence of quantum release. While minor difficulties still remain, the vesicular hypothesis is strikingly successful in explaining the morphological, electrophysiological, and biochemical data on neurotransmitter release (Kelly and Hooper 1982).

In the adrenal medulla, acetylcholine, the physiological transmitter released from the splanchnic nerve terminals, induces exocytosis of the chromaffin cells. Only under pathological conditions have the chromaffin cells been reported to release their granule contents by other modes. For example, exposure of isolated perfused cow cortex-free adrenal medullae or cow whole adrenal glands to acetaldehyde in concentrations of 1 mmol/l or greater caused a dose-related secretion of catecholamines (Schneider 1971). Catecholamine secretion induced by acetaldehyde was not accompanied by release of chromaffin granule protein or by other cell constituents such as lactate dehydrogenase or lysosomal acid deoxyribonuclease. Furthermore, reserpine has been claimed to induce the release of catecholamines but not of other granule constituents (Viveros et al. 1968). Hillarp (1960) discussed the existence of an extragranular store of catecholamines. He suggested that at least 5%–10% of catecholamines are stored in the cytoplasmic matrix. Therefore, it is possible that reserpine may cause diacrine release of the extragranular store of catecholamines without affecting the fraction of catecholamines in the secretory granules. However, little is known with certainty concerning the physiological significance of the diacrine secretion of catecholamines in the adrenal medulla.

4.4.5 Membrane Retrieval Mechanisms

In the process of exocytosis, after a secretory granule fuses with the plasma membrane, the granule contents are released into the extracellular space; the granule membrane becomes part of the plasma membrane. In earlier electron-microscopic studies, it was pointed out that the surface area of the plasma membrane must increase as exocytotic granule release repeatedly occurs (Bennett 1956; Palade 1959). It was an enigma why the cell membrane did not become excessively large. This question was answered by Palade (1982) and Farquhar (1982) who proposed that the components are removed from the plasma membrane at a rate almost equal to that of their incorporation by fusion.

Electron-microscopic studies have demonstrated that the Ω-shaped invaginations which appear during the process of exocytosis in paraneurons, such as adrenal chromaffin cells (Grynszpan-Winograd 1975; Kobayashi and Serizawa 1979), thyroid parafollicular cells (Nunez and Gershon 1976), and gastroenteric endocrine cells (Fujita and Kobayashi 1973; Osaka et al. 1974, 1975), are totally or partly provided with a *bristlelike coating*. Moreover, many coated pits and vesicles have been noticed on or around the exocytotic sites (Figs. 4.4, 4.5).

Kanaseki and Kadota (1969) revealed that the coating is not made up of hairlike projections but of a spherical "basket-work" which encircles the membrane pits or vesicles. They supported the idea proposed by Roth and Porter (1964) that the coating causes the infolding and fission of the plasma membrane. By this fission, the granule membrane components remain behind in the cell during exocytosis. Pearse (1976) isolated a unique protein called *clathrin* from the coated vesicle fraction. This 180 000-dalton protein exists as a trimer. The skeleton of the pentagonal, hexagonal, or heptagonal lattice of the basketwork is mainly composed of polymerized trimers of clathrin.

The clathrin basketwork dissociates from the coated vesicles in the cytoplasm and releases the central vesicle to fuse with a tubulovesicular structure in the cytoplasm. We visualized this cytoplasmic membranous structure by a zinc iodide-osmium tetroxide impregnation method (Uchida and Kobayashi, unpublished data). This structure forms a loose network throughout the cytoplasm and is continuous with the Golgi apparatus. The plasma membrane components retrieved as coated vesicles after exocytosis are believed to be incorporated into this tubulovesicular structure; they then return to the Golgi area by the membrane flow and are reutilized as the material for new secretory granule membranes. Thus, recycling of the granule membrane component is achieved.

4.5 Summary

The cell membrane of paraneurons is generally differentiated into three regions—receptive, conductile, and secretory.

On the receptive region, there are *receptors*, which are generally proteinic molecules located in or on the plasma membrane.

Binding of the receptor molecules with the stimulus (the ligand molecules) may evoke changes in the *membrane potential* (usually depolarization) by altering *ionic channels* of the target cell. Slow depolarization even generates *action potentials* in adrenal, pituitary, pancreatic, and some other paraneurons. Increase in cytoplasmic concentration of Ca^{2+} is the ordinary change in paraneurons caused by stimulation, and this change seems to suffice to induce exocytosis of the secretory granules. Inositol metabolites and cyclic AMP may be activated by stimulation of paraneurons. These second messengers serve to modulate the cytoplasmic concentration of Ca^{2+}.

In *photoreception*, a specific change in rhodopsin molecules induced by light causes closure of Na^{+} channels, leading to hyperpolarization of the visual cell. Mechanical stimuli also seem to alter the membrane potential of mechanoreceptor cells via unknown receptive mechanisms.

Release of secretory granules in paraneurons generally occurs by exocytosis, or by an Ω-shaped opening of the granule sac to the plasma membrane in the secretory region of the cell. By exocytosis, substances can be released *in quanta* (as pellets), and the rate of their release can be controlled by the intensity of stimulation to the cell. After exocytosis, the increased components of the plasma membrane are reutilized as coated vesicles.

The mechanisms of stimulus recognition, stimulus-secretion coupling, and secretion release are common to neurons and paraneurons. Moreover, neurons and paraneurons may share indentical or similar receptors for identical or similar stimuli (e.g., glucoreceptors).

5. Neurons As Secretory Cells

After dealing with the secretory aspect of paraneurons, it is now appropriate to shed light on the corresponding features of neurons. The electric conductivity of neurons has been investigated so extensively that many researchers tend to conceive it as the indispensable activity of neurons. It is, however, becoming more and more evident that the Na^+ spike commonly studied by electrophysiologists is generated only by a part of various functions of neurons (see p. 3). On the other hand, it is safe to say that *essentially every neuron is secretory in nature*. As the secretory nature of neurons is manifested in their possession of messenger-containing granules and vesicles, this chapter will commence with an introduction to the difference in their secretory granules.

5.1 Secretory Granules and Vesicles of Neurons

5.1.1 Peptidergic-Type Granules or Large Cored Vesicles

The granules of this type consist of relatively electron-dense and homogeneous-looking contents bounded by a smooth membrane sac usually of spherical shape, ranging in diameter from 80 to 300 nm. The size, fine structure, and electron density of the granules differ only within a limited extent according to the peptides and associated substances contained as well as to neuron type and animal species.

The term "peptid*ergic*" has traditionally pertained to peptide-containing granules for *synaptic transmission*. Therefore, the term "*peptidergic type*" is used in this text to signify granules released not only by synaptic but also by "neurosecretory" (in the classic sense) or hemocrine mechanism.

The typical peptidergic type granules are contained in the *classic neurosecretory neurons* comprising the hypothalamohypophyseal system of vertebrates secreting oxytocin, vasopressin, as well as several other messenger peptides recently added (see below). Essentially identical granules are found in the caudal hypophysis or urophysis (Enami and Imai 1958; Sano and Knoop 1959) in teleosts secreting urotensins (Lederis et al. 1982). Furthermore, large varieties of neurosecretory neurons in invertebrates contain numerous granules of the peptidergic type.

Early electron-microscopic studies of the hypothalamohypophyseal system showed that the neurosecretory neurons and their processes, especially their terminal varicosities, were densely filled with round granules, 100–300 nm diameter (Green and Van Breemen 1955; Bargmann and Knoop 1957; Palay 1957). These granules, called "elementary granules" by Bargmann and Knoop (1957), were soon isolated for chemical analysis and found to contain neurohypophyseal hormones, oxytocin and vasopressin (Weinstein et al. 1961; Barer et al. 1963). In their study on the teleostean urophysis, Sano and Knoop (1959) demonstrated that the elementary granules were derived from the Golgi apparatus. Later studies have elucidated that the neurosecretory substances are produced in the rough endoplasmic reticulum-Golgi apparatus system and released by exocytosis via the perivascular space into blood circulation (Bargmann 1966).

Recent studies have revealed that hypothalamic neurosecretory neurons project their fibers not only to the neurohypophysis (neurohemal organ) but also to various regions of the brain and spinal cord for synaptic transmission to neurons (Sofroniew 1980; Sterba et al. 1980; Yamashita et al. 1984; Buijs 1985). This is a good example of the same granules being partly used as a neurosecretion (hormone) and partly as a transmitter ("hormonal transmissional partition" Fujita et al. 1981) (Fig. 5.1). One such extrahypothalamic projection was found by Bargmann et al. (1967) and urged them to propose the term "peptidergic," as reviewed in Chap. 1 (p. 5).

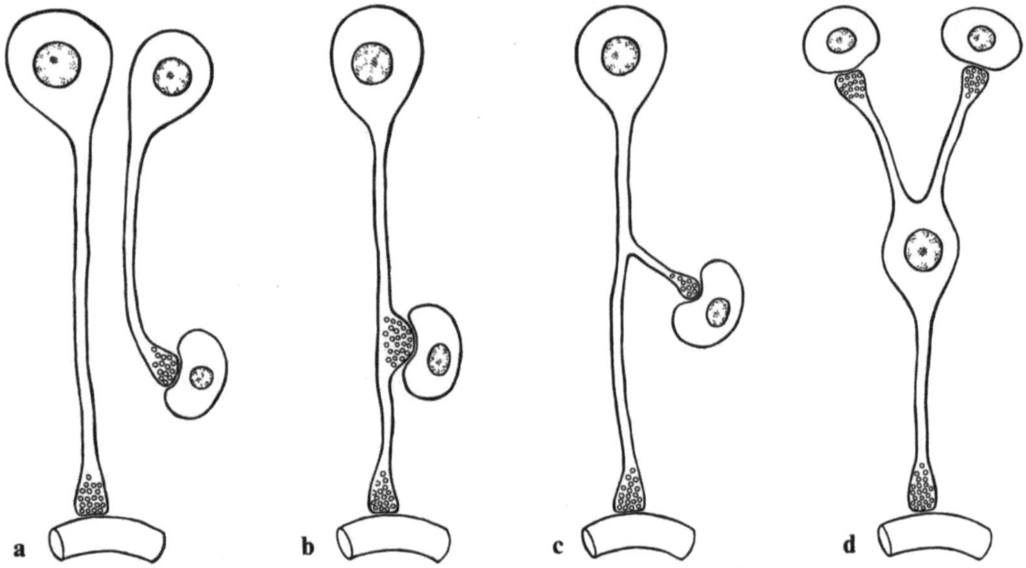

Fig. 5.1a–d. Concept of transmissional/hormonal partition. **a** Synaptocrine and hemocrine secretion by different neurons from the same identical nucleus. **b–d** Synaptocrine and hemocrine secretion by identical or different processes of a neuron

Outside the classic neurosecretory system mentioned above, granules of the same morphology as the elementary granules have been seen to occur in larger or smaller amounts in almost all varieties of neurons and have been usually called "large cored vesicles" (Fig. 5.2) In vertebrates, they tend to be densely contained in the preterminal and terminal varicosities of autonomic neurons. In invertebrates they are distributed more abundantly and extensively in the nervous system (e.g., Friend 1976).

Hager and Tafuri (1959) first demonstrated the occurrence of numerous granules in the nerves of the guinea pig myenteric plexus (Auerbach) and pointed to their resemblance to the elementary granules of the hypothalamic neurosecretory cells. However, this viewpoint was not held by later authors who were preoccupied with the idea that the (classic) neurosecretory cells were very special, atypical neurons not comparable with other systems. Thus, the large cored vesicles were claimed to be the site of monoamine storage by some researchers (Cottrell and Osborne 1970; Coggeshall 1972; Friend 1976) or the site of ATP storage by Burnstock (1972) who therefore called the granules "purinergic" (see p. 5). As detailed later (p. 63), these views on the occurrence of amines and adenine nucleotides in the granules were not erroneous given the level of the existing knowledge, but it was noteworthy that the more important constituent of the granule contents, peptide, was paid no attention.

The possibility of peptide messengers contained in the large cored vesicles was first noted by Baumgarten et al. (1970) in their study on intestinal autonomic nerves; they called these nerves and granules "polypeptide type" or "P type." Recent immunocytochemical studies especially at the electron-microscopic level have demonstrated a large variety of bioactive peptides in the large cored vesicles of different neurons (VIP—Larsson 1977b; substance P—Pickel et al. 1977b; Cuello et al. 1981). The hypothalamic neurosecretory system, on the other hand, has been known to contain, besides classic oxytocin and vasopressin, such "brain-gut peptides" as substance P (Hökfelt et al. 1975; Ljungdahl et al. 1978), VIP (Fuxe et al. 1977), the gastrin-cholecystokinin family of peptides (Vanderhaeghen et al. 1981; Yui 1983), and opioids (Sar et al. 1978). All these findings provide us with the morphological and chemical basis on which the "elementary granules" in the classic neurosecretory neurons and "large-cored granules" are identical in nature and must be investigated by the same approach (review: Scharrer 1978).

In the preceding paragraphs, we were concerned with the classic neurosecretory and some autonomic nerves where the large-cored vesicles or granules are concentrated. However, these granules are dispersed, though usually in small numbers, in essentially every type of neurons intermingled with smaller synaptic vesicles to be dealt with later.

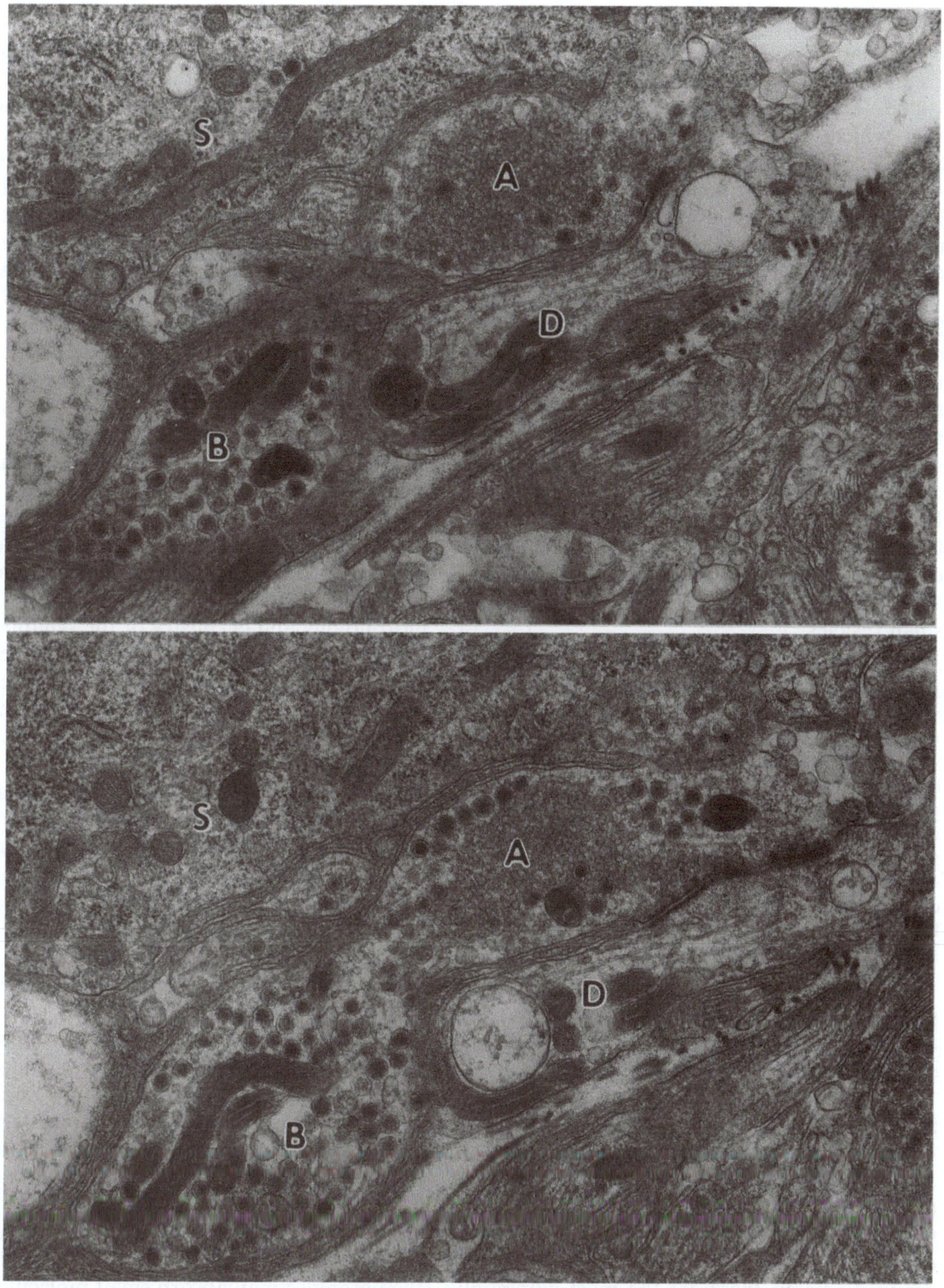

a

b

Fig. 5.2a, b. Fine structure of nerve varicosities in the guinea pig celiac ganglion as seen in two serial sections. The varicosity *A* in **a** appears to be a typical cholinergic nerve, being filled with small clear vesicles, whereas **B** appears to be a typical peptidergic type, containing numerous large cored vesicles. From **b**, however, it is evident that the two varicosities are in fact continuous with one another. This example indicates that a single nerve fiber may appear with different features depending on the section. Note that the varicosity *A* contains a few large cored vesicles and the varicosity *B* also has a few small clear vesicles. A soma (*S*) and dendrite (*D*) of postganglionic neurons are seen. Courtesy of H. Kondo, Department of Anatomy, Kanazawa University Medical School. × 32 000

With respect to "typical neurons" textbooks of neuroanatomy use the examples of Golgi type I and type II neurons in the central nervous system. The Golgi type I neurons possess a long axon which terminates either in a remote site in the brain and spinal cord or in various organs such as the skeletal muscles and exocrine glands. Large cored vesicles are found in this type of neuron, including the pyramidal neurons of the cerebral cortex (Peters and Kaiserman-Abramof 1970), giant cells of Deiters of the lateral vestibular nucleus (Sotelo and Palay 1968), and Purkinje cells of the cerebellar cortex (Herndon, 1963). On the other hand, Golgi type II neurons are characterized by short axons which break up into elaborate terminal arborizations in the neighboring gray matter. The neurons of this type, e.g., the bipolar neurons in the visual cortex (Peters and Kimerer 1981), the aspiny neurons of the striatum (Fox et al. 1971, 1972), and the stellate cells of the cerebellar cortex (Chan-Palay and Palay 1972) also contain large cored vesicles together with small vesicles.

In the above-mentioned neurons, a small number of large cored vesicles occur in the company of the small vesicles. For an unknown reason, the latter vesicles are usually condensed close to the synaptic junction, whereas the former tend to be dispersed a certain distance away from it. For example, electron micrographs of the nerve terminals of the spinal motoneurons on the skeletal muscle fibers of the frog sartorius muscle exhibit, in addition to a large number of spherical synaptic vesicles, a small number of large cored vesicles which can be found at a certain distance from the synaptic membrane (see electron micrographs by Heuser and Rees 1973).

The coexistence of the large cored and small synaptic vesicles indicates that the neurons mentioned above are peptidergic in addition to being cholinergic or aminergic (see p. 65).

5.1.2 Small Vesicles, Cored and Noncored

Attempts have been made to correlate the shape of the synaptic vesicles with the kind of transmitter substance. The first type comprises *spherical vesicles with no visible core*, measuring 40–60 nm in diameter; these are the most commonly occurring synaptic vesicles and are present in the axon terminals of the neuromuscular junction (Padykula and Gauthier 1970), in the preganglionic sympathetic nerve terminals (Coupland 1965c; Grynszpan-Winograd 1975), and in many terminals throughout the central nervous system. These synaptic vesicles contain *acetylcholine* (De Robertis 1964; Whittaker 1965). The second type of synaptic vesicle consists of flattened vesicles in the form of discs, 40–60 nm in diameter and about 30 nm in thickness. In the cerebellar cortex of the cat, Uchizono (1968) noticed that the *flattened vesicles* are specific to the inhibitory synapses, whereas the spherical vesicles occur in the excitatory nerve terminals. Although the contents of the flattened vesicles are not as yet known with absolute certainty, there is strong circumstantial evidence which indicates that they contain GABA as a transmitter substance (Uchizono 1968, 1975). The third type comprises small synaptic vesicles 40–60 nm in diameter *with an electron-dense core*. These vesicles occur at the sites where *adrenergic* nerve terminals are known to be present (Grillo 1966; Wolfe et al. 1962). They are believed to contain catecholamines such as noradrenaline and dopamine.

The spherical noncored vesicles containing acetylcholine, the flattened noncored vesicles containing GABA, and the small cored vesicles containing catecholamines are examples where morphology and chemistry of the synaptic vesicles has successfully been correlated. These examples are rather exceptional, and there are many small vesicles whose transmitter substances are still unknown. Little is known at present about the vesicular localization of identified transmitter substances in the nerve terminals. The morphology of the synaptic vesicles is only one indicator of their physicochemical situation. There are too many kinds of transmitter substances to correlate them with specific synaptic vesicles.

5.1.3 Formation, Maturation, and Release

In a neuron, the perikaryon and the axonal terminal are connected by an axon which does not contain rough endoplasmic reticulum and hence cannot synthesize peptide hormones. It is assumed that newly synthesized peptide hormones are packed into the peptidergic-type granules in the Golgi area of the perikaryon to be transported to the axonal terminal with the axonal flow as illustrated in Fig. 5.3. The peptidergic-type granules may absorb small molecules such as Ca^{2+}, ATP, and catecholamines and become mature on the way from the Golgi area to the nerve terminal; they are stored in the terminal until, responding to adequate stimuli on the receptor sites of the neuron, they discharge their contents by exocytosis (Fig. 5.4).

The origin of the (small) synaptic vesicles in the nerve terminals has been discussed by many

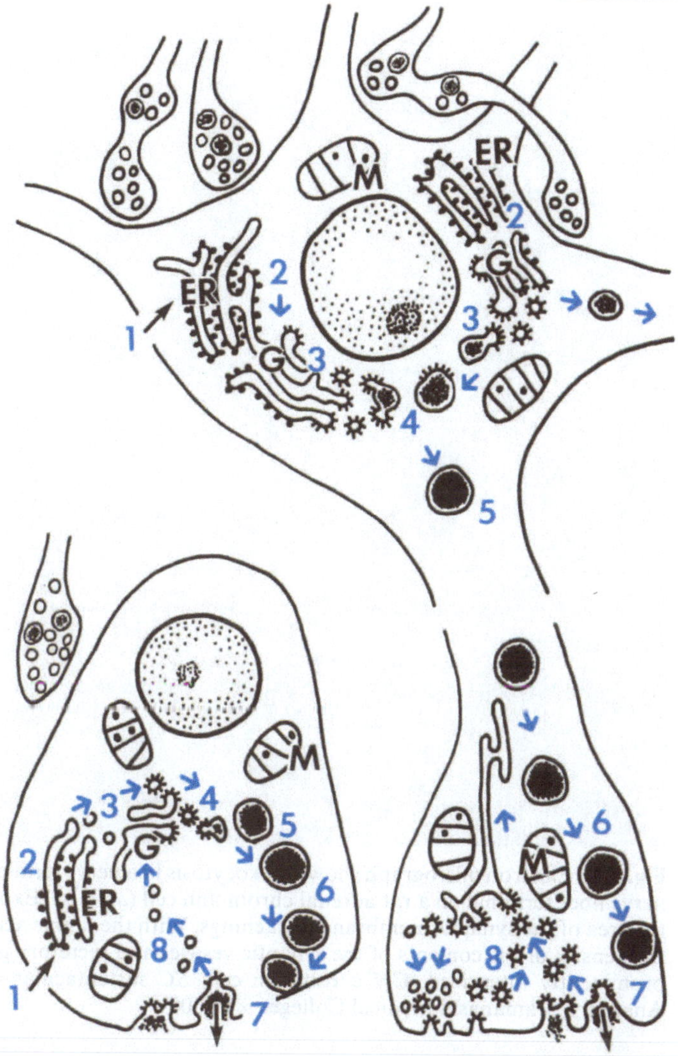

Fig. 5.3. Granule dynamics in neurons and paraneurons. The secretory activity of neurons and paraneurons takes place in the following order: *1* Uptake of material from extracellular space, *2* protein synthesis on ribosomes and subsequent segregation into cisterns of granular endoplasmic reticulum, *3* intracellular transport of proteinic material to the Golgi complex, *4* formation of new secretory granules, *5* maturation of the new secretory granules by incorporation of substances such as catecholamines, ATP, and Ca^{2+}, *6* storage of mature secretory granules, *7* exocytosis, and *8* membrane vesiculation and synaptic vesicle formation. *ER* granular endoplasmic reticulum, *G* Golgi apparatus, *M* mitochondria

authors (Pfenninger 1973; Jones 1979; Osborne 1977). It is supposed that synaptic vesicles originate in the nerve terminals by the pinching off of the coated vesicles associated with a kind of smooth endoplasmic reticulum and the subsequent loss of the clathrin coat (see p. 54).

In the nerve terminal, a *recapture mechanism* of released messenger substances (transmitters) plays the most important role in the inactivation of the transmitters in the synaptic cleft (Iversen 1977). Acetylcholine, ATP, catecholamines, and serotonin are recaptured at various rates after the exocytotic release. Incorporated transmitter substances and the metabolites of former synaptic vesicle contents are repeatedly utilized after resynthesis and repacking into synaptic vesicles (Osborne 1977).

We may thus visualize a mechanism of "*transmitter-precursor uptake and resynthesis*" (Kobayashi 1979), which may be applicable to a large variety of neurons. In the case of catecholamines and serotonin, this mechanism corre-

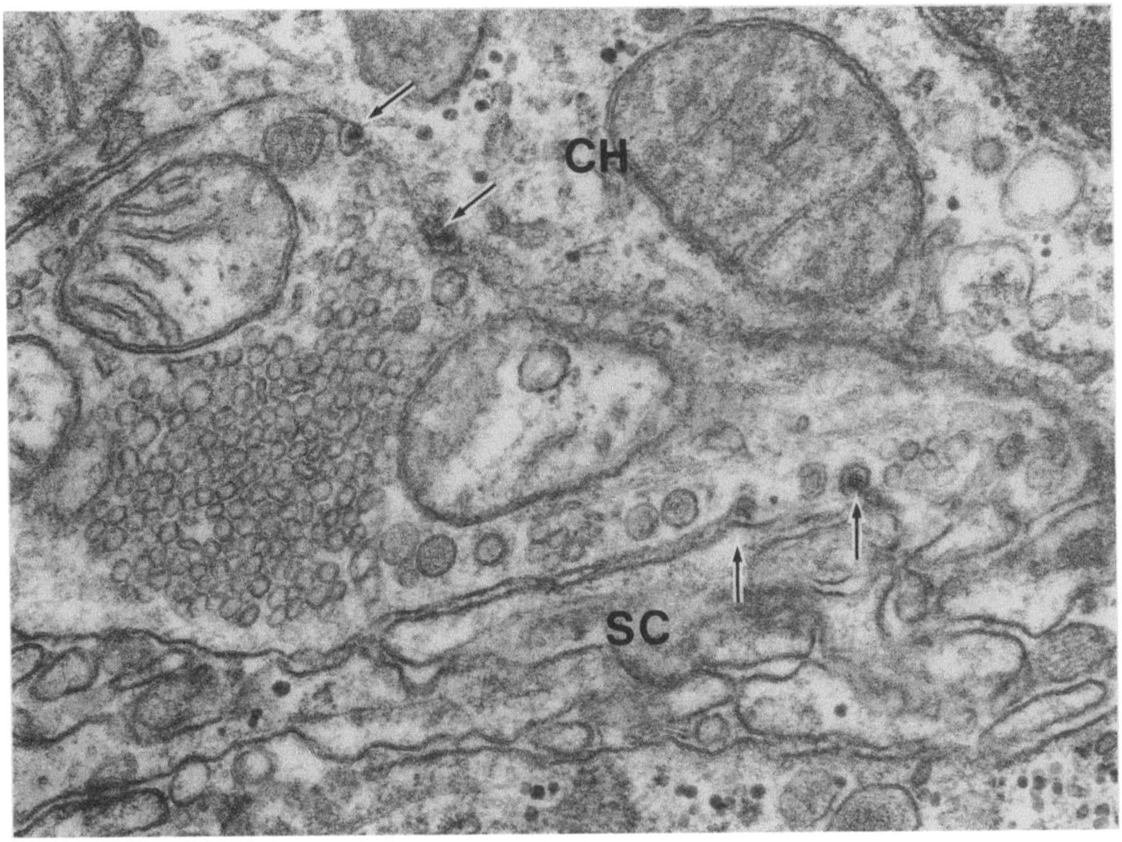

Fig. 5.4. Electron micrograph showing exocytosis in a nerve terminal. Exocytotic figures seen in a preganglionic nerve fiber terminal on a rat adrenal chromaffin cell (*arrows*). Exocytosis in nerve terminals is not restricted to the area of the synaptic membrane thickenings. With the tannic acid-glutaraldehyde-osmium tetroxide method, the density of the contents of the synaptic vesicles and secretory granules, which are undergoing exocytosis, is prominently intensified. *CH* chromaffin cell, *SC* sustentacular cell. Prepared by Y. Endo, Department of Anatomy, Yamanashi Medical College. ×65 000

sponds to the process of "amine-precursor uptake and decarboxylation" or "APUD" coined by Pearse (1966a, b, 1968, 1969). In the case of acetylcholine and ATP, these messenger substances are taken up again by neuronal terminals such as choline or adenosine, respectively, and resynthesized into acetylcholine or ATP to be reutilized as messengers. It is known that some paraneurons share the amine reutilization (APUD) system with neurons. We anticipate that the reutilization mechanisms for other messenger substances like acetylcholine and ATP might also be shared by some paraneurons, though convincing examples have not yet been demonstrated.

Both synaptic vesicles and the peptidergic-type granules of neurons are dynamic structures. They are continually formed, release their contents, and disappear. Following exocytosis, membrane substances are mostly recovered by the cytoplasm in the form of coated vesicles which subsequently move toward a specific smooth endoplasmic reticulum to fuse with it. Simultaneously, new synaptic vesicles are being formed from the smooth endoplasmic reticulum.

In the nerve terminals, each component of the contents of the granular structures shows a different speed in the recapture process (Chubb 1977). Larger molecules such as peptide messengers cannot be reabsorbed, since no effective reuptake mechanism for them seems to be available. Smaller molecules such as acetylcholine and catecholamines are quickly recaptured. Acetylcholine and ATP are metabolized into choline and adenosine, respectively, in the synaptic cleft, but most of the catecholamines and Ca^{2+} remains unchanged (Chubb 1977). Thus, once released, the bulk of the larger molecules in the peptidergic-type granules diffuse out into the tissue space, whereas smaller molecules are mostly recaptured in the cytoplasm and stored in the synaptic vesicles until the next discharge. The contrast between the distribution of the peptidergic-type granules being dispersed away from the synaptic membrane and that of the small vesicles densely gathered to it seems to be accounted for by the above-mentioned difference in metabolism between larger and smaller messenger molecules.

5.2 Contents of Granules and Vesicles

As in the case of paraneurons, neurons are believed to produce a set of different kinds of substances (review: Fujita 1983a, b; Fujita et al. 1984). These include (1) peptides (and their precursors), (2) amines or choline, (3) ATP and other adenine nucleotides, (4) chromogranins and other glycoproteins, (5) protein-processing and amine-converting enzymes, (6) presumably, acid polysaccharides, (7) lipids, and (8) calcium and other cationic ions. The peptidergic-type granules are believed to contain all of these substances, whereas small synaptic vesicles seem to lack the substances (1, 4–7).

5.2.1 Peptides and Their Precursors

Over 20 peptides have been identified as the messengers or their candidates of neurons. Substance P, VIP, NPY, GRP, CGRP, somatostatin, and hypothalamic releasing hormones as well as the classic oxytosin and vasopressin are some of the most important representatives (reviews: Hökfelt et al. 1984; Krisch 1986). These peptides are stored in the secretory granules of neurons, which may be called large cored vesicles, peptidergic-type granules or, in some neurons, neurosecretory (or elementary) granules (p. 55). This has been evidenced by many immunocytochemical studies at the electron-microscopic level (VIP—Larsson 1977b; substance P—Pickel et al. 1977b; Cuello et al. 1981; amphibian gastrin neurons—Doerr-Schott and Jean-Claude 1981). A recent protein A-gold labeling technique has more clearly localized immunoreactive neuropeptides in the large cored vesicles (Kobayashi and Uchida 1983; Uchida et al. 1985) (Fig. 5.5).

The same mechanism as that in paraneurons is presumed to occur in the neurons with regard to the processing of preprohormones into messenger and inactive peptide fragments. This view is supported by immunocytochemical studies using antisera against amino acid sequences which belong to the precursor molecules of a messenger peptide (i.e., somatostatin 28—Iwanaga et al. 1983; preproenkephalin A—Uchida et al. (1985).

As in paraneurons, many neurons have been claimed to produce two or more messenger peptides, which are derived from either one or multiple prepropeptides. This problem will be dealt with later (p. 66).

No positive data are available concerning the occurrence of messenger peptides in small vesicles which are known to be the sites of storage of the classic transmitters (amines or choline) and adenine nucleotides (see p. 63).

5.2.2 Amines, Acetylcholine, and Other Messenger Substances

Noradrenaline, adrenaline, and dopamine as catecholamines and serotonin as indolamine are used by neurons as neurotransmitters. The position of histamine as a neurotransmitter is undetermined.

Noradrenaline in adrenergic nerves is believed to be contained in the small cored vesicles (diameter—50 nm) on the basis of circumstantial evidence (Richardson 1964), false-transmitter experiments (Richards and Tranzer 1970), and chemical analysis of vesicle fractions isolated from adrenergic nerves (Euler et al. 1963; Stjärne and Lishajko 1966; de Potter et al. 1970). A recent immunocytochemical technique using an antiserum against noradrenaline indicates the localization of this amine in the small cored vesicles of adrenergic nerves (Kobayashi and Uchida 1985, unpublished). Biochemical analysis of granule fractions (Lagercrantz and Stjärne 1974) and ultrastructural cytochemical studies (Tranzer and Richards 1976) suggest that noradrenaline is also stored in sympathetic large cored vesicles (100 nm), but there is a lack of data to support a generalization of this finding (see also Castel et al. 1984).

Dopamine is believed to be stored in certain synaptic vesicles, which are still difficult to identify. Neurons which contain granules and vesicles ranging from 40 to 50 nm in diameter have been claimed to be dopaminergic (rat—Pickel et al. 1976).

Serotonin has been demonstrated by FIF histochemistry and, more specifically, by immunocytochemistry in numerous neurons in, for example, the myenteric plexus in the colon (Gershon et al. 1965; Nada and Toyohara 1987), mammalian brainstem (Takeuchi et al. 1982), and submam-

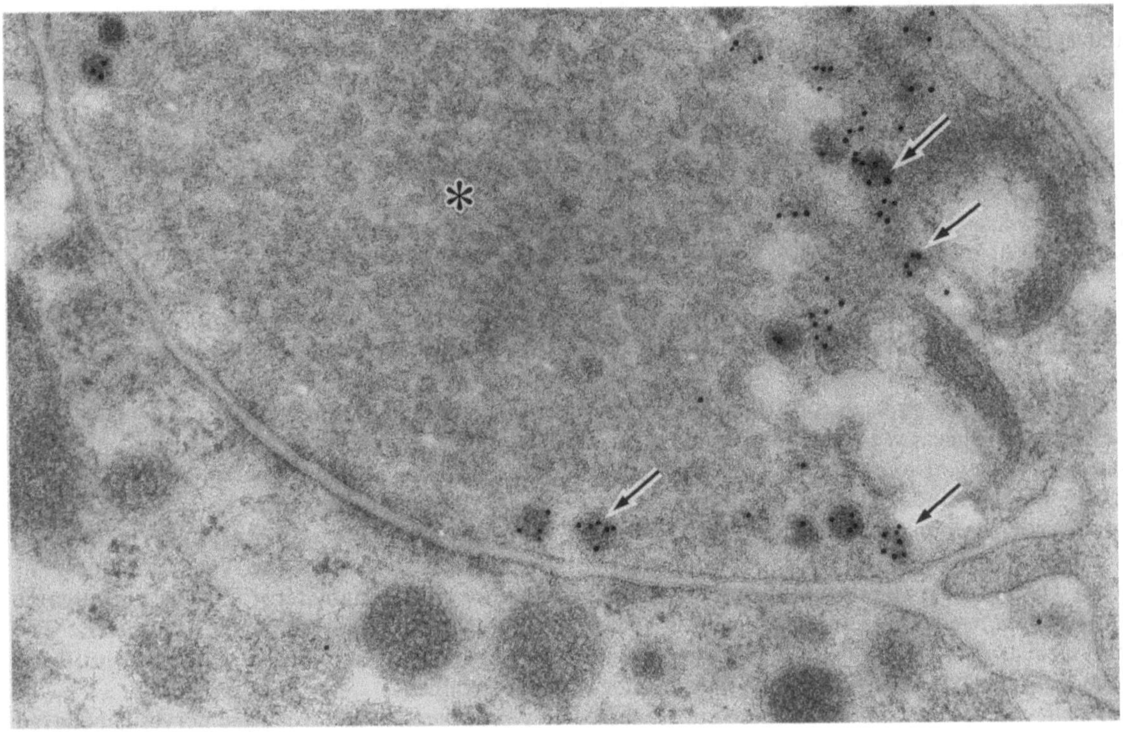

Fig. 5.5. Localization of the immunoreactivity for Met-enkephalin-Arg-Gly-Leu in a nerve terminal. Protein A gold particles are seen on the large cored vesicles (*arrows*), and not on the small clear vesicles (*asterisk*) filling the major part of the terminal. Adrenal medulla of the guinea pig. Prepared by T. Uchida, Department of Anatomy, Yamanashi Medical College. × 57 000

malian hypothalamus and paraventricular organ (Sano et al. 1983; Shimizu et al. 1983).

The electron-microscopic localization of serotonin seems unclear as yet. Using the immunocytochemical test for tryptophan hydroxylase in the rat brain, Pickel et al. (1977a) suggested that serotonin-containing neurons possess both large cored vesicles (60–80 nm in diameter) and small clear vesicles (40–60 nm). In the cerebral ganglion of the mollusc *Aplysia*, Shkolnik and Schwarz (1980) suggested that, besides the large cored vesicles (95 nm), small clear vesicles may be involved in the release of serotonin (see also Castel et al. 1984).

Histamine has been demonstrated to be contained specifically in large neurons (called C2) in the cerebral ganglion of *Aplysia*. Histamine produced from its precursor, histidine, moves by axonal transport to the terminal varicosities (Gotoh and Schwartz 1982). A study by electron-microscopic autoradiography has suggested that histamine is contained in large cored vesicles (100 nm in diameter) whose core is only moderately electron dense; small clear vesicles (70 nm) have been also recognized in the same terminals (Schwartz et al. 1986).

Neurons specifically containing histamine have been found also in the rat. Using antisera against histamine, Wilcox and Seybold (1982) demonstrated histamine neurons in the lateral hypothalamus, although the electron-microscopic nature of vesicles containing histamine has not been explored.

Acetylcholine, the most extensively distributed neurotransmitter, is stored in small noncored (clear) vesicles of the spherical type, as extensively reviewed by Kelly and Hooper (1982). With respect to the synthesis and metabolism of this substance, the reader is referred to p. 35.

As typical and classic transmitters, noradrenaline and acetylcholine have long been believed to be exclusive of one another in their occurrence. Recent studies by Landis and Keefe (1983) have evidenced that, at least in rat sympathetic neurons, adrenergic and cholinergic natures may occasionally coexist and it is possible that the former may change into the latter (for literature and details see p. 66).

Gamma-aminobutyric acid (GABA) is widely accepted to be contained in the small clear vesicles of the flattened type (Uchizono 1968, 1975) (see p. 58). Other amino acid transmitters (glutamic acid, glycine, etc.) are presumed to be contained in synaptic vesicles whose morphological types await identification.

5.2.3 ATP and Other Adenine Nucleotides

ATP is stored in nerve terminal vesicles together with neurotransmitters and released after nerve stimulation (review: Ribeiro 1978). Data are accumulating to suggest that ATP itself might be a neurotransmitter or modulator (see Burnstock 1972; Ribeiro 1978; Sneddon and Burnstock 1984). Recent studies by the research group of Burnstock demonstrated that the contractile response of the guinea-pig vas deferens is biphasic, the first phase being caused by noradrenaline, the second phase by ATP. The latter phase could be mimicked by exogenous ATP and abolished by α, β-methylene-ATP, the stable analogue of ATP. It was thus proposed that noradrenaline and ATP are cosecreted from nerve terminals and act upon the smooth muscle cells as cotransmitters (Sneddon and Burnstock 1984; Sneddon and Westfall 1984).

Within neurons, ATP is regarded necessary for the uptake and storage of catecholamines in the granules. ATP is also believed to be necessary to bind the acetylcholine in cholinergic vesicles (Whittaker 1972; Dowdall et al. 1974).

Small vesicles isolated from cholinergic nerves have been shown to contain large amounts of ATP and some GTP (Wagner et al. 1978; Kelly and Hooper 1982). That ATP is a constituent of synaptic vesicles is further supported by the finding that the peak release of acetylcholine coincides with the peak of ATP in stimulated nerves of the *Torpedo* electric organ (Zimmermann and Whittaker 1974). The vesicles in adrenergic nerves are also known to contain ATP (Euler et al. 1963; Stjärne and Lishajko 1966; De Potter et al. 1970).

Large cored or peptidergic-type granules in autonomic nerves were related to ATP by Burnstock et al. (1970). By extensive physiological and fine-structural studies on gastroenteric and some other autonomic nerves whose terminals were characterized by accumulations of large cored vesicles, they concluded that these nerves, which they called "purinergic nerves," contained in their synaptic vesicles ATP as their specific transmitter, causing relaxation of smooth muscles (review: Burnstock 1972). Studies using the quinacrine reaction for light microscopy (Olson and Alund 1976; Crowe and Burnstock 1981; Iijima and Hasegawa 1985) and the uranaffin reaction for electron microscopy (Wilson et al. 1979; Iijima 1981) supported the occurrence of stored ATP in "purinergic"-type neurons and in large cored vesicles. In the brain, "purinergic" neurons have been reported to exist and ATP, AMP, and adenosine released from them may have powerful depressant effects on central neurons (Phillis and Kostopoulos 1975; Stone and Taylor 1977).

ATP has been demonstrated in isolated neurosecretory (oxytocin and vasopressin) granules though in small amounts—several orders of magnitude less than the content of ATP in the chromaffin granules (Poisner and Douglas 1968; Gratzl et al. 1980).

As stated above, ATP is a universal constituent of secretory granules and synaptic vesicles of neurons; after the proposal of "purinergic" neurons by Burnstock, different kinds of messenger peptides with potent bioactivities have been identified in the granules of these neurons. For these reasons, Burnstock's idea that definite neurons have ATP as their specific and principal transmitter seems no longer tenable. As ATP is released from every type of neuron together with peptides and the classic transmitters, *every neuron can be said to be more or less purinergic*. It may be worthwhile to recollect that *every paraneuron is also more or less purinergic* (Fujita 1980).

5.2.4 Neurophysins, Chromogranins, and Similar Substances

Neurophysins are cystein-rich acidic proteins with a molecular weight of about 10 000 which have long been regarded as carrier proteins of oxytosin and vasopressin. The oxytocin binding "neurophysin I" and vasopressin-binding "neurophysin II" have been differentiated, and each has been shown to be produced in separate neurons (Vandesande et al. 1975), stored in neurosecretory (elementary) granules (Lösecke et al. 1982; Castel et al. 1985) and released into the bloodstream together with its partner hormone (Cheng and Friesen 1970; Robinson et al. 1971; Forling et al. 1971; McNeilly et al. 1972) (Fig. 5.6). Experimental results by Sachs et al. (1969) indirectly indicated that oxytocin and vasopressin might be formed together with their respective neurophysins from common precursors; this was, however, proven only recently by Gainer and associates who isolated and identified the precursors of neurophysins and partner

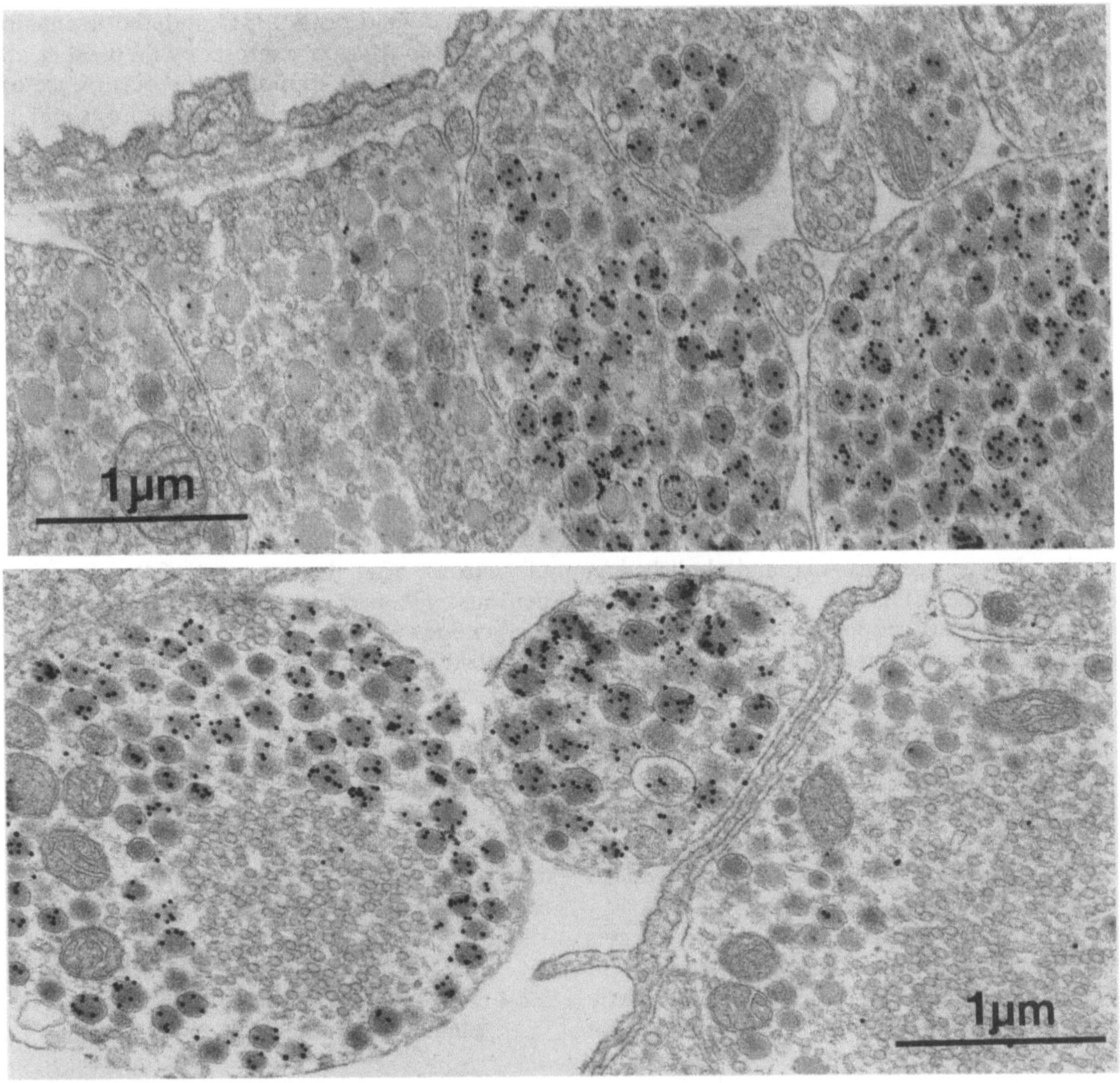

Fig. 5.6a, b. Localization of oxytocin-associated neurophysin in oxytocin-containing nerve fibers in the rat neurohypophysis. Protein A gold immunolabeling for this specific neurophysin is localized on large solid neurosecretory granules and not on small clear vesicles. Gold particles are virtually absent from probable vasopressin nerves, which can be seen on the *left* in **a** and on the *right* in **b**. **a** Axon endings close to a blood capillary (*upper left*). **b** Axonal varicosities in the neurohypophysis. Courtesy of M. Castel et al.; reproduced, with permission, from J Histochem Cytochem 33: 1015–1025 (1985)

hormones as proteins of 20 000 daltons (Gainer et al. 1977; Brownstein et al. 1977). These could be separated into a common precursor for oxytocin and its neurophysin (termed "pro-oxyphysin") with an isoelectric point of 5.4 and a common precursor for vasopressin and its neurophysin (termed "propressophysin") with an isoelectric point of 6.1 (Russel et al. 1980; see also Thorn et al. 1982).

Neurophysins resemble chromogranins in some respects. Both of them are acidic glucoproteins stored in the secretory granules in association with messenger peptides. As chromogranins, neurophysins released into the circulation are generally thought to exert little physiological activity, if any, though Robinson et al. (1974) reported that neurophysin I intravenously administered to hydropenic dogs caused a significant increase in sodium excretion. A striking difference, however, is that neurophysins are formed from one precursor molecule with their partner hormones, whereas chromogranins are apparently derived from precursors different to those of their partner peptides.

It has been expected but not demonstrated that oxytocin and vasopressin neurons possess chromogranin or similar substances besides neurophysins. Glycoproteins have been detected in vasopressin-producing neurons (Tasso et al. 1976) localized in the granules (Haddad et al. 1980). These glycoproteins may, at least partly, be derived from propressophysin, which has been shown to be glycosylated (Russel et al. 1980; see also Thorn et al. 1982). A recent immunocytochemical study by Nolan et al. (1985) demonstrates that chromograninlike immunoreactivity is contained in "neurons in the hypothalamus," which, according to the photomicrograph, probably include the magnocellular, classical neurosecretory neurons.

Chromogranin A has been detected in the fraction of "noradrenaline-containing particles" of the sympathetic (splenic) nerve (De Potter et al. 1970). Smith et al. (1970) reported that stimulation of the splenic nerve caused release of chromogranin A together with noradrenaline from the nerve terminal. In his biochemical analysis of large cored vesicles from the bovine splenic nerve, Lagercrantz (1976) reported that these vesicles contained only traces of chromogranin A, the soluble proteins of the vesicle contents being mainly dopamine β-hydroxylase (DBH; see below).

Recently, Nolan et al. (1985) reported that a variety of neurons in the central and peripheral nervous systems contained immunoreactivity for chromogranin (75 000-dalton acidic protein derived from bovine adrenal chromaffin granules). This finding supports the view that chromogranin might be a granule substance common to neurons and paraneurons.

5.2.5 Enzymes for Amine Conversion and Peptide Processing

Among the enzymes related to the production of biogenic amines, dopamine-β-hydroxylase (DBH) converting dopamine to noradrenaline, has been found to be contained in the noradrenaline vesicle fractions of the sympathetic (splenic) nerve (De Potter et al. 1970; Lagercrantz 1976; see also Burnstock and Costa 1975) and released when the nerve is stimulated (Smith et al. 1970). Immunocytochemical studies have shown that DBH is localized in the small and large granules of neurons in the rat locus coeruleus nucleus (Cimarusti et al. 1979).

Enzymes for the processing of prepropeptides have been presumed to occur within the granules of peptidergic neurons, though the evidence for this is rather meager as yet. A proopiocortin-processing enzyme cleaving the prohormone at paired basic amino acids has been demonstrated in the isolated granules of neurosecretory neurons (Chang et al. 1982; see also Castel et al. 1984).

5.2.6 Other Granule Contents

The occurrence of *glycosaminoglycans* in cholinergic vesicles has been reported (Stadler and Whittaker 1978). The occurrence of the same substances in paraneuronal granules (p. 28) makes this finding reasonable (Giannattasio et al. 1979), but no reliable evidence for it is available as yet (Kelly and Hooper 1982).

There are very few reports on lipids in neuronal granules and vesicles. Sawada et al. (1979) demonstrated that neurophysins isolated from the bovine neurohypophysis contained phospholipid molecules in a noncovalent complex.

Ca^{2+} has been detected in isolated neurosecretory granules in a concentration comparable with that in the adrenal chromaffin granules (Thorn et al. 1975). Analysis of isolated *Torpedo* acetylcholine vesicles has indicated that Ca^{2+} is very likely part of the vesicle contents, whereas evidence is weak that Mg^{2+}, Na^{+}, and K^{+} specifically occur there (Schmidt et al. 1980; see also Kelly and Hooper 1982).

5.3 Multiple Secretions and Dale's Law

It has long been believed that *one neuron produces and releases only a single transmitter*. This idea, which was associated with Dale by later researchers without due reason (for criticism see: Potter et al. 1981; Eccles 1986), has been useful in modern neurology, but it is no longer tenable as more and more evidence has accumulated to indicate that neurons may secrete two or more bioactive agents (reviews: Hökfelt et al. 1980, 1986; Potter et al. 1981; Fujita 1983a, b). As detailed above (pp. 61, ff), neuronal secretions, as a rule, comprise different categories of substances: peptides, amines or acetylcholine, adenine nucleotides, chromogranins, neurophysins and similar proteins, Ca^{2+}, etc. This phenomenon may be called *heterologous coexistence* of substances. The component which exerts the principal bioactivity differs from neuron to neuron: It may be a peptide, amine, choline, or possibly even ATP. In some neurons, two or more substances with potent bioactivities are cosecreted. For instance, it has been demonstrated that somatostatin (Hökfelt et al. 1977), opioid peptides (Wilson et al. 1980; Fried et al. 1986), or NPY (Lundberg et al. 1982; Sundler et al. 1986; Fried et al. 1986) is secreted by mammalian peripheral noradrenergic neurons, and VIP by intestinal and urogenital cholinergic nerves (Lundberg et al. 1979b; Schultzberg et al. 1979). Even cholinergic neuromuscular junctions have been demonstrated to contain CGRP (Rodrigo et al. 1985). At least at present, it is difficult to suggest which—the peptide or the classic transmitter—is predominant and which is subsidiary in function.

Just as in the case of paraneurons, Dale's law is further affected by the fact that a neuron often cosecretes two or more substances belonging to the same chemical category, either peptides, amines, or nucleotides. This phenomenon may be called *homologous coexistence* of substances. In the case of cosecreted peptides, they may be derived from a single prepropeptide. VIP and PHI, for instance, were immunocytochemically demonstrated to be costored in the same cholinergic neurons in the rat intestine (Yanaihara et al. 1983), and this has been accounted for by the result of cDNA analysis that VIP and PHI are formed by the processing of a common prepropeptide (Itoh et al. 1983).

More difficult to interpret is the fact that two or more neuropeptides derived from different precursors may be contained in the same neuron. Among the numerous instances of this category of homologous coexistence, the most intensely investigated are probably the hypothalamic neurons producing oxytocin, which simultaneously contain enkephalins (Martin and Voigt 1981; Vanderhaeghen et al. 1983) and CCK (Vanderhaeghen et al. 1983). Using the protein A-colloidal gold immunocytochemical method, Adachi et al. (1985) demonstrated that Met-enkephalin and oxytocin coexist in individual secretory granules within the nerve terminals in the rat posterior pituitary. Substance P coexists with CCK in rat dorsal root ganglia (Hökfelt et al. 1982) and with CGRP in vasodilator neurons (Uddman et al. 1986). Many other instances of this category of peptide coexistence in neurons have been reported recently, while completely the same situation has been revealed in different kinds of paraneurons (p. 37) (Fig. 5.7).

Chronological alteration in the messenger substances produced by one and the same cells has also recently been evidenced in the immunocytochemical study by Gesser and Larsson (1985) in snail neurons. The "light green cells" of *Lymnaea*, which are known to produce a body growth-stimulating hormone, contain enkephalins in spring and gastrin/CCK-like immunoreactivity in autumn; during winter, the neurons contain both enkephalin- and gastrin/CCK-like immunoreactivities. In another group of neurons, the caudodorsal cells, known to produce an ovulation hormone, an age-dependent alteration in peptides has been revealed. In young snails, these neurons contain enkephalins; in animals of intermediate age they contain both enkephalins and gastrin/CCK, and in older animals only the gastrin/CCK-like immunoreactivity can be detected in these neurons.

Multiple secretion of the classic transmitters has been shown by combined electrophysiological and electron-microscopic analyses of cultured single neruons. Adrenergic and cholinergic dual function and structure have been demonstrated in neurons dissociated from neonatal and adult rats (Furshpan et al. 1976). Chronological alteration of transmitters has been shown in cultured neurons. Sympathetic neurons from perinatal rats changed from an initial, immature adrenergic state to a cholinergic state after culture for 1–2 months, apparently by induction of certain nonneuronal cells (Landis 1980; Potter et al. 1980; review: Furshpan et al. 1982). In their in vivo observation in the sympathetic nerves of rat sweat glands, Landis and Keefe (1983) demonstrated that the transition from adrenergic to cholinergic nature of neurons actually occurs during development of the animals (see also Potter et al. 1981). Schultzberg et al. (1986) reported that in the rat transplants of encephalic or medullary raphe cells into the striatum gradually attained immunoreactivity for substance P or TRH in addition to serotonin. As the production of these peptides is normally lacking in the raphe cells and striatum, it has been postulated that plasticity of the neurons with regard to their transmitter phenotype is reflected in their attitude while transplanted into a different environment.

To sum up, the single transmitter status is no longer applicable to neuron cell biology. Neurons usually secrete multiple bioactive (messenger) substances of the same and different chemical categories. They may also change their messenger substances during development, aging, or according to seasonal and biological cycles. The latter condition may be called "chronological cosecretion" of transmitters.

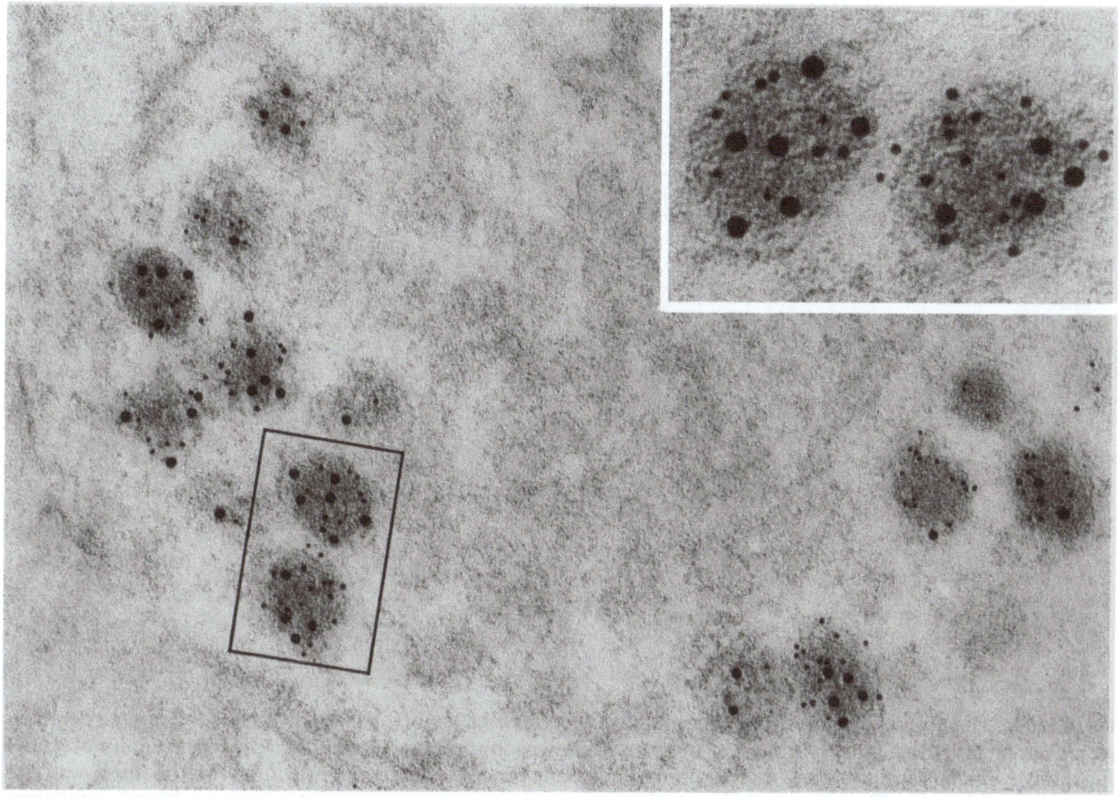

Fig. 5.7. Coexistence of two different peptides, Met-enkephalin-Arg-Gly-Leu and substance P, in the same granules in a nerve fiber in the guinea pig duodenum. By the protein A gold method, the enkephalin immunoreactivity is labeled with large gold particles, while the substance P immunoreactivity is labeled with small particles. Both particles are intermingled on every large cored vesicle. The *inset* is a magnification view of the indicated area. Prepared by T. Uchida, Department of Anatomy, Yamanashi Medical College. × 110 000, *inset* × 230 000

5.4 Multipolar (Dendritic) Secretion

5.4.1 Physiological Data and Views

Neurons receive information in the form of impulses from many other neurons, presynaptic neurons, and in turn transmit information to postsynaptic neurons. Incoming impulses may cause excitatory postsynaptic potentials (EPSP) or inhibitory postsynaptic potentials (IPSP) at the input portion of the neuron; when the net electrogenesis of the input is depolarizing, it can initiate spikes in the conductile, spike generating portion (see Chap. IV). For decades, the anterior horn cell, or the cell of the Deiters type has been presented as a typical neuron. It was said that a typical neuron has one neurite or axon, one soma or perikaryon, and many branching dendrites. The work of Golgi, Cajal, and others has shown that the majority of neurons are cells with elaborate but rather short branches and that in many instances processes cannot be easily assigned to either the axon or dendrite category. Horizontal and amacrine cells in the retina and intramural ganglion cells in the intestine are good examples (Bennett 1977). From the diversity of neuron type in the central and peripheral nervous systems, it is apparent that the pictures of "typical" neurons are a fiction.

Contrary to the conventional concept of the all or none impulse propagated along the entire length of the neuron being synonymous with the neuron in action, we must now accomodate the following facts (Ralph 1978): (1) Many neurons cannot respond in an all or none fashion and therefore cannot propagate an impulse without a decrement. (2) Input to a neuron can occur at several different kinds of loci in various parts of the neuron, in addition to the dendritic synapses. (3) Responses to afferent excitation do not automatically give rise to firing but often only help determine the rate of firing of impulses,

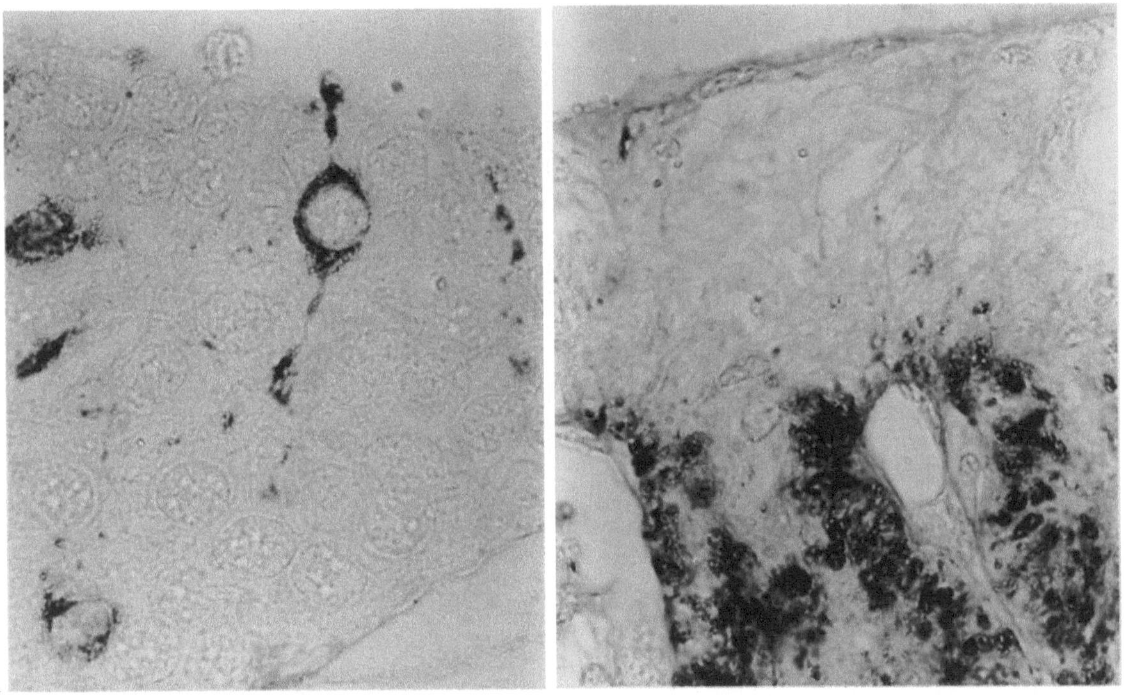
a b

Fig. 5.8a, b. Calcitonin-containing CSF-contacting neurons in the bullfrog as an instance of suspected dendritic secretion. **a** Hypothalamus. A bipolar neuron is shown containing a granular calcitonin-immunoreactive substance, which is accumulated in the bulbous end of the process (dendrite) projecting into the third ventricle. **b** Median eminence. The axons of the calcitonin-containing neurons terminate around blood capillaries, showing marked accumulation of the immunoreactive substance. Prepared by R. Yui, Department of Anatomy, Niigata University Medical School

which are initiated at a more distal site—commonly near the base of the axon. (4) Neurons sometimes have two or more axons and therefore can deliver two nonidentical, pulse-coded outputs at the same time in different directions (Fig. 5.8).

This complex behavior and diversity of neuron types has been conspicuously demonstrated in the *enteric neurons* (review: Wood 1984), which may be regarded as neurons with some paraneuronal features. Neurotransmission at the cell bodies of enteric neurons involves slow and fast synaptic mechanisms. Fast synaptic potentials have durations of less than 50 ms, and the slow potentials last for several seconds. These synaptic events may be EPSPs or IPSPs, either of which can be evoked experimentally by electrical stimulation of presynaptic fibers. Presynaptic inhibition is another form of neurochemical transmission that occurs at both the fast and slow synapses. All the fast EPSPs in enteric ganglion cells of the small intestine appear to be mediated by nicotinic-cholinergic receptors, and the slow EPSPs seems to be transmitted by the release of substance P and serotonin. Several other brain-gut related substances, such as histamine, cholecystokinin (CCK), gastrin releasing peptide (GRP), acetylcholine, and VIP, also mimic slow synaptic excitation, and at least some of them are presumed to function in the paracrine or hemocrine mode. Putative transmitters for slow IPSPs are the opioid peptides, somatostatin, serotonin, noradrenaline, and CCK. Putative transmitters for presynaptic inhibition are noradrenaline, serotonin, and acetylcholine (Wood 1984). If some enteric neurons possess paracrine and hemocrine functions, these cells can be regarded as being more paraneuronal than otherwise. Adopting the knowledge concerning paraneurons, it may be reasonable to assume that in these paraneuronal neurons, multiple information may be transmitted via receptors in the plasma membrane, cytosol, mitochondria, and nucleus.

The paraneuronal functions of neurons in the *central nervous system* were also predicted by Llinas (1983). It is now agreed that the triggering of synaptic transmitter release is produced by an inward Ca^{2+} influx across the presynaptic terminal (Rubin 1970, 1974; Douglas 1975a, b). In addition to the role of Ca^{2+} as a charge carrier in Purkinje cells in the cerebellum, Ca^{2+} may be regarded as triggering the release of intracellular substances from dendrites; it thus becomes a central issue in the problem of local circuit function. In addition, it is possible that Ca^{2+} plays a role in the release from dendrites of substances other than the classic transmitters, for example, neuropeptides and adenine nucleotides. The brain can be considered an endocrine system capable of producing peptides important to the overall maintenance of the organism.

It is also quite possible that Ca^{2+} influences the metabolic level of the cell, which would be important to the maintenance of dendritic release and to the possible plastic effects subserved by electrical activity. Ca^{2+} probably plays a role in dendrites similar to the role in axons with respect to axoplasmic flow. The voltage-dependent Ca^{2+} entry into dendrites seems to be an important factor in modulating the long-term excitability of cells, since electric activity may possibly produce neuronal changes in order to generate, for instance, memory (Llinas 1983). These paraneuronal features of neurons may be correlated with the initiation of spikes in the conductile, spike generating portion not being the sole mode of transmission of information in and among neurons.

Spikeless communication among neurons has recently been proposed for the neural mechanism underlying the facilitatory interaction of behavior in crayfish locomotion (Hisada 1984 Hisada et al. 1984a, b). Intracellular recording and staining revealed that the interaction was mediated, at least in part, by newly found local nonspiking interneurons. These interneurons, unlike the "classic" neurons, can effectively modulate the activity of uropod motoneurons by their own membrane potential change, without generating the spikes. These interneurons lack an axon and are entirely confined to the terminal ganglion, where the uropod motoneurons are present.

5.4.2 Morphological Data and Concepts

Bipolar paraneurons, like basal-granulated (endocrine) cells in the gastroenteric and bronchopulmonary epithelia as well as gustatory, olfactory, and other sensory cells, show two distinct poles—receptive and secretory. This corresponds to the concept that neurons have receptive or input and secretory or output regions, the former generally covering the dendrites and sometimes the perikaryon, the latter the axon terminal portions. This classic view of the polarity of neurons, however, should be extensively revised as recent results suggest that dendrites may be engaged in the release of bioactive substances in many kinds of neurons.

Cerebrospinal fluid (CSF)-contacting neurons may be suitable examples in this respect because of their simple bipolar structure (Fig. 5.7). These neurons extend a slender process with a swollen end to the ependymal surface, while they extend another, much longer process to different parts of the brain. The former process is generally regarded as the dendrite and the latter the neurite or axon for this type of neuron. Based on this concept, the dendritic process has generally been accepted to be a device for detecting information in the CSF. Recent studies, however, strongly indicate that the dendritic process of the CSF-contacting neurons is engaged in secretion release. Various kinds of neuropeptides and amines have been reported to be contained in the CSF of various animals, and some researchers believe that these are, at least partly, due to the secretory activity of the CSF-contacting neurons into the ventricles (Leonhardt 1980).

In the wall of the third ventricle of the frog, *Rana catesbeiana*, many CSF-contacting neurons can be found by peptide immunocytochemistry, because they contain substance P, calcitonin, gastrin, and other peptides. We found that some of these neurons, especially those with calcitoninlike immunoreactivity possessed the accumulated immunoreactive substance both in the dendritic knob projected into the CSF and in the axonal terminal on the blood capillaries in the median eminence (Yui et al. 1981; Yui 1983). Under the electron microscope, the dendritic end is swollen by densely packed granules (Fig. 5.9). Although we have not yet obtained evidence of exocytotic release of these granules, it seems most likely that granule release takes place here.

If the CSF-contacting neurons are detectors of intraventricular information, the dendritic terminal of the neurons may be engaged simultaneously in reception and secretion. A mosaic differentiation of the cell surface for these two functions can be postulated.

In neurons of more complicated structures, *dendritic secretion* also seems to occur fairly extensively. Immunocytochemical observations of neuropeptides usually show that the peptides fill almost all the dendritic branches, tending sometimes to be concentrated in the preterminal and terminal varicosities. The same is the case in the immunocytochemically demonstrated serotonin and fluorescence-cytochemically demonstrated

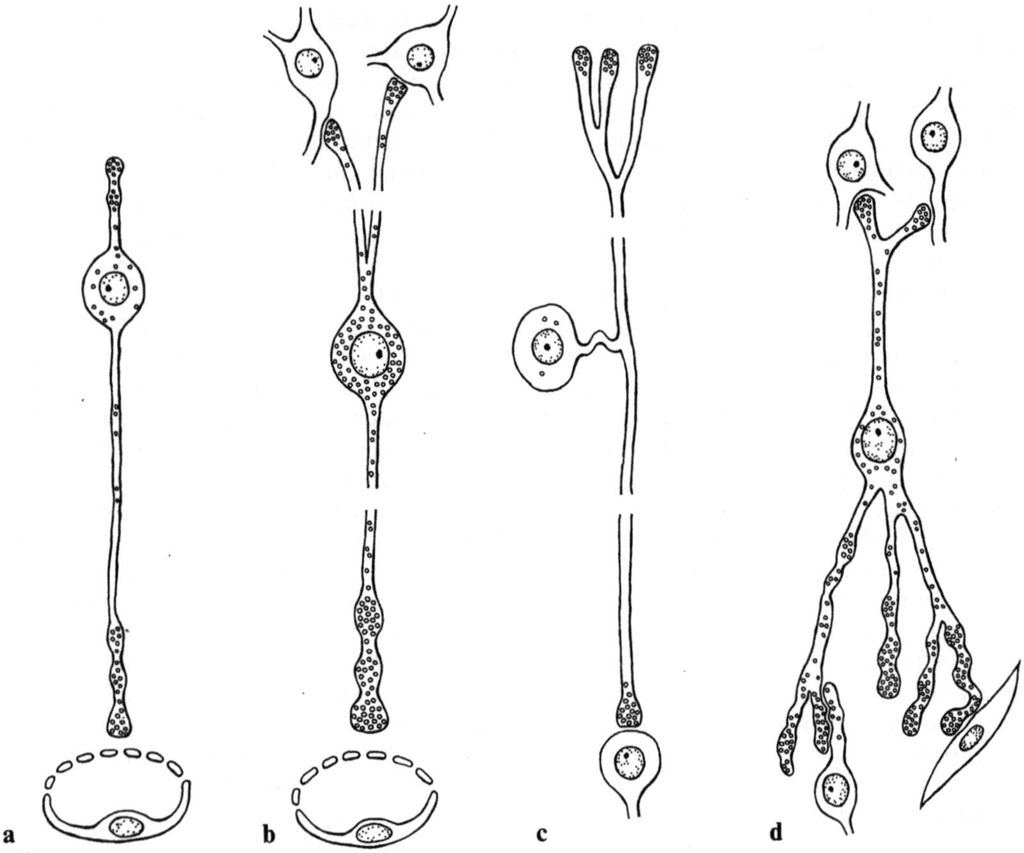

Fig. 5.9a–d. Multipolar (dendritic) secretion of neurons. **a** CSF-contacting neuron. **b** Hypothalamic neurosecretory neuron. **c** Primary sensory neuron. **d** Intramural neuron in the intestine

monoamines.

It was reported that more than 90% of the substance P-immunoreactive substance produced in sensory neurons in the nodose ganglion was transported into the dendrites (Brimijoin et al. 1980). Some nerve terminals in the taste buds of the cat have been shown to contain substance P (Lundberg et al. 1979c). Whether these nerves are dendrites from the sensory neurons of the petrosal ganglion and nucleus tractus solitarii remains unsolved. It is also unknown whether substance P serves as the transmitter of these primary sensory neurons. However, it is certain that substance P is transferred centrifugally to the nerve endings. When released here, substance P may exert an action different to that of sensory transmitter. A vasodilator function, for example, has been suggested for dendritic substance P (Morris et al. 1978).

Capsaicin, the pungent principle in red pepper, causes local pain and violent inflammatory reactions, i.e., hyperemia and edema, when applied to an area of skin. Hungarian researchers discovered that repeated application of capsaicin to the same dermal area in the rat and man (and also to the rat eye) totally inhibits the chemogenic pain and tissue reactions due to selective impairment of primary sensory neurons (Janscó et al. 1967, 1977; Janscó 1968). Janscó et al. (1977) propose a hypothesis that the violent inflammatory reactions, including increased vascular permeability, are caused "by neurohumoral substance released from the sensory nerve endings which transmit chemogenic pain" and repeated capsaicin treatment may cause exhaustion of this substance, leading to the insensitivity to chemical stimuli. Now that the primary neurons in question have been established to be substance P-like neurons (Otsuka et al. 1982), it seems reasonable to extend the hypothesis by Janscó et al. (1977) as follows: The primary sensory neurons, when stimulated by capsaicin, release substance P at their axonal ends and simultaneously also at their dendritic ends. The axonal release of substance P assists synaptic relay of the excitation, while its *dendritic release* induces elevated vascular permeability and related reactions of inflammation.

Electron microscopy has often shown synaptic vesicles and granules of different types in dendrites, especially their terminal portions. Even the dendritic terminals, which are believed to be purely afferent in nature, may contain rather numerous vesicles and/or granules. For example, the terminal associated with Merkel cell is occasionally filled with small clear vesicles (Tachibana, personal communication). The apparently sensory nerve terminals in the odontoblastic and predentinal layers of the tooth often contain small clear or cored vesicles, large-cored vesicles (granules), or a mixture of both (Avery 1981; Gunji 1982). Many electron micrographs of these and other sensory terminals containing synaptic vesicles have not been published, as they were thought to be erroneous images since they were at odds with the generally accepted doctrine of neurology. It now seems necessary to hypothesize that these terminals are reciprocal in function, i.e., they are sensory but also occasionally secretory in nature. The messenger substances thus released there may presumably regulate the sensitivity of the receptor cells.

Classic neurosecretory neurons containing oxytocin and vasopressin have also been shown to perform dendritic secretion. Besides the axons forming the hypothalamohypophyseal tract, the neurons project a second or third process, which should be called a dendrite according to classic terminology (Fig. 5.9). Immunocytochemical and more conventional stainings indicate that these processes contain as much messenger peptide as the axon (Sofroniew et al. 1979; Fujita 1985). Immunocytochemical and horseradish peroxidase injection studies have shown that the classic neurosecretory neurons project to various parts of the brain, medulla oblongata, and spinal cord (Weindl and Sofroniew 1976; Swanson 1977; Sofroniew and Weindl 1978; Hosoya and Matsushita 1979; Sofroniew 1980; review: Sterba et al. 1980). Electrophysiological studies indicate that some of the neurons simultaneously project to the neurohypophysis and to the medulla oblongata (Zerihun and Harris 1981; Kannan and Yamashita 1983). Recently, Buijs and Van Heerikhuize (1982) demonstrated that Ca^{2+}-dependent release of vasopressin and oxytocin actually occurs at the extrahypothalamic terminals of neurosecretory neurons.

5.5 Summary

The essential and universal feature of the neuron is not electric conductivity but secretory activity. This criterion will be of assistance in the detection of unknown aspects of neuronal functions, and also in the visualization of the continuity of neurons with paraneurons.

Besides the synaptic vesicles, neurons possess larger granules with solid contents which contain messenger peptides. These granules or "large cored vesicles," which are abundant in the classic neurosecretory neurons and in "peptidergic" neurons, are contained, though in smaller numbers, in almost all kinds of neurons, intermingled with the small vesicles.

The contents of the granules of neurons are largely common to those of paraneurons. Peptides coexist with amines, acetylcholine, or other messenger substances in the granules; ATP and carrier proteins (neurophysins, etc.) are also included. Essentially the same coexistence of substances also occurs in small vesicles, although these seem to be lacking in proteinic substances. Dale's law is no longer tenable since the different messenger substances coexisting in the granules and vesicles are simultaneously released from a single neuron.

There is increasing evidence suggesting that neurons may release their messenger substances not only at the axonal but also at the dendritic terminals. The secretory activity of neurons thus may occur independently of their electric conductivity.

6. Transport of Messenger Substances to Targets

6.1 Routes and Ranges of Messenger Transport

A typical *neurotransmitter* is released from a neuron and transported to its target across a *synaptic cleft* of 10- to 40-nm width. We call this mode of messenger secretion "synaptocrinia" (the term proposed by Fujita 1983b). A typical *hormone*, on the other hand, is released from an endocrine cell—or endocrine paraneuron, if we restrict the subject to the peptide/amine producing system) to be conveyed to its remote target via the *general circulation*. This type of messenger secretion can be called "telecrinia" (the term proposed by Leibson 1979), "hemocrinia" (proposed by Feyrter 1962), or, more precisely "macrohemocrinia" (Fujita 1983b). Between these two extreme cases of the typical neurotransmitter and hormone, every gradation can be found. In other words, a large variety of intermediate routes and ranges of messenger transport may occur, and these will now be classified.

6.1.1 Synaptocrinia

In this case, a messenger substance released from its source cell runs to an adjacent target cell by diffusion across a narrow intercellular space. The space, which is called the synaptic cleft usually measures only 10–40 nm in width. Typical synaptocrinia is seen in a neurotransmitter which is released from a neuronal terminal and conveyed by diffusion across synaptic cleft to an adjacent target, i.e., another neuron or a glandular, muscular or other kind of cell. At the connection of the neuron with a skeletal muscle fiber, i.e., the neuromuscular junction, the synaptic cleft is widened (about 60 nm) by the intervention of a basement membrane.

Sensory paraneurons, such as olfactory, gustatory, bronchial neuroepithelia, Merkel, and inner ear hair cells, give their messengers to the adjacent afferent nerves by synaptocrinia.

Endocrine paraneurons may also show morphological features suggesting synaptocrinia to their adjacent targets. Somatostatin-producing D cells in the fundic mucosa of the mammalian stomach have been reported to extend a process to HCl secreting parietal cells (dog—Kusumoto et al. 1979; rat—Larsson et al. 1979; see also Larsson 1979; Fujita et al. 1982) (Fig. 6.1). This finding has been interpreted as the morphological manifestation of the local inhibition by somatostatin of acid release by the parietal cells (Kusumoto et al. 1979; Larsson et al. 1979; Larsson 1979).

This interpretation seems to account for only a small part of the control of acid secretion by D cells, as only a few percent of the parietal cells are in contact with D cells (Fujita 1983b). It seems likely that somatostatin released by D cells may be distributed to parietal cells partly by this direct cell contact but largely by diffusion over a greater distance (paracrinia) and perhaps also by the microcirculation (microhemocrinia, see below).

It is notable that we have here a clear instance of endocrine paraneurons which are most probably engaged in synaptocrinia. Their slender and nodular processes filled with secretory granules and touching the surface of a target cell over a very thin cleft correspond exactly to the attitude of a neuronal terminal.

Certain other endocrine cells, including TG cells in the rat ileum, EC cells in the mouse and rat ileum and colon and human fetal urethra (see Larsson 1979), and D cells in the chicken proventriculus (Fig. 6.1) (Fujita et al. 1982), have been reported to display more or less conspicuous cell processes to their presumed target cells.

Synaptocrine control or modulation, probably mixed with paracrine actions, may presumably also be exerted by paraneurons gathered in an endocrine organ. Pancreatic islet cells, especially D cells, often extend long processes to other cell types (p. 175). Thyroid parafollicular cells may show conspicuous processes among themselves or towards follicular cells (p. 145). Here again, possible synaptocrine distribution of so-

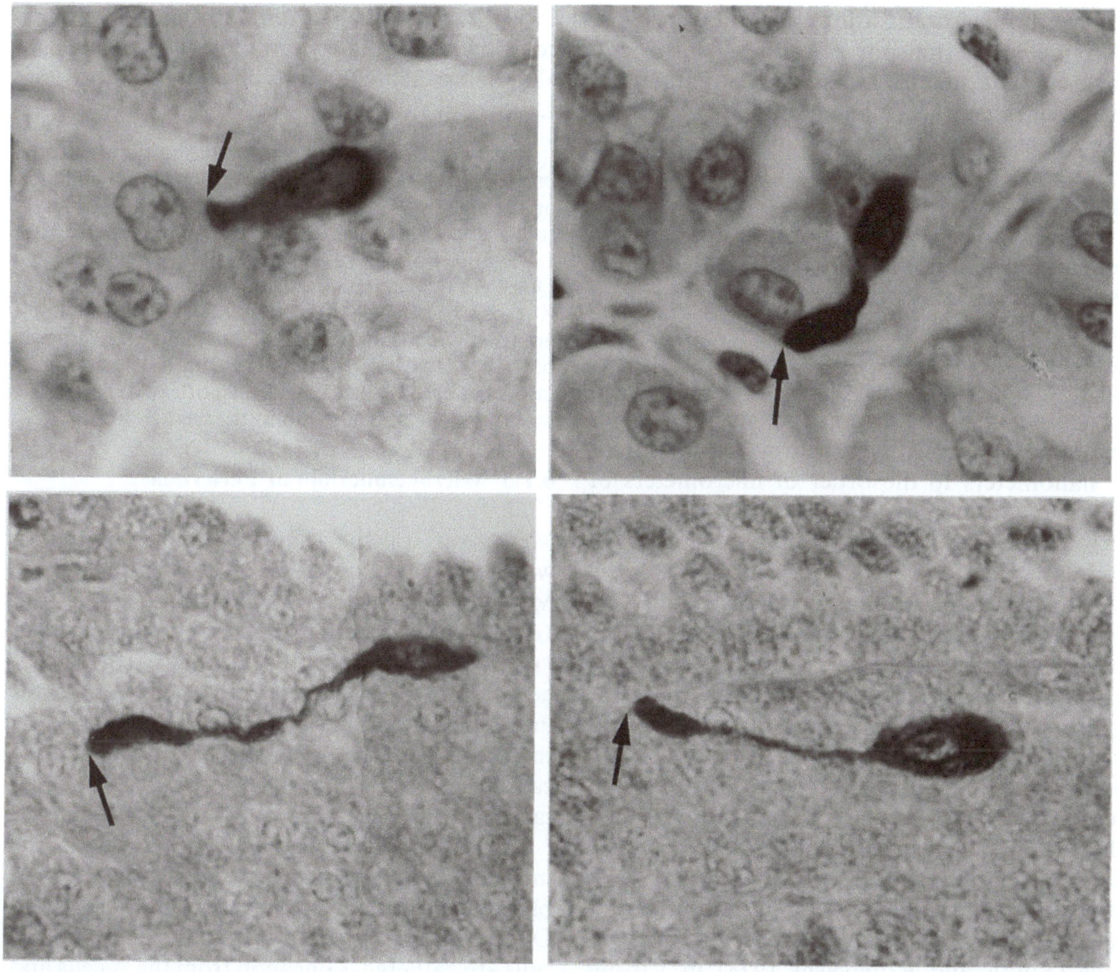

Fig. 6.1a–d. Images suggesting a synaptic regulation of secretory cells by somatostatin-containing paraneurons. **a, b** Somatostatin-immunoreactive cells in the gastric fundus of the dog extending a knobbed process filled with the immunoreactive substance in contact with (*arrows*) the side or base of a parietal cell. **c, d** Somatostatin-immunoreactive cells in the chicken proventriculus. A long knobbed process of the cells terminates (*arrows*) at the side or base of a proventricular gland cell. Indirect immunoperoxidase method with counterstaining of hematoxylin and eosin. Reproduced, with permission, from Fujita T et al. (1982) in Farner S, Lederis K (eds) Neurosecretion, Plenum

matostatin or other messenger substances to neighboring targets is suggested.

6.1.2 Paracrinia

The term paracrinia was originated by Feyrter (1953), a Viennese pathologist who proposed that the endocrine cells dispersed in the gastroenteric and bronchial epithelia ("helle Zellen" or clear cells) might distribute their secretions to nearby targets. The "helle Zellen" which Feyrter (1938, 1946, 1952, 1953) extensively investigated by light microscopy correspond to the basal-granulated endocrine cells, which are now known to release various peptidic, aminic, and other messengers whose importance as "local hormones" has recently been stressed (Fujita and Kobayashi 1973, 1977). In this respect, the term "paracrinia" has been widely accepted by current researchers and serves to signify the short range transfer of hormonal messengers (Polak et al. 1976; Pearse et al. 1977; Karlson 1982).

As pointed out elsewhere (Fujita 198?b), however, neither Feyrter himself nor recent advocators of his concept have precisely defined "paracrinia." Feyrter (1962) wrote that paracrinia is a "humorale Beeinflussung anliegender Zellen und Gewebe" (humoral influence of closely disposed cells and tissues). Besides messenger transfer by diffusion, he occasionally, and apparently without a careful and systematic approach, also included transport by the intravisceral microcirculation into the category of paracrinia. Synapselike contact of an endocrine cell with its nearby target cell in the vicinity has repeatedly been referred to as "paracrinia" (e.g., Larsson et al. 1979; Larsson 1979; see above).

In order to remove this confusion, we would define "paracrinia" as follows (see also Fujita 1983b): Paracrinia is the transfer of a chemical messenger to a neighboring target by *intercellular diffusion*, whereby the target is not located so close to the source cell of the messenger as to form a synapse. The distance between the source and target cells roughly measures from hundreds of nanometers to the order of micrometers or even tens of micrometers. Naturally, we cannot draw a clear boundary between synaptocrinia and paracrinia.

Although experimental evidence is still meager, most endocrinologists now believe that a variety of endocrine paraneurons may exert paracrine effects upon other cells of the same and/or different categories and that these effects might be so significant that the structure, distribution, and function of endocrine cells and glands could not be well understood without their elucidation. Various endocrine paraneurons, on the other hand, seem to receive paracrine regulation by nearby nerves. Among gut endocrine cells, G cells (Lechago and Barajas 1981) and EC cells (Fujita and Kobayashi 1978) in various mammals seems to be related more closely to nerve terminals, which are often at a distance of 600–1200 nm from the cells. This morphological relationship seems compatible with the physiological findings that gastrin release by G cells (Blair et al. 1977; Matsuo and Seki 1978; Grossman 1979) and serotonin release by EC cells (Tobe et al. 1976) are enhanced by vagal stimulation, whereas nervous effects upon the release of other gut hormones are indistinct. Inversely, we assume that gut endocrine cells play the roles of sensory paraneurons, although they form no synapses with the neuronal terminals (Fig. 6.2). Network formation of the bundles of neuronal processes ensheathed by enteroglial cells in the lamina propria mucosae (see p. 278, ff.) may facilitate the paracrine activity of gut endocrine cells on the enteric and other nervous system.

The relationship of a nerve terminal to its target described above corresponds to what has long been known by neurologists as a "synapse à distance" (Fig. 6.3). Typical examples of this are seen in the autonomic groundplexus (Hillarp 1959) surrounding arteriolar smooth muscle cells, the distance between the two elements usually being 250–500 nm or more (Sato 1966; Lever et al.).

Paracrine transport of neuronal and paraneuronal messengers may be facilitated by intercellular channels or canaliculi, which have been reported to occur in some endocrine glands (adenohypophysis—Foster 1971; Vila-Porcile 1972; Quay 1975; pineal body—Quay 1974; pancreatic islet—Fujita et al. 1981; Kataoka et al. 1982). As these intercellular channels are usually continuous with pericapillary spaces, they probably also assist in the permeation of messengers into the blood. Vascular transport of messengers, however, will be dealt with in the next section.

6.1.3 Microhemocrinia or Pericrinia

Messenger transport by the blood circulation was termed "hemocrinia" by Feyrter (1962). Messenger transport by the microcirculation, i.e., within a small tissue area may be called "microhemocrinia" (Fujita 1983b). In microhemocrinia, a chemical messenger permeates into blood capillaries and, after being carried by them for a certain distance, emerges from the capillaries to act upon extravascularly located targets.

In the pancreatic islets of some mammals, the microcirculation is designed in such a way that blood flows first through the region where A and D cells are arranged and then to the area where B cells are gathered. Glucagon from the A cells and somatostatin from the D cells are thought to be transported to the B-cell area by this microcirculation in addition to simple paracrinia in order to stimulate and inhibit, respectively, the release of insulin from the B cells (Fujita 1973; Fujita et al. 1976; Ohtani et al. 1986).

The precise examination of serial sections of human islets of Langerhans by Grube and Bohn (1983) suggests that somatostatin, after release from the D cells, reaches other islet cells mainly by the intrainsular microcirculation and partly, presumably, along the perivascular space. A similar mechanism of microhemocrine hormone transport is believed to exist in various endocrine glands, but little evidence is available for this.

The tunica propria mucosae of the gut is characterized by unusually long capillaries running closely beneath the epithelium. These capillaries are without doubt involved in the microhemocrine transport of various messengers released from endocrine cells (paraneurons) in the epithelium and from a pericapillary network of beaded nerves (pp. 278, ff.). The messengers are presumed

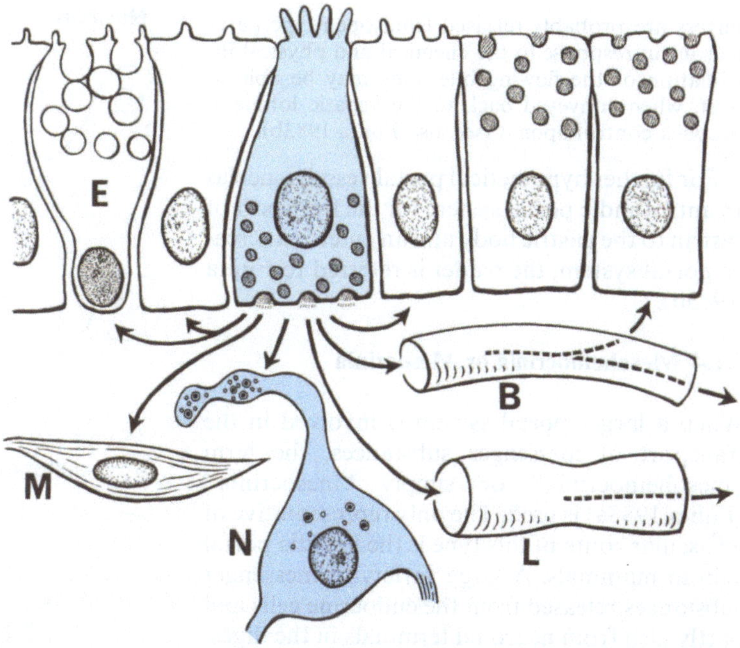

Fig. 6.2. Different routes of transport and targets of actions of messenger substances released from a gut basal-granulated cell, as an example of endocrine and sensory paraneurons. The messenger substances enter both blood vessels (*B* hemocrinia) and lymphatics (*L* lymphocrinia). The substances are transported to a considerable extent by diffusion (paracrinia) to regulate nearby targets, including neurons (*N*), muscular cells (*M*), and epithelial or glandular cells (*E*)

to reach their targets (epithelial cells, nerves, muscles, etc.) located within the same propria and may exert secretory, contractile, or trophic effects (see also Fujita 1983b).

A one-step advanced type of hemocrine messenger is that mediated by small portal vessels. The messenger substance permeates the blood capillaries. From this capillary network, the substance is transferred by a small portal vessel or vessels to a second capillary network located at a different site, either within the same or in an adjacent organ (Henderson and Daniel 1978; Ohtani et al. 1983; Fujita 1983). The well-known example of this category of microhemocrinia is the transport of hypothalamic neurohormones (pituitary hormone-releasing and release-inhibitory hormones from the median eminence and neurohypophysis to the adenohypophysis by the hypophyseal portal vessels (review: Krisch 1986).

In the pancreas, islet hormones permeate into islet capillaries and are conveyed by insuloacinar portal vessels to the capillary bed in the exocrine pancreas (Henderson 1969; Henderson and Daniel 1978; Fujita 1973; Fujita and Murakami 1973). Here, insulin and other islet hormones are believed to exert stimulatory or inhibitory effects on the exocrine functions of the pancreas. The function of insulin in potentiating the acinar cells to produce pancreozymin has, in particular, been well established (Kanno and Saito 1976, 1979; Kanno et al. 1976; Korc et al. 1981; Otsuki and Williams 1983). Moreover, numerous nerve fibers with vesicles of the cholinergic, adrenergic, and peptidergic (especially VIP containing) types terminate in the perivascular spaces of islet capillaries in some mammals (Fujita and Kobayashi 1979a). The neuronal secretions released here are thought to be transported to the exocrine pancreas to stimulate it. The acinar tissue supplied with insuloacinar portal vessels thus can be neuronally controlled without direct innervation. The islet can be regarded as the "neuroparaneuronal control center" of the pancreas (Fujita et al. 1981) from which transmitters and hormones are distributed to the exocrine pancreas by microhemocrinia.

Another instance of microhemocrinia may be found in the periductular portal vessels, which originate from the capillary plexus surrounding hepatic ductules and pour into the lobular sinusoids (Murakami et al. 1974; Ohtani et al. 1983). These vessels possibly convey messengers released from endocrine cells dispersed in the ductular epithelium to the hepatic lobule. As the messengers are probably released from open-type paraneurons in response to the chemical and physical information of the flowing bile, they may be able to exert, when conveyed back to the hepatic lobule, a feedback control upon it (see also Fujita 1983b).

For further hypothetical portal vessels, such as an antrofundic portal system for the transport of gastrin to the gastric body and an enteropancreatic portal system, the reader is referred to Fujita (1983b).

6.1.4 Mesohemocrinia or Mesocrinia

When a larger portal system is involved in the transport of messenger substances, the term "mesohemocrinia" or simply "mesocrinia" (Fujita 1983a) is used. The only representative of a vascular route of this type is the hepatic portal vein in mammals. A large variety of messenger substances released from the endocrine cells and partly also from neuronal terminals in the digestive organs are conveyed by the portal vein to the liver. Only a few of the actions of these substances upon the liver are known. Among them, the glycogenolytic and glycogenetic effects of glucagon and insulin, respectively, upon the hepatocytes and the secretagogous effect of secretin upon the bile ductules have been established. Substance P has been shown to cause increased blood flow through the liver sinusoids (Erspamer et al. 1977). Moreover, the trophic effects of some GEP hormones, especially insulin and glucagon, upon the liver parenchyma have been demonstrated (Bucher and Weir 1976; Starzl et al. 1976; Hasegawa and Koga 1977).

6.1.5 Macrohemocrinia or Telecrinia

The classic transport of messenger substances (hormones) via the general circulation can be called macrohemocrinia (Fujita 1983b). As this is a messenger transport to remote targets, the term telecrinia coined by Leibson (1979) is also suitable. Messengers with long half-lives circulate through the body many times in a bioactive form, whereas those with short half-lives are inactivated or absorbed so quickly that they hardly go round the body once. Not only the substances secreted by endocrine paraneurons but also those released from neurons may be transported by macrohemocrinia. Besides the well-known neurohormones oxytocin and vasopressin, some peptides, especially VIP, have been suggested as being released in considerable amounts into the circulation after the nerves containing them are stimulated (Fahrenkrug et al. 1978).

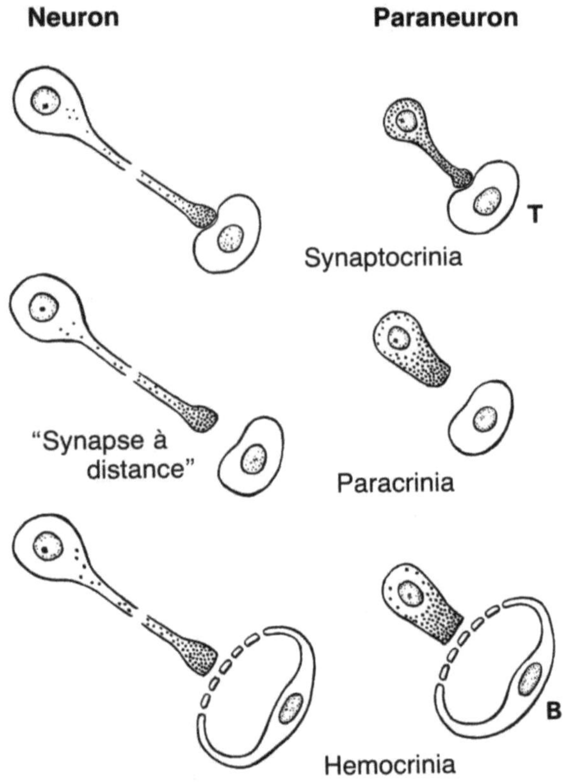

Fig. 6.3. Gradation in the range of transport of neuronal and paraneuronal messengers. Note that neurons and paraneurons are equal with regard to their relation to the target cell (*T*) and blood capillary (*B*)

6.1.6 Other Routes of Messenger Transport

Ventriculocrinia. Numerous kinds of neuronal peptides and amines are found in the cerebrospinal fluid in considerable concentrations (Post et al. 1982; Scharrer 1969). Histological studies have demonstrated bulbous endings containing peptidergic, adrenergic, or cholinergic vesicles protruding into the ventricle (Scharrer 1969). Beaded axons have been shown on the ependymal surface of the ventricle (Scott et al. 1977). In the hypothalamus of lower vertebrates, numerous CSF-contacting neurons (pp. 273, ff.) extend their dendrite to the ependymal surface of the third ventricle and show morphological features suggesting the release of their peptidic and aminic secretions (Leonhardt 1980; Yui et al. 1981).

The messenger substances released into the cerebrospinal fluid are thought to exert a variety of their actions upon different systems in the brain without being hindered by the blood-brain barrier. These actions can be mimicked by intraventricular injections of the substances. In his recent classification of messengers, Epple (1982) called the bioactive substances released into the cerebrospinal fluid "CSF hormones."

Lymphocrinia. Some messenger substances may be partly transferred by the lymphatics. The thyroid contains well-developed lymphatics (Shimada 1981), which are involved in the transport of thyroid hormones (Kotani et al. 1968). Messenger transport by the lymphatics, however, seems less important in peptide/amine-producing tissues, as lymph capillaries are usually distributed at a much greater distance from the mucosal epithelium containing the secretory elements than the blood capillaries are (Fig. 6.2)

A *terminological note* may be useful here. In his valuable review, Leibson (1979) proposed the terms "parahormone" and "telehormone" for messengers conveyed by paracrinia and telecrinia, respectively. As Epple (1982) noticed, these terms are very useful. Karlson (1982) recently suggested the shorter term "paramone" for "parahormone." His term is not really logical because "*mone*" has no meaning. However, as we have adopted the term phero*mone*, we should have no strong reasons for objecting to "para*mone*." Messengers conveyed by the microcirculation (microhemocrinia or pericrinia) may be called "perihormones," while those carried by an intervisceral portal vessel (mesohemocrinia) may be termed "mesohormones."

6.2 Continuity of Transmitters and Hormones

In Chaps. 3 and 5 we have seen that neuronal secretions, including typical neurotransmitters, are largely common to paraneuronal secretions, including local hormones (parahormones) and typical hormones or telehormones. It is impossible to distinguish neurotransmitters, parahormones, and telehormones, either by their chemical nature or by the mechanism of their secretion (Fig. 6.3).

In the preceding section, we have shown that the manner of transport of messengers to targets is gradational from synaptocrinia to macrohemocrinia or telecrinia. Moreover, we have suggested the possibility that a given messenger may exert its effects upon targets disposed at different grades of distance. A synaptocrine or paracrine messenger may partly enter the circulation to be transferred to both nearby and remote targets.

With respect to the receptors, it seems most unreasonable to postulate that receptors for neurotransmitters or synaptocrine messengers might differ in their essential nature and location from those for hormones or telecrine messengers.

For these reasons, we conclude that (neuro) transmitters and (tele)hormones are continuous in nature.

6.3 Summary

The routes and ranges of messenger substances are highly variable. They may reach the target by diffusion over a minute intercellular gap (synaptocrinia). They may also diffuse to the target over a longer distance (paracrinia). They may enter the circulation to be transported by the blood to close or remote targets (micro- and macrohemocrinia).

There is every gradation between these varieties, and every range and route of transport is found for both neuronal and paraneuronal messengers. In this respect also, neurons and paraneurons as well as "transmitters" and "hormones" are continuous with each other.

7. Effects of Messenger Substances

7.1 Continuity Between Modes of Actions of Messenger Substances Released from Neurons and Paraneurons

In the preceding chapters, it was emphasized that paraneuronal hormones and neurotransmitters *overlap* and are *continuous* with each other both in their chemical nature and in the route and range of their transport to the target cells. The continuity is also recognized between the responses of various target cells to paraneuronal hormones and to a variety of chemical substances released from neurons, but this continuity cannot easily be expressed following the rigid criteria of classic neurotransmitters and hormones. In this chapter, by amending the classic criteria of neurotransmitters, a new mode of chemical transmission of information will be proposed to unify a large variety of actions of messenger substances released from neurons and paraneurons. The term "messenger substance" is thus used to deal collectively with the chemical substances released from neurons and paraneurons, including the systemic, local, and paracrine hormones, neurotransmitters, and neuromodulators.

Messenger substances can induce diverse responses in various target cells, such as *secretory, contractile, metabolic, trophic*, and *genetic* responses. The *time scale* of these responses ranges from microseconds to years: The quickest response is induced via membrane receptors to change ion transport, while the slowest one may be induced by transcription in the nucleus.

7.2 Critical Account of Neurotransmitters and Neuromodulators

The most profound and concise knowledge of molecular mechanism of action of messenger substances has been developed in the study of synaptic transmission. Evoking *excitatory or inhibitory postsynaptic potential (EPSP or IPSP)* has been believed to be the sole action in common with all of the classic neurotransmitters.

The initiation of EPSP and IPSP corresponds to the sixth of the *eight criteria of a classic neurotransmitter* defined by Barchas et al. (1978) as follows: (1) The substance must be present in presynaptic elements of neuronal tissue, possibly in an uneven distribution throughout the brain. (2) Precursors and synthetic enzymes must be present in the neuron, usually in close proximity to the site of presumed action. (3) Stimulation of afferents should cause release of the substance in physiologically significant amounts. (4) Direct application of the substance to the synapse should produce responses which are identical to those of stimulating afferents. (5) There should be specific receptors present which interact with the substance; these should be in close proximity to presynaptic structures. (6) Interaction of the substance with its receptor should induce changes in postsynaptic membrane permeability leading to EPSP or IPSP. (7) Specific inactivating mechanisms should exist which stop interactions of the substance with its receptor in a physiologically reasonable time frame. (8) Interventions should exist at postsynaptic sites or through inactivating mechanisms. The effects of stimulation of afferents or of direct application of the substance should be equally responsive.

The advance in knowledge and understanding of postsynaptic electric responses have been brought about largely as a result of the microtechniques: electric investigations by microelectrodes, particularly with intracellular electrodes and microinjection procedures. We now know that the initial event caused by a neurotransmitter is *depolarization* (EPSP or end-plate potential) or *hyperpolarization* (IPSP) of the postsynaptic membrane of various excitable cells (neurons and muscle cells), although the final events of transmitter-induced responses in these target cells are quite different. The final event in neurons is acceleration or deceleration of release of proper neurotransmitter, whereas that in muscle cells is contraction or relaxation.

EPSP or IPSP in the postsynaptic cells is induced initially by binding of the neurotransmitter with the recognition subunit of the receptor, which, in turn, causes opening of an ionic-gating subunit (ionic channel) and accelarating passive cation or anion transport. The production of EPSPs by acetylcholine at synapses where acetylcholine is an excitatory transmitter provides a good example of the operation of these mechanisms. The amino acids GABA (γ-aminobutyric acid) and L-glumamate may be the principal inhibitory and excitatory neurotransmitters used for fast point-to-point transmission in the central nervous system (Iversen 1984).

Fast synaptic signals, ordinary EPSP, endplate potential, and IPSP, lasting for a small fraction of a second have been found at many synapses in the central nervous system of all the vertebrates that have been examined. In contrast to these fast responses, some neurons in the autonomic and central nervous systems produce electric responses lasting for many seconds or minutes. The responses produced in postganglionic neurons by stimulating their preganglionic innervation include not only a fast depolarization (fast EPSP) that generates action potentials but also a *prolonged inhibitory postsynaptic potential* (*slow IPSP*), a *prolonged excitatory postsynaptic potential* (*slow EPSP*), and a *late slow EPSP* (Kuffler and Sejnowski 1983). These three responses apparently modulate and regulate transmission through the sympathetic ganglia. The fast and slow EPSP are both produced by acetylcholine, but the former is mediated by a nicotinic receptor and the latter by a muscarinic receptor. The slow IPSP is probably produced by *dopamine*, which is secreted by an interneuron within the ganglion. The interneuron is excited by activation of a muscarinic receptor. The interneurons that secrete dopamine are SIF cells in the ganglia (Williams 1967; see also Ganong 1985). The late slow EPSP is produced by a *peptide* resembling LHRH.

The first *peptide* proposed as a neurotransmitter was *substance P* for primary afferents in the spinal cord.

Main reasons for this view are as follows: (1) Substance P is present in the primary afferent terminals in the superficial layers of the dorsal horn. (2) Substance P-like immunoreactivity is released from the spinal cord of the rat and the cat in response to electric stimulation of sensory nerves in a Ca^{2+}-dependent manner. (3) When synthetic substance P was applied to the isolated spinal cord of the neonatal rat and the response recorded from the motoneurons, substance P was found to exert a powerful excitatory action on spinal neurons. It is also shown that substance P is a neurotransmitter released from peripheral branches of primary afferents and producing a noncholinergic slow EPSP in the inferior mesentric ganglia (Otsuka and Konishi 1983).

The classic concept of the neurotransmitter, however, cannot embrace a wide variety of actions of a large number of neuropeptides or "brain-gut peptides." The concept of *neuromodulation* has thus been proposed to express the actions of a substance that alters neuronal activity (Barchas et al. 1978). In general, a synaptic transmission appears to be modulated in a variety of ways by the action of many kinds of endogenous modulators. Four types of modulatory actions of these endogenous modulators are proposed (Koketsu 1984): (1) Modulation of the amount of the transmitter released from presynaptic neurons; (2) modulation of the sensitivity of the receptor sites of postsynaptic neurons; (3) modulation of the resting membrane potential or conductance of postsynaptic neurons; and (4) modulation of the configuration of the action potential of postsynaptic neurons.

Koketsu (1984) recently reviewed the experimental data supporting the modulatory actions of endogenous substances in the nicotinic cholinergic transmission in bullfrog sympathetic ganglia. In frog spinal sympathetic ganglia, a pituitary peptide, LHRH causes a prolonged and slow depolarizing response, which resembles a late slow EPSP elicited by stimulation of preganglionic inputs (Jan and Jan 1982, 1983: Kuffler and Sejnowski 1983). Jan and Jan (1983) suggest that the peptide-mediated response shares a common ionic basis to the slow depolarizing effects elicited by acetylcholine acting on muscarinic receptors on the ganglion cells: namely inhibition of a voltage-sensitive K^+ channel (the "M" current). An LHRH-like peptide probably coexists with acetylcholine in the same nerve terminals, and although the same terminals that form conventional morphological synaptic contacts with sympathetic neurons, the peptide can diffuse a far greater distance than acetylcholine and can act on cells nonsynaptically (Jan and Jan 1983). The mode of this peptide transmitter thus can be said to be *paracrinia*. The excitatory effects of the LHRH-like peptide are very slow in onset and prolonged in duration, and by themselves cannot cause sufficient depolarization to excite the cells. The effect, however, is to render the ganglionic neurons more readily excited by other excitatory afferent inputs, and thus is regarded as a clear example of neuromodulation (Iversen 1984).

Other examples of neuromodulatory actions of neuropeptides are also reported. The *opioid peptides* (enkephalins and endorphins) can influence the release of monoamines or other neuropeptides by acting on presynaptically located opioid receptors; this includes inhibitory control of the release of noradrenaline, acetylcholine, dopamine, vasopressin, oxytocine and substance P (Bloom 1983; Jessel 1982; Hökfelt et al. 1986; Siggins and Gruol 1986). The precise mechanisms involved remained unclear, but recent studies by North and Williams (1983) suggest that opioid peptides may activate voltage-sensitive K^+ channels, which in turn may reduce Ca^{2+} fluxes and suppress release of amine or peptide.

Investigations of how transmitters modulate ionic channel function at the molecular level have greatly advanced by the patch-clamp technique for single-channel recording. Examples of such modulatory transmitter actions on single-channel currents are recently reviewed by Siegelbaum and Tsien (1983) and Sakmann and Neher (1984).

The concept of the neuromodulator is still based on the assumption that the information across a chemical synapse is solely transmitted by EPSP or IPSP. This assumption, however, should be broadened since studies in cellular neurobiology have profoundly advanced in three major fields as follow: First, countless new interneuronal connections have been revealed by new cellular tracing methods. Second, new methods of electrophysiological analysis have revealed a whole new range of ionic flux mechanisms which can be influenced by neurotransmitters to enrich their signaling capacity. Third, modern methods of chemical analysis have provided an ever-increasing list of new neurotransmitter candidates (Bloom 1984b). Thus, the term "neuromodulator" seems to be inadequate to express the recent advances in these fields.

7.3 Diverse Actions of Messenger Substances

Individual neurons may receive converging inputs from different neurons and generate several types of postsynaptic response with a wide range of time scale (Sejnowski 1982). For example, the interaction of four different signals that differ in their duration by a factor of up to 10 000 can be studied in single sympathetic ganglion cells of the frog (Kuffler 1980); interactions in the inferior mesenteric ganglion of the guinea pig may be even more complicated (Konishi et al. 1980).

The postsynaptic responses with longer time scales seem to be mediated by intracellular *second messengers*. The effects of activating the muscarinic receptor appear to be mediated via cyclic GMP (Ganong 1985). As the time scales of responses of target cells and the ranges of the transport of messenger substances are extended, the modes of actions of messenger substances become much more complex than the opening of an ionic channel.

The multiple and complex effects of messenger substances of neurons and paraneurons can, thus, be rearranged according to the *sites of the receptors*, although we are far from being able to localize the receptors for each of the diverse actions of messenger substances (Fig. 7.1). Messenger substances generally exert their effects via *membrane receptors*: (1) To increase cytoplasmic free Ca^{2+}; (2) to increase cyclic AMP in the target cells; (3) acting via *cytoplasmic receptors* to alter the transcription of mRNA; (4) acting directly on the *nucleus* to change the transcription of mRNA (Ganong 1985); or (5) acting to alter energy metabolism in the target cells (Bloom 1984a, b; Kanno 1984; Kanno and Saito 1985).

7.3.1 Effects Via Membrane Receptors to Change Ion Transport and Intracellular Ca^{2+} Concentration

Effects of messenger substances via membrane receptors to induce electric responses in the target cells have been mentioned in the preceding section. The origin of the electric responses in various secretory cells including paraneurons (see Part II; Poisner and Trifaro 1985) and in skeletal, cardiac, and smooth muscle cells seems to be the same. The molecular structures of *receptors on muscle cells* have not been demonstrated except for the acetylcholine receptor. Both in striated and smooth muscle cells, the coupling of a messenger substance with its receptor causes depolarization of the plasma membrane. The depolarization brings about an increase in the cytosolic Ca^{2+} concentration, which in turn causes contraction of the elements of myofilaments.

One of the first established cases of chemical regulation of an action potential channel was the *increase in* Ca^{2+} *conductance* by β-adrenergic agents (Siegelbaum and Tsien 1983). The β-adrenergic modulation of cardiac Ca^{2+} channels leads to increased Ca^{2+} influx during the action potential by a direct effect on Ca^{2+} conductance. The Ca^{2+}-channel modulation is mediated by cyclic AMP and protein phosphorylation.

Motile response to messengers may also be found in nonmuscular cells. In lower vertebrates, the paling reaction of the *melanocyte* to melatonin, an indoleamine derivative released from the pineal body, is a remarkable example. The receptor for melatonin is located in the plasma membrane of the melanocyte and it has been proposed that its molecular conformation stereospecifically fits the molecule of melatonin (Heward and Hadley 1975). The occurrence of second messengers and the mechanism causing the movement of melanin granules toward the cell center have yet to be determined.

In many *secretory cells*, the secretory response is started by a specific messenger substance acting on the membrane receptor of a specific target cell to promote either Ca^{2+} entry or the intracellular release of bound Ca^{2+}, which in turn mediates the secretory response (Poisner and Trifaro 1985).

The secretory response of the *pancreatic acinar cell* induced by *cholecystokinin* (CCK) at physiological concentrations (20 pmol or lower) is initiated by the binding of CCK with the receptor on the plasma membrane and maintained by a continuous increase in Ca^{2+} influx. The coupling induced by CCK at pharmacological concentration is maintained both by transient increase in Ca^{2+} release from Ca^{2+} storage sites in the aciner cell, and a continuous stimulation with CCK at pathological concentrations may cause a great increase in $[Ca^{2+}]_i$ (Kanno et al. 1983b, 1984b, 1985; Kanno and Saito 1978).

Secretin and *VIP* are believed to bind with a specific receptor on the plasma membrane of the intercalated duct cells of the pancreas to elevate the level of intracellular cyclic AMP, which mediates the secretion of pancreatic juice containing a large amount of $NaHCO_3$ (Case 1978). It is noteworthy that the secretagogous effect of these two peptides upon the pancreas is mimicked by a typical aminic messenger, dopamine (Hashimoto et al. 1971). This is remarkable evidence that peptides and amines may exert the same or a similar action.

Insulin couples with its receptor in the membrane of target cells. By this coupling, insulin increases the membrane potential of cells in skeletal muscle, adipose tissue (Ganong 1985), and the exocrine pancreas (Kanno and Saito 1976, 1979; Kanno et al. 1976), probably by potentiating the electrogenic Na^+-pump.

Changing active ionic transport may also occur via a membrane receptor. It is well known that the resting membrane potential of mammalian skeletal muscles decreases within a few hours following denervation. To explain this depolarization after denervation, an increase in the Na^+-permeability of the membrane and inactivation of the Na^+-K^+-pump have been suggested (review: McArdle 1983).

A decrease in the Na^+-K^+-electrogenic pump activity in denervated muscles was first inferred from the reduced effectiveness of ouabain in decreasing the resting membrane potential of denervated rat muscles compared with that in nondenervated muscles; however, the pump-inactivation hypothesis has been criticised on various grounds. McArdle (1983) proposed that denervation may result from a decrease in the sensitivity of the Na^+-K^+-pump to $[Na^+]_i$. Lorković and Hörr (1985) recently reported evidence compatible with the hypothesis of McArdle (1983). They used thin preparations of mouse soleus muscles, which were denervated for 3–11 days and bathed in Cl^--free solutions before the experiments. On application of acetylcholine, the muscle fibers were depolarized. Following the washout of acetylcholine, the muscle fibers were hyperpolarized. The hyperpolarization observed presumably reflects the *activation of the electrogenic Na^+-K^+-pump* by the Na ions, which entered the cells during exposure to acetylcholine; the hyperpolarization was suppressed by ouabain. Such an activation of the Na^+-K^+-pump is the second action of messenger substances of neurons.

7.3.2 Effect Via Membrane Receptors to Increase Cyclic AMP

Many of the messenger amines, peptides, and various other ligands have been known to act via the membrane receptors of their target cells to increase *intracellular cyclic AMP*, which mediates various effects. It has become evident that there is a synergic regulation of cell function by two intracellular messengers: cyclic AMP and Ca^{2+}. These two intracellular messengers are involved in coupling stimulus and response in a wide variety of cell types when those cells are called on to perform their specific function. In regulating cellular function, Ca^{2+} and cyclic AMP nearly always function together. Their interactions are plastic rather than stereotyped in character (Rasmussen and Barrett 1984).

7.3.3 Effects Via Cytoplasmic Receptors

Many messenger substances of neurons and paraneurons are known to increase protein biosynthesis; the stimulatory effect of growth hormone on protein biosynthesis appears to be exerted at the *ribosomal level*, where it has been postulated to affect either ribosomal attachment or translation. It also increases the transport of neutral and basic amino acids into cells (Ganong 1985). It is suggested that growth hormone stimulates the liver and perhaps other tissues to produce somatomedin, which has a variety of effects on cartilage, including stimulation of DNA and RNA synthesis and collagen formation. ACTH, on the other hand, acts upon membrane receptor to activate adenylate cyclase and protein kinase, which phosphorylate a component of ribosomes, leading to increased biosynthesis of a protein that facilitates the conversion of cholesterol to pregnenolone (Ganong 1985).

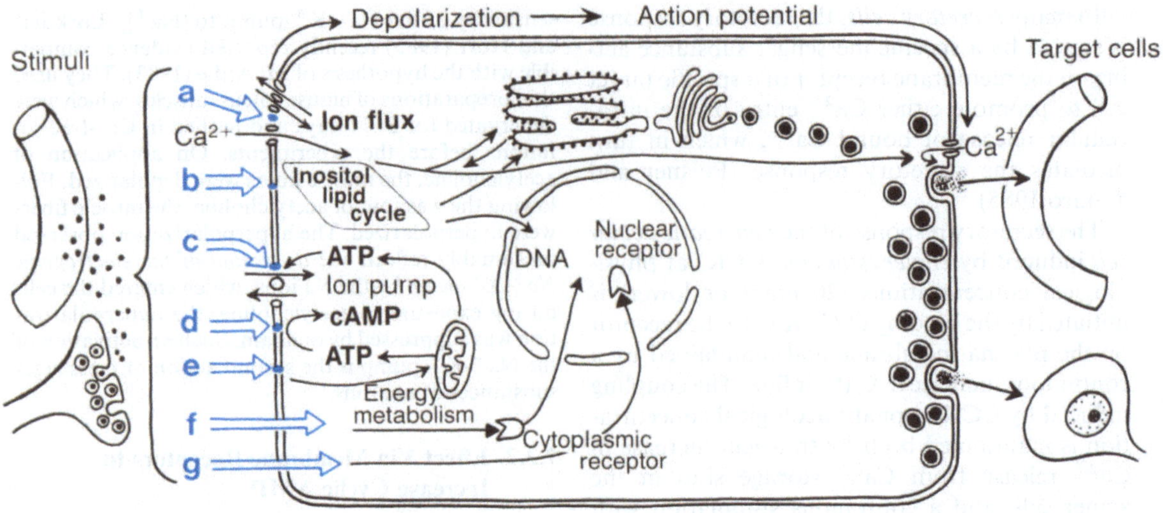

Fig. 7.1. Variety of mode of action upon a target cell. Messenger substances released from paraneurons and neurons may induce a variety of actions as follows. *a* Changes in ion fluxes generate slow depolarization in some cases. A slow depolarization may occasionally develop into an action potential, which is propagated to the secretory portion and opens the voltage-dependent Ca^{2+} channel at the portion. Opening of a receptor-activated Ca^{2+} channel at the receptor portion can also increase Ca^{2+} influx. *b* Activation of the inositol lipid cycle may induce various cellular responses, including release of Ca^{2+} from the intracellular Ca^{2+} store sites and cell growth. *c* Potentiation of ion pumps requires ATP supply and is responsible for electrogenesis. *d* Activation of cyclic AMP formation triggers various cellular responses in concert with Ca^{2+}-dependent processes and turnover of the inositol lipid cycle. *e* Activation of oxidative phosphorylation in mitochondria and the glycolysis system supplies ATP for various cellular responses. *f* Activation of cytoplasmic receptors alters the transcription of mRNA. *g* Action directly on the nucleus increasing the transcription of mRNA

The mechanism of the *trophic effect* is poorly known. As seen in the actions of gastrin upon parietal cells, CCK upon pancreatic cells, and NGF upon sympathetic neurons, the trophic effect is induced by potentiation of protein biosynthesis. When the target is a secretory cell, this effect is usually linked with a secretory effect. The trophic effect, therefore, is presumed to be an integration of multiple actions induced by messengers and second messengers, partly common to or associated with the protein-producing and secretion-releasing mechanisms and partly independent of them. The trophic action thus occurs both at the cytoplasmic (especially ribosomal) level and partly at the nuclear (gene and transcription) level.

The trophic action of gastroenteric hormones (Solomon et al. 1979) and insulin (Goldfine and Williams 1984) has been closely studied in pancreatic acinar cells. CCK and insulin interact with specific receptors on the cell membrane. Protein synthesis is initiated either through a second messenger generated by the coupling of the messenger with the receptor or through a subsequent direct interaction of the incorporated messenger with intracellular organelles, including the endoplasmic reticulum-Golgi system. A series of events then brings about stimulation of cell growth, increase in the amylase content, and elevation of sensitivity to CCK (Goldfine and Williams 1984).

7.3.4 Effect at the Nuclear Level

It was mentioned above that the *trophic effect* probably involves the messenger's action at the nuclear level. Protein synthesis depends upon the commands of the nucleus through mRNA. It is believed that the messenger substances may, directly or indirectly, alter the transcription of mRNA. Trophic effects are further based on the *DNA-level reaction* of the nucleus, especially its *division*.

Greenberg et al. (1986) recently demonstrated that cholinergic agonists rapidly and transiently induced transcription of the c-*fos* proto-oncogene and one or more actin genes in neuronally differentiated PC12 cells. Transcription was activated within minutes after stimulation of the nicotinic acetylcholine receptor and resulted in an influx of extracellular Ca^{2+} ions through voltage-sensitive calcium channels. Nicotine activation proceeded by a different pathway from activation by NGF, whose stimulation of these genes is independent of extracellular Ca^{2+} ions. These findings suggest that neurotransmitters may rapidly activate specific gene transcription in nondividing neuronally differentiated cells. The findings also suggest a functional role for neurotransmitter induction of c-*fos* and actin expression in the nervous system.

In neurons, protein synthesis has been demonstrated to be involved in the processes responsible for *memory* (references: see Rosenzweig and Bennett 1976; Nagata 1984). The neuronal activity during learning could lead to changes that enhance the synthesis of mRNA and hence the synthesis of particular proteins. These proteins could modify synaptic transmission by affecting transmitter synthesis, membrane permeability, or some other neural process. It is, however, a long way from such speculation to an understanding of the relationship between synthesis and the formation of a stable memory trace (Ganong 1985). Because both calcium and protein kinase C have been implicated in the acquisition of long-term potentiation (see 4.1, Chap. 4), there is some experimental evidence for the involvement of such messenger systems in memory (Berridge 1986).

7.3.5 Effects on Energy Metabolism

Messengers of neurons and paraneurons may alter energy production in the target cells ("energy domain," Bloom 1984a). Adrenaline and noradrenaline both cause *glycogenolysis*. They act via β-adrenergic receptors and increase cyclic AMP to activate phosphorylase and via α-adrenergic receptors to accelerate glycogen breakdown by increasing the level of intracellular Ca^{2+}. Insulin facilitates the entry of glucose into cells by promoting diffusion across the cell membrane in muscle, adipose tissue, and connective tissue. Insulin does not affect the movement of glucose across the membranes of hepatic cells directly, but it brings about facilitation of glycogen synthesis and a decrease in glucose output (Ganong 1985).

According to one theory of the intracellular action of insulin, which accounts for all of its anabolic effects on administration to whole animals or isolated cells or tissues, insulin acts by connecting hexokinase to the mitochondria of susceptible cells, thereby furnishing a respiratory control stimulus (Bessman et al. 1986). The anabolic processes, which take place adjacent to the stimulated mitochondria, receive an increase in energy (ATP) supply and are accelerated. Glucagon stimulates adenylate cyclase and increases cyclic AMP, which leads to the activation of phosphorylase and, therefore, to increased glycogen breakdown in liver cells. Glucagon increases gluconeogenesis from available amino acids in the liver and elevates the metabolic rate.

In the brain, the messengers' actions on the "energy domain" have been investigated by Bloom (1984). Among the various known cortical messenger substances of neurons tested, only VIP and noradrenaline share the ability to activate cyclic AMP synthesis and at least one step that can be initiated by cyclic AMP, namely glycogenolysis (Bloom 1984a, b). Glycogen is the single largest energy reserve of the central nervous system (Watanabe and Passonneau 1973; Cammermeyer and Fenton 1981). Enzymes for glycogen metabolism in neurons are regulated by intracellular messengers such as cyclic AMP and free Ca^{2+} (Breckenridge and Crawford 1961; Breckenridge and Norman 1965; Nelson et al. 1968; Drummond and Bellward 1970). Both VIP and noradrenaline promoted concentration-dependent glycogenolysis with an EC_{50} of 26 nmol (Magistretti et al. 1981) and 500 nmol (Magistretti et al. 1981), respectively. This action was rapid: Approximately 50% of the maximal effect was reached within 1 min (Magistretti et al. 1984). When both messenger substances were tested in combination (10 μmol/l noradrenaline and 1 μmol/l VIP), cyclic AMP levels increased far in excess of those expected through simple addition (Magistretti and Schorderet 1984). This synergism corresponds with the geometric arrangement of these two types of neuron as follows. VIP seems to affect glycogen levels by regulating the availability of energy substances locally within partially overlapping cortical columns of 100 μm in diameter (Magistretti 1985). In contrast to this radially oriented local action of VIP, the noradrenergic system seems to promote glycogenolysis throughout a vast rostrocaudal expanse of neocortex, simultaneously passing across adjacent columns (Magistretti et al. 1981). Thus VIP- and noradrenaline-containing neuronal circuits within the cerebral cortex may constitute an example of a "neuronal system with a hormonal function" (Magistretti 1985). Such systems probably operate spatially and temporally in parallel with those circuits that directly and primarily modify neuronal excitability (Magistretti 1985).

A fresh clue to the mode of action of messenger substances on the *mitochondria* in the target cells has been provided in the studies made by recording the redox response of the electron transfer system. Cholecystokinin octapeptide (CCK-8) has been shown to act on the respiratory chain and to induce an immediate and parallel reduction in cytochromes, aa_3, b, and $c+c_1$, followed by an increase in protein output in the isolated perfused rat pancreas (Asada and Kanno 1985; Kannó et al. 1981; Shibuya and Kanno 1985). CCK-8 at physiological concentrations caused little change in redox state and continuous pancreatic protein output; at pharmacological and pathological concentrations it caused a cytochrome reduction and secretory responses. The CCK-induced cytochrome reduction may reflect the excess increase in $[Ca^{2+}]_i$ (Shibuya and Kanno 1985). Acetylcholine also induced a reduction of cytochromes with concomitant increase in protein output (Kanno et al. 1983a).

In the perfused bullfrog brain it has been shown that CCK-8, neuropeptide Y (NPY), and afferent activation induced a parallel reduction of cytochromes aa_3, b, and $c+c_1$ (Kanno and Saito 1982; Saito et al. 1983; Kanno et al. 1985). A recent study showed that VIP pretreatment nullified the parallel reduction of the cytochromes induced by afferent activation in the perfused bullfrog brain (Kanno and Saito 1986).

7.4 Summary

Diverse effects are exerted by "hormones" or messenger substances of paraneurons—secretory, motile, trophic, metabolic, etc. These effects are common to those exerted by "transmitters" or neuronal messengers.

Paraneuronal messengers, like neuronal ones usually have their *receptors* on the cell surface of their targets. Besides these membrane receptors, there is the possibility of intracytoplasmic—especially ribosomal and mitochondrial—and nuclear receptors, although it is at present unclear whether actions at these sites are due to the messengers themselves or *second messengers*. The second messengers of paraneuronal messengers include Ca^{2+}, cyclic AMP, and inositol metabolites; these are also shared by neuronal messengers.

Peptidic and aminic messengers may exert similar effects upon the target cells. The mechanisms of their actions via receptors and second messengers seem essentially identical.

The concepts of *neurotransmitter* and *neuromodulator* have been based on the belief that the effect of a chemical messenger appears in terms of an excitatory or inhibitory postsynaptic potential (EPSP or IPSP). This electric phenomenon, however, represents only a small part of the diverse reactions of target cells induced by chemical transmission. Rigid usage of the terms "neurotransmitters" and "modulators" will hinder a sound insight into the continuity between transmitters and hormones.

8. Ontogeny and Kinetics of Paraneurons

8.1 Concepts on the Origins of Paraneurons

8.1.1 Unifying Theory of Common Origin

A hypothesis of a *common origin* of peptide/amine producing endocrine cells was proposed by Pearse (1966a, b, 1968). He conceived that all amine precursor uptake and decarboxylation (APUD) cells might be derived from the *neural crest*. By examining dopa-provoked formaldehyde-induced fluorescence (FIF) in mouse embryos, Pearse and Polak (1971a) reported that thyroid parafollicular cells are derived from the neural crest since the neural crest cells, cells migrating from there, and the cells which have settled in the thyroid were all FIF positive. The same conclusion was reached by Pearse and Polak (1971a, b) concerning the endocrine cells in the pancreas, gut, and respiratory tract.

The *unifying concept* proposed by Pearse that the all APUD cells are of neural crest origin proved highly appealing, especially among pathologists. This concept could clearly account for the frequent involvement of different types of endocrine cells and even of neurons in simultaneous neoplastic changes like multiple endocrine adenomatosis (MEA) or multiple endocrine neoplasm (MEN), and the occasional ectopic occurrence of peptide hormones in neoplastic endocrine cells (Weichert 1970; Pearse and Welbourn 1973; Pearse and Polak 1974; Pearse 1975a).

Objections against the Pearse concept of the neural crest origin of APUD cells were made by Le Douarin and Teillet (1973) and Andrew (1974, 1975, 1976a). *Experimental embryological studies* by these researchers have indicated that important members of the APUD series and of the paraneuron family, such as the gastroentero-pancreatic (GEP) and bronchopulmonary endocrine cells are not derived from the neural crest (reviews: Andrew 1981, 1985; Ayer-Le Lievre and Fontaine-Perus 1982).

Pearse (1975, 1977; Pearse and Takor Takor 1979) meanwhile changed his notion of the "neural crest" being the origin of the APUD cells to a "neuroectoderm" and later to a "neuro-endocrine-programmed ectoblast" or "epiblast" origin. Andrew (1985) recently pointed out that the latter two terms make no sense from an embryological point of view.

When we proposed the term and concept of the paraneuron (Fujita 1976), we were preoccupied with the idea that all paraneurons were neuroectodermal in origin. As the studies advanced, however, we withdrew this view and came to support the notion that the cells have a multiple origin.

8.1.2 Concept of Multiple Origin

Recent studies indicate that paraneurons are derived from *different germ layers*. Some of them are neuroectodermal in origin, while others are nonneural. Thus, paraneurons can be divided into the following groups according to their origin (Table 8.1).

Paraneurons of neural tube origin. Visual cells in the retina and pinealocytes in the pineal body originate from the neural tube.

Paraneurons of neural crest origin. Adrenal chromaffin cells, SIF cells in the autonomic ganglia, carotid body chief cells, and melanocytes are paraneurons that have been established as being derived from the neural crest. It should be borne in mind that many autonomic neurons also originate from the neural crest. The origin of thyroid parafollicular cells and their homologous elements in lower vertebrates, ultimobranchial body cells, has not been fully established, though a neural crest origin has been supported by experimental-embryological studies (Le Douarin et al. 1974). Excellent reviews of the neural crest and its derivatives have been provided by Weston (1970) and Le Douarin (1982).

Table 8.1. Origins of main paraneurons

Cell type	Origin
Pinealocyte	Neuroectoderm (neural tube)
Visual Cell	Neuroectoderm (neural tube)
Adrenal chromaffin cell	Neuroectoderm (neural crest)
Carotid body chief cell	Neuroectoderm (neural crest)
Parafollicular cell	Neuroectoderm (neural crest)? Ectoderm?
Inner ear hair cell	Ectoderm ("placode")
Olfactory cell	Ectoderm ("placode")
Gustatory cell	Ectoderm (lingual epithelium)
Merkel cell	Ectoderm (epidermis, mucous epithelium)
Adenohypophyseal cell	Ectoderm (Rathke's pouch)
Parathyroid cell	Endoderm (branchial pouches III and IV)
GEP endocrine cell	Endoderm (gut epithelium)
Bronchopulmonary paraneuron	Endoderm (air tract epithelium)
Urogenital paraneuron	Endoderm (epithelium of cloacal origin?)
Mast cell	Mesenchyme (bone marrow stem cell)

According to Rovasio et al. (1983) events in the migration of the neural crest cells occur in the following four stages:
1) *Phase of individualization.* Neural crest cells become free of the neural epithelium. They exhibit a multipolar shape after a transient rounding up during mitosis.
2) *Phase of migration.* Neural crest cells migrate along fixed routes which have an acellular three-dimensional fibrillar matrix.
3) *Phase of arrest.* When neural crest cells encounter barriers such as the developing endoderm, their migration ceases. They lose their filopodia, start to produce specific substances for adhesion, and become fixed.
4) *Phase of aggregation.* Neural crest cells proliferate and regroup into a structure with a specific arrangement.

Paraneurons of (nonneuronal) epithelial origin. Many kinds of sensory and endocrine paraneurons are derived from nonneuronal epithelia. As sensory paraneurons, gustatory cells in the taste buds and Merkel cells in the skin originate from the epithelia *in situ*. The former cells are ectodermal and partly endodermal (pharyngeal gustatory cells), while the latter are all ectodermal. Olfactory cells in the nasal mucosa and hair cells in the inner ear and lateral line system have been described as originating from specific *placodes*. However, "placodes" are mere thickenings in the ectodermal epithelium and there are no reasons for ascribing a special neuroectodermal nature to them.

Among endocrine paraneurons, the origin of the GEP endocrine cells has been disputed most intensively. Although they were claimed to be derived from the neural crest by Pearse and Polak (1971a, b), the classic view of their endodermal origin has been confirmed by experimental embryological studies (see p. 180). Endocrine cells in the respiratory tract and urogenital epithelia are also believed to be differentiated in the epithelia *in situ*, i.e., they are of endodermal and possibly also mesodermal origin.

Parathyroid cells are of endodermal origin, being derived from the epithelium of the third and fourth pharyngeal pouches. Adenohypophyseal cells are of ectodermal origin; they are derived from the epithelium of the Rathke's pouch, an outgrowth of the stomodeum. The origin of the parafollicular cells in the thyroid has been controversial. The classic view is an epithelial origin but recent experimental embryological studies suggest a neural crest origin (see p. 148).

All the paraneurons derived from the epithelium proliferate in the epithelial tissue and do not positively migrate. They only passively change their position by the movement of the cellular elements around them.

Paraneurons of mesenchymal origin. The data introduced above may indicate that paraneurons are cells of epithelial origin (Ayer-Le Lièvre and Fontaine-Perus 1982). However, the mast cell, although it is arbitrary whether to include or exclude this cell from the paraneuron, is a cell of mesenchymal origin. It originates from the bone marrow. Mast cells are derived from the pluripotent hematopoietic stem cells, circulate throughout the body, leave the bloodstream, and settle and proliferate in the connective tissue (Kitamura et al. 1977).

8.2 Cell Kinetics of Paraneurons

It is rather surprising that there was no concept of a cell life cycle (cell cycle) until Howard and Pelc in 1952 discovered, using autoradiography, that radioisotope-labeled DNA precursors were incorporated into the cell only during a limited part of the intermitotic phase (interphase). The concept of cell cycles now is regarded as one of the central concepts in cell biology.

Howard and Pelc (1952) divided the interval between mitoses, which was then misleadingly called the "resting" phase, into three phases. The G1 (gap 1) phase follows mitosis and proceeds into DNA synthesis; the S (synthesis) phase is the period of DNA synthesis; the G2 (gap 2) phase occupies the period between the S phase and the period of the next mitosis the M phase. The four phases, G1, S, G2, and M, make up the cell cycle of all eukaryotes. It is known in many tissues and cells that the morphological, biochemical, and physiological characteristics greatly change with the cell cycle (Pardee et al. 1978; Leblond 1981; Baserga 1981).

Most tissues throughout the body contain three types of cells: (1) Proliferating or dividing cells, which repeat the cell cycle and proliferate, e.g., generative cells in the crypt of the small intestine divide repeatedly (Leblond 1981). (2) Quiescent or resting cells, which ordinarily do not divide, but can be stimulated to divide under various conditions, e.g., liver cells are stimulated to divide after partial hepatectomy (Baserga 1981). (3) Nondividing or postmitotic cells, which leave the cell cycle at mitosis and do not divide again, e.g., erythrocytes and spermatozoa remain nondividing and perform their specific functions (Till and McCulloch 1980).

It is important that all the typical neurons are nondividing or postmitotic cells. They are diploid, having two sets of chromosomes, and are therefore arrested in the G1 phase. The occurrence of tetraploidy, the possession of four sets of chromosomes, is also reported in large neurons such as Purkinje cells of the cerebellum (Lapham 1968, Lentz and Lapham 1969). It is likely that a few large neurons are arrested in the G2 phase. However, these large neurons have never been reported to undergo mitosis.

Concerning the paraneurons, most of them possess the ability to perform *mitosis*. Many electron micrographs have been published which show the mitosis of such paraneurons as the pancreatic islet cells (Like and Chick 1969), adrenal chromaffin cells (Grynszpan-Winograd 1969), carotid body chief cells (Kondo 1971), and SIF cells of the autonomic ganglion (Yokota and Burnstock 1978). These paraneurons divide when abundant secretory granules are present in the cytoplasm. There is strong evidence that granule formation progresses during the mitosis (Fujita 1976; Kurosumi 1971). In an adenohypophyseal cell, Kurosumi (1971) demonstrated that exocytotic release of the granule contents may take place during the mitotic phase.

Mitosis of some paraneurons can be induced by certain experimental conditions. Yanatori and Fujita (1976) demonstrated a mitotic proliferation of B cells of the pancreatic islet in rats which were fed raw soybeans and given trypsin inhibitor or repeatedly injected with cholecystokinin (CCK). Kurosumi (1971) showed many mitotic figures in the rat anterior pituitary cells after thyroidectomy (thyroid-stimulating hormone (TSH)-producing cells) and after long-term administration of estrogen to the female (prolactin-producing cells). Kameda (1970) demonstrated increased mitotic activity of the thyroid parafollicular cells of the dog in experimentally induced hypercalcemia by long-term treatment with $CaCl_2$ and vitamin D_3. Thus, most, if not all, kinds of paraneurons possess the ability to divide and proliferate. Extreme stimulation of the specific function of a certain paraneuron by experimental procedures, such as the removal of the target organ, often induces the mitotic proliferation of the paraneuron concerned (Kurosumi 1971).

In the developing central nervous system of higher vertebrates, the neurons originate from the neuroblasts, which are incapable of further division. Thus, the neuroblast is the young neuron itself: The neuroblast, once differentiated, is programed to give rise only to a nondividing neuron by exposing specific features, such as dendrites and axons. On the other hand, some paraneurons or their blast cells repeat cell cycles and can proliferate throughout life. Mitosis of paraneurons is frequently encountered in the embryonic and fetal stages. However, in adult animals, mitotic figures in paraneurons are extremely rare. It seems likely that the cell generation time of paraneurons increases with increasing age of the animal. Since in other proliferating systems the time spent in the S, G2, and M phases remains constant (Pardee et al. 1978; Baserga 1981), the elongation of the G1 phase may be the most significant factor in the lengthening of the generation time of paraneurons in adult animals.

Thus, in adult animals, most paraneurons are in the *G0 phase* and not in the cell cycle. Appropriate stimuli are required for them to reenter the cell cycle.

In most tissues, the production of parenchymal cells is in equilibrium with their loss. Thus, a kind of "structural homeostasis" is maintained and the uniformity of the individual tissues continues throughout life. It seems reasonable to suggest that tissues containing paraneurons are generally under structural homeostasis and that a certain proportion of these paraneurons continue to be renewed. However, convincing data on the turnover and renewal of paraneurons are few, except for some reports on adrenal chromaffin cells (p. 89).

8.3 Aging of Paraneurons

All the features of paraneurons, like those in ordinary neurons, continually change with time. Old paraneurons (a long time after the final mitosis) possess characteristic features which do not exist in young paraneurons (soon after mitosis). Our knowledge on the aging of paraneurons is still very incomplete.

In ordinary neurons, which represent postmitotic cells, typical signs of aging are an accumulation of the *lipofuscin granules* and *Alzheimer's neurofibrillar change* (Alzheimer 1907). The lipofuscin granules are a kind of secondary lysosome, i.e., residual bodies. They contain indigested remnants of auto- and heterophagocytized cellular components. On the other hand, Alzheimer's neurofibrilar change is characterized by the increase in filaments and other cytoskeletal proteins, such as neurofilaments, intermediate filaments, actin filaments, and microtubules.

The accumulation of lipofuscin granules is frequently reported in some paraneurons such as human pancreatic islet cells (Shibasaki and Ito 1969). We have the impression that there is a tendency for human paraneurons to contain more lipofuscin granules than the paraneurons of experimental animals. This may be due to the fact that the laboratory animals we use in electron-microscopic investigations provide younger paraneurons than those we observe in human material.

The accumulation of enormous amounts of *filamentous material* is also reported in human paraneurons such as the duodenal endocrine cells (Kobayashi et al. 1970). It is likely that they represent paraneurons which correspond to the aged neurons with Alzheimer's neurofibrilar change.

In adult animals, many paraneurons remain in the G0 phase of the cell cycle and perform their specific functions. Therefore, signs of aging accumulte mainly during the G0 phase. If appropriate stimuli let the paraneurons reenter the G1 phase, the S and G2 phases follow according to the gene-coded programing, and they divide into daughter cells. The fate of the lipofuscin granules and of the filamentous material mentioned above during the mitosis remains to be demonstrated. It is likely that these structures hinder the smooth progress of the mitosis.

8.4 Ontogeny of Some Paraneurons

8.4.1 Visual Cells and Pinealocytes

Photoreceptor cells in the retina originate from the *neural tube*. Melatonin secreting pinealocytes in higher vertebrates as well as their photoreceptive ancestors in lower vertebrates are also derived from the neural tube.

The neural tube consists of the neuroepithelial cells whose nuclei change position during the cell cycle. The cell division occurs only in the zone nearest to the lumen of the central canal of the neural tube. Between the mitoses, the neuroepithelial cells extend peripheral processes to the outer surface of the neural tube, and the newly formed nuclei move to the periphery. The neuroepithelial cells then retract their peripheral processes and the next mitosis takes place. Since the neuroepithelial cells are held tightly together by junctional complexes at the luminal surface, they passively shrink in the innermost layer, when connections of their peripheral processes and the basement membrane of the neural tube are cut.

Retinal visual cells and pinealocytes are regarded as a kind of cerebrospinal-fluid contacting neuron (Vigh-Teichmann and Vigh 1983). Their receptor sites represent dendrites which are originally in direct contact with the cerebrospinal fluid. Differing from other paraneurons, the visual cells and pinealocytes, once differentiated as the blast cells within the neural epithelium, do not divide again and do not change their position throughout later life, as ordinary neurons of the central nervous system.

8.4.2 Adrenal Chromaffin Cells

Adrenal chromaffin cells originate from the *neural crest* cells which form the sympathetic chain (Balfour 1877). In the mouse, developing chromaffin cells invade the adrenal cortical tissues on the 11th–12th days of gestation. This immigration is completed on the 14th day (Theiler and Müntener 1974). The differentiation between the two types of adrenal chromaffin cells, the adrenaline-storing (A) and the noradrenaline-storing (NA) cells, starts in the prenatal period (Elfvin 1967; Theiler and Müntener 1974). In the rat, the adrenal medulla 4–8 days after birth exhibits a structure resembling that of adult animals (Elfvin 1967). The mitotic activity of the adrenal chromaffin cells in newborn mice is about ten times higher than that in adult animals (Jurecka et al. 1978).

Adrenal chromaffin cells of the adult rat and mouse incorporated ^{3}H-thymidine at a labeling index of about 1% (Malvaldi and Viola-Magni 1972; Jurecka et al. 1978). Although not all the DNA labeling represents the premitotic synthesis, it is certain that the fully differentiated adrenal chromaffin cells turnover with a mean life span of shorter than 1338–3125 days (Malvaldi and Viola-Magni 1972).

8.4.3 Carotid Body Chief Cells

In chick embryos with quail rhombencephalic grafts, Le Douarin et al. (1972) recognized that the carotid body consisted mainly of quail cells with their characteristic nucleoli. This result suggests that the carotid body chief cells arise as migrants from the anlage of the rhombencephalon, *probably from the neural crest*. Fluorescence histochemical studies have demonstrated that catecholamine-containing cells migrate into the carotid body along a process connecting it with the sympathetic chain (Korkala and Hervonen 1973).

Little is known of the cell cycle and kinetics of the carotid body chief cells. Kondo (1971) reported occasional mitosis in the cells of the guinea pig.

8.4.4 Melanocytes

The *neural crest* origin of melanocytes has been established by transplantation experiments in amphibia (du Shane 1935, 1943) and mammals (Rawles 1947, 1955) (see p. 259).

8.4.5 Parafollicular Cells

Parafollicular cells of the mammalian thyroid gland are disseminated among the thyroid follicles as isolated cells or as a small cell group. Together with their submammalian homologue, the ultimobranchial body, these cells were long regarded as endodermal origin, being derived from the fourth pharyngeal pouch. Pearse and Polak (1971c), however, proposed a neural crest origin of the cells, based on their observations of amine-fluorescent cells in mouse embryos, which appeared to migrate from a position between the epidermis and the neural tube, via the fourth pharyngeal pouch, down to the thyroid gland. Le Douarin and Le Lièvre (1970) transplanted the caudal portion of the brain from quail to chick embryos and found that the majority of cells in the ultimobranchial body identifiable as paraneuronal cells by electron microscopy and by monoamine fluorescence contained nuclei characteristic of the quail. Later, a joint study of the groups of Pearse and Le Douarin (Polak et al. 1974) demonstrated in chick embryos with quail rhombencephalic transplants that calcitonin-immunoreactive cells mostly contained quail nucleoli. These investigations indicate that, at least in birds, the parafollicular and ultimobranchial cells are of neuroectodermal origin.

The turnover of parafollicular cells in mammals may presumably be more active than that of chromaffin cells. Mitotic proliferation of the cells can be provoked by hypercalcemia induced by repeated injections of $CaCl_2$ and vitamin D (Kameda 1970) (see p. 149).

8.4.6 Olfactory Cells

Olfactory cells in the nasal mucosa are derived from a thickened epithelium, called the *olfactory placode*, forming the roof of the nasal groove. They are therefore epithelial and ectodermal in origin. By differentiation, they extend an axon to the part of the telencephalon which is destined to be the olfactory bulb.

Although olfactory cells have previously been categorized as neurons, they are *constantly renewed* in the nasal epithelium (Graziadei and Monti-Graziadei 1978). Yamamoto et al. (1976) showed that an immature olfactory cell assumes the form of a basal-granulated cell. Radioautographic studies have indicated that immature cells in the olfactory epithelium differentiate into mature bipolar cells in 7 or 8 days and that the labeled elements degenerate and disappear from the epithelium in no longer than 30–35 days (Graziadei et al. 1978) (see p. 217).

8.4.7 Gustatory Cells

Gustatory cells in the taste buds have long been known to be of *epithelial origin.* The lingual gustatory cells are ectodermal, whereas the pharyngeal and laryngeal cells are endodermal in origin. In the rat, gustatory cells differentiate on day 15 of gestation. It is likely that nerves reaching the epithelium induce the differentiation of the cells. The gustatory cells depend on the nerves for survival, as they degenerate rapidly after the nerves are sectioned (see p. 210).

8.4.8 Merkel Cells

Markel cells, the mechanoreceptor cells dispersed in the basal portion of the epidermis, have been thought to originate from the neural crest. This hypothesis, however, seems to be disproved by the fact that Merkel cells are derived from the basal cells of the epidermis. Tweedle (1978) removed the entire presumptive neural tube and crest from urodele embryos and found no effect on the development of Merkel cells. Cells intermediate in morphological characteristics between Merkel cells and keratinocytes have been recognized (English 1977a), a finding supporting the epidermal origin of Merkel cells.

Merkel cells have been presumed to differentiate and survive under the effect of the sensory nerves associated with them. Evidence, however, has accumulated indicating that the nerves exert no essential influence upon the differentiation of Merkel cells. Both in the labial mucosa of adult rabbits (Tachibana et al. 1983) and epidermis of adult salamanders (Scott et al. 1981), new Merkel cells appeared in the regenerative tissue which had been wounded and denervated. After denervation of rabbit labial mucosa, Tachibana et al. (1983) recognized no degenerative signs in Merkel cells. Nafstad (1987) studied a life cycle of mammalian Merkel cells and presented evidence that they originate from the epidermal cells and turnover continually.

8.4.9 GEP and Related Endocrine Cells

Gastro-entero-pancreatic or *GEP endocrine* cells have been much disputed with regard to their origin. Classically, these cells were believed to be endodermal in origin, though a few studies have supported their nervous origin (Danisch 1924). More recently, Pearse and Polak (1971a) renewed the concept of the nervous origin of GEP endocrine cells. In mouse embryos, they observed that the monoamine-fluorescent cells first occurred in the neural crest and later in the stomach, duodenum, dorsal pancreatic bud, and air tract epithelium, and concluded that both GEP and bronchopulmonary endocrine cells are of neural crest origin.

Experimental embryological studies, especially experiments transplanting the neural primordium between chick and quail embryos by Le Douarin and Teillet (1973) and chorioallantoic culture of presumptive gut by Andrew (1974), however, have indicated that the neural crest and even the neuroectoderm are not involved in the ontogeny of GEP endocrine cells. The *endodermal origin* of GEP endocrine cells has thus become the generally accepted view. Fujii (1979) demonstrated in the rat that the initial differentiation of endocrine cells (glucagon cells) occurs, on day 11 of gestation, in the duodenal wall, including the just forming dorsal pancreatic bud. The gastroenteric and pancreatic endocrine elements are thus formed on the base of the same, continuous endoderm from the beginning. The endocrine cells both in the pancreas and gastroenteric epithelium mitotically proliferate and differentiate into mature forms (Fig. 8.1). Their turnover rate has not been convincingly determined but is generally presumed to be much longer than that of ordinary cells in the intestinal epithelium, which live only a few days (see p. 182).

It has been demonstrated that both the gastroenteric and pancreatic endocrine cells may be associated with nervous tissue (neuronal somata or nerve bundles) in the sub- or interepithelial connective tissue (Fujita 1959; Osaka and Kobayashi 1976). These histological features might reflect possible induction of the endocrine cells by the nervous elements. If this is not the case, the close association of the endocrine and nervous cells must indicate a specific affinity between both elements. The embryological relationship may persist in adult tissues as particular structures called *neuroinsular complexes* in the case of the pancreas or, more generally, *neuroendocrine complexes* (Fujita and Kobayashi 1977) (see p. 93).

An excellent review is available on the changes of concepts on the origin of GEP endocrine cells (Andrew 1981).

Parathyroid cells have been regarded as being of endodermal origin since they are derived

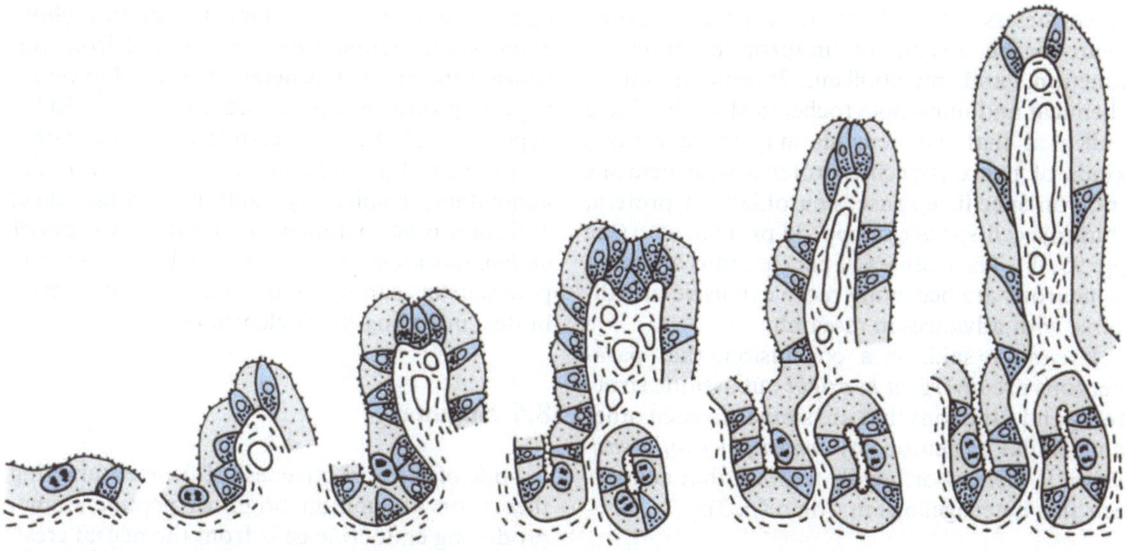

Fig. 8.1. Ontogeny and kinetics of gut paraneurons in the human duodenum. Gut paraneurons appear in the intestinal epithelial tissue at 3 months of gestation. They increase rapidly in number in the period between 3 and 5 months of gestation. During the period between 5 and 9 months of gestation, Segi's cap, a large aggregation of gut paraneurons on the tip of the intestinal villi, is formed. Segi's cap disappears shortly after birth. In adult humans, mitosis of gut paraneurons takes place in the intestinal crypt. The paraneurons move toward the tip of the intestinal villi, where they degenerate and exfoliate into the intestinal lumen

from the third and fourth pharyngeal pouches (see p. 154). Although Pearse and Takor Takor (1976) attempted to find evidence for a neural or placodal origin of the cells, no convincing data were obtained.

Adenohypophyseal cells have long been considered ectodermal in origin since they are derived from Rathke's pouch, which is an outgrowth of the stomodeum. As reviewed by Andrew (1981), some recent studies, especially by Takor Takor and Pearse (1975), have supported the hypothesis of the neural crest origin of adenohypophyseal cells, but this idea has met with objections. The generally accepted view today is that the adenohypophysis is of nonneural ectodermal origin (Andrew 1981), though a possible influence (induction in a broad sense) of the hypothalamus upon the differentiation and/or growth of the adenohypophysis cannot be excluded.

The endocrine cells first detectable by immunocytochemical means in the rat adenohypophysis are adrenocorticotropic hormone (ACTH) cells, which appear on day 14.5 of gestation. Interestingly, ACTH-containing neurons can be identified one day earlier in the hypothalamus (Daikoku et al. 1983c). Pituitary cells containing thyroid-stimulating hormone (TSH: day 15.5), luteinizing hormone (LH: day 16.5), and other peptides subsequently appear, while prolactin-containing cells are only found at the end of fetal life.

Endocrine cells mitotically proliferate in the fetal and postnatal periods, but after certain cell populations are formed they become inactive. Their life span is unclear. Mitosis can be experimentally induced in adult animals (Kurosumi 1971) (see also p. 162).

8.4.10 Mast Cells

Mast cells are not of epithelial but of mesenchymal origin. Although their ontogeny had been disputed, Kitamura et al. (1977) showed by using irradiated and bone marrow-grafted mice that they are derived from stem cells in the bone marrow (see p. 266).

8.5 General Considerations

8.5.1 Convergence Differentiation

As detailed above for individual cases, paraneurons are derived from different germ layers.

Nevertheless, the cells to be classed as paraneurons are essentially uniform in structure, function, and metabolism. Recent immunochemical and immunocytochemical studies have indicated that paraneurons may share various kinds of neuron-specific proteins with neurons (neuron-specific enolase, neurofilament protein, Purkinje cell-specific or spot 35 protein, etc.) (see pp. 95, ff). The features of paraneurons common to neurons are becoming more extensive and diverse with advances in research.

It can be said, as a conclusion, that paraneurons are *divergent in origin* but assume structures and functions that are similar to each other and similar to neurons by *convergence differentiation*. It seems worthwhile to point that neurons are also heterogenous in origin (p. 3).

8.5.2 Phylogenetic Account: Archineurons and Archiparaneurons

Investigators of vertebrate embryology are apt to believe that the neural tube and crest are indispensable for the existence of neurons. However, it must be borne in mind that the neural tube and crest are parts which have been added to the structure of protochordate organisms. Neurons exist in lower invertebrates lacking a neural tube and crest. Even coelenterates have neurons, mainly in the mesoderm around the gut and in the tentacles. It may, therefore, be possible to differentiate between *neuroectodermal neurons* or "*neoneurons*," derived from the neural tube and crest, and *primitive neurons* or "*archineurons*," which are independent of them. There is then the possibility that in vertebrates archineurons survive as "native" structures in some parts of the body, especially around the alimentary tract, after the neural crest has evolved to send out "nonnative" neoneurons. The origin of autonomic neurons, especially the intramural neurons in the gut, has been disputed. The possibility of their surviving in the vertebrate gut as archineurons is an attractive concept.

Essentially the same pattern may apply to paraneurons. In animals higher than protochordates, some paraneurons are derived from the neural tube or crest as detailed above. However, peptide-producing receptosecretory cells exist independent of the neuroectoderm. It is possible that the basal-granulated (endocrine) cells in the alimentary, respiratory, and urogenital tracts and some other paraneurons are the survivors of *archiparaneurons*. As detailed in Chap. 11, enteric paraneurons can be found in almost all invertebrates, including the coelentrates.

8.6 Summary

In spite of the attractiveness of Pearse's unifying theory on a common origin of peptide/amine-producing endocrine cells from the neural crest, recent results in experimental embryology indicate that these cells are derived from different germinal layers. Thus, *paraneurons are multiple in origin.*

The pineal and visual cells are derived from the neural tube. The paraneurons of the chromaffin system originate from the neural crest, and thyroid parafollicular cells are suspected of having the same origin. Other endocrine and sensory paraneurons are nonneural-epithelial in origin, either ectodermal or endodermal. If one counts the mast cell as a paraneuron, it is mesenchymal in origin. Whatever origins they may have, neurons and paraneurons have acquired their common features by *convergent differentiation.*

The majority of paraneurons are capable of mitotic proliferation. Their cell kinetics and life span are still poorly known.

From a phylogenetic viewpoint, it is hypothesized that cells called "*archineurons*" and "*archiparaneurons*" regulate the body at the level of coelenterates through the lower invertebrates and that at the level of protochordates the neural tube and crest are a new addition; from the latter, "neoneurons" and "neoparaneurons" are derived.

9. Subsidiary Neuronlike Features of Paraneurons

9.1 Association with Neurons

9.1.1 Involvement in Neuronal Chains

Most paraneurons have either the manifest or potential tendency to be involved in neuronal chains, forming *neuroparaneuronal chains*. Sensory paraneurons, like visual, statoacoustic, olfactory, gustatory, and Merkel cells, are synaptically connected with sensory nerves (partly also with efferent nerves). Some bronchial paraneurons are connected with nerve fibers. Internuncial paraneurons, such as SIF cells and carotid body chief cells are naturally involved in neuron chains by some synapses.

Endocrine paraneurons are highly variable with regard to their *relation to nerves*. The same paraneurons may conspicuously vary in their attitude to nerves among different animals (Bargmann 1972). Cells which most constantly have synaptic contact with nerves are adrenal chromaffin cells and pinealocytes. Less regular innervation, though conspicuously rich in certain mammals, can be found in pancreatic islet cells (Kobayashi and Fujita 1969; review Unsicker 1984). Gut endocrine cells, especially EC cells, are approached by a nervous network (autonomic ground plexus; Hillarp 1959) but usually not directly contacted. These paraneurons in the gut can be regarded as the "primary sensory neurons" in the afferent system in the gut. They recognize the luminal chemical stimuli and transduce them into chemical messengers, which may stimulate secondary neurons in the lamina propria (see p. 173; Fig. 6.2).

The adenohypophyseal cells in mammals are only rarely contacted by a nerve terminal (review: Unsicker 1984), as the cells receive hypothalamic messengers, i.e., releasing and inhibitory hormones, via the hypophyseal portal vessels. In some teleost fishes, however, nerve fibers conveying hypothalamic neurohormones do not terminate in the median eminence but enter the pituitary pars distalis to make direct contact with endocrine cells (Vollrath 1969; Kaul and Vollrath 1974). Parathyroid cells and parafollicular cells have only been rarely found in direct contact with a nerve terminal (review: Unsicker 1984).

9.1.2 Neuroendocrine Complex

Endocrine paraneurons may form a cell complex by association with neurons, and the occurrence of this neuroendocrine complex or more exactly, *neuroparaneuronal complex* clearly indicates the continuity and inseparability of neurons and paraneurons.

Neuroinsular complex. The most typical structure in this context is the neuroinsular complex (Simard 1937) of the pancreas (Fig. 9.1). This can be found constantly, though not very frequently, in the pancreas of man, the mouse, rabbit, dog, and cat. The complex consists of islet cells and neuronal somata and/or fibers closely adjacent to each other and supported by glial (Schwann) cells (Fujita 1959; Serizawa et al. 1979). The complex may occur separately ensheathed by a perineurium of nerve bundles (Böck 1986), or it may be contained in an intrapancreatic ganglion. It also comprises an islet containing neuronal somata and fibers. In the mink, essentially every islet of Langerhans contains several neurons which are adjacent to islet cells and, therefore, every islet may be termed a neuroinsular complex (Fujita et al. 1981). The islet of the nutria (*Myocastor coypus*) also frequently contains neurons forming a complex (Kern et al. 1971). In the dog, large numbers of nerve fibers gather in the islet to form a kind of complex (Fujita and Kobayashi 1979a). Thus, there are gradations between the pure ganglion, neuroendocrine mixture (representing the complex), and pure islet (Serizawa et al. 1979; Böck 1986).

Electron-microscopic studies indicate that in the neuroinsular complex a neuron (either soma or axon) and an islet cell of any cell type are directly adjacent to each other (space 30 nm) or,

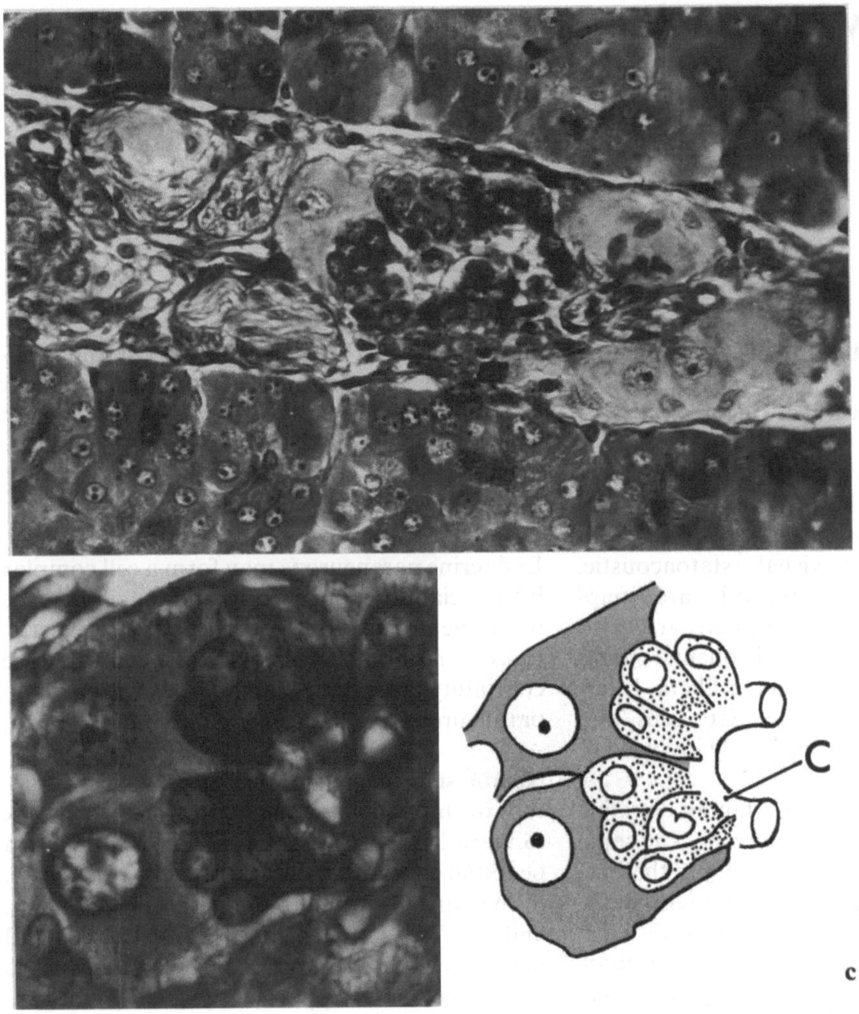

a

b

c c

Fig. 9.1a–c. Neuroinsular complexes in the dog pancreas. Nerve cells and nerve fibers are juxtaposed with islet cells (mainly B cells which are darkly stained) within an interlobular ganglion. **a** Overview. **b, c** Closer view of the neuroinsular conjugation as schema and corresponding micrograph. Two neurons are in direct contact with islet cells. Note that the islet cell granules shift toward a blood capillary; the neurons also extend short processes to the capillary (probable hemocrinia). Aldehydefuchsin-trichrome staining. Reproduced, with permission, from Fujita T (1959) Z Zellforsch 50:94–109

less typically, there is the intervention of thin glial cytoplasm (Serizawa et al. 1979; Böck 1986). Islet cell granules do not accumulate toward the associated neuron but toward the blood capillaries distributed in the complex (Fujita 1959); this may deny the possibility of the synaptocrine effect of islet cells upon the nervous elements in the complex. There are synapselike associations between the neuronal and insular elements in the complex (Fujita and Kobayashi 1979a; Radke and Stach 1986) (Fig. 9.2). However, the synapses do not seem to be more numerous in the complexes than in ordinary islets, as pointed out by Kern et al. (1971) in the nutria.

On the other hand, our immunocytochemical and electron-microscopic studies have indicated that both in complex and simple islets of the dog (Fujita and Kobayashi 1979a), mouse, mink (Fujita et al. 1981), and snake (Fujita et al. 1982) neuronal axons extend to the perivascular spaces of blood capillaries, showing an accumulation of synaptic granules/vesicles in their swollen terminals. Thus, it is most likely that the neurons, besides regulating islet cells in the vicinity, release their secretions (VIP, acetylcholine, catecholamine, etc.) into the bloodstream (Fujita et al. 1981). It seems that in the neuroinsular complex, the neuron and islet cell are disposed not so much in series as *in parallel*; both concur in the release of their secretions into the blood capillaries

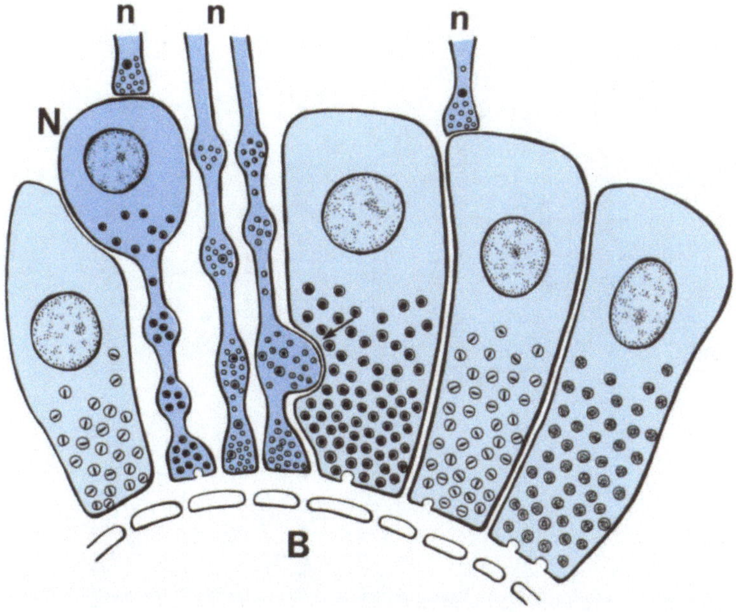

Fig. 9.2. Relationship of neurons and paraneurons (endocrine cells) in pancreatic islet. Intrinsic (*N*) and extrinsic (*n*) nerves gather around a blood capillary (*B*) for hemocrine release of their messengers (insular neurosecretion). Some of the nerve fibers are in contact with the endocrine cells. There is contact between a neuron and endocrine cell (arrow) as is usually found in neuroinsular complexes.

(Fig. 9.2). In this sense, both elements, neurons and paraneurons, are equivalent to each other.

Neuroinsular complexes are formed during the fetal period, apparently by a special affinity between the nervous and insular elements (Fujita 1959).

Neuroendocrine complex in the intestinal mucosa. In human fetuses (4–5 months), neuroendocrine complexes in the duodenum have been demonstrated by Osaka et al. (1975) and Osaka and Kobayashi (1976). Groups of endocrine cells located in the lamina propria mucosae were in close association with nerve fibers of Meissner's plexus.

Although these authors believed at the time that the endocrine cells mights be differentiated in relation to the nervous tissue and then migrate into the epithelium, this hypothesis of a neurogenic origin of gut endocrine cells no longer seems tenable. An alternative and more feasible view is that endocrine cells which have differentiated in the epithelium might temporarily migrate out of it to contact the nervous tissue. It is possible that in this contact neurons may induce differentiation of the gut endocrine cells.

In the proventriculus of the finch, *Uroloncha striata* var. *domestica*, Kataoka (1974) demonstrated that numerous endocrine cells occurred in the lamina propria and many of them formed neuroendocrine complexes. Some endocrine cells surrounded nerve axons either directly or with intercalated Schwann cell cytoplasm.

9.2 Neuron-Specific Substances Detected in Paraneurons

The view that paraneurons, even though derived from different germ layers, have cell-biological features in common with neurons is impressively supported by recent studies detecting neuron-specific proteins in paraneurons.

9.2.1 Neuron-Specific Enolase (NSE)

NSE is an isozyme of the glycolytic enzyme enolase (review: Zomzely-Neurath 1982) and has been shown immunocytochemically to be localized in neurons (Pickel et al. 1975). Not all neurons, however, are immunoreactive for NSE. A considerable variety of neurons in the central and peripheral nervous systems, including pyramidal cells in the spinal anterior horn and Purkinje cells in the cerebellum, do not seem to contain NSE-like immunoreactivity. Thus, NSE is a good marker of certain kinds neurons (Fig. 9.3).

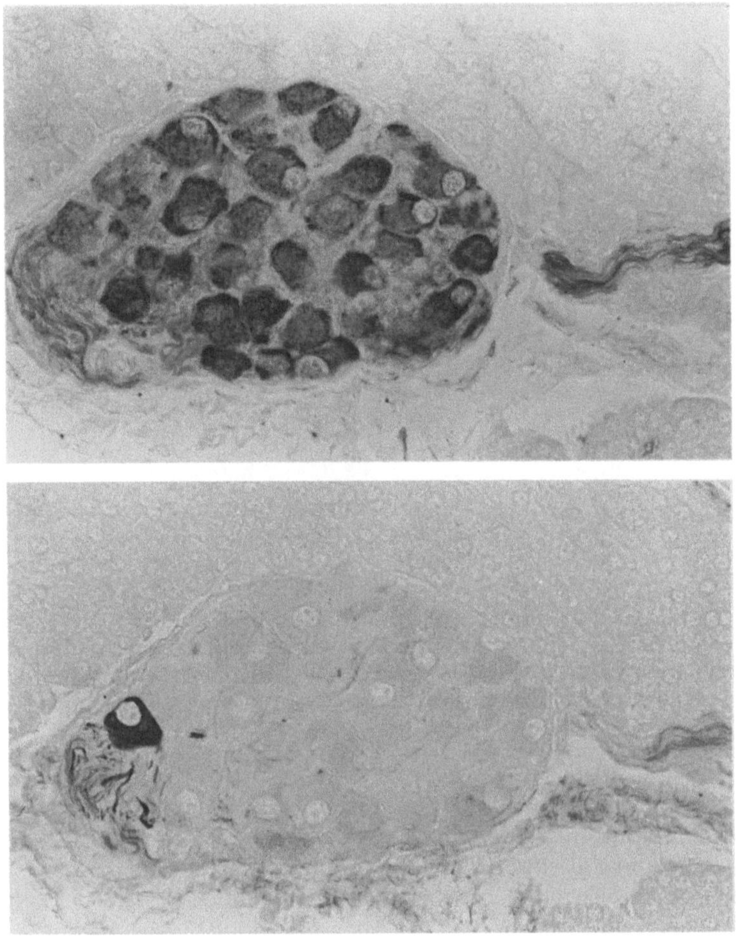

a

b

Fig. 9.3a, b. Comparison of the distribution of **a** NSE and **b** NEP in adjacent sections of an intrapancreatic ganglion of the dog. Almost all neurons are immunoreactive for NSE in this autonomic ganglion, whereas only a very restricted number of them are immunoreactive for NFP. These micrographs visualize the heterogenous distribution of neuron-specific proteins. NSE is also heterogenous in distribution in other nervous tissues. Prepared by T. Iwanaga, Department of Anatomy, Niigata University Medical School

Schmechel et al. (1978) and Marangos et al. (1982) revealed occurrence of NSE-like immunoreactivity in various "neuroendocrine" cells (paraneurons according to our terminology), including adrenal chromaffin cells, thyroid parafollicular cells, anterior and intermediate lobe cells of the pituitary gland, pancreatic islet cells (Fig. 9.4), gut endocrine cells, bronchial basal-granulated cells (see also Wharton et al. 1981), carotid body chief cells, and pinealocytes (see also Fujita et al. 1983a). Furthermore, NSE-like immunoreactivity has been demonstrated in typical sensory cells, including the gustatory cells in the taste bud (Fujita et al. 1983a), Merkel cells in the skin (Gu et al. 1981), hair cells in Corti's organ (Altschuler et al. 1985), and olfactory cells (Takahashi et al. 1984) (Table 9.1; Fig. 9.4).

In summary, many, if not all, endocrine, internuncial, and sensory paraneurons may contain NSE, supporting the idea that certain unknown mechanisms in the cellular metabolism might be shared by neurons and paraneurons. It often occurs, for unknown reasons, that a certain population in a paraneuronal cell mass is immunostained while the remainder remains immunonegative. Moreover, the immunoreactivity of the same cells may differ considerably among animal species. Such heterogeneity and species differences in the immunoreaction can also be found in the NSE immunostainability of neurons.

9.2.2 Neurofilament Protein (NFP)

Neurofilament protein (NFP) is characteristic of the intermediate filaments in neurons (review: Shelanski and Liem 1979) and can be differentiated from the proteins of glial filaments and intermediate filaments in fibroblasts and other cells (Yen and Fields 1981; Liem 1982). Mammalian NFP consists of three subunits with approximate molecular weight of 200 000, 145 000, and 68 000 (Hoffman and Lasek 1975; Liem et al. 1978; Mori and Kurokawa 1980).

As stressed in the review by Shelanski and

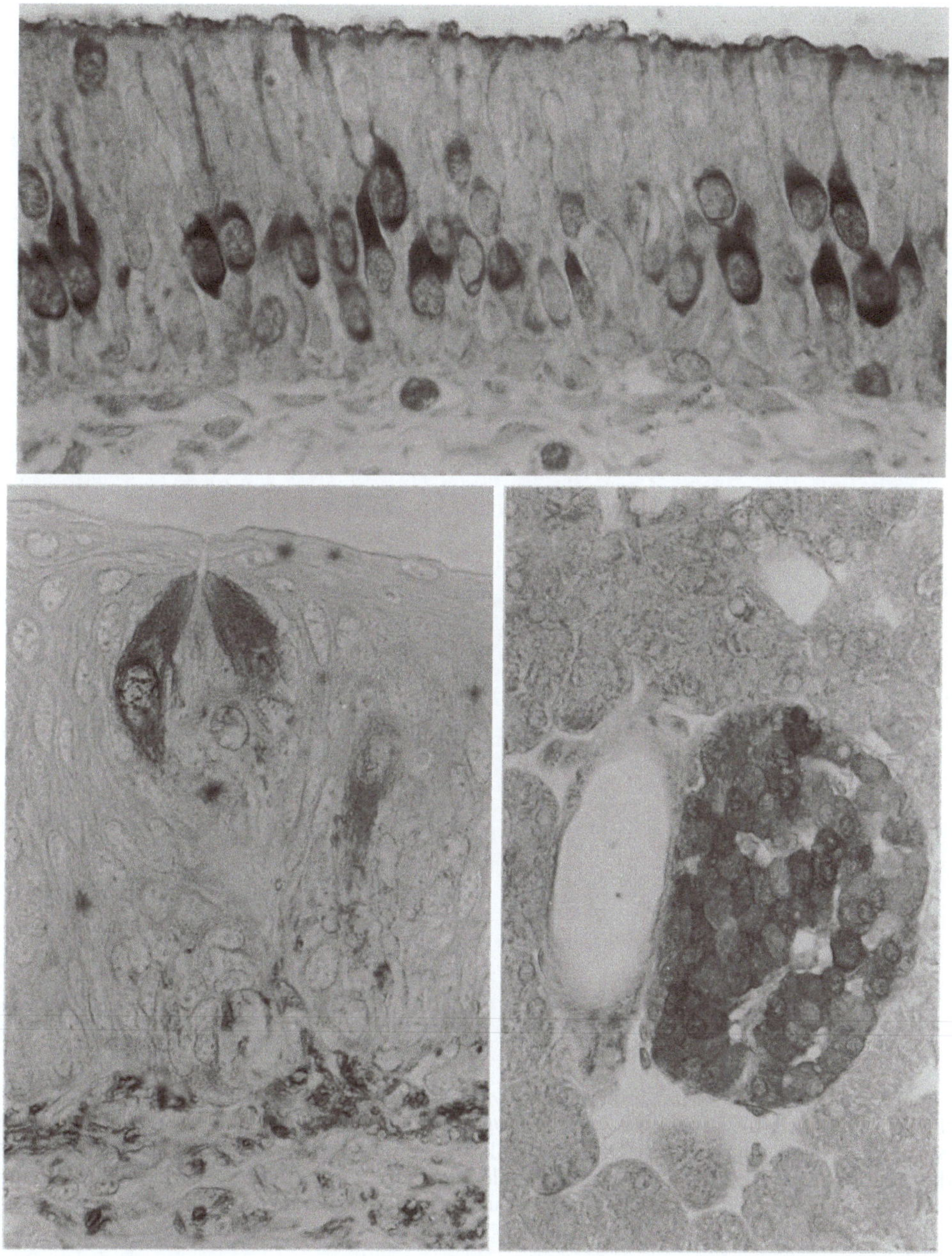

a b c

Fig. 9.4a–c. Distribution of NSE-like immunoreactivity in various paraneurons. **a** Olfactory cells in the nasal mucosa of a human fetus. **b** Presumable gustatory cells in the taste buds of the dog tongue. Subepithelial nerves are also immunostained. **c** Islet cells in the human pancreas. Prepared by T. Iwanaga, Department of Anatomy, Niigata University School of Medicine

Liem (1979), neurofilaments are considerably species-specific, tissue-specific, and cell specific in their localization and chemical nature. Probably for this reason, immunocytochemical methods using antisera against the NFP subunits can stain only part of the neurons and their processes in the nervous system (Schlaepfer and Lynch 1977). A part of the neurons in the spinal ganglion has been shown to be immunoreactive for the NFP subunits both in the human ganglion *in situ* (Schlaepfer and Lynch 1977) and in rat tissue in culture (Shaw and Weber 1981). In various mammals, we have also experienced that only a small number of the different kinds of

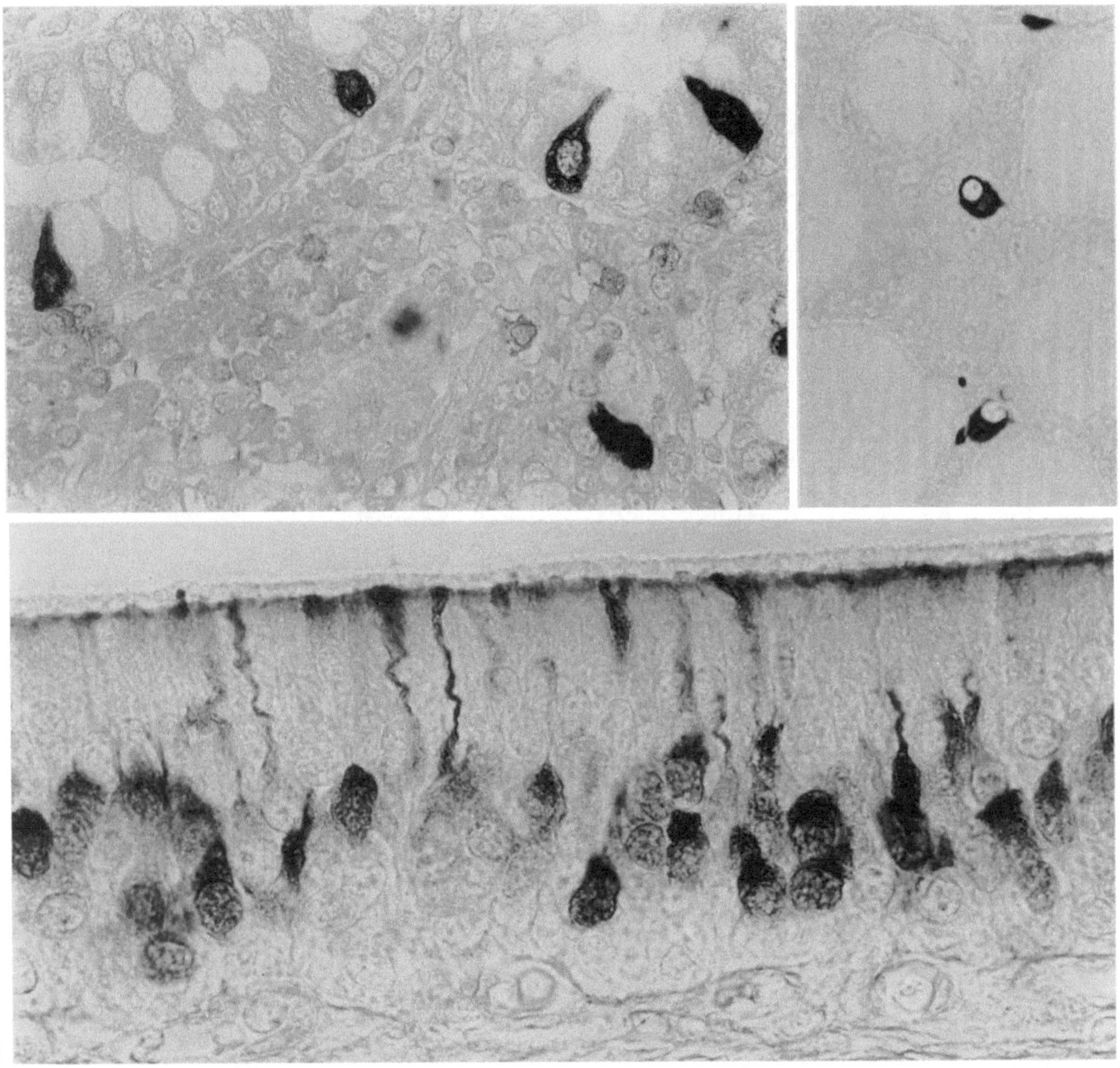

a b c

Fig. 9.5a–c. NFP (145 k)-like immunoreactivity in some paraneurons. **a** Duodenal mucosa of the pig. Some of the endocrine cells are immunoreactive. **b** Dog thyroid. A few of the parafollicular cells are immunopositive. **c** Nasal mucosa of a human fetus. Olfactory cells are immunoreactive. The NFP-like substance is concentrated in the apical part of the cells and is undetectable in the basal part and axon. Indirect immunoperoxidase (PAP) method. Prepared by T. Iwanaga, Department of Anatomy, Niigata University Medical School

neurons and, even within the same kind, only some of the neurons are immunoreactive for NFP in both the central and peripheral nervous systems. In rat myenteric ganglia, for instance, only a few percent of neurons react intensely to the antisera, while the rest are completely negative.

Murine neuroblastoma cells in culture were demonstrated immunocytochemically to contain NFP-like immunoreactivity (Jørgensen et al. 1976).

We examined various kinds of paraneurons in humans (adult and fetal) and mammals, including the dog, cat, pig, and rat, with regard to the occurrence of NFP-like immunoreactivity, using antisera against the three subunits derived from rat brain (Fujita et al. 1983a). It was found that some paraneurons in some animals, were selectively and intensely immunoreactive for NFP. Usually, a small population in a group of paraneurons is immunostained, as is the case for neurons in the myenteric ganglion mentioned above. Immunoreactivity for the 145 000 subunit could be more clearly and constantly demonstrated than that for the 68 000 and 200 000 subunits.

In gut endocrine cells of the pig, NFP immunoreactivity is found only in cells of unidentifiable type (Fig. 9.5). In the pancreas of the dog, very small numbers of islet cells (probably D cells) are immunoreactive. Adrenal chromaffin cells and thyroid parafollicular cells are partly immunopositive. There are fewer positive cells than NSE-immunoreactive elements in these tissues. As for sensory paraneurons, olfactory cells are immunoreactive for NFP and the reaction is especially intense in their apical processes (Table 9.1; Fig. 9.5) (Fujita et al. 1983a).

Paraneuromas may also reveal NFP-like immunoreactivity, though in occasional cases and usually in a small proportion of the cells (Fujita et al. 1983a).

In summary, NFP-like immunoreactivity may appear in paraneurons, and this provides further support for the neuronlike nature of paraneurons. The immunopositive cells are few in number in both the nervous and paraneuronal systems. The immunoreaction is cell-specific and species-specific, though neurofilaments and intermediate filaments are fairly universal in distribution both in neurons and paraneurons.

It is not known on what cell metabolism and function the occurrence or nonoccurrence of NFP may depend. Neurofilamens or intermediate filaments with apparently the same fine structure are present essentially in all neurons and paraneurons, and yet they are heterogeneous with regard to their NFP immunoreactivity.

9.2.3 Brain Tubulin

Tubulin, the molecular unit of microtubules, consists of two distinct subunits, α and β (Bryan and Wilson 1971). Heterogeneity occurs among tubulins derived from different tissues and it is especially marked in brain tubulin (Gozes and Littauer 1978; Dahl and Weibel 1979; Strocchi et al. 1981). Furthermore, the tubulin subunits are associated with particular proteins, which may differ according to the source tissue including the brain (Sloboda et al. 1976; Dustin 1978). It is, therefore, possible that antisera raised against brain tubulin may react to the nerve-specific parts in the tubulin subunits and/or the tubulin-associated proteins.

In the course of our immunocytochemical survey of brain-derived substances, we found that an antiserum against brain-derived tubulin reacted neuron-specifically (Iwanaga et al. 1982a). The antiserum was raised by S. Ito (Department of Internal Medicine, Niigata University School of Medicine) in a rabbit after injections of a tubulin preparation from the porcine brain by Y. Kanazawa (Department of Internal Medicine, Tokyo University School of Medicine).

In the small intestine of human fetuses, in the 4th-6th month of gestation, we found that a small population of intraepithelial endocrine cells (cell type unidentified) selectively and intensely immunoreacted to the brain tubulin antiserum. The same section revealed strongly immunoreactive nerve plexuses of Meissner and Auerbach. We found furthermore that the pancreatic islet cells in the chicken were intensely positive to the brain tubulin antiserum (Fig. 9.6).

It would be worthwhile to carry out a more systematic study on the distribution of brain tubulin immunoreactivity in neurons and paraneurons; chemical identification should also be performed on the neuron-specific component in the tubulin preparation used as the antigen.

9.2.4 Purkinje Cell Specific Protein (Spot-35 Protein)

In two-dimensional gel electrophoresis of soluble proteins extracted from rat brains, Yoshida and Takahashi (1980) noticed a 35th spot only in extracts of the cerebellum, suggesting the existence of a cerebellum-specific protein. The protein, called spot-35 protein, was soon isolated from the bovine cerebellum (Yamakuni et al. 1984). Immunocytochemical methods using an antiserum against this protein revealed that Purkinje cells were specifically stained in the rat and human cerebellum (Yamakuni et al. 1984) (Fig. 9.7). Further immunocytochemical studies, however, made it clear that spot-35 protein is contained in some other neurons, including multipolar neurons in the amygdala and habenula, periglomerular cells in the olfactory bulb (rat—Hozumi et al. 1986), horizontal cells in the retina (guinea pig—Iwanaga et al. 1985a), intraadrenal nerve cells (rat—Kondo et al. 1985), and a certain population of intramural neurons in the gut (Fig. 9.7).

In contrast to the very limited occurrence of spot-35 protein in neurons, a wider distribution of this substance can be seen in paraneurons (Table 9.1) (Hozumi et al. 1986). In the pituitary anterior lobe of the guinea pig and dog, scattered endocrine cells with rich cytoplasm are immunopositive. Thyroid parafollicular cells are strongly immunoreactive in various mammals

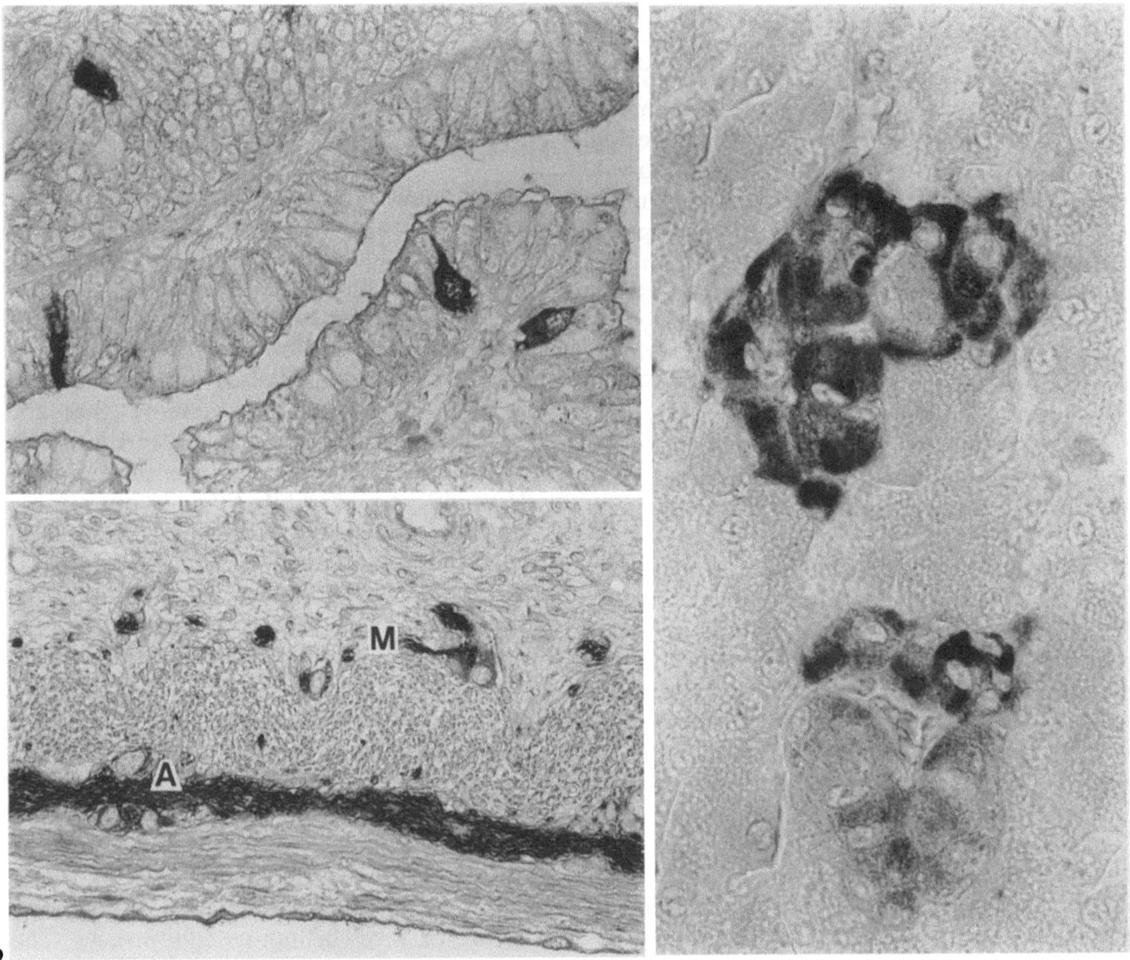

a b c

Fig. 9.6a–c. Neurotubulin (brain-derived tubulin)-like immunoreactivity in some neurons and paraneurons. Human fetal duodenum showing the immunoreactivity in **a** some endocrine cells in the epithelium and **b** neurons and their fibers in Auerbach's (*A*) and Meissner's (*M*) plexuses. **c** Pancreatic islets of the finch, *Uroloncha striata* var. *domestica*. Islet cells are immunoreactive in different intensities. Indirect immunoperoxidase (PAP) method

(Fig. 9.7). More or less restricted cell types, according to species, are immunoreactive in pancreatic islet and gastroenteric endocrine cells. Intensely and weakly immunopositive cells are intermingled in the adrenal medulla in the cat and dog (Hozumi et al. 1986). In the rat, noradrenaline (NA) cells are strongly immunopositive for spot-35 protein, whereas adrenaline (A) cells are negative (Kondo et al. 1985).

Characteristically, spot-35 protein is localized in a variety of sensory paraneurons. In the taste buds of the guinea pig, several gustatory cells are immunoreactive (Fig. 9.7). In the guinea pig cochlea, all hair cells are selectively immunoreactive, the outer hair cells showing a stronger reaction than the inner ones. The olfactory cells, though in only a very small population of them, are intensely immunostained (guinea pig) (Fig. 9.7). It is unknown why a few cells are so clearly immunoreactive, while others with probably the same function remain completely unreactive. Visual cells, contrary to expectations, could not be immunostained in the retina, where horizontal cells contain intense spot 35-like immunoreactivity (Table 9.1).

Recent biochemical studies have shown that spot-35 protein is a kind of Ca^{2+}-binding protein (Yamakuni et al. 1985) and its similarity to a vitamin D-dependent calcium-binding protein purified by Wassermann et al. (1968) from the chick intestinal mucosa (Wassermann et al. 1968) has been claimed (Iwanaga et al. 1985a). It is worthy of mention that the immunoreactivity for this intestinal calcium-binding protein is de-

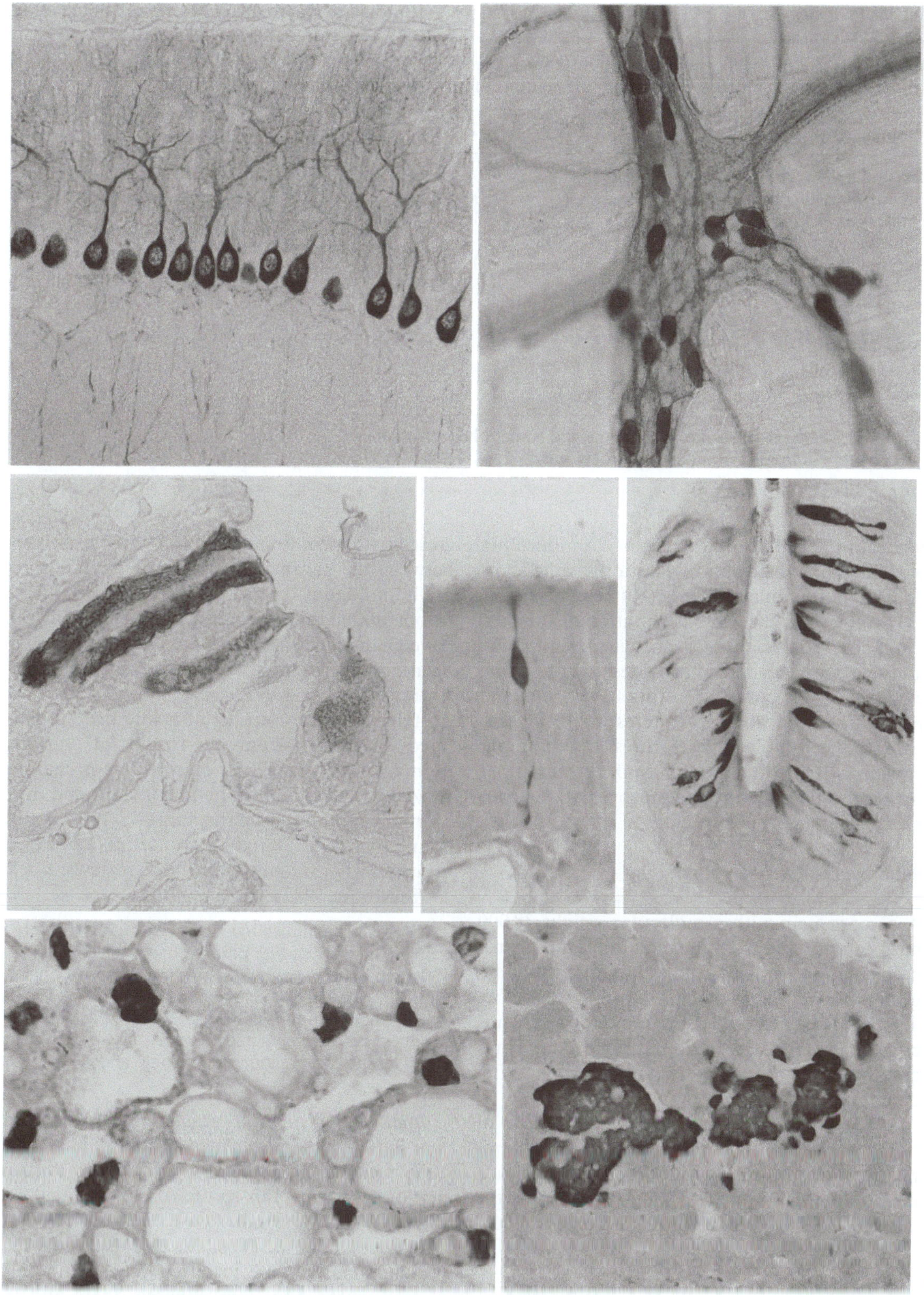

Fig. 9.7a–g. Immunoreactivity for Purkinje cell specific protein (spot 35 protein) in some neurons and paraneurons. **a** Purkinje cells in the rat cerebellum. **b** Neurons in Auerbach's plexus of the guinea pig jejunum (whole-mount preparation). **c** Inner ear hair cells of the guinea pig. Note the more intense reactivity toward the outer row of cells. **d** Olfactory cell of the guinea pig. Only a small population of the cells is immunoreactive. **e** Taste buds of the guinea pig. Probable gustatory cells are immunoreactive. **f** Parafollicular cells of the cat thyroid. **g** Islet cells of the cat pancreas. Prepared by T. Iwanaga, Department of Anatomy, Niigata University Medical School

Table 9.1. Occurrence of immunoreactivities for NSE, NFP(145,000 subunit), spot 35(Purkinje cell-specific protein), and S-100 protein in different paraneurons of endocrine and sensory types

Paraneuron	NSE	NFP(145K)	Spot 35	S100
Gut endocrine cell	(+)	(+)	(+)	—
Pancreatic islet cell	+	(+)	+	+ Schwann cell
Parafollicular cell	+	(+)	+	—
Anterior pituitary cell	+	—	(+)	+ Stellate cell
Adrenomedullary cell	+	(+)	+	+ Satellite cell
Pinealocyte	+	—	+	+ Glial cell
Visual cell	+	—	—	+ Müller cell
Corti hair cell	+	—	+	?
Olfactory cell	+	+	(+)	+ Schwann cell
Gustatory cell	+	—	+	—
Merkel cell	+	—	—	—

The data are based on studies in humans and some other mammals.
(+) positive in limited numbers (types) of cells

tected in the cochlear hair cells of the guinea pig with the same distributional pattern as the spot-35 protein immunoreactivity (Rabié et al. 1983).

In any event, the wide distribution of the Purkinje cell specific spot-35 protein in endocrine and sensory paraneurons provides us with strong evidence of the common nature of neurons and paraneurons. It would be interesting to investigate the cell-biological features which might explain why Purkinje cells and certain other neurons and endocrine and sensory paraneurons share this particular protein.

9.2.5 Appendix: Synaptophysin

A glycoprotein with a molecular weight 38000 has been purified from the membranes of neuronal synaptic vesicles and adrenal chromaffin granules and designated synaptophysin (Wiedenmann and Franke 1985). Immunocytochemical studies have suggested that synaptophysin is distributed in a variety of paraneurons (Wiedenmann et al. 1986).

Synaptophysin seems to be an example of neuron-specific substances shared by paraneurons which are derived from the granule membrane, in contrast to the substances as the contents of the granules or those in the cytosol.

9.3 Sustentacular Cells of Glial Nature

9.3.1 Detection of S-100 Protein

Paraneuronal cells may be supported by sustentacular cells which are usually stellate or crescent-shaped, extending attenuated processes between the paraneuronal cells. If paraneurons are homologous counterparts of neurons, it can be postulated that their sustentacular cells might have glialike features. This problem has recently been explored immunocytochemically for the detection of a *glia-specific substance* called S-100 protein. S-100 protein is an acidic protein isolated originally from the brain of various mammals (Moore 1965). This protein is characterized by its solubility in 100% saturated ammonium sulfate (hence the name "S-100" at pH 7) and by its Ca^{2+}-binding properties (Isobe and Okuyama 1978; Isobe et al. 1983; reviews: Bock 1978; Moore 1982).

Biochemical studies have revealed that S-100 protein is a mixture of two predominant components, S-100a and S-100b, having the subunit composition α-β and β-β, respectively (Isobe et al. 1978, 1981, 1983). The α- and β-subunits comprise 93 and 91 amino acid residues, respectively, and there is 58% identity in their amino acid sequences (Isobe and Okuyama 1981).

Immunocytochemical studies have revealed that S-100 protein is located primarily in neuroglial elements of nervous tissue. In the central nervous system, this protein is contained in astrocytes, oligodendrocytes, and ependymal cells (Ludwin et al. 1976; Cocchia 1981). In the peripheral nerves, it is localized in Schwann cells of myelinated and unmyelinated fibers and in satellite cells of sensory and autonomic ganglia (Stefansson et al. 1982; for further literature on the distribution of S-100 see Girod et al. 1985). The occurrence of S-100 protein in nonnervous cells will be discussed on p. 104).

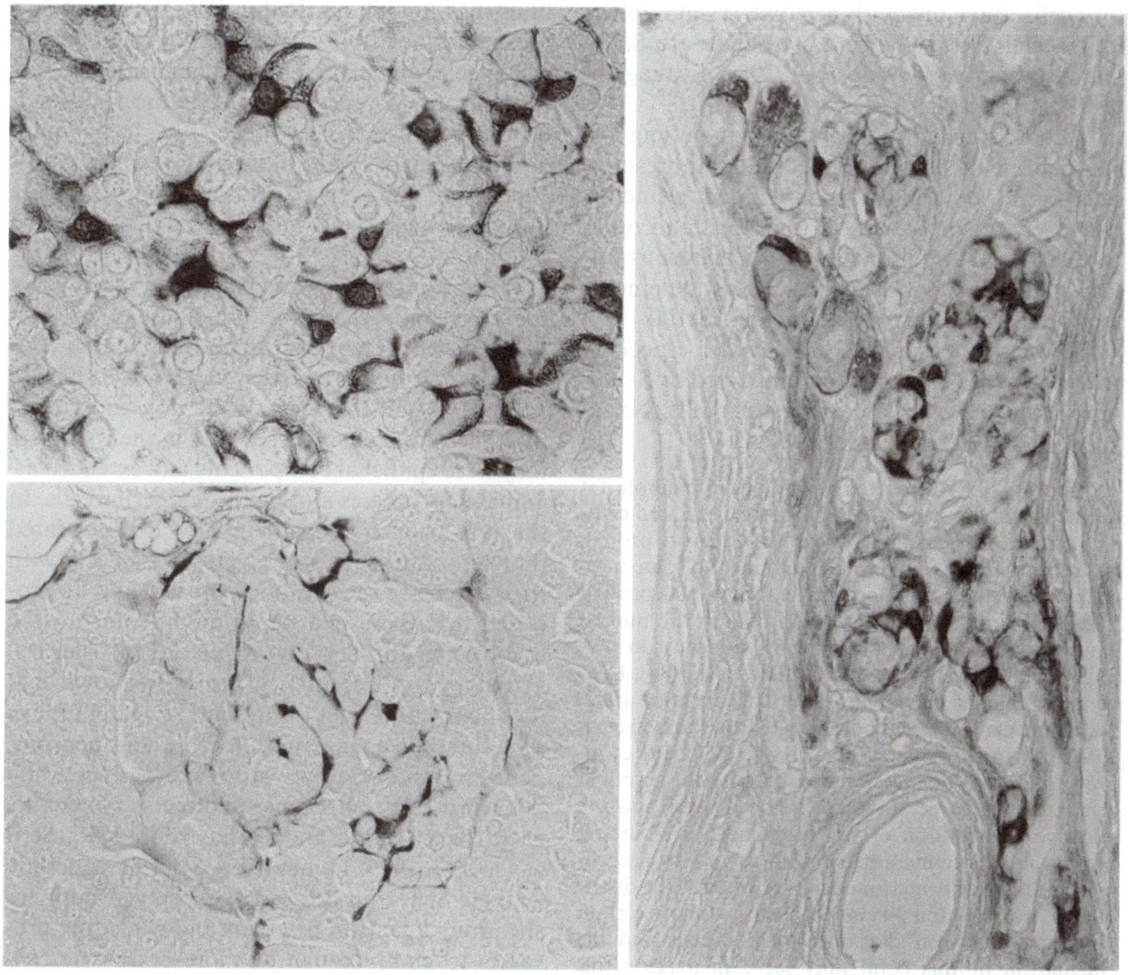

a b c

Fig. 9.8a–c. Distribution of S-100 protein immunoreactivity in sustentacular cells of paraneurons. **a** Rat anterior pituitary showing immunoreactive folliculostellate cells. **b** Cat pancreatic islet containing attenuate sustentacular cells with immunoreactivity. **c** Rat carotid body containing immunoreactive glial elements. Prepared by T. Iwanaga, Department of Anatomy, Niigata University Medical School

The application of immunocytochemical methods to various paraneuronal tissues has revealed that sustentacular cells in some of them were immunoreactive for S-100 protein. In the adrenal medulla, S-100-like immunoreactivity is selectively contained in sustentacular cells which extend laminar processes to envelop chromaffin cells (rat—Cocchia and Michetti 1981; human fetus—Iwanaga and Fujita 1984). In the carotid body, sustentacular cells embracing the chief cells were immunostained together with Schwann cells surrounding nerves (rat—Kondo et al. 1982) (Fig. 9.8). In the pineal body, S-100 immunoreactivity occurs in interstitial cells among pinealocytes (Møller et al. 1978). In the pituitary anterior lobe, supporting cells often called the folliculostellate cells were immunoreactive for S-100 protein (rat—Nakajima et al. 1980; Cocchia and Miani 1980; human—Girod et al. 1985) (Fig. 9.8). The pancreatic islets may contain stellate S-100 immunoreactive cells, whose population varies among species. In the cat and mink, S-100-immunoreactive sustentacular elements are widely distributed in the islets (Fig. 9.8), while in the rat only islets forming the neuroinsular complex (see pp. 93, ff.) are distributed with the immunopositive cells (Fujita et al. 1983a). Girod et al. (1987) recently demonstrated S-100-immunopositive cells in the islets of the monkey, though they did not identify these cells as glial elements.

Investigations into the possibility that S-100-positive cells are in the position to support sensory paraneurons, such as gustatory and olfactory cells, have met with negative results. In the retina, cells of Müller and astrocytes are S-100 immunoreactive (Kondo et al. 1983; Iwanaga et al. 1985b) (Table 9.1).

In the bronchial paraneurons, including the neuroepithelial bodies, no S-100-immunoreactive cells could be found as associated elements. However, in some cases of tumorlets and other neoplasms of the bronchiolar paraneurons, attenuated cells supporting the neoplastic elements are found to be immunoreactive for S-100.

Different types of *paraneuromas*, especially carcinoids (p. 113), are often supported by S-100-immunopositive stellate or crescent cells (Nakajima et al. 1982a; Sano et al. 1984).

The story of S-100 protein would be quite simple if it occurred only in neuroglial and possibly related cells. The case, however, is more complicated as the protein has recently been found to be contained in some mesenchymal cells. Immunoreactivity for S-100 protein has been found in interdigitating cells in the human lymph node (Takahashi et al. 1981) and human thymus (Nakajima et al. 1982b; Ushiki et al. 1984), Langerhans' cells of the human skin (Cocchia et al. 1981; Nakajima et al. 1982c), chondrocytes (Stefansson et al. 1982b), fat cells (Suzuki et al. 1982; Michetti et al. 1983), and reticular cells in the rat lymph node germinal center (Iwanaga et al. 1982b). The diverse and yet selective distribution of this peculiar Ca^{2+}-binding protein must be investigated with regard to its cell-biological significance for each type of cell. The distribution of the α- and β-subunits seems to show certain differences between glial and nonglial cells and the elucidation of this point may serve in making this complex story somewhat clearer.

9.3.2 Electron-Microscopic Observations

Electron-microscopic studies, especially those combined with S-100 protein immunocytochemical investigations, have confirmed the glial nature of the sustentacular cells in the adrenal medulla (rat—Cocchia and Michetti 1981) and carotid body (rat—Kondo et al. 1982). In the anterior lobe of the pituitary gland, the S-100-immunoreactive elements are agranular (devoid of secretory granules) stellate cells, some of which form pituitary follicles and others embrace secretory (hormone producing) cells (rat—Cocchia and Miani 1980). Concerning the identity of the so-called folliculostellate cells and their possible glial nature, however, there are controversial reports (e.g., goat—Shirasawa et al. 1984), and extensive studies would appear to be necessary using different species of animals to reach a broad view.

Gut endocrine cells are usually dispersed in the intestinal epithelium. Rubin (1972) noticed that the endocrine cell is surrounded by an attenuate cell or cells which are reminiscent of satellite cells for neurons. We can also confirm the occasional occurrence of these cells (e.g., Fig. 16 in Fujita and Kobayashi 1977). Examination with anti-S-100 serum, however, could not demonstrate any immunoreactive cells surrounding the endocrine cells.

More fruitful advances have been made using the electron-microscopic morphology of *pancreatic islets*. The occurrence of *Schwann cells* was first demonstrated by us (Kobayashi and Fujita 1969) in the islets of the dog and guinea pig. The cells were incorporated in the islets as if they were epithelial components and extended attenuated processes among the endocrine cells. In the dog pancreas, Smith (1975) demonstrated that Schwann cells cover groups of islet cells with their extensive sheaths of cytoplasm and he proposed that these glial septa might somehow be involved in the nervous regulation of islet cell activity. More recently, it has been shown that the pancreatic islet represents a ganglionlike structure consisting of endocrine paraneurons, neurons (and/or their various processes), and Schwann cells, all elements directly contacting each other (Fujita and Kobayashi 1979a; Serizawa et al. 1979; Donev 1984). In this structure, a single Schwann cell can be observed both ensheathing a neuron or axons and extending processes to embrace one or more endocrine cells (Fig. 9.9). This convincingly visualizes the equivalent relation of neurons and endocrine cells as paraneurons. It seems apparent that the glial (Schwann) cell cannot differentiate between the cell surface requirements of the neurons and paraneurons and takes care of both at the same time.

9.4 Summary

In addition to possessing the neuronlike features introduced in the preceding chapters, paraneurons reveal further properties common to those of neurons.

Paraneurons may be involved in *neuronal chains* either with synaptic connections or with relays leaving a distance. GEP endocrine paraneurons may form neuroendocrine complexes in which neurons and paraneurons are closely associated with each other as if they were derived from the same origin. In neuroinsular complexes, neurons mimic endocrine cells by extending their

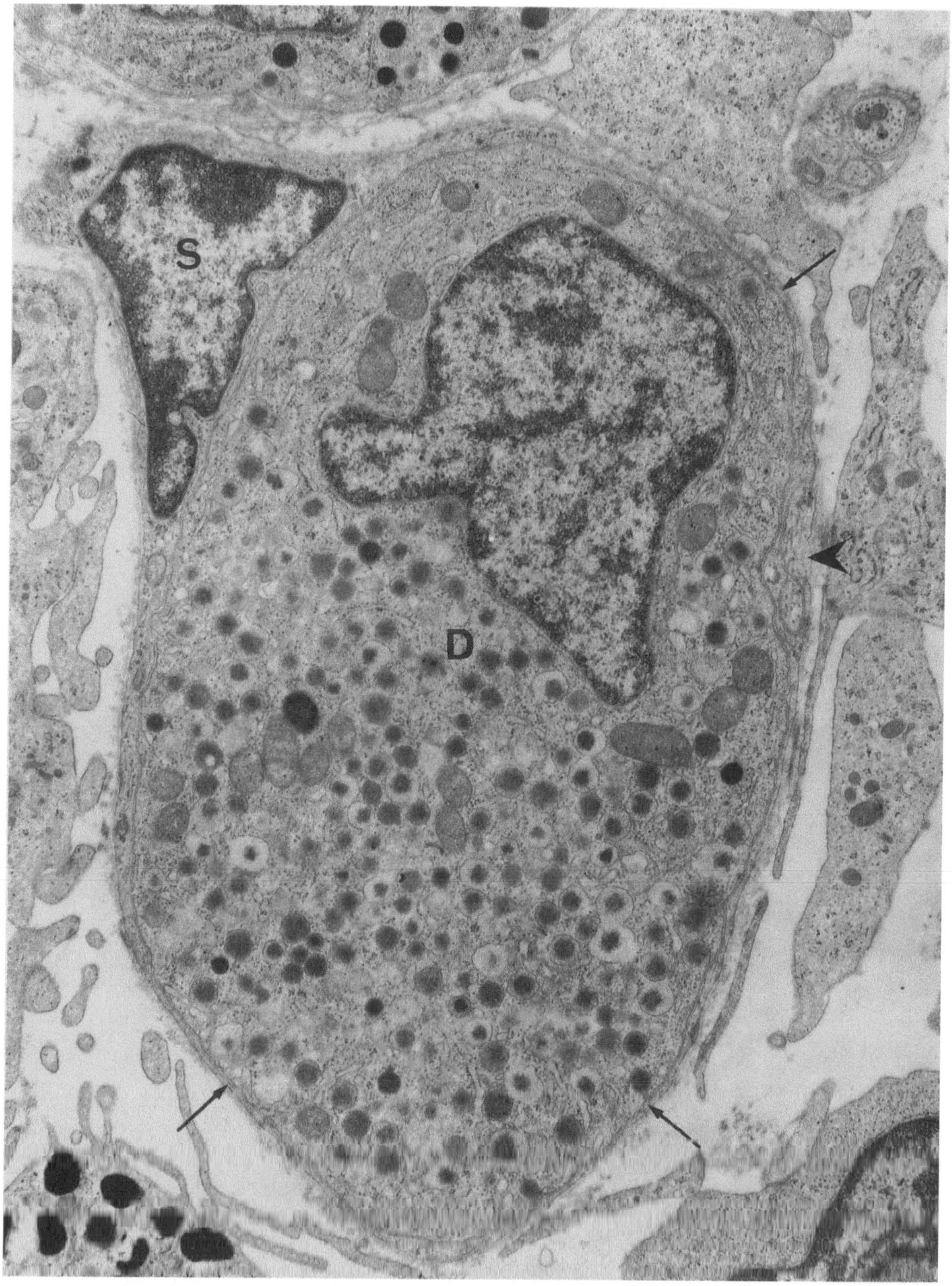

Fig. [illegible]. A pancreatic islet cell ensheathed by a Schwann cell in the dog pancreas. A D cell (*D*) is ensheathed by the attenuated cytoplasm (*arrows*) of a Schwann cell (*S*), which also invests a few axons (*arrowhead*). This micrograph visualizes the neuronlike nature of the islet cell. Reproduced from Fujita T, Kobayashi S (1979a) Arch Histol Jpn 42:277–295. ×13000

processes to blood capillaries for hemocrine release of their messenger substances, whereas the endocrine paraneurons mimic the neurons in their attitude to neuronal and glial elements.

Neuron-specific proteins, such as neuron-specific enolase, neurofilament protein, brain-derived tubulin, Purkinje cell specific protein (spot-35 protein), are immunocytochemically detected in a large variety of paraneurons. Though their functional significances are largely unknown, occurrence of these nonsecretory proteins in common to neurons and paraneurons strongly supports that cell-biological characteristics are extensively shared by both cell groups.

Sustentacular cells in some paraneurons are immunoreactive for (glia-specific) S-100 protein, supporting the idea that paraneurons may need nurses of glial nature and that these two types of cells cooperate in metabolic aspects as the neurons and their glial cells do.

10. Neoplasms of Paraneurons: Paraneuromas

10.1 General Aspects

Neoplasms are masses of cells which proliferate without normal control and without any useful function. Neoplasms of paraneurons, like those of other categories of cells, are designated by the suffix "-oma" (adapted from the Greek word "onkoma," meaning tumor). Thus, the term *paraneuroma* indicates neoplasms of the paraneurons.

"Typical" neurons and such paraneurons as inner ear hair cells are generally lacking in the ability of cell division (see p. 87). Therefore, neoplasms of these elements, such as central neurocytomas and ganglioneuromas, are extremely rare tumors which grow slowly. The term "neuroma" means a tumor largely made up of neuronal processes and glial overgrowth. The neuroblasts, embryonic cells which develop into neurons, form neoplasms which are called *neuroblastomas*. Cytological features of paraneuromas and neuroblastomas are identical in many cases. In the retina of children, there occur *retinoblastomas* whose counterparts in the central and peripheral nervous systems are neuroblastomas. Furthermore, the olfactory epithelium in the nasal cavity is the site of the *olfactory neuroblastoma*. However, it is yet to be established whether the tumor cells do in fact come from the blast cells of the olfactory neurons (Choi and Anderson 1985). Gustatory cells of the taste bud are under continual replacement (see p. 210) and, therefore, are potential sources of neoplasms. However, to our knowledge, no case report of *gustatory cell neoplasm* is available.

Paraneuromas may be *benign* or *malignant*. Benign paraneuromas usually grow slowly, do not form new foci in distant parts by metastasis, and consist of neoplastic cells retaining more or less amply the features of the normal cells. On the other hand, malignant paraneuromas frequently metastasize, grow rapidly, and are composed of neoplastic cells of more irregular and deviant appearance. The distinction between the benignity and malignancy of paraneuromas is not always clear, as there are many intermediate forms.

Paraneuromas show more or less immature cytological features, yet they usually display such a degree of differentiation that the cell types of their sources are identifiable. One of the most characteristic features of paraneuromas is the presence of the membrane-bound *secretory granules*. The neoplastic cells contain peptide hormones and their precursors together with aminic substances, which are identical to those produced by normal paraneurons. Since they secrete their substances, paraneuromas usually represent what are called *functioning tumors*. The metabolic activity of the tumor cells may be slightly or greatly different from that of normal cells. The chemical composition and physical state of the secretory granule contents in the same type of paraneuron differ from tumor to tumor. This is reflected in the markedly variable fine structure of the secretory granules.

We ourselves examined 16 cases of *insulinoma* using the electron microscope (Kobayashi et al. 1979). Secretory granules with a crystalloid core, which are specific to the islet B cells, were recognized in all 16 tumors. However, the ratio of the granules with a crystalline core to those with a noncrystalline core in the neoplastic B cells differed conspicuously from case to case. There was a tendency for well-differentiated insulinoma cells to possess more secretory granules with a crystalloid core than normal B cells (Fig. 10.1).

From a pathological point of view, well-differentiated paraneuromas have a better prognosis than undifferentiated tumors. However, paraneuromas, if benign in nature, may cause clinical symptoms due to the *hypersecretion* of the secretory substances from them. In the 16 insulinoma cases mentioned above, we noticed the tendency that clinical symptoms caused by the excess insulin secretion were more prominent in patients whose tumor cell granules revealed an appearance closer to that of normal B-cell granules.

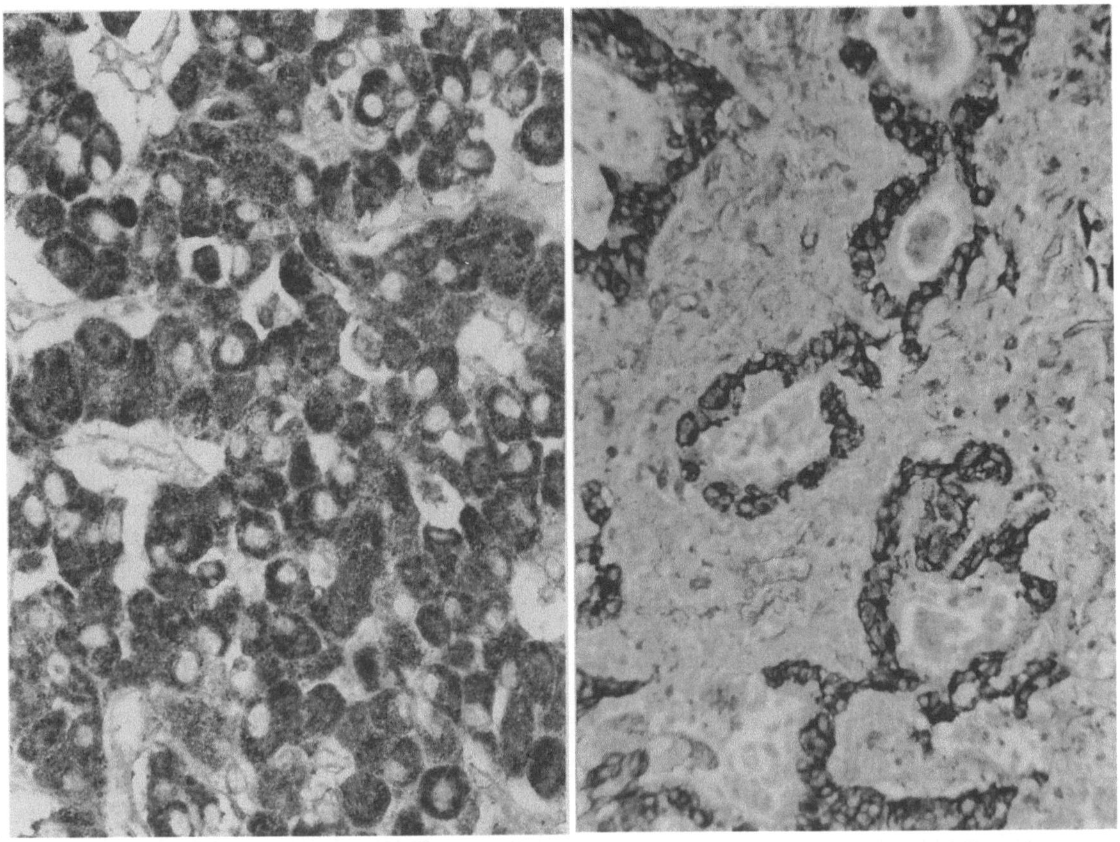

a b

Fig. 10.1a, b. Thyroidal medullary carcinoma and pulmonary tumorlet as typical examples of malignant and benign paraneuromas. **a** Human medullary carcinoma immunostained with an anticalcitonin serum. The secretory granules of most of the tumor cells are immunoreactive for calcitonin. Courtesy of Y. Kameda, Department of Anatomy, Fukuoka University Medical School. **b** Human pulmonary tumorlet stained with Grimelius silver impregnation. Hyperplastic paraneurons are blackened in and beneath the bronchiolar epithelium. Prepared by T. Iwanaga, Department of Anatomy, Niigata University Medical School

10.2 Cellular Origin

Views on the origins of the neoplasms of the peptide/amine-producing endocrine cells have changed along with those on the origins of normal cells. Weichert (1970) suggested that primitive neuroendocrine cells migrate from the neuroectoderm into the developing alimentary system (see also Gammil and Weichert 1973) and proposed that peptide-secreting endocrine tumors are of *neuroectodermal origin*. A similar unifying concept of the origin of endocrine tumors has also been proposed by Pearse and associates (Pearse and Welbourn 1973; Pearse and Polak 1974; Pearse 1975a). Szijj et al. (1969) used the term *apudomas* to designate tumors originating from the endocrine cells of the APUD series (see also Pearse and Welbourn 1973). Since Pearse and his associates believed in a *neural crest origin* of the apudoma cells as well as the normal cells of the APUD series, they proposed the concept and term of *neurolophoma* (Greek "lophos," crest). Neurolophomas are, according to Pearse and Polak (1974), tumors produced by cells having their origin in the neural crest and include, besides apudomas, Schwannomas and melanomas. This unifying concept of the neuroectodermal or neural crest origin of peptide/amine-producing cells was welcomed by pathologists, since they had been looking for a theory which would account for the pathogenesis of *multiple endocrine neoplasms* (MEN). However, the neuroectodermal origin of various peptide-producing endocrine cells was disproved in later embryological (see p. 90) and histopathological studies (Sidhu 1979). Thus, the unifying concept on the origin of the tumors we term "paraneuromas" has lost its basis.

10.3 Markers of Paraneuromas

The stainability of the granules and/or cytoplasm of cells using empirical techniques, especially silver impregnation, has been the main marker of paraneuromas, whereas the chromaffin reaction and argentaffinity have given information on the occurrence of amines (both catecholamines and indolamines). More direct methods, especially immunocytochemical techniques, can be used to detect the occurrence of specific secretory substances, amines and peptides, of paraneuroma cells (see Solcia et al. 1986).

Although the secretory substances and other marker substances of paraneuromas are not essentially different from those of normal paraneurons, the tumor cells do not reflect exactly what is seen in normal paraneurons. One of the reasons is that the substances may deviate in molecular structure from the normal state. Another reason is that *ectopic* tumors frequently occur. In this context, present knowledge of the ectopia of hormone-producing tumors seems incomplete. The concept of "ectopia" was apparently established on the basis of the "one endocrine cell-one hormone theory" (see p. 36). However, it has recently been demonstrated that normal paraneuromas simultaneously produce more than one amine and peptide in a certain proportion (see below). It is, therefore, likely that some "ectopic hormone-producing tumors" represent "eutopic" neoplasms, in which the tumor cells produce specified messenger substances in a ratio greatly different to that in the normal condition.

The markers of paraneuromas can be divided into secretory or intragranular substances and nonsecretory or extragranular substances.

10.3.1 Secretory Substances

Amines. The presence of *catecholamines* was formerly believed to be specific to pheochromocytoma and related tumors. Smaller amounts of catecholamines, however, are detectable by the formaldehyde-induced fluorescence method in many other paraneuromas. *Serotonin* has long been known to be the most characteristic substance of carcinoid tumors. The fluorescence cytochemical and immunocytochemical techniques have enabled this amine to be detected in various kinds of paraneuroma (Schimke 1976; Solcia et al. 1986).

Peptides. Immunocytochemical detection of messenger peptides is of essential importance in the diagnosis of the type of paraneuroma. *Ectopic tumors* may produce peptides which are not detectable in the normal tissue at or near the site in question. Neurosecretory neurons of the hypothalamus, such as oxytocin- and vasopressin-producing neurons of the supraoptic and paraventricular nuclei, do not give rise to neoplasms. However, there are occasional cases where their hormones are produced by oat-cell carcinomas of the lung, insulomas of the pancreas, and carcinoid tumors of the alimentary canal (Bartter and Schwarz 1967; Kameya et al. 1977). Some oat-cell carcinomas, carcinoid tumors, and islet cell tumors (insulomas) produce pituitary hormones such as ACTH, prolactin, GH, TSH, and LH (Yamaguchi et al. 1977; Solcia et al. 1986). Both gastrin and secretin, which are usually localized in the alimentary canal, can also be produced by insulomas. Gastrinoma occurs in the pancreas where no gastrin cells occur except in the fetal period (Solcia et al. 1979) (Table 10.1).

Chromogranins. Among the acidic carrier proteins in the secretory granules, chromogranins have been used as markers of paraneuromas. In particular, chromogranin A-like immunoreactivity has been demonstrated in a large variety of paraneuromas (Lloyd et al. 1984; see also Solcia et al. 1986). As in normal paraneurons, not all cells in a tumor are immunoreactive for chromogranin A and only a certain population of the neoplastic cell mass is usually immunostained.

10.3.2 Nonsecretory Substances

Neuron-specific proteins are often recognized in the extragranular cytoplasm, or cytosol, of the neoplastic paraneurons as in normal cells. *Neuron-specific enolase (NSE)* has been best investigated. Its immunoreactivity can be found in a large variety of paraneuromas (Fujita et al. 1983a; Carlei and Polak 1984; Lloyd et al. 1984; Monaghan and Roberts 1985). We have further demonstrated that immunoreactivities for *neurofilament protein (NFP)* may be contained in some paraneuromas (Fujita et al. 1983a).

S-100 protein occurs in sustentacular cells in neuronal and paraneuronal tissues, as detailed on pp. 102, ff. Its immunoreactivity is recognized in the satellite cells in neuroblastoma and ganglioneuroblastoma; it has also been shown in the satellite or sustentacular cells in pheochromocytoma, chemodectoma, carcinoid, and some other paraneuromas (Nakajima et al. 1984). It is inter-

Table 10.1. Neoplasms of main paraneurons and their secretory products

Paraneuron	Paraneuroma	Site of occurrence	Products (important substances in *italics*)
Chromaffin cell	Pheochromocytoma (b/m), pheochromoblastoma (m)	Adrenal medulla and other paraganglia, autonomic ganglia	*Adrenaline, noradrenaline, enkephalins, NPY,* somatostatin, VIP, ACTH
Parafollicular cell	Medullary carcinoma (m)	Thyroid	*Calcitonin, CGRP, GRP, 5-HT,* somatostatin
Parathyroid cell	Parathyroid adenoma (b)	Parathyroid	*PTH*
Adenohypophyseal cell	Pituitary adenoma (b)	Anterior pituitary	*ACTH, GH, prolactin,* α-subunit of HCG
Gut endocrine cell	Carcinoid (b/m)	Stomach, intestine	*Serotonin, gastrin, PP,* glucagon, somatostatin, VIP, PYY
Pancreatic islet cell	Insuloma (b/m)	Pancreas, gut	*Insulin* (insulinoma), *gastrin* (gastrinoma, Zollinger-Elisson syndrome), *VIP* (vipoma, WDHA syndrome), glucagon (glucagonoma)
Urogenital paraneuron	Carcinoid (b/m)	Urethra, prostate, urinary bladder	*Serotonin*
Bronchopulmonary paraneuron	Carcinoid (b/m), tumorlet (b), small-cell carcinoma (m)	Lung	*GRP, serotonin, calcitonin, CGRP,* PP, somatostatin, ADH, α-subunit of HCG
Carotid body chief cell	Chemodectoma	Neck	*Adrenaline, noradrenaline,* enkephalins, NPY, VIP, ACTH
Merkel cell	Cutaneous neuroendocrine carcinoma	Skin	*Calcitonin, enkephalins,* GRP, ACTH, somatostatin, VIP
Pinealocyte	Pinealocytoma	Pineal body	Melatonin mostly undetectable (see p. 112)

b benign, *m* malignant, *NPY* neuropeptide Y, *VIP* vasoactive intestinal polypeptide, *ACTH* adrenocorticotrophic hormone, *CGRP* calcitonin gene related peptide, *GRP* gastrin releasing peptide, *5-HT* 5-hydroxytryptamine, *PTH* parathyroid hormone, *GH* growth hormone, *PP* pancreatic peptide, *ADH* anti-diuretic hormone

esting that such dispersed paraneurons as those in the bronchial and intestinal epithelia, although they are not accompanied by sustentacular cells in normal states, become associated with such elements when they form tumors (Sano et al. 1984). The positive S-100 immunoreactivity of these sustentacular cells indicates their glial nature and, consequently, the neuronlike nature of the chief neoplasmatic cells.

10.3.3 Synaptophysin

Synaptophysin is a membrane glycoprotein (molecular weight 38 000) occurring in neurons and adrenal chromaffin cells (Wiedenmann and Franke 1985). Wiedenmann et al. (1986) demonstrated immunocytochemically that this substance is widely distributed in paraneurons and also represents a marker substance for such neuronal and paraneuronal neoplasms as pheochromocytomas, paragangliomas, ganglioneuroblastomas, insulomas, enteric and pulmonary paraneuromas, and medullary carcinomas of the thyroid.

10.4 Electric Activities

Cultured cells of paraneuromas have often been used by physiologists studying the electric activities of nonneuronal cells. Electric activities were recorded from clonal anterior pituitary cells of GH_3 and GH_4 strains. The GH_3 strain was established from a transplantable rat anterior pituitary tumor; it secretes both prolactin and growth hormone (Tashijian et al. 1968). The GH_4 strain was developed by serial propagation of GH_3 cells and produces little or no hormone (Ozawa 1985). Kidokoro (1975) first found that GH_3 cells generate action potentials. Biales et al. (1977) and Ozawa and Miyazaki (1979) were able to show that the action potentials of GH_3 cells have both Na^+ and Ca^{2+} components. GH_4 cells also generate action potentials, which may be as a result of increases in Na^+ and Ca^{2+} permeability (Dufy et al. 1979; Taraskevich and Douglas 1980). The Ca^{2+} component shows little variability in different cell strains, whereas the Na^+ component varies markedly (Kidokoro 1975; Taraskevich and Douglas 1980; Hagiwara and Ohmori 1982; Ozawa 1985). The possibility that the ability of GH cells to generate action potentials is caused by pathological changes in the neoplastic cells has been excluded, since electric excitability has also been observed in cells from disaggregated rat anterior pituitary lobes (Taraskevich and Douglas 1977), in slices of these lobes (Ozawa and Sand 1978), and in prolactin cells of teleostean adenohypophyses (Taraskevich and Douglas 1978). However, it is known that the hormonal content is low and more rapidly renewed in the clonal pituitary cell than in the normal cell (Bancroft et al. 1969; Tashijian et al. 1970).

The human small-cell carcinoma of the lung is a tumor probably derived from pulmonary paraneurons. The tumor cells may secrete a large variety of brain-gut peptides, including ACTH, GRP, vasopressin, enkephalins, calcitonin (see p. 114). Extensive work has been done on DMS53 cells, a long-established cell line of human small-cell carcinoma of the lung. The resting membrane potentials of these cells have two characteristics—relatively low resting potential levels and switching between two potential levels. It is possible that an inward Na^+ leak exists in these cells and that this leak is a transient event, increasing and decreasing as switching occurs. Experiments with the patch-clamp technique on DMS53 cells have revealed the presence of single-channel voltage-sensitive conductance as well as acetylcholine-induced channel activity (McCann et al. 1985).

A clonal cell line (designated PC12) was established from a transplantable rat adrenal medullary pheochromocytoma. PC12 cells express many properties characteristic of noradrenergic adrenal chromaffin cells and sympathetic neurons, including uptake, synthesis, storage, and release of catecholamines (see Green and Rein 1977). In addition, the cells can also synthesize, store, and release acetylcholine (Green and Rein 1977). In response to electric stimulation or acetylcholine, the cells produce action potentials similar to those in sympathetic neurons (Tischler et al. 1977a).

10.5 Different Types of Paraneuroma

Paraneuromas are classified into different types by the site of the tumor and the cells of origin. (Table 10.1).

10.5.1 Pineal Paraneuroma

Since the classic studies with special reference to the pathogenesis of pubertas praecox, many kinds of "pinealoma" have been described (Eberts and Ransburg 1979; Herrick and Rubinstein 1979). However, modern exact investigations have shown that only a small number of them are "true" pinealomas, i.e., neoplasms of the pinealocyte itself and/or other parenchymal cells such as glial cells, whereas others (e.g., germinomas and teratomas) are derived from nonparenchymal cells of the pineal organ or surrounding tissues (Rubinstein 1981). Few cases can be identified as neoplasms of the pinealocyte —*pinealocytoma* and *pinealoblastoma*. Features of pinealocytoma cells can be equated with those of the mature pinealocyte, whereas the pinealoblastoma is composed of undifferentiated cells.

The ultrastructure and cytochemical nature of the tumor cells are markedly different from case to case even in neoplasms classified as "*true*" *pinealomas* (Herrick and Rubinstein 1979). In an autopsy case of a 69-year-old female reported by Okeda et al. (1984), a number of true pinealoma cells contained crowded small clear vesicles and large cored vesicles in the electron microscope and showed a positive immunocytochemical reaction to NSE. The cells may be derived from pinealocytes. These "pinealocytoma" cells were surrounded by astrocytes containing "glia-specific" markers such as S-100 protein and glial fibrillary acidic protein. However, Okeda et al. (1984) could not demonstrate melatonin and related indoleamine derivatives. It is possible that the pinealocytoma cells reported by Okeda et al.

(1984) have differentiated into neurons and lost the ability to synthesize and store aminic substances specific to normal pinealocytes.

Normal pinealocytes contain hydroxyindole-*o*-methyl transferase, the enzyme responsible for the synthesis of melatonin (see p. 250). However, so far as we know, there is no case report of a pinealocytoma possessing a large amount of this enzyme. Since true pinealomas are very rare neoplasms, it seems that future studies will be required to demonstrate whether all pinealocytomas are devoid of the ability to synthesize melatonin.

10.5.2 Pituitary Paraneuroma

Neoplasms of the anterior pituitary gland cells include benign adenomas and malignant carcinomas. Ultrastructural aspects of the pituitary tumor cells were reviewed by Horvath and Kovacs (1976). Comprehensive studies of the messenger and nonmessenger products of pituitary paraneuromas have been published by Martinez et al. (1980) and Solcia et al. (1986).

Three main types of adenoma are distinguished according to the staining properties—acidophil, basophil, and chromophobe. This distinction is only for the convenience of description, because adenomas of intermediate or mixed type frequently occur. The *acidophil adenomas* are derived from the two kinds of acidophil cells—growth hormone (GH)-producing and prolactin-producing cells. Functioning (i.e., hormone-secreting) GH cell tumors are accompanied either by gigantism in young patients or by acromegaly in adult patients. On the other hand, functioning prolactin-cell tumors cause impotence, oligospermia, and gynecomastia in men and amenorrhea and galactorrhea in women. There are three kinds of *basophil adenomas*, corresponding to the three types of basophilic elements in the anterior pituitary gland. ACTH-cell tumors cause Cushing's syndrome through hyperfunction of the adrenal cortex. TSH-cell tumors are associated with hyperthyroidism, characterized by emotional lability, tachycardia, weight loss, perspiration, and fine tremor in the hand. Gonadotropin-cell tumors secreting FSH and LH are accompanied by disturbed gonadal functions. Finally, the *chromophobe adenomas* are inactive tumors; the tumor cells contain no secretory granules and thus produce no hormones.

Craniopharyngioma arises from the remnant of Rathke's pouch, from which the anterior pituitary gland develops in early fetal life. This benign tumor possesses no endocrine cells with secretory granules; hence it is not a paraneuroma (Ghatak et al. 1971).

Electric activities were recorded by Kidokoro (1975) in a prolactin- and GH-producing strain of anterior pituitary adenoma cells; detailed investigations have followed (see p. 111).

10.5.3 Medullary Carcinoma of the Thyroid Gland

Medullary carcinoma of the thyroid gland consists of small cells which grow in solid clusters (Pearse and Polak 1974; Schimke 1976) (Fig. 10.2). The tumor cells arise from parafollicular cells and so secrete calcitonin and somatostatin in addition to catecholamines and serotonin (see Suzuki et al. 1985). It is notable that almost all calcitonin-containing cells in the medullary tumor contain gastrin-releasing peptide (GRP; Kameya et al. 1983a, b). Moreover, calcitonin gene related peptide CGRP) is also recognized in these cells, although only small amounts of this peptide are contained in normal human parafollicular cells (Sikri et al. 1985; Kameya, personal communication). ACTH may occur in neoplastic calcitonin cells in such large amounts as to cause Cushing's syndrome (see Schimke 1976).

10.5.4 Parathyroid Adenoma

Most parathyroid adenomas arise from the chief cells of the gland (Nilsson 1977). The main clinical symptoms are loss of calcium from the bone (osteoporosis) and elevation of blood calcium level. These are simply accounted for by the hyperproduction of parathyroid hormone from the tumor cells.

In a case of a parathyroid adenoma (23-year-old female) which one of us (Kobayashi 1974, unpublished data) examined using electron microscopy, the tumor cells possessed numerous secretory granules with neurosecretion-like structures. This finding was one of the reasons for our including parathyroid cells among the paraneurons (see Fujita 1976). At that time, Pearse (1976) excluded parathyroid cells from the APUD group, because normal parathyroid cells showed no demonstrable APUD ability. We explain this lack of APUD ability by the very small number of secretory granules in normal parathyroid cells. Parathyroid tumors are sometimes associated with other paraneuromas such as those in the pituitary, pancreatic islet, and

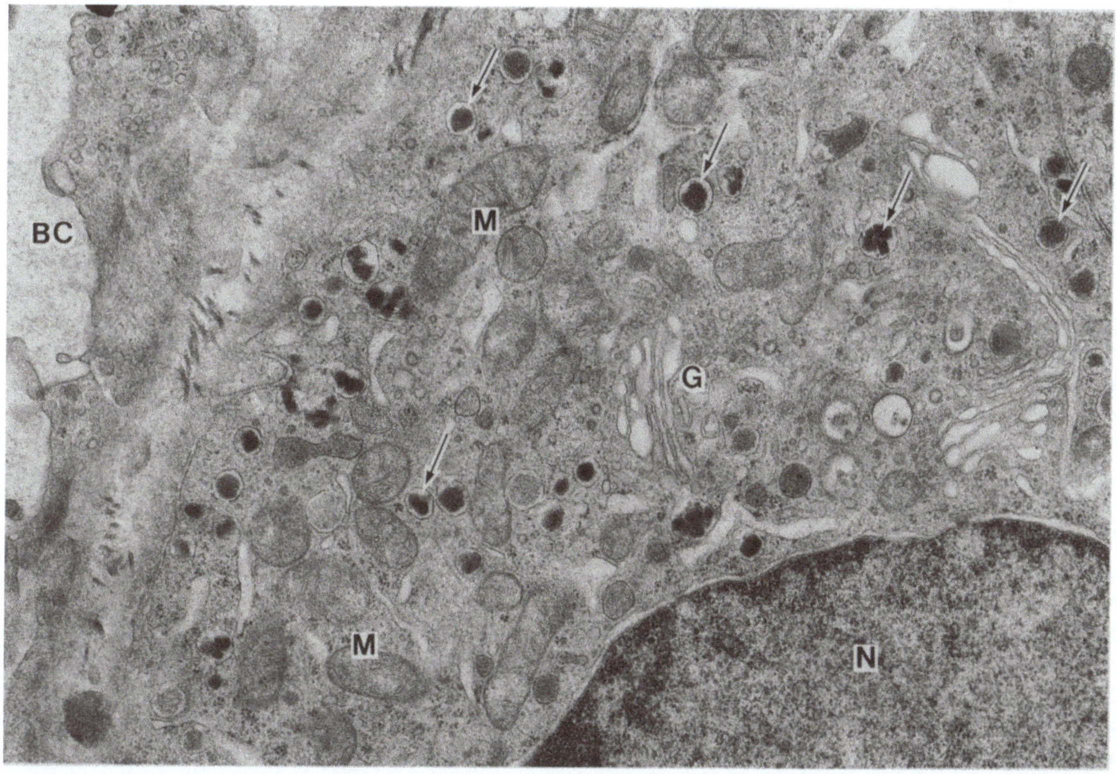

Fig. 10.2. Electron micrograph of an insulinoma cell as an example of a paraneuroma cell. In this "well-differentiated" insulinoma cell, there is a relatively large number of secretory granules with a crystalloid core, which is characteristic of insulin-producing B cells of the pancreatic islet. In this insulinoma cell, a well-developed Golgi apparatus (*G*) is shown vigorously producing secretory granules, but, unlike a normal B cell, mature granules do not show a clear tendency to accumulate at the vascular pole facing the blood capillary (*BC*). *N* nucleus, *M* mitochondria. ×20 000

chromaffin cell system (Schimke 1976) (p. 116). Therefore, if a parathyroid tumor is diagnosed, scanning of the patient for other endocrine tumors is required.

10.5.5 Gastroenteric Paraneuroma, Carcinoid Tumors

Functioning neoplasms of the alimentary tract and related organs have been traditionally called carcinoids. Oberndorfer (1907), who originated this term, defined carcinoids as small tumors of undifferentiated cells often showing a glandular structure and never metastasizing. Later, "carcinoids" came to indicate *serotonin-producing* tumors in the gut and more recently it has been shown that serotonin-producing carcinoids are but one type of a large variety of endocrine tumors. The concept of carcinoids has thus conspicuously changed (review: Soga 1973, 1976). Carcinoids, or more precisely gut paraneuromas, occur throughout the alimentary tract, from the esophagus to the rectum, and in the gallbladder and bile duct. The cells of origin are the basal-granulated cells, and therefore identical tumors also occur in the bronchopulmonary and urogenital systems. In contrast to the original idea of Oberndorfer (1907), only some carcinoids are benign in nature; more often they are malignant and cause metastasis (reviews: Soga and Tazawa 1971; Pearse et al. 1974a; Solcia et al. 1986).

Serotonin-producing (hence argentaffin) carcinoids have been known since the study of Masson (1920). Recent investigations, however, have revealed the occurrence of carcinoids producing other amines and a large variety of peptides. VIPoma, consisting of cells producing VIP, causes severe diarrhea, while other types, such as somatostatinoma, produce no marked symptoms. It is stressed that in most cases the tumor does not consist of a single cell type but comprises a mixture of multiple cell types (see Solcia et al. 1986).

10.5.6 Bronchopulmonary Paraneuroma

Three main kinds of neoplasm occur from bronchopulmonary paraneurons—carcinoid, small cell carcinoma, and tumorlet. All of them contain electron-microscopically discernible secretory granules and immunocytochemically detectable amines and peptides.

Carcinoids in the respiratory tract are essentially identical with those in the alimentary tract. The majority of them produce serotonin and GRP, reflecting the products of their source cells. ACTH and calcitonin may be detected in some cases (Kameya et al. 1983a, b; see Solcia et al. 1986).

Small cell carcinomas are extremely malignant tumors consisting of small, spindle-shaped or fusiform cells with scanty cytoplasm and numerous mitotic features. Typical forms characterized by oval nuclei with a very small amount of cytoplasm are called *oat cell carinomas*. Some oat cells possess cell processes of considerable length; electrophysiologically, these cells have been shown to generate action potentials (Tischler et al. 1977). Electron-microscopic studies have shown that the oat cells possess large cored granules (100–200 nm), even down to the tip of the cell processes. In an oat cell carcinoma of the larynx reported by Bishop et al. (1985), immunoreactive ACTH, GRP, β-endorphin, calcitonin, and keratin were demonstrated. Serotonin is not detected in many cases of oat cell carcinoma (Kameya et al. 1983a, b). Cultured cell lines from human small cell carcinoma have been demonstrated to generate electric activities (see p. 111).

Tumorlets (MacMahon et al. 1967) are small assemblies of paraneurons in the bronchopulmonary epithelium or immediately beneath it (Fig. 10.2). Their structure closely resembles that of the neuroepithelial bodies (pp. 192, ff). They are generally believed to be exaggerated structures of the neuroepithelial bodies. The tumorlets never produce malignancy. The cells forming the tumorlets usually contain GRP, calcitonin, ACTH, and serotonin (Kameya et al. 1983; see also Solcia et al. 1986).

10.5.7 Insuloma

Benign and malignant tumors arise from each type of endocrine cell of the pancreatic islet. Cell tumors are associated with hypersecretion of glucagon and its precursors; they are called *glucagonomas*. B-cell tumors produce excess insulin and its precursors and are thus termed *insulinomas*. D-cell tumors are also present; they secrete excess somatostatin and its precursors and are called *somatostatinomas*.

In addition to the above-mentioned three kinds of functioning tumors of the islet endocrine cells, there occur many kinds of peptide hormone-producing cell tumors in the pancreas. The best known are the tumors which cause *Zollinger-Ellison syndrome* (Zollinger and Ellison 1955) and *Verner-Morrison syndrome* (Verner and Morrison 1974). The Zollinger-Ellison syndrome, characterized by hypersecretion of gastrin and ulceration in the stomach and duodenum, is produced by a *gastrinoma*, the neoplasm of gastrin-producing cells in the pancreas or occasionally in the duodenum, stomach, and other organs. In contrast, the Verner-Morrison syndrome, also referred to as pancreatic cholera, is characterized by *w*atery *d*iarrhea, *h*ypokalemia, and *a*chlorhydria, hence the acronym *WDHA*-syndrome. This syndrome is ascribed to excess secretion of VIP from a *vipoma*, which in many cases occurs as a pancreatic tumor (Bloom et al. 1973).

10.5.8 Tumors of Adrenal Chromaffin and Related Cells

Adrenal and extra-adrenal chromaffin cells, including the carotid body chief cells, give rise to benign and malignant neoplasms. Tumors arising from the adrenal chromaffin cells are called *pheochromocytomas* (Manger and Gifford 1977), whereas those from the carotid body are *chemodectomas* (*chemoreceptomas*; Mulligan 1950; see also Stanulla and Wöckel 1969; Glenner and Grimley 1974) (Fig. 10.3a, b). There are a series of neoplasms arising from paraneurons in the paraganglia (chromaffin and nonchromaffin paraganglia). Reflecting the close relation to neurons of the cells of these tumors, there is every gradation between them and such neuronal neoplasms as neuroblastoma and sympathogonioma. All these neuronal and paraneuronal neoplasms are slightly different from each other, as reflected in the term, which indicates the specific cell of origin. All of these tumors, including also the "nonchromaffin" tumors, produce catecholamines, hence we doubt whether the division of these tumors into subgroups is a useful and valid exercise. We believe that it is less confusing if all these tumors be regarded as a single category. Tumors consisting of very immature cells are called *pheochromoblastomas*. How-

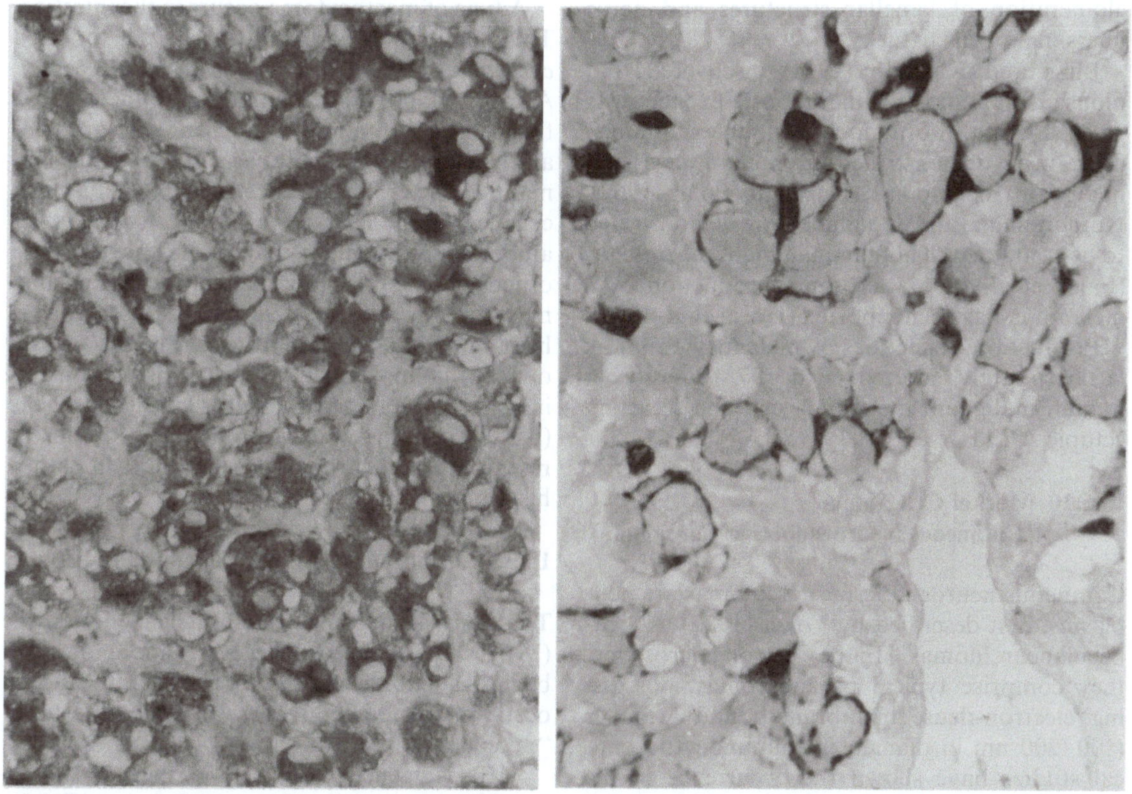

a b

Fig. 10.3a, b. Carotid body tumor or chemodectoma as immunostained with **a** anti-NSE and **b** anti-S-100 antisera. A case of human chemodectoma (provided by E. Okazaki, Niigata City Hospital) shows that the tumor cells contain NSE-like immunoreactivity and are supported by sustentacular (glial) cells with S-100-like immuno-reactivity. Prepared by T. Iwanaga, Department of Anatomy, Niigata University Medical School

ever, again it must be pointed out that there are no sharp distinctions between pheochromocytomas and pheochromoblastomas with respect to the cytological features.

Tumors of adrenal chromaffin and related cells always contain large amounts of catecholamines. However, the content of catecholamines differs from tumor to tumor. There is a tendency for the amount of adrenaline to be limited, whereas storage of noradrenaline and dopamine is predominant in the tumor cells. The occurrence of serotonin and histamine in pheochromocytoma and related tumors seems likely, because it has been demonstrated that normal adrenal chromaffin cells contain these substances (see p. 23).

Evidence is accumulating that pheochromocytoma cells may produce various neuropeptides, such as ACTH, enkephalins (mainly derived from preproenkephalin A), substance P, VIP, NPY, and somatostatin, as their normal counterparts do (see Kakudo et al. 1984). The cells forming *chemodectoma* are very similar to those of pheochromocytoma in every cytological feature (Capella and Solcia 1971). We propose that the only essential difference between pheochromocytoma of the adrenal medulla and chemodectoma of the carotid body is the site of the tumor, and therefore the category of pheochromocytoma will suffice to describe the tumor in question.

10.5.9 Melanoma

Melanomas are tumors of melanocytes and they are of neural crest origin. Therefore, melanoma cells produce melanin pigment to a greater or lesser extent. Melanomas are usually very malignant tumors. Benign tumors of melanocytes represent *nevus pigmentosus* of various types.

Different kinds of neuronal and related elements may be contained in melanocyte tumors. This may be due either to induction by the melanocytes or to a common factor causing the growth of both melanocytes and nervous sturctures. Suzuki et al. (1986) demonstrated a conglomeration of lamellar structures and nerve fibers in a nevus tissue.

Blois and Kallman (1964) discovered that mouse melanoma cells incorporate exogenously given L-dopa. Formaldehyde-induced fluorescence was demonstrated in human nevi and malignant melanomas (Rost and Polak 1969). Noradrenaline and dopamine were demonstrated to be contained in human malignant melanomas (Falck et al. 1966). Serotonin production by a melanoma has been reported (Horai et al. 1979).

Production of peptides by melanocyte tumors has been rarely reported. Orth (1973) described a human malignant melanoma cell line secreting ectopic ACTH.

10.5.10 Merkel Cell Tumor (Cutaneous Neuroendocrine Carcinoma)

Cutaneous neuroendocrine carcinomas are neoplasms, first described by Toker (1972) as "trabecular carcinomas." In the electron microscope, they comprise typical paraneuromas, possessing electron-dense membrane-bound granules (100–200 nm in diameter). Immunocytochemical studies have shown the occurrence in the neoplastic cells of such neuropeptides as calcitonin GRP, enkephalins, ACTH, somatostatin, and VIP (Gould et al. 1985).

Many authors identify cutaneous neuroendocrine carcinomas as Merkel cell tumors (e.g., Yoshida et al. 1984). However, "there is no evidence that these tumors are really derived from Merkel cells or their precursors" (Warner et al. 1983). In this respect, there was an interesting finding by Warner et al. (1983) that some of the tumor cells show straight, filament-rich cytoplasmic processes, similar to the antennalike processes in normal Merkel cells.

10.5.11 Mastocytoma

Neoplasms of mast cell origin are mastocytomas, which include both benign and malignant tumors. *Urticaria pigmentosa* is systemic mastocytosis characterized by proliferation of mast cells in the skin. There is every gradation between this state of mastocytosis and malignant mastocytomas.

The cells of mastocytomas contain large metachromatic granules as normal mast cells do. Besides heparin and other glycoseaminoglycans, the granules contain variable amounts of histamine, which may cause ulceration of the intestinal mucosa (literature: see Alroy et al. 1975).

A type of mastocytoma occurs in the intestine. This tumor was originally believed to be a *carcinoid of enterochromaffin cells*. Recently, Alroy et al. (1975) demonstrated in the cat that this tumor, containing metachromatic granules and lacking argyrophil, argentaffin, and formalin-induced fluorescence reactions, is a mast cell neoplasm. Histamine is detected in lower amounts in this tumor than in mastocytomas in other regions, reflecting the feature that normal mast cells in the intestinal mucosa contain less histamine than those in other sites. It is worthy of note that in spite of this recent distinction and in spite of the mesenchymal origin of mast cells (p. 266) certain neoplasms of these cells closely resemble enterochromaffin cell tumors in pathohistological appearance.

10.5.12 Multiple Endocrine Neoplasia (MEN)

The term "multiple endocrine adenopathy" (MEA) is also used. However, since not only benign adenomas but also malignant cancers occur in this rare disease, we prefer the term "neoplasia."

MEN is a genetic disorder (Wermer 1954). The sequential occurrence of neoplasms in different endocrine organs is considered to be mostly inherited (Steiner et al. 1968).

There are two types of MEN. Type I is called Werner's syndrome and involves the anterior pituitary, parathyroid, and pancreatic islet (Schimke 1976). Type II, Sippl's syndrome, is the simultaneous occurrence of medullary carcinoma of the thyroid gland and pheochromocytoma of the adrenal medulla (Sippl 1961; see also Schimke 1976).

The reason why different endocrine tumors simultaneously occur in various organs is unknown at present. However, it seems likely that the cause of MEN is related to the gene or genes which control the differentiation of the involved cells into peptide/amine-producing endocrine cells. Furthermore, the sequential origin of different neoplasms from various sites is at odds with the theory that cancers arise from an acquired mutation of a single cell.

10.6 Summary

The neoplasm of a paraneuron is called a paraneuroma. Its counterpart in the nervous system is a neuroblastoma. The cytological features of neuroblastoma cells, such as the presence of secretory granules, are shared by paraneuroma cells.

Paraneuromas show various degrees of differentiation. In well-differentiated paraneuromas (often called "functioning tumors"), identification of their progenitor cells is not difficult since they possess secretory substances and granules similar to those of their progenitors. In many other cases of paraneuromas, the cellular origin is difficult to identify.

The coexistence of aminic and peptidic messenger substances is a common phenomenon in both neuroblastomas and paraneuromas. Thus, the "ectopic" hormome-producing neoplasm may actually be a "eutopic" tumor, in which tumor cells produce specified messenger substances in a ratio greatly different from that in the progenitor cells.

Studies of the cells of paraneuromas are sometimes useful, since the neoplastic cells may reveal structural, cytochemical, or electrophysiological features which are not manifest in their normal progenitors and yet are important for the analysis of their paraneuronal nature. Similarities in the cytological features between neuroblastomas and related neuronic neoplasms and paraneuromas provide support for the idea that peptide/amine-producing endocrine/sensory cells form a continuous spectrum with "typical" neurons in the central and peripheral nervous systems.

11. Phylogeny of Paraneurons

Originating from the common ancester, Coelenterata, the evolution of the Metazoa has proceeded along two animal series, Protostomia and Deuterostomia. The *protostomian series* includes Platyhelminthes (planarians, tapeworms, etc.), Annelida (earthworms, clamworms, etc.), and, as more advanced forms, Mollusca and Arthropoda. The *deuterostomian series* includes Echinodermata (seaurchins, starfish, etc.), Protochordata (seasquirts, amphioxus, etc.), and Vertebrata (Fig. 11.1).

In this chapter, the Deuterostomia, comprising the evolutionary series of mammals, will be dealt with first, Protostomia second, and Coelenterata, as the common ancester of both series, last. A short review will be added on the occurrence of metazoan messenger substances in unicellular organisms.

An excellent review of the morphological, biochemical, and physiological aspects of invertebrate neuroregulatory systems is available (Haynes 1980).

11.1 Deuterostomia

Except in vertebrates, some protochordate species have been investigated with regard to the occurrence of paraneurons in the deuterostomian series.

11.1.1 Protochordata

The brain-gut peptides in the Ascidiacea (seasquirts) are relatively well examined. In the cerebral ganglion of *Ciona intestinalis*, neurons containing immunoreactivities for adrenocorticotropic hormone (ACTH) (Georges and Dubois 1979), luteinizing hormone releasing hormone (LH-RH) (Georges and Dubois 1980), VIP, substance P, PP, somatostatin, bombesin, secretin, neurotensin, motilin, β-endorphin, and (human) prolactin (Fritsch et al. 1982) have been reported. Using electron microscopy, Fritsch et al. (1982) demonstrated typical peptidergic-type neurons containing large-cored granules which could be classified into at least five types. In the gut epithelium, endocrine cells containing all but ACTH, LH-RH, VIP, motilin, and β-endorphin of the cerebral peptides were demonstrated; serotonin-containing endocrine cells were found in groups (Fritsch et al. 1982). As is also clear from immunocytochemical studies in two other ascidean species, *Styela clava* (Bevis and Thorndyke 1978, 1979) and *Styela plicata* (Pestarino 1982), endocrine cells in the digestive tract of seasquirts are mostly of the typical open type. In *Styela plicata* and *Halocynthia papillosa*, secretin-containing cells in the esophagus simultaneously stored serotonin, whereas those in the gastric epithelium revealed only secretin (Pestarino 1982; Pestarino and Tagliafierro 1983). The cerebral ganglion of *Styela plicata* was shown to contain prolactin-, secretin- and β-endorphin-immunoreactive neurons (Pestarino 1983, 1984a, b, 1985b). Further, in the same species, the neural gland, located dorsal to the cerebral ganglion, was shown to contain prolactin- and α-melanocyte-stimulating hormone (MSH)-immunoreactive cells; the nature of these cells and their resemblance to the vertebrate adenohypophysis have been discussed (Pestarino 1984a, 1985a).

Studies on the functions of brain-gut peptides in invertebrates are still very few, but there is a paper by Bevis and Thorndyke (1981), reporting that an injection of porcine cholecystokinin (CCK) caused an increased output of gastric enzymes in *Styela clava*.

In the most evolved protochordates, Cephalochordata, represented by the lancelet (amphioxus), no immunocytochemical studies on the nervous system appear to have been performed. As to the digestive tract, Kataoka and Fujita (1974) demonstrated by electron microscopy that very tall endocrine cells of the open type were dispersed in the gut epithelium of *Branchiostoma japonicum*. After some attempts to reveal the peptides in these cells in *Branchiostoma lanceolatum* (Van Noorden and Pearse 1976;

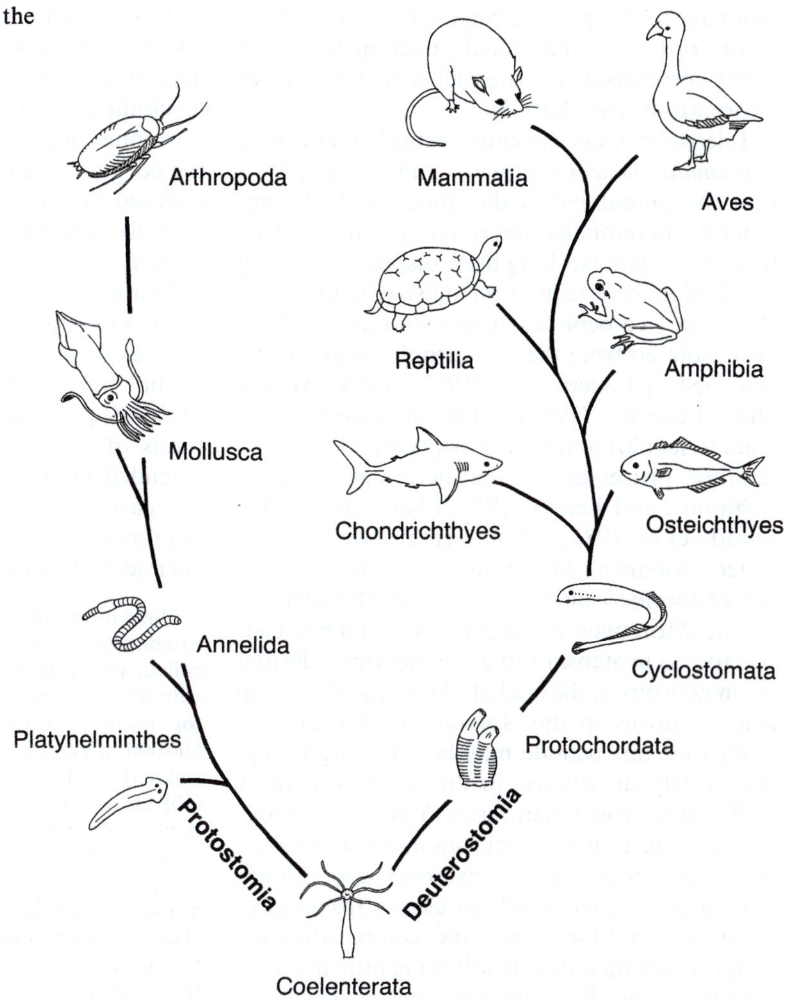

Fig. 11.1. Simplified diagram of the phylogeny of animals

Reinecke et al. 1980), Reinecke (1981) recently reported the occurrence of eight types of cells corresponding to the immunoreactivities for insulin, glucagon, somatostatin, PP, CCK/gastrin, VIP, secretin, and neurotensin.

It is now noteworthy that the insulinlike immunoreactivity in amphioxus occurs only in the open-type endocrine cells in the intestinal epithelium and no tissue corresponding to the pancreatic islets of vertebrates can be found at this stage of evolution.

11.1.2 Vertebrata

In *Ammocoetes*, the larval form of the lamprey (Cyclostoma), which represents the lowest vertebrate, insulin-producing cells are gathered in the connective tissue surrounding the anterior part of the intestine. In the adult lamprey, *Lampetra reisneri*, somatostatin cells join this insulin cell mass (Yui and Fujita unpublished data). In Atlantic hagfish, *Myxine glutinosa*, the insulin producing cell mass occurs surrounding the distal end of the bile duct (Falkmer and Matty 1966; Östberg et al. 1975, Östberg 1976), and this tissue has been shown to contain also somatostatin cells (Van Noorden et al. 1977b). The most primitive "islet" in Cyclostoma does not merit the name *pancreatic* islet since the pancreas as an exocine gland is completely lacking. The pancreas appears as an associate of the islet first at the evolutionary stage of the holocephalian and elasmobranchial fish (reviews: Fujita et al. 1980; Falkmer 1980; Van Noorden and Falkmer 1980; Fujita et al. 1981). In teleosts, pancreatic islet cells are gathered into one or more independent organs called Brockmann bodies. These are, however, usually associated with a rim of the exocrine pancreas.

Even after the pancreatic islet has been established, numerous endocrine cells remain in the epithelium of the gut throughout the evolutionary scale up to mammals. In their peptide/amine products, they partly overlap with and partly compensate for the islet cells. For further information, see Chap. 16.

Teleosts possess a peculiar neurohemal organ, the *caudal neurosecretory organ* or *urophysis* near the caudal end of the spinal cord. Resembling the mammalian neurohypophysis in structure, this organ has long been suspected of being involved in osmoregulation in fish (Enami 1956). Recently, two peptides, urotensin I and II, have been isolated from the carp urophysis and characterized by Lederis et al. (1982) and McMaster and Lederis (1983). Immunocytochemical studies localized the two peptides in neurons sending processes to the neurohemal tissue (Ishimura and Lederis 1983; Fisher et al. 1984; Owada et al. 1985). More recently, we have detected urotensin I-like immunoreactivity in some gut endocrine cells (open type) in the carp intestine. Moreover, as we detected endocrine cells reactive to urotensin I in gut endocrine cells and brain neurons in the cricket (Arthropoda) and in some neurons in the *Aplysia* (Mollusca), it is likely that this peptide may be a brain-gut peptide widely distributed in the protostomian as well as deuterostomian series (Yui et al. 1985).

The basic designs of neurons and paraneurons and their messenger substances are similar among the different vertebrate classes, from Cyclostoma up to Mammalia, and comparative descriptions of their details will bring little of essential importance from an evolutionary viewpoint. We will now move to the protostomian series.

11.2 Protostomia

For the convenience of description, the protostomian phyla will be dealt with in decreasing order of organization—from advanced to primitive forms.

11.2.1 Arthropoda

Hemolymph, which contains trehalose for our "blood glucose," streams in the body of insects. Hormones regulating the levels of trehalose have been postulated. Tager and Markese (1976) isolated glucagonlike and insulinlike peptides from the corpus cardiacum-alatum complex of the tobacco hornworm, *Manduca sexta*, and a few other species. These substances immunologically cross-reacted with mammalian glucagon and insulin, respectively, and also exerted clear effects in bioassays in terms of glycogen contents in insect adipose tissue and trehalose levels in hemolymph. Thus, it has been suggested that insulinlike and glucagonlike peptides as neurohormones might be conveyed from the brain to the corpus cardiacum and corpus alatum, to be released into hemolymph. These peptides were actually detected in insect hemolymph (Kramer et al. 1980).

Using Japanese silkworm larvae, *Bombyx mori*, Yui et al. (1980) demonstrated insulinlike immunoreactivity in eight large cells in the medial neurosecretory neuron group in the brain. Supporting the above-mentioned hypothesis of Kramer et al. (1980), these neurons extended immunoreactive processes to the corpus cardiacum and corpus allatum. Glucagon-immunoreactive neurons, however, have not been identified in the silkworm brain.

Quite independently, a great efforts have been made to characterize the insect "brain hormone" recently called, prothoracicotropic hormone (PTTH), which stimulates the prothoracic gland to release ecdysone for molting (ecdysis) and metamorphosis. From 600000 silkworm heads, PTTH-I, II, and III were isolated and characterized by Nagasawa et al. (1984a, b). Contrary to the presumption that PTTH might resemble hypothalamohypophyseal hormones of mammals, it turned out to be composed of insulinlike peptides. PTTH-I, II, and III contain amino acid sequences closely similar to the sequence of the A chain of human insulin. Now the problem remains to be solved whether the insulinlike immunoreactivity in the neurosecretory system of the silkworm and other insects is due to PTTH; if so, in addition to its ecdysone-releasing effect, PTTH might be involved in trehalose regulation as an "insect insulin."

In the *silkworm brain*, Yui et al. (1980) demonstrated gastrin and PP-immunoreactive neurons of smaller size in the medial neurosecretory cell group. The prethoracic ganglion contained two large neurons with gastrinlike reactivity. Several neurons containing α-endorphin-like immunoreactivity were demonstrated by Rémy et al. (1979) in the subpharyngeal ganglion of the silkworm. Four neurons containing Met-enkephalin-like immunoreactivity have been reported in *Locusta migratoria* (Rémy and Dubois 1981). In the subpharyngeal ganglion of this insect, Rémy and Girardie (1980) revealed two large, aldehyde-fuchsin-positive neurons containing immunoreactivities for mammalian vasopressin and neurophysin (p. 63). El-Salhy et al. (1980) detected neurons containing immunoreactivities for insulin, glucagon, somatostatin, PP, secretin, gastrin/CCK, substance P, enkephalin, and endorphin in the brain of *Eristalis aeneus* larvae (Diptera) (see also El-Salhy 1981). Fairly extensive distribution of gastrin/CCK-immunoreactive neurons was demonstrated in the brain and ganglia of the blowfly *Calliphora vomitoria* (Duve and Thorpe 1984). Distribution of similarly numerous neuropeptides has been reported in the brain of an adult lepidopteran insect, *Manduca sexta* (L) (El-Salhy et al. 1983). Veenstra et al. (1984) demonstrated that neurons in the Colorado potato beetle, *Leptinotarsa decemlineata*, contained immunoreactivities for vasopressin/vasotosin, oxytocin, BPP, and FMRFamide, the last two being contained in the same neurons. GRP/bombesin-immunoreactive neurons have also been found in *Locusta migratoria* and other insects (Veenstra and Yanaihara 1984).

The cockroach contains numerous peptidergic neurons in the brain, ganglia, and visceral nerves. PP-immunoreactive neurons are numerous, including those in the medial and lateral neurosecretory cell groups in the pars intercerebralis (Endo et al. 1982). A similar number of neurons contain somatostatinlike immunoreactivity, and a few neurons VIP-like immunoreactivity (Fujita et al. 1981).

The predominant occurrence of PP-immunoreactive neurons in the brain and PP-immunoreactive endocrine cells in the gut was stressed by Duve and Thorpe (1982) in their investigation of *Calliphora vomitoria* (Diptera).

The occurrence of *basal-granulated paraneurons* in the *insect digestive tract* has been studied quite extensively following the electron-microscopic observations of Nishiitsutsuji-Uwo and Endo (1981) in the midgut of the cockroach, *Periplaneta americana*. Numerous PP-immunoreactive cells, smaller numbers of somatostatin- and glicentin-immunoreactive cells (Iwanaga et al. 1981), and recently also PHI and urotensin I-immunopositive cells (Yui et al. 1985; Iwanaga et al. 1986) have been shown to be open-type basal-granulated cells. Silkworm larvae also have PHI immunoreactive cells (Yui et al. 1985), while the cricket *Gryllus bimaculatus* reveals numerous urotensin I-immunoreactive paraneurons in the midgut epithelium (Iwanaga et al. 1986) (Fig. 11.2). Electron-microscopic studies in cockroach and butterfly larvae have demonstrated that the endocrine cells, even exceeding 100 μm in length, possess a tuft of *microvilli* at the top of the cell and contain round secretory granules at the base (Nishiitsutsuji-Uwo and Endo 1981; Endo and Nishiitsutsuji-Uwo 1981) (Fig. 11.3). Moreover, typical images of *exocytosis* of these granules have been demonstrated (Endo and Nishiitsutsuji-Uwo 1982; see also Fujita et al. 1981) (Fig. 11.4).

11.2.2 Mollusca

Extensive studies on protostomian neuropeptides have been made in the pondsnail, *Lymnaea stagnalis* (Gastropoda). Boer et al. (1979) first demonstrated ACTH-like immunoreactivity in two huge neurons (130 μm in diameter), one located in the splanchnic ganglion and the other in the right parietal ganglion. Another 15 peptides were then localized in different neurons in the central nervous system and clear diagrams indicating their distribution were published by Schot et al. (1981). On the basis of their morphological observations, Schot et al. (1981) conceive that some of the peptidergic nerves end in the connective tissue and release their secretions into the hemolymph, while others synaptically regulate other neurons. FMRF-amide, known as the molluscan cardioactive peptide, has been shown to be widely distributed in the neurons of the pondsnail and is presumably secreted as a neurohormone at the end of their processes (Schot and Boer 1982).

In *Aplysia* (Gastropoda), bag-cell neurons, gathered in bilateral cell clusters and sending axons to a terminal area with a typical neurohemal structure (Haskins et al. 1981), have been shown to produce egg-laying hormone, a neuropeptide with 36 amino acid residues (Chiu and Strumwasser 1981). The cephalic ganglia of *Aplysia californica* contain immunoreactivity for arginine vasotocin (Moore et al. 1981), which is known to have certain physiological effects, including suppression of the gill withdrawal reflex (Thornhill et al. 1981). In both cephalic ganglia and peripheral nerves (intestine) of this animal, we recognized neurons containing PP-, PHI-, and urotensin I-like immunoreactivity (Yui et al. 1985). In *Fusitriton oregonensis* (Cymatiidae, Gastropoda), we found numerous nerves mainly immunoreactive for PHI and partly for PP and urotensin I at the epithelial base and in muscular layers of the gut, especially the foregut.

Monoamines have been found, probably as transmitters, in the central and peripheral neurons of molluscs. In the pond snail, histamine (Osborne and Patel 1984), dopamine, and serotonin (Boer et al. 1984) have been detected in neurons. The last two amines together with a peptide vasotocin are costored in the cerebral

a

b

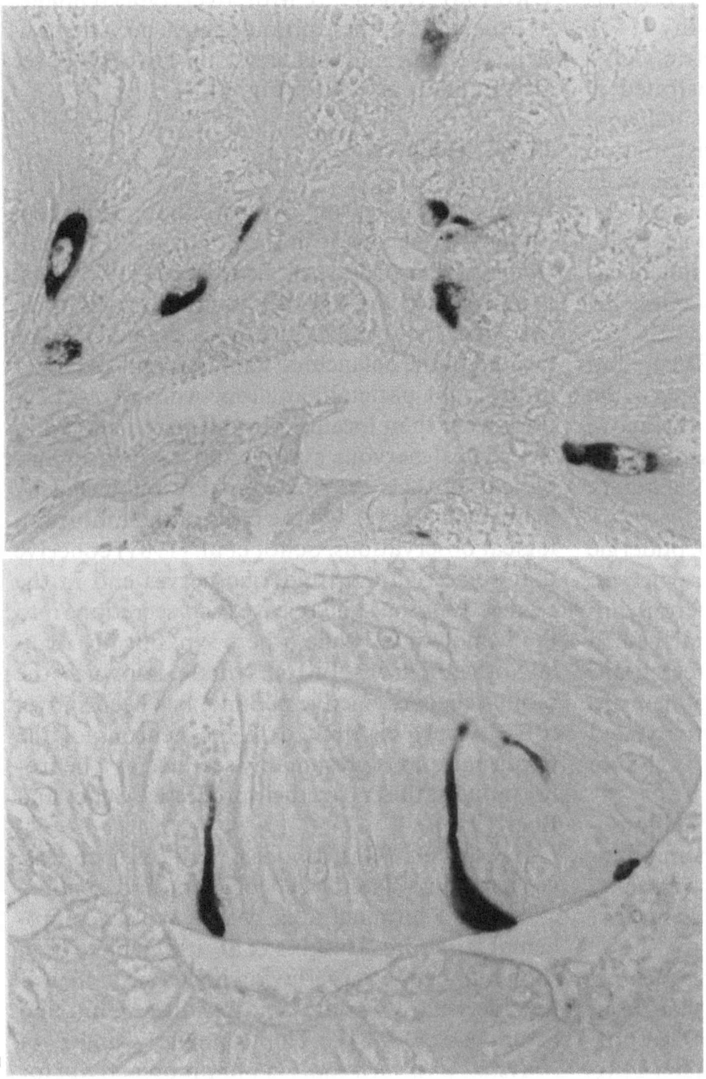

Fig. 11.2a, b. Endocrine cells in the midgut of the cricket, *Gryllus bimaculatus* (insect), containing **a** BPP-like and **b** urotensin I-like immunoreactivities. The epithelium contains dispersed immunoreactive cells, which are mainly open in type. Indirect immunoperoxidase (PAP) method

giant neuron (Boer et al. 1984). In *Aplysia*, specific neurons contain histamine (Weinreich et al. 1975; Turner and Cottrell 1977; Bailey et al. 1982). By fluorescence histochemistry, dopamine and noradrenaline/adrenaline have been detected in the nerves and intraepithelial cells of the albumen gland duct of freshwater and land snails (Hartwig et al. 1980), while serotonin and dopamine have been found in neurons in *Aplysia* (Tritt et al. 1983).

Brisson and Collin (1980) pointed out that intraepithelial bipolar cells containing catecholamines occur in the carrefour region (albumen gland duct, fertilization pocket, and oviduct sac) of *Bulinus truncatus* (Gastropoda) and represent typical paraneurons, presumably of the sensory type.

11.2.3 Annelida

The nervous system of the earthworm, *Lumbricus terrestris*, was examined by Sundler et al. (1977), who demonstrated VIP-immunoreactive neurons in the subpharyngeal ganglion and PP-positive neurons in the cerebral, subpharyngeal, and more posterior ganglia. Rémy and Dubois (1979) demonstrated in *Dendrobaena subrubicunda* that four neurons in the infraesophageal ganglia contained α-endorphin-like immunoreactivity (β-endorphin being negative). According to Alumets et al. (1979), the cephalic ganglion of *Lumbricus terrestris* contains many neurons immunoreactive to Leu-enkephalin and a few positive for β-endorphin, both extending numerous fibers.

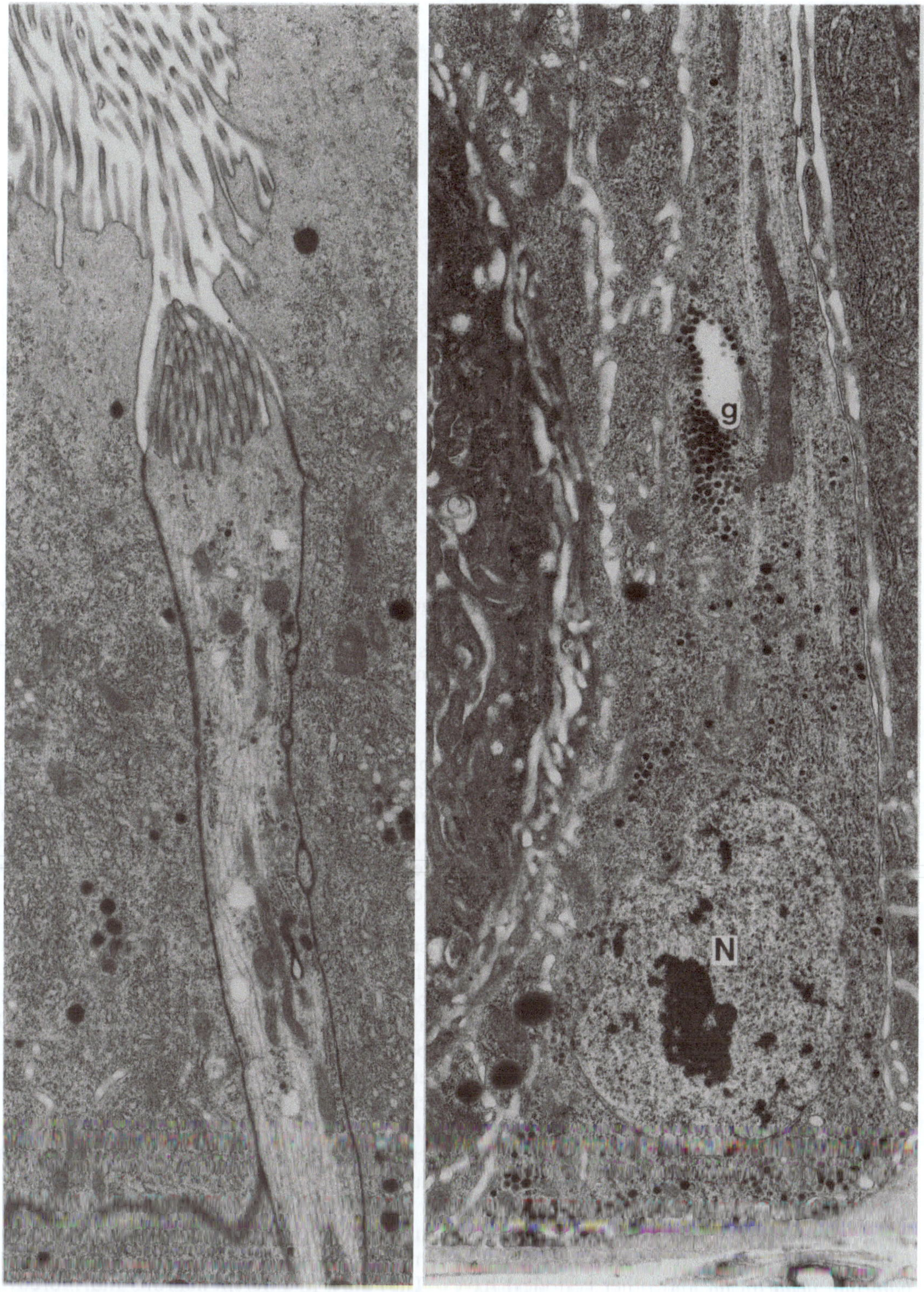

Fig. 11.3a, b. Electron micrographs showing the apical and basal parts of an endocrine cell in the midgut of a butterfly larva, *Papilio xuthus*. **a** The cell apex is provided with a tuft of microvilli, the probable receptive site of the cell. The cytoplasm contains a neurotubule-like system and some mitochondria. **b** The basal part of the cell contains a nucleus (*N*) and secretory granules mainly gathered in the infranuclear portion. *g* glycogen particles. Courtesy of Y. Endo and J. Nishiitsutsuji-Uwo; reproduced, with permission, from Biomed Res 2:270–280 (1981)

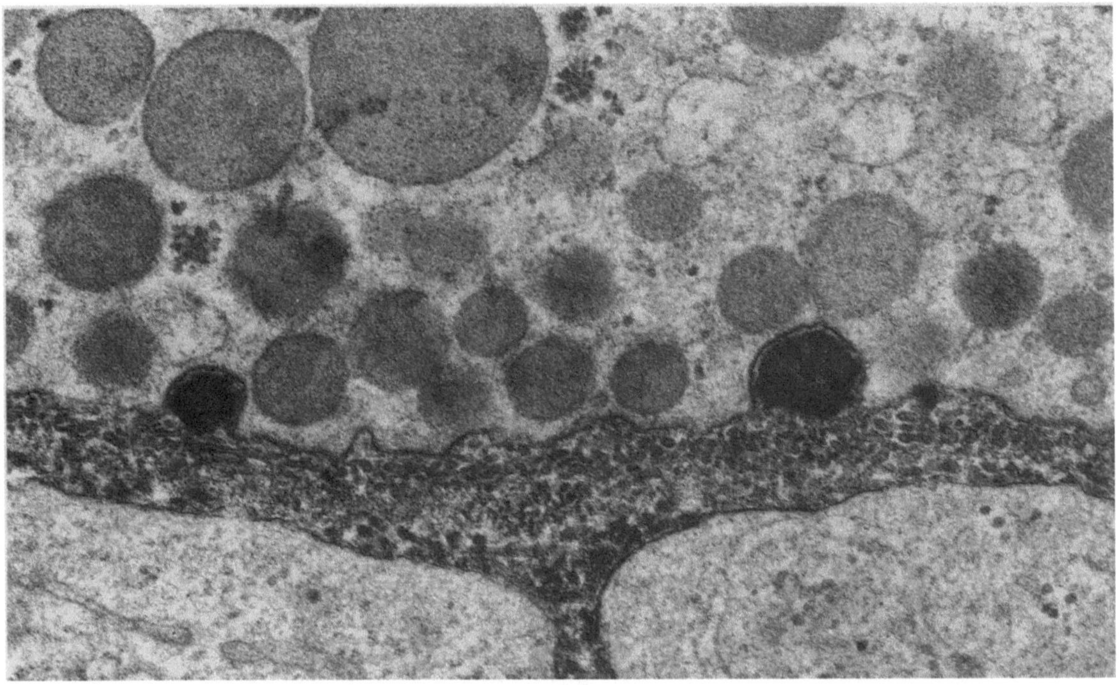

Fig. 11.4. Basal portion of an endocrine cell in the midgut of the cockroach *Periplaneta americana* showing exocytotic granule release. The specimen was prepared by the tannic acid-glutaraldehyde-osmium method; the contents in the two open granules and the intercellular substances are stained dark with tannic acid. Prepared by Y. Endo and J. Nishiitsutsuji-Uwo; reproduced, with permission, from Peptides Suppl 2: 123–131 (1981)

In the leech, *Maemopis marmorata* (Hirudinea), Leu-enkephalin-immunoreactivity has been reported by Zipser (1980) to be localized in one of the 400 cells of each posterior midbody ganglion. Serotonin has been shown to occur in many neurons of *Hirudo medicinalis* (Leake et al. 1985).

11.2.4 Platyhelminthes

In *Planaria gonocephala* (Turbellaria, Platyhelminthes), we found BPP-like immunoreactivity extensively in the nervous system; it was especially rich in the ladder-shaped nerve cords and submucosal nerve plexus. The subepithelial nerve plexus was free of BPP and contained serotonin-like immunoreactivity (Iwanaga et al. unpublished data, recorded in Yui et al. 1985).

The flatworms (*Stenostomum leucops* and other turbellarians) and tapeworms (*Diphyllobothrium dendriticum* and other cestodes) have been investigated immunocytochemically by Finnish zoologists (Reuter et al. 1984; Gustafsson et al. 1985, 1986; Wikgren and Reuter 1985; Reuter et al. 1986). Besides FMRF-amide, immunoreactivities for β-endorphin, Leu-enkephalin, vasotocin, growth hormone releasing factor (GRF), substance P, neurotensin, BPP, and the amine serotonin were recognized in neurons in the central and peripheral nervous systems, though the occurrence of each peptide differed among the species studied.

It is of note that Reuter et al. (1984) recognized endocrinelike cells in the intestinal epithelium of the flatworm *Microstomum lineare*, which appeared to contain FMRF-amide-, vasotocin-, and BPP-like immunoreactivities (Reuter 1985, personal communication). A diagram by Reuter et al. (1984) clearly indicates that these endocrinelike cells (paraneurons) are dispersed in the intestinal epithelium, partly closed in the epithelial base and partly open to the intestinal lumen.

Secretory granules and synaptic vesicles in the nervous system of platyhelminthes are of the multiple type, which is the rule in vertebrates. They differ in different neurons and, moreover, a neuron mostly contains two types of granules/vesicles. In her electron-microscopic study of the nervous system of the tapeworm *Diphyllobothrium dendriticum*, Gustafsson (1984; see also Reuter for *Microstomum*) clearly demonstrated the occurrence and combinations of four different granule types: (1) dense-cored vesicles, 60–120 nm in diameter; (2) small clear vesicles, 30–50 nm; (3) large dense (peptidergic type) granules, 80–320 nm; and (4) large lucent vesicles, 60–120 nm.

11.3 Coelenterata

It is widely accepted that there has been a bifurcation in the evolution of the metazoa with the development of Protostomia and Deuterostomia. If, then, recent findings, as reviewed above, indicate that the most evolved animals in the protostomian series insects and molluscs possess messenger peptides closely similar in structure to those in deuterostomian vertebrates, it implies that these peptides might be found in the Coelenterata, which represents the origin of both series of evolution. And this assumption is indeed correct. In the primitive coelenterate, *Hydra attenuata*, Grimmelikhuijzen and associates (1980, 1981a, b, c, 1982) recognized "nerve cells" containing immunoreactivities for gastrin/CCK, bombesin/GRP, oxytocin/vasopressin (the latter two coexisted in the same cells, Grimmelikhuijzen 1983), substance P-, neurotensin-, FMRF-amide-like immunoreactivities. The immunoreactivities were located mainly in the subepidermal network of nerve cells and partly in the perpendicular cells ("sensory cells", though Grimmelikhuijzen does not use this term) in the ectoderm, especially over the tentacles and in the basal disk. A small part of the neurotensin and oxytocin/vasopressin immunoreactivity could be found in the cells (again, presumably "sensory cells") in the endoderm.

Besides these vertebrate peptides, coelenterates possess the molluscan cardioexcitatory peptide, FMRF-amide, or similar substances in numerous nerve cells and their processes as has been clearly shown by immunocytochemistry on whole-mount preparations of the medusa *Polyorchis penicillatus* (Grimmelikhuijzen and Spencer 1984) and several species of *Hydra* (Grimmelikhuijzen 1985).

Histochemical studies indicate the occurrence of adrenaline, noradrenaline, serotonin (Wood and Lentz 1964), and monoamine oxidase (Lentz 1966) in the nervous system ("ganglion cells" and "sensory cells," see below) of *Hydra*. Acetylcholinesterase was demonstrated in most of these cells (Lentz 1966). However, biochemical investigations on the occurrence of acetylcholine in

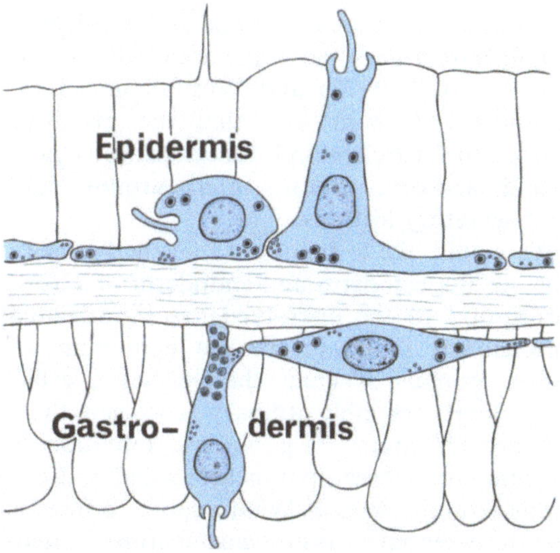

Fig. 11.5. Structure and distribution of neurons and paraneurons in *Hydra*. Both the epidermis and gastrodermis contain bipolar, perpendicularly extended cells, which are regarded as the prototypes of paraneurons. The epithelial base also contains neurons extending their processes horizontally. The cells are loosely connected with synapselike structures

Coelenterata have produced conflicting results (Lentz 1968), and it remains to be settled whether these primitive animals possess this signal substance in their nervous system.

As reviewed by Lentz (1968), *Hydra* possesses two types of neuron: One is a bipolar or multipolar "ganglion" cell lying at the base of the epitheliomuscular cells, while the other is a perpendicularly elongated "sensory" cell in the epidermis (ectoderm) and, less numerous, in the gastrodermis (endoderm) (Fig. 11.5). Electron-microscopic studies by Westfall (1973) and Westfall and Kinnamon (1978) in *Hydra littoralis* have demonstrated that both cell types possess synaptic vesicles, both of the cored and noncored types. The cored vesicles (granules) measured 100–140 nm in diameter and contained a dense core, 90–100 nm in diameter. The clear vesicles measured 100–120 nm in diameter. Payne and Cromey (1987) demonstrated two types of cored granules in the neuroendocrine cells in *Chlorohydra viridissima*. One type was small (110 nm) with a central dense core surrounded by a clear halo, the other was larger (180 nm) with an eccentric dense core. The vesicles of different types or their combinations gather in synapselike areas adjacent to nematocytes, epitheliomuscular cells, or other neurons; in addition, accumulations of cored vesicles are found in nonsynaptic elevations of the cell, suggesting endocrine or paracrine release of the vesicles. Westfall (1973) and Westfall and Kinnamon (1978) point out that the primitive neurons in *Hydra* are multifunctional cells, possessing sensory, motor, interneuronal, and neurosecretory features.

We would like to point out that the "sensory cells" in the ectoderm and endoderm of *Hydra* correspond to the open-type basal-granulated paraneurons found in the gut epithelium of higher animals, whereas the "ganglion cells" more closely resemble ordinary neurons in these animals. Of course, it is not our intention to discriminate between paraneurons and neurons in this primitive animal. What we wish to stress is that the essential morphological features of neuronal and paraneuronal cells are complete at the stage of *Hydra*. Incorporating the above results with immunocytochemical data, it may be concluded that neuroparaneuronal cells as well as their messenger substances are developed in *Hydra* in, respectively, their essential cell organization and molecular structure.

11.4 Evolution of Neurons and Paraneurons and Their Messenger Substances

11.4.1 Prototype of Paraneurons: Archiparaneuron

Hydra (Coelenterata), the most primitive metazoan, possesses a well-differentiated nervous system. Since Schneider's work (1890), "ganglion cells" and "sensory cells" have been differentiated in the nervous system of *Hydra*. The "sensory cells," which are perpendicularly extended elements in the epidermis and gastrodermis (Lentz 1968), may be regarded as the prototype of paraneurons—*archiparaneurons*. As typical receptosecretory cells, they strikingly resemble in fine structure and other cell-biological aspects the basal-granulated cells in the intestines and bronchioles of vertebrates, including man. Archiparaneurons may be said to persist in vertebrate intestines (Fig. 11.6).

In invertebrates of both the protostomian and deuterostomian series, the intraepithelial bipolar (basal-granulated) paraneurons have been identified by recent immunocytochemical and electron-microscopic studies. Numerous cells of this type have been demonstrated in the gut epithelium of deuterostomian protochordates (p. 118). In protostomian invertebrates, the flatworm *Microstomum lineare* (Platyhelminthes) has been shown to possess peptide-immunoreactive cells in the intestinal epithelium in addition to the nervous system (Reuter et al. 1984; Reuter 1985, personal communication; p. 124). In *Bulinus truncatus* (Gastropoda), intraepithelial bipolar cells containing catecholamines have been shown in the carrefour region (Brisson and Collin 1980; p. 122). The basal-granulated cells in the gut epithelium of the cockroach and some other insects are described in detail on pp. 120, ff. Their fine structure and the peptides they contain astonishingly coincide with those of mammalian gut endocrine cells.

It is to be expected that more and more instances of intraepithelial bipolar paraneurons will be found in the gut and other organs of invertebrates. In this respect, we are against the view of Falkmer and his colleagues that intestinal endocrine cells developed at an advanced (protochordate) stage of animal evolution by a dispersion of neuropeptides producing elements in the nervous system (El-Salhy et al. 1980; Falkmer et al. 1985). Intraepithelial granulated cells (paraneurons) already exist in the ectoderm (epidermis) and endoderm (gastrodermis) of *Hydra* as the prototype of paraneurons. These archiparaneurons persist in mammals, the most advanced animals in the deuterostomian series, and in insects, the most advanced forms in the protostomian series. Paraneurons are thus as ancient as neurons in their descent.

It is noteworthy that the primitive neurons of *Hydra*, called "ganglion cells' by Schneider (1890), also persist as *archineurons* in the intramural plexuses of vertebrates. Our intestine is thus regulated by archineurons and archiparaneurons, the direct descendants of the elements which regulated the primitive gut of *Hydra* some 500 million years ago (Fig. 11.6).

11.4.2 Differentiation and Diversification of Cells

The "ganglion cells" (archineurons) and "sensory cells" (archiparaneurons) of *Hydra* are both multifunctional. Portions suggesting neurosecretory (paracrine) release of secretory granules and synaptic sites facing adjacent cells coexist in a single cell (Westfall and Kinnamon 1978). During evolution, two kinds of these multifunctional cells differentiate into a large variety of neurons and paraneurons, each with a conspicuously narrowed repertory of functions. In other words, the diversification of cell types is caused by the specialization of multifunctional cells into unifunctional ones.

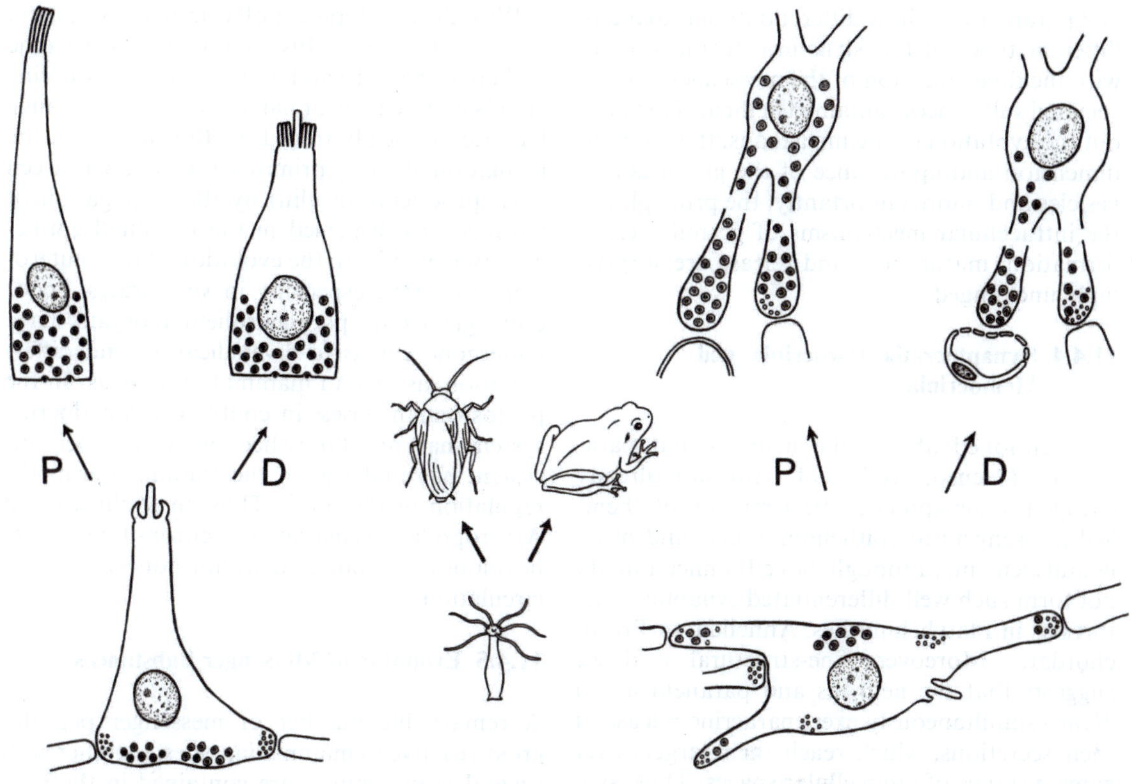

Fig. 11.6a, b. Evolution of enteric neurons and paraneurons through the protostomian (*P*) and deuterostomian (*D*) series. The primitive **a** paraneuron ("sensory cell") and **b** neuron in *Hydra* (Coelenterata) have been inherited over 500 million years to the Arthropoda and Vertebrata without essential changes in structure and function

11.4.3 Phylogeny of Granules and Vesicles

The granules or vesicles containing messenger substances are essential structures for neurons and paraneurons throughout the evolution of the metazoa. Cored or endocrinelike vesicles and clear synaptic vesicles are already distinguished at the stage of *Hydra* (p. 125). The former are gathered in swollen portions of the "ganglion" and "sensory" cells, suggesting the release of their contents toward separate targets (paracrinia); the latter, partly intermingled with the former accumulate in synapselike contact sites against closely adjacent cells (Westfall and Kinnamon 1978). Although there are claims that the cored vesicles contain no peptides (Heynen 1980), we believe that they correspond well with the peptidergic granules in higher animals and are thus in accordance with the immunocytochemical findings on the occurrence of peptides. This view was recently strengthened by Payne and Cromey (1987), who not only stressed the morphological similarity between mammalian neurosecretory granules and the cored vesicles in neuroendocrine cells in *Hydra*, but demonstrated that both share a positive uranaffin reaction. The coexistence of ATP and other nucleotides with peptidic and/or aminic secretions within granules and vesicles thus seems to be derived from the most primitive neurons and paraneurons in coelenterates.

The coexistence of peptidergic-type granules and clear vesicles both within a cell and within a narrow synaptic region of the cytoplasm has been demonstrated in a variety of invertebrates (e.g., Turbellaria *Gyratrix*—Reuter and Lindroos 1979; Polychete *Eulalia*—Whittle and Golding 1974; Polychete *Nephthys*—Zahid and Golding 1975; review Heynen 1980) as well as vertebrates. This peculiar phenomenon of granule/vesicle coexistence persists in most neurons and some paraneurons in mammals, although some cells have become so specialized that they produce a single type of granule or vesicle.

In both protostomian and deuterostomian evolution, the granules and vesicles of neurons and paraneurons show differentiation into many different types of fine-structural features along with the diversification of the messenger and associated substances contained in them. Throughout the evolution of the metazoans, the essential dimension and appearance of the granules and vesicles and, more importantly, the principles of the intracellular mechanisms of granule/vesicle formation, maturation, and release are surprisingly unchanged.

11.4.4 Synaptocrinia, Paracrinia, and Hemocrinia

As mentioned above, the neurons and paraneurons ("sensory cells") of *Hydra* are already engaged in synaptocrine transmission of chemical messengers to epitheliomuscular and other neural elements, although the cell connections do not form such well-differentiated synaptic structures as in Platyhelminthes, Annelida, or Protochordata. Moreover, fine-structural evidence suggests that the neurons and paraneurons of *Hydra* simultaneously exert paracrine release of their secretions, which reach their targets over some distance of intercellular spaces. Thus, synaptocrinia and paracrinia are both the oldest modes of chemical transmission in the multicellular animal body.

With animal evolution, a circulatory system develops containing blood or hemolymph. Some neurons send their axons to the perivascular spaces to release their secretions into the circulation (hemocrinia), and thus neurohemal organs (classic neurosecretory systems) are established. The corpus cardiacum-corpus allatum system in insects (Protostomia) and hypothalamohypophyseal and urophyseal systems in vertebrates (Deuterostomia) are the most differentiated forms.

It is interesting to note that in invertebrates, however, one may find instances of intermediate forms between paracrinia and hemocrinia. In polychaetes (Annelida), for example, numerous neurons extend their axons to the neural lamella and release their secretions into the tissue spaces extended along the lamella. The spaces are continuous with a hemolymphatic vessels and may be able to convey the nerual substances into it (Friend 1976; Golding and Whittle 1977; Tomben 1977; review: Haynes 1980). In *Aplysia*, neurosecretory endings are located below the ganglion sheath (lamella) and the neural substances, such as egg-laying hormone, are believed to be conveyed along the lamella into the hemolymph on one hand and to remote target neurons on the other (Frazier et al. 1967; Branton et al. 1978; see Haynes 1980).

With the development of a circulatory system, some endocrine diffuse paraneurons become gathered around capillaries to form endocrine glands in the classic or narrow sense. This change has been typically traced in *Ammocoetes* in the formation of a very primitive islet (endocrine cell mass producing insulin) by the aggregation of paraneurons dispersed in the intestinal epithelium (see p. 183). In the evolution of the deuterostomian series, especially in vertebrates, endocrine glands or paraneurohemal organs have undergone extensive diversification and differentiation, as seen in mammals and birds. In the protostomian series, in contrast, the endocrine system has not flourished, while the nervous system has undertaken the leading role in the regulation of the body. Thus, in Mollusca and Arthropoda, it is mainly the neurons that release hormones (therefore neurohormones) into the circulation.

11.4.5 Evolution of Messenger Substances

A remarkable number of messenger peptides cross-reactive to mammalian ones, together with several monoamines, are contained in the neurons and paraneurons in *Hydra*; these molecules become distributed during evolution to the diversified cells. The messenger substances also become diversified to a certain extent. Several kinds of peptides are known to be added by diversification of a single peptide due to gene mutation. For instance, gastrin, CCK, and caerulein and VIP, secretin, and glicentin/glucagon are probably derived from the same ancestor hormone presumably by modified genes. Opioid peptides, on the other hand, have apparently diversified by duplication within the gene (repetitive genes) (review: Acher 1984). During evolution, however, diversification of messenger peptides is far less conspicuous than the tremendous differentiation that has taken place in the cells that produce them.

In this chapter, we have concentrated on the occurrence of vertebrate peptides in invertebrates, more precisely, the occurrence of deuterostomian peptides in protostomia. A brief note is necessary on those neuropeptides which are believed to be native to protostomian invertebrates. In their recent review, Greenberg and Price (1983) counted 14 "native invertebrate neuropeptides," including head activator hormone, a cardioexcitatory peptide discovered in *Hydra* (Schaller 1979; Schaller and Bodenmüller 1981), FMRF-amide, a cardioexcitatory peptide isolated from the ganglia of *Macrocallista* (Mollusca) (Price and Greenberg 1977), egg-laying hormone isolated from the bag cells of *Aplysia* (Mollusca) (Chiu and Strumwasser 1981), and adipokinetic hormone (AKH) purified from the corpus cardiacum of *Locusta* (Insecta) (Stone et al. 1976; Broomfield and Hardy 1977). Recent radioimmunoassay and immunohistochemical studies have revealed the possibility that these "native invertebrate peptides" may be shared by vertebrates. Head activator hormone or a similar peptide has been found in the mammalian brain (hypothalamus) and intestine (Bodenmüller and Schaller 1981). FMRF-amide-like immunoreactivity has been reported in *Hydra* (Grimmelikhuijzen 1982) and in fish and mammalian brains (mouse—Boer et al. 1980; guinea pig—Triepel and Grimmelikhuijzen 1984). Furthermore, Kubben et al. (1986) reported that mammalian GEP endocrine system contains a few cells immunoreactive for FMRF-amide. There is the possibility that other peptides which have been thought characteristic of protostomians may be identified in deutrostomian animals, especially vertebrates.

It is worthy mentioning here that the heterologous and homologous coexistence of secretions (including peptidic and aminic messenger substances) in neurons and paraneurons can be recognized at the stage of *Hydra*. This can be regarded as the most conservative feature of the secretions of neurons and paraneurons in the course of evolution.

We have seen that an important part of the regulatory system of the human body derives from the Coelenterata, which 500 million years ago possessed archineurons and archiparaneurons; these exerted their communicative functions by means of messenger peptides and amines. However, it seems likely that protozoans possess substances identical with or similar to vertebrate messenger peptides and amines.

11.5 Unicellular Organisms

Le Roith et al. (1980, 1982) and Berelowitz et al. (1982) demonstrated somatostatin, ACTH, and β-endorphin in extracts of *Tetrahymena pyriformis*, a ciliated protozoan. Insulin or a closely related molecule was detected in *Escherichia coli* (Le Roith et al. 1981). Earlier studies demonstrated the occurrence of serotonin in unicellular organisms (Janakidevi et al. 1966).

Little is known concerning the functions of the substances in question in unicellular organisms. Data, however, are being accumulated supporting the view that the substances might be used for intercellular communication in microorganisms (review Roth et al. 1982; Le Roith et al. 1984).

11.6 Ectopic Secretion of Amine/Peptide Messengers

The above account indicates that the ability to produce peptide/amine messenger molecules, whether they are used as messengers or not, originates in unicellular organisms and is handed on to the metazoa, where these molecules are produced and stored in the neuroparaneuronal cell system. It is possible, however, that the potential ability to produce these molecules may be retained in every cell in a metazoan organism. This latent ability of cells seems to become manifest in certain nonneuroparaneuronal cells in some animals.

11.6.1 Cutaneous Gland of Frogs

The skin of frogs contains, according to the species, different neuropeptides, such as caerulein, bombesin, and sauvagine, as well as serotonin and other monoamines in the cutaneous exocrine glands (review: Erspamer 1983). The immunocytochemical study by Yoshie et al. (1985) even showed the *coexistence* of bombesin and serotonin in the same exocrine cells in the skin of *Bombina orientalis*.

Zaccone et al. (1986) demonstrated serotoninlike immunoreactivity in *sacciform gland* cells in *fish skin* (*Lepadogaster candollei*). Zaccone (1986, personal communication) further found bombesin- and GRP-like immunoreactivities in the sacciform gland. The possible coexistence of serotonin and bombesin in the exocrine cells of the sacciform gland parallels that in the amphibian cutaneous gland cells.

The significance of these aminic and peptidic substances released from amphibian and piscine skin (pheromonal, repellent, or other functions) provide interesting subjects for future studies.

11.6.2 Murine Salivary Gland

The *granular duct of the submandibular gland* in rats and mice has been reported to contain *nerve growth factor* (NGF), *epidermal growth factor* (EGF), *renin*, angiotensin, neurotensin, and many bioactive peptides (reviews: Barka 1980, Murphy et al 1980). These substances are secreted, at least predominantly, by exocrine fashion into the saliva (Nexø et al. 1984). The possibility that these substances serve as pheromones is an attractive hypothesis, but no experimental data supporting it seem to be available.

11.6.3 Atrial Muscle

Atrial muscle cells in mammals have long been known to possess round, membrane-limited secretorylike granules (Kisch 1956; Jamieson and Palade 1964). In the frog and other submammalian vertebrates, the granules may also be found in ventricular muscle cells (Yunge et al. 1980). Recently, strongly bioactive peptides causing diuresis and natriuresis as well as vasodilatation have been extracted from mammalian atria by many research groups (review: Cantin and Genest 1985) and designated differently: *auriculins* (Atlas et al. 1984), *atriopeptins* (Currie et al. 1983, 1984), *cardiodilatin* (Forssmann et al. 1983, 1984), *atrial natriuretic factors* (Thibault et al. 1983), and *atrial natriuretic peptides* (ANP; Kangawa and Matsuo 1984; Kangawa et al. 1984). Kangawa and Matsuo (1984) revealed that the major circulating form of the atrial peptides is a 28-amino-acid peptide, α-ANP. Immunocytochemical studies have evidenced that the atrial peptides are specifically localized in secretorylike granules in the atrial muscle cells (Cantin et al. 1984; Forssmann et al. 1984; Morinaga et al. 1985). Electron-microscopic observations support the view that the granules are formed in the Golgi apparatus and, after shifting to vascular poles of the muscle cell, are released by exocytosis in response to adequate stimuli (Sugawara 1987). Expansion of the atrium due to increased blood volume induces elevated plasma levels of the atrial peptides (Lang et al. 1985) and, correspondingly, the incidence of exocytotic images in the electron microscope is increased after atrial expansion (Sugawara 1987).

As detailed above, the atrial muscle cell seems to qualify as an *endocrine* cell producing and releasing a potent messenger peptide. It is sensitive to mechanical extension and also to chemical stimulation, such as by adrenaline (Sonnenberg and Veress 1984). In these respects, the atrial muscle cell is fully comparable with paraneurons. An important finding is that ANP-like substances have been localized in some neurons and paraneurons. Forssmann and Mutt (1985) reported that they found immunoreativity for cardiodilatin in certain neurons in the hypothalamus of the tupaia (primate). Inagaki et al. (1986) demonstrated ANP immunoreactivity in adrenomedullary cells and in SIF cells in sympathetic ganglia of the rat.

11.6.4 Juxtaglomerular Cells

The juxtaglomerular cells are epitheloid (not epithelial) cells derived from *smooth muscle cells* of an arteriole, the vas afferens in the kidney. These cells produce a regulatory peptide, renin. *Renin* is an enzyme changing angiotensinogen into angiotensin, thus producing elevated blood pressure. Lowered pressure (ischemia) in the afferent arteriole is recognized by the juxtaglomerular cells, which then proliferate and secrete larger amounts of renin (Goormaghtigh 1940). Renin is produced in the juxtaglomerular cells by the processing of a larger precursor molecule, prorenin. This relationship has been confirmed by an immunocytochemical study using antisera against prorenin and renin (Yanaihara et al. 1987). The juxtaglomerular cells contain large round basophil granules limited by a membrane sac. The granules, which are the sites of renin storage, are formed in the Golgi apparatus and released by exocytosis, as evidenced by Taugner et al. (1978) following electric stimulation of renal nerves.

On the other hand, it has been pointed out that renin is a carboxyl protease structurally related to other acid proteases and that renin granules have certain features as modified lysosomes (Taugner et al. 1985).

Renin is produced in the granular ducts of the murine submandibular gland (p. 129). Moreover, renin has been shown to be contained in the brain, and this suggests the possibility that it is produced by certain *neurons* (Inagami et al. 1978; review: Ganten et al. 1981).

11.7 Summary

The most primitive metazoan *Hydra* (Coelenterata) has prototype neurons and prototype paraneurons. The former are multipolar neurons whose descendants persist as "archineurons" in the intramural plexuses of the vertebrate gut; the latter are bipolar cells in the ectoderm and endoderm of *Hydra* and survive as "archiparaneurons" in the gastroenteric and bronchopulmonary epithelia of vertebrates, as well as in the midgut of insects. In both the protostomian and deuterostomian series of animal evolution, more differentiated varieties of neurons and paraneurons have been added to these archineurons and archiparaneurons.

The intracellular mechanisms in the production, storage, and release of secretory substances seem to have remained unchanged throughout metazoan evolution. Only the mode of messenger transport has been modified by the appearance of the vascular system.

In addition to some bioactive amines, *Hydra* possesses a considerable number of messenger peptides that are cross-reactive to mammalian peptides. During evolution, only inconspicuous changes have occurred in their molecular structures and few new peptides have been added.

Although the great majority of peptide/amine messenger substances are restricted to the neuron-paraneuron system, some of them may be produced "ectopically." The dermal gland of frogs secretes gut hormones and amines; the duct cells in the murine submandibular gland secrete NGF and other peptides; atrial muscle cells produce natriuretic peptides; the smooth muscle-derived cells in the kidney produce renin. It is notable that for each of these "ectopically" secreted substances it has been either evidenced or suggested that they are also produced "eutopically," i.e., in the neuron-paraneuron system in some animals.

II. Special Part

12. Adrenal Chromaffin Cells

12.1 Historical and General Aspects

The term *chromaffin cells* is derived from the properties of the parenchymal cells of the adrenal medulla whereby they are stained from yellow to brown after treatment with chromate- or dichromate-containing fixatives (Bennett 1941; Wood 1963; Coupland 1965a). It was Kohn (1903) who named this phenomenon the *chromaffin reaction*. The chromaffin reaction was regarded as a specific histochemical test for adrenaline and related substances. Thus, adrenal chromaffin cells have been regarded as occupying a special position among endocrine cells as the only adrenaline-secreting cells.

In his pioneering monograph, Coupland (1965a) characterized a chromaffin cell as "an element developed from the neuroectoderm, innervated by preganglionic sympathetic nerve fibers, capable of synthesizing and secreting catecholamines, and storing them in sufficient quantities to give a positive chromaffin reaction with aqueous solutions of potassium dichromate or chromate." This characterization is still valid. However, during the past two decades, the chromaffin reaction has become less important in the histochemical demonstration of biogenic monoamines, because of the development of more sensitive and specific methodologies such as *fluorescence histochemistry* (Eränkö 1952; Falck et al. 1962) and immunocytochemistry (Verhofstad et al. 1980, 1983). It has been established that catecholamine synthesis and storage is not a specific ability of the adrenal chromaffin cells but is a more or less common feature of the "neuroendocrine relatives." The only characteristic feature of adrenal chromaffin cells is that they contain larger amounts of catecholamines than any other cells in the body.

12.2 Structure and Ultrastructure

The mammalian adrenal medulla is a *modified sympathetic ganglion* (Kohn 1903; Langley 1921). Its parenchyma contains numerous chromaffin cells and a few ganglion cells. There are ample data of various kinds which support the theory that adrenal chromaffin cells are endocrine varieties of the sympathetic ganglion cells. Thus, they are typical paraneurons (Kobayashi 1977).

Adrenal chromaffin cells contain catecholamines such as *adrenaline* (Takamine 1901), *noradrenaline* (Euler and Hamberg 1949), and *dopamine* (Goodall 1951). Recent immunohistochemical and biochemical studies of Verhofstad and Jonsson (1983) have shown that *serotonin*, an indolamine, is present in rat adrenal chromaffin cells. The presence of histamine was also demonstrated (Panula et al. 1985).

The *ultrastructure* of adrenal chromaffin cells has been investigated by many authors (Yates et al. 1962; Wood 1963; Wood and Barnett 1964; Coupland 1965b, c, 1971, 1972; Benedeczky and Smith 1972; Unsicker 1973a–c; Grynszpan-Winograd 1975). The most prominent cytoplasmic organelles are *secretory granules*, first described by Lever (1955). Biochemical studies on the subcellular distribution of catecholamines in the adrenal medulla showed that the majority of these monoamines are located in the secretory granules (Hillarp et al. 1953; Blaschko and Welsch 1953; Winkler and Hörtnagl 1973). This was confirmed by recent immunocytochemical studies using specific antisera against adrenaline and noradrenaline (Verhofstad et al. 1980, 1983; Kobayashi and Uchida 1983). Catecholamines form a storage complex with glutamate rich, acidic glucoproteins called chromogranins (Blaschko et al. 1967; O'Connor 1983; see p. 24) and adenine nucleotides (Blaschko and Welsch 1953; Hillarp et al. 1953). *Opioid and other peptides* are contained in the secretory granules (Schultzberg et al. 1978, Kimura et al. 1980, Kobayashi et al. 1983a, b; review: Kondo 1985). All these granule contents are released by exocytosis (DeRobertis and Vaz Ferreira 1957; Grynszpan-Winograd 1975) on stimulation of the splanchnic nerve and other secretagogous stimuli (Douglas 1968; Nagasawa 1977).

12.3 Types of Adrenal Cells

12.3.1 Adrenaline- and Noradrenaline-Containing Cells

In the mammalian adrenal medulla, adrenaline and noradrenaline are stored in different types of chromaffin cells—*adrenaline-storing (A)* and *noradrenaline-storing (NA) cells*. This was first deduced from the observation that the ratio of A cells to NA cells coincides with that of adrenaline to noradrenaline in the extract of the adrenal medulla under different conditions and in various animal species (Bänder 1950; Eränkö 1952; Hillarp and Hökfelt 1953).

The histochemical characteristics of A and NA cells are markedly different. The A cells contain a large amount of basophilic substances, hence they stain intensely with basic dyes such as toluidine blue (Costa 1968; Hopwood 1971; Kobayashi and Watanabe 1971) and lead hematoxylin (Solcia et al. 1969; Takaya 1970). On the other hand, NA cells possess strong chromaffinity (Wood 1963), give a positive iodate reaction (Hillarp and Hökfelt 1953), stain intensely with various silver impregnation methods (Tramezzani et al. 1964; Chang and Bencosme 1968; Gorgas and Böck 1976), and exhibit weak, green fluorescence after formalin fixation, forming so-called noradrenaline cell islets (Eränkö 1952). Recent studies have shown that NA cells also exhibit yellow to orange fluorescence after freeze-drying followed by formaldehyde treatment (Falck et al. 1962; Williams and Morris 1970; Kobayashi and Watanabe 1971).

The existence of the two kinds of chromaffin cells demonstrated by light microscopists was confirmed by electron microscopists when Wood and Barnett (1964), with glutaraldehyde/dichromate-fixed specimens, and Coupland et al. (1964), with glutaraldehyde/osmium tetroxide-fixed material, found that A cells possess rounded secretory granules with fine-granular contents, whereas the NA cell granules are irregular in shape and are provided with solid, electron-dense contents. Coupland and Hopwood (1966) demonstrated that noradrenaline reacts with glutaraldehyde to produce a water-insoluble precipitate, whereas adrenaline does not react with glutaraldehyde. Autoradiographic studies performed by Coupland and co-workers confirmed this (Coupland and Kobayashi 1976; Coupland et al. 1976a, b)

12.3.2 Small-Granule Chromaffin Cells

Small-granule chromaffin (SGC) cells, the third type of adrenal chromaffin cells, were first described by Unsicker (1973a) in the avian adrenal gland. Similar cells were later described in amphibian (Yamamoto et al. 1983), reptilian (Unsicker 1976), and mammalian adrenal glands (Gorgas and Böck 1976; Kobayashi and Coupland 1977; Kajihara et al. 1978).

In the mouse, the diameter of the secretory granules is in the range of 100–230 nm in SGC cells, 170–350 nm in A cells, and 185–495 nm in NA cells (Kobayashi and Coupland 1977). In the mouse adrenal medulla, SGC cells comprise less than 4% of the parenchymal cells, hence it seems unlikely that they play an essential role in adrenomedullary function.

SGC cells possess one or more long, slender *cytoplasmic processes*, at least one of which has been traced for up to 30 μm in an electron micrograph (Kobayashi and Coupland 1977). Similar axonlike projections of the SGC and like cells have been reported in reptiles (Unsicker 1976; Watari et al. 1977), birds (Unsicker 1973a), guinea pigs (Unsicker et al. 1978), cultured rat tissues (Unsicker and Chamley 1977), and homografts in the rat anterior eye chamber (Olson and Malmfors 1970).

Nerve synapses are more frequent on SGC cells than on A and NA cells (Kobayashi and Coupland 1977). So far, only the efferent (i.e., from nerve to cell) type of synaptic complex has been found on SGC cells as well as A and NA cells (Kobayashi and Coupland 1977).

The SGC cell comprises a third and independent type of adrenal chromaffin cell. Morphological features of this cell such as the long cytoplasmic projections and rich innervations seem to imply that it occupies an intermediate position between the sympathetic nerve cells and the A or NA cells of endocrine function.

Synaptic-type vesicles in adrenal SGC cells are worthy of particular mention. Sympathetic ganglion cells contain in their terminal varicosities at least two kinds of membrane-bound vesicles—the large cored vesicles 80–120 nm in diameter and the small granular vesicles 40–60nm in diameter (Grillo 1966).

SGC cells of the mouse adrenal medulla constantly contain two populations of microvesicles: one, the *secretory granules with a dense core*; the other, small *synaptic-type vesicles*, 40–70 nm in diameter (Kobayashi and Coupland 1977). These synaptic-type vesicles contain dense pre-

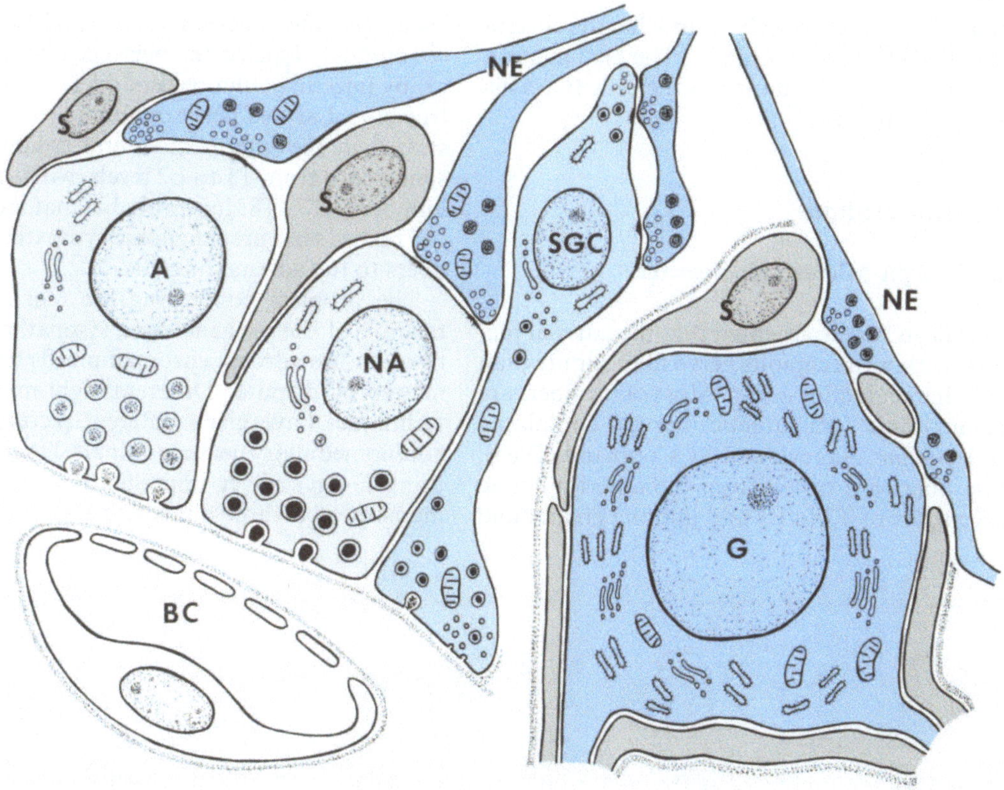

Fig. 12.1. Cellular elements of the adrenal medulla in mammals. Adrenaline (*A*) and noradrenaline (*NA*) cells are the main chromaffin cells. Note the small granule-containing (*SGC*) cell showing a morphology intermediate between that of the chromaffin cell and ganglion cell (*G*). *NE* nerve endings, *S* Schwann cells, *BC* blood capillary

cipitates along the limiting membrane and are scattered throughout the cytoplasm, forming occasional clusters (Kobayashi et al. 1978). The synaptic-type vesicles have been demonstrated by Unsicker and Chamley (1977) in cultured rat adrenal chromaffin cells and nerve cells. Furthermore, Unsicker (1976) reported the occurrence of small clear vesicles in addition to the large cored vesicles in the reptilian adrenal chromaffin cells.

Figure 12.1 illustrates parenchymal cells of the adrenal medulla. In addition to the three types of chromaffin cells, ganglion cells occur in the adrenal medulla.

12.3.3 Ganglion Cells

The adrenal medullary tissue in a variety of animals from elasmobranchs to man has been shown to possess considerable numbers of ganglion cells (Coupland 1965a; Unsicker 1973a, 1976). Dogiel (1894) described the bipolar and multipolar nature of these medullary neurons. The neurons are regarded as *sympathetic ganglion cells* which send postganglionic nerve fibers; it has not been possible to trace the termination of these fibers. Some adrenal medullary neurons contain neuropeptides such as VIP, somatostatin, and enkephalins (see Kondo 1985).

Electron-microscopic studies of adrenal medullary neurons were performed by Unsicker et al. (1978) in the guinea pig. Two types of neuron were distinguished: large multipolar cells, probably corresponding to the principal neurons of sympathetic ganglia, and small neurons, as a heterogeneous population. There are transitional forms between these small neurons and the SIF cells in the autonomic ganglia.

12.3.4 Sustentacular Cells

The chromaffin cells are intercalated by sustentacular cells, which are regarded as the neuroglial component of the adrenal medulla. The cells are small and irregular in shape; they extend attenuated processes closely embracing the chromaffin cells and nerve terminals. The glial marker, S-100 protein, is contained in the cells (Cocchia and Michetti 1981).

12.4 Innervations

12.4.1 Light-Microscopic Structure

It has been widely accepted that the adrenal medulla has a more copious nerve fiber supply than any other endocrine organ. Most of the fibers are preganglionic sympathetic nerves containing *acetylcholine*. As reviewed by Coupland (1965), these fibers are traced from the anterior spinal nerve roots of T7 to L3 and join the greater and lesser splanchnic nerves. Holets and Elde (1982) retrogradely labeled the nerve cell bodies in the rat by injecting a fluorescent dyes, fast blue, and true blue. Labeled nerve cell bodies were demonstrated in the intermediolateral column of the spinal cord from T1 to L2 levels, with the largest number at T7–T8. It is probable that these nerve cell bodies send preganglionic sympathetic nerve fibers to the adrenal medulla.

The problem concerning the morphological relation of the preganglionic sympathetic nerve fibers to the adrenal chromaffin cells has been a matter of dispute. Different light-microscopic techniques have given different pictures of the adrenomedullary nerve terminals. Dogiel (1894), using Golgi's silver impregnation method, demonstrated a three-dimensional nerve network

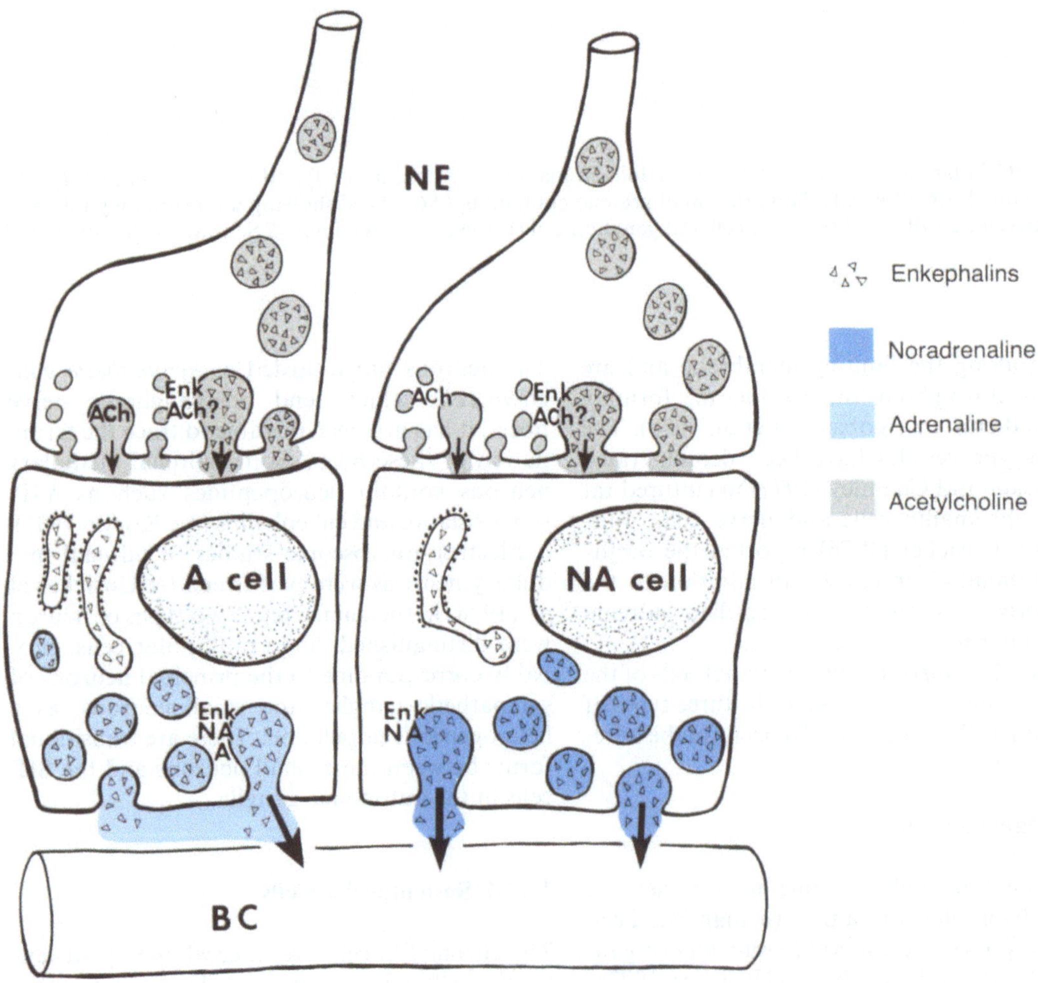

Fig. 12.2. Secretion release by chromaffin cells and by nerve endings (*NE*) supplying them. *Enk* enkephalins, *A* adrenaline, *NA* noradrenaline, *Ach* acetylcholine, *BC* blood capillary

in the adrenal medulla, but he did not show the synapse formation between the neuronal terminals and the chromaffin cells. Stöhr (1935) and Sato (1952), using Bielschowsky's silver impregnation method, described a terminal anastomosing net of neurofibrils called the "*terminal reticulum*." Hillarp (1959), using a methylene blue technique, found no terminal boutons. He stated that the nerve fibers run in fine bundles ensheathed in a coarse three-dimensional network composed of Schwann cell cytoplasm, the tangled mass forming the "*autonomic groundplexus*." Furthermore, Willard (1936), using Cajal's chloral hydrate method, described nerve terminals in the form of platelets or boutons. Thus, the structure of the nerve terminals innervating the chromaffin cells has remained enigmatic until quite recently.

The occurrence of *opioid peptides* in the nuclear area of the spinal cord containing sympathetic preganglionic neurons was reported by Romagnano and Hamill (1984). These authors described in horizontal sections of the rat spinal cord a pattern of the opioid peptide-containing structures resembling the rungs of a ladder. Thus, the term "enkephalin ladder" was proposed. It is possible that opioid peptides synthesized in the nerve cell bodies of the enkephalin ladder are carried by the axonal flow through the splanchnic nerves to the adrenal medulla.

The adrenal medulla also receives *adrenergic innervations* (Prentice and Wood 1975); these are sympathetic postganglionic neurons and distributed mostly in the perivascular connective tissue space.

The most recent immunocytochemical studies have shown that both chromaffin cells and nerve terminals in the adrenal medulla possess immunoreactivities for opioid peptides such as Met-enkephalin (Schultzberg et al. 1978a, b) and Met-enkephalin-Arg^6-Gly^7-Leu^8 (Kobayashi et al. 1983b, 1985) (Fig. 12.2). In a few animal species where the immunoreactivity of the chromaffin cells is not very strong, the structure of nerve terminals on the chromaffin cells becomes visible after immunocytochemical procedures. Thus, Schultzberg et al. (1978a, b) reported a sparse network of varicose nerve terminals which correspond to the autonomic groundplexus of Hillarp (1959). This was confirmed by Kobayashi et al. (1983b, 1985). Nerve terminal swellings of variable size and shape reported by Kobayashi et al. (1983b, 1985) correspond to Willard's end-boutons and boutons de passage (Willard 1936).

12.4.2 Electron-Microsopic Structure

The ultrastructure of the nerve terminals on adrenal chromaffin cells has been investigated by many authors in various animals (rat—Coupland 1962, 1965; hamster—Grynszpan-Winograde 1974; domestic fowl—Unsicker, 1973a, b). In all the animals examined, the nerve terminal varicosities on the adrenal chromaffin cells were found to contain two types of vesicular structure. One is the small synaptic-type vesicles; there are ample data supporting the idea that these synaptic-type vesicles contain *acetylcholine* (see p. 62). The large cored vesicles are the second type of vesicular structure.

Ultrastructural localization of *opioid peptides* in the adrenal chromaffin cells and their innervation were investigated using electron-microscopic immunocytochemical methods (Kobayashi and Uchida 1983). It was demonstrated that opioid peptides coexist with catecholamines in adrenal chromaffin cells. On the other hand, opioid peptides coexist with acetylcholine in sympathetic preganglionic neurons (Kobayashi et al. 1984a, 1985). The cellular mechanisms by which the preganglionic neurons produce opioid peptide with acetylcholine and the chromaffin cells costore the same peptides with catecholamines remain to be investigated (Fig. 12.3).

12.5 Secretory Aspect of Chromaffin Cells

12.5.1 Uptake of Catecholamines and Related Substances

Adrenal chromaffin cells possess a typical APUD ability in the literal sense. Effective uptake of exogenous catecholamines and related substances by adrenal chromaffin cells have been examined by autoradiographic techniques. Small laboratory animals were used in the experiments for reasons of economy and safety associated with the use of isotope-labeled substances.

Both A and NA cells of the adrenal medulla specifically incorporate exogenous catecholamines and related substances such as dopa (Elfvin et al. 1966; Forssmann et al. 1971; Coupland and Kobayashi 1978; Coupland et al. 1978a, b), dopamine (Hirano et al. 1977; Hirano and Kobayashi 1978), noradrenaline (Kent and Coupland 1984), and serotonin (Litzén et al. 1965; Gershon and Ross 1966; Csaba and Baráth 1974; Kent and Coupland 1984). A cells take up labeled-dopa, dopamine, and noradrenaline more

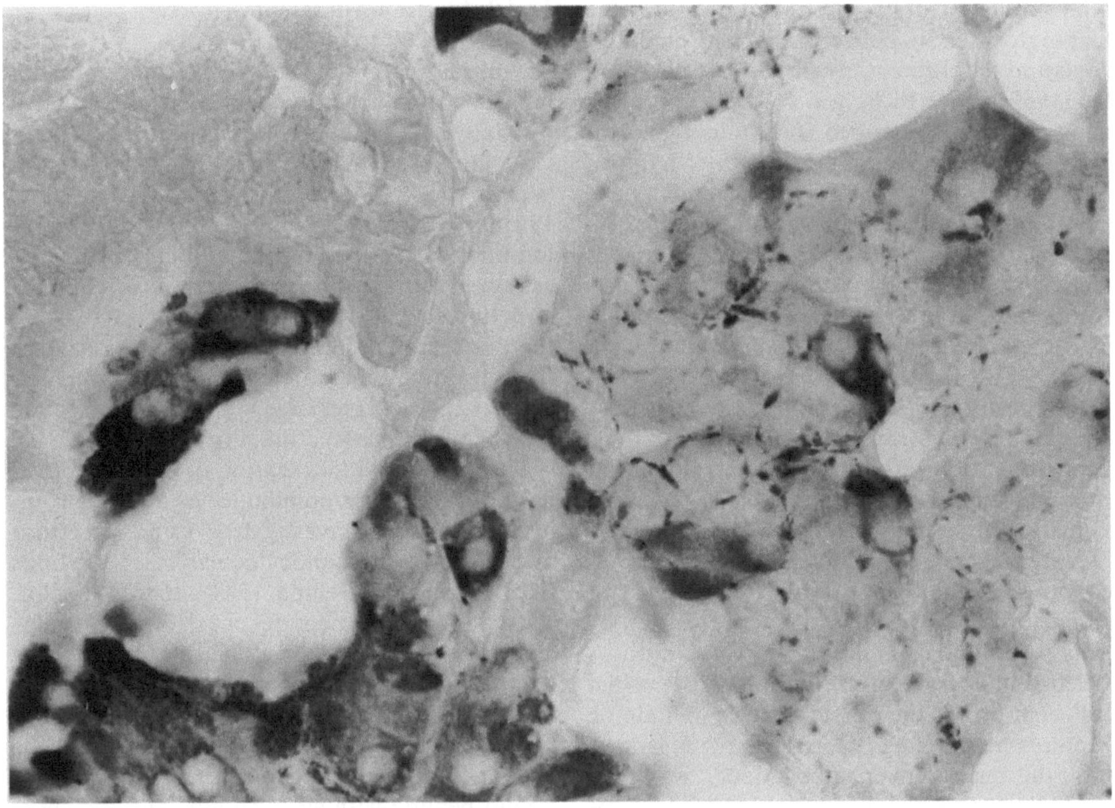

Fig. 12.3. Enkephalin-immunoreactive cells and nerves in the adrenal medulla of the guinea pig. Immunoreactivity for Met-enkephalin-Arg-Gly-Leu is demonstrated by the PAP method. The chromaffin cells are partly intensely and partly faintly immunoreactive; some appear negative. The dotlike immunoreactive structures dispersed around the cells are enkephalin-containing nerve fibers with beaded swellings

rapidly and lose newly synthesized adrenaline more rapidly than NA cells incorporate these substances and release noradrenaline. The turnover of adrenaline in A cells seems to be more rapid than that of noradrenaline in NA cells (Elfvin et al. 1966; Coupland and Kobayashi 1976; Coupland et al. 1976a, b).

In the mouse, the radioactivity in the adrenal medulla shows a characteristic zonal distribution after injection of ^{3}H-dopa, ^{3}H-dopamine, and ^{3}H-noradrenaline. The highest radioactivity is regularly seen in the zone immediately beneath the corticomedullary junction (Coupland et al. 1976b; Hirano et al. 1977; Kent and Coupland 1984). Hirano and Kobayashi (1978) showed that this zonal distribution of radioactivity after ^{3}H-dopamine injection was obscured by hypophysectomy and reappeared after ACTH treatment. They proposed that adrenal chromaffin cells possess a system for trapping and concentrating exogenous dopamine and that this uptake system is reduced by the removal of pituitary ACTH. It is likely that a similar uptake system is present for catecholamines and related substances other than dopamine. It has been hypothesized that higher concentrations of cortical hormones in the microenvironment of the medullary tissue may accelerate the uptake of amines and amine precursors into the chromaffin cells (Hirano and Kobayashi 1978).

The uptake of catecholamines and related substances by adrenal chromaffin cells is also controlled by sympathetic preganglionic nerve fibers. Hirano (1982) examined the stress effect on the distribution of ^{3}H-dopamine in the denervated adrenal medulla of the mouse. Cutting the splanchnic nerve increased the uptake into A cells but not into NA cells. The characteristic distribution of ^{3}H-dopamine (high radioactivity along the corticomedullary junction) was amplified after denervation procedures. Hirano (1982) concluded that the splanchnic nerve activity suppresses the uptake of ^{3}H-dopamine by the chromaffin cells. This view was supported by

Role and Perlman (1983) who demonstrated that acetylcholine not only stimulates catecholamine secretion from isolated guinea pig chromaffin cells, but also causes a striking inhibition of catecholamine uptake into these cells. Role and Perlman (1983) suggested that acetylcholine released from the splanchnic nerve terminals may prevent the reuptake of catecholamines released from the chromaffin cell during physiological secretory excitation.

12.5.2 Coexistence and Cosecretion of Peptides and Amines

A high concentration of *opioid peptides* exists in the adrenal gland. This was discovered by Schultzberg et al. (1978a, b) using immunocytochemical techniques. Later, Udenfriend and coworkers chemically isolated many kinds of enkephalin-containing peptides (Stern et al. 1979; Kimura et al. 1980; Kilpatrick et al. 1981). Noda et al. (1982) established the whole nucleotide sequence of cloned cDNA for *preproenkephalin A* from the bovine adrenal medulla. This precursor protein contains four copies of met-enkephalin and one copy each of leu-enkephalin, met-enkephalin-Arg^6-Phe^7, and met-enkephalin-Arg^6-Gly^7-Leu^8. Proenkephalin A thus represents a multihormone precursor for the adrenal opioid peptides.

There is accumulating evidence that newly synthesized opioid peptides in the adrenal chromaffin cells are transported via the Golgi apparatus to the secretory granules. This intracellular pathway for opioid peptide is deduced from the results of cell fractionation (Winkler et al. 1972) and autoradiographic studies (Coupland and Kobayashi 1976; Kobayashi et al. 1978; Benchimol and Cantin 1978).

Opioid peptides and other peptides such as *chromogranins* (Blaschko et al. 1967; Winkler 1976) are costored with catecholamines in the secretory granules. *Catecholamines* directly enter the secretory granules without passing through the Golgi apparatus (Coupland et al. 1976b, Benchimol and Cantin 1982). Thus, the amine/protein ratio increases in the secretory granules as they migrate from the Golgi area to the cell periphery. When the chromaffin cells are stimulated, all the contents of the secretory granules, including opioid peptides and catecholamines, are released to the exterior of the cell by an *exocytotic* mechanism (Livett et al. 1981; Asada et al. 1983).

Immunocytochemical studies were performed to solve the problem of whether the adrenal opioid peptides are produced and stored by A cells, NA cells, or both. Results of the comparative studies of Kobayashi et al. (1983c, 1984a) in the rat, dog, cat, and hamster indicated that both A and NA cells can produce and store opioid peptides. However, a remarkable species specificity was noticed with the immunoreaction for opioid peptides: In the rat, a limited number of chromaffin cells showed immunoreactivity for opioid peptides; in the dog, cat, and hamster, opioid peptide immunoreactivity was demonstrated in both A and NA cells. It is hypothesized that the immunocytochemically demonstrable species differences in the distribution of opioid peptides in chromaffin cells may reflect possible differences in the storage conditions of the peptides and other secretory substances in chromaffin granules. It is possible that catecholamines and other granule substances mask the immunoreactivity of peptides in the secretory granules during the immunocytochemical procedures (Kobayashi et al. 1983b, c, 1984a).

Immunochemical and immunocytochemical studies on the occurrence and localization of neuropeptides in the adrenal medulla have recently been performed by many authors in the rat and other mammals (review: Kondo 1985). *In addition to enkephalins*, chromaffin cells have been identified containing immunoreactivity for a variety of neuropeptides, such as NPY (Kuramoto et al. 1986; Lundberg et al. 1986), substance P (Kuramoto et al. 1985), VIP (Holzwarth 1984), somatostatin, neurotensin (Lundberg et al. 1982), and calcitonin gene-related peptide (CGRP) (Rosenfeld et al. 1983). Immunocytochemical studies, mainly in the rat, indicate that these peptides derived from different prepropeptides partly occur in different and partly in the same cells (*homologous coexistence* of messenger substances, p. 37). The functional significance of the variety of peptides in the adrenal medulla presents a large area for future studies.

Atrial natriuretic peptide (ANP)-like immunoreactivity has recently been demonstrated in rat adrenal chromaffin cells. The A cells contained larger amounts of the substance than the NA cells (McKenzie et al. 1985; Inagaki et al. 1986). This information both extends the list of peptide secretions of the adrenomedullary cells and supports the view that the ANP, originally found as the product of atrial muscle cells, belongs to the brain-gut peptides.

12.6 Electrophysiological Properties

The electrophysiological properties of adrenal chromaffin cells, as already reviewed on pp. 44, 45, are in full accord with the features of these neural crest-derived cells common to sympathetic neurons. The electric properties of adrenal chromaffin cells are as follows:

1) The cells are *depolarized by acetylcholine*, the neurotransmitter of the nerve innervating the cells (Douglas et al. 1967a).
2) The depolarizing effect is mainly due to the inward movement of Na^+ and, to a much smaller extent, to the inward movement of Ca^{2+} (Douglas et al. 1967b).
3) The cells generate *all-or-none action potentials*, which are due to the inward movement of Na^+ and Ca^{2+} (Brandt et al. 1976; Fenwick et al. 1982a, b).
4) A cloned line of rat pheochromocytoma cells (PC12), which responds to nerve growth factor (NGF) by extension of neuritelike processes, developed electric excitability and showed greatly increased acetylcholine sensitivity in the presence of NGF (Dichter et al. 1977).

It has also been pointed out that adrenal chromaffin cells develop morphological and functional characteristics in common with those of adrenergic neurons when the former cells are kept in culture (Trifaró 1982):

1) Chromaffin cells in culture grow one or more processes displaying varicosity-like and growth cone-like structures (Trifaró and Lee 1980).
2) Chromaffin cells show the presence of a high-affinity uptake mechanism for catecholamines (Kenigsberg and Trifaró 1980) similar to neuronal uptake.
3) In contrast to adrenal medullary cells *in situ*, chromaffin cells in culture are sensitive to 6-hydroxydopamine (Kirpekar et al. 1983).
4) Chromaffin granules move along the processes from the cell body to the periphery, a movement stopped by colchicine treatment (Bader et al. 1981).
5) Similar to neurons, chromaffin cells in culture show the presence of neurofilaments (Bader et al. 1984a, b).

12.7 Ontogeny

12.7.1 Origin and Differentiation

The origin and development of adrenal chromaffin cells has previously been reviewed (Coupland 1965a; Weston 1970; LeDourarin 1982). It is well-established that they derive from the *neural crest* similar to sympathetic ganglion cells. A recent fluorescence histochemical study of Polak and co-workers (1971) provides further support for the neural crest origin of adrenal chromaffin cells.

The *adrenal cortex* originates by a thickening of the *mesoderm* near the root of the dorsal mesentery. Two cortical anlagen, one on each side, grow in this region to be located close to the developing kidney. Neuroectodermal stem cells, meanwhile, migrate from the neural crest to form the *celiac ganglia.* Some of the cells leave the ganglia to advance further and take a position in the central portion of the cortical anlage.

According to Coupland (1965a), the stem cells of the adrenal medulla begin to invade the cortical anlage in 14-mm human embryos. In human embryos of 125–200 mm crown-rump length, the stem cells are gathered in folliclelike masses surrounded by glial cells which are immunopositive for S-100 protein (Iwanaga and Fujita 1984). The stem cells, which persist until a crown-rump length of 270 mm (Coupland 1965a), actively *proliferate* and *differentiate* into different types of chromaffin cells and into intra-adrenal neurons.

Levi-Montalcini and co-workers injected nerve growth factor (NGF) into the peritoneal cavity of newborn rats and showed that adrenal chromaffin cells were transformed into nerve cells by the NGF injection (see Levi-Montalcini and Aloe 1980, 1983). On the other hand, Eränkö and Eränkö (1972) administered a large amount of glucocorticoids to newborn rats and showed that ganglion cells change into chromaffin cells. These experiments indicate the "continuity" of adrenal chromaffin cells and sympathetic ganglion cells. The chromaffin cells and nerve cells of the perinatal period are interchangeable depending upon extrinsic (environmental) factors.

Adrenomedullary cells and sympathetic ganglion cells are thus twin elements that develop from the sympathetic stem cells or sympathicogonia.

In birds and lower vertebrates, the linkage of the cortical and medullary anlagen does not occur and both form independent glands. This linkage is attained at the final evolutionary level of vertebrates.

12.7.2 Corticomedullary Relations

In mammals, the medulla is almost completely enclosed by the cortical tissue. The reasons for this corticomedullary linkage have been investigated by many authors. The microvascular arrangement between the adrenal cortex and medulla probably assumes a particular importance in the differentiation and function of adrenal chromaffin cells (Coupland 1975). Thus, a detailed account of adrenal vasculature is required.

One of the most important differences between adrenaline-secreting (A cells) and noradrenaline-secreting chromaffin cells (NA cells) is the presence or absence of phenyl-ethanolamine-*N*-methyl-transferase (PNMT), which converts noradrenaline to adrenaline. Wurtman and Axelrod (1966) reported that high concentrations of glucocorticoids activate the synthesis of PNMT in adrenal medullary tissues. Thus, Coupland (1965a) suggested the possibility that a high concentration of glucocorticoids induces the differentiation of NA cells into A cells.

The adrenal gland of mammals receives a rich blood supply from the neighboring large arteries such as the abdominal aorta. In the capsule, adrenal arteries divide into many arterioles and form a capsular plexus. The adrenal medulla receives a blood supply from two sources: One is the medullary arteries (arteriae medullae), which originate in the capsular plexus, traverse the cortical tissue without division, and branch into capillaries in the medulla; the other consists of the cortical capillaries, which lie between the capsular plexus and the medullary blood vessels (Coupland 1975). Chromaffin cells which receive cortical venous blood containing a high concentration of glucocorticoids may be A in type, whereas other chromaffin cells supplied by the blood of the medullary arteries are probably NA in type. However, the studies by Coupland and Selby (1976) and Kikuta and Murakami (1982) indicate that the special vascular arrangement of the adrenal gland is not directly related to the distribution of A and NA cells in all the mammals examined (rat, mouse, cat, dog, cattle, guinea pig).

It must further be stressed that steroid hormone production from the adrenal cortex is essential for the chromaffin cells to maintain their endocrine nature (Wurtman and Axelrod 1966)

12.8 Phylogeny

As mentioned above, the adrenal medulla is a modified sympathetic ganglion and chromaffin cells originate from the same precursor cells as those of the ganglion cells. Phylogenic studies show a similar connection. Comparative anatomy shows a close relationship between the adrenal medulla and sympathetic ganglia. However, the most important point may be the relationship between the chromaffin cells and the steroid hormone-producing cells.

The origin of the adrenal chromaffin cells can be traced back with certainty, to the elasmobranchs. As reviewed by Coupland (1965a), in the elasmobranchs, peculiar structures consisting of chromaffin cells occur in association with sympathetic ganglia which have no internodal nerve strands: No sympathetic trunk chain exists in the elasmobranchs. Gorgas (1987) demonstrated that at least three types of chromaffin cell are present in elasmobranchs, among them adrenaline-producing (A) and noradrenaline-producing (NA) cells. There is a steroid hormone-producing interrenal body in the prevertebral region. However, no topographic specialization has been demonstrated between the chromaffin cells and the interrenal body.

In bony fishes and amphibians, chromaffin cells show different associations with both the sympathetic ganglia and interrenal bodies. In anurans, there are adrenal glands consisting of chromaffin cells and interrenal cells.

In reptiles and birds, masses of chromaffin cells intermingled with steroid hormone-producing cell groups form discrete adrenal glands. The formation of an adrenal gland consisting of a medulla and cortex is characteristic of mammals.

As mentioned above, the differentiation of adrenaline and noradrenaline cells can be seen already at the evolutionary stage of elasmobranchs. The occurrence of SGC cells in lower vertebrates can be traced to the amphibia (Yamamoto et al. 1983). Further studies are required to determine whether SGC cells occur in fish and lower animals. Studies on the peptides produced by chromaffin cells have only been made in frogs. In *Rana ridibunda*, Leboulenger et al. (1983) demonstrated many VIP-immunoreactive cells and some Met- and Leu-enkephalin-reactive cells. Kondo and Yui (1984) and Kuramoto (1987) recognized that about half the chromaffin cells in *Rana catesbeiana* are immunopositive for enkephalins. Kuramoto (1987) further showed the occurrence of serotonin in the NA cells of the frog.

12.9 Summary

The adrenal medulla is an *endocrine* variety of the sympathetic ganglion. Chromaffin cells and sustentacular cells occur. The chromaffin cells correspond to ganglion cells, whereas the sustentacular cells are of glial nature. The chromaffin cells make synapses with preganglionic nerve fibers, which form an autonomic groundplexus with a sustentacular (or Schwann) cell sheath inside the adrenal medulla.

Three types of chromaffin cell are present: the adrenaline-secreting (A) cell, the noradrenaline-secreting (NA) cell, and the small-granule chromaffin (SGC) cell, which has been discovered quite recently and whose secretions are not yet known. Morphological evidence supports the view that the SGC cell is an intermediate form between the classic chromaffin (A and NA) cells and neurons. Small numbers of ganglion cells also occur in the adrenal medulla. These parenchymal cells produce, in addition to catecholamines, enkephalins and several other *neuropeptides*. ATP and chromogranins are co-stored with these substances in the secretory (chromaffin) granules. The coexistence of acetylcholine and neuropeptides is demonstrated in the preganglionic nerve terminals which innervate chromaffin cells.

Chromaffin cells originate from the *neural crest*. The differentiation of A and NA cells takes place during intrauterine life. In the newborn rat, the administration of NGF induces *transformation* of chromaffin cells into ganglion cells, whereas glucocorticoid administration changes ganglion cells into chromaffin cells.

These facts and also electrophysiological data indicate that adrenal chromaffin cells are paraneuron that are very closely related to neurons.

13. Parafollicular Cells

13.1 General Aspects

The thyroid gland consists of follicles producing triiodothyronine (T_3) and tetraiodothyronine or thyroxine (T_4). As a second endocrine element, the thyroid contains clear cells dispersed solitarily or in small groups in and between the epithelial walls of the follicles. These cells, called parafollicular cells or C cells (from *clear* cells), are the source of the blood calcium level-lowering hormone, *calcitonin* (reviews: Nunez and Gershon 1978b; Ericson and Sundler 1984).

Parafollicular cells are easily definable in certain mammals, such as the dog and guinea pig, while they are more difficult to identify in other species, including humans. As a general rule, the cells are larger in number and size in young animals. Parafollicular cells are generally described as being oval or polyhedral, but they may project thick or slender cytoplasmic processes, extending between neighboring cells or to the pericapillary space.

The occurrence of secretory granules in the cytoplasm of parafollicular cells has long been established through silver impregnation (Nonidez 1932; Carvalheira and Pearse 1967; Kameda 1968) and other light-microscopic techniques, especially lead-hematoxylin staining (Solcia et al. 1969). The granules react metachromatically to such basic dyes as toluidine blue (Solcia et al. 1969) and pseudoisocyanine (Kameda 1970), suggesting an acidic nature of the substances in the granules.

The *stimulus* for activity of parafollicular cells is elevation of Ca^{2+} levels in the blood, i.e., hypercalcemia. Calcitonin released from the thyroid parafollicular cells works in mammalian calcium homeostasis and promote bone formation and calcification (review: Munson 1976).

13.2 Fine Structure

Electron-microscopic observations indicate that parafollicular cells contain a dense accumulation of *secretory granules*, which possess all the morphological features of paraneuronal granules. They are round granules, 100–200 nm in diameter, and consist of a dense homogeneous material ensheathed in a membrane sac (Ekholm and Ericson 1968; reviews: Nunez and Gershon 1978b; Ericson and Sundler 1984).

The organelles involved in the production of proteinic secretions are well developed in parafollicular cells. The rough endoplasmic reticulum may show rather modest amounts of flattened cisternae, but in some populations of the cells they may be swollen with abundant contents and extend in large areas of the cytoplasm (dog, Kameda 1973). The Golgi apparatuses are prominent in the paranuclear regions. Small new granules are produced in the Golgi areas, showing the same electron-microscopic images as those observed in other paraneurons (Fig. 13.1).

The *release* of the secretory granules can be induced by hypercalcemia after administration of $CaCl_2$ and/or vitamin D. Electron-microscopic images have indicated the typical exocytotic processes (Matsuzawa 1966; Ericson 1968). Without such experimental treatment, parafollicular cells may undergo active granule release in the neonatal period. In their observations on dog thyroids, Nunez and Gershon (1978a) demonstrated that prior to birth parafollicular cells became heavily loaded with secretory granules, which partly took the appearance of intracisternal granules. Thereafter, abundant images indicating exocytotic granule release were observed at birth and during the neonatal period; such release was hardly evident in adult thyroids.

Nervous connections. Besides the well-known adrenergic and cholinergic innervation, the thyroid is also supplied with a few VIP-, substance P-, and cholecystokinin immunoreactive nerves (mouse, Ericson and Sundler 1984). Under the electron microscope, nerve terminals can be found in close proximity to parafollicular as well as follicular epithelial cells (Fig. 13.1). Involvement of the autonomic nerves in the regulation of parafollicular cell functions has been proposed.

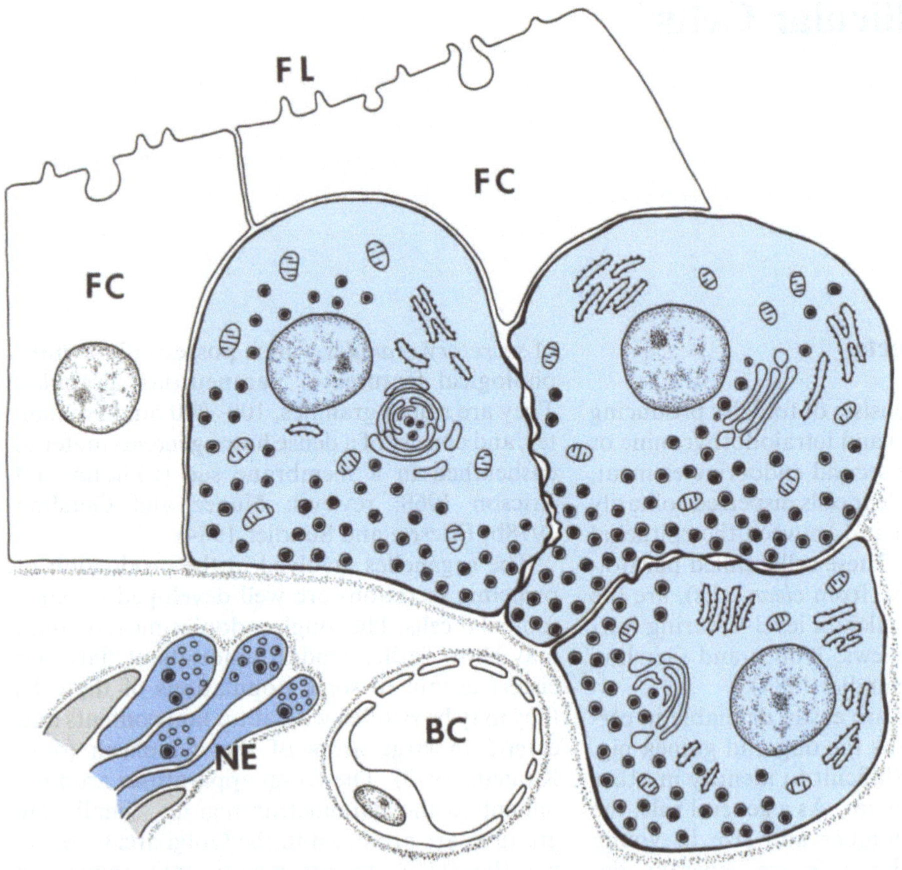

Fig. 13.1. Structure of mammalian parafollicular cells. *FC* follicular cells, *FL* follicular lumen, *BC* blood capillary, *NE* nerve endings, which do not regularly occur at this site

This possibility is supported by the findings of physiologists that gastrin, acetylcholine, and noradrenaline, which all occur in thyroid nerves, can influence calcitonin release (Munson 1976).

13.3 Secretory Products

13.3.1 Peptides

Copp et al. (1962) discovered a blood calcium-lowering hormone, *calcitonin*, and Hirsch et al. (1963) determined that it is secreted from the thyroid. The assumption that the second endocrinelike cell type, the parafollicular cell, might be the source of calcitonin was supported by immunocytochemical studies using anticalcitonin antisera (Bussolati and Pearse 1967; McMillan et al. 1974; Peng et al. 1975). The density of calcitonin immunoreactivity is quite uniform in the parafollicular cell population in most mammals, but in some species it differs conspicuously from cell to cell, including occasional immunonegative cells (dog, Kusumoto 1980).

Somatostatin is the second peptide contained in parafollicular cells as evidenced by immunocytochemical studies (Van Noorden et al. 1977; Yamada et al. 1977). In contrast to the universal occurrence of calcitonin in mammalian parafollicular cells, the distribution of somatostatin is highly variable among species and even among cells in the same thyroid. In the guinea pig and rabbit, essentially all calcitonin cells contain somatostatin, whereas in many other animals including the rat, mouse, cat, pig, and monkey, only part of the calcitonin cells possess somatostatin (Kameda et al. 1982; see also Zabel and Schäfer 1986). In the dog, both peptides tend to occur in different parafollicular cells, though the distribution is partially overlapping (Kusumoto 1980).

The possible functions of somatostatin released from thyroid parafollicular cells will be mentioned later. An interesting problem is whether calcitonin and somatostatin coexist within the same granules or are separated in different granule populations. Ericson and Sundler (1984) state that both peptides are contained in the same granules from their observations in the mouse.

Recent study by Larsson (1985) shows occurrence of immunoreactivity to a C-terminal gastrin/CCK antiserum (CCK-4-like immunoreactivity) in a majority of rat parafollicular cells.

13.3.2 Amines

Serotonin has been demonstrated in the parafollicular cells of several species, including the horse, sheep, goat (Falck et al. 1964; Solcia and Sampietro 1968), and bat (Gershon and Nunez 1970). In the bat, localization of serotonin in the calcitonin granules has been evidenced by high-resolution radioautography (Nunez and Gershon 1972).

The adult dog thyroid contains little serotonin in its parafollicular cells, but the serotonin content increases in the cells just before birth; during the 1st postnatal week, the serotonin content in the cells falls precipitously to the low adult level (Gershon et al. 1971). Taking into account the electron-microscopic finding mentioned above (Nunez and Gershon 1978a), it is likely that serotonin is abundantly released together with calcitonin at birth and in the neonatal period. The physiological significance of this peculiar phenomenon requires investigation (see also Nunez and Gershon 1978b).

The parafollicular cell was one of the most useful models for Pearse (1966a, b) in his concept of the APUD series. Pearse (1966b) demonstrated that 5-hydroxytryptophan administered to a dog is taken up by thyroid parafollicular cells and decarboxylated into 5-hydroxytryptamine, serotonin, which was detected by the Falck-Hillarp fluorescence histochemical technique. In the same way, administration of L-dopa induces the production of dopamine in parafollicular cells (Pearse 1966a, 1968).

The fractionation study of the sheep thyroid by Atack et al. (1972) indicated that intracellular monoamines, together with peptides, are almost exclusively present within the secretory granules.

13.3.3 Other Secretions

Chromogranin A is contained in the secretory granules of parafollicular cells, as recent immunocytochemical studies indicate (O'Conner et al. 1983; Lloyd and Wilson 1983).

A high-affinity *serotonin-binding protein* (SBP) has been demonstrated to be a granular component in the sheep thyroid using the cell fractionation technique (Bernd et al. 1979). This protein, which binds serotonin with high selectivity, is known to occur in serotoninergic neurons in the brain and gut (Tamir and Huang 1974; Tamir and Kuhar 1974). It seems reasonable to suggest that SBP is involved in the intragranular storage of serotonin and, therefore, shared by serotoninergic neurons and serotonin-producing paraneurons. The chemical and functional relations of SBP with chromogranins are still obscure.

13.4 Relations to Thyroid Follicular Cells

13.4.1 Comments on Thyroid Follicular Cells

As the parafollicular and follicular cells are in a "symbiotic" relationship in mammals, it now seems appropriate to discuss briefly the nature of the thyroid follicular cells. The biological position of the thyroid in the endocrine system has not hitherto been greatly discussed.

The hormonal products of the thyroid follicular cells are derivatives of amino acids but quite different in chemical construction from the peptides produced by neurons and paraneurons. Both T_3 and T_4 consist of two tyrosines. These are iodinated, but more importantly the two amino acids are combined not by a peptide bond but by an ether bond (-o-). T_3 and T_4 have a particular composition, which is categorized as diphenylether.

The mechanism of secreting T_3 and T_4 in the follicular cell is also unique. The cell first synthesizes a protein, thyroglobulin, and then releases it from the cell apex into the follicular lumen. This can be regarded as an exocrine activity of the cell, although thyroglobulin cannot be excreted by this ductless organ; it is stored in the follicle. During this process, iodination of thyroglobulin occurs. The cell then reabsorbs this iodinated large-molecule protein and breaks it up by lysosomal enzymes into small molecules, among which are exposed T_3 and T_4. The cell finally releases these from the cell base and they enter the blood circulation. Thus, the final process is the endocrine or hemocrine activity of the cell (review: Fujita 1975).

On the basis of these peculiar biochemical and cytological features of the thyroid follicular cells, we propose that these elements should be given a specific category in the endocrine regulatory system called the "diphenylether-producing system" (designation by Kobayashi 1983 unpublished; Fujita 1985).

The *ontogeny* of the thyroid essentially indicates that the follicular parenchyma is endodermal in origin, arising by a median downgrowth of epithelium at the base of the tongue.

The *phylogeny* of the thyroid indicates that it is a direct descendant of the foregut epithelium. The evolutionary precursor of the vertebrate thyroid can be traced in protochordates, such as amphioxus and *Ciona intestinalis*. Specialized cells in a thickened outgrowth of the pharyngeal epithelium, called the endostyle, produce iodinated proteins, including T_3 and T_4, and release them into the gut lumen by an exocrine fashion. The substances are later hydrolyzed and reabsorbed through the gut epithelium. Separation of the primitive thyroid organ, endostyle, from the digestive tract occurs at the evolutionary stage of Cyclostomata. The ammocetes, larval lampreys, still possess the endostyle, but during metamorphosis into the adult form the endostyle becomes isolated from the gut epithelium, thus forming a follicular tissue independent of and closed from the gut lumen (Gorbman 1959; Fujita and Honma 1968; Fujita 1975).

13.4.2 Functional Relationship

The close topographic relationship of parafollicular cells to thyroid follicular cells suggests that these cell types are also functionally related. Both cell types are supplied with a common blood capillary network (Kameda 1977) and it is very probable that their secretions exert a mutual influence by microhemocrinia as well as by paracrinia (intercellular diffusion).

Physiological experiments have provided little indication of the possible effect of follicular secretions upon parafollicular cell activity (review: Melander et al. 1975). On the other hand, some studies show that effects are exerted upon follicular cells by the peptides and amines released from parafollicular cells. No positive data appear to be available concerning the effects of calcitonin upon follicular activity. When administered in man and laboratory animals, somatostatin has been reported to inhibit thyroid hormone release (Ericson and Sundler 1984). Serotonin and dopamine are known to stimulate endocytosis and thyroid hormone release in follicular cells (review: Melander et al. 1975). In spite of these data, it is important that a systematic exploration into the significance of the "symbiosis" of follicular and parafollicular elements in the mammalian thyroid be carried out in future studies.

13.5 Ontogeny

Parafollicular cells were long believed to be of endodermal origin. The stem cells bud from the epithelium of the fourth branchial pouch, forming a cell mass called the ultimobranchial body. This body then fuses with the downgrowing thyroid primordium on the 16th day of gestation in the rat. On the 17th day, parafollicular cells become differentiated in the ultimobranchial body and are incorporated into the thyroid; they proliferate and scatter throughout the thyroid gland (Stoeckel and Porte 1970).

The endodermal origin of parafollicular cells was opposed by Pearse and Polak (1971), who proposed a neural crest origin of the cells. They reported that in the mouse embryo the last branchial pouch was colonized by APUD cells, which migrated from the neural crest. Although the report by Pearse and Polak (1971) was not in itself sufficient evidence, a neuroectodermal origin of the avian ultimobranchial body was supported by Le Douarin and her associates using the quail-chick chimera technique. They transplanted quail rhombencephalic tissue to chick embryos and recognized that parafollicular cells appearing by differentiation were mostly of quail, i.e., transplanted, origin (Le Douarin and Le Lièvre 1970; Polak et al. 1974; see p. 89). More recently, Fontaine (1979) investigated the problem in the mouse. Various fragments of the pharynx of mouse embryos were grafted onto the chorioallantoic membrane of chick embryos, and parafollicular cell differentiation occurring in the explants was evaluated by electron microscopy and immunocytochemistry. The results were interpreted as follows. The parafollicular cell precursors first appear in the fourth branchial arch *mesenchyme*, which is known to be mainly derived from the neural crest; they are later found in the fourth pouch *endoderm*, from which the ultimobranchial body arises. Thereafter, the differentiating C cells become incorporated in the thyroid as the ultimobranchial body fuses with it. Thus, Fontaine (1979) suggests that in mammals, parafollicular cells are neuroectodermal in origin and undergo "multistep migration" before settling in the thyroid.

The following features of parafollicular cells common to follicular cells deserve attention as unfavorable data for the neuroectodermal origin theory: (1) Parafollicular cells may form follicles with the same structure as those formed by follicular cells (Kameda 1982a). (2) Immunochemical and cytochemical studies have shown that parafollicular cell granules contain immunoreactivity for a 19S component of thyroglobulin (Kameda and Ikeda 1978, 1979). (3) Long-term administration of antithyroid drugs in dogs not only causes hypertrophy in follicular cells but also hypertrophy and degranulation in parafollicular cells (Kameda 1982b). (4) Embryological studies suggest that ultimobranchial cells are partly differentiated into thyroid follicular cells (Calvert 1972a, b; Kameda and Ikeda 1980). None of this evidence, however, seems sufficiently powerful to undermine the neuroectodermal origin of parafollicular cells. It appears to be difficult to differentiate between the cell's inherited features and features which have possibly been obtained by local induction or influence.

Yoshimura et al. (1972) explored a technique of culturing purely isolated parafollicular cells and purely isolated follicular cells from the rat thyroid. In culture, both kinds of cell were found to construct follicles which were morphologically very similar to each other. The follicles of parafollicular cells contained calcitonin; these cells did not show any uptake of ^{131}I nor did they show any sign of thyroxin production. Likewise, follicular cells in culture showed no functional features of parafollicular cells. This result supports the view that no functional shift occurs between follicular and parafollicular cells, at least after differentiation.

The *turnover rate* of parafollicular cells is unknown. Mitotic proliferation of the cells can be easily induced by hypercalcemic stimuli. After repeated injections of $CaCl_2$ and vitamin D in the dog, Kameda (1970) found numerous mitotic figures (mitotic index 10%) parallel with marked morphological signs of degranulation. After 1 month of repeated stimulation, the entire thyroid appeared to be filled with parafollicular cells.

13.6 Phylogeny

13.6.1 Submammalian Vertebrates

In birds, reptiles, amphibians, and fish, the embryonic ultimobranchial body does not fuse with the median thyroid anlage, and it remains into adult animals as an independent organ. The *ultimobranchial body in submammals* consists of endocrine cells, which are usually gathered in

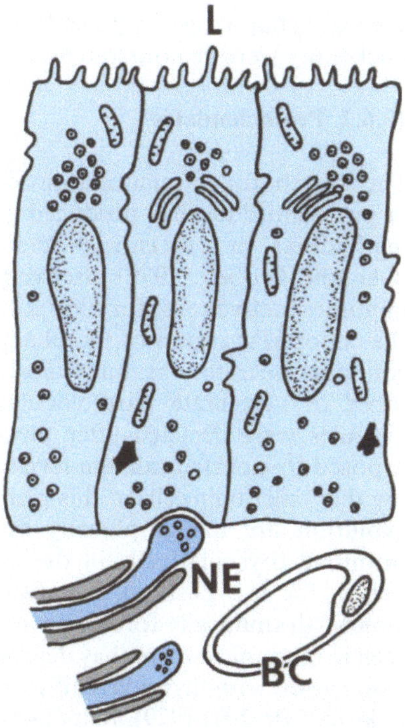

Fig. 13.2. Structure of amphibian ultimobranchial body cells. *L* lumen, *BC* blood capillary, *NE* nerve endings, part of which may enter deep into the epithelium. Based on the findings by Robertson (1967, 1968) in *Rana pipiens*

small follicles (Fig. 13.2). The individual cells are strikingly similar in fine structure to those of parafollicular cells in mammals. They produce *calcitonin* and store it in secretory granules which accumulate on the basal side of the cell. Somatostatin does not seem to be produced at least in the chicken (Ito et al. 1986). The secretory granules undergo depletion by hypercalcemic stimuli (frog, Robertson 1968; chick, Melander et al. 1971). The cells also contain monoamines, especially dopamine at least in some birds (Takagi et al. 1984). After administration of amine precursors, the cells show typical APUD ability (literature: see Melander et al. 1971).

The submammalian ultimobranchial body generally receives heavier innervation than its mammalian counterpart (Fig. 13.2). Robertson (1967, 1968), for instance, demonstrated numerous nerve terminals at the base of the ultimobranchial cells in *Rana pipiens*. The terminals contained small-cored vesicles (adrenergic) and also numerous large-cored vesicles (peptidergic).

Typical synapses accompanied by a membrane thickening were demonstrated.

13.6.2 Protochordates

The search for a calcitonin-producing system in protochordates has produced a few interesting findings. In *Styla clava* (Ascidiacea), Thorndyke and Probert (1979) discovered calcitonin-immunoreactive cells, forming a small group in the endostylar region of the pharynx. The cells contained secretory granules closely resembling those of vertebrate parafollicular cells; these granules were depleted after the animals were exposed to elevated calcium levels. It is surprising that calcitonin cells at this primitive stage of evolution are topographically associated with primitive thyroidal cells in the endostyle. The reason for this position of parafollicular cells, which "dramatically foreshadows the condition seen in mammals where they develop an intimate association with thyroidal follicle cells" (Thorndyke and Probert 1979), remains to be settled by future studies.

Calcitoninlike immunoreactivity has been recorded in certain paraneuronal cells in the gut epithelium and neurons and their fibers in the subepithelial connective tissue in *Ciona intestinalis* (Fritsch et al. 1980).

13.7 Summary

The parafollicular cell in the thyroid is a typical paraneuron secreting calcitonin, the blood calcium-lowering hormone. The release of the hormone is principally regulated by the Ca^{2+} levels in the circulating blood.

The cell also produces somatostatin, which is then costored with calcitonin in the same granules, although in some species somatostatin may occur only in some of the calcitonin-containing cells.

Amines, among others serotonin, are costored with the peptides in the secretory granules. As a carrier protein, chromogranin A is immunocytochemically localized in the granules.

The origin of the parafollicular cell has been disputed. Its neuroectodermal origin has recently been supported by transplantation studies using the quail-chick chimera technique. The parafollicular cells are derived from their primordial organ, the ultimobranchial body, which fuses with the thyroid primordium. The reasons for the adjacency of the parafollicular and follicular cells in the thyroid remain to be explored.

In submammalian vertebrates, the parafollicular cells remain separated from the thyroid. Calcitonin-producing paraneurons have been found in some protochordates.

14. Parathyroid Cells

14.1 Historical and General Aspects

The parathyroid glands (epithelial bodies) were discovered by the Swedish anatomist Sandström (1880; see Kohn 1900). Sandström thought that they belonged to the thyroid gland. It was Kohn (1895) who first recognized the parathyroids as special organs independent of the thyroid gland (see Appendix I of Böck 1982). It is interesting that Kohn (1895, see also 1900), who introduced the concept of "paraganglia" and "chromaffin cells," did not consider the parathyroid glands as paraganglia. The main reasons for this may have been that no chromaffin reaction could be demonstrated in the parathyroid cells and that the branchial epithelial origin of the cells had by then been established.

Nineteenth century surgeons noted that removal of the thyroid glands from dogs and cats caused convulsive states or death of the animals. Gley (1891) performed an experiment in the rabbit, which has an external pair of parathyroid glands besides those within the thyroid. He found that removal of either the thyroid or the external parathyroid glands caused little effect but that removal of both the thyroid and external parathyroid glands induced tetany. Vessale and Generali (1896) showed that thyroidectomy was not necessary to cause tetany; complete removal of the parathyroid tissue resulted in the tetanic states of the animals. MacCallum and Voegtlin (1909) showed that the tetany after removal of the parathyroid glands is associated with a low blood calcium level and that the tetany is relieved by administering calcium salts. Hanson (1925) and Collip (1925) prepared an extract from the parathyroid glands which raised the blood calcium levels in normal animals. Since then, the physiologically active parathyroid extract has been believed to contain parathyroid hormone (PTH, parathormone).

In 1959, two groups of researchers independently isolated and purified bovine PTH in a crystalline form (Auerbach 1959; Rasmussen and Craig 1961). Auerbach et al. (1971) determined the amino acid sequence of PTH and subsequently succeeded in the first total synthesis of this polypeptide hormone, which has 84 amino acid residues. The PTH precursor, prepro-PTH, was found by Kemper et al. (1974). It is thus believed that the parathyroid cells produce this prepropeptide and its product by processing through the secretory machinery consisting of the granular endoplasmic reticulum, Golgi apparatus, and secretory granules (see Habener et al. 1979, 1984).

As detailed below, we regard the parathyroid cell as a paraneuron. We support the view of Kohn (1900) that the parathyroid cells are distinct from the adrenal "chromaffin cells," but we would like to point out that they are similar in cytological features to the gut paraneurons which previously were collectively termed "enterochromaffin cells".

The parathyroid gland is made up of a well-vascularized parenchyma consisting of epithelial cell cords. In most mammalian species, a single cell type forms the gland. The parathyroid cells of man, monkeys, cattle, the horse, and bat, however, are classified into two types—*oxyphil cells* and *chief cells*. Oxyphil cells contain many mitochondria and hence are stainable with acidic dyes such as eosin. However, there is every transitional form between the oxyphil and chief cells (Munger and Roth 1963; Weymouth and Sheridan 1966; Nilsson 1977; review: Thiele 1977). Therefore, oxyphil cells probably represent a functional stage of the chief cells.

Parathyroid cells respond to hypocalcemia by releasing PTH. Thus, they play a vitally important role as a component of the controlling system for Ca^{2+} homeostasis.

14.2 Cytological Features

The parathyroid cells in man and various mammals are polyhedral and do not project any cytoplasmic processes. They are attached to one

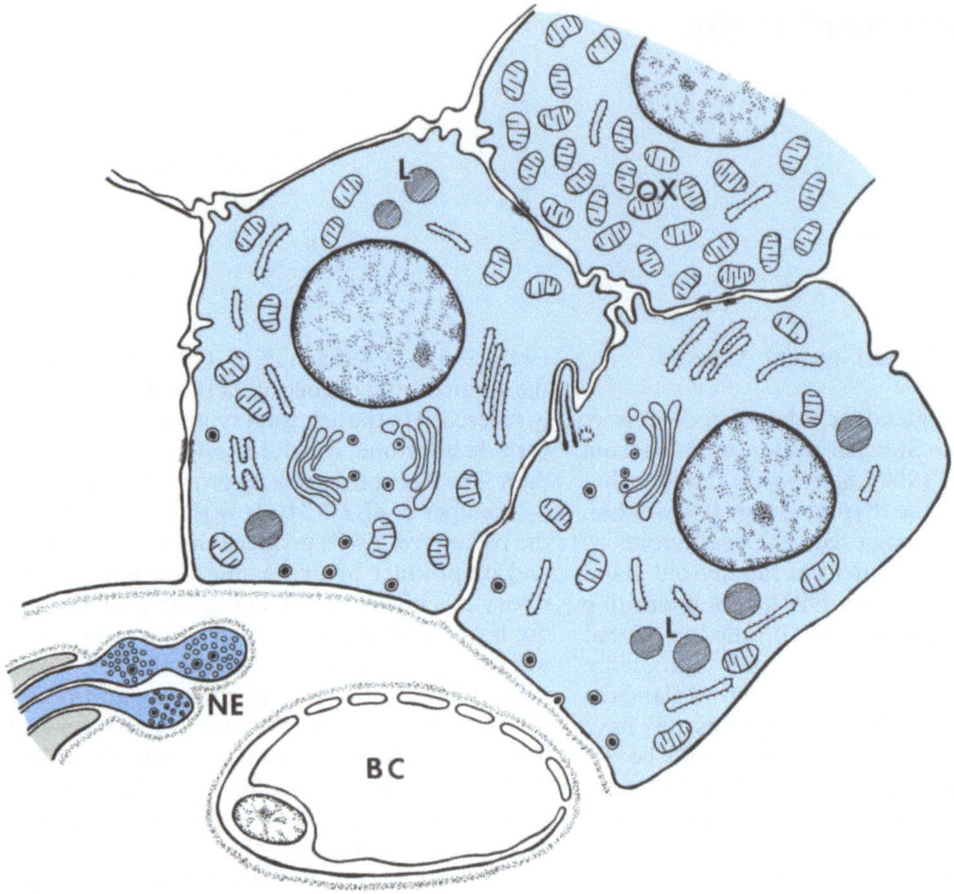

Fig. 14.1. Structure of mammalian parathyroid cells. The oxyntic cell (*OX*) is characterized by numerous mitochondria. *L* lipid droplets

another by demosomes. No gap junctions are present in the parathyroid glands and this may explain why each cell has an independent secretory cycle (Thiele 1977). A *single cilium* with the 9+0 pattern of microtubules is frequently recognized. However, its location on the cell surface does not seem to be particularly convenient for a receptive function. An increase in the number of cilia in the cells of the parathyroid adenoma is reported (Polyzonis 1970). The functional significance of these cilia remains unknown.

Parathyroid cells possess cisterns of granular endoplasmic reticulum, Golgi apparatus, and secretory granules. They are thus provided with all the indispensable organelles for the production and storage of secretory peptides and amines (Fig. 14.1). Furthermore, electron micrographs convincingly indicating the *exocytotic release* of secretory granules have been published by many authors, among them Munger and Roth (1963), Fujii and Isono (1972), Isono and Shoumura (1973), and Capen and Roth (1973) (review: Thiele 1980). Membrane fusion and recycling associated with the exocytosis have been demonstrated by freeze-fracture electron microscopy (Capen 1971; Thiele 1977). These ultrastructural features are the main basis for our identifying the parathyroid cells as paraneurons.

The secretory granules have a dense core and a pale halo; they are spherical and in man measure 200–400 nm in diameter (Shannon and Roth 1974; Thiele 1977). In the rat numerous small granules (100–250 nm) and a few large mature granules (300–600 nm) have been identified (Setoguti et al. 1981). The existence of PTH in the secretory granules has been established both by biochemical cell fractionation studies (L'Heureux and Melius 1956) and by electron microscopic immunocytochemical investigations (Setoguti et al. 1985; Inoue and Setoguti 1986).

14.3 Secretion Control and Innervation

The parathyroid cells synthesize and release PTH. This hormone activates osteocytes and osteoclasts in the bone and a consequent resorption of bone matrix, thereby increasing the blood Ca^{2+} and phosphate levels. The activity of parathyroid cells is mostly controlled by the blood Ca^{2+} level by a negative feedback mechanism: Lowered blood Ca^{2+} levels provoke an increased release of PTH from the parathyroid cells, and *vice versa* (Brown et al. 1976; Morrissey and Cohn 1979). Electron-microscopic studies by Setoguti et al. (1981, 1985) demonstrated that experimental hypocalcemia caused decreased granule number while hypercalcemia caused increased granules and their degradation by lysosomes. Setoguti et al. (1985) accounted for this phenomenon by crinophagy.

The connective tissue of the parathyroid gland contains abundant blood capillaries and fat cells. The blood capillaries are lined by endothelial cells with thin cytoplasm and numerous fenestrations. The capillaries are in close proximity to the parathyroid cells and so the exchange of substances between the bloodstream and parathyroid cells is greatly facilitated. In the space between the capillaries and parathyroid cells, there are neuronal processes ensheathed by Schwann cells. Although there is no frequent synapse formation between the parathyroid cells and neuronal processes, it seems possible that they influence each other by a paracrine mechanism. There has only been one report on direct contact between the neuronal process and the parathyroid chief cell (Altenähr 1971).

Stimulation of autonomic nerves innervating the parathyroid glands considerably modulates parathyroid function (Morii et al. 1963; Kukreja et al. 1976). This seems partly due to the changes in the parathyroid blood flow and is not necessarily due to direct control of the parathyroid cells. Setoguti et al. (1984) showed that administration of adrenaline caused a prompt decrease in the number of large (mature) secretory granules in the rat.

14.4 Parathyroid Cells As Paraneurons

Pearse (1966a, 1968, 1969, 1976) did not include parathyroid cells in the APUD series. Parathyroid cells show no specific fluorescence after histochemical treatment for catecholamines and serotonin. No uptake of amine precursors is demonstrable by the available histochemical methodologies. We, on the other hand, do regard the parathyroid cells as paraneurons (Fujita and Kobayashi 1975, 1977, 1979b; Fujita 1976).

We do not consider the amine-handling ability of parathyroid cells to be as important as Pearse does. As pointed out above (p. 6), there are numerous neurons and paraneurons which do not exhibit this ability. There are very few, secretory granules in parathyroid cells, hence it is likely that, even if individual granules do contain monoamines, the amount of monoamines is too small to give a distinct fluorescence reaction at the cellular level. Detailed and reliable fluorescence histochemical or autoradiographic studies are necessary to determine whether monoamines are present or absent and to clarify the uptake mechanisms in parathyroid cells. The possibility that the cells contain nonaminic transmitter substances of a small molecular size can not be excluded.

Attempts to demonstrate ATP and other adenine nucleotides which might coexist with PTH in parathyroid granules have been unsuccessful. Böck (1980c) applied the quinacrine fluorescence method to parathyroid glands and could not detect any fluorescence. This result may also be ascribed to the small number of secretory granules in the parathyroid cells.

Coexistence of gastrin or gastrinlike peptides with PTH has been demonstrated in some cases of hyperparathyroidism (Fabri et al. 1986). An attractive interpretation of this finding is the possibility that gastrin or a related peptide is present in a minute amount in normal parathyroid cells and produced in a large amount under certain pathological conditions. At any rate, the potentiality of parathyroid cells to produce a gut hormone strongly supports their relation to gastroenteric paraneurons.

The nature of the *carrier protein* in the secretory granules further supports the paraneuronal property of parathyroid cells. The acidic, soluble protein obtained from parathyroid granules, which is called secretory protein (SP 1) or parathyroid secretory protein (PSP), closely resembles chromogranin A (pp. 24, ff) in its chemical and immunochemical characteristics (Cohn and Elting 1983; see also Habener et al. 1984).

The final, equally important reason for regarding parathyroid cells as paraneurons is that they are involved and often take the leading role in general neoplastic disorders of the neuroparaneuronal cells, called multiple endocrine adenomatosis (MEA) and multiple endocrine neoplasia (MEN) (see p. 116).

14.5 Chemical Structure, Biosynthesis, and Release of PTH

The chemistry, biosynthesis, secretion, and metabolism of PTH have been reviewed by Habener and Potts (1976) and Habener et al. 1984. The complete amino acid sequence of bovine PTH was determined by Brewer and Ronan (1970) and Niall et al. (1970), followed by porcine (Sauer et al. 1974) and human PTH. All the identified forms of PTH are a single-chain polypeptide consisting of 84 amino acid residues. There is a free amino group at the amino terminal end of the molecule, but the significance of this moiety in the hormonal action is as yet unknown. No cysteine is contained in the molecule.

Biosynthesis of PTH has been investigated both in vivo and in vitro. Nakagami et al. (1971) studied the time-course of the secretory cycle *in vivo*, i.e., in rat PTH cells by electron-microscopic autoradiography using ^{3}H-tyrosine as a label. They found the label in the Golgi apparatus 2 min after injection of the label, in the secretory granules in 25–30 min, and out of the cells in 45–60 min. This time-course of the hormone synthesis is not greatly different from that in other paraneurons such as adrenal chromaffin cells (Kobayashi et al. 1978a). In a series of autoradiographic studies, Nakagami et al. (1971) demonstrated that ^{3}H-galactose is built into the glycoprotein component of the secretory granule membrane. The development of an *in vitro* system to study the biosynthesis has resulted in a significant advance in our knowledge of the mechanims. PTH is formed from a precursor, prepro-PTH, which consists of 115 amino acids (Kemper et al. 1974; Habener et al. 1975, 1976, 1984). During the synthesis of the hormone in the granular endoplasmic reticulum, the signal peptide, or "pre-" segment, containing 25 amino acids is removed from the *N*-terminal portion followed by the six amino acids in the "pro-" part of the pro-PTH molecule. This processing takes place during the intracellular transport of the molecule to the secretory granule through the Golgi apparatus (Habener et al. 1976).

Recent chemical studies of the parathyroid glands do not seem to have been directed toward analysis of the composition of the purified secretory granules. However, earlier studies using the differential centrifugation of parathyroid tissues provided sufficient biochemical evidence that PTH is localized in the secretory granules (L'Heureux and Melius 1956).

Immunocytochemical investigations at the light- and electron-microscopic levels were performed by many authors using antisera raised against various segments of the PTH molecule or against the whole molecule (Roth and Capen 1974; Ravazzola et al. 1978; Limacher et al. 1979; Setoguti et al. 1985). The immunoreaction product was demonstrated light-microscopically in the cytoplasm of the parathyroid cells and electron-microscopically in their secretory granules. The possibility has been proposed that pro-PTH may also be released into the circulation (see Habener and Potts 1976; Habener et al. 1984); this implies that the storage of the secretory granules in the cytoplasm of the parathyroid cells is too brief to complete the processing of the PTH molecule. Once released into the tissue space and blood-stream, PTH is cleaved into smaller molecules by various peptidases (Fischer et al. 1972). It is known that there are fragments of PTH showing the same activity as that of the hormone with the full 84 amino acid sequences (Niall et al. 1974).

14.6 Electrophysiology

No electrophysiological studies have been published on the parathyroid cells to the authors' knowledge.

14.7 Ontogeny and Kinetics

Mammalian parathyroid cells arise from the *endodermal branchial pouches* in early fetal life (Norris 1937, 1946). They originate from the upper dorsal portion of the third and fourth branchial pouches. The area lined by parathyroid cells protrudes into the subjacent connective tissue. The cells are then separated, as a mass, from the epithelial sheath and become the endocrine gland. Nakagami et al. (1968) examined the parathyroid cells of the human fetus by electron microscopy and found abundant glycogen in the cytoplasm of the cells.

It is interesting that in urodele amphibians the glands first appear after metamorphosis, i.e., after loss of the gills. The primordial parathyroid cells multiply by *mitosis* and form the glandular parenchyma. They are then separated from the branchial pouch epithelium and move to the due position in accordance with the growth of adjacent structures.

At least some parathyroid cells in adult mammals possess the ability to proliferate by *mitosis*. There is a tendency for mitosis to take place more frequently in young animals (see Roth and Schiller 1976).

14.8 Phylogeny

Histological and ultrastructural studies on parathyroid cells have been performed in numerous species of all vertebrate classes. A comprehensive review on the comparative morphology of the parathyroid glands was published by Roth and Schiller (1976). Parathyroid glands are present in all amphibians, reptiles, birds, and mammals. Fish do not have parathyroid glands, although they have ultimobranchial bodies which secrete calcitonin (see p. 149). Thus, the mechanism of the control of Ca^{2+} homeostasis in fish must be greatly different from that in other vertebrates.

Phylogenetically, the simplest animals to possess parathyroid glands are urodelen amphibians. Electron-microscopic studies by Setoguti et al. (1970) and Isono et al. (1971) indicate that the parathyroid cells in the newt *Triturus pyrrhogaster* possess essentially the same intracytoplasmic structures as those in mammals. Distinct seasonal changes in the cytoplasmic organelles have been demonstrated in anuran amphibians (Brehm 1963).

Most reptiles have two pairs of parathyroid glands with similar histological and ultrastructural characteristics to those in mammals. In several species of freshwater turtles, the parathyroid cells form follicles (Clarke and Khairallah 1972). These follicles are similar to those occasionally seen in amphibian, avian, and mammalian parathyroid glands. Whether or not the follicular lumen corresponds to the remnant of the primordial intestinal lumen needs further confirmation.

Birds usually have two separate parathyroid glands located caudal to each thyroid lobe; but in some species, the glands on each side tend to fuse. In small birds, such as *Uroloncha domestica*, the parathyroid glands partially envelop the carotid body and become associated with it (Watzka 1943; Kobayashi 1969a, b). The significance of this peculiar topographic relationship between the two different organs consisting of different paraneurons remains to be established. Avian parathyroid cells have similar histological and ultrastructural features to those of mammalian cells (Fujii and Isono 1972; Gould and Hodges 1971). No morphological changes were found in association with the mobilization of large amounts of Ca^{2+} for eggshell formation.

14.9 Summary

Parathyroid cells produce parathyroid hormone, which is made up of 84 amino acid residues and controls the homeostasis of the blood Ca^{2+} level. The cells have very little innervation, hence their secretory activity is predominantly controlled by the blood Ca^{2+} level.

Parathyroid cells are regarded as endocrine paraneurons. They contain typical paraneuronal secretory granules, which are, however, extremely few in number. The secretory granules are continuously formed and released, since they cannot be stored for a long time. The sparsity of the secretory granules in the cell has hindered studies on the nonpeptidic elements that they contain. Thus, it has not been settled whether monoamines and ATP occur in the granules. An acidic carrier protein called SP-1 or PSP was obtained from the granules and when characterized it was shown closely to resemble chromogranin A. There are cases of hyperparathyroidism in which parathyroid hormone coexists with gastrin and/or related peptides in the parathyroid cells.

Parathyroid cells originate from the epithelium of the third and fourth branchial pouches. The cells can proliferate by mitosis, even in adults.

Phylogenic studies indicated that parathyroid glands are absent in invertebrates and fish.

15. Adenohypophyseal Cells

15.1 General Aspects

The hypothysis, an appendage of the brain on its basal aspect, is also called the pituitary gland, since Andreas Vesalius (1543) in his "Fabrica" interpreted this organ as the production site of pituita (phlegm). The pituitary gland, which weighs about 0.5 g in man, is divided into two main components: the *neurohypophysis*, which embryologically is derived from the floor of the diencephalon, and the *adenohypophysis*, which originates from Rathke's pouch, the invagination of the roof of the buccopharyngeal cavity. In this chapter, we will concentrate on the adenohypophyseal cells as the paraneuronal component of the pituitary gland; the neurohypophysis was dealt with in relation to the classic neurosecretory system (p. 55).

The *adenohypophysis* is divided into three portions: *pars distalis*, *pars tuberalis*, and *pars intermedia.* The pars distalis occupies the largest part of the adenohypophysis and consists of anastomosing cords of parenchymal cells. The pars tuberalis is a small portion of the adenohypophysis extending upward from the pars distalis around the infundibulum. The pars intermedia is wedged in the space between the pars distails and the neurohypophysis. Each portion of the adenohypophysis possesses specific endocrine cells. Several cell types have been differentiated in mammals and other classes of vertebrates by the cell size and shape and, in particular, by the stainability of their secretory granules with different dyes (Romeis 1940; Baker 1974). More recently, the use of *immunocytochemical* methods has made possible the reliable identification of the cellular sites of adenohypophyseal hormones (Nakane 1970; Girod 1977; Kurosumi 1986). Furthermore, the *fine structure* of the secretory granules of each cell type can now be correlated with the principal secretory products of the cell.

15.2 Parenchymal Cells of the Adenohypophysis

Among the secretory products of the adenohypophysis, those of a *proteinic* (and *glucoproteinic*) nature are the most important and correspond to what are traditionally called (anterior) pituitary hormones. The established hormones are *g*rowth *h*ormone (somatotropin, GH), *pro*l*actin (mammotropin, PRL), *a*dreno*c*ortico*t*ropic *h*ormone (corticotropin, ACTH), *g*onado*t*ropic *h*ormones [gonadotropins, GTHs, i.e., *f*ollicle *s*timulating *h*ormone (FSH) and *l*uteinizing *h*ormone (LH)], and *t*hyroid *s*timulating *h*ormone (thyrotropin, TSH). Chemically, GH, PRL, and ACTH are simple proteins, whereas GTHs and TSH are glycoproteins. These pituitary peptides were previously believed to occur exclusively in the adenohypophyseal endocrine cells, but recent immunoassay and immunocytochemical studies have revealed that some of them are *shared by neurons* in the hypothalamus. It is widely accepted that these pituitary peptides include ACTH, MSH, β-endorphin (and some other peptides derived from corticotropin-β-lipotropin precursors), and prolactin (review: Imura 1985).

Adenohypophyseal endocrine cells occur as the follows (review: Kurosumi and Inoue 1986) (Fig. 15.1).

15.2.1 Pars Distalis

According to Kurosumi (1968, 1974), there are at least five types of structurally and functionally differentiated endocrine cells in the parenchyma of the pars distalis in the rat (see also Kurosumi and Inoue 1986).

GH cells (somatotrophs). These cells coincide with the acidophil α-cells observable with light microscopy. Electron microscopically, they are characterized by round, dense, secretory granules 300–350 nm in diameter.

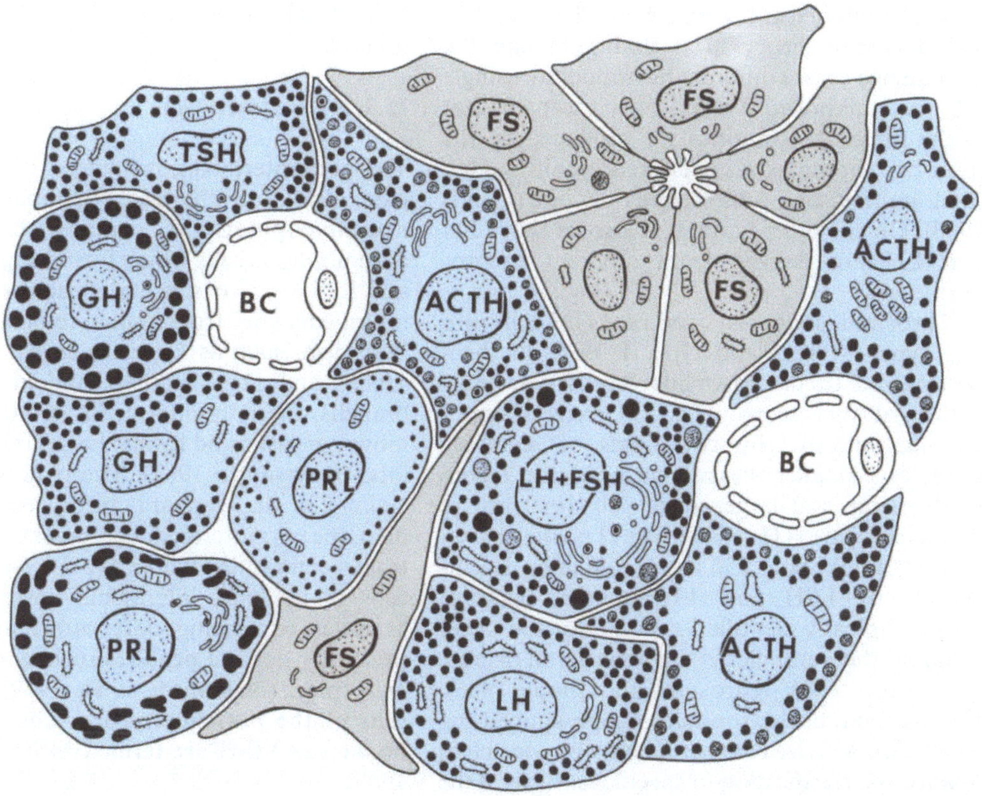

Fig. 15.1. Anterior pituitary cells in mammals. The different types of endocrine cells are labeled with the names of the hormones they produce. *FS* folliculostellate cells, *BC* blood capillaries. Based on and modified from Kurosumi (1986)

PRL cells (mammotrophs). The PRL cells are the acidophill ε cells detectable under light microscopy. The secretory granules of the PRL cells are the largest of those of the adenohypophyseal cells and measure about 700 nm in maximum diameter. The granules are irregular in shape. Elliptical, dumbbell-shaped, or cylindrical profiles of the secretory granules are frequently encountered. The granules are abundant in the female but sparse in the male. The size and number of PRL cells greatly increase during pregnancy and lactation.

ACTH cells (corticotrophs). The ACTH cells are large chromophobe cells evident under light microscopy. Nakanishi et al. (1979) deduced from the nucleotide sequence analysis of cloned cDNA for the bovine ACTH precursor that this polypeptide, composed of 39 amino acid residues, is *cosynthesized* with γ-melanotropin and β-lipotropin. The latter contains small component peptides such as γ-lipotropin, β-endorphin, β-melanotropin, and Met-enkephalin. ACTH itself contains α-melanotropin and corticotropinlike intermediate lobe peptide (CLIP). Nakanishi et al. (1979) demonstrated the whole amino acid sequence of the "corticotropin-β-lipotropin precursor" (preproopiomelanocortin, Cox 1982). However, ACTH is the oldest-known hormone among these peptides derived from the corticotropin-β-lipotropin precursor and so the term "ACTH cell" is still widely used in cytology.

Kurosumi et al. (1984) classified ACTH cells of the rat adenohypophyseal cells into two types. The typical ACTH cells (type II) are small cells with slender cytoplasmic processes and contain small secretory granules about 150–200 nm in diameter aligned in a single row along the cell surface. On the other hand, the atypical ACTH cells (type I) are polygonal or oval and contain many secretory granules, which fill the cytoplasm.

GTH cells (gonadotrophs). It had been believed that FSH and LH were produced by different cells, but Nakano (1970) immunocytochemically demonstrated that both hormones are secreted by the same cell type in the rat. This view was strongly disputed at the time when the "one cell, one hormone theory" was dominant and the coexistence of two different hormones in a single cell was unknown. Today, it is accepted that in the rat both hormones are produced by the same cell (homologous coexistence), although a small number of δ-cells secrete only one (Childs 1986). On the other hand, it is known that FSH and LH are derived from different preprohormones.

Kurosumi (1974) classified rat GTH cells into two types—type I and type II. Both of them correspond to the basophile δ-cells detectable with light microscopy. The GTH cells of type I are characterized by two kinds of secretory granules—smaller ones measuring about 200 nm in diameter and larger ones 700–1000 nm in diameter. The GTH cells of type II possess secretory granules about 250 nm in diameter. Localization of FSH and LH in the electron-microscopically identified subtypes of the GTH cells and their granules is still a matter of dispute (see Girod 1984; Kurosumi and Inoue 1986). Careful immunoelectron-microscopic study using a well-characterized antiserum may provide clues toward the clarification of specific features of the subtypes of the GTH cells.

TSH cells (thyrotrophs). Under light microscopy, the TSH cells are large basophil cells giving a positive periodic acid-Schiff (PAS) reaction. This reaction is ascribed to the glycoproteinic nature of the TSH molecule. They have abundant secretory granules with a solid core, diameter 100–150 nm, and are grouped in the periphery of the cell.

Our impression is that exocytotic figures of TSH cells are more frequently reported than those of other types of adenohypophyseal cells (see also Kurosumi 1974), though the reason for this phenomenon is unknown. Thyroidectomy induces such drastic morphological changes as the accumulation of intracisternal granules in the granular endoplasmic reticulum (Farquhar and Rinehart 1954; Kurosumi and Oota 1966).

15.2.2 Pars Tuberalis

The pars tuberalis, also called the pars infundibularis, is a small portion of the adenohypophysis extending from the pars distalis and surrounding the infundibulum. It contains a few basophil cells, but practically no acidophil cells. There are indeed many endocrine cells with round secretory granules, diameter 100–200 nm (Stoeckel and Porte 1984), but no special hormone has been demonstrated so far in the pars tuberalis.

15.2.3 Pars Intermedia

The pars intermedia is wedged in between the pars distalis of the adenohypophysis and the neurohypophysis. It contains basophil and chromophobe cells which form irregular strands and follicles. The size and structure of the pars intermedia vary from animal to animal. In some bony fish, such as *Cyprinus carpio*, *Parasilurus asotus*, and *Fugu rubripes*, and in the elasmobranchs, the pars intermedia is mixed with the lobus nervosus and forms the neurointermediate lobe (Kawamoto 1967). The pars intermedia is lacking in the elephant, birds, and whales. In man, the pars intermedia occupies only 2% of the total weight of the whole hypophysis. Endocrine cells of the pars intermedia possess secretory granules containing corticotropin-β-lipotropin precursor-derived peptides such as MSH and β-endorphin (Nakanishi et al. 1979). Endocrine cells of the pars intermedia containing immunoreactive MSH are termed MSH cells (Pearse 1969).

15.2.4 Folliculostellate Cells

These parenchymal cells universally occurring in the above-mentioned three portions of the adenohypophysis contain no secretory granules and hence are nonparaneuronic in nature. Folliculostellate cells occasionally line follicles, which resemble thyroid follicles (Yamamoto et al. 1982; Barberini and Correr 1984). Some folliculostellate cells extend long cytoplasmic processes on the opposite side of the follicular lumen. These processes are frequently in contact with the endocrine-granulated cells and terminate on the surface of the adenohypophysis or on blood vessels. Immunocytochemical studies have shown in some mammals that folliculostellate cells possess S-100 protein (Nakajima et al. 1980; Shirasawa et al. 1983; Girod et al. 1986), a calcium-binding protein specific, though not absolutely, to glial elements in both the central and peripheral nervous systems (see p. 102). Intense interest has thus been shown in the reevaluation of the supportive cell theory (Kagayama 1965; Vila-Porcile 1972) and helper cell theory (Shiotani 1980) of folliculostellate cells. If folliculostellate cells have some glialike features, the neuronlike (paraneuronal) nature of the adenohypophyseal endocrine cells they support is strengthened.

15.3 Cytological Features

A comprehensive review on the cytological features of adenohypophyseal cells is given by Baker (1974). The secretory aspects of the cells are in complete conformity with the features known in other endocrine paraneurons.

15.3.1 Secretory Pathway

The secretory pathway of adenohypophyseal cells in the rat has been investigated both morphologically by electron-microscopic autoradiography (Racodot et al. 1965; Hopkins and Farquhar 1973; Salpeter and Farquhar (1981) and biochemically by cell fractionation methods (Costoff and McSchan 1969). It has been established that polypeptides are synthesized at the ribosomes of granular endoplasmic reticulum, segregated into the cisterns of the reticulum, and transported to the Golgi apparatus, where they are packed into the secretory granules. Thus, the adenohypophyseal cells have the basic secretory machinery as other peptidergic and neurosecretory neurons (p. 17). According to Racodot et al. (1965), newly synthesized polypeptides occur in the endoplasmic reticulum within 10 min after the injection of ^{3}H-leucine, in the Golgi apparatus after 30 min, and in the mature secretory granules after 90 min. On the other hand, Salpeter and Farquhar (1981) showed that in rat PRL cells, the granular endoplasmic reticulum is removed with newly synthesized polypeptides by 30 min, the Golgi apparatus by 1 h, and the young secretory granules by 2 h.

With respect to the biosynthesis of the side-chain polysaccharide moieties of some pituitary hormones such as GTHs and TSH, autoradiographic studies by Neutra and Leblond (1966) using ^{3}H-glucose demonstrated that these moieties are added to polypeptide moieties of the hormones in the Golgi apparatus.

15.3.2 Exocytotic Granule Release, Membrane Retrieval

Since the first convincing demonstration of exocytotic granule release by Ichikawa (1959) in rat GH cells, ultrastructural data have accumulated indicating that the granule release in different types of adenohypophyseal cells takes place by *exocytosis* (reviews: Kurosumi 1968, 1974; Kurosumi and Inoue 1986). Various stimuli induce granule release from specified or nonspecified cell types of the adenohypophysis. Striking images of the exocytotic release were presented by Fujita et al. (1983) in adenohypophyseal cells incubated in Krebs' medium containing the calcium ionophore A23187; the latter has been utilized in studying the effect of calcium influx upon the secretory activity of various paraneurons, including mast cells (Kanno et al. 1973), adrenal chromaffin cells (Ricci et al. 1975), and carotid body chief cells (Grönblad et al. 1980), as well as nonparaneuronal acinar cells of the pancreas (Kanno and Saito 1978). The electron micrographs of Fujita et al. (1983) suggest that the anterior pituitary glands are a good model for the morphological and physiological analysis of the Ca^{2+}-dependent exocytotic granule release in paraneurons and other secretory cells.

The mechanism of the *disposition of the surplus membrane* after the exocytotic granule release is one of the most important topics in the cytochemical aspects of adenohypophyseal cells. Farquhar (1978) demonstrated in an electron-microscopic study using dissociated rat hypophyseal cells that cationized ferritin, but not native (anionic) ferritin, binds to the surface of the endocrine cells and is incorporated via clathrin-coated vesicles into the lysosomes, Golgi cisterns, and condensing vacuoles (p. 54). It is likely that the intracellular transport of cationized ferritin more or less represents the pathway of the membrane for reutilization in the recycling events (see also Farquhar 1983). However, successful visualization of the intracellular transport of the membrane retrieved after exocytotic release of the granule contents has not yet been achieved.

The concept of *crinophagy* was introduced by Farquhar and co-workers on the basis of their electron-microscopic studies on adenohypophyseal cells (Smith and Farquhar 1966). Excess secretory granules are inactivated and degraded by fusion with lysosomes; for example, if suckling baby rats are taken away from the lactating mother many secretory granules of the PRL cells fuse with lysosomes. Thus, the number of the secretory granules in a single adenohypophyseal cell is kept constant.

15.3.3 Monoamines in the Adenohypophyseal Cells

The existence of monoamines has been demonstrated in certain types of adenohypophyseal endocrine cells of some mammals (ACTH- and MSH-cells) using the FIF method (see p. 6) by Pearse and McGregor (1964) and Dahlström and

Fuxe (1966). The uptake of amine precursors such as L-dopa into the adenohypophyseal cells has been also demonstrated.

Pearse (1966a, 1968, 1969) originally included only ACTH and MSH cells of the anterior pituitary gland in his APUD series. Thus, he originally made a sharp distinction between pituitary endocrine cells with and without the APUD ability. Indeed, the APUD characteristics, i.e., the positive FIF reaction after the loading of L-dopa, are very distinct in these two types of pituitary endocrine cell. However, Pearse (1981) later modified his APUD concept and included all the other endocrine cell types of the adenohypophysis notwithstanding their negative APUD ability.

There exist some data which are at odds with the view that the amine-handling ability is an important feature of adenohypophyseal cells. Partanen et al. (1975) demonstrated that of the cells accumulating L-dopa in the adult rat adenohypophysis 96% were folliculostellate cells (chromophobes), 2.0% GH plus PRL cells (acidophiles), 0.9% ACTH cells, and 1.2% GTH plus TSH cells (glycoprotein-containing cells). It is striking that exogenous L-dopa is mainly incorporated into the nonendocrine folliculostellate cells in the adenohypophysis. The significance of this phenomenon in pituitary function is not as yet known.

Using the quinacrine method, Böck (1980c) showed that in some mammals adenohypophyseal cells exhibit a histochemical reaction for *adenine nucleotides*.

Chromogranins also seem to be components of the granule substances in the adenohypophyseal cells, although their occurrence has been evidenced immunocytochemically only in a certain population of endocrine cells in the human (Wilson and Lloyd 1984), bovine (Nolan et al. 1985), and rat adenohypophysis (Fischer-Colbrie et al. 1985). In several mammalian species, Rindi et al. (1986) demonstrated occurrence of chromogranins A, B, and C in certain adenohypophyseal cells including TSH, FSH, and LH cells.

15.4 Innervation of the Adenohypophyseal Cells

The secretory activity of adenohypophyseal endocrine cells of mammals is mainly controlled by humoral factors such as the releasing and inhibiting hormones from the hypothalamus. Thus, direct neuronal control of adenohypophyseal function is incidental. However, distinct nerve fibers and terminals are present in the adenohypophysis, though there are marked quantitative and qualitative differences during their phylogenetic development. As pointed out by Bargmann (1972) and Unsicker (1984), what applies to one vertebrate species does not necessarily hold true for a different vertebrate class or species.

Bargmann et al. (1967) found nerves apparently terminating on the endocrine cells in the pars intermedia of the cat pituitary gland. They classified these nerve terminals into three types—cholinergic, adrenergic, and peptidergic. The cholinergic terminals contained mainly small noncored vesicles. The adrenergic terminals were characterized by small vesicles with a dense core. The third type of nerve terminal contained an accumulation of large cored vesicles, 150–200 nm in diameter, which resembled the neurosecretory granules in the neurohypophysis. To designate this last type of nerve, Bargmann et al. (1967) invented the term "peptidergic" (see also p. 5).

No innervation can be found in the pars distalis of the pituitary gland of the mouse, rat, hamster, cat, and pig (Unsicker 1971); in the pars tuberalis of these mammalian species, Unsicker (1971) described two types of nerve terminal—cholinergic and peptidergic. The scarcity of the nerve elements in the pars distalis is in keeping with the fact that the function of the endocrine cells in these animals is mostly controlled by nonneuronal factors conveyed by the *hypothalamohypophyseal portal vessels*.

In poikilothermic animals, especially in teleost fish, neuronal control seems to be essential for adenohypophyseal function. As reviewed by Vollrath (1969) and Kaul and Vollrath (1974), in many fish, nerve terminals bringing hypophyseal releasing and inhibiting hormones into the adenohypophysis do not terminate in the median eminence but directly enter the pars distalis, where they come into close contact with the endocrine cells. However, the species difference in the degree of innervation in the adenohypophysis is significant even in teleosts. For example, in the tench *Tinca tinca*, the whole parenchyma of the adenohypophysis is innervated, whereas in the eel *Anguilla anguilla* nerve fibers are restricted to the dorsal portion of the adenohypophysis (Unsicker 1984).

15.5 Electrophysiology

Certain differences have been shown in the electrical characteristics of adenohypophyseal cells not only between different species but also between different cell types in different regions of the same species. Nevertheless, it is evident that adenohypophyseal cells, at least many of them, have resting membrane potentials that are as high or higher than those found in neurons. It is also clear that this potential depends largely on the transmembrane gradient of K^+, that the cells are electrically excitable and often show spontaneous action potentials, and that certain hypophysiotropic substances known to stimulate or inhibit secretion have corresponding stimulant or inhibitory effects on the electrical activity. This electrical activity involves to a greater or smaller degree an influx of Ca^{2+} through voltage-dependent Ca^{2+} channels. The involvement of inward Ca^{2+} currents during basal electrical activity and the stimulant or inhibitory effects of the hypophysiotropic substances suggests that electrical events may be important factors in the physiological control of adenohypophyseal secretion (Douglas and Taraskevich 1985; Ozawa 1985).

Israel et al. (1985) recorded electrical activity from isolated human prolactin-secreting adenoma cells and concluded that the dopamine modulation of electrical activity involving the Ca^{2+} current may be an early important step in the mechanism by which dopamine inhibits prolactin release.

15.6 Ontogeny and Kinetics

The adenohypophysis originates from a small *ectodermal* diverticulum, *Rathke's pouch*, which occurs in the roof of the primitive mouth (*stomodeum*) adjacent to the buccopharyngeal membrane. Rathke's pouch enlarges and subsequently becomes constricted at the pharyngeal roof. In the human fetus, Rathke's pouch loses its connection with the stomodeal epithelium by the end of the 2nd month and then comes into close contact with the infundibulum, which originates at the bottom of the third ventricle of the developing brain. The epithelium of Rathke's pouch is not provided with the granule-containing cells. During further development of the embryo, the cells of Rathke's pouch undergo a great modification; they rapidly increase in number and differentiate into granulated parenchymal cells of the adenohypophysis.

There have been many studies on the developmental relationship between the *hypothalamus* and *adenohypophysis*. Evidence has accumulated showing that the hypothalamus affects the development of the adenohypophysis (Ishikawa et al. 1977; Dubois et al. 1978; Gross and Baker 1979). Daikoku et al. (1982) cultured Rathke's pouches of fetal rats with and without various other fetal organs. They demonstrated that Rathke's pouches incubated with tissues of the diencephalon were eight times larger than those in single culture or in combined culture with the telencephalon or liver anlage. No chronological difference in immunoreactive ACTH or PRL cells between single and combined cultures was found. Daikoku et al. (1982) suggested that differentiation of the type of adenohypophyseal cell has already been determined at the stage of Rathke's pouch and that the fetal diencephalon contains factors which promote the growth of Rathke's pouch. It is argued that the cavity of Rathke's pouch becomes the residual lumen in the hypophysis of adult animals.

A small portion of Rathke's pouch consistently persists in the wall of the pharynx of adult animals (Boyd 1956). Some cells of the remaining part of Rathke's pouch acquire the ability of hormone production and form a kind of diffuse endocrine organ. The ectopic aggregation of adenohypophyseal cells in the roof of the nasopharynx is termed the pharyngeal hypophysis.

The neoplasm craniopharyngioma originates in the remnant of Rathke's pouch in the nasopharyngeal wall, but this tumor produces no bioactive peptides and amines. The epithelium of Rathke's pouch itself is not provided with endocrine cells. It is likely that the tumor cells of craniopharyngioma, which contain no secretory granules, represent the cells of the undifferentiated adenohypophysis.

On the basis of their observation of avian embryos, Takor Takor and Pearse (1975) emphasized that Rathke's pouch is not derived from the stomodeal ectoderm but from the *neuroectoderm* of the ventral neural ridge. Thus, they proposed that adenohypophyseal cells are of neuroectodermal origin, thereby giving support to their unified concept of the neuroectodermal origin of cells of the APUD series (see also Pearse and Takor Takor 1976). This view of Takor Takor and Pearse on the neuroectodermal origin of the adenohypophysis, however, has not been widely accepted. The argument on the basis of studies using classic methods on sections of embryos is hardly convincing with regard to whether the primordial cell group of the adenohypophysis is derived from the *neuroectoderm* or stomodeal ectoderm.

The group of LeDouarin examined the ontogeny of the adenohypophysis using the quail-chick chimera technique. Interspecific grafts of mesencephalic and prosencephalic neural primordia revealed that neural crest cells involved in the formation of meninges do reach the Rathke's pouch region (Le Lièvre 1976). The cells contribute connective tissue components to the forming adenohypophysis but "have never been seen to differentiate into one of the glandular cell types" (Ayer-Lièvre and Fontaine-Perus 1982; see also Le Douarin 1982).

Concerning the *kinetics* of adenohypophyseal cells, Heidenhain (1911) assumed that such fully differentiated cells were not capable of further division. However, electron microscopists demonstrated mitotic division in various types of adenohypophyseal cells (Kurosumi and Inoue 1986). Thus, at least some adenohypophyseal cells have the potential to proliferate by *mitosis*. There is a tendency for each type of adenohypophyseal endocrine cell to increase selectively in number as well as size after removal of the specific target organs. For example, castration in the male rat induces the proliferation of GTH cells (Inoue et al. 1985).

15.7 Phylogeny

The adenohypophysis is present only in those animal species which possess a central nervous system of neural tube origin. No equivalent of this organ is known in invertebrates, which have no neural tube-derived central nervous system. The evolutionary reason for the simultaneous occurrence of the adenohypophysis and the neural tube is unknown.

In higher vertebrate classes, the adenohypophysis differentiates from Rathke's pouch. However, in some anuran amphibians, the hypophyseal primordium occurs outside the mouth as a wedge of tissue just beneath the forebrain, forming a placode (Eakin and Bush 1957). This thickened plate of ectoderm is completely homologous to Rathke's pouch in higher vertebrates. It is interesting that in placode formation the amphibian adenohypophysis resembles such sensory organs as the olfactory epithelium and acoustic and lateral line organs.

There is phylogenic evidence supporting the hypothesis that adenohypophyseal cells, like gut paraneurons (p. 182), develop from a kind of basal-granulated cell. Olsson (1973) found that the larval teleost milk fish *Chanos chanos* retains the embryonic orohypophyseal duct and the pituitary gland contains a canal continuous with the oral cavity. The canal is lined by basal-granulated cells which are prolactin-immunopositive cells (Fig. 15.2). Electron-microscopic studies by Olsson (1973) showed that the apices of the prolactin-immunopositive cells extend to the pharyngeal lumen and extend one cilium and one rodlike cytoplasmic projection. Pores in the special flattened epithelial cells covering these prolactin cells in the milk fish adenohypophysis are reminiscent of the taste pores, which connect the taste bud with the oral cavity (see p. 207). The prolactin cells possess many secretory granules in the basal cytoplasm, and structures evident of the exocytotic release of the granule contents are frequently seen at the cell base. The prolactin cells of the milk fish adenohypophysis thus resemble the open type basal-granulated cells of the gut (Fig. 15.2). One of the important functions of prolactin in lower vertebrates is the regulation of osmolarity in the body. Olsson (1973) proposes that the prolactin-containing pituitary cells may detect water osmolarity through the orohypophyseal duct and accordingly release prolactin into the circulation.

In another teleost fish *Engraulis japonica*, Olsson and Fujita (1976) demonstrated that the apical pole of the prolactin-immunopositive cells is almost completely overlaid by a special layer of flattened epithelial cells with a thick glycocalyx. This teleost fish has one large follicle in the adenohypophysis, but there is no persistent orohypophyseal duct in postmetamorphic stages. Secretory granules of the prolactin-immunoreactive cells show a tendency to gather in the basal cytoplasm of the cell. It is not known whether the hypophyseal function is controlled by osmolarity in the oral cavity in this teleost fish.

Yamamoto et al. (1982) examined both light and electron microscopically the fine structure of the follicles in the ironfish, a natural hybrid between *Carassius carassius* and *Carassius auratus*. They found that the follicles of the ironfish adenohypophysis consist of two kinds of cell—prolactin and folliculostellate cells (Fig. 15.3). The prolactin cells are basally granulated and release their granules by exocytosis. Like open-type paraneurons in the gut, they extend the apical process bearing microvilli and a cilium into the follicular lumen. Folliculostellate cells extend among the prolactin cells and face the follicular lumen. ACTH cells and other endocrine cells and glial cells are distributed around the cells forming the follicles. There are considerable numbers of nerve terminals in the parenchyma of the ironfish adenohypophysis, and some

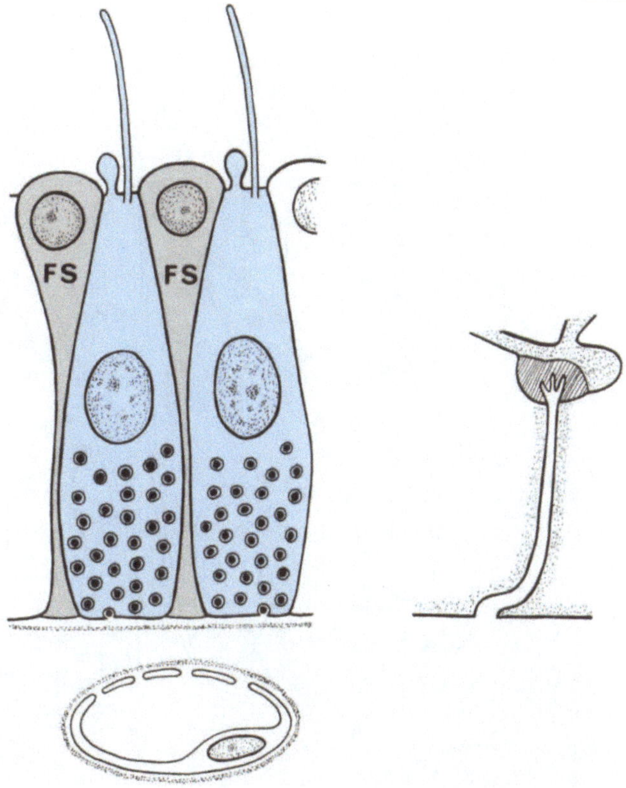

Fig. 15.2. Open-type prolactin cells (eta cells) in larval *Chanos chanos* (teleost). On the *right*, the topographic relationship of the pituitary gland is shown. A duct extends from the oral cavity to the anterior lobe, where it forms a canalicular system. Prolactin cells open to the lumen and extend a cilium and a bulbous cytoplasmic projection. It is presumed that the cells detect osmotic changes in the environment through the canalicular system and respond to them by releasing prolactin, which is the principal hormone for osmoregulation in fish. *FS* presumable folliculostellate cells. Based on Olsson (1973)

of them are in close contact with the endocrine cells.

It seems likely that there is a gradation in adenohypophyseal endocrine cells between the open-type, apparently sensory cells in the milk fish and the endocrine cells closed from the circumstances in higher animals. The duct which makes a direct connection between adenohypophyseal cells and the external environment disappears at an early phylogenic stage, leaving follicular structures as a remnant in fish and higher vertebrates (Yamamoto et al. 1982). We propose that the primitive adenohypophysis represented by that of *Chanos chanos* converts, through the intermediate forms represented by that of the *Engraulis japonica* and the ironfish, to the discrete ductless organ with clustered endocrine cells as seen in the higher vertebrate species.

15.8 Summary

The endocrine cells of the adenohypophysis contain membrane-bound secretory granules and release their contents by *exocytosis* followed by typical processes of membrane retrieval. Among the secretions of the adenohypophyseal cells proteinic components known as *pituitary hormones* have been the most extensively studied. It is stressed that at least some of them are *shared by neurons* such as those in the hypothalamus. *Amine* storage or the amine-handling ability can be recognized in only some of the cells. It is most likely that *adenine nucleotides* as well as *chromogranins* are costored in the secretory granules. Folliculostellate cells are immunoreactive for S-100 protein in some mammals, favoring the idea that pituitary paraneurons are supported by glia-like cells.

The *innervation* of the adenohypophysis greatly differs from species to species. In some teleost fish, the adenohypophysis is richly supplied with nerve terminals making close contact with the endocrine cells. In the mammalian pars distalis, nerve terminals are rare, although distinct nerve terminals are present in the pars tuberalis and pars intermedia of some mammalian species such as the cat.

Spontaneous action potentials and potential changes caused by thyrotropin-releasing hormone have been studied in tumor cells of adenohypophyseal origin.

Ontogenetically, the adenohypophysis develops from Rathke's pouch and is, therefore,

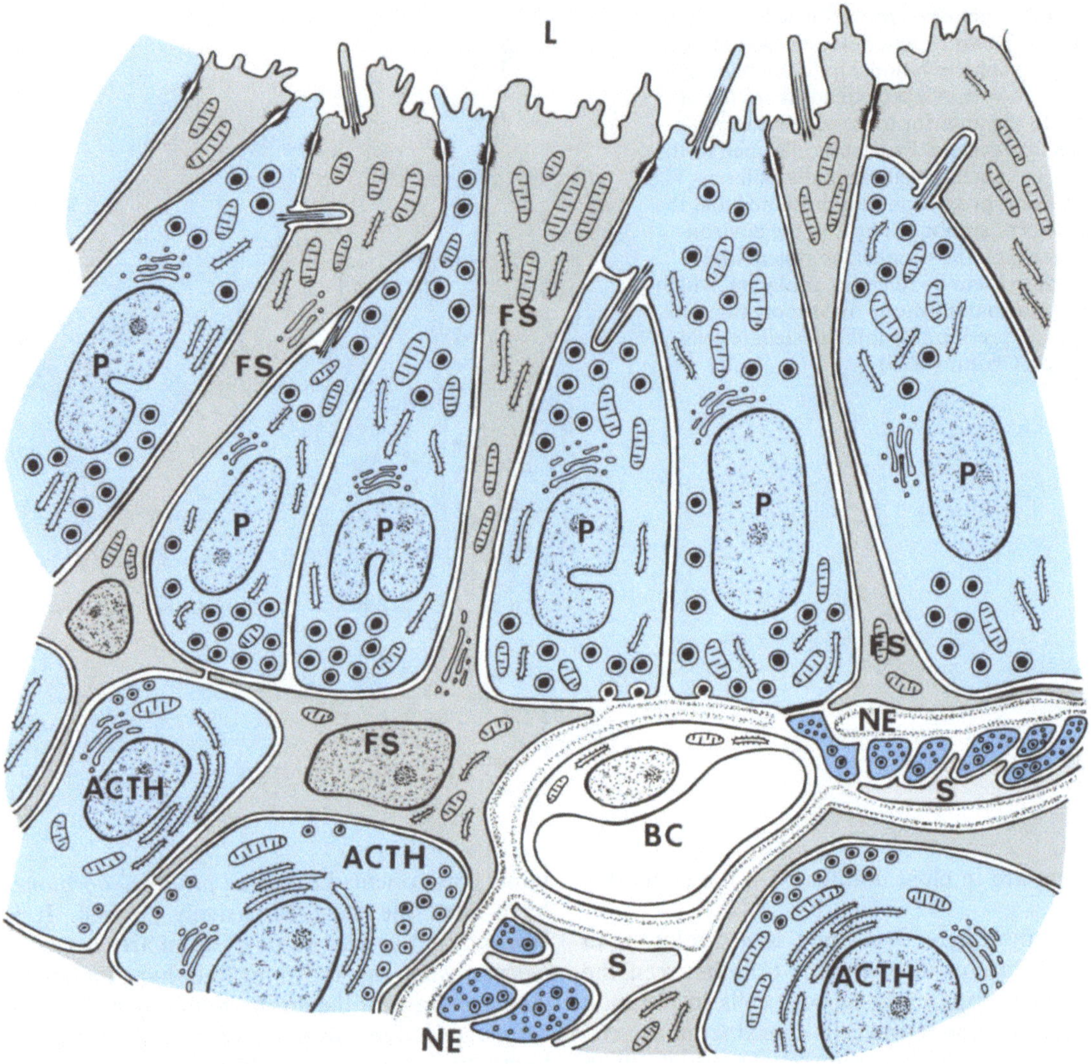

Fig. 15.3. Structure of the adenohypophysis in the iron fish, a natural hybrid between *Carassius carassius* and *Carassius auratus*. Follicles and prolactin cells (*P*) open to the lumen (*L*). Heavy innervation, including occasional synapses, characterizes this tissue. *ACTH* cells are not involved in the follicle lining. *FS* folliculostellate cell, *BC* blood capillary, *NE* nerve ending, *S* Schwann cells. Redrawn, with permission, from Yamamoto T et al. (1982) Arch Histol Jpn 45:355–364

derived from the stomodeal ectoderm. It has been proposed that Rathke's pouch originates from neuroectodermal elements, but this view has been challenged recently following experimental studies using avian interspecific grafts.

Phylogenetically, the hypophysis is present in animals with a neural tube. We regard the adenohypophysis of fish, especially that reported in the larval milk fish *Chanos chanos*, as a prototype form. The prolactin cells in this fish are open-type basal-granulated cells lining the intrahypophyseal canaliculi, which are continuous with the oral cavity through the orohypophyseal duct. It is hypothesized that these "open-type endocrine cells" secrete prolactin in response to stimuli to the oral cavity from the environmental water, e.g., changes in osmolarity. This prototype hypophysis is presumed to evolve into the ductless hypophysis of higher vertebrates through the follicular hypophysis, which is seen in some teleosts.

16. Gastroenteropancreatic Endocrine System

Endocrine cells distributed in the gastroenteric mucosa and those located in the pancreas mainly forming islets of Langerhans are collectively termed the gastroenteropancreatic (GEP) endocrine cells. (Fig. 16.1). The gastroenteric endocrine elements are mostly bipolar basal-granulated cells reaching the gut lumen which receive chemical information from the food, whereas their counterpart cells in the pancreas are mainly gathered around the blood capillaries to detect chemical information in the blood. In spite of these differences, both endocrine cell groups share many structural, functional, and metabolic features. The same cell types are even found to produce the same hormones in both groups.

The ontogeny of the gastroenteric and pancreatic endocrine cells makes it clear that it is impossible to deal with the two separately (see p. 180). Phylogenetic studies have indicated that the pancreatic islets originate from the endocrine cells in the intestinal mucosa, which leave to become associated with the pancreas at an early stage of vertebrate evolution (see pp. 182, ff).

16.1 Gastroenteric Endocrine Cells

16.1.1 Historical and General Aspects

It has been known for over a hundred years that cells containing granules in the basal cytoplasm, hence today collectively called *basal-granulated cells*, are dispersed in the gastroenteric epithelium of mammals (Heidenhain 1870; Schwalbe 1872; Nicolas 1890; Kultschitzky 1896, 1897). These cells came to be designated *enterochromaffin cells*, since their granules produced a yellow reaction deposit when treated with a fixative containing dichromate (Schmidt 1905; Ciaccio 1906, 1907). This reaction was ascribed by Erspamer and Asero (1952) to the occurrence of serotonin (5 HT) in the granules. It thus came to be generally believed that the enterochromaffin cells might produce the monoamine and store it in the cytoplasmic granules.

Some early light microscopists noticed the occurrence of one or a few other types of basal-granulated cells in addition to the enterochromaffin cells (Kull 1912; Tehver 1930), but most researchers were of the opinion that these represented a different functional phase or immature stage of the enterochromaffin cells (Hamperl 1925; Kull 1925; Ratzenhofer and Leb 1965). Thus, the view that all basal-granulated cells in the gut are serotonin-storing cells or represent their nonfunctional states came to prevail.

A few investigators proposed the hypothesis that the gut basal-granulated cells are the sources of gut hormones, especially of secretin (Parat 1924; Wermel and Kacharova 1948), but this view was paid little attention as the cells had been connected only with serotonin by most scientists.

Electron-microscopic studies in the 1960s gradually provided convincing evidence that the gut basal-granulated cells are *heterogenous in cell type* (Ito and Winchester 1963; Toner 1964; Orci et al. 1968; Forssmann et al. 1969; Forssmann 1970). As in the cells of the adenohypophysis and pancreatic islets, the basal-granulated cells could be clearly classified into different types on the basis of the fine-structural features of the basal granules, and no gradations could be found between them.

In 1965, the Italian pathologists Solcia and Sampietro (1965b) published a light-microscopic study on the basal-granulated cells in the pyloric antrum and proposed that gastrin, which was known to be concentrated in this portion of the gut, might be contained in a specific cell type, which they called *G cells*. This view was strengthened by later electron microscopic observations in which Solcia and his associates (1967) demonstrated the characteristic fine structure of the granules of the G cells. A few years later, immunohistochemical studies showed that G cells in the pyloric antrum contained gastrinlike immunoreactivity (McGuigan 1968; Bussolati and Pearse 1970). Since then, over a dozen cell types

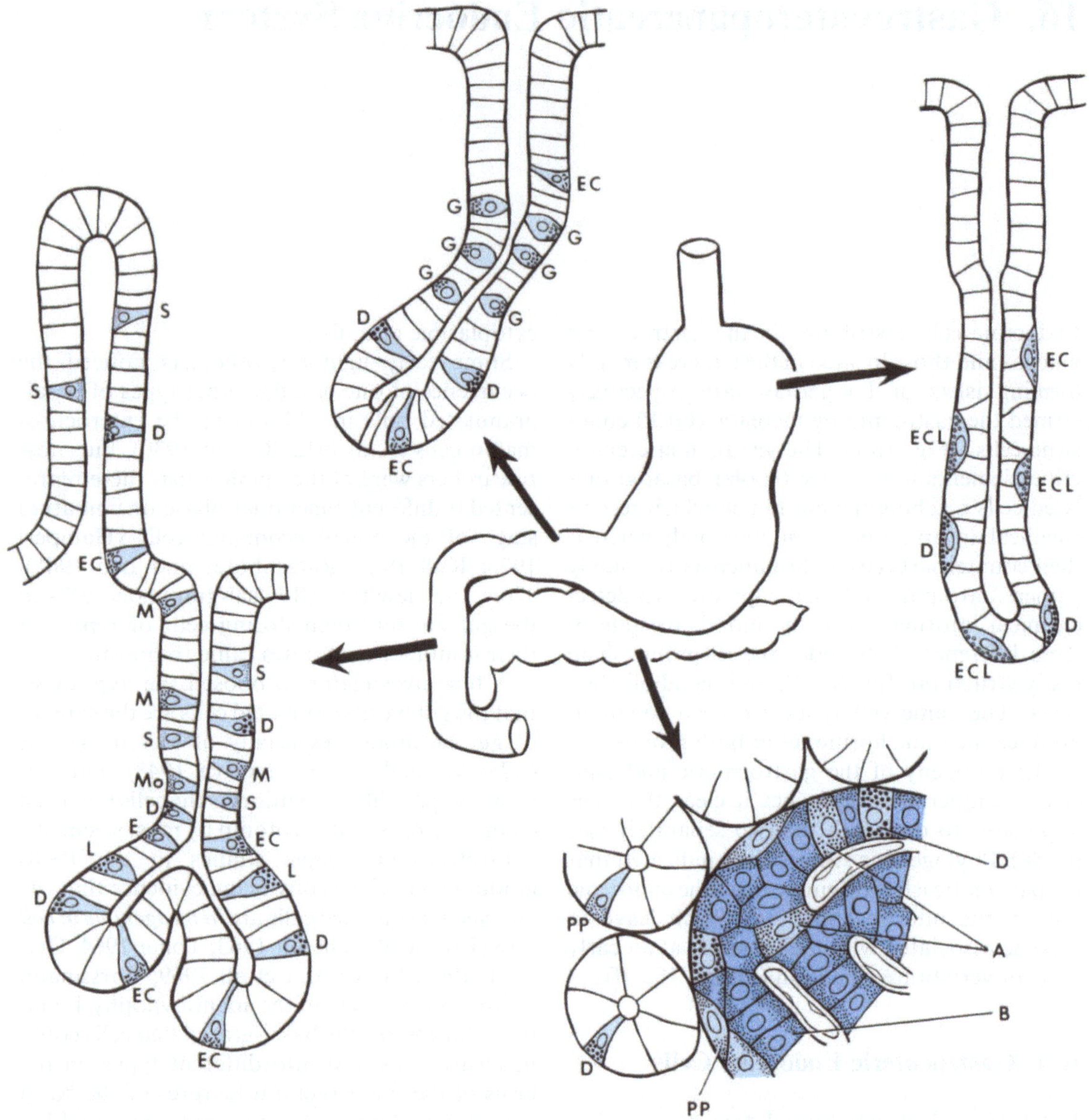

Fig. 16.1. Distribution of GEP endocrine cells. Note that the cells are open to the gut to detect chemical information of food, except in the fundic gland region of the stomach. The pancreatic islet cells are separated from the gut wall and face the bloodstream to detect chemical information

have come to be identified under the electron microscope and there was a lot of competition to correlate each of these cell types with known and partly also unknown peptide hormones. The immunohistochemical detection of peptides at the light- and electron-microscopic levels promoted this kind of investigation (reviews: Grube and Forssmann 1979; Solcia et al. 1978, 1981).

Through this trend in the studies it has become the general view that the basal-granulated cells in the gut are really endocrine in nature, secreting peptides and amines in combination. Some non-chromaffin types of cells were demonstrated by Pearse and his associates (Carvalheira et al. 1968) to contain dopamine after the animal was given an injection of L-dopa (see also Aures et al. 1970; Grube 1976a, b). That the basal granules are actually released, by exocytosis, at the base of the cell was evidenced by our electron-microscopic studies (Fujita and Kobayashi 1971; Kobayashi and Fujita 1973).

16.1.2 Cell Types and Their Hormones

Table 16.1 shows the generally accepted cell types of gut endocrine cells, fine structure of their granules, their distribution in portions of the gut, and the hormones contained in them. As useful reviews are available dealing with the cell types (Sasagawa et al. 1973; Grube and Forssmann 1979; Solcia et al. 1978, 1981), we will comment only on those associated with particular problems.

EC cell (*enterochromaffin cell*). This cell is defined by its possession of serotonin, since this amine causes the chromaffin reaction on which the name of the cell is based. The basal granules are electron dense and distorted in shape; their size varies conspicuously among different populations of the cells even in the same individual (Böck and Gorgas 1976). For this and other reasons, the occurrence of several subtypes of EC cells has been proposed (Solcia et al. 1978).

We have suggested that a bioactive peptide is contained in the basal granules of EC cells together with the amine (Fujita and Kobayashi 1973). An immunocytochemical study by Pearse et al. (1974b) reported motilin to be the EC cell peptide, but this was later disputed by several authors (Forssmann et al. 1976; Kobayashi et al. 1980b). Substance P was thereafter proposed to be the peptide (Nilsson et al. 1975; Pearse et al. 1975), but this is also controversial. Alumets et al. (1978) and our research group (Nihei et al. 1984) demonstrated enkephalinlike immunoreactivity in EC cells of the pig duodenum and pyloric antrum. Enkephalin-like immunoreactivity was shown in EC cells in the small intestine of human fetuses (Iwanaga and Fujita 1987, unpublished data). It is suggested that the negative results in EC cells in many species and portions of the intestine may not necessarily indicate the absence of enkephalins but may possibly be ascribed to the peptide being masked within the granules.

EC-like cell (*ECL cell*). This cell is characterized by vacuolar granules with an electron-dense core and distributed only in the gastric body. In the mouse and rat, this cell is known to contain histamine (Håkanson et al. 1970), but little information is available concerning a peptide secreted from this cell. Håkanson et al. (1969) proposed that this cell could be the storage site of the intrinsic factor, but this view has not been accepted widely.

D_1 cell. Cells containing round, gray-looking granules which are smaller (150–200 nm) than those of D cells (350 nm) and show no space between the granule sac and contents have been called D_1 cells. Gastric inhibitory peptide (GIP) was localized in the D_1 cells by Polak et al. (1973), but recently this category has included different types of cells corresponding to different peptides. Motilin-containing cells, today called Mo cells, were previously regarded as D_1 cells.

S and M (or I) cells. Cells characterized by small round granules (100–150 nm) have been termed S (small) cells, whereas cells with granules larger than the granules of the S cell but smaller than those of L (large) cells have been designated M (medium size) or I (intermediate size) cells. Immunocytochemical studies, mainly in the dog, by the research group of Solcia reported that the S cell was the source of secretin, while the M cell was that of CCK (Bussolati et al. 1971). Recent investigations of the same group, however, have indicated that at least in the dog secretin is not produced by the S cell but by a cell which possesses larger granules and has previously been described as the "K cell" (Usellini et al. 1984), while CCK is produced by a cell which belongs to the category of the S cell (Usellini et al. 1985). As this confusion shows, the correspondence between cell types and hormones in the gut mucosa determined in the 1970s must be reinvestigated.

16.1.3 Receptive Aspects of the Cell

The endocrine cells in the gut are dispersed in the epithelium of the gastroenteric mucosa. In the regions from the pyloric antrum down to the rectum, the cells lie on the basement membrane and extend a cytoplasmic process to the mucosal surface. This type of basal-granulated cell is called the "*open type*" (Fujita and Kobayashi 1971, 1973; Fujita et al. 1971). The other type of cell is the "*closed type*," which is seen in the gastric body (oxyntic area) in man and many mammals (Kobayashi et al. 1971). The cell lies flat on the basement membrane and does not reach the mucosal surface (Fig. 16.2).

It has long been known that gut hormones are released by chemical stimuli in the gut lumen. For instance, secretin is released when 0.1 *N* hydrochloric acid is introduced into the intestine, and gastrin is released when meat extract or ethanol is introduced into the pyloric antrum. Since we recognized that the endocrine cells were regularly open to the lumen in the pyloric antrum and the entire intestine and we found a peculiar tuft of microvilli at the apical end of these cells we hypothesized that the cells recognize stimuli in the food with the cell apex and respond by releasing their secretions (Fujita and Kobayashi 1971, 1973) (Fig. 16.3). Only this hypothesis seems to account for the diverse gastroenterological phenomena.

According to the previous view on the recognition of the luminal stimuli, represented by the hypothesis of Grossman (1963), a *nerve terminal* reaching the mucosal surface recognizes the luminal stimuli. This nerve was thought to transmit the stimuli to an endocrine cell over some synaptic relays belonging to the vagal system (Fig. 16.3).

Table 16.1. Main GEP endocrine cells, their distribution and hormonal products

	Distribution	Granules (diameter in nm)	Peptides	Amines
EC	Stomach, intestine	200–400 according to cell subtype; distorted, very dark	Enkephalins[a]	Serotonin
ECL	Gastric fundus	400; vacuolated with an eccentric core	?	Histamine (mouse and rat)
G	Pyloric antrum *duodenum*	300; round sac with flocculent contents	Gastrin	Dopamine (small amounts)
D	Stomach, small intestine, pancreas	300–450; round, grey	Somatostatin	
D_1	Stomach, intestine	150–200; round, grey	GIP, etc.?	
S	Small intestine	100–150; small, round	Secretin[b] CCK (dog)[c]	
M(I)	Small intestine	150–250; dark, round	CCK[b]	
K	Small intestine	350; dark, slightly distorted	GIP? Secretin (dog)[c]	
L	Intestine	300–400; dark, round	Glicentin	
Mo	Duodenum	?	Motilin[d]	
N	Small intestine	300; dark, round	Neurotensin[e]	
A	Pancreas, Stomach (dog), *small intestine*	200–400; round with dark core and clear halo	Glucagon	
B	Pancreas, *fetal intestine*	200–400; round sac containing an angular core	Insulin	
PP	Pancreas, *stomach, small intestine*	150–170; round grey	PP	

Data are from human studies if not otherwise specified in *parentheses*. *Italics* indicate an infrequent occurrence
[a] Alumets et al. (1978), Nihei et al. (1984); there are several subtypes in this cell
[b] Bussolati et al. (1971)
[c] Usellini et al. (1985)
[d] Solcia et al. (1981)
[e] Polak et al. (1977)

Throughout our morphological study, however, we were unable to find a nerve terminal in or on the gut epithelium. Neither could we find a nerve synapse to an endocrine cell in the gut, although we recognized more or less dense networks of nerve fibers in the lamina propria. We furthermore made some *pharmacological experiments* to examine the possible intervention of nervous elements between the luminal stimuli and hormone release. Using anesthetized dogs, the amount of cholecystokinin (CCK) release from the duodenum stimulated by luminal administration of amino acids (mixed solution of tryptophan and phenylalanine) was measured in terms of the rate of protein output in the pancreatic juice. *Atropine* (1 mg/kg) administered intravenously did not affect the CCK release by stimulation of the amino acids. Also, an intravenous injection of *tetrodotoxin* (5 µg/kg) did not inhibit but enhanced the reaction for

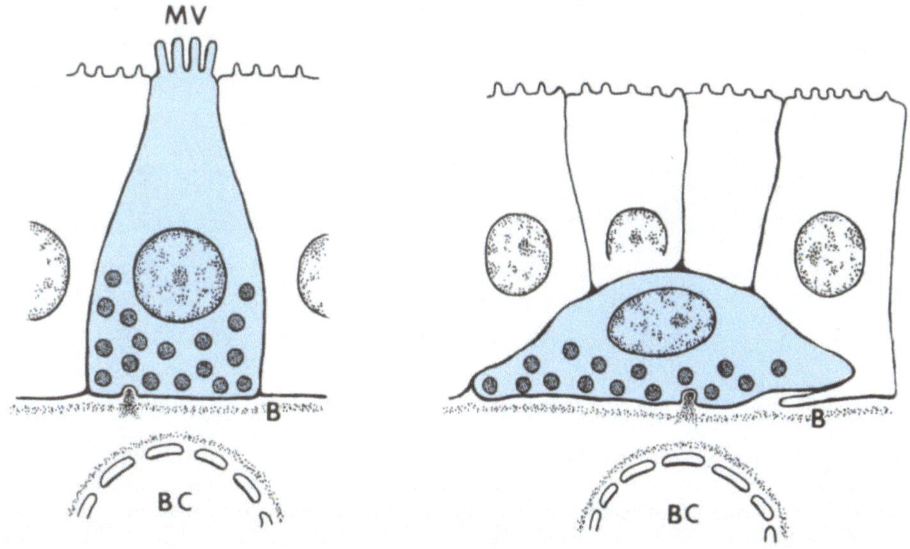

Fig. 16.2. Open- and closed-type endocrine cells in the gut epithelium. *MV* microvilli which represent the chemoreceptive site of the cell, *B* basement membrane, *BC* blood capillary

unknown reasons. These experiments support our view that it is not nervous elements but the endocrine cell itself that recognizes and conducts the luminal stimulus (Fujita et al. 1980).

It has long been known that *local anesthetics* administered intraluminally suppress the hormone release in response to adequate luminal stimuli (Schapiro and Woodward 1965; Slayback et al. 1967; Hong and Magee 1970; Fujita et al. 1978). This finding provided important evidence for luminal chemoreception by nerve terminals (Slayback et al. 1967; Harper 1972; Shaw and Heath 1973). According to our interpretation, however, this simply indicates that the open type endocrine cells of the gut are sensitive to local anesthetics (Fujita et al. 1978, 1979).

The open-type endocrine cells can be called the "*taste cells in the gut*" (Fujita and Kobayashi 1973, 1977). They differ from the taste cells on the tongue in that they are not connected with nerves and therefore could not transmit their excitation to the brain.

It is interesting that paraneurons both in the gut and on the tongue are easily affected by local anaesthetics, while both are resistant to tetrodotoxin.

Studies in isolated perfused rat duodenum. Stimulus-secretion coupling in gut endocrine cells has been examined in our system of the isolated perfused duodenum with attached pancreas in the rat (Kanno et al. 1980). After infusion of a synthetic trypsin inhibitor, FOY 305 (Ono Pharmaceutical Co., Osaka), into the duodenal lumen, the vascular perfusate was collected from the portal vein for bioassay using isolated rat pancreas. Results showed that the trypsin inhibitor caused CCK release in the duodenum and that the release was nullified when the blood vessels were perfused with the Ca^{2+}-deficient solution, although a considerable amount of $CaCl_2$ was contained in the duodenal lumen. This indicates that the *Ca^{2+} influx* in the stimulus-secretion coupling occurs at the cell base, i.e., at the *secretory pole* of the cell, and not at the apex or the receptive pole. Experiments using this isolated perfused system also showed that the CCK release was *not* inhibited by intraduodenal and intravenous application of *tetrodotoxin*, suggesting that the tetrodotoxin-sensitive Na^+ spike may not be involved in the coupling. The intraduodenal infusion of a high-K^+ solution caused dose-dependent CCK release without any other stimulus in the duodenum. This suggests that a voltage-dependent Ca^{2+} influx may be involved in the coupling.

Receptor molecule. On the basis of our experiments on the dog and rat, we proposed that trypsin or chymotrypsin molecules are involved in the reception of certain amino acids in the CCK-producing cell (Fujita and Yanatori 1976). A detailed and revised account of this hypothesis is given in Fujita and Kobayashi (1977).

Significance of the closed type. The closed-type cells in the gastric body have been postulated to

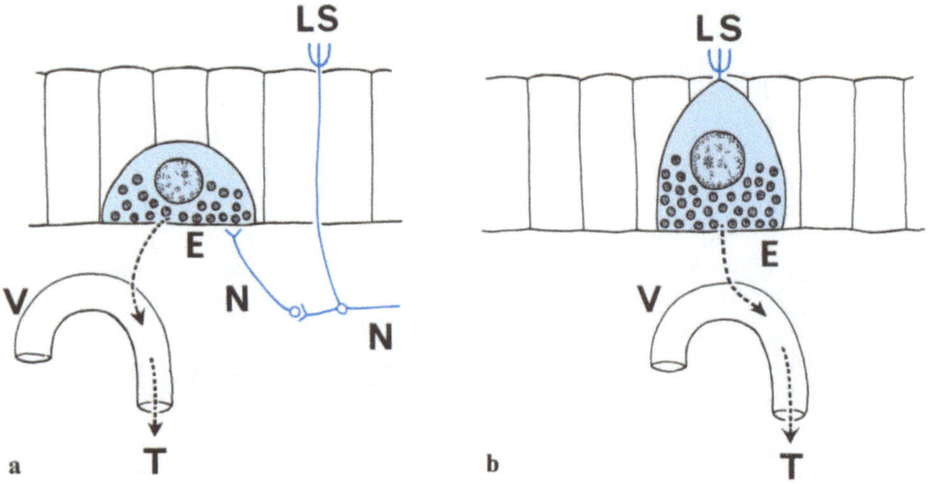

Fig. 16.3a, b. Concepts of reception and transmission of luminal information to the endocrine cell in the gut. **a** The view which prevailed until recently. The luminal stimuli (*LS*) are received and transmitted by nerves (*N*) to the endocrine cell (*E*), which releases its secretions and regulates a target (*T*) via blood vessels (*V*). **b** Our opinion, which neglects the intervention of nerves in the reaction to the luminal stimuli. The endocrine cell, with its apical microvilli, can directly respond to the luminal information

respond to *mechanical* stimuli to the mucous membrane (Fujita and Kobayashi 1973). In the proventriculus of the chicken and Japanese quail, endocrinelike cells extend long dendritic processes; the presence of those structures favors the hypothesis that the cells are sensitive to stress and strain in the gastric wall (Iwanaga, unpublished data). The peptidic and aminic secretions contained in these dendritic cells have not yet been determined.

16.1.4 Cell Structure and Function

Gut endocrine cells appear more or less clear in routine light-microscopic staining, hence they were called "helle Zellen" (clear cells) by Feyrter (1952). The basal granules of some cell types can be stained by acidic aniline dyes, e.g., azocarmine and acid fuchsin. Heidenhain's hematoxylin and, more effectively, *lead hematoxylin* (Solcia et al. 1969) can be used to stain the granules of every cell type. Basic aniline dyes are effective for staining the granules. Solcia's *metachromatic reaction* after hydration of the sections in warmed HCl (Solcia et al. 1968) specifically demonstrate the basal granules of every cell type (see p. 26).

Silver impregnation techniques, especially those of Grimelius (1968a, b) and Hellman and Hellerström (1960) are useful for staining certain cell types. Both the granules and the entire cytoplasm are blackened in these methods. Moreover, silver impregnation methods may demonstrate well-developed argyrophil fibrils in the cytoplasm, extending into the apical cell process. In appearance, they resemble neurofibrils in neurons.

In contrast to these argyrophil reactions, an *argentaffin reaction* is revealed by the basal granules of EC cells. This reaction is known to be based on the silver-reducing ability of serotonin in the cells.

Fine structure. The gut endocrine cells are provided with cytological organelles characteristic of peptide hormone-producing cells. The granular endoplasmic reticulum is more or less marked, sometimes forming a stack of lamellar cisternae. The Golgi apparatus is well developed near the nucleus and either occupies a large area or is separated into a few smaller areas. Lamellar, vacuolar, and vesicular components constitute the apparatus; some vesicles are coated with bristles (coated vesicles). Immature secretory granules are found in and near the Golgi apparatus. Their sacs are often coated with bristles. (Fig. 16.4)

Secretory granules, gathered in the infranuclear and partly in the paranuclear cytoplasm, are wrapped in a membrane sac which derives from the Golgi membrane. The size of the granules ranges from 50 to 500 nm in diameter. The granules in most cells are round but EC cell granules have a peculiar tortuous shape. The contents of the granules are variable in ap-

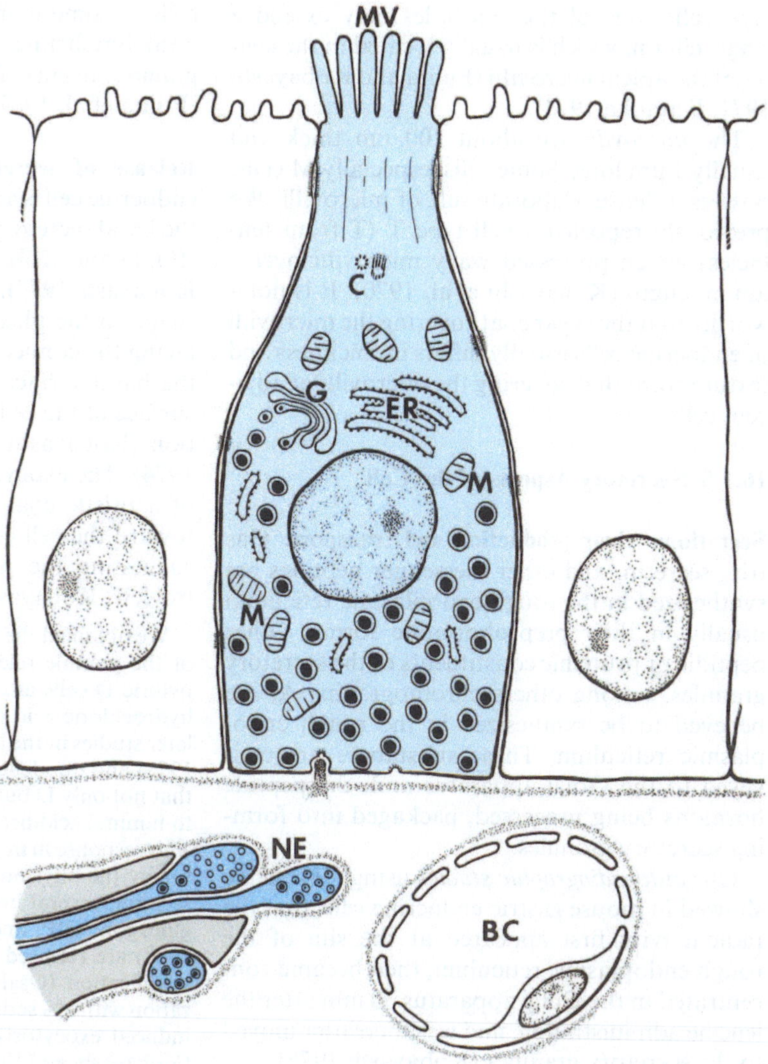

Fig. 16.4. Structure of the gut endocrine cell. *MV* microvilli, *C* centrioles, *ER* rough endoplasmic reticulum, *G* Golgi apparatus, *M* mitochondria, *NE* nerve endings, *BC* blood capillary

pearance among different cell types. Some cells are characterized by a clear halo beneath the sac. In other cells the sac is filled with homogeneous or fine-granular substances.

In some cells, such as G cells, the granules appear vesicular, containing only a cloudy substance. Such an appearance of the granules seems to be due to insufficient fixation in the routine specimen preparation procedures. Mortensen and Morris (1977) demonstrated that G cell granules maintain a dark solid appearance using acidified fixatives (pH 5).

The cytoplasm contains moderate numbers of elongate and oval mitochondria. The EC-like (ECL) cell in the gastric body is characterized by numerous, conspicuously small mitochondria (Kobayashi et al. 1971). The significance of these peculiar mitochondria is unknown, though they are reminiscent of the small, abundant mitochondria in some sensory nerve terminals.

Lysosomes of different sizes and shapes may be contained in the cytoplasm. Microtubules and microfilaments may be found to run partly in random bundles, partly surrounding the nucleus. Some of them extend into the apical process. Some cells, such as S cells, are conspicuously rich in microfilaments (Iwanaga et al. 1979). The functions, for instance contractility, of these fibrous structures in the gut endocrine cells have not been investigated. Their occurrence attracts attention as immunocytochemically neurofilament protein and neurotubulin-derived protein have been detected in small populations of gut endocrine cells (Iwanaga et al. 1982a; Fujita et al. 1983a).

Centrioles forming a diplosome are common in the supranuclear portion of the cell. In open-type cells, one of the centrioles may extend a single cilium, which is usually located in the center of the apical microvilli (Fujita and Kobayashi 1971; Kataoka 1973).

The *microvilli* are about 100 nm thick and usually 1 μm long. Some cells, especially M cells, possess a dense, elaborate tuft of microvilli. We previously reported a cell type T (T from tentacles) which possessed wavy microvilli over 2 μm in length (Kobayashi et al. 1970). It is noteworthy that the sugar coat covering the microvilli of endocrine cells usually differs in thickness and texture from that covering the microvilli of adjacent cells.

16.1.5 Secretory Aspects of the Cell

Secretions, their production and transport. Gas trin, secretin, and other messenger peptides are synthesized in the rough endoplasmic reticulum usually in their preprohormone forms. Other peptidic or proteinic constituents of the secretory granules, among others chromogranins A, are believed to be synthesized in the rough endoplasmic reticulum. These substances are conveyed to the Golgi apparatus and, the peptide hormons being processed, packaged into forming secretory granules.

Our *autoradiographic studies* using ^{3}H-leucine showed in mouse gastric endocrine cells that the radioactivity first appeared at the site of the rough endoplasmic reticulum, then became concentrated in the Golgi apparatus 30 min after the leucine administration, and was thereafter moved to the secretory granules (Kobayashi 1975).

As detailed in Chap. 3, ATP and other adenine nucleotides as well as monoamines (serotonin, noradrenaline, dopamine, and histamine) are taken up into maturing secretory granules. In our autoradiographic study using the mouse stomach, we observed that after ^{3}H-dopa administration the radioactivity was quite evenly dispersed in the cytoplasm without being concentrated in the endoplasmic reticulum or Golgi apparatus; 30 min after dopa administration, the radioactivity appeared to be concentrated in the secretory granules (Kobayashi 1975).

Recent immunocytochemical studies have indicated that *chromogranin A* and other related substances are contained in the secretory granules of mammalian gut endocrine cells (Wilson and Lloyd 1984; Nolan et al. 1985; Grube et al. 1986). Rindi et al. (1986) demonstrated different localizations of immunocytochemically detectable chromogranins A, B, and C in the gut endocrine cells in some mammals. We could demonstrate weak but distinct immunoreactivity for chromogranin A in gut endocrine cells of *Rana catesbeiana* (Fujita et al. 1987).

Release of secretory granules. When the gut endocrine cell is adequately stimulated, release of the basal secretory granules occurs by *exocytosis* (Fujita and Kobayashi 1971; review: Fujita and Kobayashi 1977). The granule sac opens in an Ω-shape to the plasma membrane at the cell base facing the connective tissue. In the EC cell, both the basal surface and lower parts of the lateral surface of the cell are involved in the degranulation (Kobayashi and Fujita 1973; Fujita et al. 1974). The exocytosis is followed by attachment of a bristle coat and separation and transport toward the cell center of the coated membrane, suggesting the process of membrane retrieval (review: Kobayashi and Sasagawa 1976).

We obtained the first electron-microscopic evidence of the granule release of gut endocrine cells in dog pyloric D cells after luminal acidification with 0.1 *N* hydrochloric acid (Fujita and Kobayashi 1971). Our later studies in the human gastric antrum (Osaka et al. 1973) and duodenal bulb (Osaka et al. 1974) showed that not only D but also D_1, L, M, and S cells reacted to luminal acidification by exocytotic granule release. This response in multiple cell types could be accounted for by the physiological evidence that in addition to secretin, several hormones such as CCK-PZ, enteroglucagon, GIP, and somatostatin (Shusdziarra et al. 1978) are released into the bloodstream by mucosal acidification (Osaka et al. 1974). Luminal alkalinization with 3% sodium bicarbonate, on the other hand, induced exocytosis of G-cell granules in the canine (Kobayashi and Fujita 1973) and human (Osaka et al. 1973) pyloric antrum. In their studies on humans, Miyagami et al. (1977) demonstrated that electron-microscopic findings of granule release induced by alkalinization with sodium bicarbonate or by administration of meat extract corresponded well with the blood gastrin levels as measured by radioimmunoassay. In the human duodenum, both phenylalanine solutions and magnesium sulfate (known to induce CCK-PZ release) were shown to cause exocytosis in a cell type resembling the D cell (Kobayashi and Sasagawa 1976).

EC cells seem to react to harmful or toxic stimuli in the intestinal lumen and to cause diarrhea by their secretions, especially with serotonin. Thus, hypertonic glucose in the canine duodenum (Kobayashi and Fujita 1973) and cholera toxin in the intestine of young rabbits (Fujita et al. 1974; Osaka et al. 1975) were demonstrated to induce severe degranulation in the EC cells.

Exocytotic granule release in gastroenteric endocrine cells apparently is not restricted to mammals, though evidence in lower animals is slight. Typical exocytotic images have been recorded in the intestinal endocrine cells in the frog (Kataoka 1973) and insects (Endo et al. 1982).

Problem of bipolar secretion. Feyrter (1952) was of the opinion that the gut endocrine cells are engaged in endocrinia and exocrinia simultaneously. This hypothesis of "Endokrinie-Exokrinie" has been neglected by us on the basis of morphological recognition that the messenger-containing granules gather at the basal pole of the cell and exocytosis occurs exclusively at this pole (Fujita and Kobayashi 1971, 1973, 1977). Recently, this principle has been challenged by Nilsson et al. (1987): In the rat EC cell, both serotonin-immunoreactive material and electron-microscopic secretory granules were concentrated toward the basal pole of the cell, but a small portion gathered at the apical pole. Although they could not find evidence of exocytosis, they proposed a "bipolar" release of serotonin. This morphological observation may possibly account for the enigmatic phenomena collectively termed "exocrine-endocrine partition" (Fujita 1983b); these include the previous findings that serotonin (for literature: see Nilsson et al. 1987) and gastrin (literature: Fujita 1983b and Grube 1986) are also released into the gut lumen when their release into the circulation occurs. It remains to be elucidated whether this luminal release of messengers can be ascribed to apical secretion in the endocrine cell concerned as Nilsson et al. propose or to "leakage" in the junctions between epithelial cells (reviews: Fujita 1983b; Grube 1986). If the former possibility is true, the apical release of messengers in bipolar paraneurons could be compared with the dendritic release of messengers in bipolar neurons (p. 56). At any rate, the apical release of messengers in gut endocrine cells, if it occurs at all, must be very small in amount compared with the basal release.

16.1.6 Fate and Functions of Secretions

The functions of such major gut hormones as gastrin, CCK, and secretin have been established as being exerted via the general circulation. Radioimmunoassay can detect the blood levels of these hormones. On the other hand, it is becoming more and more widely accepted that a considerable portion of the messenger substances released from the gut endocrine cells is conveyed to vicinal targets by *paracrinia* without entering the circulation (reviews: Fujita 1983b; Grube 1986). In fact, such common aminic and peptidic messengers as serotonin and somatostatin are barely able to circulate once through the body and thus their main mode of function would appear to be paracrine.

"Gut hormones" are thus involved in the regulation of muscular movement (e.g., elevated peristaltic movement by CCK), secretion by intestinal and associated glands (e.g., enhanced release of intestinal fluid by enteroglucagon, release of mucus by serotonin), and microcirculation (e.g., elevated capillary permeability by serotonin) in the local tissue.

The paracrine effects of the messengers of the gut endocrine cells *upon the nervous system* deserve special attention. It is most probable that the open-type endocrine cell transduces the luminal stimuli into chemical messengers, and these in turn stimulate the underlying nerve fibers which form a subepithelial terminal reticulum (Kobayashi et al. 1986). Physiological data supporting this function are available.

Mei (1978) perfused the rat duodenum with glucose and found an electric response in the afferent nerve fibers in the vagal nerve. Niijima (1980, personal communication) obtained a similar electrophysiological response in nerve fibers from the rat duodenum perfused with a glucose solution.

In electron-microscopic observations of human and canine intestines, we noticed that the serotonin-secreting EC cells were located more closely than other cells to the nervous network. We thus suggested that serotonin and possibly other secretions of EC cells may exert some effects upon nerves (Fujita and Kobayashi 1978). In an effort to test this notion, Niijima (1980) examined the effect of serotonin in the rat intestine upon the afferent nerves. An intraduodenal infusion of serotonin caused a prompt increase in the discharge rate in the nerve, whereas an intra-arterial infusion of serotonin caused a 10- to 15-min delayed response in the nerve. Niijima (1980) interpreted this result as substantiating the above-mentioned notion. Of great interest in this context is the report by Ádám-Vizi and Vizi (1978) who, using a strip preparation of guinea pig ileum demonstrated that serotonin caused acetylcholine release from the intramural neurons, leading to contraction of the muscles.

In summary, the gut endocrine cell or paraneuron can be regarded as the "*primary sensory neuron*" for the detection of food chemical information. It transmits the information to a "secondary neuron" in the propria mucosae by messenger substances by paracrinia, i.e., over a synapsis à distance. Inversely, the nerves exert certain efferent effects upon the endocrine cells.

This regulatory system of the gut composed of neurons and paraneurons is inherited from Hydra, the most primitive metazoan, as detailed on pp. 125, ff.; its cellular elements, therefore, may be called *archineurons* and *archiparaneurons*.

16.2 Pancreatic Islet Cells

16.2.1 General and Historical Aspects

The islet of Langerhans consists of endocrine cells of at least four or five different types supplied with blood capillaries and unmyelinated nerves. In contrast to its counterpart dispersed in the gastroenteric mucosa, the pancreatic islet has a long history of intense study because it comprises a compact cell mass which is easy to approach and, especially, because the islet is related to the pathogenesis of diabetes mellitus (reviews: Wellmann and Volk 1977a; Epple and Brinn 1986).

Early light microscopists succeeded in differentiating A and B cells in mammalian islets (Lane 1907; Bensley 1911). Bensley (1911) described a third type, the C cell (clear cell), in the guinea pig islet, which was not stained by various kinds of dye. Later, a fourth type, the D cell, was recognized by Bloom (1931) in the human islet as an element which was tinged blue in azan staining. The existence of the D cell, however, was neglected by most later researchers until it became clear that the cell is a normal and constant constituent of the endocrine pancreas in vertebrates and that the C cell of Bensley in the guinea pig is identical with the D cell in other animals (Fujita 1964b, 1968). Furthermore, the previously known argyrophil cell of the islet was identified with the D cell (Epple 1964; Fujita 1964a, b, 1968) and the D cell was demonstrated to increase in number in the pancreas of some diabetic patients (Fujita 1966). Electron-microscopic studies later indicated that D cells are also shared by the gastroenteric endocrine system.

By the middle of the 1950s, insulin and glucagon were established as hormones produced in the B and A cells, respectively (review: Wellmann and Volk 1977a). In 1975, the long-pursued hormone of the D cell in the pancreas and gut was identified as somatostatin (Arimura et al. 1975; Hökfelt et al. 1975; Orci et al. 1975). The finding that somatostatin, a growth hormone release-inhibiting peptide produced by neurons in the hypothalamus, is also produced by the D cells in the GEP endocrine system was instrumental in the extensive discoveries of the *brain-gut peptides*.

A fourth hormone of the pancreas, pancreatic polypeptide (PP), was isolated (Kimmel et al. 1968, 1975; Langslow et al. 1973) and immunocytochemically localized in an independent endocrine cell type, the PP cell (Larsson et al. 1975, 1976). This peptide, differentiated into avian (APP), porcine (PPP), and human pancreatic polypeptide (HPP), has also been identified as a typical brain-gut peptide; it is also produced in neurons and is especially abundant in the nervous system of invertebrates (see Chap. 11)

16.2.2 Cell Types and Their Hormones

Since many textbooks and reviews (Lacy and Greider 1972; Wellmann and Volk 1977b; Munger 1981) are available on the islet cell types and their features, only brief notes on the main cell types in the mammalian islet will be given here.

B cell. The B cells in mammals always constitute the major population of the islet. Their secretory granules are selectively stained by chrome-hematoxylin or aldehyde fuchsin. Electron-microscopically, the granules consist of a crystalloid core loosely contained in a round membrane sac. The core may be bar-shaped in profile as in the dog, irregularly angular as in man, or more or less rounded as in rodents. Insulin is produced in the B cells and stored in the granules. Impaired function of the cell causes diabetes. Alloxan specifically destroys this cell, causing alloxan diabetes.

A cell. The A cells in mammals represent a minor population in the islet cells. Their secretory granules are stained with such acidic, red aniline dyes as acid fuchsin, azocarmine, and phloxin. Electron-microscopically, the granules are characterized by a round core of very high electron density loosely surrounded by a membrane sac. *Glucagon* is produced in the A cells and stored in the granules.

D cell. The D cells usually constitute a still smaller population in the islet than the A cells. Their secretory granules can be stained with such blue acidic aniline dyes as light green or aniline blue. They are more clearly blackened by some silver impregnation methods (Fujita 1968). Under the electron microscope, the granules consist of a round fine-granular mass of moderate electron density tightly surrounded by a membrane sac. The D cells produce somatostatin and store it in the granules.

PP cell. Immunocytochemical studies indicate that in humans and other mammalian species the PP-like immunoreactivity is contained in a small population of endocrine cells, which are partly located in the peripheral portion of the islets and partly dispersed in the exocrine pancreas. Within the pancreas, the cells are generally condensed on the side close to the duodenum. PP cells (designation proposed by Solcia et al. 1978) correspond to the F cells in the dog and some other mammals (for literature, see Solcia et al. 1978.)

The *F cell* was first described by Munger et al. (1965) in the right lobe (so-called uncinate process) of the dog pancreas. It is characterized by secretory granules, 180–300 nm in diameter, with irregular profiles; the granule contents are of moderate electron density and close to the membrane sac. These granules are labile to aldehyde fixation and often swell into clear vesicles, thus making the identification of this cell type controversial (Munger 1981). Another problem is that the PP-immunoreactive cells in the human pancreas possess granules markedly different in appearance from those of the F cells in animals; their granules are much smaller (150–170 nm) and more regular in profile and may be difficult to differentiate from the granules of D_1 cells (for references, see Solcia et al. 1978).

16.2.3 Cell Structure and Function

Cell shape. The A and B cells are usually uni-, bi-, or occasionally multipolar cells with a columar or polyhedral shape. The "pole" here signifies the secretory pole represented by a cytoplasmic area, often elongated, facing a blood capillary and containing an accumulation of secretory granules (Fig. 16.5).

The D cells conspicuously differ in configuration from the A and B cells. They are elongated, extending one, two, or more cytoplasmic processes. Grube and Bohn (1983) examined the three-dimensional extensions of somatostatin-immunoreactive (D) cells in human islets using serial sections and revealed that most of the cytoplasmic processes with accumulations of somatostatin were destined to reach the blood capillaries, suggesting microhemocrine regulation of other islet cells by the D cells. Grube and Bohn also demonstrated that D cells tend to be contiguous with each other.

The tendency of D cells to extend long cytoplasmic processes becomes manifest when the cells are isolated and cultured. Matsuba et al. (1982) cultured human pancreatic islet cell clones and showed that D cells extended dendritelike processes, resembling neurons in appearance, whereas B cells remained rounded and A cells adopted a spindlelike or stellate shape.

It has been claimed that *GABA* is contained in B cells (Taniguchi et al. 1979). A recent immunocytochemical study in the rat pancreas suggests that GABA might be secreted with insulin and act as a paracrine modulator of islet cells of other types (Sakaue et al. 1987). Although more precise localization and the release mechanism of GABA in the B cell remain to be elucidated, the occurrence of this neurotransmitter substance in the pancreatic B cell seems to provide us with an additional neuronlike feature of this cell.

Secretory function and secretions. Insulin, glucagon, and other messenger peptides in the islet are produced in the rough endoplasmic reticulum-Golgi system in each endocrine cell. An account of the preprohormones and their processing in this system will not be repeated here (see pp. 29, ff).

The islet paraneuron also produces acidic carrier proteins, among others *chromogranin A* and similar substances, which have been localized in the islet cell granules by immunocytochemical techniques (O'Conner et al. 1983; Nolan et al. 1985; Ehrhart et al. 1986). Grube et al. (1986) examined islet cells in ten mammalian species including man and concluded that "basically all pancreatic endocrine cell types are chromogranin A-immunoreactive; however, every species has its own pattern of the immunoreactive cell types." This heterogeneity in the immunoreactivity due to cell types and animal species was accounted for by Grube et al. (1986) either by the heterogeneity of the antigen molecules or by the possibility of the antigens being "masked" by combination with aminic, peptidic, or other substances.

While they mature in the cytoplasm, the secretory granules take up ATP and other adenine nucleotides and amines, as is the case with other paraneurons. Dopamine is a common amine in mammalian islet cells; it is increased in the cells after L-dopa is administered to the animal (Cegrell 1968). There are also a few EC cells that contain serotonin (Cegrell 1968; Inage 1973).

When the islet cell is stimulated, the granule substances are released by exocytosis (see p. 49).

The secretions released from the islet cell enter the blood through the fenestrated capillary endothelium located close to the secretory pole of the cell. The targets of these secretions are not only remote organs and cells but also other endocrine cells in the same islet and exocrine cells in the surrounding exocrine tissue.

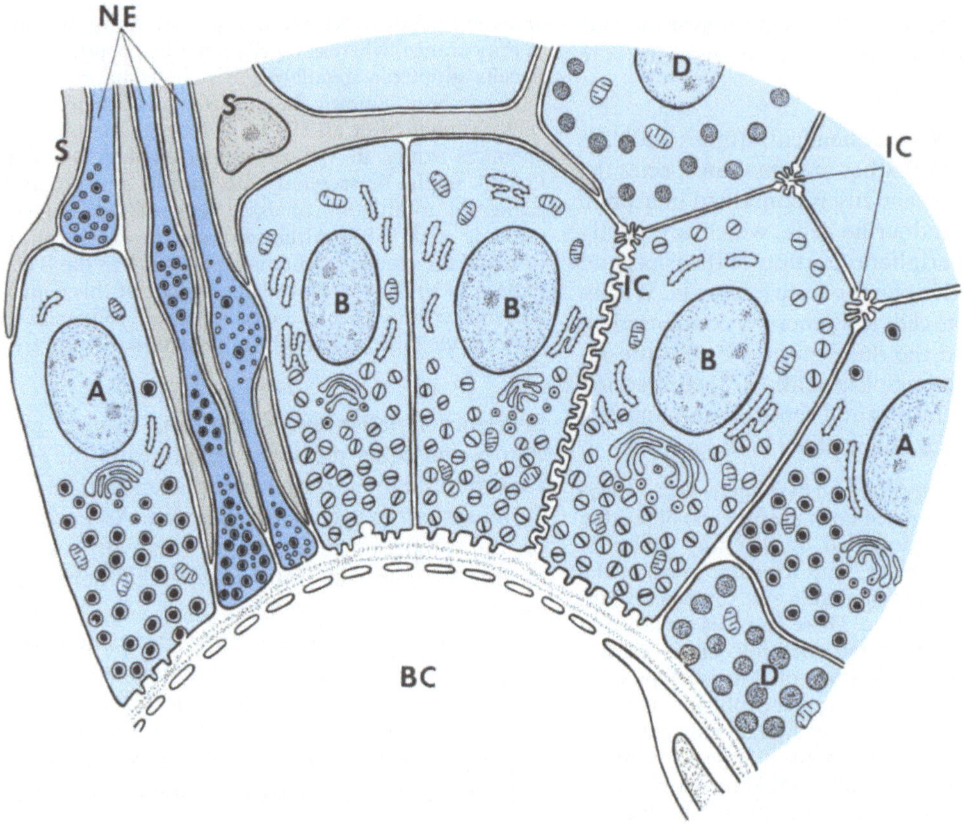

Fig. 16.5. Cellular construction of the pancreatic islet in the dog. The islet in the dog and some other species receives numerous nerve endings (*NE*), which partly contact endocrine cells (*A*, *B*, and *D* cells) but partly terminate close to the blood capillaries (*BC*). Schwann cells not only invest the nerves but extend their attenuated processes between the endocrine cells. Intercellular canaliculi (*IC*) are developed in this species. The endocrine cells project irregular microvilli, which possibly are receptive sites of the cells, into the canaliculi and pericapillary space

It is an accepted hypothesis that glucagon released from the A cells stimulates insulin release from the B cells, while somatostatin from the D cells inhibits it. These actions are presumed to be exerted by microhemocrine transport of the hormones, i.e., through the local capillaries. It has been demonstrated that the islet cells and blood capillaries are arranged in such a way that the blood perfuses first the A and D cells and then the B cells (Fujita 1973; Fujita and Murakami 1973; Fujita et al. 1976). In addition, the hormones of the A and D cells may be given to the B cells by paracrine (diffusion) and synaptocrine (directly adjacent) mechanisms.

The effects of islet hormones upon the secretory activities of the exocrine pancreas have been investigated (Fujita and Watanabe 1973). In particular, we demonstrated the effects of insulin potentiating the activities of acinar cells (Kanno and Saito 1976). The blood capillaries in the islet drain to the capillary bed in the surrounding exocrine tissue via radiating efferent vessels, which are called the *insuloacinar portal vessels*. By this small portal system, the secretions of the islet cells are directly conveyed to the exocrine cells in very high concentrations to exert their actions (Fujita 1973; Fujita and Murakami 1973; Fujita et al. 1976).

Receptive function. The B cell releases insulin following stimulation by glucose (and some other sugars) and certain amino acids. Conversely, glucagon release is suppressed by these stimuli. The islet cells are adjacent to blood capillaries with fenestrated endothelium and respond to glucose and other chemicals in the blood. Our scanning and transmission electron-microscopic observations revealed irregular-shaped microprojections (up to 1 μm in length) on the vascular pole of the cells (Fujita et al. 1981). We propose that *glucoreceptors* and other receptors of the islet cell are located on those microprojections. The vascular pole of the islet cell is presumably composed, in a *mosaic* fashion, of receptive and secretory areas, the receptors being located on the projecting portions and granule release occurring on the flat portions of the plasma membrane.

Of special interest in this context is the occurrence of an *intercellular canalicular system* in the islet of some animals which is considered to be a device to extend the receptive area of plasma membranes. Thus, in the dog, mouse (Fujita et al. 1981b), and Mongolian gerbil (Kataoka et al. 1982), narrow canalicules 0.5–1 μm in caliber are extended from the pericapillary space deep into the islet cell mass. Numerous microvilli of irregular shape (up to 1 μm in length) and a cilium issue from the islet cells surrounding the canalicules (Fig. 16.5). The canalicular system does not occur in human and rat islets.

In collaboration with Y Kanazawa, we observed the monolayer culture of rat islet cells (mainly B cells) by scanning electron microscopy and found oval patches densely covered by microvilli, frequently with a cilium at the center (Fujita and Kanazawa, unpublished data). It appears possible that these microvillous patches may represent the receptor sites of the islet cells and correspond to the microvillous surface of the islet cells *in situ* facing the pericapillary space and intercellular canaliculi. It may be further proposed that the microvillous patches on the cultured cells are an atavistic occurrence of the microvillous apex in the gastroenteric endocrine cells.

As mentioned above, the pancreatic A and B cells have receptors recognizing glucose. By perifusion of isolated rat islets with α- and β-anomers of D-glucose, Niki et al. (1974) and Niki and Niki (1976) demonstrated that it was the α-anomer that caused insulin release. This result indicates that the *glucoreceptor* of the B cell is stereospecific for the α-anomer of D-glucose. It further favors the view that the glucoreceptor is located on the plasma membrane (Niki et al. 1974). It is likely that the A cell has the same or similar glucoreceptor on its plasma membrane. Furthermore, the intestinal enteroglucagon-producing cell, lingual gustatory cell, and pituitary somatotroph have been demonstrated to share the α-anomer specific receptor (see pp. 42, 43). The satiety center neuron in the hypothalamus, on the other hand, has been suggested as being susceptible to the β-anomer of D-glucose. Niki and Niki (1976) and Asplund (1976) have proposed that *diabetes mellitus* (of the second type) may be a *general glucoreceptor disturbance* in these paraneurons and neurons, which are more or less simultaneously affected in their sensitivity to glucose.

16.2.4 Electrophysiology

The electric activity of the B cell is outlined in Chap. 4 (p. 45).

In the absence of glucose or in the presence of a low concentration of sugar, the membrane potential of the B cell is stable and usually between −60 and −70 mV (Henquin and Meissner 1984a, b).

Stimulation with D-glucose depolarizes the B-cell membrane and triggers typical electric activity (Henquin and Meissner 1984a, b; Matthews 1985), the characteristics of which are illustrated in Fig. 4.2 (Henquin and Meissner 1984a). Raising the glucose level from 3 to 15 mmol/l is followed by an *initial depolarization* of 10–15 mV, which brings the membrane potential to a threshold at which the electric activity starts. The latter exhibits a biphasic pattern. An initial long phase of continuous spike activity is followed by a *partial repolarization* without spikes and finally by the development of *slow waves*. Each slow wave is characterized by a fast depolarization from a threshold potential to a plateau potential onto which spike activity is superimposed. At the end of the burst of spikes, the membrane repolarizes to a level slightly more negative than the threshold potential. During the interval, the membrane slowly depolarizes until the threshold is reached again.

When the concentration of the sugar is elevated further, the absolute values of the membrane potential are little affected, but the slow waves with activity lengthen and the intervals shorten. Finally, when the glucose level reaches approximately 18 mmol/l, the B-cell membrane remains persistently depolarized at the plateau potential and exhibits continuous spike activity. It is thus evident that each B cell can adapt the magnitude of its response to the prevailing concentration of glucose (Henquin and Meissner 1984a, b).

The dose-dependent correlation of insulin release and electric activity strongly suggested a causal relationship between the two phenomena (Dean and Matthews 1970a, b; Matthews and Dean 1970), a conclusion substantiated by subsequent observations (Meissner and Schmeltz 1974; Meissner and Atwater 1976; Scott et al. 1981).

The resting membrane potential of the B cell is

dominated by electrodiffusion of K^+; activity of the electrogenic Na^+-pump may contribute less than 5%–10% (i.e., 2–7 mV) of the measured resting membrane potential, and it tends to hyperpolarize the membrane (Henquin and Meissner 1984a, b; Matthews 1985).

There seems to be general agreement that the depolarization between the resting and threshold potential is brought about by a *decrease in K^+ permeability* of the plasma membrane (Henquin and Meissner 1984a). It is well established that the metabolism of glucose by the B cell is necessary for the sugar to decrease the rate of K^+ (and $^{86}Rb^+$) efflux and to induce depolarization and electric activity. However, the coupling factor between metabolic events and K^+ permeability of the plasma membrane remains elusive. A number of possibilities have been envisaged (Henquin and Meissner 1984a). There is some debate concerning the ionic mechanisms underlying the slow waves of membrane potential triggered by glucose concentrations of between 7 and 18 mmol/l. The prepotential is the progressive depolarization that precedes the fast depolarization of the slow wave. Measurements of membrane resistance suggest that it is due to the gradual decrease of an outward K^+ current (Henquin and Meissner 1984a). The *voltage-dependent entry of extracellular Ca^{2+}* is believed to be primarily responsible for the appearance of spikes and the increase in $[Ca^{2+}]_i$ required for the glucose-evoked secretion of insulin; some other channels also exist for the *inward transport of Na^+* across the B-cell membrane. Tetrodotoxin-sensitive Na^+-channels in the cell membrane appear to open for both Na^+ and Ca^{2+} entry (Matthews 1985). The mechanisms which control the relative durations of the active plateau and the silent interburst phases are key processes governing stimulus-secretion coupling of glucose-induced insulin release (Matthews 1985). The repolarizing phase involves an activation of K^+-channels, but inactivation of Ca^{2+}-channels cannot be ruled out (Henquin and Meissner 1984a).

As cited by Henquin and Meissner (1984a), there is now strong evidence that, under physiological conditions at least, the electric events—and the underlying modifications of ionic permeabilities and fluxes—play a causal role in stimulus-secretion coupling in the B cell. It also seems clear that certain accompanying ionic fluxes have no direct stimulatory role, but may be important in maintaining cellular homeostasis. Recent experimental evidence has also shown that the electric activity in *B cells is not an all-or-none stereotypic response:* Not only can its intensity be adjusted to the magnitude of the stimulus, but its characteristics can also be modulated by potentiators. Our knowledge of stimulus-secretion coupling has markedly progressed over the past few years, but elucidation of several important steps remains a challenging goal.

Recent studies indicate that there is electric coupling between islet B cells. Two microelectrodes were used to measure membrane potentials simultaneously in pairs of islet cells in the mouse. Synchronized electric excitation could be detected in both the glucose-stimulated and nonstimulated cells, less than 35 µm apart from each other (Eddlestone et al. 1984). This electric coupling of excitation among vicinal cells is apparently ascribed to the morphological coupling of the cells by gap junctions. A similar mechanism of triggering excitation of cells as a mass may reasonably be explored in other gland-forming paraneurons.

16.2.5 Relation to Nerves

Innervation of islet cells. Rich innervation of pancreatic islet in mammals as well as in lower vertebrates has been detected by silver impregnation, monoamine fluorescence, and other light-microscopic methods. It has thus been demonstrated that both adrenergic and cholinergic fibers form a network in the islet, apparently ending among the endocrine cells and on the wall of capillaries, though the density and distribution of the nerves conspicuously differ among species (review: Cegrell 1968; Woods and Porte 1974). The dog, cat, hamster, bat, and mouse have densely innervated islets, while the islets of humans and some other species have a moderate or poor supply of nerves.

Electron-microscopic studies have confirmed the existence of the nerve terminations mentioned above and have further revealed the occurrence of several different fibers as indicated by the structure and size of the synaptic vesicles/granules, some of which apparently represent peptidergic types (Watari 1973; Fujita and Kobayashi 1979a; Radke and Stach 1986). In the dog and some other animals, either the A, B, or D cell receives nerve terminals of, often simultaneously, different types with adrenergic, cholinergic, or peptidergic features. Thus, there is no correspondence between the type of nerve terminal and the type of endocrine cell innervated (Esterhuizen et al. 1968; Kobayashi and Fujita 1969; review: Woods and Porte 1974). Between the closely adjacent endocrine cell and the nerve fiber, which may be swollen with synaptic vesicles and/or granules, there is a slit with a width of

30 nm. A beaded nerve fiber may, with its side, protrude into an indentation of an endocrine cell. The existence of "en distance" type terminals a few microns from an endocrine cell has also been suggested (Kobayashi and Fujita 1969).

Immunocytochemical studies have shown that many VIP-containing and some enkephalin-containing nerves supply the pancreatic islet.

Pericapillary nerves: insular neurosecretion. In the dog, mouse, mink, and snake, we have demonstrated many nerve fibers running to the pericapillary space and terminating there. Swollen axons filled either with electron-microscopically discernible synaptic vesicles/granules or with immunocytochemically detectable messenger substances (among others VIP) are exposed from their Schwann sheaths and face the base of the fenestrated endothelium over a cleft of 300–500 nm (Fig. 16.5). The cleft contains the basement membrane of the nerve and that of the endothelium—the same situation as in the cleft between the islet cells and the endothelium (Fujita and Kobayashi 1979a; Fujita et al. 1981a; see also Radke and Stach 1986). We propose that here the neuronal secretions are released into the bloodstream (*hemocrinia*) together with the secretions of the endocrine cells, which are located parallel to the nerves. The neuronal secretions, together with the islet hormones are distributed to the exocrine pancreas in high concentrations via the *insuloacinar portal vessels*. The secretin-like effect of VIP upon the exocrine pancreas is known; the possible actions of other neurosecretions in the islet, such as acetylcholine, noradrenaline, and dopamine upon the exocrine pancreas have been reviewed (Fujita and Kobayashi 1979a). With its typical *neurohemal structure*, the islet in some mammals, especially the dog and mouse, represents a *nervous control center* of the pancreas. This view accounts for the fact that in many animals the exocrine portion of the pancreas has a strikingly sparse nerve supply compared with the endocrine portion. The neurohormones are distributed by blood vessels and need not be conveyed by individual nerve fibers.

Neuroinsular complex. Pancreatic islets of various vertebrates may contain neuronal somata. Conversely, intrapancreatic ganglia may contain islet cells (Fig. 9.1). In either case, the neurons are in direct contact with islet cells over a cleft of 15–20 nm, thus forming a *neuroinsular complex of type I* (Fujita 1959; see p. 93). The islet of the mink regularly contains several neuronal somata, most of which contain VIP. The neuronal processes in the islet of this and other animals mainly terminate in the pericapillary space, suggesting neurosecretory activity besides their apparent role of regulatory endocrine cells in the vicinity (Fig. 16.5).

Islet cells may be closely associated with a bundle of nerve fibers, thus forming a *neuroinuslar complex of type II* (Fujita 1959). Islet cells may be enclosed by thick perineurium and thus completely isolated from the pancreatic parenchyma (Böck 1986).

Studies in animal fetuses suggest that both types of neuroinsular complex are formed by islet cells and nervous elements approaching and attaching to each other. The occurrence of the neuroinsular complexes is evidence that the islet cells, though not nervous in origin, have a special *affinity with neurons*. In the complex, the islet cells and neurons are similar in their attitude to each other and to Schwann cells (see p. 93) as well as in their position to the blood capillaries.

It is furthermore suggested in this context that the pancreatic islet is *a kind of paraganglion* (Serizawa et al. 1979; Donev 1984). It is made up of islet cells, the monoamine-producing paraneurons, and variable amounts of neuronal elements, which are tightly associated and cooperate closely.

Schwann cells. Stellate or crescent cells of apparently sustentacular nature are known to extend between groups of endocrine cells in the islets of some mammals (Fig. 16.5). These cells are mostly, if not exclusively, Schwann cells, since electron-microscopic observation discloses that their processes ensheath axons on the one hand and on the other cover endocrine cells (Kobayashi and Fujita 1969). The glial nature of the sustentacular cells is supported by their positive immunoreactivity for S-100 protein, the marker substance of glial cells, though this reaction can be proved only in restricted species (Iwanaga and Fujita 1982, unpublished data; Fujita et al. 1983a; Girod et al. 1987) (Fig. 9.8). In the dog (Smith 1975) and sand rat (Donev 1984), a thin discontinuous sheet of Schwann cell cytoplasm has been demonstrated to enclose the islet and separate the endocrine cell groups or even individual cells. Smith (1975) presumed that these Schwann-cell sheets provided insulation for the electrically active endocrine cells, but nothing is clear in this context. What is clear is that, at least in some mammals, islet cells may be ensheathed by glial cells as if they were genuine neurons.

16.3. Ontogeny of GEP Endocrine Cells

16.3.1 Origin of the Cells

Although it has long been vaguely believed that gastroenteric endocrine cells are of *endodermal* origin, some early research workers held other opinions. Finding cells with chromaffin granules in the lamina propria of fetal intestines, Kull (1925) and Dias-Amado (1925a, b) suggested a *mesenchymal* origin of the cells. In his study of human fetal intestines, Danisch (1924), on the other hand, proposed a *neurogenic* hypothesis of the endocrine cells. He wrote that the cells were derived from the anlagen of sympathetic ganglia and migrated into the intestinal epithelium through Meissner's plexus. The French pathologist Masson (1924) proposed that enterochromaffin cells and intramural neurons may share the same origin on the basis of his observations on neoplastic (carcinoid) cells in the appendix.

More recently, Pearse (1969) observed that gastroenteric as well as pancreatic endocrine cells have the ability of amine precursor uptake and decarboxylation (APUD) and suggested a *neural crest* origin (see p. 86). Pearse and Polak (1971) reported that the formaldehyde-induced fluorescence (FIF) reaction indicating the APUD ability could be found in neural crest cells, cells "migrating" from there, and the endocrine cells which have "settled" in the intestinal epithelium and in the pancreas. This view of the neural origin of GEP endocrine cells seemed to be supported by the findings of neuroendocrine complexes in the mammalian pancreas (Fujita 1959; Fujita 1976; see pp. 93, ff.) and avian proventriculus (Kataoka 1974).

More recent embryological experiments using the *quail and chicken chimera* (Le Douarin and Teillet 1973) and other techniques (Andrew 1974), however, have provided no evidence supporting a neural origin of GEP endocrine cells (see p. 90). The general opinion has thus returned to the classic view, i.e., an endodermal origin of GEP endocrine cells.

While the origin of endocrine cells in the gut and pancreas in vertebrates is difficult to determine, studies on the midgut endocrine cells in *insects* have provided clearer information in this context. In insects, the foregut and hindgut are ectodermal in origin; the midgut, however, is endodermal and this is the only portion to contain endocrine cells (Nishiitsutsuji-Uwo and Endo 1981; Endo 1984). Endo and Nishiitsutsuji-Uwo (1982) demonstrated by means of electron microscopy that in the midgut of the cockroach endocrine cells mitotically divide within nidi, the nests of epithelial cell regeneration, and, moving outside the nidi, differentiate into open-type bipolar cells. This finding provides evidence that in this protostomian animal the endocrine cells are derived from endodermal epithelial cells.

16.3.2 Possible Nervous Induction

Close contact between endocrine cells undergoing differentiation and neurons has been observed in the interstitial tissues in the pancreas of mammalian fetuses (Fujita 1959) and in the subepithelial connective tissue of the human fetal duodenum (Osaka 1975; Osaka and Kobayashi 1976). Some of these cell associations seem to develop into neuroendocrine complexes (p. 93), while in others it is presumed that the endocrine cells may possibly dissociate from the nervous elements and return to the epithelial region. Thus, it is possible that GEP endocrine cells receive nervous induction for their differentiation. Further precise studies on the attitude of the endocrine and nervous elements in fetal tissues are necessary to clarify this matter.

16.3.3 Cell Kinetics

The chronological and topographical data on the differentiation and proliferation of GEP endocrine cells in different species will be omitted from this text, and references only are given (review: Pictet and Rutter 1972; man—Moxey and Trier 1977; rat—Larsson 1977a; Fujii 1979; chick—Andrew 1976a, b, c).

As Fujii (1979) demonstrated in his immunocytochemical study on rat fetuses, the first differentiation of endocrine cells (glucagon cells) occurs on day 11 of gestation in the ventral and dorsal portions of the foregut epithelium, where pancreatic buds are still rudimentary (Fig. 16.6). On day 12, some endocrine cells are incorporated into the growing pancreatic buds and then into the pancreas, whereas others remain in the foregut epithelium and become either open or closed forms of gut endocrine cells. This ontogenic relationship implies a continuous and overlapping distribution of gastroenteric and pancreatic endocrine cells. In fact, Larsson (1977) demonstrated that GEP endocrine cell types in fetal and neonatal rats are distributed less site-specifically than in adult animals. Thus, gastrin cells constantly occur in the pancreatic islets, while sporadic insulin (B) cells and PP cells are found

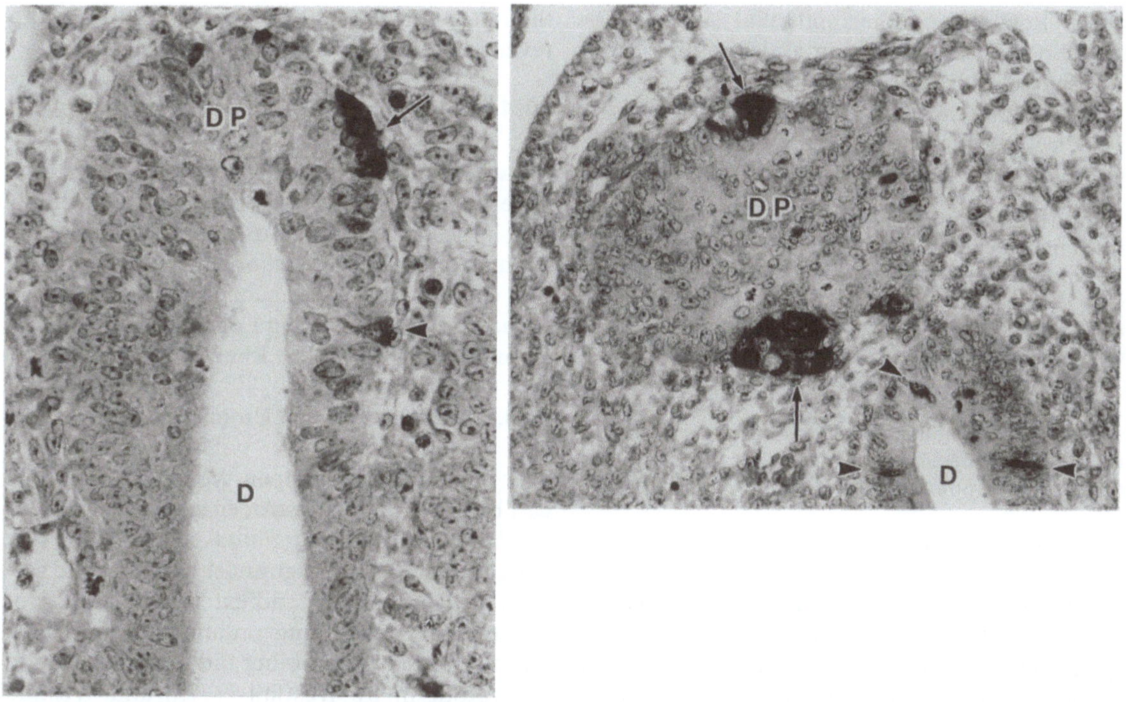

Fig. 16.6a, b. Initial stages of the differentiation of GEP endocrine cells in rat fetuses. **a** As the first endocrine cells to be recognized, glucagon-immunoreactive cells appear in the epithelium of the duodenum (*D*) on day 11 of gestation. Some of them (*arrow*) are in the primordium of the dorsal pancreas (*DP*), which is now quite an inconspicuous cell mass, whereas others (*arrowhead*) show the shape and location of gut endocrine cells. **b** On day 12 of gestation, glucagon-immunoreactive cells are increased in number. They are partly incorporated in the now growing dorsal pancreas (*DP*, *arrows*) and partly dispersed in the epithelium of the duodenum (*D*, *arrowheads*). Courtesy of S. Fujii; reproduced, with permission, from Arch Histol Jpn 42:467–479 (1979)

in the duodenum and antrum, respectively. These cells represent "transitory cell populations only occurring during a restricted fetal and/or neonatal period" (Larsson 1977).

The endocrine cells are generally believed to differentiate from "immature precursor cells" in the generative zone of the epithelium (review: Inokuchi et al. 1983). It is not settled as to whether the cells normally undergo proliferation after differentiation or are directly derived from the precursor cells.

In their autoradiographic study on mouse antral G cells, Lehy and Willems (1975, 1976) proposed that the large majority of G cells originate by mitotic proliferation of other gastrin cells. On the other hand, Fujimoto et al. (1980) reported that in the hamster the majority of G cells are derived from the precursor cells and only a small percentage are formed by proliferation. Mitotic proliferation of EC cells has been reported in the human rectum (Deschner and Lipkin 1966), though this has been refuted by more recent authors (for references, see Inokuchi et al. 1983). Odartchenko et al. (1970) pointed out that the proliferation of EC cells in the mouse duodenum occurred at such an early stage of differentiation that they were hardly recognizable as belonging to the EC-cell series. Judgment as to whether a given cell is proliferative or not may depend on the sensitivity of the staining methods employed for identification of the cell types. Here, Inokuchi et al. (1983) applied a highly sensitive immunostaining technique for serotonin and were able to recognize some EC cells showing mitotic activity.

Mitotic proliferation seems to be a common occurrence in endocrine paraneurons, at least when they are stimulated; mitotic proliferation can be induced in islet B cells by an elevated blood level of cholecystokinin (Yanatori and Fujita 1976) and in antral G cells after vagotomy (Inokuchi et al. 1983).

After differentiation and proliferation in the generative cell zone, gastroenteric endocrine cells migrate toward the site of epithelial cell degeneration, i.e., up toward the villous tip in ordinary gut regions and, in the antrum, down toward the bottom of the pyloric gland. The endocrine cells are believed by many researchers to migrate more slowly than other epithelial cells (Odartchenko et al. 1970; Chang and Leblond 1971; Tsubouchi and Leblond 1979).

Concerning the turnover of gastroenteric endocrine cells, estimations in previous autoradiographic studies are very divergent.

A turnover rate of 35–100 days has been proposed for human rectal EC cells (Deschner and Lipkin 1966), 2–4 months for mouse antral G cells (Lehy and Willems 1976), 23.3 days for mouse colon EC cells (Chang and Leblond 1971; Tsubouchi and Leblond 1979), 15 days for hamster pyloric EC cells (Hattori et al. 1977), 10–15 days for hamster antral G cells (Fujimoto et al. 1980), 4 days (Ferreira and Leblond 1971; Cheng and Leblond 1974b) and 3 days (Odartchenko et al. 1970) for mouse duodenal EC cells, and 2 days for rat duodenal S cells (Inokuchi et al. 1983).

16.3.4 Segi's Cap

Segi (1935, 1936) described a large aggregation of basal-granulated cells *on the top of every villus* in the upper small intestine of human fetuses with a gestational age of 5 months and over. The villous top is concave and contains endocrine cells, which directly face the intestinal lumen with their cell apices. This peculiar structure was rediscovered 45 years later and given the name of Segi's cap (Kobayashi et al. 1980a). Both electron-microscopic and immunocytochemical techniques have confirmed that the endocrine cells in Segi's cap are still active in the secretion of bioactive peptides and amines (Iwanaga et al. 1980; review: Yamada et al. 1983). Besides human fetuses, those of cattle, pigs, and monkeys have been shown to possess Segi's caps (Yamada et al. 1981; Kuramoto et al. 1983; review: Yamada et al. 1983; Kobayashi and Segi 1984).

The mechanism of formation of Segi's cap is not well understood as yet. It has been proposed that the endocrine cells live longer than ordinary enterocytes and, therefore, survive them after arriving at the villous top (Iwanaga et al. 1980; Kobayashi and Segi 1984). Moreover, it has been suggested that the endocrine cells have a stronger affinity to the basement membrane than the enterocytes and cannot be exfoliated from it so easily (Tsubouchi 1981). Segi's caps disappear during the perinatal period, but the exact term and fashion of their deterioration are unknown.

The function of Segi's caps is poorly understood. It has been postulated that the endocrine cells in the caps release large amounts of hormones needed for the perinatal period of life when the substances in the intestine become subject to great and sudden change, especially at the initiation of lactation (Iwanaga et al. 1980; review: Kobayashi and Segi 1984).

16.4 Phylogeny of GEP Endocrine Cells

Basal-granulated cells of a paraneuronal nature have been found dispersed in the gut epithelium of a large variety of protostomian and deuterostomian animals. The islet organ, i.e., an aggregation of these cells in the subepithelial connective tissue or pancreas, occurs only in the deuterostomian series above the level of cyclostomes.

16.4.1 Coelenterata and Protostomian Animals

The prototype of the open-type endocrine cells can be found in the gastrodermis (endoderm) of *Hydra* (Coelenterata) (Fujita et al. 1980). The cells contain secretory granules and synaptic vesicles in the basal part and extend an apical process provided with a single cilium to the epithelial surface, closely resembling the cells observed in the ectoderm by Westfall (1973) and Westfall and Kinnamon (1978). Although it is still not known for certain which of the neuropeptides in *Hydra* are contained in the gut paraneurons, the reports by Grimmelikhuijzen (1981, 1982) suggest that at least neurotensin-like and oxytocin/vasopressin-like immunoreactivities are contained in the endodermal paraneurons.

The flatworm *Microstomum lineare* has been reported by Reuter et al. (1984) to possess endocrinelike cells of both the open and closed type dispersed in the intestinal epithelium. These cells seem to contain FMRFamide-, vasotocin-, and BPP-like immunoreactivities (Reuter 1985, personal communication).

In molluscs, it is not known whether intestinal paraneurons are present. However, intraepithelial bipolar cells containing catecholamines have been identified by Brisson and Collin (1980) in the albumen gland duct, fertilization pocket, and oviduct of *Bulinus truncatus* (Gastropoda).

Among arthropods, insects have been investigated with regard to intestinal paraneurons. The midgut of the silkworm (Kobayashi 1971), cockroach, and butterfly larvae (Nishiitsutsuji-Uwo and Endo 1981; Endo and Nishiitsutsuji-Uwo 1981) have been shown by electron microscopy to contain basal-granulated cells with a conspicuous tuft of microvilli at the cell apex. Cells containing immunoreactivities for PP, glicentin, somatostatin, and urotensin I have been identified in the cockroach (Iwanaga et al. 1981, 1966). The silkworm and cricket have been

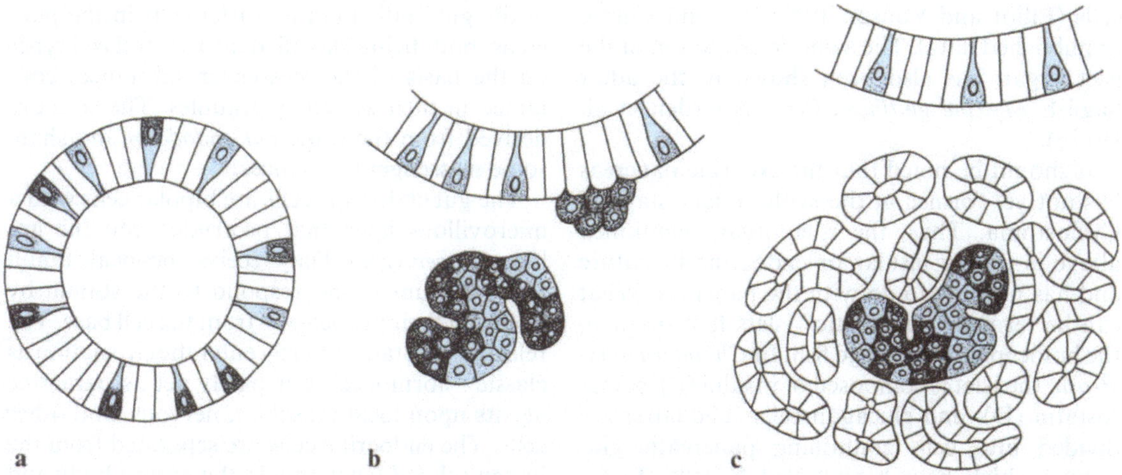

Fig. 16.7a–c. Stages of evolution of pancreatic islets. **a** Invertebrates. Endocrine cells secreting insulin and other hormones are all contained in the gut epithelium as open-type basal-granulated cells. **b** Cyclostome. In the evolutionary stage of *Ammocoetes*, insulin-producing cells migrate into subepithelial connective tissue to form a primitive islet tissue. In the adult lamprey, some somatostatin cells join the islet. **c** From holocephalian fish (*Chimaera*) up to mammals, an exocrine pancreas is associated with the endocrine islet, which contains glucagon and some other endocrine cell types. Note that the gut epithelium still contains endocrine cells of the open type to detect directly the chemical information of food

shown to contain PP-, PHI-, and urotensin I-immunoreactive gut paraneurons (Yui et al. 1985).

16.4.2 Deuterostomia

The gut epithelium of the sea-squirt (Ascidacea, Protochordata) contains numerous basal-granulated cells, mostly of the typical open type (Bevis and Thorndyke 1978, 1979; Pestarino 1982). In *Ciona intestinalis*, Fritsch et al. (1982) demonstrated gut endocrine cells containing immunoreactivities for somatostatin, gastrin, bombesin, secretin, PP, and neurotensin as well as serotonin.

In the lancelet or amphioxus (Cephalochordata), the occurrence of endocrine cells of the open type has been shown by electron microscopy in the gut epithelium of *Branchiostoma japonicum* (Kataoka and Fujita 1974). In *Branchiostoma lanceolatum*, eight cell types containing immunoreactivities for insulin, glucagon, somatostatin, PP, CCK/gastrin, VIP, secretin, and neurotensin have been demonstrated (Reinecke 1981).

It is noteworthy that insulin-containing B cells are open to the intestinal lumen in amphioxus (Van Noorden and Pearse 1976). In higher animals, the B cells are gathered outside the intestinal epithelium and form the islet organ; they can no longer be recognized in the epithelium. Many other types of GEP hormone-producing cells, however, remain in the gut epithelium up to the level of mammals (Fig. 16.7).

16.4.3 Phylogeny of the Islet Organ

The distal end of the bile duct in the hagfish and the most oral portion of the intestine in the lamprey have long been known to form a whitish swelling called the "pancreaslike" organ. This structure, however, does not contain any acinar elements but represents a lobulated mass of insulin-producing B cells (Falkmer and Matty 1966; Winbladh 1966; Östberg et al. 1975; for earlier literature, see Östberg 1976).

More recent immunocytochemical studies have revealed that the most primitive islet organ occurs in the larval form (ammocete) of the lamprey as a budding of B cells into the subepithelial connective tissue (Fig. 16.7). Thus, lobulated cell cords immunoreactive for mammalian insulin have been recognized to form the islet organ in the ammocete of *Lampetra fluviatilis* and *L. planeri* (Van Noorden et al. 1972), *Petromyzon marinus* L. (Elliott and Youson 1987), and *L. reissneri* and *L. japonica* (Yui and Fujita, unpublished data). In the adult lamprey, somatostatin-immunoreactive D cells join the insulin cells; the islet organ comprises a mixture of the B and D cells (Elliot and Youson 1987; Yui and Fujita, unpublished data). The same construction of the islet organ has also been shown in the adult hagfish *Myxine glutinosa* (Van Noorden et al. 1977b).

It should be noted that the exocrine pancreas has not yet formed at the evolutionary stage of Cyclostoma. Thus, the islet organ mentioned above is purely insular or endocrine in nature and has nothing to do with the pancreas. What can be termed "pancreatic" islets first occur in the holocephalic cartilage fish. In *Chimaera monstrosa*, the islet is composed of insulin (B), somatostatin (D), and glucagon cells. The latter are divided into cells containing pancreatic glucagon, which have been called X cells (Fujita 1962), and those containing glicentinlike peptide, which have been called A cells (Yui and Fujita 1986).

In Elasmobranchii, islet cells are mainly represented by the outer layer of the duct system of the pancreas (Thomas 1940). In Teleostei, the islet cells form one or two Brockman bodies. Each of these bodies, however, is surrounded by a layer of acinar tissue. Immunocytochemical studies have shown that both elasmobranchial and teleostean islet tissues contain the four main islet hormones—insulin, glucagon, somatostatin, and pancreatic polypeptide (PP) (Klein and Van Noorden 1980; Sekine and Yui 1981). Islet tissue producing the four main hormones and associated with the exocrine pancreas is thus established in fish and inherited by mammals via amphibians, reptiles, and birds, all of which possess typical islets in the pancreas (Review: Epple and Brinn 1986) (Fig. 16.7).

16.5 Summary

The gastroenteropancreatic (GEP) endocrine system consists of basal-granulated cells dispersed in the gut epithelium and islet cells in the pancreas, both being classified into several cell types on the basis of the messenger substances contained in their secretory granules. The cells are derived from the same *gut endoderm* and share some messenger substances.

The gut endocrine cells are bipolar cells with a microvillous apex that protrudes into the gut lumen (*open type*). They receive chemical stimuli from the lumen and respond to the stimuli by releasing their messengers from the cell base. The released substances partly enter the circulation as classic "hormones" but partly act as paracrine agents upon local muscular, nervous, and other cells. The endocrine cells are separated from the lumen (*closed type*) only in the gastric body and probably respond to mechanical stimuli.

Phylogenetic studies have revealed that the gut basal-granulated cells and their messenger substances are a feature throughout metazoan evolution and can be traced back to *Hydra*.

The pancreatic islet cells are so disposed that they do not receive direct stimulation from the gut lumen but detect the effects of nutrient absorption in the blood. The islet represents a center regulating the exocrine pancreas; endocrine cells and nerve terminals are concentrated in the islet. The peculiar relations of islet cells to nervous elements, among others the formation of *neuro-insular complexes*, reflect the paraneuronal nature of the cells.

Phylogenetically, the islet forms by the sinking of the gut endocrine cells into the connective tissue at the stage of cyclostomes and by connection with the exocrine pancreas at the stage of fish.

GEP endocrine cells secrete *peptide* messengers together with *aminic* secretions; *ATP* and *chromogranins* coexist with them in the secretory granules. Clear *action potentials* have been detected in pancreatic A and B cells stimulated by glucose.

17. Urogenital Paraneurons

17.1 General Aspects

Feyrter (1951, 1953) described the occurrence of *clear* and *argyrophil* cells dispersed in the urethra and certain other parts of the urogenital epithelia. These cells, which he regarded as typical elements of his "Helle-Zellen-System," were impregnated in the Bielschowsky-Gros method (argyrophilia) and most, if not all, showed *argentaffinity* and *chromaffinity*. Although these reactions indicated that the cells contained monoamines (p. 23), this point was emphasized neither by Feyrter himself nor by other researchers until modern fluorescence histochemical methods for monoamines came to be applied to urogenital tissues (Forsberg et al. 1964; Håkanson et al. 1974).

Since the cells now in question are the urogenital counterpart of gastroenteric paraneurons, as discussed below, they will be referred to as urogenital paraneurons.

17.2 Distribution

Feyrter (1951, 1953) published extensive and precise observations on the distribution of the "helle Zellen" (which correspond to what we call paraneurons) in *human* urogenital epithelia, and his descriptions are very useful even today.

The *male urethra*, according to Feyrter, contains paraneurons dispersed throughout its entire extent, i.e., from the interior to the exterior ostium; the cells tend to decrease distally and completely disappear in the fossa navicularis (Feyrter 1951, 1953). Later, Lendon et al. (1976) claimed that in the male urethra no paraneurons could be found distal to the orifices of the ejaculatory ducts. Quite recently, however, we were able to confirm by the immunocytochemical demonstration of serotonin that numerous paraneurons are dispersed in the cavernous (penile) part of the urethra (Iwanaga et al. 1987). The lacunae and urethral glands contain rather more paraneurons than the flat portions of the urethral epithelium (Feyrter 1951, 1953). The *prostate* also is richly distributed with paraneurons, which occur both in the acini and ducts and in the prostatic urethra (Pretl 1944; Feyrter 1953; Kazzaz 1974; di Sant'Agnese and de Mesy Jensen 1984b). Similarly, the bulbourethral gland of Cowper contains some paraneurons (Feyrter 1951, 1953).

The human *female urethra* has paraneurons distributed along its entire length. *Bartholin's gland* was declared by Feyrter (1951, 1953) as lacking paraneurons, but Fetissof et al. (1985) demonstrated that Bartholin's and paraurethral glands constantly contain paraneurons; these authors reported that the cells which were argentaffin and serotonin-immunoreactive markedly increased in number during inflammation of Bartholin's glands.

The *urinary bladder*, both in males and females, has been described by Feyrter (1951, 1953) as containing some argyrophil cells, which increased markedly in chronic cystitis. The ureter, pelvis, and kidney lack paraneuronal elements (Feyrter 1951, 1953).

The above-described sites of the distribution of paraneurons correspond to the derivatives of the endodermal cloaca (sinus urogenitalis) (Feyrter 1951; Lendon et al. 1976). The mesodermal organs (derivatives of the Müllerian and Wolffian ducts), such as the vagina and uterus in females as well as the seminal tract in males, were held to be devoid of paraneuronal elements (Feyrter 1951, 1953). This concept, however, has encountered contradictory findings, as discussed on p. 189).

Putting this theoretical problem aside, we will give an overview on the findings in *laboratory animals*. Owman et al. (1971) demonstrated flask-shaped serotonin-fluorescent cells dispersed in the urethral epithelium of female cats. Dixon et al. (1973) made an extensive study in rats, rabbits, and guinea pigs. They recorded the occurrence of serotonin-fluorescent cells in all regions of the male and female urethra but not in the bladder (Dixon et al. 1973). Håkanson et al. (1974) noted that in the urethra of the guinea pig the serotonin-fluorescent cells were most numerous in the middle portion; in males, the distal two-thirds of the cavernous part contained considerably fewer cells than the remainder of the urethra. A similar pattern of distribution of serotonin-immunoreactive paraneurons was confirmed in male dogs by our research group (Hanyu et al. 1987). Håkanson et al. (1974) demonstrated that the prostate in guinea pigs contains cells which are non-argentaffin and show green fluorescence (not yellow like serotonin cells) after L-dopa administration. The female canine urethra also contains numerous serotonin-immunoreactive paraneurons and it is noteworthy that a smaller number of them are also distributed in the epithelium of the vestibulum vaginae (Hanyu et al. 1987). In their pioneering work in the rabbit, Forsberg et al. (1964) reported that serotonin-fluorescent cells are dispersed in the *vestibulum* and *vagina* (see p. 189).

17.3 Light-Microscopic Structure

The shape of urogenital paraneurons is highly variable from cell to cell as well as among different organs and species. In *man*, the cells are neuronlike in shape, either bipolar or multipolar. The cells mostly extend a slender apical process to the luminal surface of the epithelium, thus they are *mostly open and partly closed* in type (Fetissof et al. 1985; Iwanaga et al. 1987) (Fig. 17.1). The basal processes are unusual in that they may be single but are mostly double or multiple; they branch like neuronal dendrites (Feyrter 1951, 1953; di Sant'Agnese and de Mesy Jensen 1984b; Iwanaga et al. 1987). The processes are irregular in thickness; a very thin thread with beaded swellings is often found. In closed-type cells lacking an apical process, the basal processes are also well developed. The processes may occasionally extend over a long distance in oblique or horizontal directions (Fig. 17.1). In any case, the basal processes end at the epithelial base and this terminal portion is characterized by a buttonlike thickening. In serotonin cells, the serotonin-immunoreactive substance is seen condensed at the basal part of the cell, especially at the buttonlike endings (Iwanaga et al. 1987). The variable profiles of the cells are also caused by the position of the nuclei. In most cells, the nucleus occurs in the basal half of the epithelium, but in others it is in a higher position, sometimes even close to the epithelial surface.

In *laboratory animals*, the urogenital paraneurons appear simpler in shape than in man; they are mostly flask-shaped with an apical process extending to the epithelial surface and with a single, fairly straight basal process (Owman et al. 1971; Dixon et al. 1973). Håkanson et al. (1974) noticed in the guinea pig urethra and prostate that the nuclei of the cells were located at highly variable levels in the epithelium, causing variable cell shapes.

17.4 Cell Types and Fine Structure

Careful light-microscopic observations in human urogenital tissues have indicated that argyrophil cells exceed argentaffin cells in number (Feyrter 1951, 1953; Kazzaz 1974). Just as in the case of gastroenteric paraneurons, this finding has been interpreted by some authors as signifying that the argyrophil but not argentaffin cells represent a different functional phase of the argentaffin cells (Kazzaz 1974). For other authors, however, the fact that the argyrophil cells outnumber the argentaffin cells implies that *more than one cell type* exists (Håkanson et al. 1974; di Sant'Agnese and de Mesy Jensen 1984a, b). In their combined silver and fluorescence histochemical studies on paraneuronal cells in the urethra and prostate of the guinea pig, Håkanson et al. (1974) made it clear that there are at least two cell types: The predominant type is argentaffin (and chromaffin) and stores serotonin, whereas the other less numerous type is argyrophil, nonargentaffin (and nonchromaffin), devoid of serotonin, but can be induced to store dopamine if supplied with L-dopa.

Electron-microscopic observations by Casanova et al. (1974) in the human urethra and by Fetissof et al. (1983) in the human prostate and urethra demonstrated the occurrence of two types of paraneuron characterized by different shapes and sizes of granules.

Di Sant'Agnese and de Mesy Jensen (1984b), also in the human prostate and prostatic urethra, have demonstrated nine cell types with different fine structures of the secretory granules. Two or three of these types possessed dark granules of distorted shapes and were classed in a single category of enterochromaffin (EC) cells. Other cell types were characterized by small round granules containing a loose material of moderate electron density, resembling the granules in gastroenteric S or D_1 cells. Di Sant'Agnese and de Mesy

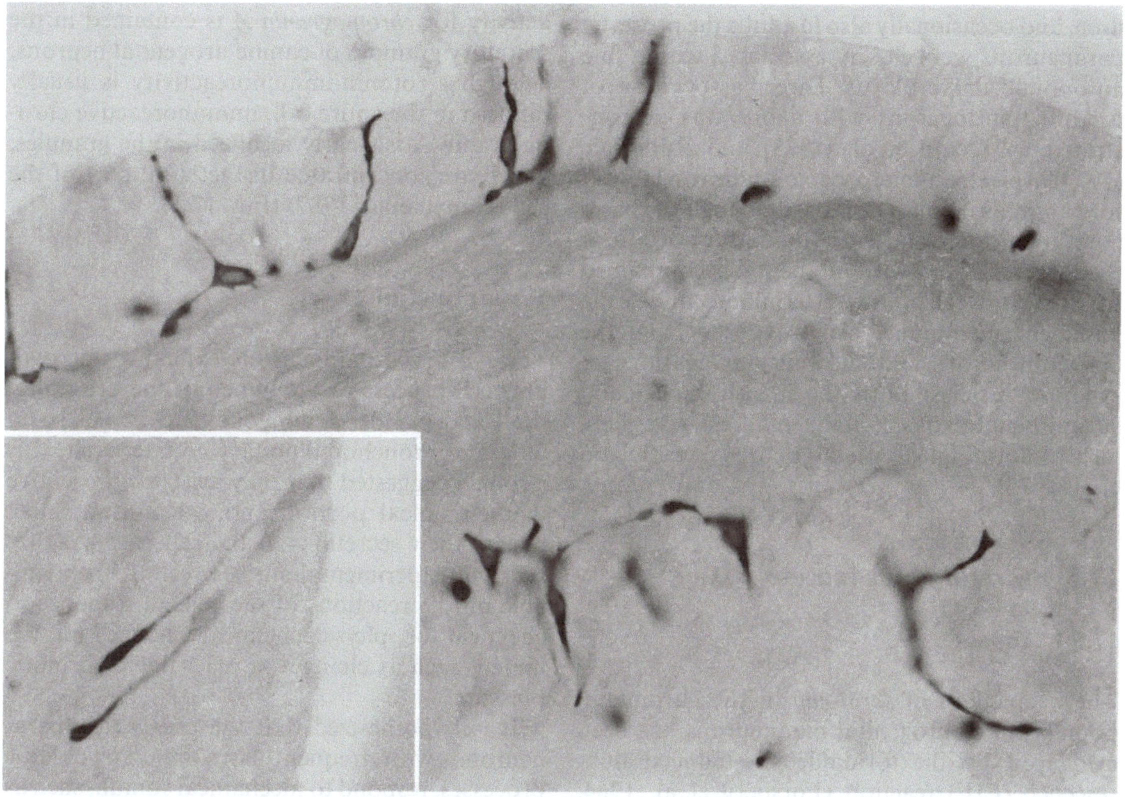

Fig. 17.1. Paraneurons in human (male) urethral epithelium as immunostained with an antiserotonin serum and an antichromogranin A serum (*inset*). Note the irregular cell shape and dendritelike pocesses. The serotonin immunoreactivity is seen in the whole extent of the cell processes, probably due to the diffusion of the amine, whereas the chromogranin immunoreactivity shown in the *inset* is concentrated in the basal process, indicating the endocrine nature of the cells

Jensen (1984b) believed that the cell types they identified "were not discrete but formed a continuum" as "many cells with intermediate or mixed granules were found." Looking at the electron micrographs of the nine granule types of these authors, we are of the opinion that although some of the micrographs seem to represent different functional phases of the same cell type, the above-mentioned two cell types, at least, appear to exist independently of each other.

In urethral paraneurons in the rat, rabbit, and guineapig, Dixon et al. (1973) recorded the occurrence of small round granules containing a granular material of moderate electron density. These granules seem to correspond to the small round granules mentioned above in human cells. The fine-structural cell typing in urogenital paraneurons seems to be more than 15 years behind that of the gastroenteric counterparts. As with the latter (from the one-cell-type theory to the multiple type theory), sooner or later precise knowledge of the cell types of urogenital paraneurons and their distribution correlated with granule morphology and secretory products will be achieved.

The cell organelles, especially those involved in the production of secretory substances, are essentially the same in development and appearance as those in gastroenteric paraneurons, and details of these structures will be omitted here.

The *apical process* of urogenital paraneurons, in the case of the open type, reaches the lumen and is covered with a tuft of microvilli, again in the same way as in the gut paraneurons (Dixon et al. 1973; Ramsdale 1974; di Sant 'Agnese and de Mesy Jensen 1984b). Like these authors, we hypothesize that this microvillous cell apex is the site of *chemoreception* of urogenital paraneurons, but little is known as yet concerning the nature and type of stimuli which may occur in urine, prostatic secretions, etc.

A *close relationship* between urogenital paraneurons and *nerves* was noted by Feyrter (1951, 1953). He observed that in human fetuses, children, and occasionally also in adults the prostatic paraneurons were closely associated with a fine autonomic nerve plexus. Direct association of urethral paraneurons with axons was demonstrated by Dixon et al. (1973) and Ramsdale (1974) in electron-microscopic studies on laboratory animals. Some axons pass through the basement membrane and enter the basal portion of the paraneurons, without being accompanied by Schwann cells. Others terminate under the basement membrane, being exposed from the Schwann sheath and showing aggregations of synaptic vesicles. Both the intraepithelial and subepithelial axons contain small clear vesicles, partly intermingled with large cored ones (Dixon et al. 1973).

17.5 Secretory Substances

17.5.1 Amines

The occurrence of *serotonin* in the chromaffin population of urogenital paraneurons has been evidenced by the formaldehyde-induced fluorescence (FIF) reaction (Forsberg et al. 1964; Håkanson et al. 1974) and by immunocytochemical techniques using antiserotonin sera (Fetissof et al. 1983; Iwanaga et al. 1987; Hanyu et al. 1987) (Fig. 17.1).

The nonchromaffin paraneurons lack serotonin but they take up extrinsic L-dopa and decarboxylate it into dopamine (Håkanson et al. 1974).

17.5.2 Peptides

Peptide messengers are believed to be contained in the granules of EC and other cell types. Attempts using antisera against a variety of brain-gut peptides have given negative results (Fetissof et al. 1985; Iwanaga et al. 1987), except for the report by di Sant 'Agnese and de Mesy Jensen (1984a), who were able to demonstrate some *somatostatin*-immunoreactive cells in the human prostate.

17.5.3 Other Substances

It seems most likely that urogenital paraneurons also possess other secretory substances, including adenine nucleotides and carrier proteins, though no reports appear to have dealt with this subject. Recently, we recognized that immunoreactivity for *chromogranin A* is contained in the secretory granules of canine urogenital neurons. Though serotonin-immunoreactivity is usually diffused in the entire cell, immunoreactive chromogranin A is clearly localized in the granules, which are concentrated in the basal part of the cell (Hanyu et al. 1987) (Fig. 17.1).

17.6 Functions

The *open-type* paraneurons extend a microvillous cell process to the lumen of the urogenital tract and glands. Analogous to their gastroenteric and bronchopulmonary counterparts, it is strongly suggested that they are chemoreceptive in their apical portion and, when stimulated, release their secretions at the cell base. At present, no experimental studies seem to be available on the reactions of the cells in response to chemical or physicochemical changes in the lumen, such as changes in pH value or osmotic pressure.

It is hypothesized that the *closed-type* paraneurons, with frequent horizontal or oblique processes, respond to mechanical stimuli, such as expansion of the lumen. No experimental data, however, are available supporting this idea.

The administration of *reserpine* was found to cause depletion of serotonin-fluorescence from the vaginal paraneurons of the rabbit, although many of the cells were resistant to reserpine (Forsberg et al. 1964).

The targets and effects of the secretory substances released, mainly in the paracrine and partly in the hemocrine fashion, from urogenital paraneurons are very poorly understood. It has been suggested that the effects of serotonin enhance the secretion of mucus, elevate vascular permeability, and cause contraction of certain smooth muscles. Nothing definite, however, has been proved in the urogenital regions.

Urogenital paraneurons are at least partly supplied with synaptic nerve terminals. These terminals contain synaptic vesicles suggesting their efferent nature, but afferent fibers may possibly be intermingled; reciprocal terminals may also be involved. It has been suggested that urogenital paraneurons represent sensory devices for detecting information about the luminal contents and, via the afferent nerves associated with them, lead to local reflexes, such as muscular contraction (Dixon et al. 1973; Ramsdale 1974).

17.7 Ontogeny and Phylogeny

Paraneurons are dispersed throughout the gastroenteric epithelium, which is derived from the endoderm. The tissues that formed by budding of the gut epithelium—especially the bronchopulmonary and pancreatic epithelia—and which are therefore of the same endodermal origin contain similar paraneurons. Is it then possible to account for the occurrence of paraneurons in the urogenital system in the same way? Feyrter (1951, 1953) and Lendon et al. (1976) gave a clear answer to this question. They defined paraneuronal elements as being distributed in the urethra, urinary bladder, prostate, and some accessory genital glands (see above), which are all endodermal in origin since they are derived from the *sinus urogenitalis*, a ventral protrusion of the *cloaca*.

This cloacal origin theory seems to be compatible with the finding by Forsberg et al. (1964) that serotonin-fluorescent paraneurons were numerous in the lower half of the rabbit vagina. The lower half of the vagina covered by stratified cuboidal epithelium contained the paraneurons, while the upper half covered by simple columnar epithelium was devoid of them. Although Forsberg et al. (1964) stated that the lower part of the vagina is derived from the Wolffian duct, modern texts of embryology state that the vaginal portion lower than the squamocolumnar junction is derived from the sinus urogenitalis.

Finally, a cell distribution apparently beyond the endodermal (cloacal) territory has also been recorded: In the human uterine cervix, Fox et al. (1964) found a few argyrophil cells, though the incidence was very low. Paraneuronal elements were further found in different types of ovarian cystadenoma. Numerous serotonin-fluorescent cells were found by Fetissof et al. (1983, 1985) in well-differentiated carcinomas in the cervix and in Brenner tumors, which are mesodermal in origin.

Reviewing all these data concerning the distribution of urogenital paraneurons, it seems reasonable to conclude that they occur in the territory of cloacal or endodermal derivatives, but tissues of apparently mesodermal origin may also contain a few of these elements either ectopically or pathologically.

Phylogenetic aspects of urogenital paraneurons have not been investigated to our knowledge. It would appear to be an interesting area for future studies.

17.8 Summary

Basal-granulated cells are dispersed in the epithelia of the urethra, prostate, vaginal vestibule, and certain other urogenital organs. The cells are distributed in the derivatives of the *sinus urogenitalis*, the ventral part of the *cloaca*, and can be regarded as comprising a division of the gastroenteric paraneurons.

Urogenital paraneurons consist of two or more cell types. The fact that the predominant cell type secretes *serotonin* is the only established finding concerning the messengers of the cells. Immunoreactivity for *chromogranin A* is contained in the secretory granules.

Urogenital paraneurons are more complex in shape than their counterpart cells in the gut. They are mostly categorized as belonging to the *open type*, extending their apical processes to the epithelial surface. It is thus apparent that the cells release their messenger substances in response to chemical or physical stimuli from the passing urine or other excreta.

18. Bronchopulmonary Paraneurons

18.1 Historical and General Aspects

The epithelium of the air tract contains basal-granulated cells, which comprise the respiratory division of the gastroenteric paraneurons. The paraneurons of the air tract can be divided into *solitary* and *aggregated* types. The former cells are dispersed among the epithelial cells, whereas the latter form specifically innervated corpuscles called neuroepithelial bodies (Lauweryns and Peuskens 1972), which are thought to be *chemoreceptors* for the partial pressures of O_2 and CO_2 in the inhaled air.

The paraneurons dealt with in this chapter are distributed in the larynx, trachea, bronchi, bronchioli, and alveoli. The term "bronchopulmonary" is used for convenience in this text to include all these parts of the airway.

18.1.1 Solitary Paraneurons

Clear cells ("helle Zellen") dispersed either solitarily or in small groups in the bronchopulmonary epithelium were described by Fröhlich (1949) in mammals (rabbits and cats) and by Feyrter (1953) in humans. Fröhlich (1949) recognized that the cells were argyrophil and that they were connected to fine nerve fibers. He suggested a chemoreceptive function of these cells, which he regarded as neurogenous in origin. Feyrter (1953), on the other hand, believed in an epithelial nature of the cells and ascribed an endocrine (paracrine) function to them. He included the cells in his "diffuse endokrine epitheliale Organe" and furthermore stressed their pathological significance as the sources of bronchial carcinoids.

Light- and electron-microscopic studies in the 1960s and early 1970s made it clear that the bronchopulmonary clear cells are typical *basal-granulated cells.* In their electron-microscopic study on the bronchial epithelium in human adults. Bensch et al. (1965) characterized the cells as the "bronchial counterpart of intestinal Kultchitzky cells." Based on the same idea, the cells have also been called "enterochromaffin-like cells" (Ericson et al. 1972), "endocrine-like cells" (Cutz and Conen 1972), or simply "endocrine cells" (Hage 1972). In this book, we will use mainly the last term together with "paraneurons".

Meanwhile, Pearse (1969) proposed that the bronchopulmonary endocrine cells belong to his APUD series (p. 4). His group (Ewen et al. 1972) and, more unequivocally, Ericson et al. (1972) demonstrated that the cells take up amine percursors (L-dopa and 5-hydroxytryptophane) and decarboxylate them into monoamines (dopamine and serotonin).

Pathologists have been interested in identifying the source cells of oat cell carcinomas in the lung and carcinoids in the bronchi. The endocrine cells they demonstrated in human fetal (Hage 1972, 1973a, b; Cutz and Conen 1972; Capella et al. 1978) and adult (Terzakis et al. 1972; Hage et al. 1977) lungs contained membrane-bounded "neurosecretory-type granules" (Cutz and Conen 1972), which were shared by oat cell carcinomas and bronchial carcinoids (Bensch et al. 1965a, b, 1968; Terzakis et al. 1972).

Recently, immunocytochemical studies mainly in fetal and neonatal human lungs have revealed that the endocrine cells contain immunoreactivities for bombesin (Wharton et al. 1978), more exactly gastrin-releasing peptide (GRP) (Iwanaga 1983; Tsutsumi et al. 1983) and calcitonin (Cutz et al. 1981), peptides which are usually detected in bronchopulmonary neoplasms (Tsutsumi et al. 1983).

18.1.2 Neuroepithelial Bodies

The possible existence of *intrapulmonary chemoreceptors* detecting changes in the composition of lung gases has long been suggested (review: Dejours 1962; Comroe 1968; see also Laros 1971). It has been postulated by physiologists that these receptors provoke a local constriction of the pulmonary vessels in response to hypoxia in the inhaled air (Daly and Hebb 1966; Dawes and Comroe 1954). Hypoxemic blood itself did not cause a pulmonary vasoconstriction (Nisell 1951; Hauge and Staub 1969), while blood made hypoxemic through hypoxically aerated lungs produced that effect (Duke 1954; Daly and Wright 1957).

Twenty years after the early unconvincing suggestion of a chemoreceptive function of the bronchopulmonary argyrophil cells by Fröhlich (1949; see above), an *electron-microscopic* study by the British anatomists Cook and King (1969) revealed the existence in the chicken bronchial epithelium of a tall granulated cell associated with a nerve terminal. They pointed out that the structure resembled the carotid body cell and Merkel cell and that it may represent either a chemoreceptor for monitoring the gases in the airway or a mechanoreceptor in the bronchial wall. Following this breakthrough, the Belgian pathologists Lauweryns and associates more clearly and systematically clarified the morphological substance of the intrapulmonary chemoreceptors in mammals, using the term "*neuroepithelial bodies*" (human infant—Lauweryns and Peuskens 1972; rabbit and other mammals—Lauweryns et al. 1972, 1973). The neuroepithelial bodies comprised masses of basal-granulated cells supplied with unmyelinated nerve fibers (Lauweryns and Cokelaere 1973b) (Figs. 18.1–3). The chemoreceptive function of the bodies was supported by the results that *exocytotic granule release* was induced in the cells after the rabbit lung was hypoxically aerated (Lauweryns et al. 1973). Serotonin (Lauweryns et al. 1973) and gastrin-releasing peptide (GRP) (Tsutsumi et al. 1983) have been localized in the neuroepithelial body cells and these substances are believed to be released by luminal hypoxic stimulus.

18.2 Distribution, Cell Types, and Structure

The bronchopulmonary endocrine cells are largely divided into solitary or single cells and aggregated or corpuscular cell groups (neuroepithelial bodies), although there seem to be intermediate types consisting of only a few cells gathered together.

18.2.1 Solitary Paraneurons

The solitary cells are argyrophil but show no or inconspicuous argentaffinity (Lauweryns et al. 1972). They are predominant in the air tract proximal to the bronchi.

The epithelium of the larynx has been reported to contain numerous solitary endocrine cells in the rat (Ewen et al. 1972; Taira 1985). The cells are abundant in the trachea of man, lamb, rabbit, armadillo (Cutz et al. 1975), and mouse (Ericson et al. 1972). Segmental bronchi of man (Terzakis et al. 1972) and some animals also contain considerable populations of the cells.

The majority of the solitary paraneurons are pyramidal in shape and *open in type*. They extend a slender process to the epithelial surface, and the end of the process is covered by microvilli as in the case of gastroenteric endocrine cells (Ericson et al. 1972; Hage 1973a, b; Taira 1985).

Some of the paraneurons are *closed in type*, as our observations in human fetuses (Iwanaga 1983) and rabbits indicate. The cell lies flat on the basement membrane, extending long horizontal processes filled with secretory substances.

In the cytoplasm of the solitary cells of both the open and closed types, mitochondria are relatively numerous. A well-developed Golgi apparatus is seen near the nucleus, and a cilium may be encountered in this vicinity. Small secretory granules are observed to be formed in the vesicles of the Golgi apparatus (Hage 1973a).

Membrane-bound granules measuring 100–300 nm depending on the cell type (human—Cutz and Conen 1972) gather in the lower half of the cell. They are round and contain either dense, moderate, or flocculent contents, according to the cell type and method of fixation and staining. In fetal and adult human lungs, Capella et al. (1978) differentiated three cell types according to the size and fine structure of the granules. Electron-microscopic images indicating *exocytotic* release of the basal granules have been demonstrated in the mouse trachea (Taira and Shibasaki 1978).

A review of the controversial reports on the innervation of the bronchopulmonary endocrine cells suggests that only part of them are directly innervated, while the rest are free of nerves. Thus, synaptic association of a nerve fiber with the endocrine cell can be recognized (Hage et al. 1977; Taira 1985). The fine-structural features of the nerve terminals do not clearly indicate whether they are efferent or afferent (sensory) in nature.

18.2.2 Brush Cells

A probable subtype of the bronchopulmonary paraneurons comprises the so-called "brush cell" discovered and designated by Rhodin and Dalhamn (1956) and more precisely observed by Luciano et al. (1968); both groups studied the rat trachea. The brush cells are at present accepted to be a widely distributed cell type both in the tracheobronchial (Watson and Brinkman 1964; Rhodin 1966; Ishida 1977; Taira and Shibasaki 1978; Taira 1985) and alveolar epithelia (Mayrick and Reid 1968; Hijiya et al. 1977). Brush cells of almost identical morphological features have been found to be dispersed in the epithelia of the gallbladder and bile duct (Luciano and Reale 1969; Luciano et al. 1981).

Distributed solitarily or occasionally paired, the brush cells are barrel-shaped and apically extend a dense tuft of strong, straight microvilli of uniform size (150 nm in thickness). Characteristically, the axial filaments of the microvilli penetrate deep into the cytoplasm without ending in the terminal web (Luciano et al. 1968).

Unmyelinated nerve fibers from the propria mucosae contact the brush cells to form synapses. The nerve terminals were elegantly demonstrated by Luciano et al. (1968) to contain both assemblies of mitochondria and synaptic vesicles (numerous small clear vesicles and some large cored ones); however, Luciano et al. (1968) were preoccupied with the concept of unidirectional transmission and declared that the nerves were solely afferent (sensory) in nature. This view influenced later researchers. It seems appropriate now to hypothesize a *reciprocal* function of the synapses.

Although the possible occurrence of *secretory granules* in brush cells was long neglected (e.g., Ishida 1977) or overlooked, recent studies have demonstrated their existence. In the mouse trachea, Taira and Shibasaki (1978) reported that the basal part of the brush cells contained membrane-bound round granules.

18.2.3 Neuroepithelial Bodies

Neuroepithelial bodies comprise corpuscular assemblies of basal-granulated cells, which are believed to be specialized for monitoring hypoxic and hypercapnic changes in the inhaled air. In contrast to the solitary paraneurons, which occur mainly in the proximal portion of the air tract, neuroepithelial bodies are found in the distal portion, especially in the bronchioles. In the adult, infant, and fetal human lung, neuroepithelial bodies are distributed in the bronchi, bronchioles, and even in the alveoli (Lauweryns and Goddeeris 1975).

In the rabbit, cat, monkey (Lauweryns et al. 1972), and mouse (Hung and Loosli 1975), the bodies are distributed throughout the entire bronchioli, extending as far as the most peripheral part of the terminal bronchioles and especially occurring at the bifurcations of the bronchioles (Lauweryns et al. 1972; Hung et al. 1979).

Typical neuroepithelial bodies extend from the base to the luminal surface of the epithelium. All the specialized cells of the bodies may open to the lumen as it is typically seen in the rabbit (Lauweryns et al. 1972) (Fig. 18.1), while only a portion of the cells may reach the lumen in the case of the mouse (Hung and Loosli 1975) and probably also in man (Lauweryns and Goddeeris 1975). In their scanning electron-microscopic study of the neonatal mouse lung, Hung et al. (1979) showed that modified Clara cells cover most of the swollen surface of the neuroepithelial body, leaving small oval areas for the open-type chemoreceptive cells.

In contrast to the open cell groups described above, groups of *closed-type* paraneurons are also found as indicated by our observations in the fetal human lung (Iwanaga 1983). Their significance is unknown, but it is possible that these bodies represent mechanoreceptors responding to stress and strain in the bronchiolar wall.

The fine structure of the endocrine cells forming the neuroepithelial bodies essentially corresponds to that of the solitary cells described above. Lauweryns et al. (1986) propose that there are two types of secretory granules (dense cored vesicles) in rabbit neuroepithelial body cells: One has a pleomorphic profile, measuring 130 nm in long axis, while the other is more rounded in shape and measures 110 nm in diameter. Immunocytochemical study at the electron-microscopic level indicates that serotonin is located only in the first type of granules.

The neuroepithelial bodies are constantly supplied with *nerves*. The electron-microscopic studies by Lauweryns and Cokelelaere (1973b) and Lauweryns and Lommel (1982) show that unmyelinated nerves in the lamina propria penetrate the body at different levels to form synapses with the chemoreceptive cells. In the rabbit, some profiles of the nerve terminals contain a dense accumulation of small mitochondria suggesting an afferent nature, while others include aggregations of synaptic vesicles (small clear and a few

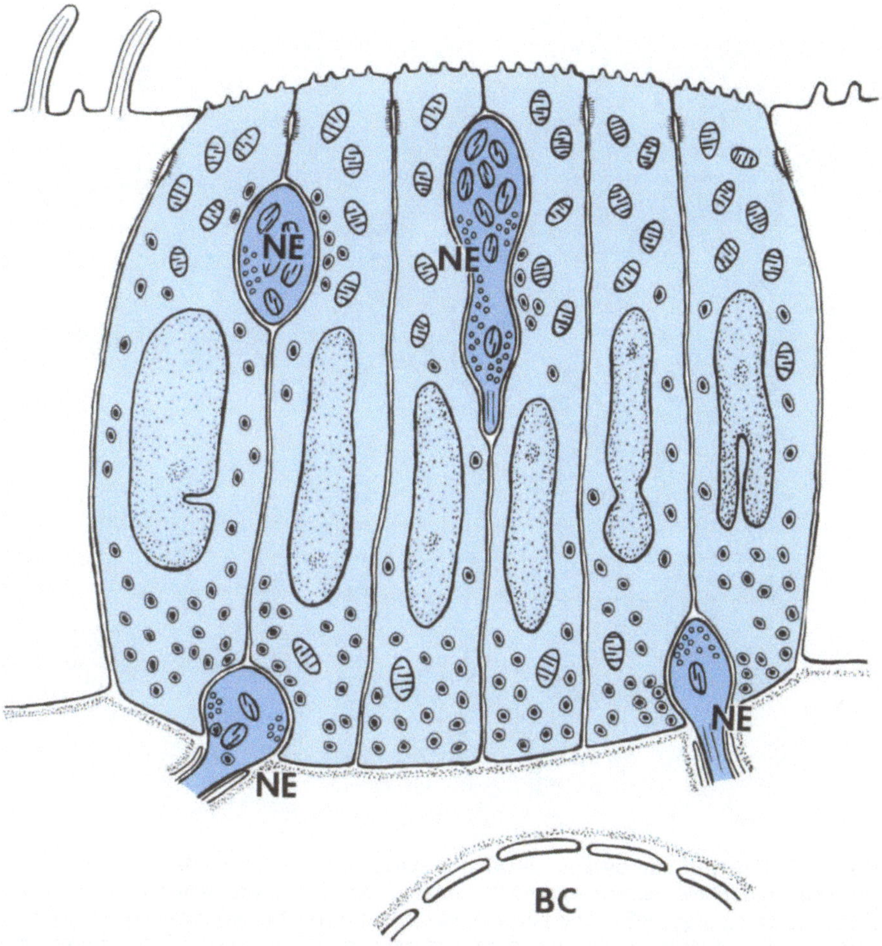

Fig. 18.1. Structure of a neuroepithelial body in the respiratory epithelium of the rabbit. *NE* nerve endings, *BC* blood capillary. Modified from Lauweryns and Lommel (1982)

large cored vesicles) suggesting an efferent function (Lauweryns and Cockelaere 1973b; Lauweryns and Lommel 1982). In the rat, laminar expansions of the *afferent-type terminals* supply the endocrine cells apparently independently of efferent-type terminals (Luciano et al. 1981). However, in other mammals including the rabbit, the majority of the axonal profiles contain separate aggregations of mitochondria and synaptic vesicles, suggesting *reciprocal transmission* of impulses. More convincing images of reciprocal synapses can be seen in the electron micrographs of neuroepithelial bodies in the toad (Rogers and Haller 1978) (Figs. 18.2, 18.3). In the rat and mouse, nerve terminals have been reported to contain numerous mitochondria and no synaptic vesicles (Hung and Loosli 1975; Wasano 1977). The discrepancy among animals must be further investigated (Review: Scheuermann 1987).

18.3 Neuroepithelial Bodies As Chemoreceptors

Following the systematic investigations of Lauweryns and associates, it now appears most probable that the neuroepithelial bodies represent chemoreceptive organs for the O_2 and CO_2 content of the inhaled air. By exposing neonatal rabbits to *hypoxia* (low O_2) or to *hypercapnia* (high CO_2), these authors succeeded in inducing *exocytotic granule release* at the base of the neuroepithelial body, which is faced by a fenestrated capillary (Lauweryns and Cokelaere 1973a, b; Lauweryns et al. 1977). The granule release is caused only by the inhaled gas and not by hypoxic blood circulation, as evidenced by cross-circulation experiments in rabbits (Lauweryns et al. 1978). It has been further reported that during hypoxia exocytotic figures of the secretory granules of the neuroepithelial body cells were

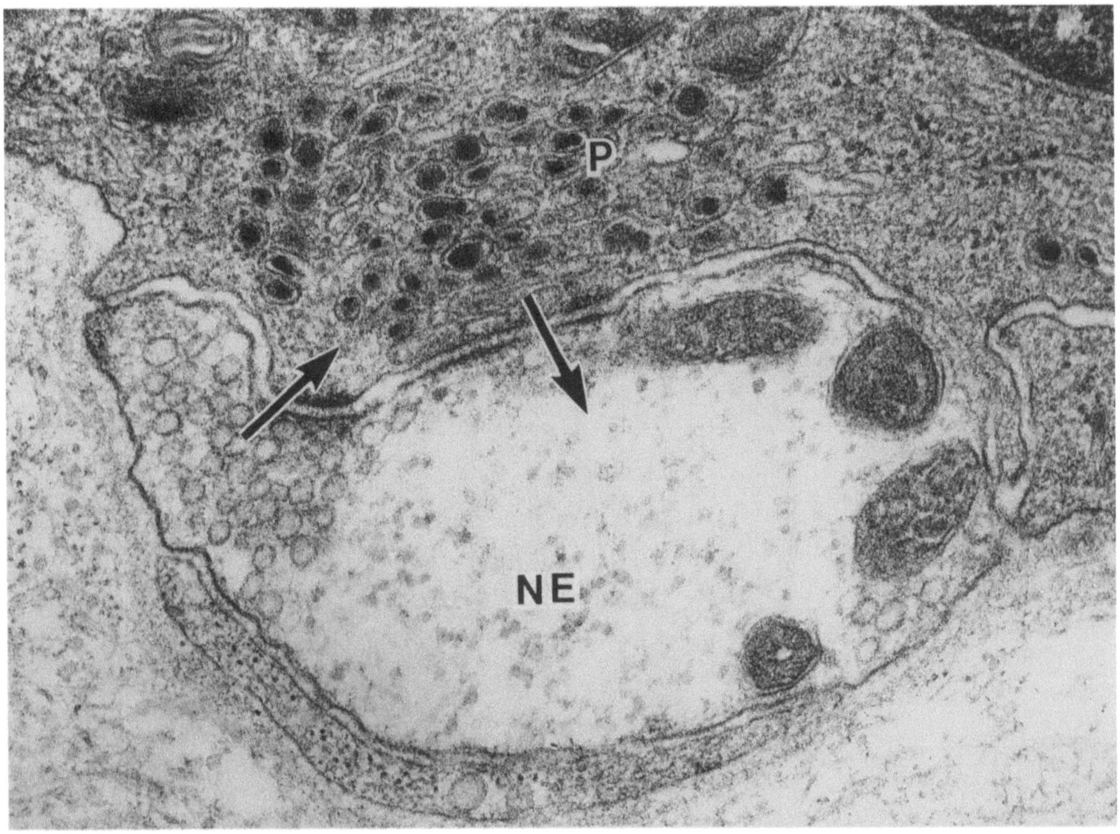

Fig. 18.2. Electron micrograph showing a synapse in a neuroepithelial body of the toad *Bufo marinus*. A paraneuronal (chemoreceptive) cell (*P*) contains secretory granules accumulated toward the synaptically associated nerve ending (*NE*). The nerve appears empty in the cytoplasm behind this synapse but contains an accumulation of clear vesicles in another synaptic region on the left. Thus, the synapse shown here appears to be reciprocal in nature (*arrows* indicate the presumed directions of transmission). Courtesy of Rogers DC, Haller CJ (1978) Cell Tiss Res 195: 395–410 (The labels and interpretation of the micrograph were added by the present authors)

increased at the site facing the afferent-like zone of the synapses (Lauweryns and Lommel 1982). On the other hand, the synaptic vesicles in the efferent zone of the synapses were depleted during hypoxia, suggesting an efferent effect on the chemoreceptive cells.

These results lead to the following hypotheses:

1) Neuroepithelial body cells release their secretions in response to hypoxia and hypercapnia in the inhaled air; these secretions may locally cause (by paracrinia and microhemocrinia) vasoconstriction in hypoxically or hypercapically aerated areas of the lung. Serotonin, which is one of the principal products of the cells (see below) is known to have this vasoconstrictive effect (Sjoerdsma 1959; see also Lauweryns et al. 1977). It has been postulated that the neuroepithelial bodies may also modulate mucosal secretion and smooth muscle tone in the bronchioles (Lauweryns et al. 1973a, b).
2) The possibility that the secretions of the neuroepithelial body cells partly enter the circulation and exert some effects in remote targets may not be excluded.
3) Hypoxia and hypercapnia may also cause release of the secretions of the neuroepithelial body cells to the nerve endings, generating afferent (centripetal) nerve impulses.
4) At the same time, the neuroepithelial body cells are presumed to be modulated by their efferent nerve endings, as is suggested by the depletion of synaptic vesicles in the nerves supplying the hypoxia-stimulated cells.

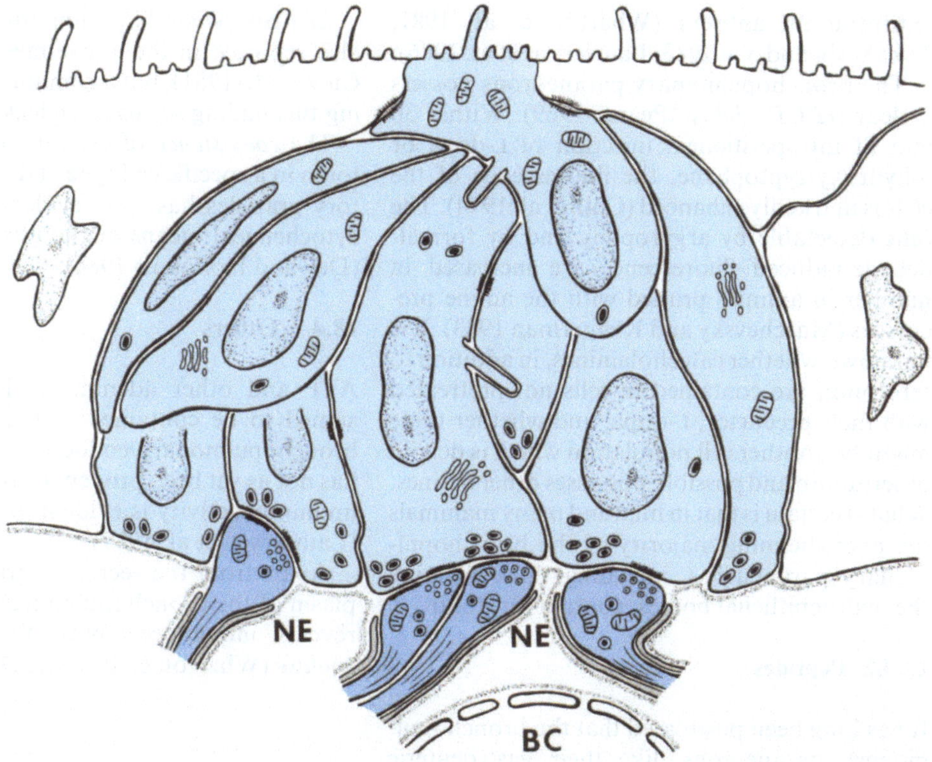

Fig. 18.3. Structure of a neuroepithelial body in the respiratory epithelium of the toad. Note the characteristic cell construction of the body and synapses which are reciprocal in nature. *BC* blood capillaries, *NE* nerve terminals. Modified, with permission, from Rogers DC, Haller CJ (1978) Cell Tiss Res 195:395–410

It now seems appropriate to compare neuroepithelial bodies with the *carotid body*. The former are open to the inhaled air to monitor hypoxic or hypercapnic changes, while the latter is enclosed in the tissue and detects the same information in the blood. The chemoreceptive cells in both organs are typical paraneurons and have the same peptide/amine-producing devices in the cytoplasm, especially the membrane-bound secretory granules. Afferent and efferent nerves supply both organs, predominantly showing reciprocal synapses. Enhanced chemoreceptor activity of both organs causes a depletion of synaptic vesicles in these nerve endings, suggesting an afferent nervous effect on the chemoreceptive cells (McDonald 1977; Morgan et al. 1981; Lauweryns and Lommel 1982). The messengers secreted from the cells, however, differ between the carotid and neuroepithelial bodies: In the former they are predominantly enkephalins and catecholamines, while in the latter, they are GRP and serotonin, as detailed in the following section.

18.4 Secretory Products

It seems reasonable to conceive that in the bronchopulmonary endocrine cells all the categories of secretory substances of paraneurons mentioned on p. 19 are stored in the secretory granules, although studies to substantiate this idea are still at the early stage.

18.4.1 Amines

Serotonin is the main, if not exclusive, monoamine of mammalian bronchopulmonary paraneurons of both the solitary and aggregated types. With the fluorescence histochemical method of Falck and Hillarp, the cells emit an intense yellow fluorescence characteristic of serotonin (Lauweryns et al. 1972, 1973; Wasano 1977). The serotonin fluorescence is recognized also in amphibian neuroepithelial bodies (frog and toad; see below). Serotonin in the bronchopulmonary paraneurons has recently been evidenced more directly by immunocytochemical methods using antiserotonin antisera (Wharton et al. 1981; Takahashi and Yui 1983; Lauweryns et al. 1986).

The bronchopulmonary paraneurons possess a clear *APUD ability* (Pearse 1969). Within 30 min of intraperitioneal injection of L-dopa or 5-hydroxytryptophane, the fluorescence of the cells is markedly enhanced (Cutz et al. 1974). The cells detectable by argyrophily and by formaldehyde-induced fluorescence are increased in number in animals primed with the amine precursors (Marchevsky and Kleinerman 1983). It is unknown whether catecholamines, in addition to serotonin, are contained in cells not pretreated with their precursor, L-dopa, and whether there might be another cell population which is devoid of serotonin and possibly possesses other amines. What is certain is that in man and many mammals the overwhelming majority of the bronchopulmonary paraneurons, including those forming the neuroepithelial bodies, contain serotonin.

18.4.2 Peptides

It has long been postulated that the bronchopulmonary paraneurons, like their gastroenteric counterpart, produce a peptidic messenger besides amines. Pearse (1969) proposed the term "pneumokinin" for this hypothetical substance (see also Fujita 1976). In their immunocytochemical studies, Polak et al. (1976) and Wharton et al. (1978) revealed that the "pneumokinin" was represented by an immunoreactivity for *bombesin*, a tetradecapeptide isolated from amphibian skin. More recent studies, however, support the view that *GRP* and not bombesin is present in the bronchopulmonary paraneurons (Tsutsumi et al. 1982, 1983a, b; Iwanaga 1983). GRP, comprising 27 amino acid residues, contains the bioactive sequence of bombesin in its C-terminal portion. As GRP/bombesin has been shown in mammals to cause smooth muscle contraction (Mayer et al. 1982), it seems reasonable to assume that the peptide released from the cells may induce contraction of the airway smooth muscles and possibly also blood vessels.

Calcitonin-like immunoreactivity, meanwhile, was demonstrated by Becker et al. (1980), Cutz et al. (1981), and Marchevsky and Kleinerman (1983). In their study of fetal and adult human lungs, Tsutsumi et al. (1983) reported that calcitonin is contained in a portion of the GRP cells, that calcitonin appears much later than GRP in the fetal lung, and that the cells forming neuroepithelial bodies contain GRP but are free of calcitonin immunoreactivity.

Leu-enkephalin-like immunoreactivity was also reported in the above-mentioned study by Cutz et al. (1981), but a communication confirming this finding seems to be lacking.

The *coexistence* of GRP/bombesin and serotonin in a specific cell type and in identical secretory granules has been evidenced by immunocytochemical means at the ultrastructural level (Dey and Hoffpauir 1986).

18.4.3 Others

ATP and other adenine nucleotides are presumed to be contained in the granules of the bronchopulmonary endocrine cells, though this has not as yet been proven. *Chromogranin* A-like immunoreactivity is evident in the cell granules (Lauweryns et al. 1987).

Apart from the secretory granules, the cytoplasm of the bronchopulmonary endocrine cells reveals immunoreactivity for *neuron-specific enolase* (Wharton et al. 1981; Dayer et al. 1983).

18.5 Electrophysiology of Neoplasms

Neoplasms of different categories, including carcinoid, small-cell carcinoma, and oat cell (undifferentiated) carcinoma, are known to be derived from the bronchopulmonary paraneurons or their precursors. Their types and features have been described in the chapter on neoplasms (p. 114); here, electrophysiological studies using some of these tumor cells will be introduced.

Human small-cell carcinoma of the lung has been shown to secrete as many as 13 different peptides, including ACTH, vasopressin, bombesin, Leu-enkephalin, calcitonin (Tischler et al. 1977a, b; McCann et al. 1985). It is possible that an inward Na^+ leak exists in these cells and that this leak is a transient event, increasing and decreasing as switching occurs. Preliminary experiments with the patch-clamp technique on DMS 53 cells, a long-established cell line of human small-cell carcinoma of the lung, have revealed the presence of single-channel voltage-sensitive conductance as well as acetylcholine induced channel activity (McCann et al. 1985). Cultured tumor cells from two oat cell carcinomas of the human lung were demonstrated to generate all-or-nothing, short-duration action potentials similar to those previously reported in neurons, bronchial carcinoid cells, and some endocrine cells (Tischler et al. 1977a, b).

18.6 Ontogeny

Although Fröhlich (1949) postulated a neurogenous origin of bronchopulmonary endocrine cells, no evidence is available to support this view. Neuroepithelial bodies are differentiated directly from the epithelium of the lung at an early stage of fetal development when the organ is a glandular structure. The bodies differentiate on a proximal-to-distal gradient along the forming bronchi; the solitary paraneurons do not seem to differentiate earlier than the neuroepithelial bodies (rat—Carabba et al. 1985). In the rabbit, Hung (1980) reported that the bodies were detectable at 21 days of gestation, while Lauweryns and Bock (1981) reported 18 days. In the rat, Carabba et al. (1985) noted the earliest detectable bodies at 15 days of gestation and well-differentiated bodies at 21 days. In human fetuses, Cutz et al. (1984) recorded the first morphologically identifiable endocrine cells at 8 weeks of gestation. Shortly after, grouped endocrine cells appeared as primitive neuroepithelial bodies. The endocrine elements differentiated more numerously in the distal than in the proximal part of the airway system.

The bodies undergoing differentiation are slightly swollen, dense aggregations of tall cells, which are, from the earliest stage of differentiation, slightly argyrophil and positive in the fluorescent (FIF) reaction for monoamines (without amine-precursor loading) and in the acetylcholinesterase reaction (Hung 1980).

Penetration of nerves into the bodies has been observed in fetal rabbits at a very early stage of their differentiation—19 days of gestational age (Lauweryns and Bock 1981). It is an interesting problem whether nerves approaching the bronchial epithelium induce differentiation of the paraneuronal cells at this site or whether the cells under differentiation induce growth and invasion of the nerves. In their organ culture study of the lung, however, Carabba et al. (1985) stressed that nervous elements did not appear to be necessary for the differentiation and growth of the neuroepithelial bodies. The opinions are thus controversial concerning the significance of nerves in the differentiation and maintenance of bronchopulmonary paraneurons.

The development of bronchopulmonary paraneurons *in man* has been not so precisely studied as in the rabbit. By immunocytochemical staining of GRP in the human lung, Iwanaga (1983) reported that the immunoreactive cells both of neuroepithelial bodies and of the solitary type were first detected at 12 weeks of gestation and increased in number, reaching a peak at 14–16 weeks, as indicated by the cell count per area of the microscopic sections. Iwanaga (1983) stated that the GRP-immunopositive elements decreased thereafter and could only be rarely encountered after birth and in adults. Yet there is evidence that neuroepithelial bodies and solitary endocrine cells are not uncommon in the respiratory tract of adult humans (Lauweryns and Goddeeris 1975; Tsutsumi 1983) and adult rabbits (Lauweryns et al. 1973).

It remains to be settled whether the high population of the endocrine cells in the fetal lung and the rare occurrence in adults are solely due to the growth of the organ, causing dispersion of the cells, or to an actual postnatal decrease of these cells. Furthermore, the question arises concerning the dense distribution and apparently completed secretory machinery of the neuroepithelial cells during late fetal life, as pointed out by Carabba et al. (1985). What luminal information do these cells detect when no air enters the lung and what local or general effects do the secretions of the cells exert in the fetal body?

18.7 Phylogeny

Bronchopulmonary paraneurons in submammalian vertebrates have not been studied extensively and our knowledge in this field is only fragmental. It is interesting to note that neuroepithelial bodies are developed in the primitive lung of amphibia even more elaborately than in the mammalian respiratory tract.

18.7.1 Amphibia

In the lungs of the toad, *Bufo marinus* (Rogers and Haller 1978), and the frog, *Hyla alborea* L (Goniakowska-Witalińska 1981), the neuroepithelial bodies are domelike swellings located on the luminal surface of the primary septa (*Bufo*) or the primary and secondary septa (*Hyla*). The bodies consist of up to 100 columnar cells (*Bufo*), which rest on the basal lamina, beneath which a blood capillary is located. Only a small portion of the cells reach the lumen of the lung. Other, more basically located cells are covered by the columnar cells and by flattened pneumocytes. The apical end of the open-type cells is provided with microvilli and a cilium (*Bufo*). Both the open and closed cells of the neuroepithelial bodies contain a number of cored vesicles (granules), consisting of a dark substance loosely contained in a polymorphic membrane sac. The granules tend to gather at the sites of synapses (see below).

Neuroepithelial body cells in the toad and frog show yellow fluorescence after formaldehyde treatment, indicating a serotonin content (Rogers and Haller 1978; Wasano and Yamamoto 1978).

In the lung of *Rana nigromaculata*, Wasano and Yamamoto (1978) claimed that the neuroepithelial bodies were located in the basal part of the epithelium isolated from the lumen of the lung. They suggested a possible mechanoreceptive function of the bodies. Further studies using serial section observation or scanning electron microscopy would appear to be necessary to determine whether the bodies are really separated from the lumen. Yet it seems reasonable to hypothesize that amphibian neuroepithelial bodies may have a mechanoreceptive function either in addition to the chemoreceptive role or more specifically, as Wasano and Yamamoto (1978) suggested for *Rana*. The neuroepithelial body of the toad consists of a complex pseudostratified assembly of specialized cells which insert single cilia into adjacent cells (Rogers and Haller 1978). Together with the intraluminally protruded cilium, the intercellular cilia may possibly serve in the perception of touch, pressure, and strain in the lung.

The neuroepithelial bodies of the toad and frog are supplied by unmyelinated nerves at their base. Synapses with distinct membrane thickenings are numerously observed. The nerve terminals comprise an adrenergic type, containing small cored vesicles, and a cholinergic type, containing small clear vesicles; both include some peptidergic-type granules (Rogers and Haller 1978). Compartmentalization of a nerve terminal into efferent areas rich in synaptic vesicles and afferent areas free of them and with mitochondria are clearly seen in Fig. 18.2 from the paper of Rogers and Haller (1978), strongly suggesting the reciprocal nature of the synapse.

18.7.2 Birds

Granulated cells associated with nerve terminals, which seem to correspond to the solitary paraneurons, were recorded by Cook and King (1969) and King et al. (1974). The cells were high-pyramidal basal-granulated cells, though their apical end to the bronchial lumen could not be confirmed. The nerve terminals to the base of these cells contained numerous slender mitochondria and aggregations of small clear vesicles (30–100 nm in diameter) and larger cored ones (60–100 nm). In spite of the existence of the synaptic vesicles, these nerves were regarded by those authors as being purely afferent or sensory in nature.

No studies seem to be available on the occurrence in the avian and reptilian airways of a corpuscular assembly of endocrine cells comparable with the mammalian and amphibian neuroepithelial body.

18.8 Summary

The basal-granulated cells dispersed in the epithelia of the trachea, bronchi, bronchioli, and alveoli comprise the bronchopulmonary division of the gastroenteric paraneurons, as the air tract is ontogenetically derived from the alimentary tract. Bronchopulmonary paraneurons are partly solitary cells(including brush cells) and partly aggregated in corpuscles, the neuroepithelial bodies. These bodies are well developed in some mammals (rabbit, mouse, etc.) and in amphibians.

The solitary cells are mostly *open* to the lumen but some cells are *closed*, suggesting chemoreceptive and mechanoreceptive functions, respectively. The cells forming the neuroepithelial bodies are open in type and are selectively sensitive to *hypoxia* or *hypercapnia* in inhaled air, responding by exocytosis of their granules (rabbit—Lauweryns and associates).

Nerves synaptically supply the neuroepithelial body cells and some of the solitary cells. Their fine structure suggests a *reciprocal* nature of the synapses. It is stressed that the bronchopulmonary paraneuron as the sensor of hypoxic *air* and the *carotid body* paraneuron (chief cell) as that of hypoxic *blood* share the reciprocal pattern of innervation as well as many cell-biological features.

Serotonin as a monoamine and GRP and calcitonin as peptides are established secretions of bronchopulmonary paraneurons. *Neoplasms* of bronchopulmonary paraneurons in small-cell carcinomas or oat cell carcinomas reveal more numerous kinds of peptides.

By means of electrophysiological techniques on cultured cells from these tumors neuronlike *action potentials* have been demonstrated.

19. Carotid Body Chief Cells

19.1 Historical and General Aspects

The human carotid body is a small structure located at the upper end of the common carotid artery. It consists of nerve fibers, connective tissue, and clusters of parenchyma which contain two kinds of cells—*chief* and *sustentacular* (Schwann) cells. In addition to these two kinds of cells, there are enlarged *nerve terminals* in the parenchyma so that the structural unit of this organ is composed of three elements: (1) chief cells, (2) sustentacular cells, and (3) nerve terminals.

The carotid body has been variously called a ganglion, gland, vascular glomus, paraganglion, and chemoreceptor; its principal cell has been termed a ganglion, glandular, glomus, paraganglion, chemoreceptor, type I glomus cell, and so forth. Opinions concerning the nature and origin of this organ and its chief cells have greatly varied from author to author. To avoid any misconception, the terms "carotid body" and "chief cell" are used in this book. A brief historical survey on the concept of this enigmatic organ may clarify the complex situation concerning the terminology.

19.1.1 Early Studies

The carotid body was discovered in 1743 by Taube, one of the pupils of von Haller, a famous Swiss physiologist (Adams 1958; Pick 1959). At first, this organ was thought to be one of the sympathetic ganglia ("ganglion intercaroticum"), owing to its morphological similarity to autonomic ganglia. However, microscopic study was not then available. Around the middle of the nineteenth century, with the advance of light-microscopic techniques, Luschka (1862) revealed that this organ was not a simple ganglion but a *glandular structure* rich in nerves and blood vessels ("glandula carotica"). Luschka (1862) included the carotid body in the "Nervendrüsen" (nervous glands). Some nineteenth century anatomists, among them Stieda (1881) and Prenant (1896), thought that the carotid body was a branchial organ derived from the pharyngeal pouch. Schaper (1895) thought that the carotid body was simply the parathyroid III.

In 1865, against the glandular theory of Luschka (1862), Arnold (1865) proposed a hypothesis that the carotid body was an organ consisting of highly convoluted blood vessels or glomeruli. Arnold (1865) thought that the chief cells of the carotid body were *derived from vascular endothelial cells* ("glomeruli arteriosi intercarotici"). The vascular hypothesis of Arnold (1865) was widely accepted in the latter half of the nineteenth century.

19.1.2 Carotid Body As a Paraganglion

Stilling (1892) first described the positive *chromaffin reaction* in the carotid body. He found the development of a brown color in some chief cells of the carotid body when they were treated with dichromate or chromic acid. Since then, the presence or absence of the chromaffin reaction in the carotid body has been one of the most disputed problems in the study of this organ. Kohn (1903) demonstrated that the carotid body was derived from the *neuroectoderm*, being innervated by preganglionic sympathetic neuronal terminals, and showed a positive chromaffin reaction. Kohn (1903) emphasized the chromaffinity of the carotid body and homologized it with the adrenal medulla and other *paraganglia* ("paraganglion intercaroticum"). Kohn did not call the chief cells "paraganglion cells" but "chromaffin cells". However, it is doubtful whether Kohn really recognized a distinctly positive chromaffin reaction in all the chief cells of the carotid body (Kobayashi 1971a). Later, Kose (1907) gave the name "farblose chromaffine Zellen" (colorless chromaffin cells) to the chief cells of the carotid body, since he recognized that most of them did not become brown upon treatment with chrome salts.

19.1.3 Carotid Body As a Chemoreceptor

In the latter half of the 1920s, De Castro (1929) published a series of papers concerning the histological structure of the carotid body. These studies had a great influence upon subsequent interpretations of this organ. De Castro (1929, 1951) proposed the hypothesis that the carotid body was not a paraganglion but a *chemoreceptor*, detecting the concentration of oxygen and carbon dioxide in the bloodstream. He showed that severing of the sinus nerve supplying the carotid body below the petrosal ganglion caused degeneration of most nerve terminals in the carotid body, whereas intracranial cutting of the glossopharyngeal nerve caused no nerve degeneration.

19.1.4 Revised Vascular Glomus Theory

In the 1930s, the vascular glomus theory of Arnold (1865) was restored by Schumacher (1938), who concluded that the carotid body was neither a chemoreceptor nor a paraganglion but an *arteriovenous anastomosis*. Schumacher (1938) supposed that the carotid body chief cells were "myoepithelial cells" secreting acetylcholine or some vasodepressor substances. One of the most important bases for the hypothesis of Schumacher (1938) was that the carotid body of some mammalian species possessed no chromaffin cells in the literal sense and in other species there were only a few of these cells. Carotid body chief cells are thus different from the adrenal chromaffin cells and resemble the cells of the arteriovenous anastomosis.

19.1.5 "Revised" Paraganglion Theory

When the chemoreceptor theory of De Castro (1929) was becoming popular, Kohn's paraganglion theory was essentially revised by Watzka (1943). He classified paraganglia into two categories: *chromaffin paraganglia* originating from the sympathetic nervous system and *nonchromaffin paraganglia* derived from the parasympathetic nervous system. According to Watzka (1943), the carotid body was a type of nonchromaffin paraganglion. It was proposed that the nonchromaffin cells secreted acetylcholine and/or some other parasympathomimetic substances which antagonized adrenaline and/or other sympathomimetic substances secreted from the chromaffin paraganglia.

19.1.6 Second Receptor Cell Theory

De Kock (1954) investigated the carotid body of several mammalian species by means of the silver-impregnation method. He classified "chemoreceptor cells" of the carotid body into two types—a *glomus cell type I* and a *glomus cell type II*. The type I cell was the chief cell; the type II cell was considered to be a second receptor cell. However, it now appears that the type II glomus cell which De Kock first described was an artefact of the silver-impregnation method (Kobayashi unpublished data).

19.1.7 Indubitable Demonstration of Monoamines

In the 1960s, the occurrence of monoamines, i.e., adrenaline or "sympathomimetic substances" of the researchers of the 1940s (see Watzka 1943), was firmly established by means of the *formaldehyde-induced fluorescence* (FIF) method (see p. 23). Muscholl et al. (1960) seem to have been the first to apply the fluorescence method for investigating the carotid body. They found that some chief cells emitted specific fluorescence after fixation with formaldehyde. Later, using the Falck-Hillarp method (see p. 23), many authors investigated the carotid body of various vertebrate species (review: Kobayashi 1971a; Böck 1982). It was demonstrated that all the chief cells in the carotid body contained monoamines such as catecholamines and serotonin. Thus, reviewing previous discussions, Kobayashi (1971a) concluded that the lack of a positive chromaffin reaction in the carotid body does not necessarily mean the absence of monoamines in the cells, but it may imply their presence in amounts smaller than that detectable by the reaction. The previous contradictory results with the chromaffin reaction in the carotid body could be accounted for by underestimation of the specificity (e.g., De Castro 1929, 1951; Schumacher 1938) or overestimation of the sensitivity of this reaction (e.g., Watzka 1943).

19.2 Structure and Fine Structure

19.2.1 Light-Microscopic Features

Cytological features of the carotid body chief cells have been reviewed by Biscoe (1971), Kobayashi (1971b), Kjaergaad (1973), Dreyer (1975), Verna (1979), McDonald (1981), and

Böck (1982). The chief cells are typical paraneurons. They share a number of cytological, cytochemical, and ultrastructural characteristics with adrenal chromaffin cells and the SIF cells of the autonomic ganglia. There is accumulating evidence supporting the view that the carotid body chief cells are specific chemosensory paraneurons which detect information in the bloodstream and transmit it to the nerve terminals.

According to Capella and Solcia (1971), secretory granules of human carotid body chief cells are positively stained with lead-hematoxylin, react metachromatically after hydrochloric acid treatment, and are silver-impregnated by Grimelius' method. These stainabilities are prominent in the human and canine carotid body, but in the rat, mouse, guinea pig, and some other commonly used laboratory animals it is difficult to detect these properties of the chief cells (Kobayashi, unpublished data). This may be ascribed to the small number of secretory granules in the carotid body chief cells of these animal species (see also p. 206).

19.2.2 Electron-Microscopic Features

The ultrastructure of the chief cells of the carotid body is similar to that of adrenal chromaffin cells. The cytoplasm contains mitochondria, rough endoplasmic reticulum, Golgi complexes, secretory granules, lysosomes, microtubules, and filamentous structures of various thickness. The cells are characterized by *secretory granules*, consisting of a round membrane sac (diameter—100–200 nm in the dog; Kobayashi 1968) and a dark irregular-shaped core loosely surrounded by the sac. The number, size, and distribution of the secretory granules differ among species (Böck 1982). In the dog, at least *two types of chief cell* have been distinguished (Kobayashi 1968). The first type possesses secretory granules whose ultrastructure resembles that of the A cells of the adrenal medulla (see p. 136) (Fig. 19.1), though no adrenalinelike immunoreactivity has been demonstrated here (Kobayashi et al. 1983b). The second type is characterized by secretory granules resembling the NA cells of the adrenal medulla (Kobayashi 1968). The second type of chief cells are distinctly positive in the chromaffin reaction. In the rat, mouse, and guinea pig, the carotid body chief cells are of one type and their secretory granules are small in both size and number compared with those in the dog (Kobayashi 1971b). In the rat, mouse and guinea pig, there is a tendency for the secretory granules to gather in a narrow zone of cytoplasm immediately below the cell membrane.

The secretory granules, as in other paraneurons, are formed in the Golgi apparatus and released by exocytosis, as has been demonstrated in various laboratory animals (Kobayashi 1976; review: Böck 1973, 1982). In addition to the secretory granules, the carotid body chief cell contains *small synaptic-type vesicles* of about 40 nm in diameter in the mouse (Kobayashi and Uehara 1970). In many laboratory animals, among them the rat (McDonald and Mitchell 1975), the vesicles show a tendency to gather in the cytoplasm immediately beneath the synaptic membrane thickenings.

19.3 Secretory Aspects

19.3.1 Aminic Substances

Carotid body chief cells possess a typical *APUD ability*. Fluorescence cytochemical methods have been used to demonstrate not only the existence of monoamines in these cells, but also the specific property that they incorporate exogenous amine precursors such as dopa and decarboxylate them to produce amines (Pearse 1968, 1969; Kobayashi 1971a). The uptake and ultrastructural localization of exogenous amine precursors have also been demonstrated in our autoradiographic study using ^{3}H-dopa (Kobayashi 1976).

Recent immunocytochemical studies have shown that *noradrenaline* (Kobayashi et al. 1983b) and *serotonin* (Grönblad et al. 1983) occur in carotid body chief cells. Characterization of the aminic substances identified in these cells will improve with advances in the immunocytochemical techniques employed.

19.3.2 Peptidic Substances

Pearse (1969) anticipated the occurrence of a specific peptide in the carotid body chief cells and proposed the term "glomin" for the presumptive peptide. Immunocytochemical studies were performed using antisera against various peptides such as ACTH, insulin, and glucagon (see Kobayashi 1977), but all these early trials were inconclusive. Later, immunoreactive *enkephalins* were demonstrated in the cytoplasm of the chief cell (Lundberg et al. 1979; Hansen et al. 1982; Varndell et al. 1982). Kobayashi et al. (1983) demonstrated immunoreactive met-enkephalin-Arg^{6}-Gly^{7}-Leu^{8} in the secretory granules of the

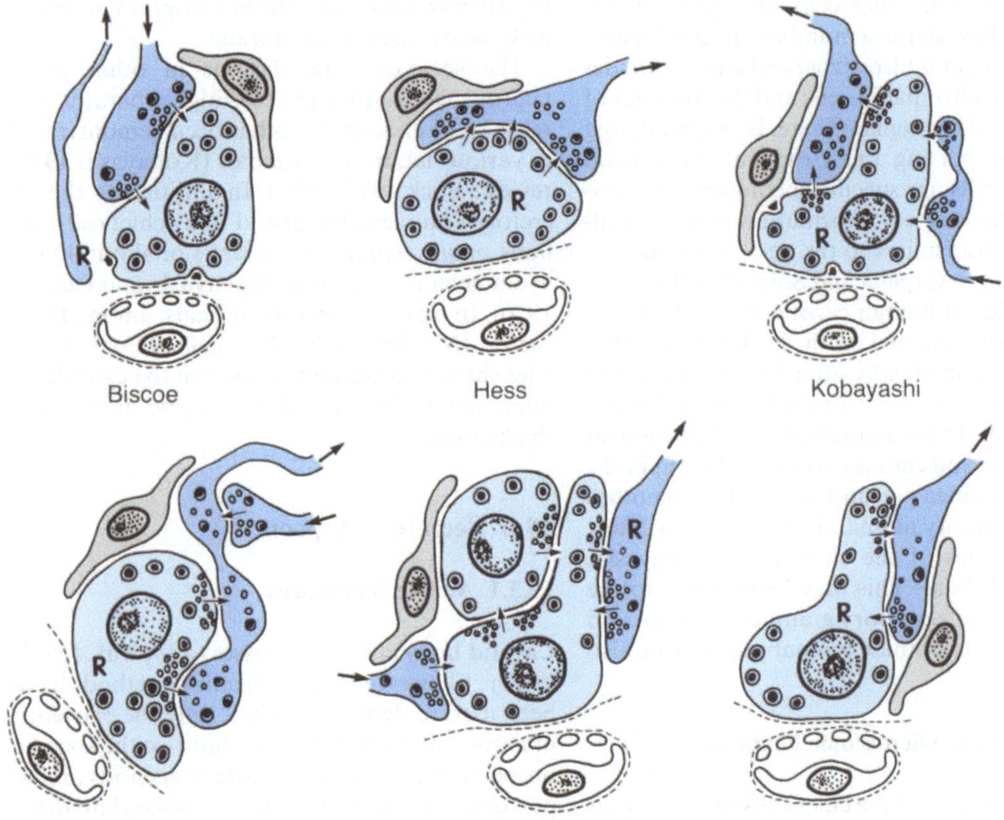

Fig. 19.1. Different views on the carotid body chemoreceptor. The models proposed by Biscoe (1970), Hess (1968), Kobayashi (1971), Verna (1973), McDonald and Mitchell (1975), and Osborne and Butler (1975) are illustrated in simplified form. These authors reached their conclusions based on electron-microscopic observations of the carotid body; all agree that the chemoreceptor end organ consists of the chief cells, sustentacular cells, and sensory nerve terminals. *R* indicates the chemoreceptor element proposed by each author. Hess (1968), Kobayashi (1971), Verna (1973), and Osborne and Butler (1975) regarded the chief cell as the chemoreceptor. Biscoe (1970) and McDonald and Mitchell (1975) proposed that the chemoreceptor is the nerve terminal around the blood vessel or adjacent to the chief cell, respectively. *Thick arrows* indicate the direction of the impulse conduction in the nerve fibers; *thin arrows* show the direction of synaptic transmission

chief cells using the protein A-colloidal gold method for electron-microscopic immunocytochemical study. This octapeptide is specific to preproenkephalin A, an opioid precursor also present in adrenal chromaffin cells. Therefore, a close relationship between carotid body chief cells and adrenal chromaffin cells has been substantiated.

19.3.3 Coexistence of Aminic and Peptidic Substances

The coexistence of aminic and peptidic substances in the secretory granules of the carotid body chief cell was suggested by Pearse (1969). We proposed that ATP and other adenine nucleotides may coexist with these substances in the same secretory granules (Kobayashi 1976). Böck (1980a, b) published cytochemical data supporting the existence of ATP in the secretory granules of the chief cell.

Immunocytochemical studies have shown that the chief cells of the carotid body, like adrenal chromaffin cells, produce and store preproenkephalin A-derived opioid peptides and that these opioid peptides coexist in the secretory granules with catecholamines such as noradrenaline (Kobayashi et al. 1983b).

The granules of carotid body chief cells are immunoreactive for chromogranin A. The same

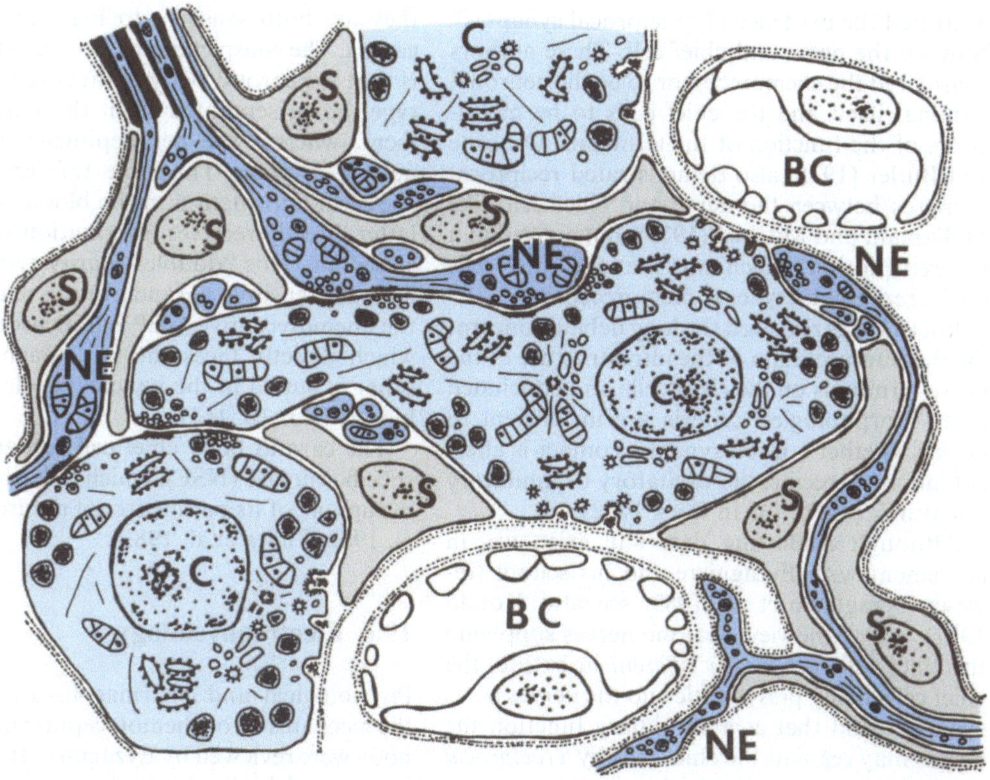

Fig. 19.2. Celluar structure of the carotid body. *C* chief cells, *S* sustentacular cells, *NE* nerve endings, *BC* blood capillaries

immunoreactivity was recognized in the monoamine-storing cells in the media of the avian thoracic aorta (carotid body-like structure introduced on p. 205).

19.4 Innervation of the Chief Cell

Sensory nerve terminals arriving in the carotid body via the sinus nerve make a reticular aggregation around the chief cells (De Castro 1929). Both the chief cells and nerve terminals are incompletely ensheathed by cytoplasmic processes of sustentacular cells, which are immunopositive for S 100 protein (Kondo et al. 1982).

Synapses are formed between the chief cells and varicose nerve terminals. However, the functional interpretation of the ultrastructural features of the synaptic relationship between the chief cell and nerve terminal is difficult. There has been a great deal of debate concerning the polarity of these synapses (Fig. 19.2).

Hess (1968) supported the opinion of De Castro (1929) that the chief cell is a chemoreceptor, although Hess could not find accumulations of synaptic vesicles on the chief cell side. Biscoe and co-workers (see Biscoe 1971), like Hess (1968), showed aggregations of small clear vesicles only on the side of the nerve terminals. However, they reached a conclusion different to that of Hess (1968): Synaptic transmission takes place in the direction from the nerve to the chief cell and there are special nerve terminals whose function is chemoreceptive. Verna (1971) found small synaptic vesicles on the chief cell side, but in his electron micrographs these vesicles showed no tendency to gather in the synaptic region. Verna (1971) further showed that the sensory nerve terminal itself receives innervations which are afferent to it. On the other hand, Kobayashi and Uehara (1970) found aggregations of the small cored vesicles on the chief cell side of the synapses and suggested that these may represent chemosensory synapses. They also described synapses which they thought to be afferent to the chief cell. McDonald and Mitchell (1975) demonstrated the existence of "reciprocal synapses" between the nerve and chief cell. These authors considered the chemoreceptor to be the neuronal terminal itself and the chief cells to be modulators of the function of the terminals. Osborne and Butler (1975) also demonstrated reciprocal synapses between the nerve and chief cell, like McDonald and Mitchell (1975), but arrived at a different notion, which was that the chief cells might represent a type of chemoreceptor.

Böck (1982) reviewed the long debate concerning the interpretation of the ultrastructure of the nerve terminals on the chief cells and concluded that "morphological criteria are insufficient to decide whether a given synaptic contact is efferent, afferent, reciprocal, excitatory or inhibitory in nature" (see p. 102 in Böck 1982).

Although conflicting data still exist, we, in agreement with Eyzaguirre and his school (review: Eyzaguirre et al. 1983; see also Kondo 1977), are of the view that the nerves supplying the chief cells are *mostly afferent* in nature, the chief cells being presynaptic and the nerves postsynaptic, and that as a subsidiary function the nerves may regulate the chief cells by a *reciprocal* mechanism of synapses. In this context, it seems worthwhile to recollect that ultrastructural evidence is accumulating which suggests the reciprocal nature of nerve terminals synaptically supplying sensory paraneurons, including Merkel, inner ear hair, gustatory, and neuroepithelial body cells. In all these cases, it is likely that the afferent nerves receive information from the peripheral paraneurons, while influencing the activities of the latter with their efferent properties.

Immunoreactivities for substance P (Lundberg et al. 1979; Wharton et al. 1980), VIP (Lundberg et al. 1979; Wharton et al. 1980), and NPY (Kondo et al. 1986) have been localized in nerves which were close to small blood vessels in the carotid body but not in the synaptic contact with the chief cell.

19.5 Carotid Body Chief Cell As a Paraneuron

With its fine structure and secretory function the carotid body chief cell qualifies as a typical paraneuron. It seems appropriate to compare the carotid body chief cell with the paraneuron in the neuroepithelial bodies in the bronchiolar epithelium (Fig. 19.3). Both secrete amines (serotonin) and share synaptic innervation, principally afferent but partly reciprocal. Above all, they are both sensitive to hypoxia and hypercapnia. The conspicuous difference between them is that the carotid body paraneuron is closed in type, being separated from the outer environment, whereas the neuroepithelial body paraneuron is open. Thus, the former detects the chemical information in the blood, whereas the latter directly receives information from the inspired air. This "double-security system" of the hypoxia-sensitive paraneurons is reminiscent of the chemoreceptive GEP paraneuron, a part of which directly faces the gut lumen, while the other is buried in the pancreatic islet facing the bloodstream (p. 167).

The carotid body chief cell contains neuron-specific enolase (NSE), which is further evidence in support of its paraneuronal nature (Kondo et al. 1982; Fujita et al. 1983).

19.6 Electrophysiology

Physiological and pharmacological studies on the mechanisms of chemoreception in the carotid body were reviewed by Eyzaguirre et al. (1983). If the carotid body chief cell is a chemoreceptor paraneuron, its electrophysiological properties must be in accordance with those of sensory transducers.

A resting membrane potential of 10–60 mV with a mean value of about 20 mV has been obtained both in vivo (Baron and Eyzaguirre 1977) and in vitro (rabbit embryo, culture—Acker and Pietruschka 1977). However, so far as we know, there is no report of the recording of action potentials. It is likely that the chief cells are too small and too labile for the insertion of a microelectrode, thereby preventing accurate electrophysiological measurement. Improvements in recording techniques are necessary to clarify whether the chief cells produce action potentials or not.

Several recent reports have stressed the importance of the *calcium ion* in the transduction process across the chemoreceptor synapse (Fidone and Gonzalez 1986). Grönblad et al. (1979) demonstrated by electron microscopy that the Ca^{2+} ionophore A23187 produced a striking increase in the exocytotic release of secretory granules of the chief cell. It is likely that a variety of natural and pharmacological stimuli induce the elevation of the intracellular Ca^{2+} level, followed by the exocytotic release of monoamines and other messenger substances within the secretory granules and synaptic type

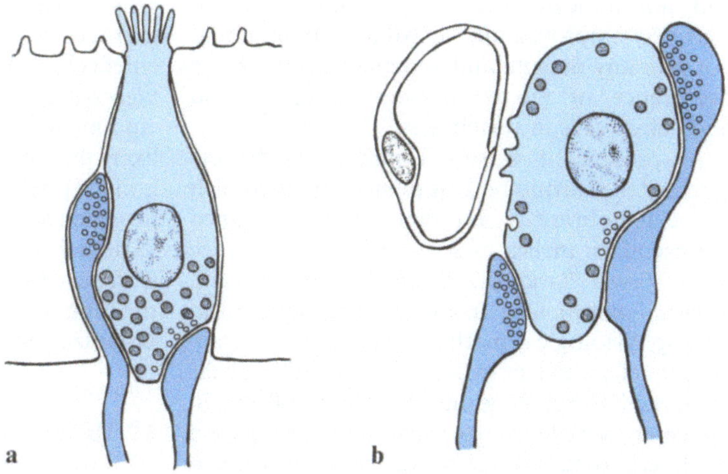

Fig. 19.3a, b. Comparison between the hypoxia-sensitive paraneuron of **a** the open type in the repiratory tract and **b** that of the closed type in the carotid body. The cell **a** directly detects the chemical information in inspired air, while the cell **b** detects that in the blood

vesicles. However, the transmitter substance of the chemoreceptor synapses has not yet been identified (Eyzaguirre et al. 1983).

19.7 Carotid Body-Like Structures

Carotid body-like small organs containing parenchymal cells similar to the chief cells have been described at various locations in the body. They include subclavian bodies, aorticopulmonary bodies, tympanojugular bodies, laryngeal paraganglia, peribronchial paraganglia, and cardiac paraganglia (for references, see Böck 1982).

The existence of monoamine-storing cells was reported in the media of the thoracic aorta (Ookawara et al. 1974; Kondo 1974) and pulmonary trunk and arteries (Taha and King 1986) of birds. Morphological features of these cells and nerve terminals on them are apparently similar to those of carotid body chief cells. However, their functional significance is unknown. Likewise, it is not known whether similar cell groups exist in the vascular wall of mammals, especially in man. It is of interest here that Le Lièvre and Le Douarin (1975) found a group of cells similar to carotid body chief cells in the large arteries arising from the heart in their quail-chick chimera system.

19.8 Ontogeny

The ontogeny of carotid body chief cells has been extensively discussed. Stieda (1881), who regarded the carotid body as a branchial organ, considered it to be derived from the pharyngeal endodermal pouch. Kohn (1903) regarded the carotid body as a chromaffin paraganglion and believed that it was a derivative of the sympathetic ganglia. In the chemoreceptor theory of De Castro (1929), a common origin of the carotid body and dorsal ganglia was suggested. Watzka (1943), who classified the carotid body among the nonchromaffin paraganglia, thought it to originate in the anlage of the parasympathetic ganglia. Arnold (1865) and Schumacher (1938) considered the carotid body to be a vascular glomus or arteriovenous anastomosis and believed in a mesenchymal origin of the chief cell. In the light of the past debate, it is clearly difficult to give an unequivocal decision on the origin of such a small organ as the carotid body based on classic techniques of embryology.

Studies by Korkala and Hervonen (1973) using the FIF method showed that the catecholamine-containing chief cells migrate from the developing sympathetic trunk to the carotid body. Le Douarin et al. (1972) using their quail-chick marker system showed that after a quail rhombencephalon was grafted into a chick embryo, the carotid body of the host contained quail chief cells. This is consistent with the theory that carotid body chief cells are derived from the *neural crest* (see also Le Douarin 1982).

19.9 Phylogeny

A comprehensive review of phylogenic aspects of the carotid body in the era of the classic histological methodologies is available in Adams (1958). In this monograph, the carotid body and related organs from elasmobranchial fish to man are critically discussed. At the time of writing, fluorescence cytochemical, electron-microscopic, and immunocytochemical methods were not routinely employed and so Adams (1958) could not give any definite and systematic data on the occurrence of monoamine-containing carotid body chief cells in vertebrates.

At present, it is still a matter of dispute whether the carotid body is present in teleost fish and elasmobranchs. Laurent (1974) examined the pseudobranches of some fish such as *Tinca tinca*, *Perca fluviatilis*, *Sander lucioperca*, and *Torpedo marmorata* using morphological and neurophysiological methods. He found characteristic cells which he called chief cells. These cells contained secretory granules, about 100 nm in diameter, sent long cytoplasmic processes, and made synapses with the dendrites of "large neurons." Laurent (1974) proposed that fish pseudobranches are arterial chemoreceptive structures.

We examined carotid bodies in amphibians (*Cynops pyrrhogaster pyrrhogaster*, *Rane nigromaculata*, *Rana catesbeiana*, *Bufo bufo japonicus*), a lizard (*Eumeces latiscutatus*), birds (*Uroloncha domestica*, *Gallus domesticus*), mammals (rat, mouse, dog), and human fetuses by the fluorescence histochemical method for monoamines. In all the vertebrate species examined, the carotid body contained FIF-positive chief cells (Kobayashi 1971a).

The carotid body in *amphibians* deserves particular mention. The common carotid artery forms a *rete mirabile* at the origin of the internal and external carotid arteries (see Noguchi and Kobayashi 1977). Thus, the carotid body of amphibians is often called the carotid labyrinth (Adams 1958). There are many monoamine-containing cells in the connective tissue space of the carotid labyrinth (Kobayashi 1971a). These cells extend two to several short cytoplasmic processes and represent the amphibian equivalent of the mammalian carotid body chief cell.

We examined the ultrastructure of chief cells and their innervations in amphibians (*Hynobius nigrescens*, *Cynops pyrrhogaster pyrrhogaster*, *Bufo bufo japonicus*, *Rana nigromaculata*), birds (*Uroloncha domestica*, *Gallus domesticus*), and mammals (cat, dog, mouse, guinea pig) (Kobayashi 1971b). All these animals and human fetuses were found to have the same fundamental composition in the carotid body, i.e., aggregation of chief cells, sustentacular cells, and nerve terminals (Kobayashi 1971b). Neuronal terminals made synaptic contacts with the chief cells, which always contained membrane-bound granules. Similarities in the ultrastructure of the chief cells of lower vertebrates, such as *Hynobius nigrescens*, and those of higher vertebrates, including human fetuses, were striking. It seems likely that the fundamental structure of the carotid body is phylogenetically complete at the stage of urodelen amphibians and that its involvement in various body functions, such as respiration control, develops in association with the evolution of the lung, vascular system, and respiratory center organization in the central nervous system.

19.10 Summary

Carotid body chief cells are intercalated between the bloodstream and nerve terminals, which transmit information concerning the pressure of CO_2 and/or O_2 in the blood to the respiratory center of the medulla oblongata. Carotid body chief cells are characterized by membrane-bound secretory granules containing monoamines (noradrenaline and serotonin), peptidic messengers (enkephalins), probably adenine nucleotides (ATP, etc.), and chromogranins. With these secretory features, the carotid body chief cell can be regarded as a typical paraneuron.

Nerve terminals make a conglomeration around the chief cells. Both the chief cells and the terminals are ensheathed by *sustentacular (Schwann) cells*, which are immunoreactive for S-100 protein. Synapses are formed between chief cells and the nerve terminals. There has been some debate on the site of chemoreception and on the mechanisms of stimulus transmission at these synapses. Recent electrophysiological and morphological data support the view that the *chemoreceptor* is represented by the chief cell itself and its excitation is conducted to the nerve terminal. It is also suggested that the synapses are reciprocal in nature. Nerves may also regulate the function of the chief cell.

In all vertebrate classes from amphibians to mammals, the carotid body contains aggregations of chief cells, sustentacular cells, and sensory neuronal terminals. A similar complex formation of cells and nerve terminals is present in fish *pseudobranches*.

Recent embryological studies using the quail-chick marker system indicate that carotid body chief cells derive from the *neural crest*.

20. Gustatory Cells in Taste Buds and Related Sensory Cells

20.1 General Aspects

Taste buds are *chemoreceptor end-organs* for taste. In man, most of them occur in the stratified squamous epithelium of the dorsal surface of the tongue concentrated in the circumvallate, foliate, and fungiform papillae. A small number of taste buds also exist in the stratified squamous epithelium of the soft palate, lower part of the pharynx, and laryngeal surface of the epiglottis. The sense of taste, however, is widely distributed in the oral cavity and pharynx. The structural and functional differentiation of taste buds corresponding to the four kinds of taste sensation, i.e., sweet, bitter, sour, and salty, is unknown. Thus, it is still to be established how important a role in taste sensation is played by taste buds.

The shape of taste buds resembles that of closed rosebuds, hence the name. A gustatory canalicule or taste pore connects the taste buds with the oral, pharyngeal, and laryngeal cavities. There are at least two kinds of cells in the taste bud in mammals:

1) *gustatory cells*, which contain membrane-bound vesicular structures and which make synaptic contacts with the *nerve terminals* on their lateral surface; and
2) *supporting* or *sustentacular cells*, which have an elongated heterochromatin-rich nucleus and a dark cytoplasmic matrix. These two kinds of taste bud cells extend from the basement membrane to the lumen of the gustatory canalicule. They have a tuft of microvilli on their apical surface; the microvilli are invested with a special glycosaminoglycan material within the small dilatation at the bottom of the taste canalicule.

20.2 Cytological Features

20.2.1 Cytochemical Properties

The cytochemical properties of taste buds were extensively investigated by light and electron-microscopy. Iwayama and Nada (1967) examined the localization of *alkaline phosphatase*, *ATPase*, and *acid phosphatase* activities in the rabbit and rat by light-microscopic cytochemical methods. They demonstrated alkaline phosphatase activity on the luminal surface, ATPase activity on the basal, lateral, and luminal surfaces, and acid phosphatase activity within the cytoplasm, especially in the supranuclear area of the taste bud cells. The occurrence of ATPase activity on the surface of taste bud cells was confirmed by electron-microscopic cytochemistry (Yamamoto and Ozawa 1977). ATPase activity was also seen on the surface of the nerve terminals (Yamamoto and Ozawa 1977).

Acetylcholine esterase activity was demonstrated around the nerve terminals and in the intercellular space between some taste bud cells (Paren and Mattern 1975). However, it has been pointed out that the localization of the acetylcholine esterase reaction and that of acetylcholine are not topographically related.

Moura and Miraglia (1978) investigated by light microscopy the localization of various substances including enzymes such as phosphorylases and alanyl aminopeptidases in the taste buds of marmosets. None of the reactions examined by these authors were specific to any particular cell type, i.e., gustatory or sustentacular cells.

Neuron-specific proteins have been demonstrated in a restricted type of taste bud cells. These cells contain NSE (Fujita et al. 1983a) and spot 35 protein or Purkinje cell-specific protein (Iwanaga et al. 1985a) in the rat and some other mammals. It was proposed that the immunoreactive cells are gustatory cells and that they are paraneurons supported by (immunonegative) sustentacular cells and innervated by the sensory nerve terminals (Iwanaga et al. 1985a). Immuno-electron-microscopic studies are required to clarify this point. The sustentacular elements of the taste bud regrettably do not reveal immunoreactivity for such glia-specific substances as S-100 protein (Fujita, unpublished data).

20.2.2 Ultrastructure

Using the transmission electron microscope, the taste buds of various mammalian species were investigated by many authors (Farbman 1965a; Murray et al. 1969; Uga 1969; Paren et al. 1975; Takeda and Hoshino 1975; Idé and Munger 1980; Farbman et al. 1985). A comprehensive review on the ultrastructure of taste bud was published by Murray (1971, 1973). There is a conspicuous species variation in the ultrastructure of taste buds even in mammals. Categorization of the cell types in the taste bud is different from author to author (Murray 1971, 1973; Idé and Munger 1980).

In all mammals, gustatory cells occur containing *large çored vesicles* and *small synaptic-type vesicles*. The large cored vesicles are dispersed in the basal cytoplasm, whereas the small synaptic-type vesicles show a tendency to gather immediately beneath the synaptic membrane thickenings on the basal and lateral surfaces. The cytoplasm contains mitochondria, granular endoplasmic reticulum, Golgi apparatus, and lysosomes between the large cored and small synaptic-type vesicles. Filamentous elements are also present throughout the cytoplasm. Thus, the ultrastructure of gustatory cells is in line with the criteria for paraneurons (Fig. 20.1).

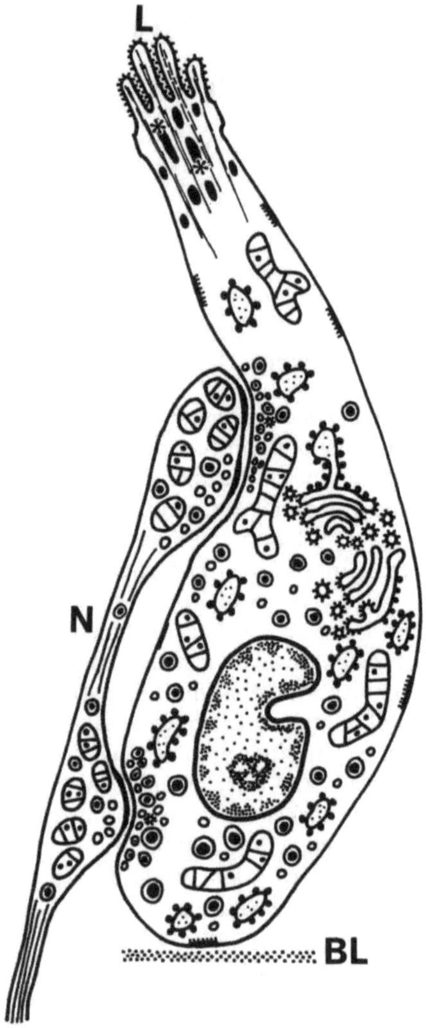

Fig. 20.1. Structure of a gustatory cell in mammalian taste bud. The gustatory cell extends from the basal lamina (*BL*) to the oral cavity (*L*). A sensory nerve (*N*) makes synaptic contact with the gustatory cell. Small cored and clear vesicles accumulate in the cytoplasm adjacent to the synapses. Large cored vesicles are scattered in the basal cytoplasm, but they show no special tendency to gather at the synaptic sites. In the apical cytoplasm, there are bundles of actin filaments intermingled with dark ellipsoidal bodies of unknown significance (*asterisks*)

20.2.3 Amine-Handling Ability

Gustatory cells contain monoamines such as *dopamine* and *serotonin* (Nada and Hirata 1975a, b). Selective incorporation of exogenously administered monoamine precursors has also been demonstrated by fluorescence cytochemical (Geerdink and Drukker 1973), electron-microscopic (Nada and Hinata 1975c; Takeda and Kitao 1980; Takeda et al. 1982), and autoradiographic methods (Takeda and Suzuki 1983). The large cored vesicles are the sites of storage of the exogenous monoamine precursors (Takeda 1977; Takeda et al. 1982; Fujimoto 1982; Uchida 1985) (Fig. 20.3). Some of the small synaptic-type vesicles also accumulate exogenous monoamine precursors (Takeda et al. 1982). More sensitive and specific methods are needed to determine whether the small vesicles possess monoamines under nonexperimental conditions.

Electron-microscopic studies have shown that the large cored vesicles contain fine granular material (Murray 1971, 1973). It is probable that this materials is mainly peptidic in nature. However, peptidic substance specific to the gustatory cell vesicles has not as yet been demonstrated.

20.3 Sensory Innervation

In man, taste sensation of the anterior two-thirds of the tongue is conducted by the facial nerve, though the number of taste buds in this area is small. Sensory nerve fibers entering the taste buds of the posterior third of the tongue mostly come from the glossopharyngeal nerve, which has two peripheral ganglia—ganglion superius et inferius. The taste buds of the pharynx and epiglottis are innervated by the vagus nerve, which, like the glossopharyngeal nerve, has two peripheral ganglia. These nerve fibers penetrate the basement membrane, lose the myelin shealth, and end on the gustatory cells in varicose, richly branching nerve terminals (Murray 1971, 1973).

Immunocytochemical investigations have shown that some of the nerve fibers in the taste bud contain substance P (Lundberg et al. 1979c). Electron-microscopic immunocytochemical studies have demonstrated that the substance P-immunopositive nerve terminals possessed accumulations of small synaptic-type vesicles and that none of them made synaptic contact with the gustatory cells (Yamasaki et al. 1984). Thus, the substance P-immunopositive nerve terminals may not be gustatory in nature, though a possible reciprocal nature of the nerves concerned may not be excluded. The peptides in the nerve terminals which make synapses with the gustatory cells have not yet been identified.

There are many monoamine-containing nerve terminals both in and around the taste buds (Gabella 1969; Geerdink and Drukker 1973; Takeda 1976). Takeda (1976) is of the opinion that some of them are related to taste sensation.

20.4 Electrophysiology

Gustatory cells in taste buds are *contact chemoreceptor cells* that respond to substances in the oral fluids bathing them. These substances appear to evoke *receptor potentials* (Kimura and Beider 1961; Akaike et al. 1976), but the manner in which taste-producing molecules in solution interact with receptor cells to produce receptor potentials is unsettled. There is evidence that taste-producing molecules such as water, NaCl, HCl, and quinine act on the membranes of the receptor cells or their processes and modulate the number of action potentials produced in the chorda tympani nerve.

Concerning the *transduction mechanism* in vertebrate gustatory cells, Kurihara et al. (1981) proposed a hypothesis that adsorption of chemical stimuli on the microvillous membranes of gustatory cells induces depolarization by disturbing the boundary potential at the membrane. Support for this view that disturbance of the boundary potential results in depolarization has been provided by experiments carried out in cultured neuroblastoma cells, which respond to several gustatory stimuli. However, so far as we know, no successful experiments along similar lines have been performed on mammalian gustatory cells.

Avenet and Lindemann (1987) recently dissected taste discs from the tongue of frog, *Rana ridibunda*, dissociated the gustatory receptor cells, and studied using the patch-clamp technique. Their whole-cell recordings did not indicate anion-conductive pathways, but they found a voltage-gated tetrodotoxin-blockable Na current and voltage-gated K currents which were blocked by Ba. Therefore *the taste receptor cells should be capable of firing action potentials*, which were indeed observed during passage of depolarizing currents under current-clamp conditions. Furthermore, they characterized the blockage of K currents by cyclic AMP and forskolin, suggesting that cyclic AMP may, in analogy to olfactory transduction (see p. 216), be one of the intracellular messengers in taste transduction. The recent report by Kinnamon and Roper (1986) that acid gustatory stimuli depolarize Necturus taste cells *in situ* by recording the K conductance is in a general way consistent with the results obtained by Avenet and Lindemann (1987) if one postulates that the acid stimuli leads to an increase of the cyclic AMP-level in intact taste cells.

The results of previous histological (DeHan and Graziadei 1971; Graziadei and DeHan 1973) and physiological (Morimoto and Sato 1975, 1982) studies are consistent with the view that a probable candidate for the *messenger substance* contained in the secretory granules in the gustatory cell is noradrenaline. Morimoto and Sato (1975, 1982) added catecholamines and their antagonists to a solution perfusing the frog lingual artery and showed that noradrenaline enhances both the spontaneous impulse discharges in the glossopharyngeal nerve and nerve responses to taste stimuli. These authors also showed that the addition of a β-blocker suppresses the taste nerve responses. Nagahama and Kurihara (1985) have examined the suppression of the frog taste responses by the injection of catecholamine-depleting agents (reserpine and guanetidine) and the recovery of the responses by catecholamines added to a solution perfusing the lingual artery. It has been shown that catecholamines added to the lingual artery do not greatly affect the spontaneous nerve discharges but enhance the nerve responses to taste stimuli by being incorporated into the taste cells. These results provide further support for the view that the transmitters in the gustatory synapse are catecholamines, especially noradrenaline.

The possibility of *Ca^{2+}-dependent release of a messenger substance by exocytosis* from the gustatory cell has also been examined. Nagahama et al. (1982) reported that the frog taste nerve responses to chemical stimuli of group 1 ($CaCl_2$, NaCl, distilled water, D-galactose, L-threonine) greatly decrease with a decrease in the Ca^{2+} concentration of a solution perfusing the lingual artery; the responses are enhanced by cyclic GMP (cGMP) and suppressed by cyclic AMP (cAMP). On the other hand, the responses to chemical stimuli of group 2 (quinine hydrochloride, theophylline, ethanol, and HCl) are practically unaffected by Ca^{2+}, cGMP, and cAMP.

These results reported in recent years are in keeping with the view that the gustatory cell in the taste bud is a receptor paraneuron bearing all the characteristics of a receptoconductile-secretory cell.

20.5 Ontogeny and Kinetics

Electron-microscopic investigations on the developing taste bud were performed by Farbman (1965b) and Takeda (1972). An extensive review was published by Farbman (1971).

In the rat, in which intrauterine life lasts for approximately 21 days, the primordial taste bud can be identified on the 15th day of gestation. The primordial taste buds possess many nerve fibers, which extend almost to the surface of the epithelium. At this stage, there are characteristic cored vesicles, 50–100 nm in diameter, in the future gustatory cells. It is striking how much the developing gustatory cells resemble such chemosensory paraneurons as carotid body chief cells (p. 205) and pulmonary neuroepithelial body cells (p. 197). It may be significant that the large cored vesicles, which are abundant in the gustatory cells during intrauterine life, are reduced in number after birth (Farbman 1965b).

Taste bud cells in the adult rat have a *life span* of about 10 days (Beider and Smallman, 1965). There is also evidence of cell death within the taste bud (Beidler and Smallman 1965; Idé and Munger 1980). Mitosis is seen in the basal portion of the stratified squamous epithelium around the taste bud. It is widely accepted that the cells in the taste bud are constantly replaced by newly differentiated cells from the surrounding epithelium (Beidler and Smallman 1965). However, mitotic cells exist within the taste bud (Toyoshima and Tandler 1986). It is not known whether these self-renewing populations of taste bud cells produce the gustatory cells.

The integrity of the taste bud is dependent on the *trophic influence* of *innervation* (Guth 1971). After sectioning the facial, glossopharyngeal, or vagal nerves, the taste buds rapidly disappear (Iwayama et al. 1967). They reappear after reinnervation (Fujimoto and Murray 1967; Zalewski 1970; Iwayama 1970). Taste bud formation does not require the presence of the original gustatory epithelium (Zalenski 1970).

Summing up, the gustatory cell represents a paraneuron which shares precursor cells with ordinary epithelial cells and whose occurrence is induced and sustained by special innervation.

20.6 Phylogeny

The size, shape, and distribution of taste buds differ from class to class (Ariens Kappers et al. 1967; Kare 1971).

No taste buds are present in amphioxus. Taste buds are sparsely distributed on the surface of the head and on the mucous membranes of the mouth in cyclostomes. Taste buds become more numerous in higher fishes. Teleosts have the most numerous taste buds among vertebrates. In fish, taste buds are not only distributed in the mouth but are widely spread over the surface of the body. They are not so numerous in amphibians and are found in still smaller numbers in reptiles. In birds, taste buds are sparse. In mammals, taste buds again become numerous (Ariens Kappers et al. 1967; Kare 1971).

The ultrastructure of taste buds has been studied in submammalian vertebrates: birds (Suzuki and Takeda 1984), reptiles (Korte 1980; Uchida 1980), amphibians (Düring and Andres 1976; Goossens and Vandenberghe 1974), and teleosts (Hirata 1966; Storch and Welsch 1970; Reutter 1971). The monoamine-handling ability of some taste bud cells in submammalian species has been extensively investigated (Nada and Hirata 1975b, c; Hirata and Nada 1975; Goossens and Vandenberghe 1974). These monoamine-handling cells are not always engaged in the taste sensation. The range of gustatory cells and taste buds seen in mammalian and submammalian phyla is illustrated in Fig. 20.2.

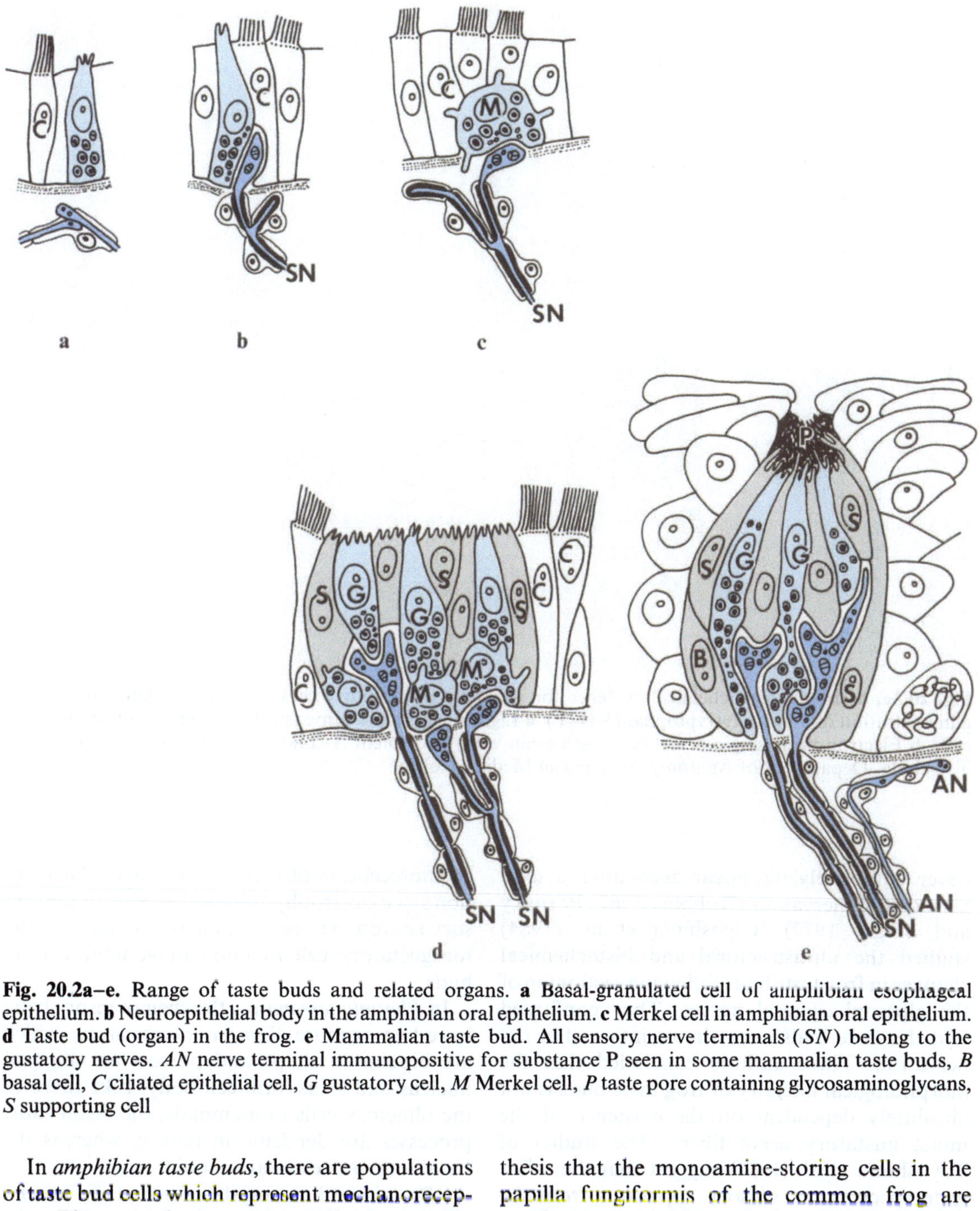

Fig. 20.2a–e. Range of taste buds and related organs. **a** Basal-granulated cell of amphibian esophageal epithelium. **b** Neuroepithelial body in the amphibian oral epithelium. **c** Merkel cell in amphibian oral epithelium. **d** Taste bud (organ) in the frog. **e** Mammalian taste bud. All sensory nerve terminals (*SN*) belong to the gustatory nerves. *AN* nerve terminal immunopositive for substance P seen in some mammalian taste buds, *B* basal cell, *C* ciliated epithelial cell, *G* gustatory cell, *M* Merkel cell, *P* taste pore containing glycosaminoglycans, *S* supporting cell

In *amphibian taste buds*, there are populations of taste bud cells which represent mechanoreceptors. These mechanoreceptor cells are usually localized in the basal portion of the taste buds and are not connected with the external environment; they cannot receive chemical information directly from the oral cavity. Furthermore, their cell surface is provided with characteristic projections which are identical to the digital projections of Merkel cells (see p. 229; Düring and Andres 1976; Toyoshima et al. 1984). Goossen and Vandenbergh (1974) proposed the hypothesis that the monoamine-storing cells in the papilla fungiformis of the common frog are touch receptors which assist in catching prey. It is possible that the taste buds of submammalian vertebrates may be used for more than one perception system. It has even been suggested that frog taste bud cells possess a paracrine function (Toyoshima et al. 1984).

As mentioned above (p. 208), mammalian taste buds are dependent on the gustatory nerves for their morphological and functional integrity (Guth 1971). However, amphibian taste buds

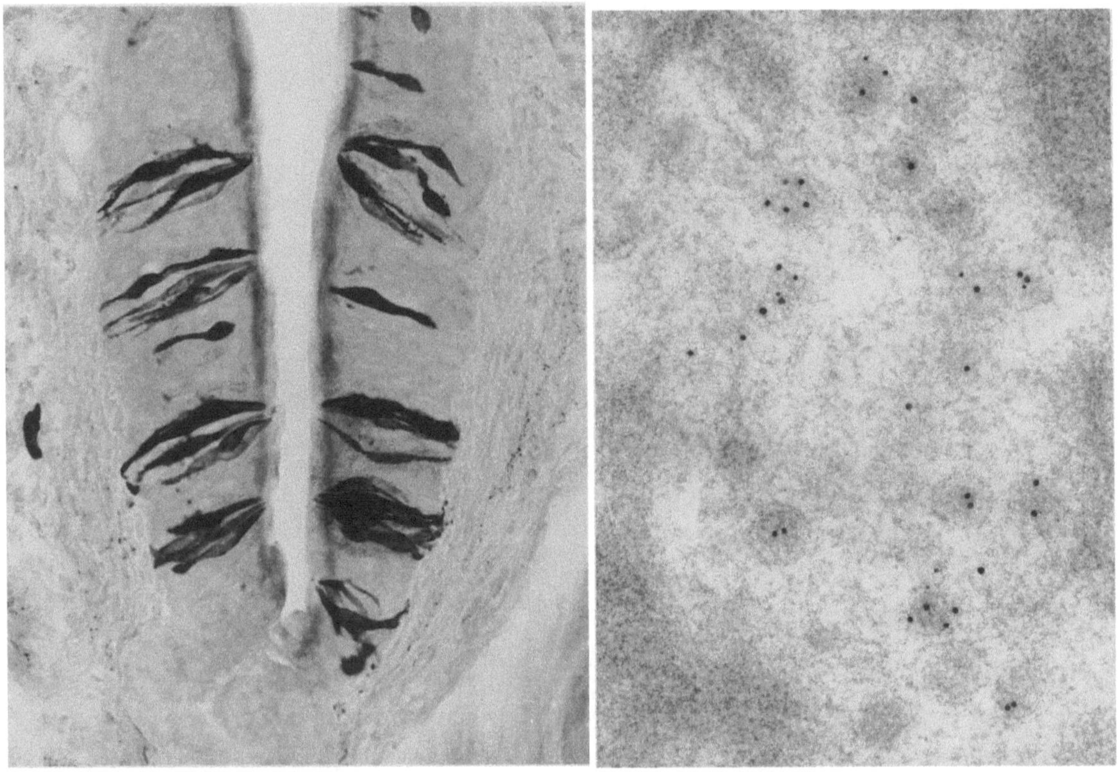

a b

Fig. 20.3a, b. Immunocytochemical evidence for serotonin in gustatory cells of mouse vallate papilla after administration of 5-hydroxytryptophan(5-HTP). **a** Light-microscopic demonstration of serotonin in taste bud cells. **b** Electron-microscopic detection of serotonin with the protein A-colloidal gold method. Prepared by T. Uchida, Department of Anatomy, Yamanashi Medical College

never completely disappear even after a long period of denervation (Robbins 1967; Poritsky and Singer 1977). Toyoshima et al. (1984) studied the ultrastructural and histochemical changes in frog taste buds following sectioning of the glossopharyngeal nerves. They confirmed that the taste buds were almost intact after denervation. These authors concluded that the morphological integrity of frog taste buds is not absolutely dependent on the existence of the intact gustatory nerve fibers. The studies of Toyoshima et al. (1984) suggest that there is a wide spectrum of sensory apparatus from the nerve-dependent gustatory organ of mammals to the self-governing gustatory mechanoreceptor-paracrine organ of amphibians.

20.7 Appendix: Labellar Chemoreceptor of Flies

Many kinds of *contact chemoreceptor* are described in invertebrates. Among them, labellar chemoreceptors of flies deserve particular mention since electrophysiological aspects of the sensory neurons of this organ may provide a model for gustatory cell function in vertebrate taste buds.

In *Phormia terranovae*, the chemosensory hairs have two or more *chemosensory neurons*, which are bipolar in shape and possess an elliptical nucleus and a fusiform cell body. Like those of the olfactory cells in mammals, the distal short processes are dendritic in nature, whereas the basal long processes are axonic.

Prominent hairs on the labellum of *Phormia* contain two, three, or four dendritic endings, the tips of which are sensitive to chemical stimulation (Ralph 1978). One of the labellar receptors responds to sugars, another to cations, and a third to water. A fourth chemoreceptor cell, present in some labellar hairs, has been shown to be an anion receptor. No specific receptors for amino acids have been demonstrated in flies, but such receptors are found in Crustacea.

The frequency of impulses in chemoreceptors

bears a logarithmic relationship to the stimulus concentration over much of the dynamic range, like other kinds of receptors. As with mechanoreceptors, it is common to find, in all types of chemoreceptor, low-frequency, spontaneous activity occurring without any application of stimuli. Generator potentials have been recorded from taste receptors of insects. However, insect chemoreceptors exhibit paradoxical features, such as depolarization when stimulated by some substances and hyperpolarization in response to others. Moreover, certain chemical substances may depolarize one receptor cell and at the same time hyperpolarize another (Ralph 1978).

The cellular constituent of taste receptors of insects resembles that of olfactory receptors of vertebrates, in which the reception of stimuli and generation of afferent impulses are processes in an olfactory cell (Chap. 21).

20.8 Summary

There are at least two kinds of cells in the taste bud in mammals: the *gustatory cell*, which contains neuron specific substances (NSE and spot 35 protein) and makes synaptic contact with the *nerve terminals* on their lateral surface, and the *sustentacular cell* of a probable glial nature.

The gustatory cell are provided with a specific tuft of microvilli on the cell apex where chemical stimuli in the oral cavity are detected. The cell contains large cored vesicles and small synaptic-type vesicles, the latter show a special tendency to gather to the synaptic site. Gustatory cells possess an amine-handling capacity, but a species difference is significant in both qualitative and quantitative aspects of monoamines in the cell. There is increasing evidence that the transmitter substance of the gustatory synapses is noradrenaline. No report is available on the action potentials from the gustatory cell.

Gustatory cell originates from the epithelial cells around the taste bud. The presence of nerve terminals is essential for the development and maintenance of the taste bud and gustatory cells. Gustatory cell thus represents a paraneuron which is *non-nervous in origin* but nerve-dependent in nature. Gustatory cells are constantly turning over and have a half-life of about 10 days.

Phylogenetically, taste buds first appear in cyclostomes, become numerous in teleost fish, decrease in number in birds, and again increase in number in mammals. Taste buds are closely related to the development of the gustatory center of the central nervous system. Their morphological and functional integrity in mammals is dependent on the gustatory nerves.

Contact chemosensory cells in the labellar organs of insects provide experimental models of the electrophysiological studies of gustatory cells of mammalian taste buds.

21. Olfactory Cells

21.1 General Aspects

The olfactory membrane is a yellowish area in the upper posterior part of the nasal cavity. The epithelium here is thicker than the surrouding respiratory epithelium, ranging from 30 μm in moles, 60 μm in humans, and 150–200 μm in frogs and turtles (Graziadei 1973b).

In the epithelium, slender *olfactory cells* are scattered among *supporting cells*, which are high columnar elements covered by microvilli. *Basal cells* which do not reach the epithelial surface are found in the basal portion of the epithelium; these are immature cells which will differentiate into the olfactory cells (see below).

In the propria mucosae under the basement membrane, there are dispersed Bowman's glands. The ducts of these glands ascend between the supporting cells and open onto the epithelial surface. Their secretions, partly proteinic and partly mucous in nature, cover the olfactory membrane and serve as the solvent for odoriferous substances.

Through the work of early histologists, especially Schultze, Babuchin, Cajal, and Retzius (review: Graziadei 1973b), the histological architecture of the olfactory membrane and the essential structure of the receptor cell were established at the end of the last century. In the 1950s, electron-microscopic techniques were applied to the study of the olfactory membrane (human—Bloom and Engström 1952; Amphibia—Bloom 1954) and this method has since become the main tool of investigation in this field (e.g., Moran et al. 1982) with scanning and high-voltage electron microscopes having been applied effectively (e.g., Menco et al. 1978).

The *olfactory cell* is the receptor cell for smell sensation and shows a typical bipolar nature. The oval nucleus of this cell is located in the middle or lower levels at the height of the olfactory epithelium. A coarse apical process (1–3 μm thick) extends to the epithelial surface, and here it swells into a knob called the olfactory vesicle. This extends more than ten olfactory hairs, which measures around 30 μm in length in mammals and up to 200 μm in Amphibia; the hairs run on the epithelial surface and are embedded in the mucous layer of the olfactory membrane.

The basal process of the olfactory cell is a long axon. After passing the epithelium, it evaginates the basement membrane and becomes invested by Schwann cells. It runs through the propria mucosae without branching. The axons thus form the nerve fibers of the fila olfactoria and reach the olfactory bulb, the extension of the brain resting on the cribriform plate of the ethmoid bone. Here, the axons form synapses with mitral cells.

The olfactory cell has been considered a primitive type of sensory cell, reflecting that olfaction is phylogenetically one of the oldest and most primitive of all the senses (Brodal 1981). It has generally been dealt with as a *typical bipolar neuron*. On the other hand, recent studies have indicated that the olfactory cell has a life span of only 30–40 days and is *constantly renewed* by cells differentiating from the basal cells (see below). For this reason, it may be more appropriate to consider the olfactory cell as a paraneuron rather than a neuron (Bennett 1977; Fujita 1977; Fujita and Kobayashi 1979).

21.2 Fine Structure

The olfactory cell, intercalated between supporting cells, is connected to the latter by a junctional complex, especially the zonula occludens, surrounding a constriction just beneath the olfactory vesicle. Two or more olfactory cells often occur in close contact to plasma membranes (Amphibia—Graziadei 1971, 1973a; man—Moran et al. 1982) (Fig. 21.1).

The *apical process* ("dendrite") comprises a columnar cytoplasm containing mitochondria of moderate size, neurotubulelike tubules, and some intermediate filaments running longitudinally. The swollen end of the process, called the *olfactory vesicle*, contains tubules and filaments, though in a less orderly arrangement. This portion also contains considerable numbers of cytoplasmic vesicles, some of which are pinocytotic in nature. The olfactory vesicle contains over ten ciliary basal bodies with the typical 9+2 set of tubules. Each basal body extends an olfactory hair, which consists of a proximal segment about 250 nm thick and a distal segment 100–150 nm thick. The former contains two central (sometimes four in man, Moran et al. 1982) microtubules and nine peripheral doublets of microtubules characteristic of a motile cilium, while in the latter, the peripheral set does not form doublets but nine single tubules.

The problem of whether the olfactory hairs are motile has not been settled (for reference, Graziadei 1973b). In their recent electron-microscopic study, Moran et al. (1982) demonstrated that the axonemes of human olfactory cilia lack dynein arms; these are known to contain Mg^{2+}-ATPase, which generates the force for ciliary motility. Moran et al. (1982) thus suggest that human olfactory hairs must be immotile.

The olfactory cell body is 5–8 µm in diameter and contains a large nucleus with one or two nucleoli. The larger portion of the cytoplasm is on the apical side of the nucleus. This supranuclear cytoplasm contains a prominent Golgi apparatus, some mitochondria, and cisternae of rough and smooth endoplasmic reticulum. There are fewer microtubules and intermediate filaments than in the dendrite.

The cell body tapers basally into the *axon*. (Fig. 21.1). Several axons from different receptor cells gather inside the epithelium in small bundles. The axons here average 0.2 µm in diameter but are often swollen in a spindle-shaped structure, which contains a large oval mitochondrion; microtubules and a few mitochondria are usually seen in these axons (Graziadei 1973b). After passing the basement membrane, the axons are enveloped by the Schwann cell cytoplasm, occasionally separately but usually as a bundle of several or tens of axons in direct contact with each other.

It is of note that olfactory cells, as mentioned above, contact each other at every level of their height, i.e., at the level of the dendrite, cell body, and axon. Distinct synapses between contiguous cell bodies have been demonstrated; the synaptic vesicles here comprise both small clear and large cored vesicles (De Lorenzo 1970). Yamamoto (1976) demonstrated the occurrence of large cored vesicles in olfactory cells of the bat (Yamamoto 1976). It was further shown that the same granules were already present in immature cells still lacking an axon (Yamamoto et al. 1976). The occurrence of these *peptidergic-type vesicles* in the olfactory cells seems worthy of special attention in connection with our view that neurons and paraneurons are generally peptide-secreting cells.

The olfactory cell axons or fila olfactoria penetrate the cribriform bone plate to enter the olfactory bulb and make synapses with the brushlike terminals of mitral cell dendrites, forming the *olfactory glomeruli* (Price and Powell 1970). The olfactory cell terminals contain small clear vesicles and very few, if any, large cored vesicles (Andres 1970). Tufted cells, smaller neurons, also send their dendrites to the glomeruli. The glomeruli and mitral cells occur in much fewer numbers than the olfactory cells, implying an extensive convergence of impulses (Allison and Warwick 1949).

21.3 Physiological Features

The olfactory receptor cells respond only to substances that dissolve in the mucous layer covering the olfactory epithelium and come in contact with the olfactory hairs. The cells are remarkably sensitive to certain substances. For example, methyl mercaptan, the substance that gives garlic its characteristic odor, can be perceived by humans at a concentration of less than 10 ng/l air. On the other hand, discrimination of differences in the intensity of any given odor is poor.

As to the mechanism of the *reception of smell*, the stereochemical theory of Amoore et al. (1964) is widely accepted. According to the theory, the molecules of odoriferous substances are accepted by "keyhole-like" dints in the receptor molecules which occur on the surface of the olfactory hairs (and possibly also on the olfactory vesicle). A dense distribution of intramembranous particles, presumably corresponding to the receptors, has been demonstrated in some electron-microscopic studies (Menco et al. 1976; Usukura and Yamada 1978).

When odoriferous molecules react with a receptor, they generate a *receptor potential*, though the mechanism by which the molecules generate the potential is unknown (review: Getchel 1986). One theory holds that odoriferous molecules inactivate enzyme systems in the epithelium and alter chemical reactions in this tissue. According to another theory, the molecules alter the surface of the receptor cells and thus change their electri-

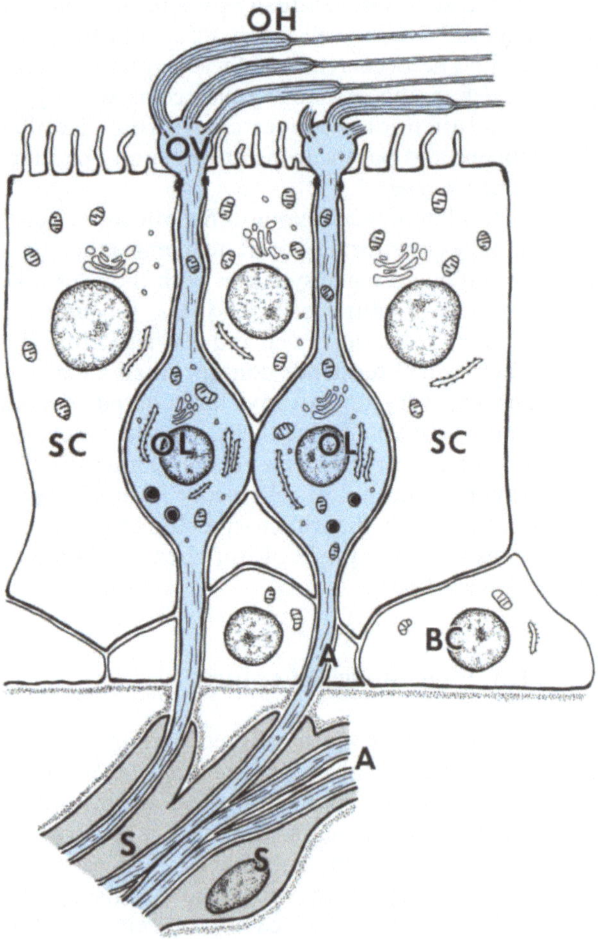

Fig. 21.1. Olfactory cell and related elements in mammalian olfactory epithelium. *OL* olfactory cells, *A* axons, *OV* olfactory vesicles, *OH* olfactory hairs, *BC* basal cell, *SC* sustentacular cell, *S* Schwann cell

cal state. A third theory holds that the molecules simply alter the Na^+ permeability of the receptor membrane (Ganong 1985).

The discovery of an odorant-sensitive guanosine triphosphate (GTP)-dependent adenylate cyclase and the presence of odorant-gated ion channels in olfactory cilia have led to current working hypotheses for the molecular pathway that may mediate signal transduction in olfaction (Anholt 1987). One model suggests that cyclic AMP generated by the olfactory adenylate cyclase causes the opening of an ion channel either directly or by phosphorylation of the channel via cyclic AMP-dependent protein kinase. Alternatively, odorants may cause activation of the olfactory receptor cell by directly opening ion channels, and phosphorylation of the channel via a cyclic AMP-dependent protein kinase would cause its inactivation.

In general, recording the intracellular potential of an olfactory cell is very difficult because the cells are so small. Recording the intracellular potential has only been done with olfactory cells of the lamprey (Suzuki 1977) and giant salamander (Trotier and MacLeod 1983). L-arginin induced slow depolarization, on which spike potentials overlapped; this coincided with an increase in the membrane conductance of olfactory cells in the lamprey (Suzuki 1977). Kurihara (1984) postulated that the depolarization is caused by stimulus-induced opening of ion channels on the olfactory hairs, which evokes a spike potential in a more proximal portion of the olfactory cell; the spike potential is then transmitted via the axon to the olfactory bulb.

Maue and Dionne (1987) recently isolated olfactory receptor cells from embryonic, neonatal, and adult mice, and studied using the patch-clamp technique. Several distinct types of ion channels were characterized in patches of membrane from the cell soma and the dendritic knob of receptor cells. The receptor cells had high input resistances, and *action potentials* could be elicited by picoampere amounts of depolarizating current. These receptor cells responded to applied odorant molecules and to forskolin with increases in membrane conductance. Increases in membrane conductance in response to increased levels of cyclic AMP at the intracellular surface of the membrane have also been recorded in isolated patches of membrane from the soma, dendrite, and cilia of bullfrog receptor cells (Nakamura and Gold, 1987). It seems thus possible that ordorants modulate the activity of ion channels via intracellular cyclic AMP, since the receptor cells often exhibit low levels of spontaneous activity (Getchell, 1974; Trotier and MacLeod, 1983), which become more regular in response to odorants (Getchell, 1974). K^+ channels are often the point at which cellular activity and responsiveness are modulated (Siegelbaum and Tsien, 1983; Levitan, 1985). This model resembles that recently, postulated in the taste receptor cells (see p. 216).

21.4 Olfactory Cell As a Paraneuron

The olfactory cell has been regarded as a bipolar *neuron*. This view is supported by the occurrence of the long cell process (axon) forming an unmyelinated nerve and by the electrophysiological property described above. However, recent information on the ability to regenerate and the short life cycle of the olfactory cell has impressively indicated that the cell is a rather unusual neuron with respect to its cell kinetics.

Ontogenetically, the cell is derived from the ectoderm and not from the neuroectoderm. It should here be noted that the immature form of the olfactory cell, lacking an axon, assumes the structure of a *basal-granulated paraneuron* as seen in the gastroenteric epithelium. In their study of the bat olfactory epithelium, Yamamoto et al. (1976) demonstrated large cored vesicles in such immature cells before and after extending an axon. It seem reasonable to propose that the olfactory cell first develops as a secretory cell before differentiating into a sensory cell.

On the basis of the above findings, the olfactory cell may be categorized either as a neuron or paraneuron. It would perhaps be more accurate to say that the olfactory cell is intermediate between neurons and paraneurons (Fig. 21.1); such a view may be useful in exploring various features of the cell.

Biochemical and cytochemical studies of the *secretory substances* of the olfactory cell have not advanced as yet. Even the transmitter in the axon terminals has not been determined, though the abundant small clear vesicles would suggest that it is acetylcholine. The occurrence of the large cored vesicles (granules) indicates that peptidic messenger(s) are produced, as is generally the case in neurons and paraneurons. Olfactory cells in the human fetus and some mammals contain immunoreactivities for neuron-specific enolase (NSE) and neurofilament protein (NFP) (Takahashi et al. 1984), which are *markers of neurons and paraneurons*. The other marker, spot 35 protein or Purkinje cell-specific protein (p. 99), has been immunostained in a few olfactory cells in the guinea pig. It is not known why particular cells are so selectively immunoreactive, while other cells remain completely immunonegative.

21.5 Ontogeny

In vertebrate embryos, a thickened patch of ectoderm, the *olfactory placode*, appears on the ventrolateral part of the head. This soon sinks to form the olfactory pit, forms the roof of the developing nasal cavity, and becomes the olfactory epithelium. The olfactory cells differentiate in this epithelium and develop axons from their basal end. These axons converge to form the olfactory nerve, which establishes synaptic connections in the olfactory bulb of the developing brain.

It has been generally assumed that olfactory cells, as neurons, are unable to regenerate, though early pioneering observations in Amphibia (Nagahara 1940; Smith 1951) and the Rhesus monkey (Schultz 1960) did indicate that they are regenerated. More recently, light- and electron-microscopic observations of autoradiographic material have indicated that olfactory cells are constantly replaced in adult vertebrates by the proliferation of basal cells. Degeneration and regeneration of olfactory cells are thus routine and continuous phenomena throughout the life of vertebrates (Andres 1965; Graziadei and Metcalf 1971; Graziadei 1973b; for reviews: Takagi 1971; Alberts 1974; Graziadei and Graziadei 1978).

The precursor cells lying at the base of the olfactory epithelium causing regeneration of the olfactory cells were called "resting cells" by Nagahara (1940). Andres (1965) described special blastema cells among the basal cells and regarded them as the precursors of olfactory cells. Graziadei and Graziadei (1978) recognized precursors of the olfactory cells which they termed "globose cells"; the cells were distributed on and between the basal cells and were characterized by being larger and having more distinct nucleoli. *Mitotic figures* are commonly seen in the basal and globose (resting) cells in mammals and lower vertebrates (Graziadei and Metcalf 1971; Graziadei and Graziadei 1978). For various reasons, especially on the basis of their radioautographic studies (see below), Graziadei and Graziadei (1978) proposed that the basal cells are not the stem cells common to the olfactory and supporting cells, but that they produce exclusively olfactory cells via the globose cells.

Radioautographic analyses of the olfactory cell *turnover* were performed by Graziadei and Graziadei (1978) in the frog, mouse, and several other vertebrates. In spite of some variability due to the different animals used, the results may roughly be summarized as follows. After the administration of ^{3}H-thymidine, the label was concentrated in the basal cells during the first 12–24 h. After 7 days, the label gathered in the globase cells and in some of the mature olfactory cells. The labeled cells disappeared from the olfactory mucosa over a period of 30 days. Graziadei and Graziadei (1978) concluded that it takes 8 days for the basal cells to differentiate into young mature neurons and that after about 30 days they are eliminated and replaced by new elements.

Severance of the fila olfactoria, i.e., section of the olfactory cell axons, is known to cause retrograde degeneration of the olfactory cells in mammals and lower vertebrates, especially frogs. The process of degeneration is maximal 8 days after axotomy. The basal cells are accelerated into mitosis and globose cells are increased. At 15 days, differentiation of young olfactory cells is marked. Maturation of the cells proceeds very fast and the olfactory mucosa regains its normal structure after 25–30 days.

21.6 Phylogeny

The essential structure of olfactory cells, the fila olfactoria, and their terminations in the olfactory bulb are identical in all classes of vertebrates, including *cyclostomes* (Andres 1970). There are, however, considerable differences in the occurrence and arrangement of the olfactory hairs even among mammalian species (Menco 1977). In *amphioxus*, a single specific pit contains olfactory cells, which send their axons to the anterior part of the neural tube (Graziadei 1973b).

The olfactory system is widely distributed in *invertebrates*. It is generally represented by chemoreceptive haired cells, extending axons to the central nervous system. *Insects* are protostomian invertebrates that are highly sensitive to odor stimuli, and their olfactory organs have been the most extensively investigated. In addition to many other receptive functions, the antennae are usually the sites of olfactory reception. Odoriferous molecules in the air reach the antennae and enter a space limited by a pored chitin layer, containing mucus. This space houses a long olfactory hair or hairs on which receptor molecules are believed to be disposed. These hairs are modified cilia, like vertebrate olfactory hairs, and stand on spindle-shaped bipolar cells in the epithelium. These olfactory cells project axons to the glomeruli in the olfactory lobe of the cerebral ganglion. The biological significance of olfactory stimuli (e.g., as pheromones and in the regulation of reproductive activities) and their electrophysiological mechanisms have been extensively studied in insects (Schafer 1971; Hawke and Farley 1971; Boeckh 1976). An excellent review by Hansen (1978) is recommended for reference.

21.7 Appendix: Vomeronasal Organ

There has been a tendency among vertebrates for a ventral segment of the olfactory epithelium to form a separate sensory organ. This origin, called the vomeronasal or Jacobson's organ, is conspicuous in amphibians and reaches its evolutionary peak in lizards and snakes (reptiles). The organ is absent in birds. Mammals generally have either a well-developed or a vestigial organ. In man and some monkeys, the organ can only be seen in the fetal period (Kent 1973; Starck 1982).

When developed, the vomeronasal organ in mammals is a *receptor for pheromones*, odoriferous sexual signals (Wysocki 1979; Wysocki et al. 1983), and smell recognition between the mother and young (Teicher et al. 1984).

The sensory cells, like the ordinary olfactory cells, of the vomeronasal organ are derived from the *olfactory placode*. The epithelium of the vomeronasal organ is a high columar epithelium consisting of three types of cell—sensory cell, supporting cell, and basal cell.

The sensory cell is a slender structure with a perinuclear and an apical thickening. Basally, it extends a long axon toward the olfactory bulb. The axons form the vomeronasal nerve, a division of the olfactory nerve. The apical portion of the receptor cell markedly differs from that of the olfactory cell; it is devoid of the olfactory vesicle and sensory cilia. The domed cell apex is covered by numerous long microvilli and a few thicker cytoplasmic processes. The apical cytoplasm contains numerous centrioles but, curiously, they do not extend cilia (turtle—Graziadei and Tucker 1970; sheep—Kratzing 1971; cattle—Adams 1986).

In their study on the turtle vomeronasal organ, Graziadei and Tucker (1970) pointed out that the vomeronasal sensory cells, like olfactory cells, are not completely isolated by the supporting cells; vomeronasal sensory cells contact each other at the apical, perikaryal, and axonal levels (see p. 214). These authors recorded *electrical responses* from the vomeronasal and olfactory nerves in the gopher tortoise. They found the same pattern of spikes in both nerves, though each nerve responded to different odors.

It is likely but not yet evidenced that the receptor cells of the vomeronasal organ are constantly renewed like olfactory cells. The cells in question have been regarded as *neurons*, but it seems reasonable to categorize them as sensory *paraneurons* if their olfactory counterparts are also included among the paraneurons.

21.8 Summary

The olfactory cell is a bipolar receptor cell for smell sensation dispersed in the olfactory epithelium. It extends an apical process (dendrite) to the nasal cavity; the apical process ends in a rounded olfactory vesicle, which extends many long olfactory hairs. Basally, the cell extends a long process (axon), which forms the olfactory nerve.

The *olfactory hairs*, which are modified cilia, carry *receptor molecules*. When odoriferous molecules contact them, a *receptor potential* (depolarization) is generated by an unknown mechanism. The spike potential is then transmitted through the axon (olfactory nerve) to the olfactory bulb.

The olfactory cell possesses a well-developed Golgi apparatus and other organelles for the production of secretory material. The cell contains numerous *small clear vesicles* gathered at synaptic sites and *large cored vesicles*, which may be found in the cell body. A peptidergic mechanism is suggested by the fact that the large cored vesicles have also been demonstrated in an immature form of the olfactory cell, which, lacking in an axon, resembles the gut basal-granulated paraneuron.

The olfactory cell is *continuously renewed* from the basal cells at a turnover rate of about a month. *Phylogenetically*, an olfactory cell with essentially the same structure and function can be traced through cyclostomes down to amphioxus. A similar olfactory mechanism widely occurs in invertebrates, especially in insects.

The vomeronasal organ, an independent part of the olfactory epithelium, is differentiated for the recognition of pheromones and sexual odors. The receptor cell of the organ has the same fundamental structure and electrophsiological activity as the olfactory cell, but its apical structure differs markedly.

The olfactory (and vomeronasal sensory) cell is in its cell-biological features intermediate between neurons and paraneurons.

22. Inner Ear and Lateral Line Hair Cells

22.1 Hair Cells of the Mammalian Inner Ear

The mammalian inner ear includes several kinds of sensory organ with distinct connections to the central nervous systems. All of the sensory organs of the inner ear belong to the membranous labyrinth, which consists of a series of fine tubes and sacs filled with the fluid endolymph and surrounded by another fluid, perilymph. The membranous labyrinth as a whole arises embryologically from the otic vesicle of ectodermal placodal origin (Rubel 1978).

The *organ of Corti* is the acoustic apparatus within the cochlear duct. The *crista ampullaris* is concerned with the sense of motion and is located in the ampulla of each semicircular duct. The *maculae of the utricle and saccule* are equilibrium receptors. They are provided with a gellylike membrane possessing calcium carbonate crystals (otolithic membrane), hence they are collectively called the *otolith organs*. Both cristae ampullares and maculae utriculi et sacculi are localized in the vestibulum of the osseous labyrinth; they comprise the *vestibular organs*.

All the sensory organs of the mammalian inner ear possess highly specialized sensory cells called *hair cells*, which are provided with characteristic hairs, i.e., microvilli and a single cilium, and are *mechanoreceptors* in function. The hair cells of Corti's organ are also called *cochlear hair cells*. They are devoid of a single cilium. In the newborn mouse, however, there is a single cilium on the cuticular plate of the cells. It is lost within a few days after birth. The basal body of the single cilium remains throughout life (Kikuchi and Hilding 1969).

Two types of cochlear hair cells are present: (1) the inner hair cells form one row along the whole length of the cochlear duct; (2) the outer hair cells are usually arranged in three rows. Since there are considerable structural differences between these two kinds of hair cells their functions are considered to be different. However, the exact physiological significance of this morphological difference remains to be established.

One of the most important common features of the inner and outer hair cells of Corti's organ is a well-developed *cuticular plate* from which characteristic microvilli project toward the tectorial membrane. The tips of the tallest microvilli are embedded in and affixed to this rooflike membrane, which consists of mucoproteins and glycosaminoglycans. Sound makes the basal membrane of Corti's organ vibrate. The hair cells lying on the membrane also vibrate such that their cuticular plate with the characteristic hairs is displaced with respect to the tectorial membrane. This mechanical distortion of the cuticular plate provokes a depolarization of the hair cell, followed by changes in the impulse activity in the afferent terminal of the acoustic branch of the eighth cranial nerve, with which the hair cells make synapases.

The crista ampullaris of the semicircular duct and the maculae of the utricle and saccule are histologically similar. Thus, they are grouped together in a single system, the *vestibular organs*. The epithelium of the vestibular organs possesses two types of hair cells, *type I hair cells* and *type II hair cells*. The type I hair cells possess an apical pole, which is occupied by the cuticular plate with a tuft of microvilli and a nonmotile single cilium. Their basal and lateral surfaces are covered by a calyxlike enlargement of an *afferent terminal of the vestibular nerve*. The apical specialization of the type II hair cells is apparently the same as that of the type I cells. The basal portion of the type II cells is innervated by numerous afferent nerve terminals of small diameter. The type I cells are also innervated by *efferent nerve terminals*, which are characterized by abundant synaptic vesicles. In the crista ampullaris, microvilli and the single cilium of the hair cells are considerably longer than those in the maculae of the utricule and saccule.

Movement of the head induces movement of the endolymph in the membranous labyrinth.

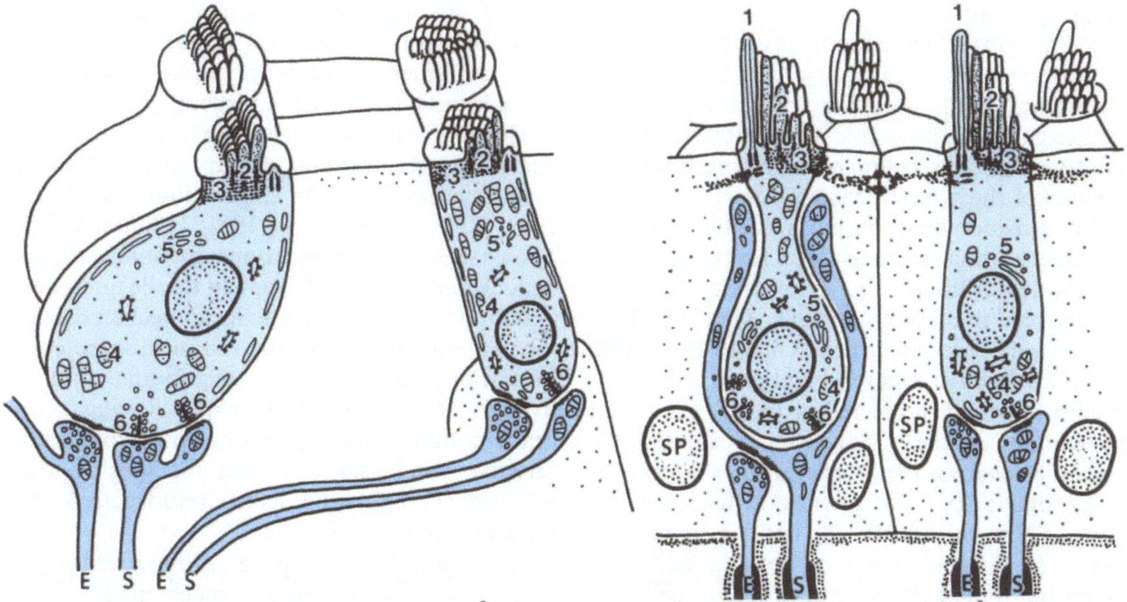

Fig. 22.1a–d. Structure of a hair cell of the mammalian inner ear. **a** Inner hair cell of Corti's organ. **b** Outer hair cell of Corti's organ. **c** Type I hair cell of the vestibular organ. **d** Type II hair cell of the vestibular organ. Efferent nerve terminal (*E*), sensory nerve terminal (*S*), supporting cell (*SP*), are partially illustrated. *1* cilium, *2* microvilli, *3* cuticle, *4* mitochondria, *5* Golgi apparatus, *6* aggregated synaptic vesicles and synaptic ribbon

This causes displacement of the cupula of the crista ampullaris and the otolithic membrane of the maculae of the utricle and saccule. The characteristic single cilium and microvilli of the two types of vestibular hair cells are affixed to the cupula in the crista ampullaris and to the otolithic membrane in the maculae of the utricle and saccule, respectively. Therefore, movement of the head stimulates the vestibular hair cells to depolarize. Impulses produced in the nerve terminals affixed to the hair cells are transmitted to the central nervous system through the vestibular branches of the eighth cranial nerve (Fig. 22.1).

22.2 Inner Ear Hair Cell As a Paraneuron

The hair cells of the inner ear are not neuroectodermal in origin but show characteristics of neurons. They undergo their final mitosis during fetal life and *never proliferate* during postnatal life (Ruben 1967; see also Rubel 1978). They are excitable cells; they *depolarize* when the base of the hairs is distorted by an adequate mechanical stimulus (see p. 222). They make synaptic contact with afferent (and efferent) nerve terminals which belong to the eighth cranial nerve. *Synaptic vesicles*, 30–50 nm in diameter, are located in the basal cytoplasm of the hair cells; they often gather around a cylindrical or platelike body. Thus, a unique structure, called a presynaptic bar or ribbon, is formed in the cytoplasm immediately beneath the cell membrane opposite the afferent nerve terminal.

The hair cells of the mammalian inner ear are not typical neurons; they are small and do not have long cytoplasmic projections corresponding to either dendrites or axons. Their nuclei are also markedly small. In the cytoplasm, there are small mitochondria, a few fragments of granular endoplasmic reticulum, an inconspicuous Golgi apparatus, a small number of lysosomes, and numerous subsurface cisterns.

We regard hair cells of the mammalian inner ear as the paraneurons for the following reason. **Secretory machinery.** The hair cells of common laboratory mammals possess a complete system of the secretory machinery (see p. 224), which consists of granular endoplasmic reticulum,

Golgi apparatus, synaptic vesicles, coated vesicles, and subsurface cisterns. It is likely that the synaptic vesicles release their contents by exocytosis and that the coated vesicles and subsurface cisterns play a role in the retrieval of the putative transmitters and of the membrane components of the synaptic vesicles.

We consider that the parts of the secretory machinery which are specialized for the secretion of peptide hormones, e.g., in gut endocrine cells, are not seen in mature hair cells. Thus, the large cored vesicles are absent here. On the other hand, the hair cells have apparently retained and developed the part of the secretory machinery which participates in the production of nonpeptidic transmitter substances. As detailed in the section on ontogeny (p. 223), premature hair cells, at least in the chick embryo, contain numerous *large cored vesicles*, which both indicate the similarity of the cells with gut endocrine cells and suggest the possible secretion of peptides.

Polarity. The functional and morphological polarities of the hair cells are similar to those of the gut endocrine cells; both are exteroceptors which are stimulated by the immediate external environment; they send external information to the internal environment by means of chemical messengers. Thus, we can regard the hair cells of the mammalian inner ear as a typical type of *receptosecretory* cell (see p. 7).

Central connections. The relationship between the hair cells of the mammalian inner ear and the sensory neurons of the spiral and vestibular ganglia is considered to be equivalent to the central connections of such typical paraneurons as the basal-granulated cells in the neuroepithelial body of the lung (see p. 192), chief cells of the carotid body (see p. 204), and Merkel cells of the skin (see p. 228). All of these cells hold a position where their supposed function as sensory neuron-associated receptor is facilitated (Fig. 22.1). The type I hair cells of the vestibular organs possess efferent innervations in addition to the sensory innervations.

22.3 Electrophysiology

The perilymph that immerses the basolateral surfaces of the inner ear hair cells is ordinary extracellular fluid and hence is similar in chemical composition to lymph and tissue fluid. On the other hand, the endolymph immersing the apical pole of the hair cells is a special fluid, which is enriched with K^+ and has reduced Na^+. The Ca^{2+} movement on the endolymphatic surface of the hair cells is low (see Hudspeth 1983).

There is increasing evidence that the transducers of vertebrate hair cells are associated with their microvilli and/or single cilium (Hillman 1969; Hillman and Lewis 1971; Hillman 1972). These specialized projections pivoting on the cuticular plate exert alternately tensile and compressional forces, and cause local mechanical deformational changes in the cell membrane conformation, thereby altering membrane conductivity to the ion flow. Whether such altering membrane conductivity to the ion flow produces a selective conductance change for specific ions has not yet been established (Hudspeth 1983; Katsuki 1985).

Hudspeth (1983) proposed a hypothesis concerning the *mechanoreceptor* mechanisms of the inner ear hair cells. When the microvilli and/or single cilium on the apical surface of the hair cells are moved, ion channels in the apical portion of the hair cells open. Positively charged ions attracted by the negative potential in the hair cells rush across the cell membrane. (The main positively charged ion in the endolymph is K^+, hence it is regarded as the carrier of charge across the membrane.) The changes in the membrane potential in the apical area may result in the exocytotic release of transmitter substances from the basal area of the hair cells, although the detailed intermediate mechanisms remain to be established.

In the basal area of the cell membrane, there are many channels that selectively admit Ca^{2+}. The activation of the Ca^{2+} channels is voltage-dependent: When the hair cells are depolarized, their Ca^{2+} channels open. The Ca^{2+} thus allowed to flow across the membrane into the cell cause the synaptic vesicles to fuse with the cell membrane and release their contents by exocytosis into the synaptic gap that separates the hair cell from the terminal of the sensory neurons.

The transmitter substance contained in the synaptic vesicles of the hair cell is unknown.

22.4 Ontogeny

22.4.1 Development of the Inner Ear

The statoacoustic sensory anlage is the *otic placode*, which in man can be recognized in an embryo of two or three somites, 2 mm in body length (review: Anniko 1983). The otic placode is a thickening of ectoderm in the hindbrain region. Invagination of this ectodermal area occurs at the 10- to 12-somite stage, forming a deep optic pit. This thereafter becomes closed from the surface and forms the *otic vesicle* or *otocyst* in embryos of about 4 mm length. The differentiation of the otocyst gives rise to the vestibular organs and to the cochlea. The ectoderm portions for presumptive sensory epithelia, either static or acoustic, differentiate into hair cells and different kinds of supporting cells.

22.4.2 Differentiation of Hair Cells

Cells passing their terminal mitosis near the epithelial surface differentiate into hair cells. Initially, the cell apex is covered with *microvilli*, which are simply columnar projections containing microfilament bundles extending into the apical cytoplasm; the constriction as seen at the base of the stereocilia in mature hair cells is not recognized (Fig. 22.2). A *kinocilium* may be identified, but it is located quite centrally in the cell apex. The kinocilium then migrates peripherally to its predestined site, while the microvilli reform into stereocilia with their peculiar morphology and arrangement. A cuticular plate is formed and the centrally located nucleus moves toward the base (review: Anniko 1983; Romand 1983; see also Tanaka et al. 1979).

As one of the earliest recognizable signs of differentiation of the hair cell in the chick embryo, membrane-limited *granules* (100–160 nm) with a dense core (80–100 nm) appear in the cytoplasm. Initially, they are dispersed in the basal half of the cell but later gather to the site where a nerve becomes synaptically associated. Subsequently, small clear vesicles also appear and presynaptic ribbons are formed (Ginzberg and Gilula 1980) (Fig. 22.2).

It is noteworthy that in their premature forms the hair cells resemble the basal-granulated paraneurons in the gastroenteric and bronchopulmonary epithelia. They contain endocrine-type granules (large cored vesicles) in the cell base, which are seldom found in mature hair cells. They are covered by rather randomly arranged microvilli and a cilium.

Yokoh (1974) noticed a very close relationship between the epithelial cells of the otocyst and the ganglionic cells in 5-mm-long human embryos. In other vertebrates, the neurons are located close to the sensory epithelium before its differentiation starts. The neurons then protrude their processes through the basement membrane and contact precursors of the hair cells. Later, the nerve fibers elongate, while the epithelium and neuronal somata are separated from each other. Afferent nerves first form synapses, and when efferent nerves occur they follow the course of the afferent nerves. The significance of nerve fiber contact for the *induction* of hair cell differentiation is controversial (Anniko 1983); Ginzberg and Gilula (1980), however, in their study using chick embryos are of the opinion that hair cell differentiation occurs only after ingrowth of nerve fibers.

22.5 Phylogenetic Aspects

22.5.1 Inner Ear Hair Cells of Submammalian Vertebrates

Anatomical features of the inner ear in submammalian vertebrates were extensively reviewed by Baird (1974).

In submammalian vertebrates, various sensory organs of the inner ear show different degrees of differentiation and development. However, in all vertebrates, the membranous labyrinth is composed of sacs and channels filled with endolymph and surrounded by perilymph. The characteristic sensory epithelium containing hair cells always exists in the form of a macula or crista.

The occurrence of vestibular organs containing specific sensory organs for equilibrium (i.e., otolith organs) and motion (i.e. semicircular canals) is phylogenetically earlier than the occurrence of the acoustic receptor (i.e., cochlea). Up to the bony fish, the inner ear consists only of vestibular organs. In the Amphibia, a new outpocketing of the vestibular organs called the lagena develops. This structure precedes the development of the primordial cochlea.

Various types of hair cell in the sensory epithelium of the inner ear of submammalian vertebrates have been shown to possess essentially the same mormophogical features as those of mammalian hair cells. Fish and frogs have only one type of cylindrical hair cell, whereas reptiles and birds have at least two types. The uniformity in the fine structure of the inner ear hair cells throughout the vertebrates is striking.

22.5.2 Lateral Line Organs

Morphological aspects. In aquatic vertebrates from cyclostomes through amphibians, a system

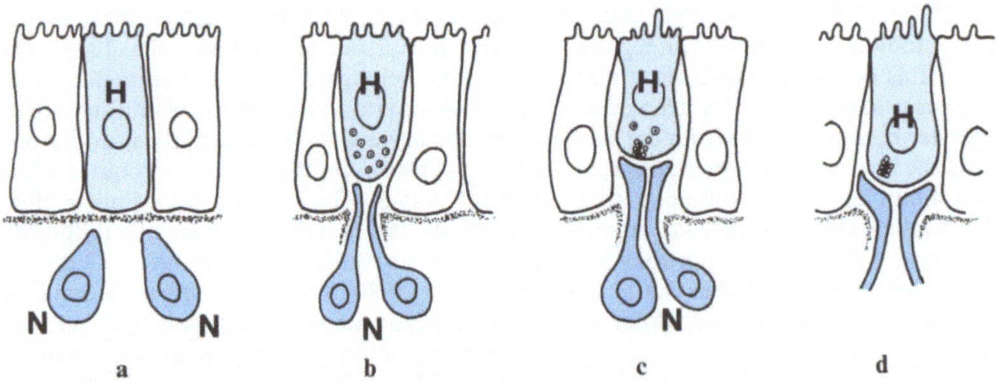

Fig. 22.2a–d. Differentiation stages of inner ear hair cells in the cat. **a** Nerve cells (*N*) start to extend processes to an undifferentiated epithelial cell (*H*). **b** When the nerve processes reach the cell, it possesses basal granules. **c** The premature hair cell (*H*) now possesses a kinocilium and microvilli, which are quite randomly arranged. **d** The kinocilium moves to a pole of the cell apex and stereocilia are formed with their characteristic shape and arrangement. The cell base now contains well-differentiated synaptic ribbons. Modified from RĐ, Ginzberg Gilula NB (1980) J Neurocytol 9:405–424

of epidermal sense organs is present in pores or canals arranged in a line down each side of the trunk. The arrangement of these sense organs is complex on the head, but the organs as a whole are called the *lateral line system.*

Lateral line organs are classified into two groups, ordinary and specialized organs. Ordinary organs are further classified into *canal organs* and *free neuromasts*. Specialized organs contain *electroreceptive ampullary and tubular organs* (p. 226). It is known that the lateral line organs of aquatic animals are homologues of various sensory organs in the mammalian inner ear (for references, see Katsuki 1982).

The sensory cells of the lateral line organs are hiar cells, which project their hairlike projections into the surrounding media. A gelatinous substance often covers the projections of these hair cells. The lateral line hair cells are stimulated by movement and vibration in the water.

Sensory nerve fiber terminals make synapses with the lateral line hair cells. Yanagisawa and Shiozawa (1980) illustrated hair cells of the free neuromast in *Xenopus laevis*. These hair cells make synapses with both afferent and efferent nerve terminals.

Physiological aspects. An electrophysiological characteristic of lateral line organs is irregular *spontaneous afferent discharge* (Harris and Flock 1967). This suggests that a messenger substance is released spontaneously from the hair cell and this excites the attached afferent nerve terminals. Stimulation of the efferent nerves *inhibits* the rate of spontaneous afferent discharge (Asanuma 1983). Endocupular potential in the cupula of *Xenopus* is +15–+50 mV, and the K^+ concentration is in the range of 24–100 mmol/l (Rusell and Sellick 1976). This situation causes sustained *depolarization* due to a continuous flux of cations. The sustained depolarization may open the voltage-dependent *Ca^{2+} influx*, which in turn causes spontaneous release of a messenger substance from the hair cells. This view was supported by Asanuma (1983) who perfused the inner and outer sides of isolated skin preparations of *Xenopus laevis* containing a small piece of the lateral line organ and a physiological saline solution containing Tris-HCl buffer or tap water. Asanuma (1983) also recorded discharges from the afferent nerves of the preparation.

Lateral line organ as a chemoreceptor. The lateral line organ as a chemoreceptor has been introduced elsewhere (see p. 11).

Chemical responses of the lateral line organs of aquatic animals have been observed mainly in free neuromasts. Stimuli for the primary responses are various metallic ions (Katsuki and Yanagisawa 1982). Although sensitivity to the ions differs from species to species, the chemical response appears to be a type of primitive *salt taste sense*, a function which may be indispensable to survival; environmental selection for food supply may be involved.

In *Xenopus*, the lateral line organ responds to certain chemical substances, i.e., salts, acids, and certain *amino acids* present in the water (Yoshioka et al. 1978). The fact that *Xenopus*, a kind of toad which has no tongue, shows rather complicated chemical responses is very suggestive of the role of the lateral line organ as a taste receptor. There are other indications that tadpoles have chemoreceptive lateral line organs but no muscular tongues and that degeneration of the lateral line organ and development of the muscular tongue proceed in parallel during metamorphosis. Kawamura and Yamashita (1981) studied the difference in responsiveness to chemical stimulation of taste buds and free neuromasts in *Pseudorasbora parva*. He reported that taste buds responded to all four types of basic taste, but free neuromasts responded only to acids and salt solutions. However, lateral line organ sensitivity to salt solutions was much greater than that of taste buds.

It is likely that mechanoreceptor hair cells in the lateral line organs also play the role of chemoreceptor. However, further studies are needed to confirm this.

22.5.3 Mechanoreceptor Cells in Invertebrates

Mechanoreceptor cells similar to the hair cells of the mammalian inner ear are present in many invertebrates.

Comparing the electron micrographs of mammalian inner ear hair cells with those of "sensory neurons" in the ectoderm of *Hydra* (see p. 125), the similarity in the general architecture is striking. It seems possible that the sensory neurons of *Hydra*, which are regarded as prototype paraneurons, liberate the contents of their secretory vesicles in response to mechanical (and perhaps also chemical) stimuli on their ciliated apical process.

Established mechanoreceptor cells are widely distributed in many invertebrate species. Among them, the sensory cell of the *statocyst* of some invertebrates is perhaps one of the best known examples. The statocyst is an equilibrium receptor; it consists of a sac, containing a statolith composed of sand grains cemented together which stimulate sensory cells. Kreidl (1893) replaced the statolith in a crab statocyst with iron filings. When the crab was exposed to a strong magnetic field it oriented itself not according to the direction of gravity but to the direction of the magnetic field. Thus, the crab was induced to live in an upside-down position.

The sensory cells of the statocyst of *Ascatus flaviatilis* have long apical and basal projections (Schöne and Steinbrecht 1968); these projections may represent the dendrite and axon, respectively. Thus, the sensory cells of the statocyst are neuronal rather than paraneuronal in morphology.

The mechanisms of the perception of gravity and angular acceleration in invertebrates have been extensively reviewed by Harkl (1974). Sensory cells bearing cilia similar to those in the crustacean statocysts are used in the perception of equilibrium and motion throughout many invertebrate species.

22.6 Appendix: Electroreceptors

22.6.1 Ampullary Organ and Its Functions

In some fish, the most important sense appears to be an electric one. Three freshwater families, the Mormyridae and Gymnarchidae of Africa and the Gymnoidae of South America, have many species with weak electric organs, which in conjunction with electroreceptors are apparently used to discern conductivity features of objects in their environment or possibly for communication between members of the species. The organs, derived from muscle tissue, are the source of low-voltage electric pulses. Many electric fish live in murky water where visibility is poor, and they often have poorly developed eyes. The value of an electric sense that permits scanning of the environment and location of prey is therefore obvious (see Bullock 1974; Ralph 1978; Schmidt-Nielsen 1975).

The receptor organs involved in electric field detection are called *ampullary electroreceptors*, which are distributed on and around the head. Here, the skin is indented in numerous places. The pore leads into an ampullary pit often filled with a jellylike substance or a loose aggregation of cells. These are electric sense organs, which contain, in their basal epithelium, receptor cells which are either solitary or grouped according to species. Between the receptor cells and the brain run thick cranial nerves (review: Szabo and Yvette 1974; Murray 1974; Ralph 1978).

Ampullary electroreceptors have been found in Elasmobranchii (Dijkgraaf and Kalmijn 1966), Chondrostei (Jørgensen et al. 1972), Dipnoi and Brachiopterygii (Roth 1973), and the teleosts Mormyriformes (Szabo 1965), Gymnotoidei (Szabo 1965), and Siluridae (Mullinger 1964). Fish groups like Brachiopterygii and the lungfish apparently have no electricity-generating organs though they possess ampullary receptors. This discrepancy has been accounted for by the hypothesis that these fish sense the distribution of the electric field and in this way, obtain infor-

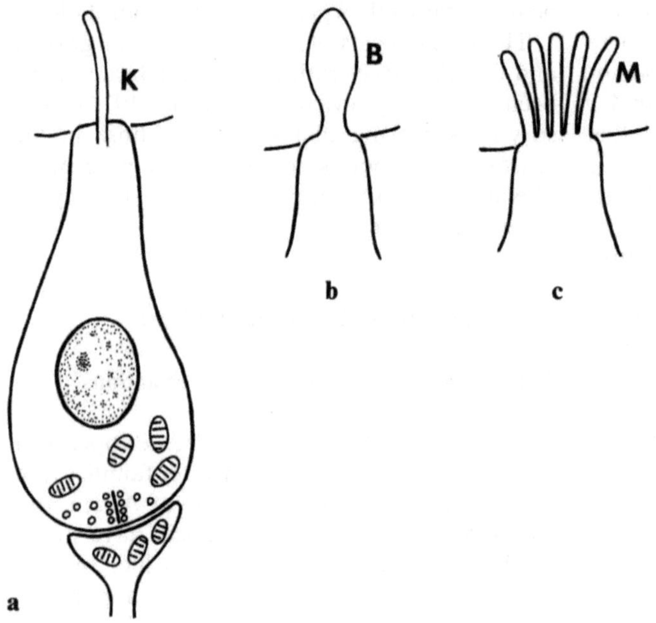

Fig. 22.3a–c. Structure of electroreceptor cells in different fishes. **a** Crossopterigii, elasmobranchii, and chondrostei. **b** Dipnoi. **c** Teleost. *K* kinocilium, *B* bulbous swelling of cytoplasm, *M* microvilli

mation about the distance from the water surface. Both lungfish and Brachiopterygii often surface in order to breathe. A distortion of the electric field surrounding the mouth would occur each time the fish approached the insulator air (water surface). Information about the water surface could then be processed by a similar active electrolocation mechanism to that found in Mormyriformes (Lissmann and Machin 1958) and Gymnotoidei (Heiligenberg 1973).

To test this hypothesis, Roth and Tscharntke (1976) made an interesting experiment. The water in which a *Protopterus dolloi* was kept was covered with a 4 cm-thick layer of liquid paraffin oil. The fish rose from the bottom of the tank to the water surface to breathe. It put its head into the oil and opened its mouth to breathe in exactly the same way as it would normally do in air. The fish even swallowed some oil before spitting it out. Apparently, the fish could not distinguish between the two insulators, air and oil, until they were in its mouth.

A detailed review of the electrophysiology of electroreceptor cells is available (Szabo and Fessard 1974).

The *tubular organ* may be distinguished from the ampullary organ. The tubular organ is also filled with a jellylike substance and provided with receptor cells at the base, though it has a more elongated pit than the ampullary organ. The tubular organ represents a quickly adapting receptor in contrast to the ampullary organ, which reacts tonically to long-lasting electric stimuli (Welsch and Storch 1973).

22.6.2 Fine Structure of Electroreceptor Cells

The electroreceptor cells are located, either solitarily or grouped, in the epithelium of the ampullary organ. They are vertically elongated cells with a swollen basal cytoplasm surrounding a large oval nucleus and a slender apical process extending to the lumen of the ampullary organ.

The apical end of the receptor cell differs in structure among different kinds of fish (Fig. 22.3). In Elasmobranchii and Chondrostei, the cell apex is provided with a *kinocilium* (Jørgensen et al. 1972) (Selachii: Waltman 1966; Szabo 1972). In Teleostei, it is covered by *microvilli* (Szabo 1965), whereas in Dipnoi it has a *club-like protrusion* of the cytoplasm (Roth and Tscharntke 1976). It is believed that all these cell projections serve to reduce the apical membrane resistance by enlarging the surface area and focusing the stimulus current to the receptor cells (Roth and Tscharntke 1976).

The cytoplasm contains cell organelles which usually occur in other paraneurons. The infranuclear cytoplasm contains numerous *clear vesicles* (about 100 nm in diameter), which are mostly round though some are irregular in shape. A few presynaptic ribbons associated with the vesicles stand perpendicularly on the cell base. According to Roth and Tscharntke (1976), the ribbon consists of two parallel membranes which are continuations of the inner leaflet of the basal (presynaptic) membrane; the vesicles appear to open to the ribbon and the transmitter is thought to be released into the ribbon space, eventually reaching the synaptic cleft.

The base of the electroreceptor cell is closely adjacent to a cuplike expansion of an *afferent nerve*. The nerve ending contains numerous mitochondria. In the lungfish, many electron-dense granules have been found in the nerve ending (Roth and Tscharntke 1976), and this may imply *reciprocal* transmission.

22.6.3 Electroreceptor Cell As a Paraneuron

The electroreceptor cell can be regarded as a typical paraneuron of bipolar shape. The occurrence of a kinocilium and synaptic ribbons in the electroreceptor cells is common to the hair cells. The identity of the transmitter(s) contained in the basal vesicles is unknown. The mechanism of electroreception as well as the electrophysiological features of the cells are poorly understood. The possible occurrence of neuron-specific proteins in the electroreceptor cell remains to be explored.

All the hitherto known electroreceptor organs are *modified lateral line organs* and are innervated by the *lateralis nerves* (Starck 1982).

22.7 Summary

Because of the distinct innervation and electrophysiological excitability, the *hair cells of the mammalian inner ear* have been widely accepted as being proper neurons. However, they deviate from typical neurons in many of their features and may reasonably be included in the family of sensory paraneurons.

The inner ear hair cells are provided with characteristic hairs, i.e., microvilli (stereocilia) and a single cilium (kinocilium), which pivot on the apical cuticular plate of the cell. The mechanical distortion of the cuticular plate provokes a *depolarization* of the hair cell.

The inner ear hair cells project neither a dendrite nor axon. There is a typical *secretory machinery*, consisting of granular endoplasmic reticulum, Golgi complex, and synaptic type vesicles. No large cored vesicles occur in mature hair cells. It has been established that Ca^{2+} *influx* induces an exocytotic release of the synaptic vesicle contents, however the transmitter substance has not yet been identified.

The cells make synapses with sensory nerve terminals which belong to the eighth cranial nerve at their basal pole. There are efferent nerve terminals on the type I hair cells of the vestibular organs.

Ontogenetically, hair cells originate from the ectodermal placode. They are devoid of the ability to undergo mitosis in adult mammals. Avian hair cells under differentiation contain *large cored vesicles* in the cell base and possess uniform microvilli and central cilium, indicating the hidden features of the hair cells to be shared with the basal-granulated paraneurons in the gut and airway.

Phylogenetically, cytological and electrophysiological characteristics of the mammalian hair cells are conserved throughout submammalian vertebrates.

According to one theory, the inner ear of higher vertebrates is a modified part of the lateral line system of aquatic animals, i.e., fish and amphibians. The *lateral line hair cells* are regarded as a suitable model for electrophysiological investigation of the inner ear hair cells. The lateral line hair cells may be *chemoreceptive*, in addition to being mechanoreceptive and thus present a continuity between chemoreceptive and mechanoreceptive paraneurons.

As modified lateral line hair cells, *electroreceptor cells* of the ampullary and tubular organs of some fish are regarded as specific sensory paraneurons. The morphological and physiological properties of these cells are similar to those of lateral line hair cells.

Mechanoreceptor cells in the form of bipolar paraneurons are presnet in a large variety of *invertebrates*.

23. Merkel Cells in the Skin

23.1 General Aspects

The cell dealt with in this chapter is distributed in the skin, either separately or in groups, mostly in the deepest layer of the epidermis. This cell is discoid, several to 10 µm in diameter. It is associated with a sensory nerve terminal and is believed by many researchers, as it was by its discoverer, Merkel (1875, 1880), to be a special mechanoreceptor. Over a hundred years ago, this German histologist gave a surprisingly precise and extensive description of this cell, which he called a "Tastzelle" (tactile cell), and associated nerves in humans and animals. In honor of his achievement, the cell is now widely called the Merkel cell (Munger 1971).

23.2 Distribution

Merkel cells are extensively distributed in the skin and part of the mucous membrane. Here, only well-established or widely studied sites will be examined; only data on mammals will be shown, while lower animals will be mentioned in the section on phylogeny.

23.2.1 Hairy Skin Epidermis, Especially Tactile Domes

Merkel cells are isolated and dispersed in the basal layer of the epidermis. In addition, larger numbers of them are condensed in the same layer in the *Haarscheiben* (*hair discs*) of Pinkus (Smith 1967; Kawamura et al. 1971). These discs, also called tactile domes, are small (0.3–0.5 mm in diameter) slightly elevated skin areas discovered by Pinkus (1905) to be located close to body hairs; they are involved in the slow-adapting mechanoreception of the skin (Iggo and Muir 1969; review: Munger 1971).

In humans and a variety of animals, the hair discs are associated with certain hair follicles, including lanugos. In cats, the discs are located cephalic to a hair, whereas in most other mammals they are caudal to a hair (Mann and Straile 1963). In any case, the hair and the disc are arranged so that the movement of the former effectively fires the latter.

A disc may contain 50 Merkel cells in its epidermal base (Iggo and Muir 1969). A thick myelinated nerve supplies a disc by issuing, in the upper layer of the dermis, unmyelinated branches to each disc (review: Winkelmann and Breathnach 1973).

23.2.2 Hairless, Touch-Sensitive Skin

The glabrous snout skin of the pig, cattle, and some other mammals have been known since Merkel (1875) to be rich in Merkel cells, which are localized in the basal layer of the epithelium. The snout of the marsupial opossum has also been studied for Merkel cells (Munger 1965). Slowly adapting mechanoreceptors, as in Haarscheiben in hairy skin, are known to be localized in the glabrous snout skin (Barker and Welker 1969).

Human fingertips and *nail matrix* have been reported to contain Merkel cells associated with sensory nerves (Hashimoto 1972).

23.2.3 Sinus Hairs or Vibrissae

Sinus hairs or vibrissae on the perioral regions of the cat, dog, rat, and many other animals represent elaborate sensory organs including slowly adapting mechanoreceptors (Andres 1966; Andres and Düring 1973). These tactile hairs contain numerous Merkel cells associated with sensory nerves in the basal layer of their outer hair sheath.

23.2.4 Oral Mucous Membrane

Merkel (1875) described his tactile cells in the *gum* of the sheep. The labial mucosa of the rabbit is known to contain numerous Merkel cells (Tachibana and Nawa 1980). In humans, Merkel

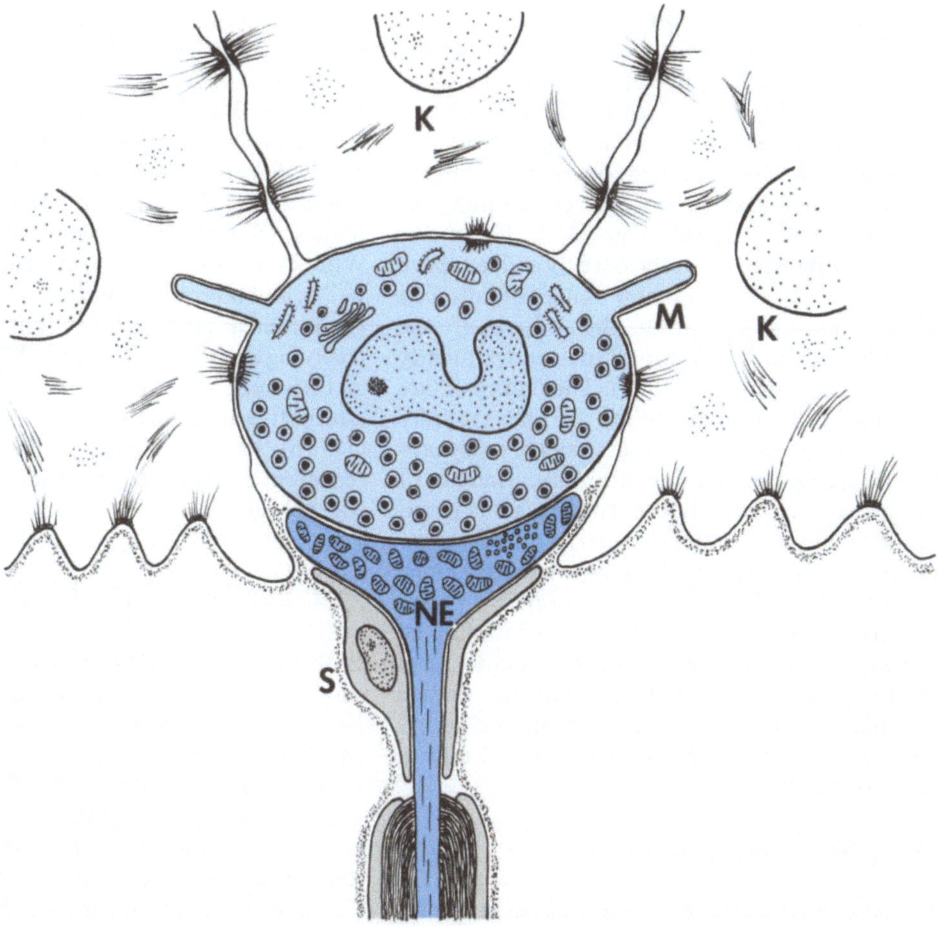

Fig. 23.1. Structure of a Merkel cell and related elements in mammalian skin. The Merkel cell containing numerous endocrinelike granules extends straight microvilli (*M*) into adjacent keratinocytes (*K*). *NE* nerve ending called a tactile meniscus which contains numerous mitochondria and occasional assemblies of small vesicles, *S* Schwann cell

cells were richly found in the *gingiva* and the vermilion border of the *lip* (Hashimoto 1972). They are dispersed in the basal layer of the epithelium and are supplied with a terminal nerve fiber.

23.3 Morphological Evidence

23.3.1 Fine Structure of Merkel Cell

A Merkel cell is located in or immediately above the stratum basale of the epidermis (epithelium in the case of the mucous membrane). For a nerve fiber to reach the cell, it frequently has to traverse a certain distance through the epidermis.

A Merkel cell is ovoid or discoid in shape. It is anchored with the surrounding keratinocytes by desmosomes, which are smaller and simpler in structure than the desmosomes between two keratinocytes (Kurosumi et al. 1979) (Fig. 23.1).

Several long and straight *microvilli* project from the cell into intercellular spaces or are inserted into adjacent keratinocytes. (Fig. 23.1). It is a widely accepted hypothesis that these microvilli are detectors of intraepidermal distortion. The projections contain numerous microfilaments, which are arranged longitudinally and are continuous with the microfilament system in the cell body.

The nucleus of the Merkel cell is deeply indented and therefore lobulated in outline. The Golgi apparatus, often divided into a few areas, is located on the side of the nucleus opposite that of the nerve ending. The cytoplasm is rich with microfilaments, and they sometimes form conspicuous bundles (Hashimoto 1972).

Endocrinelike granules, 70–100 nm in diameter (Kurosumi 1976), are dispersed in the cytoplasm of the Merkel cell and show a clear tendency to gather at the attaching site of the nerve ending. The granules consist of a smooth limiting membrane loosely containing an electron-dense material (Fig. 23.1). The membrane is separated from the core by a clear space 8 nm across (Winkelmann and Breathnach 1973). Images indicating the origin of the granules in the Golgi apparatus are frequently found (Kurosumi et al. 1976). The fine structure and intracellular distribution of the granules strongly support the view that they are presynaptic, messenger-containing granules. Images of the granules closely adjacent to the plasma membrane facing the nerve have been repeatedly demonstrated but those showing the *exocytotic granule release* have been very meager. In their electron-microscopic study on unstimulated Merkel cells in the barbels of carp, Tachibana et al. (1984) demonstrated images of omega-shaped granule release. This observation should be extended to various conditions of the cell and to various species.

23.3.2 Nerve Ending and Synapse

A myelinated nerve fiber approaches the dermoepidermal junction and issues unmyelinated branches to the epidermis. As Merkel cells are most frequently located immediately above the stratum basale, the nerve fibers must traverse the epidermis for a short distance, after having lost their Schwann sheath upon entering the epidermis.

Each nerve ends in a disc- or cup-shaped expansion called the "*tactile meniscus*," to be associated with a Merkel cell. The nerve and the cell together form a large synapselike structure. The occurrence of specialized structures such as a (probably postsynaptic) plasma membrane thickening and dark substance assembly has been demonstrated (Mihara et al. 1979).

Electron micrographs by different authors indicate the contact site between the Merkel cell and the tactile meniscus in mammals extending some 6–8 μm in profile. This means that the contact area measures 30–50 μm^2. The significance of this unusually large synaptic site seems worthy of investigation.

The tactile meniscus often contains numerous clear and some cored vesicles which cannot be distinguished from the typical synaptic vesicles (Winkelmann and Breathnach 1973; Whitear and Lane 1981). The accumulation of such vesicles toward the junction, thus suggesting the occurrence of a *reciprocal synapse* between the Merkel cell and the nerve, has been demonstrated in a variety of animals: in primates (Munger 1977), mice (Mihara et al. 1979), young rats (English 1977b), in avian Merkel and Grandry bodies (Saxod 1978), amphibians (Fox and Whitear 1978), teleost fish (Lane and Whitear 1977), and lampreys (Whitear and Lane 1981). It is an interesting possibility that the nerve, upon receiving an impulse from the Merkel cell, might in turn regulate the sensitivity of the cell. An alternative idea has been proposed by English (1977b) that the vesicles in the meniscus might represent a packaged trophic substance for the Merkel cell.

23.4 Physiological Properties

Both tactile discs (Haarscheiben) and glabrous snout skin have been known to be *slowly adapting mechanoreceptors* (Iggo 1963b; Werner and Mountcastle 1965; Barker and Welker 1969; review: Burgess and Perl 1973). The hypothesis is generally accepted that the Merkel cell is stimulated by mechanical deformation in its microvillous projections and transmits the excitation to the associated sensory nerve, making the latter adapt slowly (Iggo and Muir 1969; review: Burgess and Perl 1973).

The Haarscheibe transmits neural information to the spinal cord which is proportional to both the rate and the degree of mechanical stimulation. The frequency of discharge becomes progressively greater during at least the first half of sustained mechanical stimulation and can become as high as 1500 impulses/s. The frequency of discharge in the dynamic phase of response in the Haarscheibe is determined by both the velocity and amplitude of displacement. The velocity component of the response is related to the stimulus by a power law or $R = K S^n$, where R is the response, S the magnitude of stimulus, K the factor of proportionality, and n the exponent. There is a linear relationship between the logarithm of frequency of the discharge and the logarithm of magnitude of the stimulus. The static response of the Haarscheibe can continue for 30 min or longer if the displacement is sufficient, but usually after 20–30 s the response slowly fades away (Smith 1977; Darian-Smith 1984). Physiological experiments have mostly been confined to stimuli which last less than 30 s; the vast majority are concerned with events in the first second. A three-phase response has been described (Tapper 1965): a dynamic, displacement-rate-sensitive period; a transient, rapidly adapting period; and a static displacement-sensitive period.

The problems that have remained largely unsolved for many years are the *transduction mechanisms* of mechanoreceptors in skin and other skeletal tissues and the mechanical properties of the skin involved in the activation of these mechanoreceptors (Darian-Smith 1984). On the basis of ultrastructural evidence, it was suggested that Merkel cells could be secondary sensory cells (Andres 1966; Iggo and Muir 1969). Since the cytoplasm of Merkel cells is rich in dense-core granules, it was assumed that mechanical deformation of these cells would cause the generation of a receptor potential, which would be transmitted to the nerve endings through the release of a *chemical transmitter* from the dense-core granules at specialized zones of contact between the Merkel cells and the nerve endings. Horch et al. (1974), who studied the discharge and excitability characteristics of afferent responses elicited from touch corpuscles, concluded that the action potentials were generated through a synaptic mechanism. They considered Merkel cells to be receptor elements, and this interpretation is now generally accepted, even though it conflicts with several other findings. Met-enkephalin-like and VIP-like immunoreactivities were demonstrated in Merkel cells (Hartschuh et al. 1979, 1983, 1984), and this finding, together with ultrastructural verification of mechanoreceptor-like microvilli and of "synapses" between Merkel cells and nerve endings, supports the function of Merkel cells as receptor cells (Hartschuh and Weihe 1980).

However, since the nerve endings morphologically resemble those in other types of mechanoreceptors, a second possibility is that the mechanoelectric transduction process takes place at the nerve endings and not at the Merkel cells (Andres 1966). In fact, as pointed out by Gottschaldt and Vahle-Hinz (1981), the nerve terminal to the Merkel cell in the sinus hair follicles of the cat has structural features, especially the aggregated mitochondria, common to sensory endings, including other types of mechanoreceptors, e.g., Pacinian and Meissner corpuscles and free endings. These mechanoreceptors responded with very precise phase locking to high-frequency vibratory stimuli as well as to static hair displacements. The mechanoelectric transduction process was faster than that known for any other somatic mechanoreceptor. Thus, the nerve terminal per se seemed to Gottschaldt and Vahle-Hinz (1981) to be capable of direct reception of stimuli without the involvement of the Merkel cell. Gottschaldt and Vahle-Hinz (1981) as well as Ide (1985, personal communication) are inclined to regard the Merkel cell as trophically serving and anchoring the nerve terminal in the epidermis.

23.5 Secretory Substance and Cytochemical Data

Some researchers postulated that a *monoamine* might be contained in the Merkel cell granules because of their morphological resemblance to chromaffin granules, but every attempt to demonstrate this in mammals has failed (Smith 1967; Iggo and Muir 1969; Hartschuh and Grube 1979). From electron-microscopic autoradiographic studies in mice, Kurosumi (1976) reported that only a trace, if any at all, of administered dopa was taken up by Merkel cells. In a teleost, *Conger conger*, Zaccone (1986) demonstrated serotonin-immunoreactivity in Merkel cells of the skin.

ATP and other adenine nucleotides have been suggested to occur in the Merkel cell granules. Merkel cells are selectively stained by fluorescent quinacrine (in young frogs, *Xenopus laevis*—Crowe and Whitear 1978; sinus hair follicles—Böck 1980c). In their more extensive study of amphibian and mammalian tissues, Nurse et al. (1983) demonstrated that the quinacrine fluorescent elements are actually Merkel cells and probably correspond to their granules. Purine derivatives, especially ATP, in Merkel cell granules were presumed to bind quinacrine (Crow and Whitear 1978; Böck 1980c), though there remains the possibility that acidic substances in the granules may also bind this basic dye (Nurse et al. 1983).

Acidic *glycosaminoglycans* have been suggested as being contained in Merkel cell granules (Crow and Whitear 1978), but no detailed studies are available. A positive periodic acid-Schiff (PAS) reaction of Merkel cells has been reported by some authors (Munger 1965; Lyne and Hollis 1971).

Concerning a most interesting problem on the possible occurrence of a bioactive *peptide*, Hartschuh et al. (1979) reported that a *Met-enkephalin-like* immunoreactivity was localized in the granules of Merkel cells in rat sinus hair follicles. Hartschuh et al. (1983, 1984) further demonstrated *VIP-like* immunoreactivity in the Merkel cells of various mammals. To the authors' knowledge, no report from other research group has confirmed the important results found from the group of Hartschuh.

Among the neuronal substances contained in the cytosol, *neuron-specific enolase* (NSE) has been demonstrated to occur in mammalian Merkel cells (Gu et al. 1981). NSE has also been shown in the Merkel cells of the conger eel (teleost—Zaccone 1986).

23.6 Ontogeny and Differentiation

23.6.1 Development and Origin

Merkel cells have been recognized by electron microscopy in human fetal skin in the 16th week of gestation, already with essentially the same but simpler ultrastructural features as in the mature epidermis (Breathnach and Robins 1970; Hashimoto 1972; review: Winkelmann and Breathnach 1973). A similar development has also been observed in the sheep (Lyne and Hollis 1971). It is of note that many of the fetal Merkel cells were encountered at various levels of the subepidermal connective tissue (Breathnach and Robins 1970; Hashimoto 1972). Breathnach and Robins (1970) and Winkelmann (1977) demonstrated a Merkel cell under an "apparent passage across the epidermal-dermal junction." It has also been noticed that Merkel cells in the epidermis appeared distinctly different in morphology from keratinocytes and they were "surrounded by spaces similar to those seen in relation to melanocytes" which recently reached that location (review: Breathnach 1971).

These observations and the neurosecretory-like features of the cell gave rise to the hypothesis that the Merkel cell might be *derived from the neural crest* like a melanocyte (Breathnach and Robins 1970; Hashimoto 1972; Winkelmann and Breathnach 1973). This hypothesis has been supported by the idea that the Merkel cell could belong to the APUD series (Winkelmann 1977). Recently, Ide and Munger (1978) observed that avian Merkel cells differentiated together with Schwann cells accompanying nerves and that the Merkel cells were believed to be of neural crest origin.

Other authors, on the other hand, reached quite a different view that the Merkel cell *differentiates from the keratinocyte* of the epidermis. The occurrence of desmosomes linking the Merkel cell with surrounding keratinocytes was regarded as evidence for this view (Munger 1965; Kurosumi et al. 1969). This, however, turned out to be insufficient evidence, as desmosomes came to be found even in melanocytes, which are defined as being of neural crest origin (English 1974). Stronger support for the hypothesis of the epidermal origin of the Merkel cell has been the observations on the occurrence of cells showing an intermediate ultrastructure between that of the Merkel cell and keratinocyte (Iggo and Muir 1969; Smith 1970; Nafstad 1971b; English 1974, 1977b; chicken—Saxod; 1978; Tachibana and Nawa 1980; Ochiai and Suzuki 1981).

Only a few *experimental studies* are available pursuing the origin of the Merkel cell. In amphibian larvae, Tweedle (1978) removed regions of the neural crest and yet found that the development of Merkel cells in the corresponding skin areas was not affected. Scott et al. (1981) recognized Merkel cells in denervated and regenerating skin of the salamander. Utilizing the unusually high potentiality of would healing in mammalian oral mucosa, Tachibana and Ishizeki (1981) examined the regeneration of Merkel cells after producing a wound on the labial mucosa of adult rabbits. Ten days after the injury, immature Merkel cells appeared; after 30 days, mature cells were found in a regular arrangement in the epidermis. These studies support the view that the immature and regenerating epidermis produces Merkel cells without the contribution of the neuroectoderm.

The problem then arises as to whether or not the *contact with a nerve* is necessary for a Merkel cell precursor to differentiate. As it has been reported that every immature Merkel cells are not attached by nerves at all both in the rabbit (Tachibana and Nawa 1980) and anuran tadpole (Tachibana 1979), it seems reasonable to suggest that nerves are later elements not involved in the initial differentiation of the Merkel cell. Still another problem to be settled concerns the above observations that in fetal skin Merkel cells are frequently encountered in the subepidermal connective tissue. If it is admitted that the differentiation of the Merkel cell is initiated in the epidermis without the presence of a nerve, it is possible that the cell migrates out of the epidermis and stays in the dermis for a while so as to make contact with a nerve or receive nervous induction for further differentiation. The problem is reminiscent of similar difficulties concerning possible migration and nervous induction in gut endocrine cell differentiation.

Epidermal origin and continuous turnover of mammalian Merkel cells have recently been supported by Nafstad (1987).

23.6.2 Denervation Experiments

Early light-microscopic studies reporting the degeneration or disappearance of Merkel cells after transection of the nerves supplying them (Jalowy 1934; Brown and Iggo 1963; Palmer 1965) were followed by electron-microscopic examinations as to what extent the Merkel cells might be dependent on the presence of nerves. Mainly using cat cutaneous Merkel cells, English (1974, 1977a, b) and associates (Burgess et al. 1974) demonstrated that 20–25 days after denervation Merkel cells decreased in number, underwent degeneration *in situ*, and apparently were phagocytosed by Schwann and Langerhans cells.

A strikingly different result was produced by Tachibana et al. (1982, 1983) who examined the labial mucous epithelium of rabbits after mental nerve resection. Merkel cells showed no sign of degeneration *in situ* but shifted toward the mucosal surface to be desquamated. These changes were evident more than 3 weeks after denervation. Tachibana and associates further reported a marked increase in the number of immature Merkel cells in the denervated epithelium. It was thus suggested that differentiation of Merkel cells from keratinocytes might be induced by denervation, presumably in order to compensate the cells which move superficially to be desquamated. Kurosumi et al. (1979) examined rat cutaneous Merkel cells after denervation. Although they recognized decreased numbers of granules, lost polarity in their distribution, and some more severe degenerative changes in the Merkel cells, they never found complete disappearance of the cells even after 30 days of denervation. Kurosumi et al. (1979) concluded that the functions of Merkel cells depend upon the nerve, but the cells do not necessarily need the nerve for their survival.

23.7 Phylogeny

Merkel cells have been described in cyclostomes, fish, amphibians, and birds. The cells are strikingly similar in structure and in relation to nerves throughout the many stages of evolution.

23.7.1 Cyclostomata

Fahrenholz (1936) recorded small rounded cells in the epidermis of three European species of lamprey as "Tastzellen" (tactile cells). The electron microscopic structures of these Merkel cells were elucidated by Whitear and Lane (1981) in adult and larval (ammocoetes) stages of *Lampetra*. They are round cells provided with long rodlike microvilli filled with microfilaments projecting into adjacent epidermal cells. The synapse with a nerve meniscus has essentially the same structure as in mammals, but a cytoplasmic projection called a "spur" by the authors characterizes the cyclostomian meniscus. The spur stands at the center of the synaptic face of the meniscus and indents the Merkel cell. It is of note that the dark round granules (50–60 nm in diameter) of the Merkel cell are conspicuously gathered around the spur. On the other hand, numerous clear vesicles (about 70 nm in diameter) occur in the body of the meniscus. Thus, a reciprocal synapse is likely formed here with an afferent area represented by the spur and an efferent area represented by the remaining portion of the meniscus (Whitear and Lane 1981).

23.7.2 Teleost Fishes

Merkel cells associated with a nerve meniscus and showing essentially the same ultrastructure as in mammals have been described in two teleostean species, *Ictalurus melas* and *Phoxinus phoxinus* (Lane and Whitear 1977). The synapse also shows a vesicle distribution that suggests reciprocal transmission. The Merkel cell in the conger eel (*Conger conger*) contains electron-dense granules dispersed in the cytoplasm and is associated with the surrounding cells by a few desmosomes (Zaccone 1986). According to a personal communication (1986) of Whitear, *Anguilla anguilla* and some other teleosts also have many granules in the Merkel cell, while some others, including *Heteropneustes* and *Corydoras*, have so few granules that the cytoplasm of their Merkel cell appears rather empty.

As mentioned above, the conger eel Merkel cell has been shown by Zaccone (1986) to contain serotonin-like and NSE-like immunoreactivities.

23.7.3 Amphibians

A dozen species of Apoda, Urodela, and Anura have been shown to possess Merkel cells in their epidermis; *Xenopus laevis* in particular has been thoroughly investigated (Fox and Whitear 1978). The labial ridge of adult and larval anurans (*Rana japonica*) has been shown to contain many Merkel cells (Tachibana 1978, 1979). The fine structure of amphibian Merkel cells demonstrated by these authors is surprisingly similar to mammalian structures. A reciprocal manner of vesicle distribution may also be conspicuous (*Rana pipiens*—Nafstad and Baker 1973; *Rana temporaria*—Fox and Whitear 1978).

23.7.4 Birds

There are four main types of cutaneous mechanoreceptors in birds: Herbst corpuscles, Merkel corpuscles, Grandry corpuscles, and free nerve endings. Of these, Merkel and Grandry corpuscles will be dealt with as only these structures contain paraneuronal elements.

In nonaquatic birds, Merkel cells tend to gather in rounded bodies called *Merkel corpuscles*, which are numerously found in the dermis of the beak and palate (Anderson and Nafstad 1968; Saxod 1978). The corpuscle consists of several Merkel cells and associated nerves all flattened and alternately piled up. The fine structure of the receptor cells forming the corpuscles is essentially the same as in Merkel cells in other vertebrate classes, characterized by dense granules and ample filaments. The synapses formed by the cells against the nerve terminals are also similar in structure to those in other animals (for details, see the review by Saxod 1978).

In aquatic birds, peculiar bodies called *Grandry corpuscles* are found in the dermis of the beak and tongue. The corpuscle consists of two large hemispheric cells, with dimensions of approximately 50 × 15 μm, and a nerve terminal sandwiched between them; the whole body is surrounded by small satellite cells. Although the Grandry corpuscles conspicuously differ from the Merkel cells and Merkel corpuscles especially in the large size and the paired arrangement of the receptor cells, their essential construction and cell structures are similar; they both have dense granules and a synaptic relationship to the nerve terminal (review: Saxod 1978). It seems reasonable to regard the receptor cells of Grandry corpuscles as a modified form of Merkel cells. In support of this view, Saxod (1978) stresses that the same species never possesses both Merkel and Grandry corpuscles.

It is of note that both Merkel and Grandry corpuscles of birds are located not in the epidermis but in the dermis. In his experimental studies, Saxod (1978) demonstrated that nerves were indispensable for the differentiation of cutaneous receptor cells in birds, though he seemed to consider the cells themselves to be of dermal-mesenchymal origin. On the other hand, the electron-microscopic analysis by Ide and Munger (1978) showed that the avian cutaneous receptor cells might be derived, together with Schwann cells, from migrating cells of the neural crest.

23.8 Summary

Merkel cells distributed in the basal layer of the epidermis are heavily loaded with electron-dense *granules* derived from the Golgi apparatus. Several long *microvilli* are projected in an upper lateral direction, probably representing the detectors of intraepidermal distortion.

The Merkel cell is associated with a disc- or cup-shaped expansion (*tactile meniscus*) of a myelinated nerve fiber. It is widely accepted, though not fully evidenced, that the Merkel cell is a mechanoreceptive cell, which releases granules at the cell base to stimulate the nerve. The tactile meniscus often contains synaptic vesicles, suggesting a *reciprocal* mechanism of the synapse.

Met-enkephalin, VIP and ATP have been proposed as the *secretory products* of the Merkel cell, but the transmitter and coexisting substances of the cell remain to be explored.

The origin of the Merkel cell has been controversial. Some researchers propose a neural crest origin, while others claim a derivation from nonneural ectoderm. Some researchers believe that nerves are essential for the differentiation and survival of Merkel cells, while others suggest the converse.

Phylogenetically the Merkel cell is well developed in the epidermis of cyclostomes, fish, amphibians, and birds.

24. Visual Cells in the Retina

24.1 Historical and General Aspects

The fundamental structures of the retina and visual cells in relation to their functions were established by the endeavor of light microscopists (e.g., Hannover 1840; Polyak 1941, 1957; Walls 1942). The application of electron-microscopic techniques to this field markedly enriched this knowledge, revealing the subcellular structures of visual cells and their synaptic relations to other nervous elements (Sjöstrand 1953a, b, 1956, 1961; Moody and Robertson 1960; Tokuyasu and Yamada 1959; Nilsson 1965; for further references: see Dunn 1973; Ehinger 1982).

The retina is organized into ten layers and contains the *rods* and *cones*, which are the *visual receptors*, plus *four types of neurons*—bipolar, ganglion, horizontal, and amacrine cells. A curious feature of the vertebrate retina is that the photoreceptors lie in the deepest (outermost) layer and point toward the pigment epithelium (Fig. 24.1). Thus, light must pass through the neurons, connective tissue, and blood vessels before reaching the sensory cells, except at the fovea, the region with the most acute vision of the retina, where there are no blood vessels and no overlying retinal neurons. The rods and cones, which are next to the pigment epithelium, synapse with bipolar cells, and the bipolar cells synapse with ganglion cells. The axons of the ganglion cells converge and leave the eye as the optical nerve.

The rods are extremely sensitive to light and are the receptors for night vision (scotopic vision). The scotopic visual apparatus is not capable of resolving the details and boundaries of objects or determining their color. The *cones* have a much higher threshold, but the cone system has a much greater acuity and is the system responsible for vision in bright light (photopic vision) and for color vision.

The visual cell consists of a slender perinuclear region traditionally called the rod or cone fiber, its long outward extension (40–70 μm, varying according to the regions in the retina) is called the *rod* or *cone*, and another smaller extension on the opposite side ends in a synaptic swelling called the *rod spherule* or *cone pedicle*. The rod and cone are divided into an *outer* and an *inner segment*, which are connected by a conspicuous constriction. The *outer segment*, which is either a slender column or a stout cone, is the primary site of photoreception and contains a photosensor substance (see p. 237).

24.2 Fine Structure

24.2.1 Outer Segment

The outer segment consists of a plasma membrane covering the surface and a pile of flattened sacs called rod or cone discs. These sacs, 13–15 nm in thickness, are formed by deep invaginations of the plasma membrane (Fig. 24.2).

The rod discs are newly formed every day at the top of the inner segment and disposed by phagocytosis and lysosomal digestion in the pigment epithelial cells. In the monkey, the rod outer segment contains about 1300 discs and their turnover time is reportedly about 10 days (Young 1971).

24.2.2 Connecting Portion

The outer segment is a specialized structure of the cytoplasm and plasma membrane surrounding a cilium which grows on top of the inner segment. Reflecting this relation, both segments are connected by a cilium with a small amount of cytoplasm associated with it.

The cilium is made up of nine pairs of peripheral tubules; a central pair is lacking (9+0 pattern). The cilium, which extends deeply into the outer segment, arises from a centriole in the upper (distal) portion of the inner segment. A second centriole lies typically at right angles to the first.

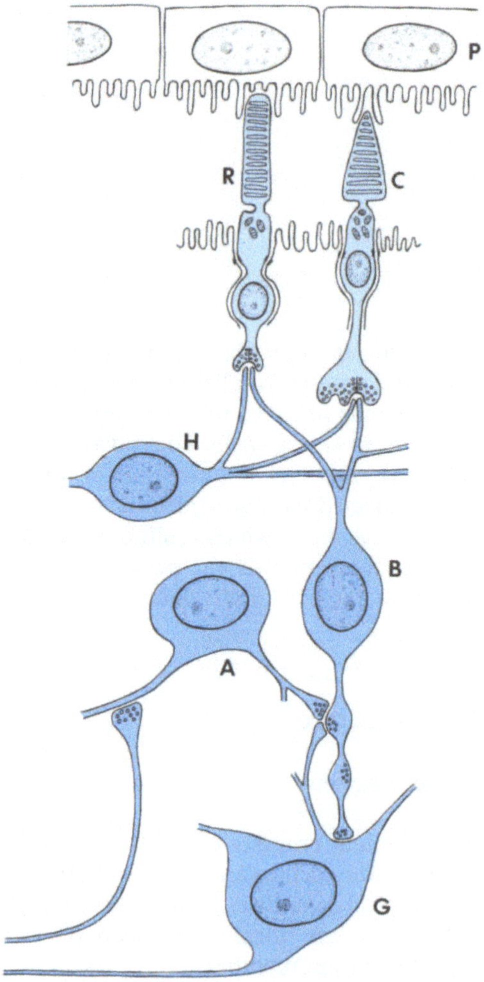

Fig. 24.1. Construction of mammalian retina. *P* pigment epithelium, *R* and *C* rod and cone cells, *H* horizontal cell, *B* bipolar cell, *A* amacrine cell, *G* ganglion cell.

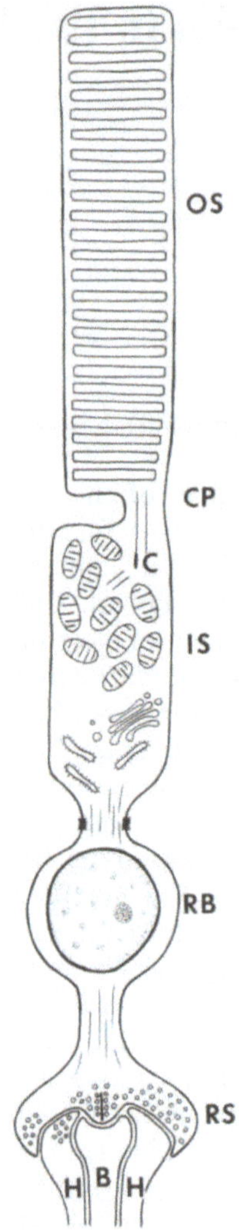

Fig. 24.2. Structure of the visual cell (rod cell). *OS* outer segment, the photoreceptive site; *CP* connecting portion with its cilium; *IS* inner segment containing numerous mitochondria; *RB* rod body with the nucleus; *RS* rod spherule, which is the synaptic portion associated with a central bipolar cell process (*B*) and lateral horizontal cell processes (*H*)

24.2.3 Inner Segment

The upper (distal) half of the inner segment, called the *ellipsoid* by light microscopists, contains a dense accumulation of mitochondria and thus represents the energy-storage site of the visual cell.

Two specialized inner segment organelles are seen in submammalian vertebrates—the paraboloid and oil droplet (review: Dunn 1973). The *paraboloid* is located in a proximal part of the ellipsoid and consists of smooth endoplasmic reticulum in a complicated network and a dense accumulation of glycogen particles in its meshes. The *oil droplets*, which conspicuously differ in number, size, and color (colorless, green, yellow, orange, and red) according to species, are contained in the distal end of the ellipsoid, usually adjacent to mitochondria. The exact functions of the paraboloid and oil droplets are unknown. The view that they may represent sites of energy storage has been proposed by many authors. As to the oil droplets, a function as a lens and/or color filter has been postulated.

The inner (proximal) half of the inner segment is called the *myoid*. It contains free ribosomes, rough and smooth endoplasmic reticulum, and Golgi apparatus. Radioautographic studies using ^{3}H-labeled amino acids as tracers have revealed that opsin, which represents the glycoproteinic part of rhodopsin, is synthesized in the rough endoplasmic reticulum-Golgi system and transferred through the connecting portion to the outer segment (Young 1967, 1976).

The myoids in lower vertebrates are contractile (hence the term "myoid") and apparently regulate the length of the cones and rods. In the cones, they contract in the light and extend in the darkness, whereas in the rods they extend in the light and contract in the darkness. The fine-structural and chemical bases of this contractility are unclear as yet. Microtubules and microfilaments have been reported to occur to a varying extent and in different amounts according to the animal species (review: Dunn 1973).

24.2.4 Rod (Cone) Fiber and Synaptic Pedicle

Proximal to the inner segment, the rod or cone extends in a slender form called the rod or cone fiber. In the middle of the fiber, the perinuclear portion is swollen and is called the rod or cone body (Fig. 24.2).

The proximal end of the rod or cone fiber is expanded into a synaptic portion called the *rod spherule* or *cone pedicle*. Numerous *clear vesicles* and a *synaptic ribbon* are contained here. The synaptic ribbon is a lamella studded with a row of clear vesicles and lies close and perpendicularly to the synaptic membrane against the bipolar cell process. The functional significance of the synaptic ribbon is unknown. Its structural details, variability among species, and the developmental process have been reviewed by Dunn (1973). The clear vesicles filling the synaptic pedicle are 20–50 nm in diameter. An opening of these vesicles to the synaptic membrane has been suggested by De Robertis (1958) but a convincing figure showing exocytosis is not available as yet. The transmitter which these vesicles are believed to contain is unknown (see p. 240).

The deeply invaginated synaptic pedicle of the visual cell is in synaptic contact with a process of the bipolar cell and, on either side, with processes of the horizontal cells. The rod possesses one set of these synapses in its spherule, whereas the cone has several sets in its pedicle (Figs. 24.1, 24.2). The synapse between the visual cell and the bipolar cell is evidently sensory or afferent in function, whereas that between the visual and horizontal cells may be partly afferent and partly efferent, as the horizontal cell may occasionally reveal aggregations of clear vesicles against the visual cell. Synaptic associations, furthermore, are formed directly between the synaptic pedicles of adjacent visual cells. Besides several mitochondria, microfilaments and microtubules are contained in the synaptic pedicles.

24.3 Physiological Findings

The rods and cones receive light, transform light energy, and most probably secrete messenger substance(s) at the synaptic pedicle. The rods and cones possess several properties that are unlike those of other receptor paraneurons. First, the initial step of photoreception in the rods and cones is absorption of light energy by photosensitive compounds in these cells. The light-induced changes in the molecular structure of the photosensitive compounds are responsible for initiating the electrophysiological activity of these cells (Fig. 24.3). The photosensitive compounds in the eyes of humans and most other mammals are made up of a protein called opsin and retinene$_1$, the aldehyde of vitamin A_1. The photosensitive pigment in the rods is called *rhodopsin* or visual purple. Its opsin is called scotopsin. Light bleaches rhodopsin by breaking the retinene-scotopsin bond; the reaction proceeds through a series of short-lived intermediates to retinene and scotopsin. There are three different types of cones in primates, and there are three different cone pigments. One cone pigment, iodopsin, has been isolated. It is found in human cones and appears to be the pigment most sensitive to red light. It is made up of retinene$_1$ and photopsin, a protein different to scotopsin. The other cone pigments also contain retinene$_1$, and the differences between them are due to differences in the structures of opsins. The photic visibility curve is the composite curve of the three cone systems (see Ganong, 1985).

The second peculiar property of rods and cones is their *receptor potentials*. In most animals, especially mammals, retinal receptor cells and neurons are relatively small. Thus, intracellular recordings are difficult, and only in a few cold-blooded animals have such recordings been made from all of the retinal cells. Recordings from one or a few of the cell types, however, have been reported for several species including some mammals (references, Dowling and Dubin

1984). The results obtained indicate that the response characteristics of retinal cells are similar across species and that generalized interpretations can be made. The view here focuses on recordings from the mudpuppy (Werblin and Dowling 1969); where relevant, results from other species are also reviewed.

In ordinary spike-generating neurons, hyperpolarization is usually associated with inhibition. But the visual cells in the distal portions of the retina, receptor (rod and cone), horizontal, and bipolar cells, respond to retinal stimulation with sustained *hyperpolarizing potentials* that are graded in intensity. Nerve impulses have never been associated with these responses. The neurons in the proximal retina—the amacrine cells and ganglion cells—respond with depolarizing potentials on which nerve impulses are superimposed. All the vertebrate photoreceptor cells recorded hyperpolarize when excited by light; this appears to be compelling evidence that excitation can be signaled by hyperpolarizing potentials in the rods and cones. In other words, the rods and cones are active when unstimulated; in the dark, the ion channels in the plasma membrane (light-sensitive channels) are opened and the continuous inflow of Na^+ gives rise to the "dark" or photosensitive current, with the consequence that the receptors are depolarized. Illumination causes the channels to close so that the whole receptor (rod outer segment, inner segment, and synaptic pedicle) hyperpolarizes and the transmitter release is reduced (references: Dowling and Dubin 1984; Altman 1985), and this generates a signal which ultimately leads to action potential in ganglion cells. The action potentials are transmitted to the brain (Fig. 24.4). The following is an experiment that directly tests this hypothesis (Dowling and Dubin 1984).

It is well known that a high level of extracellular Mg^{2+} blocks neurotransmitter release from the presynaptic terminal at chemical synapses. If photoreceptors release a messenger substance in the dark, the application of Mg^{2+} to the retina should block this release, causing the horizontal cells to hyperpolarize as they do in light. This was indeed shown to be the case: Within a few minutes of Mg^{2+} application to skate retinas, the horizontal cells hyperpolarized, and there was no longer a response to light (Dowling and Ripps 1973).

The third peculiar property of the rod is the *intracellular messenger system*. Intracellular messengers, in common with neurons and paraneurons, are intracellular Ca^{2+} and cyclic AMP (see Chap. 4). According to the long-held Ca

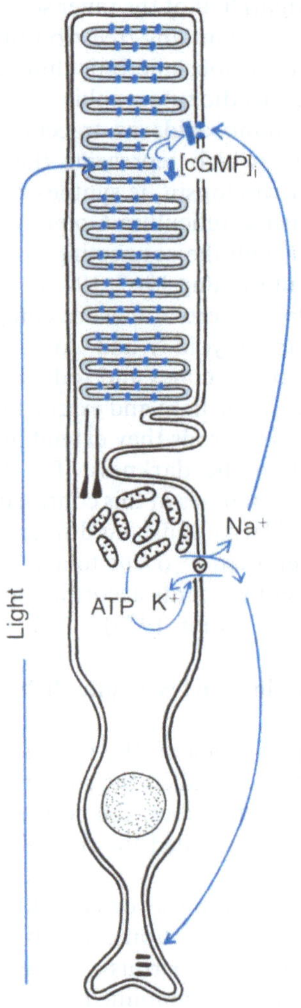

Fig. 24.3. Current flow in a rod. The outer segment is packed with disks, which are flattened membrane sacs containing rhodopsin. Mitochondria supply ATP to the Na^+, K^+-pump. A nucleus (*N*) and synapse (*S*) are found in the inner segment. In the dark, a current of Na^+ flows through the plasma membrane of the outer segment. Modified from Saibil HR (1986) News Physiol Sci 1:122–125

hypothesis of an intracellular messenger substance in the rod, light-sensitive conductance in the rod outer segment is blocked by a rise in intracellular free Ca^{2+} triggered by light. The hypothesis of a rise in internal free Ca^{2+} in the light has been supported by the findings that an induced rise in internal Ca^{2+} leads to blockage of the light-sensitive conductance and that light triggers a net Ca^{2+} efflux from the outer segment via a Na-Ca exchanger (references: Yau and Nakatani 1985). Yau and Nakatani (1985) measured both Ca^{2+} influx and efflux through the outer segment plasma membrane and found that, contrary to the Ca hypothesis, light seems to decrease rather than increase the free Ca concentration in the rod outer segment. This result implies that Ca^{2+} does not mediate visual excitation but it probably plays a role in light adaptation.

It has now been established that cyclic GMP keeps the light-sensitive channels open in the dark and that its light-activated hydrolysis allows the channels to close (for references see Altman 1985). Cobbs and Pugh (1985) used a novel technique combining measurements of membrane currents of isolated salamander rods with a suction electrode and the introduction of cyclic GMP through a whole-cell recording patch pipette. They measured the current only in the rod outer segment, all of which is light-sensitive. In the dark, only about 10% of the light-sensitive channels are open at one time (Gray and Attwell 1985), so that the increased depolarization seen when exogenous cyclic GMP is introduced suggests that cyclic GMP is a negative messenger that keeps the channel open in the dark. Fesenko et al. (1985) studied the effects of possible conductance modulators on excised "inside-out" patches from the plasma membrane of the rod outer segment in the frog; they found that cyclic GMP acting from the inner side of the membrane markedly increased the cationic conductance of such a patch (EC_{50} 30 μM cyclic GMP) in a reversible manner, while Ca^{2+} was ineffective. The cyclic GMP-induced conductance increase occurred in the absence of nucleotide triphosphates and, hence, was not mediated by protein phosphorylation but seemed rather to result from a direct action of cyclic GMP on the membrane. Thus, it seems probable that the cyclic GMP-sensitive conductance is responsible for the generation of the rod photoresponse *in vivo*. It remains to be clarified whether the cyclic GMP hypothesis in the rod can also be applied to the cone.

The messenger substances that are most probably contained in the synaptic vesicles in the rod and cone pedicles have not been demonstrated convincingly (review: Lam et al. 1982; Ehinger 1982).

Asparate and glutamate acid have been suggested as candidates for the visual cell transmitters as these substances were observed to depolarize horizontal and bipolar cells from turtle, skate, and carp retinas and also to inhibit the b-wave of the electroretinogram without directly affecting the visual cell responses (Cervetto and MacNichol 1972; Dowling and Ripps 1973; Sugawara and Negishi 1973; Murakami et al. 1975; Wu and Dowling 1978). Acetylcholine has also been suggested as the visual cell transmitter. Isolated cone cells of the turtle have been shown to synthesize acetylcholine and contain choline acetyltransferase activity (Lam 1972; Lam et al. 1982). But it remains unclear to what extent acetylcholine is a neurotransmitter in turtle photoreceptors (Gerschenfeld and Piccolino 1979), there is no indication that it might be a photoreceptor transmitter in homeotherm animals (Ehinger 1982).

24.4 Neurons in the Retina

As mentioned above, the neurons in the distal retina, horizontal and bipolar cells, share two physiological properties with rods and cones. First, these neurons respond to retinal stimulation with sustained potentials that are graded in intensity. The sustained potentials are not associated with nerve impulses. The absence of impulses corresponds to the fact that these neurons have relatively short neurites which do not transmit information over long distances. The electrotonic spread of slow potentials is sufficient for information to reach the furthest extension of these cells, and this indicates that these cells are *nonspiking interneurons* (see Chap. 5) in vertebrates.

The second property of these neurons in common with rods and cones is that they respond mostly with hyperpolarizing potentials.

Horizontal cells in the mudpuppy respond with large hyperpolarization potentials to light spots that are presented over a retinal area with a diameter of several hundred micrometers. The cells are electrically coupled to one another, increasing their receptive-field spread. In the mudpuppy, virtually all horizontal cells only hyperpolarize in response to light, regardless of stimulus intensity, wavelength, or configuration. All such horizontal cells are termed L-type (luminosity-type) cells. In other species, particularly those possessing color discrimination, horizontal cells may both hyperpolarize and depolarize, depending on the wavelength of the stimulus. Such horizontal cells have been termed C type (chromaticity-type) cells (review: Dowling and Dubin 1984).

Two physiological types of bipolar cells have been found in the mudpuppy and in all other retinas in which bipolar cell responses have been recorded. One type has a sustained hyperpolarizing response to illumination by a certain spot; the other type has a depolarizing response. In almost all species studied, an antagonism between the center and surrounding regions of the receptive field is observed at the bipolar level (review: Dowling and Dubin 1984).

Amacrine cells and ganglion cells respond with depolarizing potentials on which nerve impulses are superimposed. These cells in the proximal retina are typical neurons. In the mudpuppy, amacrine cells respond in a predominantly transient way to retinal illumination regardless of the configuration of the stimulus used. Amacrine cells are the first cell type along the visual pathway to respond in a primarily transient fashion, and they usually give on-off responses to illumination presented anywhere within their receptive fields. Nerve impulses are superimposed on the transient depolarizing responses of amacrine cells (review: Dowling and Dubin 1984).

Two types of ganglion cell response are found in the mudpuppy retina and appear to be related to the activity of one or the other of the cell types (i.e., macrine or bipolar cells) providing input to the ganglion cells. One ganglion cell type strongly resembles the amacrine cell response, giving transient responses at both the onset and cessation of stimulation. Furthermore, these ganglion cells have numerous spikes riding on the transient depolarization. The other type of ganglion cell has a receptive-field organization, which more closely resembles that of the bipolar cells. With central illumination, a sustained slow potential and a steady discharge of spikes are evoked from the cell. With some central illumination maintained, large annular illumination hyperpolarizes the cell and inhibits firing in a sustained fashion (review: Dowling and Dubin 1984).

Transmission of information by receptor paraneurons, nonspiking neurons, and spiking neurons in the retina is schematically represented in Fig. 24.4. The rod (R) on the left is receiving a flash of light that activates the receptive field center of a bipolar cell (B) on the left and the overlying photoreceptor. This same field is thus in the surrounding region of the receptive field of the bipolar cell (B) on the right. A large receptor response is observed only in the illuminated receptor. The adjacent, nonilluminated receptor shows a small response that may reflect some direct coupling between receptors.

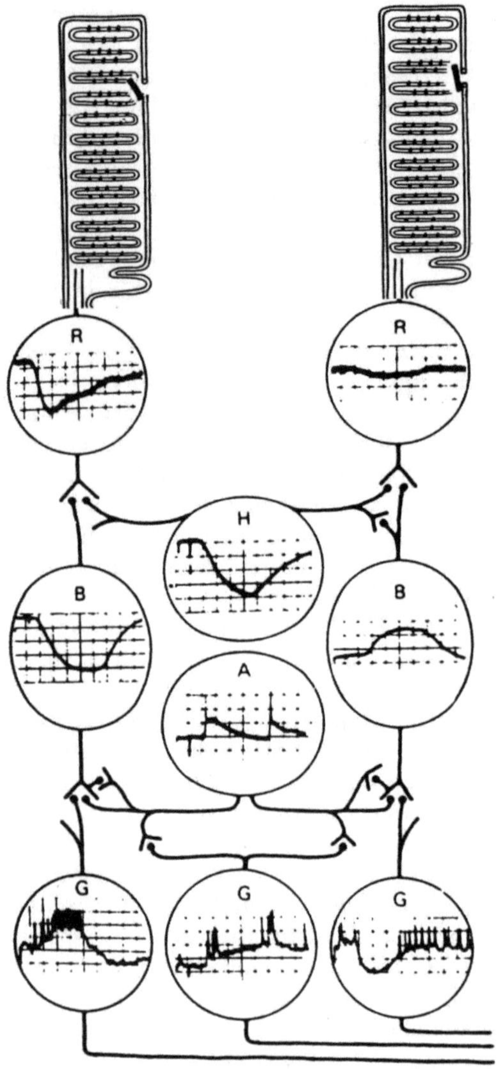

Fig. 24.4. Intracellularly recorded responses of cells in the retina to light. The synaptic connections of the cells are also indicated. The rod (*R*) on the *left* is receiving a light flash whereas the rod on the *right* is receiving steady, low-intensity illumination. *H* horizontal cell, *B* bipolar cell, *A* amacrine cell, *G* ganglion cell. Reproduced, with permission, from Dowling JE (1970) Invest Ophthalmol 9:655–680

Possible messenger substances in retinal neurons. Identifying the specific messenger substances in paraneurons and neurons in the retina has proved to be a difficult task. The possible messengers of the visual cells were mentioned on p. 239. In the inner plexiform layer, five messenger substances—acetylcholine, GABA, glycine, dopamine, and an indolamine (perhaps serotonin)—have been identified by histochemical and, in some cases, other methods. Among these substances, GABA has been suggested to be closely related to the horizontal cells and has been proposed as a possible messenger substance of these cells (see "General Discussion" in a recent monograph edited by Kaneko et al. 1982). In addition, recent studies have indicated that a number of neuropeptides also occur in this layer. All of these substances have been localized in different populations of amacrine cells and their

Table 24.1. Localization of messenger substances in the carp retina

Cell	Identified substances
Receptor cell (rod and cone)	Asparatate, glutamate
Horizontal cell (outer layer)	GABA
Horizontal cell (intermediate and inner layer)	Unknown
Bipolar cell	Unknown
Amacrine cell	Substance P, somatostatin, enkephalin, neurotensin, dopamine, GABA, acetylcholine, taurine, serotonin
Ganglion cell (in some cells)	Asparatate, glutamate

Based on a personal communication with S. Kato, University of Kanazawa, School of Medicine

processes (review: Dowling and Dubin 1984). These peptides include substance P, somatostatin, LHRH, enkephalins, B-endorphin, CCK, VIP, neurotensin, and glucagon (review: Ganong 1985).

Localization of messenger candidates in neurons and paraneurons in the retina is markedly different in different species. Their localization has most profoundly studied in the retina of fish. Messenger candidates identified in the retina of carp are shown in Table 24.1.

24.5 Ontogeny

The retina develops from the optic cup which is a frontal outgrowth of the brain. The outer thin layer of the optic cup becomes the pigment epithelium, whereas the inner thick layer gives rise to the neural layer of the retina. These two layers are separated by an intraretinal space, and only secondarily at a later stage of development do they adhere to each other.

The neural layer develops into a thick neuroepithelium by the influence of the forming lens and, consequently, the neuroblasts differentiate into visual, bipolar, horizontal, and ganglion cells. The visual cells, with their rods and cones, are in contact with the intraretinal space during the development, and after this space is closed they become covered with the pigment epithelial cells. As the intraretinal space is an extension of the ventricular cavity of the brain, the visual cells can be regarded as *CSF-contacting neurons* (Vigh-Teichmann and Vigh 1983).

The retina in amphibia contains bipolar cells which retain most of the features of the CSF-contacting neurons (p. 273). These cells extend a rodlike dendrite called *Landolt's club* to the photoreceptor space among the cones and rods. The rounded terminal of the club is provided with a 9+0 cilium and filled with mitochondria (Vigh et al. 1983). The function of Landolt's clubs is unknown.

When the visual cells are damaged by beam or ablation, they detach from the pigment epithelium and may change into cells which are identical in shape with CSF-contacting neurons (review: Vigh-Teichmann 1983).

In his electron-microscopic study on the developing cat retina, Morrison (1983) demonstrated the morphogenesis of the photoreceptor outer segment (Fig. 24.5). In the rudimentary, cilium-associated outer segment, vertically oriented lamellae are first formed, which are gradually replaced by newly formed horizontal lamellae. This sequence of formation of the vertical and horizontal lamellae appears to recapitulate the phylogenetic sequence of the appearance of both types of lamellae.

24.6 Phylogeny

About one billion years ago, a *unicellular organism* invented a sequence of nucleotides in its DNA which coded for a photoreceptive protein, consisting perhaps of the aldehyde of vitamin A conjugated with a protein carrier, opsin. The new protein, called photopigment, became embedded in the plasma membrane of an early protist in a favorable orientation to trap photons (review: Eakin 1982a).

A good model of the primitive photoreceptive organ is seen in the "eyespot" in a green alga *Chlamydomonas*, which serves the phototactic movement of this swimming protist. Unique clusters of probable photoreceptive molecules have been localized in the cell membrane covering the eyespot by a recent freeze fracture study (Melkonian and Robenek 1980; review: Eakin 1982a).

Two lines of evolution in the morphology of photoreceptive cells have been proposed (Eakin 1979). One is the *ciliary line* predominant in deuterostomes, and the other is the *rhabdomeric line* typical for protostomes (Fig. 24.6). At the first

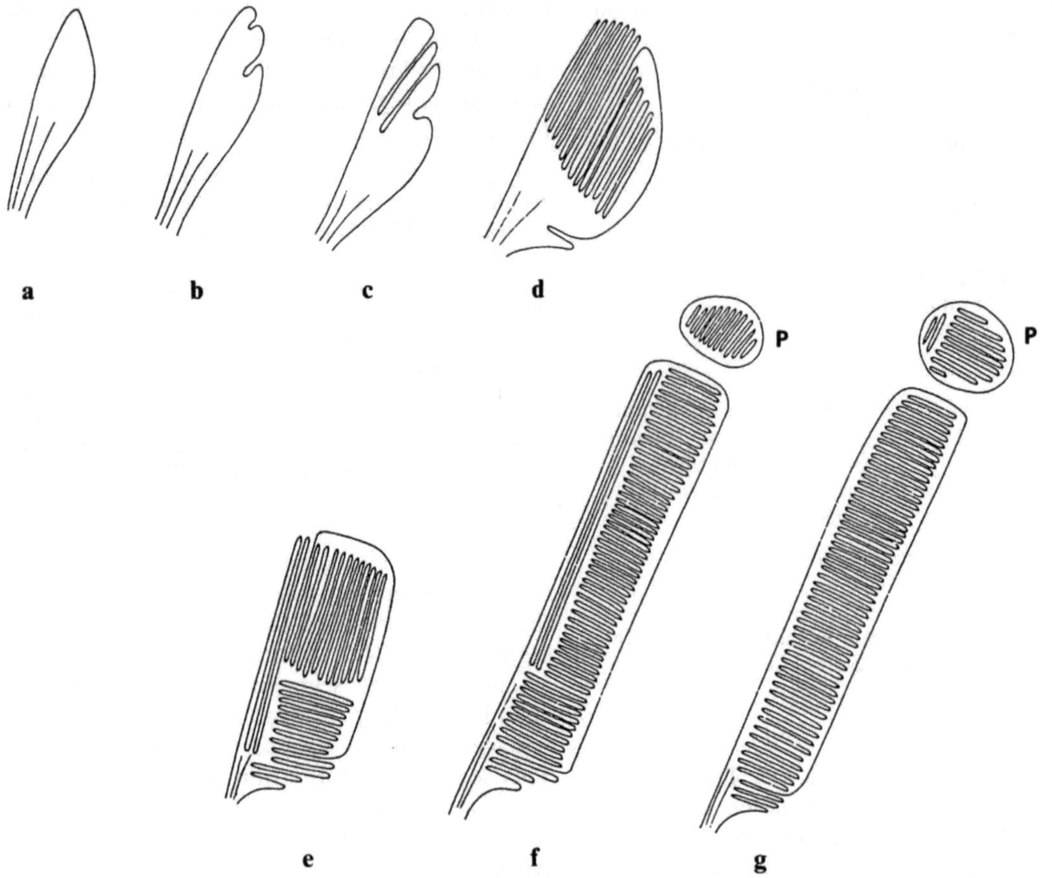

Fig. 24.5a–g. Differentiation of kitten rod outer segment. **a** Rudimentary outer segment. **b** Indentations of plasma membrane. **c** Formation of horizontal lamellae. **d** Commencement of horizontal lamellae. **e** Coexistence of vertical and horizontal lamellae. **f** Commencement of removal of vertical lamellae by pigment epithelium (*P*). **g** Outer segment with horizontal lamellae only. Courtesy of Morrison J D (1983) J Anat 136:521–533

step of the ciliary line evolution, photopigment gathers in a basal portion of the cilium. In a modern protist, for example, *Euglena spirogyra*, a swelling at the base of one of its flagella represents the photoreceptive site (Leedale et al. 1965; Grell 1968). As evolution advances, the photopigment becomes distributed more extensively on the ciliary plasma membrane. The cilia may increase their surface area by extending a large number of irregular microvilli, which are outgrowths of the ciliary membrane, as is the case in certain hydromedusans (Cnidaria) (Eakin 1982a). In other organisms, the cilia may be grouped at some strategic site; larval bryozoans possess an eyecup, which comprises a pit filled with densely gathered photosensitive cilia (Hughes and Woollacott 1982; Melkonian and Robenek 1980). The increase in the photosensitive area of the ciliary membrane may be achieved by tubular infoldings (e.g., arrowworms, Chaetognatha—Ducret 1978) or may be realized by vertically oriented lamellae (e.g., ascidian larvae, *Ciona intestinalis*—Eakin 1982a). In the evolution to vertebral lateral and pineal eyes, the ciliary outer segment developed infoldings into horizontally arranged lamellae.

A second line of photoreceptor evolution is based on the formation of a microvillous or lamellar array from the plasma membrane. The term *rhabdomere* was introduced by Grenachev (1979) for microvillous photoreceptive portions of certain arthropods and mollusks but now is widely used for nonciliary photosensitive structures of a similar nature. Elaborate, horizontally arranged labdomeres can be found at the level of Platyhelminthes. Labdomeres with horizontally, vertically, or irregularly oriented lamellae are found in every protostomian phylum. Highly strategic arrangements of rhabdomeres occur in the lateral eye of arthropods (e.g., jumping spider,

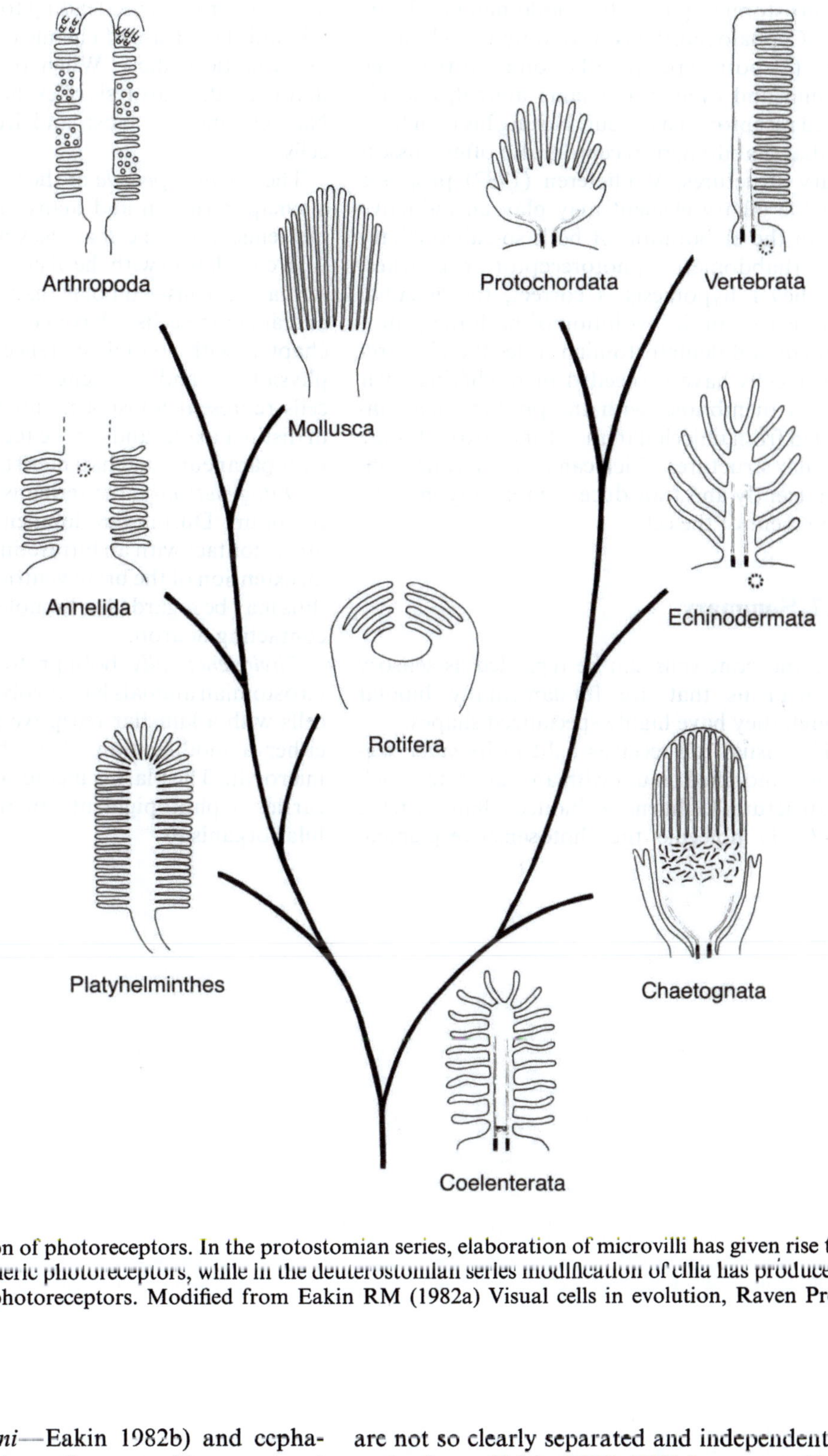

Fig. 24.6. Evolution of photoreceptors. In the protostomian series, elaboration of microvilli has given rise to a variety of rhabdomeric photoreceptors, while in the deuterostomian series modification of cilia has produced a variety of ciliary photoreceptors. Modified from Eakin RM (1982a) Visual cells in evolution, Raven Press, pp 91–105

Phidippus johnsoni—Eakin 1982b) and cephalopod mollusks (e.g., *Octopus*—Yamamoto et al. 1965; Kataoka and Yamamoto 1983).

Vanfleteren (1982) claims that the ciliary and rhabdomeric lines of photoreceptor evolution are not so clearly separated and independent of each other as Eakin (1979, 1982a) proposes. Vanfleteren points out that: (1) some protostomian phyla (Gastrotricha, Kamptozoa, Bryozoa) possess ciliary photoreceptive sites, whereas some deuterostomian phyla (Echinodermata and perhaps Cephalochordata) have only rhabdomeric ones; (2) both types may be found in the same phylum and even in the same animal; and (3) mixed receptors may occur. Noting his own finding that rhabdomeric receptors also often possess ciliary structures, Vanfleteren (1982) proposes that the ciliary element may play an inductive role in the elaboration of both so-called ciliary and rhabdomeric photoreceptor organelles. Whichever hypothesis is correct, the essential point is that in the evolution of both the protostomian and deuterostomian series the photoreceptor cells have succeeded in modifying their plasma membrane, with the photopigment inherited from unicellular ancestors, into elaborate lamellar structures which can trap photons highly efficiently and transduce their energy into the excitement of the cell.

24.7 Summary

Rod and cone cells can be regarded as sensory paraneurons that are fundamentally bipolar though they have highly specialized shapes.

The visual cell receives light in its *outer segment*, a modified cilium with an elaborate lamellar structure. Light energy induces changes in the *rhodopsin* molecule, the photosensitive pigment in the outer segment, leading to excitation of the cell and the release of chemical messengers from its synaptic pedicle. When the visual cells are illuminated, hydrolysis of cyclic GMP allows the Na^+ channels to close and hyperpolarizes the cells.

The *synaptic pedicle* of the visual cell contains a synaptic ribbon and many clear vesicles. The messengers of the cell are as yet unidentified.

In correlation with the visual cell, four kinds of retinal neurons—bipolar, ganglion, horizontal, and amacrine cells—have been dealt with in this chapter, with special reference to their electrophysiology and messenger candidates. These cells represent nonspiking interneurons lacking in distinct axons and can be regarded as neurons with paraneuronal chatacteristics.

Ontogenetically, the retina is an outgrowth of the brain. During development, the visual cells are in contact with an intraretinal space, which is an extension of the brain ventricle. The visual cell thus may be regarded as homologous to the CSF-contacting neuron.

Phylogenetically, both protostomian and deuterostomian animals have evolved photoreceptor cells with a lamellar receptive portion, which is either a modification of a cilium or modified microvilli. The plasma membrane of this portion carries a photopigment inherited from unicellular organisms.

25. Pinealocyte

25.1 General Aspects

25.1.1 Pineal Body and Pineal Complex

The pineal body in mammals is a small organ which is an ovoid evagination of the diencephalon at the posterior corner of the third ventricle. The pineal body contains *pinealocytes*, which are endocrine in nature, and numerous *glial cells*. The organ is heavily innervated by *sympathetic fibers*. It is developed throughout the vertebrate classes, but in fish, amphibia, and reptiles it is characteristically associated with the *parietal eye*, which contains *photosensitive* cells behind a primitive lens. In these animals, the pinealocytes in the pineal body are also photosensitive. Both organs, the parietal eye and pineal body, which are connected by a nervous stem, are collectively called the *pineal complex*.

25.1.2 Pinealocyte

Pinealocytes are not nerve cells but are derived from the neuroectoderm. Their structure and function completely meet the criteria of paraneurons (Fujita 1976; Ueck and Wake 1977, 1979).

The pinealocyte in lower vertebrates is photosensitive, possessing a distinct outer segment and it has *synaptic ribbons* in its contact with sensory nerves (Fig. 25.1). In phylogenetic development, this sensory nature of the cell gradually diminishes and, instead, secretory (endocrine) characteristics become more prominent (for details see pp. 246, ff., 252). The cell contains typical endocrine granules and releases *melatonin*, an indolamine derivative. Melatonin was isolated by Lerner et al. (1958) from the bovine pineal body as a factor causing the skin color of tadpoles to lighten. It causes contraction of melanin granules toward the nucleus in melanocytes in lower vertebrates. Melatonin production already occurs at the level of fish (references: Vollrath 1981).

In higher animals, melatonin is produced more and more abundantly by the pinealocytes, and in mammals it represents an important hormone suppressing a variety of sexual and reproductive functions (Figs. 25.2, 25.3). The endocrine pinealocyte in mammals lacks sensory nervous connections. Instead, it is synaptically regulated by autonomic nerves, especially sympathetic fibers coming from the superior cervical ganglion.

25.1.3 Light and Circadian Rhythm

In mammals, the center of the circadian rhythm (the biological clock) is located in the suprachiasmatic nucleus. Neurons in this nucleus send fibers to the superior cervical ganglion to relay to the sympathetic nerve, which in turn sends fibers to the pineal body. The organ is strictly but indirectly regulated in a circadian manner. The rhythm thus caused in this organ is manifested in the production of melatonin in pinealocytes.

Melatonin is produced from serotonin through N-acetyl serotonin. The enzymes necessary for this process, serotonin-*N*-acetyl-transferase (NAT) and hydroxyindol-*o*-methyltransferase (HIOMT), are produced in the pinealocyte. The activities of both enzymes are tremendously increased at night compared with during the day. Continuous darkness causes enhanced melatonin secretion by elevated activity of the melatonin-producing enzymes. General inhibition of the structure and function of the gonads is brought about. With continuous light, on the other hand, the enzymatic activities are suppressed, melatonin production is lowered, and the gonadal functions are enhanced. If the sympathetic fibers reaching to the pineal body are cut, the reactions of the organ to light stimuli cease (Wurtman et al. 1964).

In birds, the pinealocyte itself is the site of the *biological clock*. The rhythm in the production of melatonin in the cell does not disappear when the pinealocyte is isolated and cultured. Deguchi (1979), who cultured isolated chick pinealocytes under a dark and light rhythm for several days, found that the amount of NAT produced by the cells in the dark was seven to ten times larger than in the light. This study indicates that the chick pinealocyte simultaneously possesses three functions—photosensitivity, a biological clock, and melatonin production.

In reptiles and lower animals, the biological clock is located elsewhere in the nervous system. The pineal complex is involved more directly in receiving light and responds by releasing melatonin; melatonin regulates the cutaneous melanocytes so that the body color can change.

25.2 Structure and Function in Lower Vertebrates

Although the structure and function of the *pineal complex* conspicuously differ among the various vertebrate classes, its fundamental composition is that of a specialized ependymal tissue consisting of pinealocytes and supporting cells; nerve cells are usually associated and form synapses with the pinealocytes (Collin 1969; Ueck and Wake 1977) (Fig. 25.1).

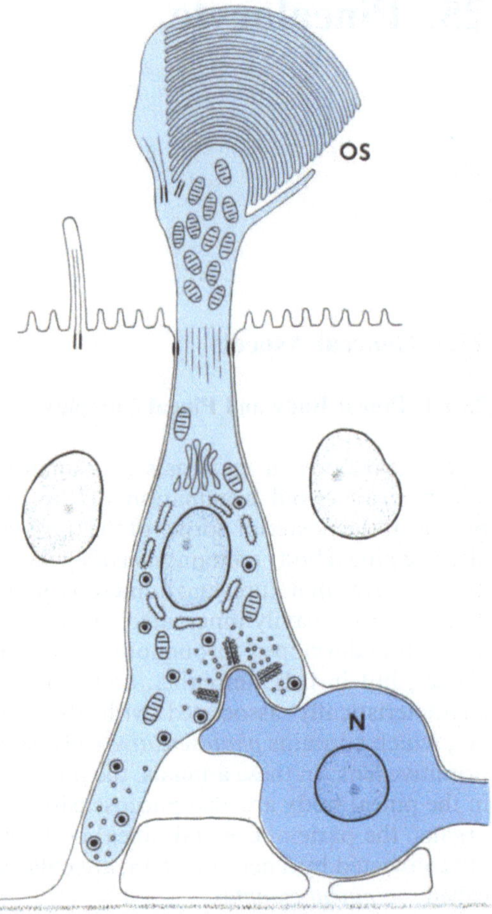

Fig. 25.1. Pinealocyte in fish and amphibians. The cell carries a photosensitive outer segment (*OS*) and is synaptically associated with a neuron (*N*). Based on Collin (1969)

Pinealocytes in lower vertebrates, including the parietal eye receptors of amphibia and reptiles, possess an *outer segment*, which is much smaller than that in the visual cells in the lateral eyes but is similarly differentiated (Fig. 25.1). The membrane stacks are originated from a modified cilium and comprise 50–150 discs characteristic of the photoreceptor site. In birds, the pinealocyte possesses a bulbous cilium or an outer segment with irregular membrane structures (Oksche and Haupel-von Harnack 1966; Fujie 1968) (Fig. 25.2).

The occurrence of *photosensitive pigments* has been demonstrated in the pineal complex in various species of lower vertebrates by electrophysiological and microspectrophotometric methods (Morita and Bergmann 1971; Hartwig and Baumann 1974; reviews: Oksche and Hartwig 1979; Vollrath 1981). Using an antiserum against opsin, the peptidic component of the rhodopsin-like photopigment isolated from the rod outer segment of the frog retina, Vigh and Vigh-Teichmann (1981) proved that the peptide was located on the outer segment of the pinealocytes in different species from fishes to birds.

A constricted connecting piece, showing a 9+0 tubular pattern of the *cilium*, combines the outer segment with the inner segment. The *inner segment* of the pinealocyte is characterized by an aggregation of mitochondria as in the retinal visual cell. The outer and inner segments represent the part of the pinealocytes that protrudes freely into the pineal lumen, while the basal part of the cell is embedded in the layer of the supporting cells. The basal part of the cell contains the nucleus, a well-developed Golgi apparatus, and lamellar cisterns of rough endoplasmic reticulum.

The basal processes of the pinealocyte contain, besides bundles of filaments and some mitochondria, *vesicles* of 40 nm diameter and a few 100-nm *granules* (large cored vesicles). They make synaptic ribbon contacts with dendrites of the neurons, whose axons reach the brain, forming the pineal tract (review: Ueck 1979).

In fish, the pineal ependium contains somata of the efferent nerves (Wake 1973). In amphibia, two different kinds of neuron are present in the pineal ependium: One is a pseudounipolar cell

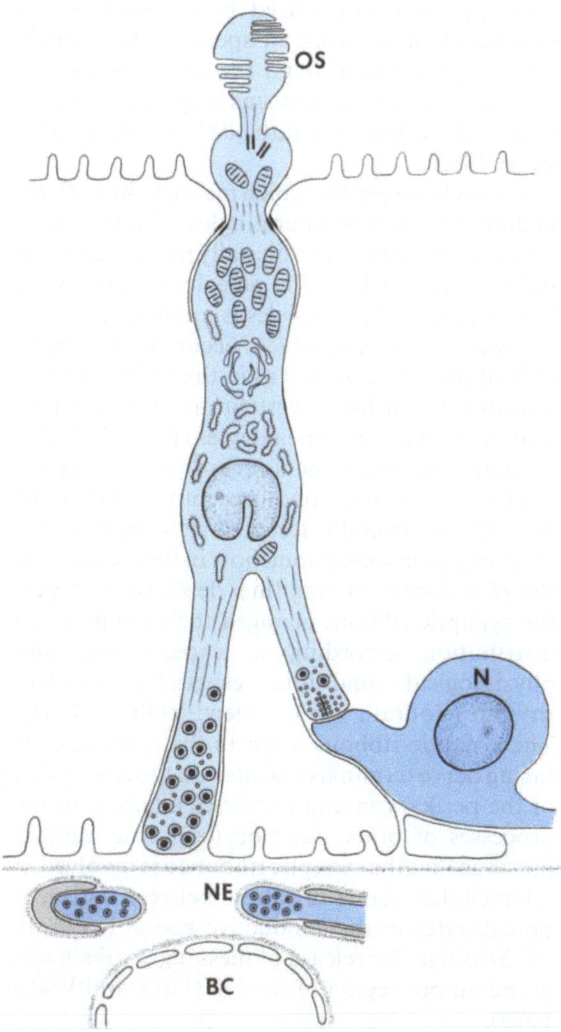

Fig. 25.2. Pinealocyte in reptiles. The cell possesses a rudimentary outer segment (*OS*). Synaptic connection with a neuron (*N*) is retained but the cell extends endocrine processes to the basement membrane, beneath which are blood capillaries (*BC*) and nerve endings (*NE*). Based on Collin (1969)

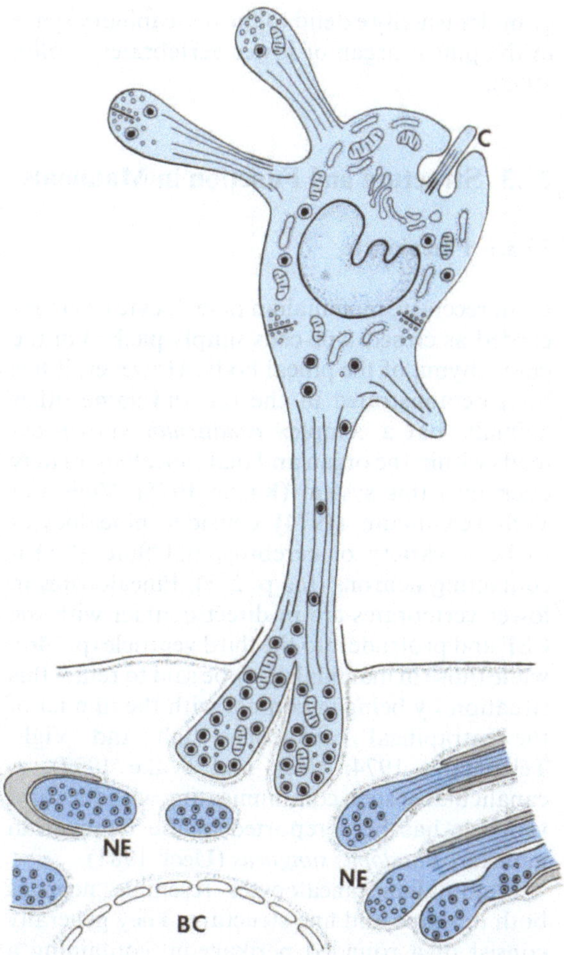

Fig. 25.3. Structure of pinealocyte in mammals. The cell resembles a multipolar neuron. One of the processes is shown to represent a secretory pole projected into a pericapillary space. Nerve endings (*NE*) are also seen in this space, suggesting their hemocrine activity. *BC* blood capillary, *C* centriole

afferent in nature, while the other is a multipolar cell, presumably an intercalated neuron (Wake et al. 1974).

The *photosensitive function* of the pineal complex has been shown electrophysiologically in different species of lower vertebrates (review: Dodt 1973). In the rainbow trout, hyperpolarized slow potentials induced by light were obtained from the pinealocyte (Tabata et al. 1975). These are the data concerning the sensory aspect of the pinealocytes in lower vertebrates.

The *secretory aspect* of the pinealocytes in lower animals has also been studied. With the fluorescence cytochemical method of Falck-Hillarp and with the microspectrofluorometric technique, *serotonin* was detected in the pineal organ in fish (Hafeez and Quay 1960) and amphibia (Owman et al. 1970). The amphibian pineal body was shown by Axelrod et al. (1965) to contain anzymes synthesizing melatonin from serotonin, i,e., NAT and HIOMT. *Melatonin* is presumably contained in the granules and may possibly be released, at least partly, into blood capillaries, as the basal processes with the granules usually extend to the pericapillary space in the pineal organ of lower vertebrates (Collin 1969).

25.3 Structure and Function in Mammals

25.3.1 Pinealocyte

Until recently, mammalian pinealocytes were regarded as closed-type cells simply packed in the parenchyma of the pineal body. However, it has been demonstrated in the rat and some other animals that a complex *canalicular system* extends within the organ and that pinealocytes may open into this system (Krstić 1975). Vigh and Vigh-Teichmann (1974) consider pinealocytes to be a variety of cerebrospinal fluid (CSF)-contacting neurons (see p. 273). Pinealocytes in lower vertebrates are in direct contact with the CSF and protrude into the third ventricle (p. 246), while those in mammals may be said to retain this situation by being in contact with the lumina of the intrapineal canaliculi (Vigh and Vigh-Teichmann 1974; Ueck and Wake 1977). A canalicular system communicating with the third ventricle has been reported in the Djungarian hamster, *Phodopus sungorus* (Ueck 1981).

Mammalian pinealocytes resemble neurons both in shape and fine structure. They generally consist of a rounded perikaryon, containing a relatively large, lobulated nucleus with a conspicuous nucleolus and some slender processes. The cell processes end in bulbous terminals either contacting other pinealocytes or in the perivascular spaces of blood capillaries (Fig. 25.3).

Mammalian pinealocytes possess the fine structures characteristic of peptide/amine-secreting cells. The rough endoplasmic reticulum and Golgi apparatus are well developed. Mitochondria are numerous, together with filaments and tubules in the cytoplasm.

Secretory granules measure about 100–200 nm in diameter and are bounded by a membrane sac. Electron-microscopic images provide evidence that they are formed in the Golgi apparatus. The secretory granules tend to be gathered in the bulbous ends of cell processes. Another secretory mechanism involves vacuoles, which are vesicles and tubules of irregular profile, containing a substance of intermediate electron density. They are not formed in the Golgi apparatus but are formed directly from the rough endoplasmic reticulum (Pévet 1977). It has been suggested that in species which have two types of pinealocyte, each type is characterized by a special secretory mechanism, while in species with a single type of pinealocyte (e.g., the mole, hedgehog, and rat) the two mechanisms may be seen in the same cell (review: Vollrath 1981; Karasek 1983; Pévet 1983).

Small clear vesicles, which measure 40–100 nm in diameter, may be intermingled with the secretory granules; they are especially numerous in the bulbous terminals of the pinealocyte processes. The nature of these vesicles is unknown.

Synaptic ribbons, which occur in the visual cells of the retina as well as in the photosensitive pinealocytes in lower vertebrates, are also present in mammalian pinealocytes (Figs. 25.3). Although some researchers regard them as a remnant of the original photoreceptive function of the cell, the generally accepted view today is that they may be involved somehow in the activities of the pinealocyte. It was thus demonstrated that the synaptic ribbons changed their number and distribution according to experimental and physiological conditions, especially circadian rhythm (Vollrath 1973; review: Vollrath 1981). The synaptic ribbons occur in the bulbous ends facing nerve terminals and also on the lateral side of the perikaryon and processes adjacent to the processes of other pinealocytes. Some authors presume that the synaptic ribbons are involved in intercellular communication between adjacent pinealocytes in the conductive region (Vollrath 1973) and in the release of messenger substances in the output region of the cell (Ueck and Wake 1979).

The mammalian pinealocyte generally possesses a *cilium* which shows the $9\times2+0$ pattern of microtubules (Figs 25.3., 25.4). This cilium corresponds to the outer segment of the submammalian pinealocyte. Vigh and Vigh-Teichmann (1981) suggested that the ciliated pinealocyte in mammals has the capacity to perceive light. Their attempts to detect opsin by immunocytochemical means have so far given negative results (literature: Oksche 1982; Vigh-Teichmann and Vigh 1983).

Heterogeneity in the structure and function of pinealocytes has been suggested by some authors:

Romijn (1975) distinguished "light" and "dark pinealocytes" in the rabbit. The light cell usually extends a cytoplasmic process into the dark cell, which is located close to a blood capillary. The rough endoplasmic reticulum-Golgi system and secretory granules formed in it are recognized only in the light cell. The dark cell possesses some pigment granules. Romijn (1975) suggests that the dark cell may take up the products of the light cell to store them or even

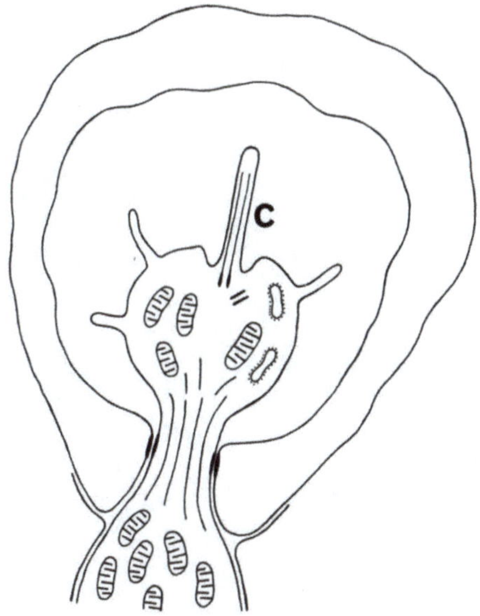

Fig. 25.4. Rudimentary outer segment in a perinatal rat. Based on Collin (1976)

"convert them into a final hormone."

In his critical review, Pévet (1977) describes that in some mammals (gerbil, macaque, mouse, chinchilla, etc.), two different populations of pinealocytes can be identified. The first type is characterized by the formation of Golgi-derived granules, while the second one is characterized by the occurrence of vesicles and vacuoles of irregular profile, which apparently are derived directly from the rough endoplasmic reticulum.

It seems reasonable to suggest that the "dark cell" and the "second type of pinealocyte" mentioned above may be variations of glial elements and should not be included as pinealocytes in the strict sense. Semm and Vollrath (1979) reported two types of pinealocyte in the guinea pig which responded differently to light and darkness. Kuwano et al. (1983) found in the bovine pineal that the majority of the pinealocytes were immunoreactive to HIOMT, while a small population were negative. The HIOMT-reactive cells were NSE-immunoreactive, while the negative cells were NSE-negative (Fig. 25.5). This finding also supports the heterogenous cell types in the pineal organ.

25.3.2 Blood Vessels

The arterial supply and venous drainage (review: Bargmann 1943; Vollrath 1981) will not be mentioned here. The pineal parenchyma is permeated by a network of capillaries. Their endothelial lining is fenestrated in the rat, mouse, and some other mammals but not fenestrated in the chinchilla (Matsushima and Reiter 1975) and many other animals (review: Vollrath 1981). A wide pericapillary space is common in most species; its development and contents conspicuously differ among mammalian species (Wartenberg 1968).

25.3.3 Nerves

Numbers of *sympathetic* nerves, derived from the *superior cervical ganglion*, end in the pericapillary space. The fibers contain numerous small cored and noncored vesicles of about 50 nm in diameter intermingled with a few large cored vesicles (Matsushima and Reiter 1977). A few cholinergic-type terminals have been demonstrated in some species (Romijn 1975; Matsushima and Reiter 1977) (Fig. 25.3). For a detailed morphology of the nerves and their vesicles, see the reviews by Ueck (1979) and Vollrath (1981).

A synapselike direct contact of nerve terminals and pinealocytes has been recorded in the rat (Wolfe 1965; further references: Vollrath 1981). Generally, however, it is accepted that there is very little, if any, contiguity between pinealocytes and nerve terminals (Matsushima and Reiter 1977). Both elements are loosely packed in a wide perivascular space, forming a kind of *synapse à distance*. It has been established that noradrenaline released from the sympathetic terminals acts upon the pinealocytes, causing inhibition of melatonin production.

On the other hand, the possibility seems to have been generally overlooked that the abundance of the nerve terminals occurring in the perivascular space exposed by the coverage of Schwann cells (this being the case in, e.g., the rat, hamster, and ground squirrel: Wartenberg 1968; Matsushima and Reiter 1977; Ueck 1979) must release their noradrenaline and other secretions into the capillary blood. This pericapillary region deserves to be called a typical *neurohemal organ*, at least in these animals (Fig. 25.3). It seems reasonable to suggest that here neurons (sympathetic terminals) and paraneurons (pinealocytes) extend toward the capillary endothelium to concur their hemocrine activity. A similar view was advanced in the light-microscopic study of the monkey pineal body by Lukaszyk and Reiter (1975).

25.4 Secretory Substances and Histochemical Data

25.4.1 Melatonin and Related Substances

Melatonin has hitherto been considered to be the main, if not exclusive, hormone secreted from the

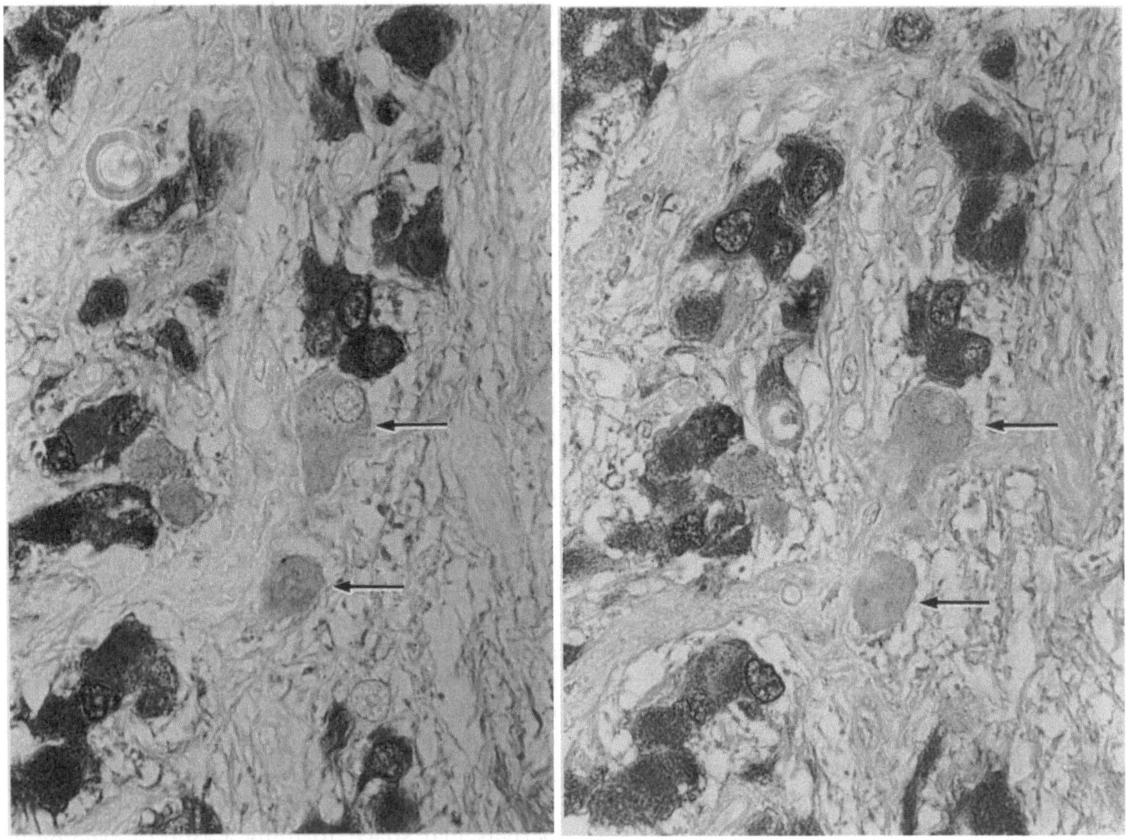

a b

Fig. 25.5a, b. NSE- and HIOMT-immunoreactivities in bovine pinealocytes. Adjacent serial sections were stained with antibodies against **a** NSE and **b** HIOMT. Heterogeneity of pinealocytes is evident as part of the cells do not react to both antisera (*arrows*). Courtesy by R.Kuwano and T. Iwanaga

pineal body. Melatonin was originally obtained by Lerner et al. (1958) from the bovine pineal as a factor which lightens the color of amphibian skin (see p. 245). Its physiological functions of inhibition of sexual and reproductive activities were later established in mammals. The effect of light on the production of melatonin in the pineal was mentioned above (p. 245).

Melatonin is one of numerous derivatives of the indolamine serotonin, which are contained in the pinealocyte. This cell first takes up the amino acid tryptophan and then converts it into 5-hydroxytryptophan (5-HTP) with tryptophan hydroxylase, which is contained in the cytosol. Thereafter, 5-HTP is decarboxylated by the enzyme L-aromatic amino acid decarboxylase, also in the cytosol, into serotonin or 5-hydroxytryptamine (5-HT). The pinealocyte thus is a typical member of Pearse's APUD series. Falck-Hillarp's fluorescence histochemical method clearly shows the occurrence of serotonin in pinealocytes (Håkanson and Owman 1966).

Serotonin is converted to melatonin in a two-step process. First, it is acetylated into *N*-acetylserotonin by the enzyme serotonin *N*-acetyltransferase (NAT). Second, *N*-acetylserotonin (NAS) is changed into melatonin by the enzyme hydroxy-indole-*o*-methyltransferase (HIOMT). NAS (Bubenick et al. 1974, 1976) and HIOMT (Kuwano et al. 1983) have been demonstrated immunocytochemically in pinealocytes.

It is likely that serotonin is taken up into secretory granules under maturation as it occurs in other paraneurons. Serotonin contained in pinealocyte granules has been demonstrated in few reports, though in avian pinealocytes Juillard and Collin (1976) showed a large amount of serotonin in the dark granules by electron-microscopic histochemical methods. The conversion of serotonin to melatonin is believed to occur within the granules. HIOMT has been immunocytochemically demonstrated by Kuwano et al. (1981) in bovine pinealocytes. We examined their preparations under the microscope and could confirm that the enzyme HIOMT was localized in the secretory granules (Iwanaga and Fujita, unpublished data).

Through the above-described process, the secretory granules of the pinealocytes contain melatonin in addition to greater or lesser amounts of serotonin, which is not immediately converted to melatonin. The amount of melatonin in the cells is conspicuously altered by the circadian rhythm and by the effect of light, and this is mediated by changes in the amounts of NAT and HIOMT (review: Reiter 1981; Vollrath 1981).

25.4.2 Peptides and Proteins

The occurrence of peptides, probably including a messenger or messengers, and proteins—carrier proteins and enzymes—is one of the prerequisites for the pinealocyte as a paraneuron. Only very few studies, however, have been based on this viewpoint. Juillard and Collin (1976) and Collin et al. (1977) clearly noted, in their studies on avian and reptilian pinealocytes that indolamines should be bound to protein(s) and/or glycoprotein(s) within the granules.

Cytochemical approach. Using mainly bovine and monkey pineal glands, Lukaszyk and Reiter (1975) reported that the pinealocytes showed stainabilities to different dyes which suggested that they are involved in the secretion of polypeptides. The methods included: (1) aldehyde fuchsin, (2) chrome hematoxylin-phloxine, (3) periodic acid-Schiff (PAS) reaction, and (4) toluidine blue staining after hydrolysis with hydrochloric acid (Solcia et al. 1968). Clear stainability with aldehyde fuchsin and chrome hematoxylin-phloxine occurred in the pinealocytes as well as in axonal terminals in the vicinity of blood capillaries. As Lukaszyk and Reiter (1975) note, this result seems to suggest the occurrence of *neurophysinlike peptides*. A positive PAS reaction was seen in some pinealocytes, which may imply the occurrence of glycoproteins. The toluidine blue treatment after hydration produced positive staining in the pinealocytes and axons. In our view, this result may be ascribed to the occurrence of *acidic carrier proteins* and possibly also more acidic glycosaminoglycans (see p. 28).

In accordance with this assumption, Reinharz et al. (1975, 1985) reported that immunoreactive neurophysins could be detected in human, bovine, and ovine pineal glands. Immunoreactivities for *chromogranins* could not be recognized in bovine pinealocytes (Nolan et al. 1985).

Lukaszyk and Reiter (1975) found that glial cells in the pineal gland contained granular structures showing the same stainabilities to aldehyde fuchsin, chrome hematoxylin-phloxine, and toluidine blue. In our opinion, these may be lysosomes (which are positive to aldehyde fuchsin and chrome hematoxylin-phloxine after oxidation) and/or possibly phagocytosed secretory material of the pinealocytes and axons.

Biochemical and physiological approach. Some researchers maintain that the hormonal effects of the pineal gland are mediated by peptides rather than melatonin. The best known of these peptides is *arginine vasotocin*, first isolated from extracts of bovine pineal bodies (Milcu et al. 1963); it has been chemically characterized (Cheesman 1970) and identified in a variety of other mammals (review: Vollrath 1981). This nonapeptide was shown to possess inhibitory effects on the reproductive system (Vaughan et al. 1976; Pavel 1978). It is, however, controversial whether this peptide is produced by pinealocytes. Pavel (1971) proposed that arginine vasotocin is produced by ependymal cells lining the pineal recess and released into the cerebrospinal fluid. Cultured pineal ependymal cells from rat fetuses have been shown to synthesize arginine vasotocin (Pavel et al. 1977). On the other hand, Pévet et al. (1979) demonstrated the biosynthesis of a vasotocinlike substance in a portion of the rat pineal body devoid of ependymal cells. If vasotocin is synthesized, possibly in parallel with the ependymal cells, in the pineal parenchyma, there remain alternative possibilities that it may be produced in neuronal elements, glial cells, or pinealocytes.

Some other peptidic substances, including hypothalamic releasing hormones (LRH, TRH, etc.), have been claimed to be contained in the mammalian pineal body (review: Vollrath 1981). However, no widely accepted information seems to be available as yet.

There has been a search for *neurophysins* in the pineal body as the possible counterpart of vasotocin. Reinharz et al. (1975, 1985) reported that they isolated proteins from extracts of the bovine, ovine, and human pineal gland which shared many characteristics of pituitary neurophysins.

25.5 Physiology

Melatonin synthesis occurs in pineal parenchymal cells; it is decreased in light and increased in darkness. The sympathetic nerves to the pineal regulate this circadian rhythm in melatonin synthesis by regulating the activity of NAT that catalyzes the formation of *N*-acetyl-5-hydroxytryptamine in the gland. The neurotransmitter *noradrenaline* from the sympathetic nerve ending is released primarily during darkness and stimulates indol production within the gland (Reiter 1981). In the hamster (Morgan et al. 1976) and gerbil (*Meriones unguiculatus*; Morgan and Reiter 1977), pineal noradrenaline levels remain relatively stable throughout the 24-h light/dark cycle. However, in the rat, the noradrenaline content of the pineal fluctuates markedly; in this species, the highest levels are measured during the day and the lowest levels during the night (Wurtman and Axelrod 1966), coinciding with the period of maximal release of the catecholamine. Following its release from the nerve terminals, noradrenaline acts on β-*adrenergic receptors* (Zatz 1978; Lipton et al. 1981) and cyclic AMP to increase the activity of the enzyme and thus increase melatonin synthesis and secretion (Axelrod 1974).

Recent studies using rat pinealocytes *in vitro* have indicated that VIP (and also PHI and secretin) stimulate adenylate cyclase and serotonin NAT activities and enhance the secretion of melatonin (Kaneko et al. 1980). These peptides have been shown to exert their effect independently of the β-adrenergic receptor system (Kaneko and Kaku 1983). On the other hand, immunocytochemical attempts have been unsuccessful in demonstrating significant amounts of VIP-containing nerves in the rat pineal body.

A comprehensive review of the physiological aspects of the mammalian pineal body is available (Ebels and Balemans 1986).

25.6 Neuronlike Features

The characteristics of the pinealocyte described above support the view that this cell shares a variety of cell-biological features with neurons. This view is further strengthened by the fact that the cell can be immunostained with antisera against neuron-specific enolase (NSE) (Schmeckel et al. 1978; Kuwano et al. 1983). Slender or stellate cells intercalated between the pinealocytes or adjacent to blood capillaries are immunopositive for S-100 protein. These cells are believed to be the glial cells in the pineal body. The pinealocytes as paraneurons seem to be supported by glial cells as neurons are.

25.7 Phylogeny

The phylogenetic aspects of the pineal organ and pinealocytes have been dealt with in some detail (pp. 246–249). The phylogeny of the pinealocyte can be summarized as follows. The primitive forms of the pinealocyte can be found as photosensitive cells in the pineal organ in lampreys and fish. Through amphibia and reptiles, the cell is a *photosensitive paraneuron* possessing a distinct outer segment. Round cored granules and synaptic ribbons with small clear vesicles are shared by the pinealocyte at every evolutionary stage (Collin 1969a, b; review: Vollrath 1981). Melatonin, its precursors, and enzymes necessary for its production, as seen in the mammalian pineal body, are present in lower vertebrates. The day-night rhythm of HIOMT activity in the pineal body can be recognized even in the lamprey (Joss 1977). The photoreceptive pinealocyte of lower vertebrates is synaptically associated with *afferent nerves*, which conduct the impulse to the brain (review: Dodt and Nauheim 1973).

From birds to mammals, the pinealocyte with its sensory features becomes more secretory or endocrine in form. The outer segment becomes rudimentary in birds, though some photosensitivity may be retained. It is almost completely lost in the mammalian pinealocyte. The cored granules containing melatonin increase in number from reptiles through birds to mammals. *Efferent*, predominantly *sympathetic nerves* replace the afferent nerves in the course of evolution. In reptiles, efferent nerves are more numerous than afferent nerves; in birds and, especially, in mammals they represent the important sympathetic route controlling the pineal body according to the light input to the lateral eyes.

25.8 Ontogeny

Pinealocytes develop, in all vertebrate classes, from precursor cells in the diencephalic ependymal area. They are thus derived from the neural tube (reviews: Bargmann 1943; Kappers 1960).

Collin (1976) proposed the term *rudimentation* for the reduction of the sensory features, especially the outer segment structure, of the pinealocyte through vertebrate phylogeny. He noticed that rudimentation apparently occurs also at the level of ontogeny. In *Lacerta vivipara* (reptile), all the pinealocytes possess a well-developed outer segment with a lamellar structure in the late embryonic stages, but around the time of hatching a population of the cells undergoes rudimentation in the outer segment, while showing progressively marked endocrine features at the cell base. Collin (1976) notes that rudimentation may also occur in mammals. In the rat and hamster (Clabough 1973), a knoblike cytoplasm with a centriole, clearly corresponding to the outer segment, protrudes into the pineal lumen in the final stage of fetal life (Fig. 25.4). Thereafter, along with the obliteration of this lumen in early postnatal days, the apical knob retracts, while the sensory features of the cell at the basal portion become more prominant. These seem to be instances of ontogeny recapitulating phylogeny.

25.9 Summary

The pineal body consists of pinealocytes, glial cells, and nerve fibers. The pinealocyte is a typical paraneuron derived from the neuroectoderm (neural tube) and sharing many cell-biological features with neurons. The pinealocyte secretes *melatonin*, a gonad-suppressing hormone. The cell first produces *serotonin* and then converts it into melatonin by means of two enzymes, NAT and HIOMT. *Bioactive peptide(s)* have been proposed but not convincingly identified to be co-secreted with melatonin. Histochemical studies suggest the occurrence of *acidic carrier proteins.*

The pinealocyte in *lower vertebrates* (from cyclostomes to reptiles) is a *photosensitive cell* possessing a distinct outer segment as in the visual cells in the lateral eyes. The pineal body in these animals is usually associated with a *parietal* eye which also contains similar photosensitive cells. Both of these photosensitive paraneurons are synaptically associated with afferent nerves, which conduct the light stimulation to the brain.

In birds, the pinealocyte is more secretory in nature and in mammals its photosensitive features are completely lost, while its cytoplasm contains numerous endocrine granules. Melatonin and other cell secretions contained in the granules are released into blood capillaries. Efferent sympathetic nerves supply the avian and mammalian pinealocyte and regulate its function according to the light input to the lateral eyes. The avian pinealocyte represents the *biological clock*, whereas in mammals the clock is in neurons of the suprachiasmatic nucleus, whose impulse is transmitted to the pinealocyte for its rhythmic function.

In its *ontogeny*, the avian and mammalian pinealocyte recapitulates the process of *rudimentation of photosensitive cell structures* in its phylogeny. The phylogeny and ontogeny of the pinealocyte evidence the continuity of the sensory and secretory paraneurons and that of their synaptocrine and hemocrine secretion modes.

26. Melanocytes

26.1 Types of Melanin-Producing Cells

26.1.1 Melanocytes As Pigment Cells

There are many kinds of melanin-producing cells in animals; the cells vary greatly in morphology and function from species to species. Care must be taken in deciding whether a specific melanin-producing cell type of one species corresponds to that of other species.

Melanocytes are only one of a variety of pigment cells (chromatophores) which are distributed throughout the animal kingdom. Table 26.1 shows the principal members of the pigment cells in vertebrates and their specific pigments. In addition to *melanocytes* (melanophores), there are *xanthophores* (erythrophores) and *leukophores* (iridophores). Experimental data indicate that all these chromatophores in vertebrates are of *neural crest origin* and possess some neuron-like characteristics (see Fujii and Oshima 1986). Invertebrates also possess a large variety of pigment cells; a homologous relationship between these cells and vertebrate melanophores has not yet been established.

In this chapter, we will concentrate on the melanocyte since it is the most extensively investigated of the various types of pigment cells. It is not our intention to make a comprehensive treatise of melanin-containing cells throughout the animal kingdom.

In medicine, especially dermatology, the term *melanocyte* is used, whereas in zoology the homologous melanin-containing cell is termed a *melanophore*. In medical research mainly homoiothermic animals, especially mammals, are used in the experiments, whereas many zoologists specialize in the study of pigment cells of poikilothermic animals. To avoid confusion in this book, we shall adopt the term *melanocyte* for both poikilothermic and homoiothermic animals. The suffix "-cyte" signifies cell, whereas "-phore" indicates bearer. International debate on the terminology for vertebrate melanin-containing cells is reported in Fitzpatrick et al. (1966).

26.1.2 Melanin: Distribution, Chemistry, and Function

Melanin is a dark endogenous pigment distributed in a wide variety of animals; in different concentrations it causes a brown to yellow coloration. It consists mainly of dihydroxyindoxylic acid. Thus, melanin is not chemically identical to any neurotransmitter substance or hormone. However, its *precursor*, *dopa* is the same as the

Table 26.1. Pigment cells (chromatophores) and their pigments in vertebrates

Chromatophore type	Specific organelles	Location in skin	Pigments
Melanocytes, i.e., melanophores (yellow to black)	Melanosomes	Epidermis, dermis	Melanin
Xanthophores[a] (yellow to red)	Pterinosomes and carotenoid vesicles	Dermis	Pteridine and/or carotenoids
Iridophores[b] (light-reflecting)	Reflecting platelets	Dermis	Purines

[a] Erythrophores are included. Both xanthophores and erythrophores are included in the lipophores
[b] Leukophores are included, although their coloring mechanism is somewhat different from that of other iridophores

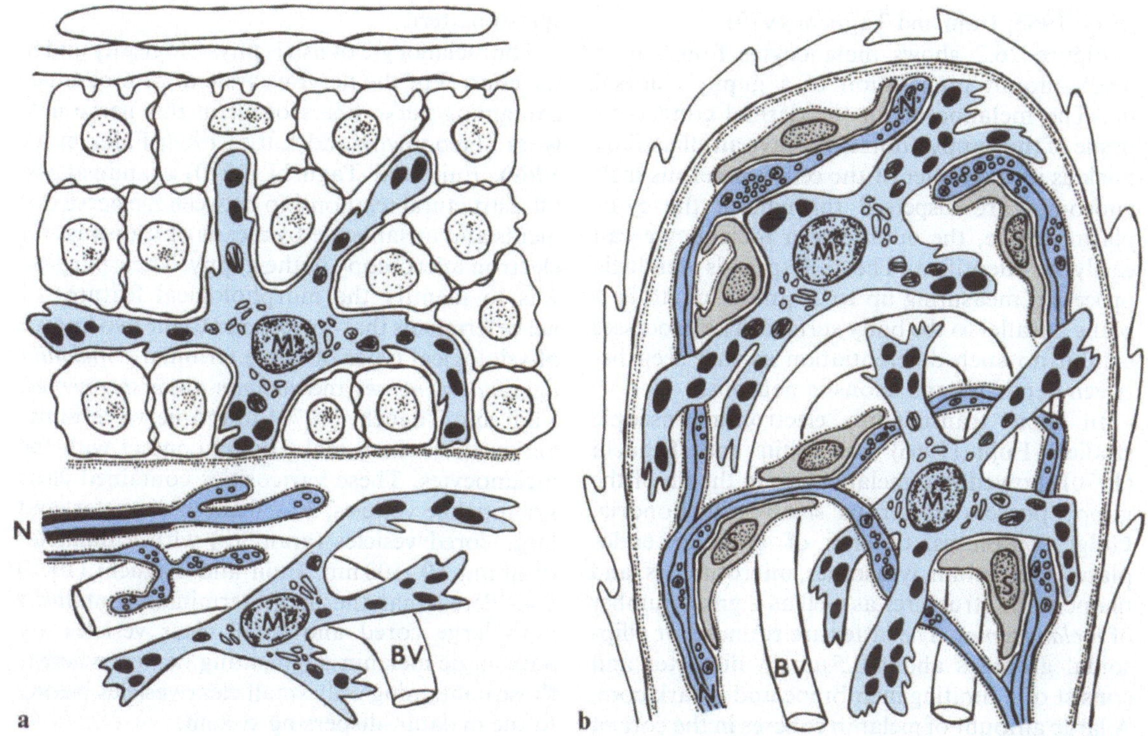

Fig. 26.1a, b. Structure of melanocyte in mammals and fish. Melanocytes (*M*) in **a** human skin and **b** the fish fin The melanocytes in the human skin have no innervation, whereas those in fish are directly innervated by autonomic nerves. *BV* blood vessel, *N* nerve fibers, *S* Schwann cells

precursor of catecholamines, and dopa often coexists with aminic and proteinic substances within the melanin-containing cells (Rost et al. 1969).

Melanin is insoluble in water, organic solvents, and dilute acids and bases. Strong oxidizers such as hydrogen peroxide and potassium permanganate are needed to decolorize melanin in the laboratory.

Melanin production gives animals the great advantage of being able *to shield themselves from ultraviolet light*, which has harmful genetic effects. The distribution of melanin in the skin, especially in the epidermis of higher vertebrates is in accordance with this purpose. Skin melanin and other pigments provide lower vertebrates and invertebrates with *protective coloration and sexual features*. Melanin is widely used in the animal kingdom to create *a dark wear around the eye*. In addition to these cases where the presence and distribution of melanin are clear, large amounts of melanin occur in some animals and the reason for its occurrence is unknown: There are large numbers of melanin-containing cells in the connective tissue deep in the body of many lower vertebrates; many kinds of fish have a surprisingly dark abdominal wall due to a dense distribution of melanocytes in the subserous layer.

Even in the skin of vertebrates, there is a *wide spectrum of melanocytes*. For clarity of explanation, first the innervated melanocytes in fish fins and then the noninnervated melanocytes in the human epidermis will be described in detail in the following sections. We take the former to be representative of melanocytes of poikilothermic vertebrates, whereas the latter are typical of those in homoiothermic animals (Fig. 26.1).

26.2 Melanocytes in Poikilothermic Vertebrates

26.2.1 Fish Fin Melanocyte: Structure and Ultrastructure

The so-called *split fin preparations from small teleosts*, such as the guppy *Lebistes reticulatus* provide a good experimental model for morphological and physiological investigations of melanocytes and their innervations. There are many melanocytes in the dermal connective tissue in the space between the blood vessels and ossicles

(Fujii 1966; Fujii and Taguchi 1970).

Figure 26.2 shows melanocytes found in a whole-mount preparation of a guppy's dorsal fin. The melanocytes in the dermal connective tissue of the guppy's dorsal fin have an ellipsoidal nucleus in the center of the cell. Numerous melanosomes are dispersed throughout the cytoplasm; hence, the outline of a melanocyte can easily be identified. The cell spreads dendritic processes measuring up to 30 μm in length in a plane parallel to the body surface. The processes exhibit no such differentiation as that seen between dendrites and axons in neurons.

In their transmission electron-microscopic studies, Fujii (1966) and Fujii and Taguchi (1970) showed that melanocytes in the fin of the guppy possess numerous small mitochondria, Golgi complexes, cisterns of granular endoplasmic reticulum, lysosomes, microtubules, and filamentous structures as well as a great number of *melanosomes*. The latter are rounded or ellipsoidal granules about 0.5 μm in diameter and consist of a limiting membrane and a dark core. A large amount of melanin coheres in the core of these melanosomes. As in mammalian melanocytes, premelanosomes are localized around the Golgi complex.

The mode of formation, maturation, and storage of melanosomes in the fish fin is essentially identical to that in human skin (see p. 258). So far as we know, no convincing electron micrographs are available indicating the possible exocytotic release of the melanosome contents.

No attachment devices such as desmosomes are seen on the surface of the melanocyte.

The fish fin melanocyte described above is only one example of the melanin-containing cells found in poikilothermic vertebrates. In their cytological features, amphibian and reptilian melanocytes resemble those of fish fin melanocytes.

26.2.2 Innervation

The melanocyte of teleost fish is partly covered by a thin cytoplasmic process of sustentacular cells, which are completely homologous to *Schwann cells*. In the space between the melanocyte and sustentacular cells, there are bundles of thin neuronal processes. Large cored vesicles and small synapticlike vesicles are present in the varicosities of the neuronal processes. It is likely that the neuronal bundles ensheathed by the Schwann cells form a network which extends throughout the dermal connective tissue. Melanocytes are inserted in this nerve plexus in a splash pattern.

The melanocyte of fish is physiologically under the control of the nervous system. Monoamine-containing nerve fibers occur in this nerve network (Jacobowitz and Laties 1968; Falck et al. 1969). Fujii and Taguchi (1970) examined the ultrastructural relationship between the nerve elements and melanocyte by means of transmission electron microscopy in the guppy. Their purpose was to identify the morphological features of nerve terminals that correspond to the two known physiological types of nerve terminal—*melanin-aggregating nerves* and *melanin-dispersing nerves*. Fujii and Taguchi (1970) found nerve terminal varicosities which make direct contact with the melanocytes. These varicosities contained clear synapticlike vesicles, 30–50 nm in diameter, and large cored vesicles (granules) with a diameter of about 60–100 nm. Fujii and Taguchi (1970) considered that the nerve terminals containing both large cored and small clear vesicles are adrenergic melanin-aggregating nerves, whereas those containing only small clear vesicles belong to the malanin-dispersing system.

26.2.3 Regulation of Melanosome Movement

The movement of melanosomes is usually controlled by light-induced activation of hormonal and neuronal systems. In the tail of the tadpole, however, the melanocyte is capable of responding *directly to stimuli.*

Blanching occurs in the larval stages of most amphibians, which is due in part to a decrease in the release of *melanocyte-stimulating hormone (MSH)* in darkness. *Melatonin*, melanin-contracting hormone, from the pineal also induces a profound aggregation of melanosomes in the melanocyte, leading to rapid blanching of skin color.

In some fish, hormonal mechanisms are thought to be predominantly responsible for the aggregation or dispersion of pigment-containing organelles within the melanocyte, whereas in others a neural mechanism seems to be dominant. There are numerous examples in fish in which both hormonal and neural mechanisms function in different proportions. In the *hormonal mechanisms*, MSH, melatonin, and other hormonal substances are involved; in the *neural mechanisms*, catecholamines (Fig. 26.2b), acetylcholines, adenine nucleotides, and neuropeptides (gastrin, VIP, substance P, neurotensin, etc.) have been recognized as the messenger substances or as possible candidates (Fujii and Oshima 1986).

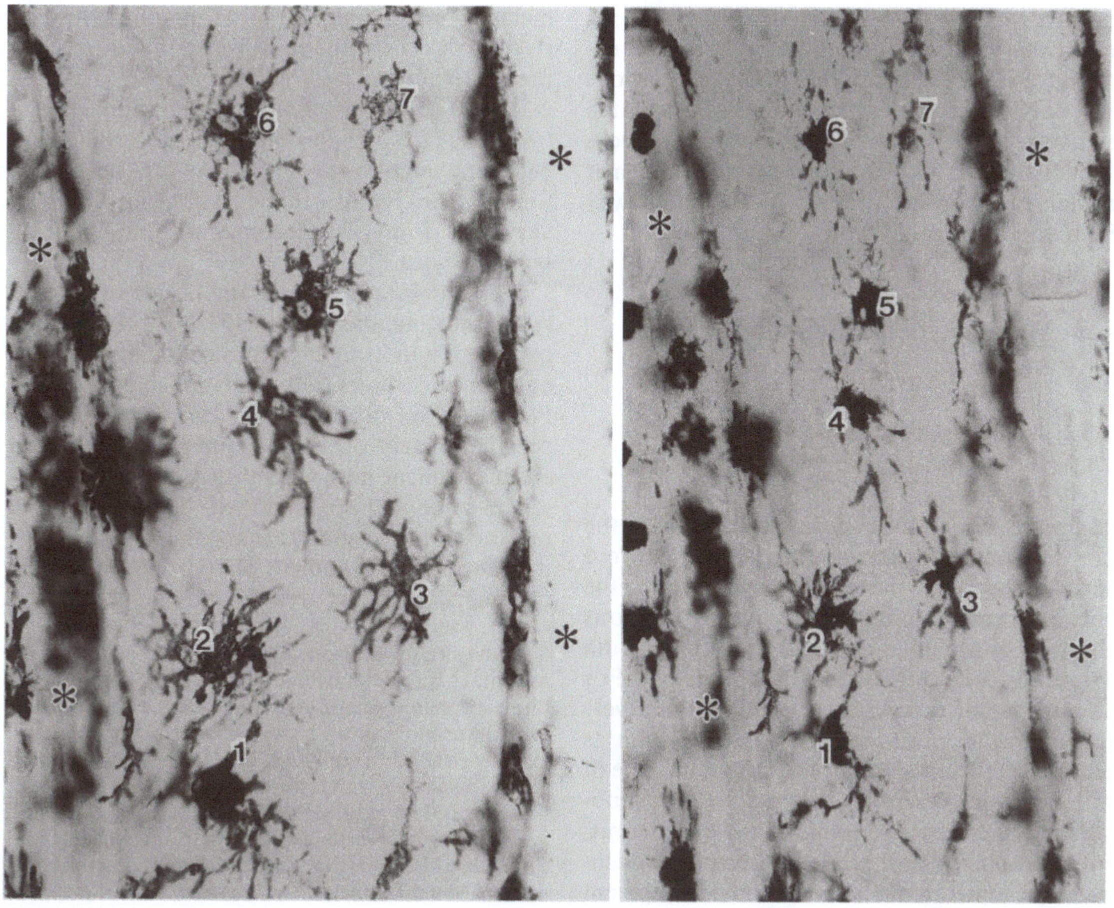

Fig. 26.2a, b. Melanocyte in a guppy's dorsal fin. **a** The outline of the melanocyte can be recognized by the localization of melanosomes. **b** After application of adrenaline, melanosomes rapidly aggregate in the perinuclear cytoplasm of the cell. Asterisks indicate framing ossicles of the dorsal fin. Each of the corresponding figures (*1–7*) in **a** and **b** points the same melanocyte before (**a**) and after (**b**) the adrenaline treatment.

26.2.4 Electrophysiological Aspects

The link between the reception of messenger substances and the movement of melanosomes in melanocytes was investigated using various electrophysiological methods (see Fujii 1969). The messenger substances may act on specific *receptors* on the cellular surface membrane. The receptors in some cases may constitute an assembly of regulatory subunits in the adenylate cyclase system. When a certain ligand interacts with its specific receptor, the catalytic speed of the catalytic subunit of the system is either accelerated or decelerated, probably leading to an increase or decrease in *cyclic AMP level* in the cytosol of the melanocyte. The increase in cytosolic concentration of cyclic AMP may cause a dispersion of melanosomes in the melanocyte.

Among the receptors on the melanocyte membrane, the MSH specific receptor is known to require extracellular Ca^{2+}. The exact sequence of the Ca^{2+}-dependence and the reason why only the MSH receptor needs Ca^{2+} remain to be clarified. It is believed that the α_1-adrenoceptors are not coupled with the adenylate cyclase system but are connected to a *Ca^{2+}-calmodulin regulatory system*, leading to the activation of several enzymes such as phosphodiesterase and C-protein kinase. Thus, for the motility of melanosomes in the fish melanocyte, Ca^{2+} is presumed to be an important intracellular messenger in addition to cyclic AMP, though the motility of the cell per se does not require extracellular Ca^{2+} (Fujii and Oshima 1986).

The action of ligands can be mediated intracellularly without the active intervention of electric forces across the melanocyte membrane, since these substances are able to exert their normal effects in K^+-rich saline in which the membrane was clamped and depolarized (Fujii and Taguchi 1969). If potential fluctuations across the melanocyte membrane could be detected following ligand action, these fluctuations may only be secondary phenomena. Although relatively little is known about pigment cells other than melanocytes, it is assumed that their cellular mechanisms operate in the same way (Fujii and Oshima 1986).

26.3 Melanocytes in Homoiothermic Animals

26.3.1 Melanin-Containing Cells in Man and Mammals

In humans, melanin-producing cells occur in the skin and hair, in the retina, iris, and choroid plexus of the eye, in the leptomeninx covering the central nervous system, around the membranous labyrinth of the inner ear, and in the substantia nigra and locus ceruleus of the brain stem. For the purpose of description, we shall tentatively classify the melanin-producing cells of man as follows: (1) melanocytes in the stratum basale of the epidermis and in the hair bulb; (2) melanocytes in the connective tissue of the eye, dermis, and leptomeninx and related tissues (i.e., membranous labyrinth of the inner ear); (3) pigment epithelial cells of the eye; and (4) melanin-producing neurons.

Keratinocytes of the epidermis and keratinized cells of hair contain abundant melanin and protect the body surface from the adverse effects of ultraviolet radiation. The amount of melanin in the cells determines the degree of pigmentation of skin and hair of the different human races. However, these melanin-containing cells are devoid of the ability to synthesize melanin. Thus, the keratinocytes of the epidermis and melanin-containing histiocytes are excluded from consideration here.

26.3.2 Human and Mammalian Epidermal Melanocytes

The human epidermal melanocyte is a dendritic cell which forms specific melanin-containing organelles called *melanosomes* (Seiji et al. 1961). This cell, like many types of paraneuron, contains FIF-positive substances, presumably catecholamines (Rost et al. 1969). In laboratory animals, such as the mouse and hamster, the melanocyte takes up amine precursors like dopa and decarboxylates them (Rost et al. 1969). There is ample circumstantial evidence that epidermal melanocytes in man and other mammals are derived from cells of the neural crest that migrate to the epidermis and hair bulb (Rost et al. 1973) (see p. 259). The epidermal melanocyte radiates long (up to 50 μm) cytoplasmic processes, which run between neighboring keratinocytes, devoid of tonofilament bundles in the cytoplasm, and form very few, if any, desmosomes with neighboring keratinocytes.

Melanosomes are membrane-bound organelles specific to melanocytes (Seiji et al. 1961); they consist of a limiting membrane and a core. Melanosomes contain both melanin and proteins. Pearse (1969) proposed the term *nigrin* for the specific protein of melanosomes. However, the exact chemical nature of this presumptive protein has not yet been identified. In any event, the melanosomes in mammalian melanocytes, like the secretory granules of paraneurons, are formed in the Golgi area (Maul 1969).

Immature melanosomes, usually called stage I melanosomes (premelanosomes), are vesicles with amorphous proteinaceous material in the center and fibrilar tyrosinase molecules at the periphery (Stanka et al. 1969; Maul and Brumbaugh 1971). They contain proteinic substances which show no tyrosinase activity at this stage. The stage II melanosomes are elongated vesicles whose interior is filled with many helicoidal tyrosinase filaments and an amorphous proteinaceous matrix. Stage III melanosomes show a strong tyrosinase activity and contain accumulated melanin together with tyrosinase filaments and a proteinaceous matrix. stage IV melanosomes represent mature melanosomes. They are filled with melanin and show no tyrosinase activity. In mature melanosomes, melanin is stored and forms a complex with proteinic substances; this situation is comparable with the storage-complex formation between aminic and peptidic substances in the secretory granules of paraneurons.

Unlike the secretory granules of neurons and paraneurons, the contents of *mature melanosomes are not extruded by exocytosis*. The melanosomes become mature in the perinuclear cytoplasm and are transported to the tip of the cytoplasmic processes of the melanocytes by a mechanism similar to axonal flow. The melanosome-containing tip then penetrates the neighboring keratinocytes and is engulfed by the latter. This process of the transport of melanosomes from the melanocytes to the keratinocytes is called *cytocrinia* (Masson 1948). The detailed mechanism of this intercellular process is not yet fully understood, although many electron-microscopic studies have been performed on the extrusion mechanisms of melanosomes (see Kurosumi 1976).

The amount of melanin in keratinocytes determines the degree of pigmentation of skin and hair. Melanosomes are frequently released in the extracellular space without being engulfed by the neighboring keratinocytes. In the epidermis of nonwhite human races, a considerable amount of melanin exists in the space between the keratinocytes.

Melanin synthesis in the melanocytes is mediated by the enzyme tyrosinase. This enzyme contains copper and converts tyrosine to dopa and then dopa to dopaquinone, which makes melanin by a complex series of spontaneous reactions such as oxidative polymerization. Thus, if melanocytes are dipped in a dopa solution, they accumulate melanin in the cytoplasm because of their tyrosinase activity (Bloch 1917). This reaction, called the *dopa test*, is an effective histochemical technique for the demonstration of melanocytes (Mishima 1960, 1964).

26.3.3 Melanocytes in the Dermis and Other Tissues

Melanocytes occur in the iris, ciliary body, and choroid plexus of the eye, leptomeninx, membranous labyrinth of the inner ear, and dermis.

Like melanocytes of the epidermis and hair bulb, melanocytes in the connective tissue are derived from cells of the neural crest and synthesize and store melanin in their melanosomes.

26.3.4 Pigment Epithelial Cells As Modified Glial Cells

The pigment epithelium of the retina is derived from the neural tube. The space between the pigment epithelium and the visual cells corresponds to the ventricular lumen of the central nervous system. Pigment epithelial cells are different from melanocytes of neural crest origin in shape; the former are columnar or cuboidal rather than dendritic and retain the epithelial arrangement of ependymal cells.

In darkness, most melanosomes of the pigment epithelial cells accumulate in the apical pole of the cells. In bright light, on the other hand, melanosomes move into the cytoplasmic processes which surround the photosensitive outer segments of the visual cells.

The pigment epithelial cells of the retina are usually regarded as *modified glial cells* and possess an active phagocytotic activity; they phagocytize and digest degenerated tips of the outer segments of the visual cells. Thus, they play a critical role in the "structural homeostasis" of the visual cells.

In addition to those of the retina, pigment epithelial cells exist in the iris and ciliary body.

26.3.5 Melanin-Containing Neurons

Melanosomes are present in some neurons of the substantia nigra and locus ceruleus of the brain stem. It has been pointed out that the melanosomes of these neurons considerably differ in ultrastructure from those of the epidermal melanocyte, but the essential points with regard to their formation and maturation are identical. Since these neurons normally contain catecholamines such as dopamine their melanin is considered to be a by-product of catecholamine metabolism.

26.4 Ontogeny

Melanocytes (melanophores) and all other chromatophores in vertebrates are derived from the neural crest. The cells that migrate out of the *neural crest* are called *melanoblasts*, or more generally *chromatoblasts*. They travel over long distances and ultimately differentiate into melanocytes presumably as a result of microenvironmental and/or hormonal influences (review: Rawles 1948). Hair bulb melanocytes, for example, become associated with developing hairs at an early stage of development and produce melanin in dark-haired individuals during the later months of gestation.

Du Shane (1935) discovered that when part of the neural crest is removed from an amphibian embryo the animal growing from that embryo will later have a white area of skin corresponding to the crest region removed. In removal and grafting experiments, the amphibian neural crest was proved to be the source not only of melanophores but also of other chromatophores (review: Du Shane 1943). The neural crest origin of melanocytes was later confirmed in higher vertebrates including mammals. Rawles (1947) transplanted different portions of the embryos of a dark-haired mouse strain into white leghorn embryos. She found that the chicks growing from these embryos showed pigmented feathers only when the graft had contained neural crest cells.

The impact brought of the experiments of Du Shane (1935) was great half a century ago when the neural crest was regarded as a mere anlage of nerve ganglia. Developmental studies on melanocytes on the basis of their inherited marker—the ability to produce melanosomes (Rawles

1948)—opened the way to the current understanding of a large variety of cells differentiating from the neural crest (review: Le Douarin 1982).

The ontogeny of invertebrate pigment cells will be discussed in the following section.

26.5 Phylogeny of Chromatophores in Invertebrates

Invertebrates of both the protostomian and deuterostomian series possess many kinds of pigment cells. Dendritic cells contain cytoplasmic structures similar to melanosomes of the vertebrate melanocytes. In some crustaceans, multipolar cells with numerous melanosome-like structures are seen in the cutaneous tissue just beneath the exoskeleton. However, the phylogenetic relationship between these invertebrate pigment cells and the vertebrate melanocyte is unknown.

Recently, the origin of pigment cells (echinochrome-containing cells) has been studied by Gibson and Burke (1985) with an immunocytochemical technique in larvae of the sea urchin *Strongylocentrotus purpuratus*. Although it was only vaguely believed that the pigment cells in echinoid larvae were of mesenchymal origin (literature: see Gibson and Burke 1985), these authors demonstrated that the precursors of the pigment cells are released from the *vegetal plate* and migrate to the entire embryo, developing pigment granules. This vegetal plate is a part of ectoderm at the margin of the archenteron and continues to the neurogenic ectoderm near the animal pole (Burke 1983).

It is an interesting point whether pigment cells in protostomian invertebrates also originate in relationship with the neurogenic germ layer.

26.6 Are Melanocytes Paraneurons?

Melanocytes (and chromatophores) in vertebrates share certain cell-biological features with neurons. They originate from the *neural crest*; they possess *Golgi-derived melanosomes*, which are regarded as equivalent to secretory granules. Besides peptidic products of rough endoplasmic reticulum, the melanosomes contain melanin, a derivative of dopa. Conversely, melanosomes and melanin also occur in certain neurons. The melanocyte extends *dendritic processes*, which are especially conspicuous in lower vertebrates. In fish, the cell receives *synaptic innervation* and represents the terminal element of the sympathetic nervous system. Moreover, the cell is associated with a typical *Schwann cell*. Taking these findings into account, it seems reasonable to categorize the melanocyte as a paraneuron.

However, in the sense that a paraneuron should be a recepto *secretory* cell, the melanocyte (or other kinds of chromatophore) cannot be included in the category of paraneurons. The main product of the cell, melanin (or another pigment), is not a messenger substance and hence is not released to regulate the function of target cells. The identity and functional significance of amines and peptides reported to be contained in the granules of melanocytes represent an interesting field of study. Furthermore, it remains to be explored whether ATP and acidic carrier proteins (chromogranins) are present in melanosomes.

26.7 Summary

There are wide varieties of melanin-containing cells throughout the animal kingdom. Melanin is not chemically identical with any neurotransmitter or hormone, but its precursor, *dopa*, is a neuroparaneuronal product.

Melanocytes and chromatophores in vertebrates are derived from the *neural crest* and contain specific organelles, *melanosomes*, which are Golgi-derived structures apparently equivalent to the secretory granules of paraneurons. However, no convincing electron micrographs of exocytotic granule release have been demonstrated. Melanocytes release no messenger substance, hence it is unlikely that they control the functions of other cells. Because of the lack of secretion release, the melanocyte cannot be included among (at least typical) paraneurons.

Melanocytes in fish fins provide a good experimental model for melanocyte studies. The distribution and movement of melanosomes in melanocytes are controlled by synaptically associated *nerves* (transmitters: catecholamines and neuropeptides) and *hormones* (pineal melatonin and pituitary MSH). The cell surface of melanocytes is provided with specific *receptors* for various messenger substances. The MSH-specific receptors require extracellular Ca^{2+}. The cyclic AMP and Ca^{2+}-calmodulin systems are present as intracellular messengers.

Melanocytes in the human skin are mainly localized in the stratum basale of the epidermis. They have no innervations and no sensitivity to specific hormones.

27. Mast Cells

The mast cell is a unicellular endocrine/paracrine element dispersed in the connective tissue. Since it is a typical receptosecretory cell producing monoamines it would seem reasonable to include this cell among the paraneurons. As discussed in the general section (p. 9), however, the mast cell also shows some cell-biological and ontogenetic features that deviate from those of paraneurons, and it thus becomes debatable whether it should be classed as a paraneuron. In this chapter, the paraneuronal features of mast cells will be presented. Experimental data concerning the secretory mechanisms in an isolated mast cell will be included as a model for paraneuron study.

27.1 Distribution and Morphology

Mast cells are ovoid cells with a rather small rounded nucleus and rich cytoplasm filled with secretory granules. These are stained reddish purple by blue basic aniline dyes such as toluidine blue. This *metachromatic reaction* discovered by Ehrlich (1877) is due to the occurrence of acidic glycosaminoglycans (mucopolysaccharides), especially heparin, in the granules of the cell.

Mast cells in mammals are dispersed in connective tissue spaces in the body. There is a marked tendency for them to be gathered around capillaries and venules. A close topographical relation of mast cells to nerve fibers in normal tissue (Heine and Förster 1975; Dyer 1978) and to Schwann cells in schwannoma (Isaacson 1976) has been reported.

In some urodeles, e.g., *Triturus pyrrhogaster*, mast cells with long *dendritic processes* occur besides ordinary ovoid mast cells (Setoguti and Yonemoto 1965) (Fig. 27.1). The dendritic mast cells are especially numerous in the brain, i.e., close to blood capillaries in the brain parenchyma, in meninges, and even within the ventricles (Inoue et al. 1968). The significance of the neuron-like shape and the affinity to nervous tissue of these cells is unknown.

In the rat and mouse, many mast cells float in the peritoneal fluid and provide useful material for cell-biological study.

27.2 Fine Structure

The surface of mast cells of mammals is densely provided with *microvilli* (0.5–1 μm long), which may partly be microplicae (microfolds) under conditions.

In the cytoplasm, dispersed mitochondria, rather inconspicuous rough endoplasmic reticulum, and a large *Golgi apparatus* are found. *Centrioles* (diplosome) are located at the center of the Golgi area. Small secretory granules are formed in the Golgi vesicles to be matured and enlarged in the membrane sac, while shifting to the cell periphery (Fujita 1965) (Fig. 27.2). Combs (1966) has demonstrated that glycosaminoglycans, the major components of mast cell granules are derived from the Golgi appara-

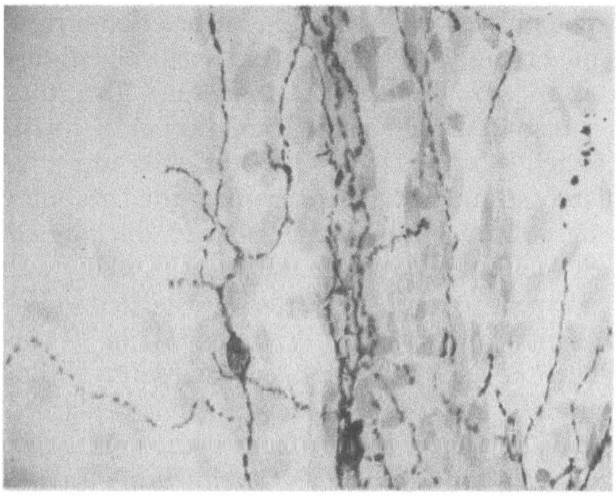

Fig. 27.1. Dendritic mast cells in Urodeles. The specimen was taken from the abdominal wall of a salamander, *Triturus pyrrhogaster* and stained in toluidine blue

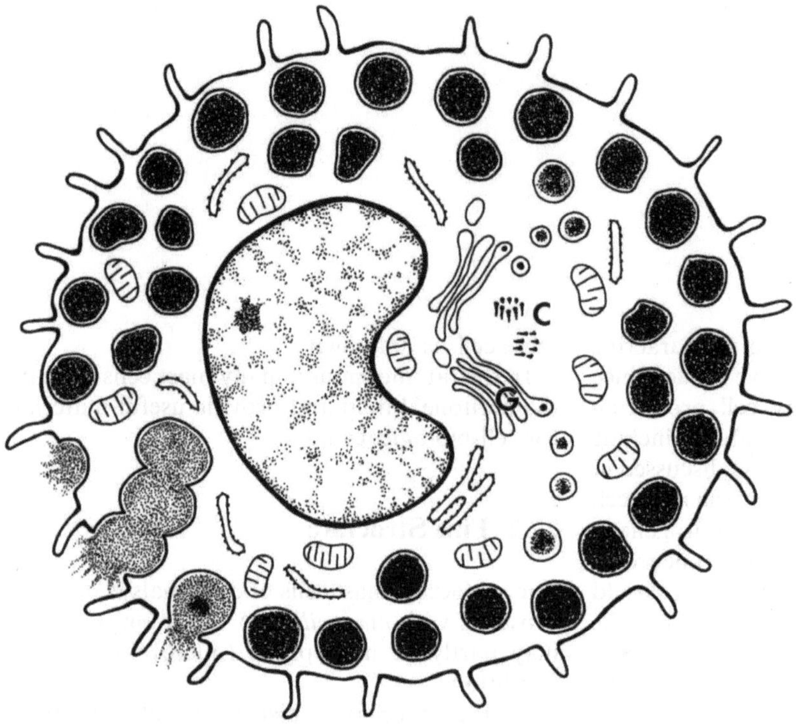

Fig. 27.2. Structure of rat mast cell. Granule release is shown in *bottom left*. *C* centriole, *G* Golgi apparatus

tus and, while they are packaged in membrane sacs, cisterns of rough endoplasmic reticulum are connected with them to add proteinic components.

The *mature granules* are around 0.8 μm in diameter (human and rat) and show very electron-dense contents surrounded by a smooth limiting membrane. The granules markedly differ in contour and fine structure among species. Human mast cell granules show a peculiar fingerprint-like structure.

The process of *granule release* has been investigated mainly in rat peritoneal mast cells stimulated *in vitro* by different histamine liberators (see below) (Smith 1958; Yamasaki et al. 1970; Röhlich et al. 1971). When the cells are adequately stimulated, peripherally located granules undergo *exocytosis*, i.e., Ω-shaped opening of their membrane sac. Often deeper located granules open to more superficially located ones, causing multigranular exocytosis (Fig. 27.3). When the cell is stimulated intensely, deeper granules may fuse together into enlarged vacuoles. Occasionally, a huge space containing denuded granules occupies the entire cell, leaving only perinuclear and marginal areas of the cell; the latter is then broken and the granule contents are extruded. In any case, the central part of the cell containing the nucleus and Golgi area is left intact for the regeneration of the cell after severe degranulation and loss of cytopolasm.

27.3 Secretory Products

Glycosaminoglycans (mucopolysaccharides). Mast cell granules consist mainly of heparin. Other acid glycosaminoglycans like hyaluronic acid and chondroitin sulphate are contained only in very small amounts (Smith 1963; Zachrisson and Schulz-Haudt 1967). These strongly acidic substances are found with basic monoamines (see below) in the granules. Green and Day (1960) demonstrated that the levels of heparin are roughly proportional to the levels of histamine and serotonin in neoplastic mast cells.

Amines. Histamine is contained in the mast cell granules and is regarded as the most important substance of the cell. Mast cells possess histidine decarboxylase, which converts histidine into histamine. If one injects animals histidine or histamine, these are taken up by mast cells, which then show elevated histamine content.

Serotonin is additionally localized in mast cell granules in the rat and mouse (Parratt and West 1957). In their fluorescence histochemical studies,

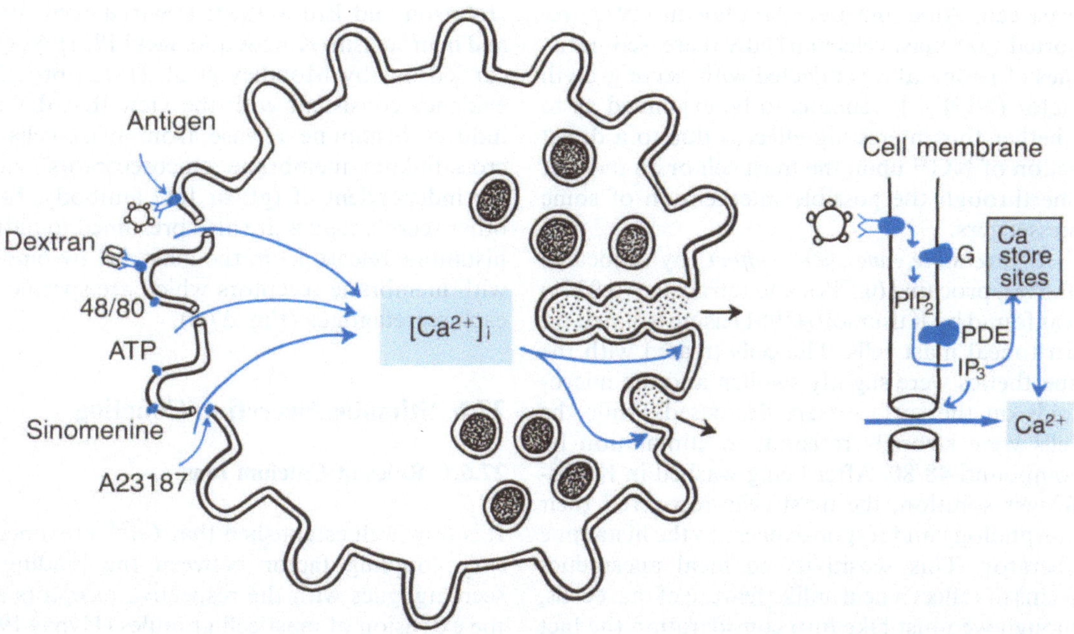

Fig. 27.3. A model for the mechanism of stimulus-secretion coupling in the mast cell. The stimuli (secretagogues) are shown acting on the proper receptor site, which when activated is considered to mediate the opening of the Ca^{2+} channel. Increase in cytosolic Ca^{2+} concentration causes extrusion of contents of granules by exocytosis. The calcium ionophore (A23187) by-passes the stimulus-operated Ca^{2+} channel. A model in concert with the inositol lipid cycle and Ca^{2+} influx via the Ca^{2+} channel is represented on the *right*. *G* guanine nucleotide binding protein, *PDE* phosphatidylinositol 4,5 bisphosphate phosphodiesterase, *PIP_2* phosphatidylinositol 4,5 bisphosphate, *IP_3* inositol triphosphate

Håkanson et al. (1970) suggest that serotonin coexists with histamine in (gastric) mast cells of a variety of mammals, including the rabbit, cat, dog, and monkey. Furthermore, dopamine-containing mast cells have been reported (Parratt and West 1957). Coupland and Heath (1961) reported the occurrence of chromaffin, dopamine-containing cells categorized as mast cells in cattle and sheep.

Although histamine is generally regarded as the essential substance for mast cells, which are often called "histaminocytes," comparative studies indicate that the histamine content in mast cells conspicuously decreases from birds to reptiles and it is undetectable in amphibia and fish (Takaya and Fujita 1967, Takaya 1969).

Neuropeptides. Cutz et al. (1978) reported that the neuropeptide VIP was contained and released by stimulation in rat peritoneal mast cells. However, later studies, including our own immunocytochemical investigations in different mammals, have not been able to confirm this remarkable report. The possible occurrence of this and other bioactive peptides in mast cell granules deserves further exploration.

Enzymes. Mast cell granules in mammals (man, rat, and several other species) have been proved to contain chymotrypsin-like and trypsin-like proteases (Lagunoff and Benditt 1961; Glenner et al. 1963; Lagunoff et al. 1964).

ATP and other adenine nucleotides. ATP and probably also other adenine nucleotides are contained in mast cell granules. ATP, together with more acidic glycosaminoglycans, has been presumed to be involved in loose linkage to basic amines in the granules. In support of this view, Green et al. (1962) demonstrated that ATP levels were roughly parallel to the levels of histamine and serotonin in cultured mastocytoma cells.

27.4 Other Paraneuronal Features

Attempts to detect either chromogranin-like immunoreactivities in mast cell granules or neuron-specific proteins (NSE, etc.) in the cytoplasm have been unsuccessful (Iwanaga and Fujita 1986, unpublished data).

In favor of the paraneuronal nature of the

mast cell, Aloe and Levi-Montalcini (1977) reported that mast cells markedly increased in tissues of neonatal rats injected with nerve growth factor (NGF). It remains to be examined as to whether this interesting effect is due to a direct action of NGF upon the mast cell or an indirect one through the possible intervention of some messengers.

A reversible anesthetic effect by lidocaine (0.5%), procaine (0.5%), and tetracaine (0.02%) was found by Kusumoto (1981) using isolated rat peritoneal mast cells. The cells treated with the anesthetics were slightly swollen and the microfolds on the surface were decreased, while the cells were strongly resistant to stimulation by compound 48/80. After being washed in Krebs-Ringer solution, the mast cells recovered their morphology and responsiveness to the histamine liberator. This sensitivity to local anesthetics seems to reflect a neuronlike feature of mast cells, though we must take into consideration the fact that many nonnervous cells may be influenced by the anesthetics to a greater or lesser extent.

The close resemblance of mastocytoma to paraneuroma has been noted by Alroy et al. (1975), who reported that a distinctive intestinal tumor in the cat hitherto known as carcinoid of enterochromaffin cells was actually a kind of mastocytoma.

27.5 Immune and Signal Recognition

Mast cells and basophil leukocytes possess *membrane receptors* which bind specifically the so-called hemocytotropic antibody now known as IgE in most species (Foreman 1980). There are about 1×10^5–5×10^5 IgE receptors molecules/cell and the equilibrium dissociation constant for the binding of IgE to its cellular receptor is about 10^{-11} mol/l. The signal to the mast cell or basophil leukocyte to secrete histamine-containing granules is the binding of the cell-bound IgE antibody on the plasma membrane with antigen present in the surrounding milieu. The IgE receptor is a glycoprotein with a molecular weight of about 80000 and is freely mobile in the horizontal plane on the plasma membrane (Metzger 1977).

In addition to IgE, a wide variety of substances can activate stimulus-secretion coupling; representatives of such substances are dextran, compound 48/80, sinomenine, ATP (Foreman et al. 1976; Kazimierczak and Diamant 1978; Kanno 1981), and bioactive peptides like substance P (Johnson and Erdös 1973; Fewtrell et al. 1982) and neurotensin (Kurose and Saeki 1981). Studies carried out by Moodley et al. (1982) provided evidence consistent with the view that dextran induces histamine release from mast cells by cross-linking membrane glucoreceptors which are independent of IgE or IgG antibody. Most other secretagogues are also presumed to induce histamine release from the mast cell by binding with membrane receptors which are specific for each secretagogue (Fig. 27.3).

27.6 Stimulus-Secretion Coupling

27.6.1 Roles of Calcium Ions

It is now well established that Ca^{2+} are a necessary coupling factor between the binding of secretagogues with the respective receptors and the extrusion of mast cell granules (Uvnäs 1971; Foreman et al. 1976; Kanno, 1981; Kanno et al. 1973; Cochrane and Douglas 1974; Kagayama and Douglas 1974; Gomperts 1976; Kazimierczak and Diamant 1978; Foreman 1980). It has been recently demonstrated that direct binding of IgE with the receptor is sufficient to increase the membrane permeability to Ca^{2+} and thereby to induce histamine extrusion (Ishizaka et al. 1979, 1980).

A model of stimulus-secretion coupling in the mast cell was proposed by Foreman et al. (1976) and this has been further revised by Foreman (1980) as follows. Immunologically triggered histamine release from mast cells has been shown to be potentiated specifically by phosphatidyl serine or its lyso-analogues; the lipid itself is not capable of stimulating histamine release. The action of phosphatidyl serine was clearly related to the role of Ca^{2+} in simulus-secretion coupling, and it has been shown that phosphatidyl serine increases the membrane permeability to Ca^{2+} in stimulated cells. The potentiating effect of phosphatidyl serine in immunologically mediated histamine release is associated with the binding of vesicles of the phospholipid to the mast cells. Binding of IgE receptors on mast cells initiates a sequence of membrane phospholipid metabolism and enzyme activation, which results in the formation of Ca^{2+} channels. Phospholipid metabolism in rat mast cells activated by antigen was further examined with reference to phosphatidyl inositol (PI) turnover by Ishizuka and Nozawa (1983). They also confirmed that antigen-induced histamine release from passively sensitized mast cells with IgE was potentiated by adding phosphatidyl serene. The addition of antigen to [^{3}H] glycerol-prelabeled and -sensitized mast cells induced a marked loss of radioactivity of PI and a concurrent accumulation of 1,2-diacylglycerol and phosphatidic acid in 5–60 s. Furthermore, this antigen-induced PI breakdown was enhanced in the presence of Mg^{2+}. Histamine release occurred parallel to PI breakdown. On the other hand, the transient Ca^{2+} influx into mast cells, as measured by uptake of $^{45}Ca^{2+}$, was found to occur quickly after the cells were activated by antigen, which was linked to PI breakdown. These results suggest that enhanced PI turnover may be an important step in the biochemical sequence of events leading to release of histamine and that both Ca^{2+} and Mg^{2+} appear to take part in stimulus-secretion coupling in rat mast cells.

The histamine released by compound 48/80 uses Ca^{2+} mobilized from both cellular and extracellular sources (Douglas and Ueda 1973; Cochrane and Douglas 1974; Douglas 1975a; Woldemussie and Moran 1979; Johansen 1980). Similar findings were obtained in the secretory response to polymixin B and bradykinin (Cochrane et al. 1982). To isolate and study the extracellular component of secretory response elicited by these secretagogues, mast cells subjected to prior incubation with Ca^{2+}-free Locke solution containing EGTA were used, which renders mast cell secretion dependent upon extracellular Ca^{2+}. In Ca-chelated mast cells "primed" by 48/80 or A23187, it was shown that the addition of Ca ions to the extracellular fluid promoted the exocytotic extrusion of secretory granules and release of histamine (Douglas and Ueda 1973; Cochrane and Douglas 1974; Kagayama and Douglas 1974). Both antigen (Foreman et al. 1977; Ishizaka et al. 1979) and compound 48/80 (Spataro and Bosmann 1976; Foreman et al. 1977; Ranadive and Dhanani 1980) stimulate increased $^{45}Ca^{2+}$ uptake by mast cells. In mast cells not depleted of Ca^{2+}, the secretagogues seem to mobilize Ca^{2+} of both extracellular and cellular (membrane) sources (Douglas 1975a; Spataro and Bosman 1976; Johansen 1980; Ranadive and Dhanani 1980; Cochrane et al. 1982).

The view that a rise in the intracellular concentration of Ca^{2+} suffices to induce exocytosis has been directly supported by injecting ions through micropipettes placed within rat peritoneal mast cells (Kanno et al. 1973). Injection of Ca^{2+} but not Mg^{2+} or K^{+} elicited extrusion of secretory granules, which indicates the general requirement for Ca^{2+} in stimulus-secretion coupling.

The increase in Ca^{2+} permeability is thought to be transient, since if mast cells are stimulated by antigen in the absence of Ca^{2+} and Ca^{2+} are added after an interval of time, histamine release declines (Foreman and Garland 1974; Foreman et al. 1973, 1977). A similar transient effect was observed in Ca^{2+}-free media with the stimulation of mast cells by secretogogues, compound 48/80, polymyxin B, and bradykinin, which are believed to act independently of IgE (Cochrane et at. 1982). These observations support the view that stimulation of Ca-chelated mast cells by these secretagogues results in a rapid increase in the permeability of the cell membrane to Ca^{2+}. Following stimulation, this permeability decreases and the addition of extracellular Ca^{2+} becomes progressively less effective in eliciting histamine release. The ionophore A23187, by circumventing this mechanism, is able to elicit a full secretory response at any time.

27.6.2 Roles of Cyclic AMP

Cyclic AMP was postulated to be involved in inactivation of the Ca channel (Foreman 1980). Cyclic AMP inhibits histamine release, and there has been evidence suggesting that a fall in intracellular cyclic AMP accompanies the activation of histamine release. As the levels of cyclic AMP return to their basal value after the release activation, inactivation of Ca channels seems to be transient. Mast cells possess a protein kinase which is activated by cyclic AMP and may be responsible for phosphorylation of membrane protein during inactivation of Ca channels.

The view that high levels of cyclic AMP inhibit histamine release by decreasing the permeability, however, has been criticized as being too simplistic. Alm and Bloom (1982) reviewed studies which implied that there was little or no correlation between histamine release and intracellular levels of cyclic nucleotides. The inhibitory effects of the catecholamines (noradrenaline and adrenaline) and the adrenergic agonist isoproterenol were studied and found to be independent of cellular cyclic AMP levels (Alm and Bloom 1983). The inhibitory effects of phosphodiesterase on the histamine released by antigen or compound 48/80 were studied and it was found that there was no correlation between the measured levels of cyclic AMP and the effect on histamine release (Sydbom et al. 1981). Although there was a relation between the accumulation of cyclic AMP in mast cells and secretion of histamine mediated by human IgE and ACTH (1–24) peptide as well as by an IgE-dependent mechanism, a simultaneous increase in cyclic AMP and release of histamine suggested that the two events may not be causally related (Burt and Stanworth 1983).

27.6.3 Roles of Inositol Lipid Cycle

The inositol lipid cycle (see Chap. 4) was postulated to be involved in antigen-activated rat mast cells by Ishizuka and Nozawa (1983) several years ago. Their view has been expanded and analyzed profoundly (Dainaka et al. 1986).

27.7 Ontogeny

The traditional view is that mast cells originate from local undifferentiated mesenchymal cells (Alfejew 1924). Some researchers, on the other hand, have suggested a possibility that mast cells might be derived from the neural crest like melanocytes (Hirano et al. 1974; see also Fujita 1977). Recent studies by Kitamura et al. (1977, 1979a, b), using bone marrow grafts in irradiated mice indicate that mast cells originate from *hemopoietic stem cells*.

The most immature mast cells appear on the 15th gestational day in rat fetuses. In the first stages of mast cell differentiation, when mitosis occurs actively, the cell granules first obtain proteases and then heparin. Histamine appears much later in the perinatal period (Combs et al. 1965, Combs 1971). This ontogenic sequence of the appearance of mast cell secretions coincides with the sequence of appearance of substances in mast cell granules in adult mammals.

27.8 Phylogeny

Mast cells identified with metachromatic granules are found in every class of vertebrate. Amphibia and fish generally possess numerous mast cells in their connective tissue. As mentioned above, mast cells in lower vertebrates contain much smaller amounts of histamine than higher animals, and it is undetectable in amphibians and fish (Takaya and Fujita 1967; Takaya 1969). This phylogenetic relation coincides with the ontogenetic one, whereby histamine appears later than glycosaminoglycans and proteins in mast cell granules.

27.9 Summary

The mast cell dispersed in connective tissue contains numerous *secretory granules* bounded by a Golgi-derived membrane sac. The contents of the granules are partly the same as those of typical paraneurons and partly different. The granules contain glycosaminoglycans (heparin, etc.), amines, especially histamine, and trypsin-like and chymotrypsin-like proteases. ATP is probably present, while the reported occurrence of VIP (and also other neuropeptides) must be reexamined.

The mast cell recognizes stimuli by means of its *membrane receptors*. The IgE receptor is the most important of these owing to the immunological reactions of the cell.

The isolated rat peritoneal mast cell has been investigated as a model of *stimulus-secretion coupling*. Intracellular injection of Ca^{2+} has proved that increased concentrations of Ca^{2+} in the cell induce exocytosis.

The mast cell is not synaptically associated by nerves, but it occasionally shows a peculiar topographic affinity to nervous elements. The urodele mast cell mimics a multipolar neuron and extends long, branched dendrites.

The mast cell shows some *neuronlike features*, including its sensitivity to NGF and local anesthetics. Nevertheless, we recognize this cell as being peripheral to the main group of paraneurons. The mast cell is *mesenchymal in origin*, being derived from bone marrow stem cells, unlike ordinary paraneurons which are epithelial in origin.

28. Small Intensely Fluorescent (SIF) Cells

28.1 General Aspects

Many ganglia throughout the body consist of two kinds of parenchymal cells—principal ganglion cells and SIF cells.

The *principal ganglion cells* are highly variable in shape. Their perikaryon is surrounded by satellite cells, whereas their long processes are mostly covered by Schwann cells. They usually possess a large ovoid or spherical nucleus, an abundance of granular endoplasmic reticulum, a well-developed Golgi apparatus, many lysosomes, and numerous large cored vesicles containing various biogenic monoamines and neuropeptides. The large cored vesicles are scattered in the perikaryon and cytoplasmic processes. They are formed in the perikaryon, transported to the axon terminals by axonal flow, and discharged into the extracellular space by exocytosis. Although the principal ganglion cells of the autonomic ganglia represent a classic and well-accepted type of neuron, a distinction between their axons and dendrites is not always possible. Moreover, amphibian sympathetic ganglia are devoid of dendrites (Huber 1900; Uchizono 1964; Nishi et al. 1967): Synapses afferent to these neurons are confined to the perikaryon and all the long processes are axonic in nature.

The second type of parenchymal cells of the autonomic ganglia are *SIF cells*. Eränkö and Härkönen (1963), in their study on the formaldehyde-induced catecholamine fluorescence of the superior cervical ganglion of the rat, demonstrated specific fluorescence not only in all the principal ganglionic cells but also in occasional small cells which appeared extraordinarily bright. These cells were subsequently called *small intensely fluorescent cells* (Eränkö and Härkönen 1965; see also Eränkö 1976). This name has been widely used since then, especially in its abbreviated form—SIF cells. SIF cells, however, were not discovered for the first time by fluorescence histochemistry; the same cells had been identified as chromaffin or nonchromaffin cells of the paraganglia by early light microscopists (Kohn 1903; Watzka 1943); electron microscopists designated them as small granule-containing (SGC) cells because electron-microscopic studies revealed that the cytoplasm contains a large number of secretory granules (Siegrist et al. 1968; Mathews and Raisman 1969).

28.2 Distribution

The occurrence of SIF cells has been reported in various autonomic ganglia, such as the superior cervical ganglion, cardiac ganglia, celiac-superior mesenteric ganglion, inferior mesenteric ganglion, thoracic and lumbar paravertebral sympathetic ganglia, and pelvic autonomic ganglia, including the paracervical ganglion of female animals and hypogastric ganglion of male animals (Partanen and Hervonen 1979; review: Böck 1982). The occurrence of SIF cells is not specific to sympathetic ganglia and nerves; the parasympathetic system also contains these cells. Thus, vagal nerves possess a considerable number of SIF cells (Mascaro and Yates 1970). Immunoreactivity for opioid peptides can be demonstrated in most, if not all, parasympathetic SIF cells like SIF cells in the sympathetic ganglia (Kobayashi et al., unpublished observation). The occurrence of somatostatin-like immunoreactivity has also been reported (Uvnäs-Wallenstein et al. 1978).

There are many monoamine-containing cells in the retroperitoneal tissue outside the autonomic ganglia and some of these cells show a positive chromaffin reaction after treatment with a dichromate-containing fixative; they are usually called extra-adrenal chromaffin cells. Their general cytological features are similar to those of SIF cells.

Prominent clusters of these extraganglionic SIF cells are represented by *Zuckerkandl's organ* and other retroperitoneal paraganglia. Species variation must be taken into account with respect

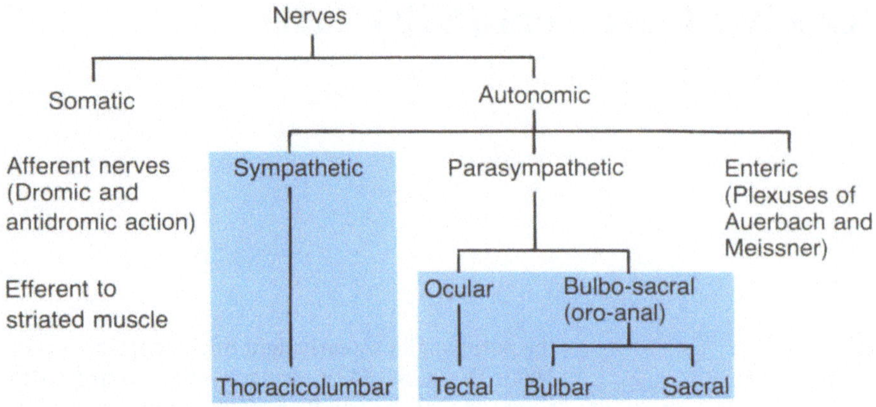

Fig. 28.1. Langley's (1921) classification of peripheral nerves. Olfactory, optic, and auditory nerves are excluded. Nerves containing the SIF cells are masked in blue

to innervation of extraganglionic SIF cells. However, in many laboratory animals, synapse formation between the preganglionic sympathetic nerve and extraganglionic SIF cells is rare.

Langley, who introduced the concept of the autonomic nervous system, classified it into three categories: (1) sympathetic, (2) parasympathetic, and (3) enteric nerves (Langley 1921). He further classified the parasympathetic nerves into ocular and bulbosacral (oroanal) groups. It seems practical to correlate the occurrence of SIF cells with Langley's classification of the peripheral nerves (Fig. 28.1). There is a distinct tendency for SIF cells to occur within or associated with the sympathetic nerves and parasympathetic bulbosacral groups.

28.3 Cytochemistry and Ultrastructure

Compared with the principal ganglion cells, SIF cells have a smaller elliptical nucleus, more condensed heterochromatin, and less-developed cell organelles, such as granular endoplasmic reticulum, Golgi apparatus, and mitochondria. The perikaryon and the cytoplasmic processes are incompletely surrounded by satellite cells (Grillo 1966; Elfvin 1968; Siegrist et al. 1968; Mathews and Raisman 1969).

SIF cells frequently form clusters of various size within the autonomic ganglia. These clusters of SIF cells are provided with a well-developed blood supply (Heym 1978; McDonald and Blewett 1981). Capillaries between the SIF cell clusters are of a fenestrated type. Cytoplasmic processes of the SIF cells, often free of satellite cell investment, are seen along the capillary endothelium 100–600 nm apart. Pericytes are located between the capillary basal lamina and SIF cell basal lamina. It is supposed that some SIF cells are *hemocrine* in nature (see pp. 72, ff.), i.e., release their granule contents into the bloodstream by an exocytotic mechanism (Siegrist et al. 1968; Mathews and Raisman 1969).

Williams and co-workers investigated the morphology and synaptology of SIF cells in the superior cervical ganglion of some mammals using monoamine fluorescence and electron-microscopic techniques (Williams 1967; Williams and Palay 1969; Williams et al. 1976; Chiba et al. 1977). They distinguished at least two types of SIF cell. The first type receives preganglionic sympathetic innervation and makes efferent synapses on the principal neurons. Thus, cells of this type represent interneurons in the autonomic ganglia. The second type of SIF cell is adjacent to the blood vessels and possesses no cytoplasmic projections among the principal neurons. Williams and co-workers (Williams et al. 1976; Chiba et al. 1977) and Heym (1978) assumed that these SIF cells might be, like the adrenal chromaffin cells, endocrine in function because their vascular pole often comes very close to the fenestrated endothelium of the blood capillaries.

28.4 Secretory Substances

The SIF cell possesses a well-developed *secretory machinery* including rough endoplasmic reticulum and Golgi apparatus. The cytoplasm usually contains numerous membrane-bound granules, roughly 70–100 nm in diameter, derived from the Golgi apparatus. It seems most probable, though not as yet evidenced, that the mechanisms of formation, transport, and costorage of secretory substances established in the adrenal chromaffin cell also hold true in the SIF cell. Cytochemical and immunocytochemical studies into this have not been numerous, but some findings are available and these will now be presented.

Amines. The SIF cell produces *noradrenaline* and stores it in the secretory granules. Dopamine-β-hydroxylase (DBH) has been localized in the granules in some SIF cells in the guinea pig (Schultzberg et al. 1983). The existence of serotonin has also been reported (Verhofstad et al. 1981).

Peptides. A few studies have demonstrated the occurrence of bioactive peptides. Immunoreactivity for *substance P* has been demonstrated in SIF cells of the guinea pig celiac ganglion (Schultzberg et al. 1983). In SIF cells of bullfrog sympathetic ganglia, Kondo and Yui (1981) demonstrated immunoreactive *enkephalins*.

ATP. The occurrence of ATP and possibly other adenine nucleotides has been supported by Böck (1980), who applied quinacrine staining to SIF cells of some mammals.

Chromagranins. Immunoreactivity for chromogranin A could be demonstrated in SIF cells of the guinea pig celiac ganglion (Kuramoto and Fujita 1986, unpublished data).

28.5 Ontogeny

There is no doubt that SIF cells, like sympathetic ganglion cells, originate from *neural crest* cells. However, direct evidence is poor, because the distinction between the principal neurons and SIF cells in the developmental stage is practically impossible (review: Le Douarin 1982).

SIF cells are not evident in cranial sensory ganglia such as the trigeminal, spiral, and vestibular ganglia, which are mostly of placodal origin. There are no SIF cells in spinal ganglia, which are of neural crest origin (Le Douarin 1972). Although a population of small neurons exists in these ganglia, no group formation associated with fenestrated blood capillaries has been demonstrated. The neural crest cells destined to differentiate into SIF cells in the sympathetic chain, celiac and mesenteric ganglia, and paravisceral ganglia, such as the cardiac and pelvic ganglia, possess the ability to form groups and induce fenestrated capillary formation around them. It is not known when and how the neural crest cells obtain this ability. It is interesting to note that in intramural ganglia, e.g., enteric ganglia, where no SIF cells exist there are no blood vascular elements (see pp. 278, ff.).

SIF cells exist in some cranial sensory ganglia such as the nodose ganglion (Grillo et al. 1974). If the principal autonomic neurons here are explained as an ectopic phenomenon, this may also apply to SIF cells. There are considerable numbers of SIF cells along the course of cranial nerves such as the vagus nerve (Chen and Yates 1970; Kummer and Addicks 1982). The presence of the SIF cells may be due to the neural crest involvement in the formation and elongation of these cranial nerves.

28.6 Phylogeny

The phylogeny of SIF cells is directly related to that of paravertebral. prevertebral, and paravisceral ganglia, which Langley (1921) included in either sympathetic ganglia or bulbosacral groups of the parasympathetic ganglia (see p. 278). It can be said with some certainty that all vertebrate species which possess these and related ganglia have SIF cells. The size, population, and electron opacity of the secretory granules are different from species to species. However, it is not our purpose here to describe species differences in the fine structure of SIF cells.

SIF cells form unique structures in a few animal species; of these the SIF cells in the avian aortic wall and those in the cyclostome heart will be briefly mentioned here.

SIF cells in the avian aortic wall. In all vertebrate classes, the cytological features of the SIF cell are similar or identical to those of the carotid body chief cell (see pp. 200, ff.). The existence of cells similar to carotid body chief cells in the avian aortic wall was detected using light microscopy (Palme 1938; Nonidez 1935). Monoamine-specific fluorescence was demonstrated in these cells (Bennett 1971; Kobayashi 1971a). The fine structure of these cells is identical to that of intraganglionic SIF cells (Kondo 1974; Ookawara et al. 1974). Synapse formation between these SIF cells and nerve terminals has been demonstrated using electron microscopy (Kondo 1974; Ookawara et al. 1974). Further studies are needed to decide whether these SIF cells act as chemoreceptors.

SIF cells in the cyclostome heart. Unique cells occur in the heart of cyclostomes; histochemically, they are chromaffin and contain catecholamines (Lignon and Le Douarin 1978). The cells have large cored vesicles and project two or more thick cytoplasmic processes, forming a discontinuous network which spreads between the myocardial cells. The fine structure of these cells is identical to that of mammalian SIF cells; however, they are devoid of innervation and have no sustentacular cell sheath. It is not known whether the occurrence of cardiac SIF cells is specific to cyclostomes. The functional significance of cardiac SIF cells remains unknown.

28.7 Methodological Improvement and Conceptual Changes

Concepts on the significance and nature of SIF cells have developed in parallel with methodological improvements in neuroendocrine cytology. It is impossible to deal with historical aspects of SIF cells independently from those of adrenal chromaffin cell and carotid body chief cell.

Kohn (1903) described the occurrence of nonprincipal parenchymal cells in and around the autonomic ganglia and nerves; he introduced the concepts of *paraganglia* and *chromaffin reaction* (Böck 1982). Watzka (1943) improved Kohn's techniques with the chromaffin reaction and obtained good preparations of paraganglia. He noticed that many of Kohn's chromaffin cells did not really show a chromaffin reaction. Watzka (1943) was also influenced by the concept of the autonomic nervous system, which was widely accepted at the time, and classified paraganglia into two categories:

1) *Chromaffin paraganglia.* These paraganglia consist of chromaffin cells and contain adrenaline and related amines. They originate from the anlage of the sympathetic nervous system and are innervated by sympathetic preganglionic nerve fibers.

2) *Nonchromaffin paraganglia.* These paraganglia consist of nonchromaffin cells and contain acetylcholine. They originate from the anlage of the parasympathetic nervous system and are innervated by parasympathetic preganglionic nerve fibers.

However, it has now been established that Watzka (1943) overestimated the selectivity and sensitivity of the chromaffin reaction for the histochemical demonstration of biogenic monoamines (Kobayashi 1968). The nonchromaffin cells which Watzka (1943) believed to contain no monoamines do in fact contain monoamines. The *fluorescence cytochemical method* for the demonstration of monoamines developed by Eränkö (1952), who used formaldehyde in solution, and improved by Falck and Hillarp (see Falck et al. 1962), who used formaldehyde vapor, has greatly helped to increase our knowledge of paraganglia, in consequence of which our ideas on the chromaffin and nonchromaffin cells of Watzka (1943) have essentially changed.

As mentioned above, the term *SIF cells* was first used by Eränkö and Härkönen (1965). They demonstrated monoamine fluorescence in the carotid body, though they were overinfluenced by the accepted contemporary view that the carotid body contains no monoamines (Watzka 1943). Thus, Eränkö and Härkönen (1965) stated: "the organ was first thought to be the carotid body but this was subsequently found to be composed of cells exhibiting a green, less intense fluorescence, therefore, the small cells apparently form a hitherto unknown endocrine gland which perhaps secretes 5-hydroxytryptamine." Thus, the definition of SIF cells was convenient from the outset. With electron microscopy, the fine structure of SIF cells and their relation to the nerve terminals and sustentacular cells were demonstrated. Recent improvements in the immunocytochemical demonstration of monoamines and neuropeptides have made it possible to obtain a Golgi-type image of SIF cells. Thus, at present, SIF cells include a wide spectrum of monoamine-containing cells associated with the peripheral nervous system.

A spectrum of SIF cells has been proposed, indicating the gradual transformation from the authentic neuron type to the pure endocrine cell type, as follows (Fig. 28.2):

1) *Multipolar neuronal type (SIF cells with dendritic projections).* This type of SIF cell occurs for example in the course of the vagus nerve. The cells have long cytoplasmic projections with arborizations.

2) *Interneuron type (SIF cells with multiple innervation).* These cells, originally described by Williams (1968), have two or more cytoplasmic processes. They make both afferent and efferent synapses with nerve terminals and are connector

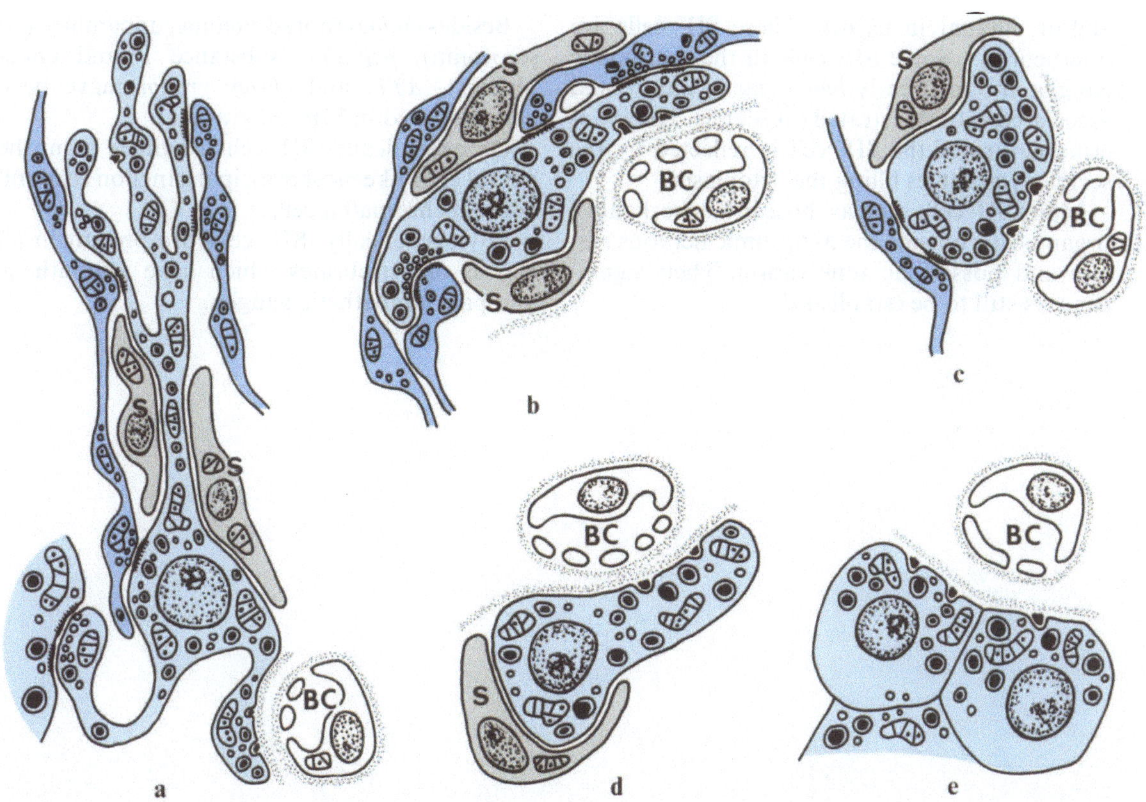

Fig. 28.2a–e. Spectrum of SIF cells. **a** Neuronal type. **b** Interneuronal type (SIF cell with multiple innervations). **c** Endocrine type with single innervation. **d** Endocrine type without innervation. **e** Extraganglionic type. Topographic relations between the SIF cell and sustentacular cells (*S*), basal lamina, and blood capillary (*BC*) are illustrated

neurons belonging to a sympathetic effector circuit. The cells are partly covered by a sustentacular cell sheath.

3) *Sensory or neurosecretory type (SIF cells with single innervation)*. These cells are similar to either carotid body chief cells or adrenal chromaffin cells. They are innervated by a nerve terminal which is either afferent or efferent to themselves. They are incompletely surrounded by a sustentacular cell sheath.

4) *Non innervated type (SIF cells without innervation)*. These cells exist inside the ganglia but have no innervation. They project a short cytoplasmic process with secretory granules and are partly covered by a sustentacular cell sheath.

5) *Extraganglionic type (SIF cells outside the autonomic ganglia)*. A typical examples of this type is seen in Zuckerkandl's organ of the human fetus. The cells have no innervation and are devoid of a sustentacular cell sheath.

28.8 Summary

Sympathetic and parasympathetic ganglia are not simple relay stations between preganglionic and ganglionic neurons: these autonomic ganglia contain considerable numbers of interneurons. The latter are usually smaller than the principal ganglionic neurons and possess catecholamine-containing secretory granules. A strong FIF reaction is evident in these interneurons; hence they are called *small intensely fluorescent (SIF) cells*.

There are different types of SIF cells with regard to the morphology, innervation, fine structure, cytochemistry, and localization. Some SIF cells, such as those along the course of the vagus nerve, project several long cytoplasmic processes, which resemble the dendrites of authentic neurons. Many of the cells, especially in sympathetic and some parasympathetic ganglia, are innervated by more than two nerve terminals, which appear in their fine structure to be afferent and/or efferent in nature. These SIF cells are interneurons. Some SIF cells in the autonomic ganglia are apparently *hemocrine* in nature and associated with fenestrated capillaries. The endocrine nature of the SIF cell is reflected by the secretory granules filling the cytoplasm.

Many SIF cells such as those in Zuckerkandl's organ occur outside the autonomic nervous system and possess no innervation. Their significance is still to be established.

Besides *amines* (noradrenaline, dopamine, and serotonin), *peptides* (substance P and enkephalins), *ATP*, and *chromogranins* have been demonstrated in SIF cell granules.

Ontogenetically, SIF cells originate from the neural crest like most principal ganglion cells and adrenal chromaffin cells.

Phylogenetically, SIF cells are present in all classes of vertebrates which have sympathetic and parasympathetic ganglia.

29. Cerebrospinal Fluid-Contacting Neurons

29.1 General Aspects

The cerebrospinal fluid (CSF)-contacting neuron is not a paraneuron. The reasons why a chapter of the present volume is devoted to this neuron are: (1) It has certain cell-biological features in common with paraneurons (Nakai et al. 1977a, b); (2) it is homologous to retinal and pineal paraneurons, as has already been discussed (see pp. 241, 248); and (3) it has recently been regarded as the prototype of neurons, possibly sharing chemoreceptive and secretory features with bipolar paraneurons (see below).

Three types of CSF-contacting neuron have been classified (Vigh-Teichmann and Vigh 1983): (1) Neurons contacting the CSF via their dendrites; (2) neurons doing so via their axons; and (3) neurons located entirely within the space of the CSF. It is mainly the cells of the first category that will be dealt with here, because these are the elements that deserve evaluation with regard to their relationship to paraneurons.

CSF-contacting neurons directly protrude into the lumen of the cerebral ventricle or of the central canal of the spinal cord, which is otherwise isolated from neurons by an ependymal boundary. These neurons occur in all vertebrates but tend to be more numerous in lower animals, i.e., amphibians, fish, and cyclostomes. In the lancelet (cephalochordate), it is a general feature of neurons that they contact the CSF (see p. 276). In vertebrates, the CSF-contacting neurons are concentrated in the hypothalamic areas facing the third ventricle and along the central canal of the spinal cord (for details of cell distribution, see Vigh and Vigh-Teichmann 1973; Vigh-Teichmann and Vigh 1974, 1983). Further phylogenetic aspects of the CSF-contacting neurons will be dealt with later (p. 276).

29.2 Structure and Fine Structure

The CSF-contacting neuron is a bipolar cell with a rather stout process, corresponding to the *dendrite* projected into the CSF, and a thinner process, corresponding to the *axon* extended deep into the parenchyma of the brain or spinal cord. The end of the dendrite is generally globular or clubbed and its cytoplasm contains smooth and rough endoplasmic reticulum, mitochondria, and some lysosomal structure (Fig. 29.1). Dense granules may be numerous and these will be dealt with later. The cytoplasm further possesses paired *centrioles*, one of which represents the basal body and gives rise to a *cilium* of different structure and to a cross-striated rootlet of various length (Vigh and Vigh-Teichmann 1973; Vigh-Teichmann and Vigh 1974, 1983). The cilium of hypothalamic CSF-contacting neurons shows the $9 \times 2 + 0$ pattern of microtubules, while that of spinal CSF-contacting neurons has the $9 \times 2 + 2$ pattern (Vigh and Vigh-Teichmann 1973).

The perikaryon contains rough endoplasmic reticulum, Golgi complexes, and mitochondria. Granules and vesicles are contained partly in association with the Golgi complexes. Besides dense granules (100–200 nm in diameter), small cored vesicles (60–100 nm) have been shown in the anuran hypothalamus (Nakai et al. 1977).

The axon contains numerous neurotubules, mitochondria, and dense granules. Its termination is variable. Some axons of the hypothalamic CSF-contacting neurons end in the pericapillary space of the neurohemal structure, while others apparently terminate in contact with other neurons. The axons of the spinal CSF-contacting neurons terminate on the pial surface of the spinal cord. In any case, the axonal end is usually swollen and contains accumulations of dense granules.

CSF-contacting neurons usually receive synaptic terminals from other neurons. The synapses can be found both on the perikaryon and dendrite. The bulbous end of the dendrite may occasionally receive a synapse of an intraventricular neuron, which usually contains small clear and some large cored vesicles (Vigh and Vigh-Teichmann 1973, 1974).

29.3 Secretory Products

29.3.1 Peptides

Peptidic neurohormones have been demonstrated in many CSF-contacting neurons. Considerable numbers of magnocellular neurosecretory cells in the preoptic and paraventricular nuclei in *submammalian* vertebrates assume the structure of CSF-contacting neurons, extending a dendrite to the ependymal surface of the third ventricle (frog—Scharrer 1933; literature: Diepen 1962; Leonhardt 1980). Immunocytochemical studies have shown the occurrence of mesotocin, vasotocin (Bons and Pérézi 1981), and somatostatin (Goossens et al. 1980; Fasolo and Gaudino 1981) in these CSF-contacting neurons in the reptilian paraventricular nucleus (review: Vigh-Teichmann and Vigh 1983).

Parvocellular hypothalamic nuclei in submammalian vertebrates possess numerous peptide-containing CSF-contacting neurons (Vigh-Teichmann et al. 1969; Vigh-Teichmann and Vigh 1974). The frog brain has been investigated especially intensively and it has been demonstrated that substance P- and somatostatin-immunoreactive neurons in the preoptic nucleus and calcitonin-immunoreactive neurons in the nucleus infundibularis ventralis in *Rana catesbeiana* are of the CSF-contacting type (Yui et al. 1981; Yui 1983).

Neuropeptides in *mammalian* CSF-contacting neurons are poorly documented. The occurrence of numerous peptidergic-type granules in CSF-contacting elements of the infundibular and arcuate nuclei in rats and guinea pigs (Vigh-Teichmann and Vigh 1974, 1983) suggests that various releasing hormones may be contained in the CSF-contacting neurons. In the rat spinal cord, Shimosegawa et al. (1986) demonstrated the occurrence of Met-enkephalin-Arg-Gly-Leu-like immunoreactivity in the CSF-contacting neurons.

29.3.2 Amines

In *submammalian* vertebrates, it has been repeatedly recorded using fluorescence histochemical techniques that the majority of hypothalamic CSF-contacting neurons either contain monoamines or possess the APUD ability (fish—Ekengren 1975; amphibians—Vigh-Teichmann 1969; Nakai et al. 1977; Shimizu et al. 1982; reptiles—Parent and Poitras 1974; review: Vigh-Teichmann and Vigh 1983). Two kinds of fluorescent neuron have been distinguished—green fluorescent cells containing catecholamines (presumably dopamine) and yellow fluorescent cells containing serotonin (literature: Vigh-Teichmann and Vigh 1983). The occurrence of serotonin has also recently been demonstrated using immunocytochemical methods (Sano et al. 1983). In addition, many spinal CSF-contacting neurons in submammalian species show monoamine fluorescence (literature: Vigh-Teichmann and Vigh 1983).

29.3.3 Acetylcholine

The occurrence of acetylcholine in some CSF-contacting neurons, both in the hypothalamus and spinal cord, has been suggested by the cytochemical detection of acetylcholine esterase (Vigh-Teichmann et al. 1970; review: Vigh-Teichmann and Vigh 1983).

It is probable that, analogous to the situation in other neurons and paraneurons, the occurrence of amines or choline in CSF-contacting neurons coincides with that of peptides, though this point does not seem to have been systematically investigated.

29.4 Functions

The functions of CSF-contacting neurons are enigmatic. As suggested by Nakai et al. (1977), it is likely that they are engaged in various functions and detect chemical and physical information from the CSF like sensory cells, conduct impulses like nerve cells, and secrete amines and peptide hormones like endocrine paraneurons.

29.4.1 Secretory Activity

As mentioned above, some axons of hypothalamic CSF-contacting neurons terminate synaptically on other neurons, whereas others terminate in the pericapillary spaces of the neurohemal tissue, i.e., median eminence or neurohypophysis (Yui 1983). The axons of spinal CSF-contacting neurons terminate on the basal membrane at the outer surface of the spinal cord or on blood vessels (Leonhardt 1980). Evidence for the release at the axonal terminals is provided by the accumulations of cytochemically or immunocytochemically detectable secretory substances in the swollen terminals (Yui et al. 1981; Yui 1983; Shimosegawa et al. 1986).

The problem is whether the *dendrite*, which is frequently filled with aminic or peptidic secretory

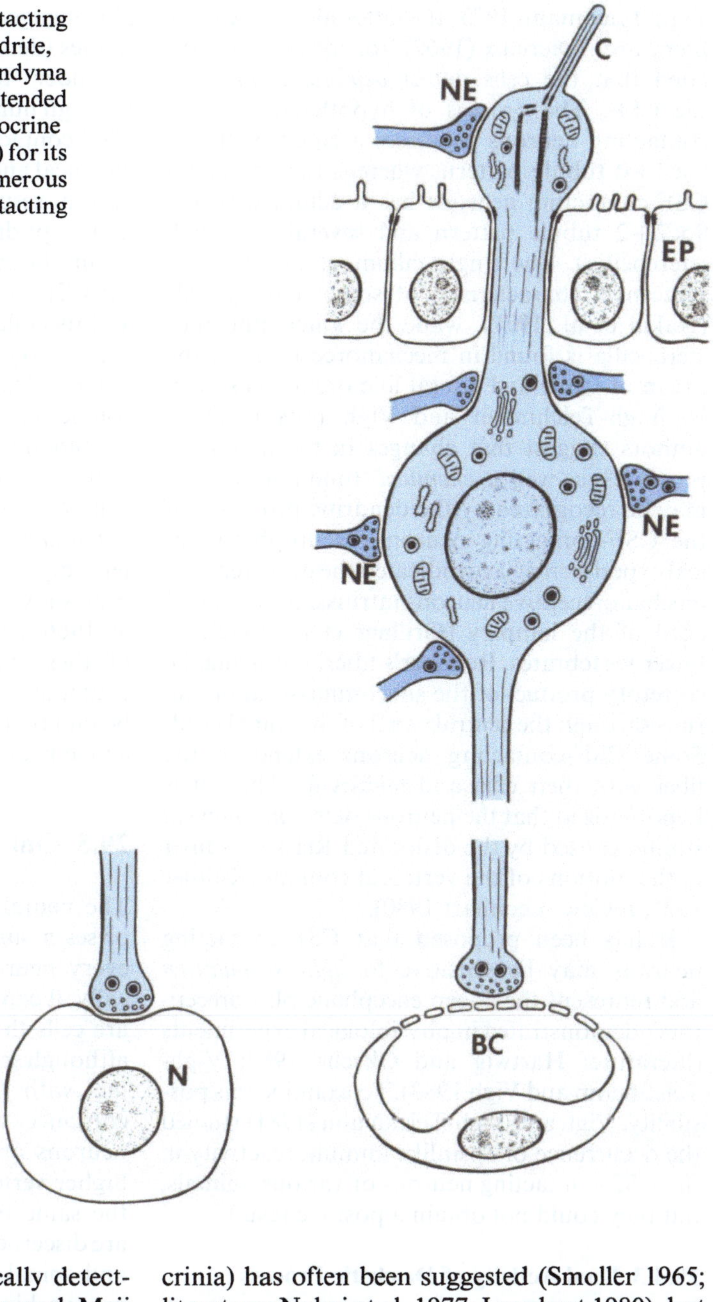

Fig. 29.1. Structure of a typical CSF-contacting neuron. The cell projects a knobbed dendrite, containing a cilium (*C*), through the ependyma (*EP*) into the brain ventricle. Its axon is extended either to a blood capillary (*BC*) for hemocrine release of its secretions or to a neuron (*N*) for its synaptic regulation. The cell receives numerous nerve endings (*NE*) even on its CSF-contacting portion

substances and electron-microscopically detectable granules (Takeichi 1967; Peute and Meij 1973; Yui 1983; Shimosegawa et al. 1986), is *secretory or sensory* in nature. Exocytotic granule release has not been evidenced by electron microscopy. However, the dendritic end-bulb of, for instance, the CSF-contacting neurons on the ventricular surface of the amphibian hypothalamus is so densely packed with secretory granules (which apparently contain substance P, calcitonin, or some other peptide) that it is most unlikely that they are not released here. The release of secretory substances of CSF-contacting neurons by their intraventricular processes (ventriculocrinia) has often been suggested (Smoller 1965; literature: Nakai et al. 1977; Leonhart 1980), but no conclusive evidence has yet been produced. Bioassay studies have indicated the occurrence of various kinds of neuropeptides and amines in the CSF (review: Leonhart 1980). These data may support ventriculocrinia of CSF-contacting neurons but do not prove it.

29.4.2 Sensory Activity

It was suggested some time ago that the dendritic process of CSF-contacting neurons is the receptor site of the cells (Edinger 1906; Vigh and

Vigh-Teichmann 1973, the latter also cite earlier literature). Dierickx (1962), for instance, postulated that the cells detect *osmotic changes* in the CSF. The process of hypothalamic CSF-contacting neurons possesses a cilium with the $9 \times 2+0$ tubule pattern, whereas that of spinal CSF-contacting neurons has a cilium with the $9 \times 2+2$ tubule pattern and several microvilli (stereocilia). The single cilium of the $9 \times 2+0$ pattern is characteristic of some sensory cells (Nakai et al. 1977), while the kinocilium plus stereocilia is found in mechanoreceptors in the organ of Corti and lateral line organ, as shown by Vigh-Teichmann and Vigh (1983). These authors suggest that changes in the *flow* (and pressure?) as well as *chemical* stimuli in the CSF may be recognized by the dendritic processes of the CSF-contacting neurons. Neurophysiological experiments have indicated the occurrence of mechanoreceptive neurons intrinsic to the spinal cord of the lamprey (Grillner et al. 1982). In lower vertebrates, Reissner's fiber, the stringlike secretory product of the subcommissural organ, runs through the central canal of the spinal cord. Some CSF-contacting neurons extend to the fiber with their cilia and microvilli. Thus, it is hypothesized that the neurons detect *mechanical* stimuli caused by the dislocated Reissner's fiber in the motions of the vertebral column (Kolmer 1921; review: Leonhart 1980).

It has been proposed that CSF-contacting neurons may be sensitive to *light stimulation* and represent the "deep encephalic photoreceptors" demonstrated in physiological experiments (literature: Hartwig and Oksche 1982; Vigh-Teichmann and Vigh 1983). To examine this possibility, Vigh and Vigh-Teichmann (1981) studied the occurrence of opsinlike immunoreactivity in the CSF-contacting neurons of various animals, but they could not obtain a positive result.

29.4.3 Dual Activity of Dendritic Process

Although the function of the intraventricular process of the CSF-contacting neurons has hitherto been postulated as being either secretory or sensory, we hypothesize, supporting the early view by Smoller (1965), that the *neurons* are *simultaneously secretory and sensory*. We posit that the dendritic process of CSF-contacting neurons is covered with a mosaic composition of receptive and secretory membranes. This hypothesis better accounts for the morphological and physiological observations and is also in accordance with the fact that neuronal dendrites are generally engaged in both receptive and secretory activities (dendritic secretion, pp. 67, ff.).

Under the regulation of inputs coming through numerous synapses from other neurons, CSF-contacting neurons are believed to receive chemical and physical information from the CSF and, in response to it, they release their aminic and peptidic products. This release probably occurs both through the dendritic process into the CSF (vetriculocrinia) and through the axon and its collaterals. *Both synaptocrine and hemocrine secretion* may take place at the axonal terminals. The effects of the secretions of CSF-contacting neurons released into the CSF or into the blood have not been identified.

It is unknown whether CSF-contacting neurons are able to generate and conduct *action potentials*. Mammalian neurosecretory cells extending their axons to the neurohypophysis have been shown to conduct electric impulses (Yagi et al. 1966; Dyball and Koizumi 1969). Since some of these neurosecretory cells represent CSF-contacting neurons, as mentioned above, it can be said that these elements at least are electrically active neurons.

29.5 Ontogeny and Phylogeny

The neural canal in early embryonic life comprises a simple epithelium, in which essentially every neuron contacts the lumen. Ontogenetically, it can be said that CSF-contacting neurons are cells that retain this primitive arrangement, although some neurons may possibly regain contact with the CSF after once losing it. In accordance with the ontogeny, CSF-contacting neurons are more numerous in lower than in higher vertebrates. *In the lamprey* (cyclostome), the same two types of CSF-contacting neuron are discerned as in vertebrates, i.e., hypothalamic and spinal types, the latter being characterized by one kinocilium and many stereocilia (Vigh-Teichmann and Vigh 1983). *In the lancelet* (protochordate), it has been pointed out that nearly all neurons contact the CSF, some of them showing kinocilia and stereocilia on their luminal surface (Vigh-Teichmann and Vigh 1983).

Deuterostomian invertebrates appear to possess neurons that are comparable with CSF-contacting neurons in protochordates and vertebrates. *In starfish* (echinoderms), the central nervous system is platelike and opens to the external environment. Neurons in the ectoneural radial cord and hyponeural nerve extend their dendrites to the cuticle that faces the sea water. They are provided with a cilium (either $9 \times 2+0$ or $9 \times 2+2$ type) associated with a long striated rootlet. These cells usually possess large cored granules, which accumulate in the swollen terminals that abut on the basal lamina lining the connective tissue of the nervous tissue, reminiscent of the neurohemal termination in the vertebrate neurosecretory system (Hehn 1970; Vigh and Vigh-Teichmann 1982; Vigh-Teichmann and Vigh 1983).

Vigh-Teichmann and Vigh (1983) propose that CSF-contacting neurons should be regarded as "protoneurons." These cells conserve the primitive characteristic of contacting the external environment presumably so as to detect chemical and physical information. After the neural plate has enclosed its lumen during evolution, the cells differentiate into various forms circumventing the brain ventricles. It is noteworthy that neurosecretory activity has been an attribute of these "protoneurons" since the earliest stage of their phylogeny.

29.6 Summary

The CSF-contacting neuron is considerably different from typical neurons and shows some paraneuronal features.

The typical CSF-contacting neuron extends a stout *dendrite* to the ventricular surface, while its long *axon* and collaterals end in *synaptocrine* and/or *hemocrine* terminals. As to the much-disputed function of the dendrite of the CSF-contacting neuron, a dual activity—*sensory and secretory*—would seem to be most likely. It has been proposed that the dendrite recognizes *physical* (osmotic changes and flow of CSF; mechanical stimuli) *and chemical information from the CSF.*

The CSF-contacting neuron generally possesses a well-developed secretory machinery in the perikaryon. Its products include *peptides*, *amines*, and, in some cells, acetylcholine. Both endocrine-type granules and synaptic-type vesicles may be found in the cell; they are concentrated in the dendritic and axonal ends.

Action potentials of the cell have not been proved.

Ontogenetically, the CSF-contacting neuron retains the arrangement of the embryonic neuron lining the neural canal.

Phylogenetically, the primitive type of the CSF-contacting neuron is seen in the platelike central nervous tissue of the starfish (echinoderm), where the neuron extends its dendrite to the external milieu (sea water). Around the closed neural canal of the lancelet, essentially all neurons contact the CSF. Cyclostomes and lower vertebrates still possess numerous CSF-contacting neurons, whereas higher vertebrates retain much fewer numbers of them mainly around the third ventricle and central canal of the spinal cord.

30. Enteric Neurons

The cells dealt with here are not paraneurons, but they would seem to deserve a specific chapter since they have certain paraneuronlike features and have close topographic and functional relations to enteric and other paraneurons.

30.1 General Aspects of Enteric Nervous System

The enteric nervous system, also called the intramural nervous system, is located in the wall of the whole of the digestive tract from the esophagus to the anus. It comprises two main assemblies of neurons and their fibers—Meissner's submucous plexuses and Auerbach's myenteric plexuses—as well as their extensive projections to the lamina propria mucosae and submucous and muscular layers. The system participates in the control of almost all functions of the gut from peristaltic movement to the secretory and absorptive functions. Many review articles have recently been published on the morphology and physiology of the enteric nervous system, among them Schofield (1968), Gabella (1979), Furness and Costa (1980), Gershon and Erde (1981), Wood (1984), and Cooke (1986).

30.1.1 Independence of Central Nervous System

The cellular elements of the enteric nervous system resemble those of the central nervous system. The parenchyma consists of enteric neurons and enteroglial cells. There is little extracellular space in the enteric nervous system, which contains no connective tissue and no blood vessels; the system is bounded by a basement membrane, which separates it from the surrounding tissues. The membrane is assumed to function as a filtration membrane. There are no such distinct perineurial and endoneurial sheaths around enteric ganglia and nerve strands as those seen in somatic ganglia and nerves.

According to Langley (1921), the enteric nervous system represents a *third division of the autonomic nervous system* (Fig. 28.1). After demonstrating that basic neuronal regulation of gastroenteric functions could be maintained *in vitro*, he concluded that this regulation is essentially intrinsic to the enteric nervous system since the central nervous system is clearly incapable of acting on the gut *in vitro*. In contrast to the sympathetic and parasympathetic divisions of the autonomic nervous system, which have their preganglionic neuron somata in the central nervous system, the enteric nervous system seemed to Langley (1921) to be more independent of the central nervous system. He assumed that there were many enteric neurons without central innervation. This was the major reason for Langley (1921) classifying this system as the third division of the autonomic nervous system.

The human enteric nervous system contains more than 10^8 neurons (Gershon and Erde 1981), a number compatible with that of spinal cord neurons (Furness and Costa 1980). On the other hand, there are fewer than 2×10^3 efferent nerve fibers in the abdominal vagus nerves (Hoffman and Schnitzlein 1969). A simple calculation that 2×10^3 extrinsic fibers cannot innervate 10^8 intrinsic neurons additionally supports the notion that the enteric nervous system is essentially independent of the central nervous system. Thus, among Langley's three divisions of the autonomic nervous system, automatism is strongest in the enteric nervous system.

30.1.2 Basic Structure of Enteric Nerve Plexuses

The enteric nerve plexuses form a continuous network consisting of three components—ganglia, nerve strands, and the autonomic ground plexus.

The ganglia, containing the somata of the enteric neurons, exist in the myenteric and submucous plexuses. In the ganglia, many synaptic con-

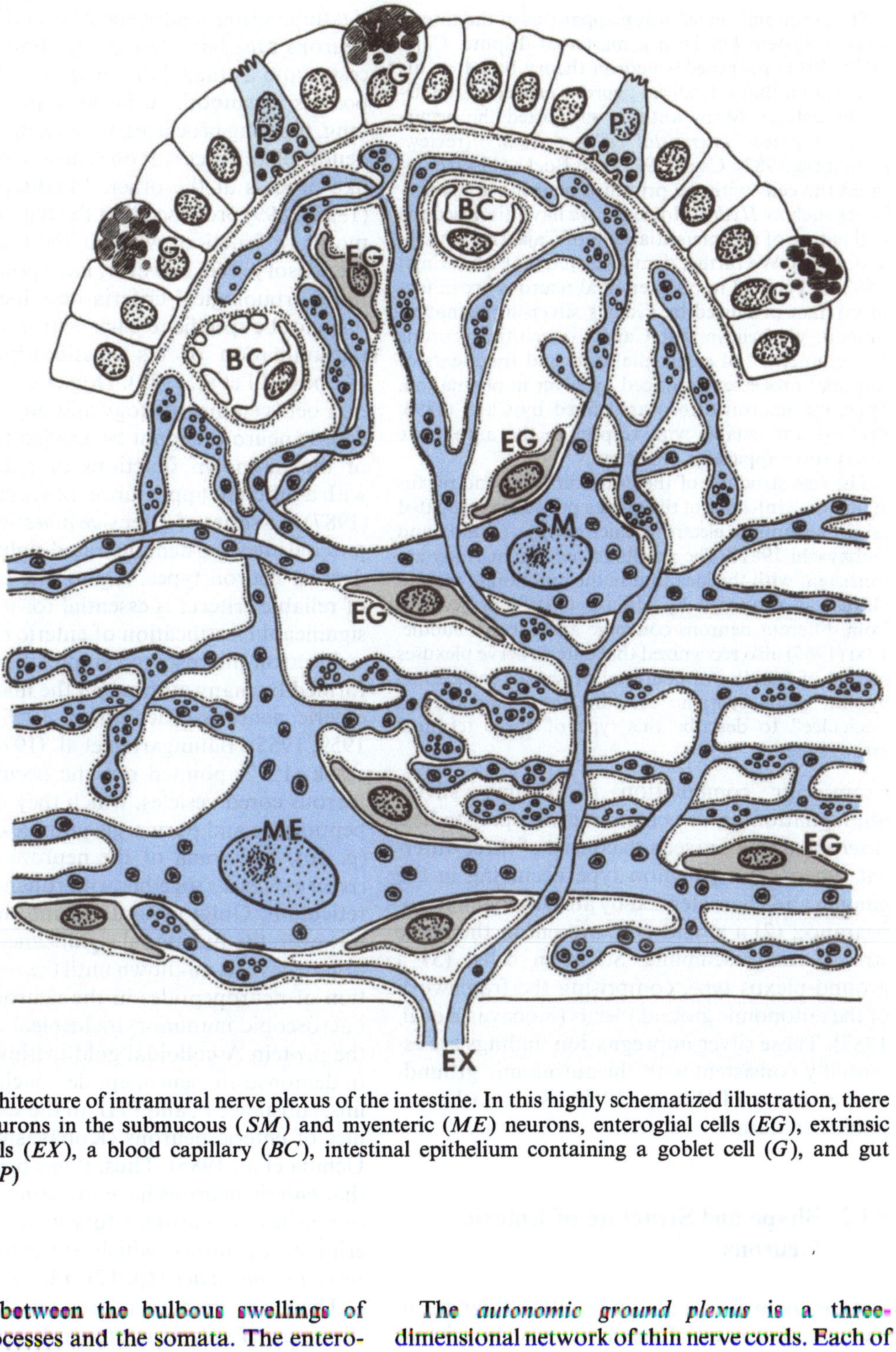

Fig. 30.1. Architecture of intramural nerve plexus of the intestine. In this highly schematized illustration, there are enteric neurons in the submucous (*SM*) and myenteric (*ME*) neurons, enteroglial cells (*EG*), extrinsic nerve terminals (*EX*), a blood capillary (*BC*), intestinal epithelium containing a goblet cell (*G*), and gut paraneurons (*P*)

tacts occur between the bulbous swellings of neuronal processes and the somata. The enteroglial cells of the ganglia project many processes which surround the neuronal elements.

The nerve strands both interconnect the ganglia and combine the ganglia with the autonomic ground plexus. The nerve strands consist of thick bundles of nonvaricose neuronal processes which are ensheathed by the enteroglial cells.

The *autonomic ground plexus* is a three-dimensional network of thin nerve cords. Each of the nerve cords consists of varicose fibers, i.e., neuronal terminals, which are incompletely ensheathed by enteroglial cells. Thus, in the autonomic ground plexus, terminals from different enteric neurons converge as a nerve cord, which divides and forms a meshwork.

The autonomic innervation apparatus of the enteric nervous system has been a matter of dispute. Cajal (1893, 1911) proposed a neuron theory based on his observation that interstitial neurons existed in the tissue interstices. Many authors supported the occurrence of these "*interstitial cells of Cajal*" (review: Thuneberg 1982). Cajal (1911) and Boeke (1949) compared the cells with the primitive neurons in invertebrates such as *Hydra*. However, we have disclosed the glial nature of the interstitial cells of Cajal (Kobayashi et al. 1986). We further demonstrated that what Cajal (1893, 1911) held to be interstitial neurons are in fact an artefact produced by Golgi's silver-impregnation method; we believe that Cajal's interstitial neurons are "chimeras" of enteroglial cells and fragments of neuronal processes produced by silver impregnation. Thus, the neuron theory as defined by Cajal (1893, 1911) is not tenable with respect to the autonomic innervation apparatus of the gut.

The fine structure of the autonomic ground plexus in the small intestine of the guinea pig was investigated using scanning electron microscopy (Endo and Kobayashi 1987). The results of our recent study are consistent with the idea that in the autonomic ground plexus, as proposed by Hillarp (1959), projections from different neurons converge and form a bundle. Taxi (1965) also recognized that enteric nerve plexuses consist of bundles of neuronal processes ensheathed by enteroglial cells; he used the term "innervation fasciculée" to describe this type of nerve terminal arrangement.

From our combination of Golgi's silver-impregnation and electron-microscopic studies, *enteroglial cells* are seen to consist of three different types: (1) a ganglion type occurring in the ganglia and characterized by an astrocytelike appearance; (2) a strand type present in the nerve strand and resembling Schwann cells; (3) a ground-plexus type, comprising the framework of the autonomic ground plexus (Kobayashi et al. 1987). These silver-impregnation findings are essentially consistent with the autonomic ground-plexus theory of Hillarp (1959) (Kobayashi, unpublished data.)

30.2 Shape and Structure of Enteric Neurons

Enteric neurons are significantly different in shape and structure from Deiters or Golgi type neurons in the central nervous system. There are no myelinated fibers in the enteric nerve plexuses. The neurons project one or more long processes, but *no axon* can be distinguished from the *dendrites*. (Fig. 30.2).

Dogiel (1899) classified the enteric neurons into three types based on the results of his investigations using methylene blue staining: Type 1 neurons are characterized by short, broad processes and a single long process; type 2 neurons possess a smooth-surfaced soma and several long, tapering processes; type 3 neurons possess a dendritelike process at one end and a single axonlike process at the other. Furthermore, Dogiel (1896, 1899) proposed that the type 1 neurons are motor in function, whereas the type 2 neurons are sensory. We, however, have pointed out that the morphological criteria described by Dogiel may not be specific to each neuron type and that reexamination of his classification is needed (Kobayashi et al. 1985). Dogiel's simple correlation between morphology and physiology of the enteric neurons cannot be applied to an analysis of the unknown functions of enteric neurons with a different appearance. More recently Stach (1987) in a series of extensive investigations in the porcine intestine demonstrated eight structurally defined neuron types. Again, the establishment of reliable criteria is essential for a functionally significant classification of enteric neurons.

Electron-microscopic studies have been performed by many authors on the fine structure of enteric neurons (Richardson 1958, 1960; Taxi 1959, 1965). Baumgarten et al. (1970) and Burnstock (1972) pointed out the occurrence of nu merous cored vesicles, which they designated as peptidergic and purinergic granules, respectively (p. 56). The soma of the neurons contains *secretory granules* together with rough endoplasmic reticulum, Golgi apparatus, and mitochondria; however, the functional significance of these cell organelles was unknown until the recent localization of neuropeptides in the neurons. Electron-microscopic immunocytochemical studies using the protein A-colloidal gold method enabled us to demonstrate neuropeptides, such as enkephalins, substance P, and PHI, in the secretory granules of enteric neurons (Kobayashi et al. 1984; Uchida et al. 1985). Thus, it has become evident that enteric neurons have the same principal cell organelles as neurosecretory neurons and endocrine paraneurons, which are provided with a *secretory machinery* (p. 17) (Fig. 3.2).

There are many synapses on the surface of the soma of enteric neurons. Accumulations of synaptic-type vesicles are usually not recognized in the neuronal soma but are found in and near the terminals. Thus, the soma of enteric neurons is regarded as the site of stimulus reception.

It is impossible to trace a true terminal of a specified neuronal process in muscular and glandular tissues. It is assumed that most neuronal

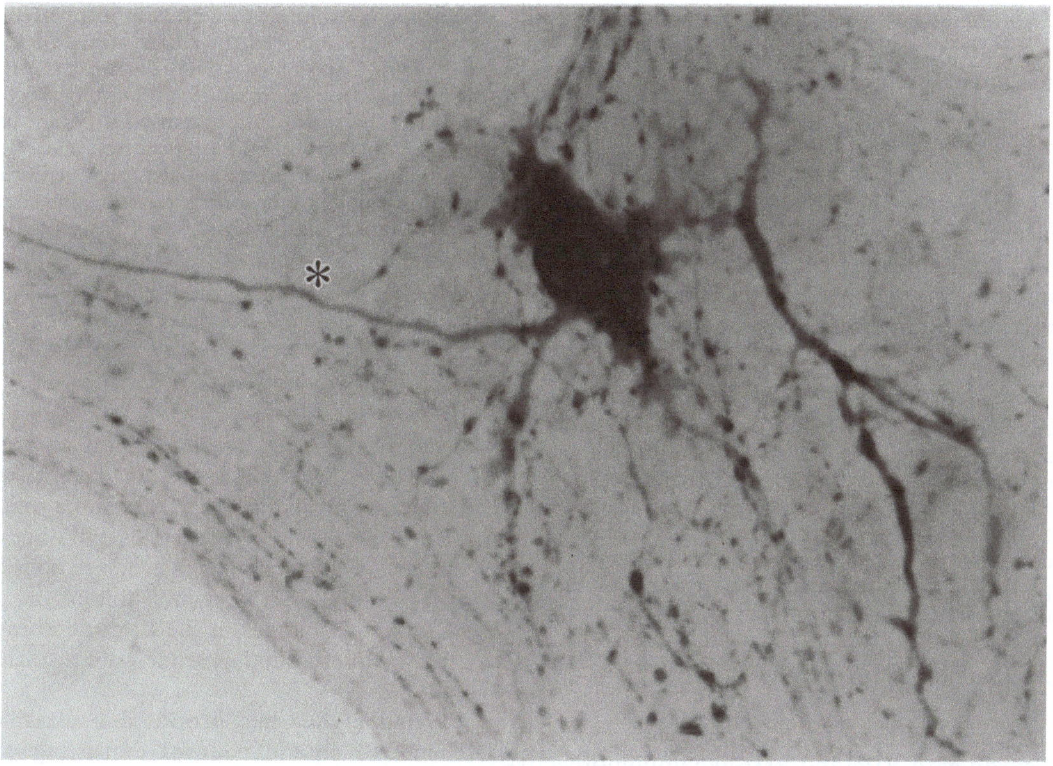

Fig. 30.2. Peptide immunocytochemical detection of intestinal neurons. A myenteric neuron with elaborate processes is immunostained by the peroxidase-antiperoxidase method using an antiserum for PHI. The axonlike process marked by the *asterisk* reached the adjacent ganglion. There are many PHI-immunopositive nerve terminal varicosities throughout the ganglion. Prepared by M. Yazawa, Department of Anatomy, Yamanashi Medical College. × 900

processes terminate in the autonomic ground plexus without making any specific end apparatus. We support Hillarp's view that the autonomic ground plexus itself represents the end apparatus of enteric neurons (Endo and Kobayashi 1987). However, direct contact occasionally exists between neuronal varicosities and smooth muscle cells (Taxi 1965). The functional significance of this direct contact between the neuronal terminals and presumptive effector cells remains unknown.

30.3 Chemical Diversity of Enteric Neurons

Recent advances in immunocytochemical techniques have greatly increased our knowledge on the chemical diversity of enteric neurons. The two classic neurotransmitters, *acetylcholine* and *noradrenaline*, have long been known as enteric transmitters; however, *various kinds of amines and peptides* have been cytochemically identified in enteric neurons, including 5-hydroxytryptamine (Furness and Costa 1982), enkephalins (Furness et al. 1983; Kobayashi et al. 1984, 1985), substance P (Costa et al. 1981), somatostatin (Costa et al. 1980; Endo et al. 1986), VIP-PHI (Costa and Furness 1983), NPY (Furness et al. 1983), and galanin (Melander et al. 1985).

Costa et al. (1980) developed a method for the immunocytochemical localization of peptides in whole-mount preparations of the gut wall. Furness and Costa (1979) introduced microsurgery techniques for the investigation of the flow of immunocytochemically detectable transmitter substances in the enteric nervous system. These technical improvements greatly facilitated the morphological analysis of the origins, pathways, and terminations of chemically different enteric neurons. More recently, we introduced a laser-photocoagulation technique for neurotomy experiments in the enteric nervous system (Kobayashi and Nishisaka 1985). Further technical improvements are expected to provide better strategies for the visualization of the elaborate design of the organization of enteric nerve plexuses.

Of the large variety of enteric neurons, we will concentrate here on enkephalins, VIP-PHI, and somatostatin.

30.3.1 Enkephalin Neurons

Varicose nerve fibers in the nerve plexuses of the muscular coat of the stomach, small intestine, and large intestine contain enkephalin-like immunoreactivities. Many somata of neurons in the myenteric ganglia are enkephalin-immunopositive.

In the guinea pig jejunum, immunoreactivity accumulated in the enkephalin neurons after laser microsurgery. The anally directed thick process of the enkephalin neurons transported a larger amount of the immunoreactivity than the orally directed thin branches. The direction of the enkephalin neuron processes may be related to neuronal control of the peristaltic and antiperistaltic movement of the gut (Kobayashi et al. 1985).

30.3.2 VIP-PHI Neurons

The term "VIP-PHI neuron" indicates an enteric neuron with VIP- and PHI-like immunoreactivity. VIP and PHI are derived from the same prepropeptide (see p. 66).

Laser irradiation combined with immunocytochemical techniques was used for the morphological investigation of VIP-PHI neurons in the guinea pig small intestine. The results obtained were in agreement with those of Costa and Furness (1983) concerning the origins, pathways, and terminations of VIP-PHI neurons. Most, if not all, VIP-PHI neurons of both the myenteric and submucous ganglia possessed a single thick process. It ran within the myenteric plexus in the anal direction. In addition to this thick process, somata of VIP-PHI neurons in the submucous ganglia projected several long tapering processes. The latter showed no directional polarity and seemed to be involved in forming the autonomic ground plexus of the lamina propria mucosae.

30.3.3 Somatostatin Neurons

Processes of somatostatin neurons are widely distributed in the digestive tract of the guinea pig. However, they are unevenly localized in the myenteric ganglia; some neuronal somata are surrounded by a basketlike formation of the somatostatin neuron processes, whereas others are only sparsely innervated.

Costa et al. (1980) demonstrated that somatostatin neurons project anally in the myenteric plexus and that they supply terminals around the neuronal somata in the more distally located myenteric ganglia. Our nerve-lesion experiments with laser irradiation (Endo et al. 1986) confirmed the observations of Costa et al. (1980). Endo et al. (1986) further demonstrated using electron-microscopic immunocytochemical methods the axosomatic synapse formation between the somatostatin-immunopositive neurons.

Somatostatin neurons are also present in the submucous ganglia of the guinea pig small intestine; they showed no directional polarity (Costa et al. 1980; Endo et al. 1986).

30.4 Electrophysiological Properties of Enteric Neurons

Electrophysiological properties of enteric neurons have been discussed in recent reviews (Holman and Hirst 1977; Gershon and Erde 1981; North 1982; Wood 1984). Using extracellular and intracellular microelectrodes, mainly in the small and large intestine of the guinea pig and cat, changes in the transmembrane potentials of individual neurons were recorded and analyzed.

Wood (1984) has proposed a classification scheme of enteric neurons consisting of four categories based on intracellular recordings of the electric behavior of the cell body—S/type 1, AH/type 2, type 3, and type 4. The electric and synaptic properties of these cells are as follows.

S (spike) neurons or type 1 neurons. These cells are characterized by: (1) Low resting membrane potential relative to the other cell types; (2) high input resistance relative to the other cell types; (3) discharge of spikes during membrane depolarization that increases in frequency as a direct function of the level of depolarization; (4) anodal-break excitation at the offset of intracellularly injected hyperpolarizing current pulses; (5) no hyperpolarizing after potentials; and (6) tetrodotoxin-sensitive somal spikes.

AH (prolonged hyperpolarizing afterpotential) neurons or type 2 neurons. The criteria of AH/type 2 cells are: (1) High resting membrane potential and low input resistance relative to S/type 1 neurons; (2) discharge of one or two spikes only at the onset of intracellular injection of prolonged depolarizing current pulses; (3) no anodal-break excitation; (4) prolonged hyperpolarizing after potentials; (5) tetrodotoxin-resistant action potentials; and (6) membrane depolarization and increased input resistance in the presence of calcium-channel blockers such as magnesium and manganese.

Type 3 neurons. These cells have high resting potentials and a low input resistance similar to AH/type 2 neurons, but *they do not discharge spikes* to intracellularly injected depolarizing current pulses. They show prominent nicotinic-cholinergic EPSPs that never trigger spikes. Unlike AH/type 2 cells, elevated magnesium or manganese does not affect the cells.

Type 4 neurons. This cell type resembles AH/type 3 neurons in having high resting membrane potentials and a low input resistance. *Neither synaptic potentials nor somal action potentials* can be evoked in these cells immediately after impalement by microelectrodes. These cells were formerly thought to be glial cells (Nishi and North 1973); however, the intracellular injection of dyes reveals them to be neurons. These initially inexcitable cell bodies behave like AH/type 2 neurons 20–30 min after impalement by microelectrodes (Hodgkiss and Lees 1978) or after treatment with calcium antagonists (Grage et al. 1980).

When the results from large numbers of preparations are tallied, all four neuronal types are seen to be impaled with about equal frequency in the cat and guinea pig myenteric plexus (Wood and Mayer 1978; Wood 1981); whereas S/type 1 neurons predominate in the guinea pig submucosal plexus (Hirst and McKirdy 1975). No definite relationship has been found between the morphology, electric behavior, and functional role of enteric neurons.

These four types of enteric neuron can be regarded as intermediates between a typical neuron and typical paraneuron: S/type 1 neurons are typical neurons; type 3 neurons are nonspiking neurons, which resemble endocrine paraneurons; and type 4 neurons fluctuate between the nonspiking and spiking states. It is also postulated that synaptic transfer of information within the enteric neurons occurs at axosomatic, axoaxonal, and axodendritic junctions and that some of the transmitter substances, for example, acetylcholine, serotonin, catecholamines, and peptides, may function as paracrine or endocrine messenger substances and influence enteric neural functions (Wood 1984).

The electrophysiological properties of enteric neurons may indicate that they represent *endocrinelike archineurons* (see pp. 126, 285) and are greatly different from *neoneurons*, which extend a well-differentiated, often myelinated axon and generate distinct sodium channel spikes.

30.5 Paraneuronal Features

The morphology, secretory functions, and electrophysiology of enteric neurons reviewed above indicate that the cells have not differentiated toward neurons so much but have remained close to endocrine cells. Many of their cell-biological features are those of endocrine paraneurons. Recently, we were surprised by the close resemblance in electron micrographs between the secretory pole of a somatostatin paraneuron (D cell) and the terminal expansion formed after laser severance of a somatostatin neuron in the gut (Endo et al. 1976).

With respect to proteins specific for neurons and paraneurons, gastroenteric neurons are immunopositive for NSE; only a small population of them are positive for NFP (rat, pig, human, etc.—Iwanaga and Fujita, unpublished data). It is of particular interest that some enteric neurons are immunoreactive for spot-35 protein (Purkinje cell-specific protein, p. 99), which is contained in paraneurons more widely than in neurons (Iwanaga and Fujita, unpublished data).

Relation to paraneurons. As already stressed in Chap. 16 on gut endocrine paraneurons, these and the neurons form an integral unit for the regulation of gastroenteric functions. There are wide varieties of neurons and paraneurons that make up this integrated system; their modes of assembly are also variable.

We recognize a gradation or spectrum from grouped paraneurons to ganglionic neurons as follows:

1) Aggregated basal-granulated (endocrine) cells in the epithelium. A typical instance is Segi's cap, occurring on the top of fetal intestinal villi (see p. 182).

2) Isolated basal-granulated cells in the epithelium. Some of these cells project fairly long cytoplasmic processes in both the luminal and basal directions (see p. 167, ff.).

3) Multipolar granulated (endocrine) cells in the epithelium of the avian proventriculus. They closely resemble neurons, at least in morphology (see p. 170).

4) Neuroendocrine complex in the lamina propria mucosae of the gut (see pp. 95, 180).

5) Solitary enteric neurons. In the submucous plexus of the guinea pig stomach, there are many solitary neurons. Similar isolated neurons also occur in the myenteric plexus.

6) Enteric neurons whose somata are aggregated in the myenteric or submucous ganglia. They are similar in all cytological features to the principal neurons of the sympathetic and parasympathetic ganglia.

The formation of *neuroendocrine complexes* well visualizes the intimate relationship and specific affinity of the neurons and paraneurons concerned. Moreover, it is interesting to note that not only neurons but also paraneurons are associated by glial cells in these complexes.

30.6 Ontogeny

A comprehensive treatise on the origin of enteric neurons is given in Andrew (1971). Theories on the origin of enteric neurons are classified into the following three categories.

30.6.1 Mesodermal Origin

Earlier investigators such as Van Campenhout (1932) suggested that enteric neurons originate from the local gut mesoderm. Keuning (1948) presented experimental support to the mesodermal origin of enteric nerve plexuses. The problem has apparently been confused by Cajal's interstitial "neuron," which is now believed to be partly made up of enteroglial cells and partly artifactual elements from silver impregnation (see p. 280).

30.6.2 Endodermal Origin

Masson and Berger (1923) and Schack (1932) were of the opinion that enteric neurons originate from the endoderm. They illustrated transitional forms between the basal-granulated cells and enteric neurons. Matsuo and Seki (1978), Osaka and Kobayashi (1976), and Kataoka (1974) reported the occurrence of neuroendocrine complexes (see p. 95) in the lamina propria mucosae of the mammalian, human, and avian gut, respectively. The occurrence and morphology of the neuroendocrine complexes were in keeping with the hypothesis that enteric neurons and gut endocrine cells have a common origin, either endodermal or neuroectodermal (see p. 90); the findings did not, however, provide actual confirmation of the hypothesis.

30.6.3 Neuroectodermal Origin

The neural crest origin of enteric neurons is the one most widely accepted. By removing the neural crest from the developing chick embryo, Andrew (1963, 1964) demonstrated that the neural crest is not the source of basal-granulated (endocrine) cells, whereas enteric neurons originate from the neural crest. Using the quail-chick marker system, Le Douarin and co-workers (Le Douarin 1982; Le Douarin and Smith 1983) demonstrated that the neural crest cells (the vagal level cells located at somites 1–7 and "lumbosacral" parasympathetic cells) migrate into the gastroenteric tract. These neural crest cells differentiate into enteric neurons but do not colonize the epithelial layer, which contains basal-granulated cells. Le Douarin and co-workers also demonstrated that the neural crest cells located between somites 7 and 28 do not participate in the formation of enteric neurons, though they form the sympathetic trunk neurons and adrenal chromaffin cells.

Enteric neurons are first identified in the 6th week of gestation in humans (Kubozoe et al. 1969; Daikoku et al. 1975), on the 25th day of gestation in the guinea pig (Gershon 1981), and on the 12th day of gestation in the rabbit (Daikoku et al. 1975). Okamoto and Ueda (1967) in humans and Cantino (1970) in rats and rabbits showed that enteric neurons first appeared in the myenteric and then in the submucous ganglia. The developmental relationship between enteric neurons and that of the enteroglial cell framework is not known.

30.7 Phylogeny

The phylogeny of enteric neurons has not as yet been established. The construction of enteric nerve plexuses as well as the structure of enteric neurons are essentially identical in all classes of vertebrates, though they have been investigated only fragmentally in submammalian animals.

The coelenteron of *Hydra* represents the most primitive gut. The endoderm and ectoderm of *Hydra* possess granule-containing sensory cells, which are regarded as the *prototypes of paraneurons* (see p. 126). These cells are loosely contiguous to a diffuse net of bipolar and multipolar neuronal cells spreading throughout the body of *Hydra*. The occurrence of large cored (endocrinelike) and small clear (synapticlike) vesicles is evident (Westfall 1973). Recent immunocytochemical studies have demonstrated that the neurons of *Hydra* contain various neuropeptides such as oxytosin/vasopressin, bombesin/GRP, gastrin/CCK, and FMRFamide (Grimmelikhuijzen 1983, 1985; Grimmelikhuijzen et al. 1980,

1981a–c). We can regard these cells as the *prototypes of enteric neurons* and throughout evolution these neurons have been inherited without undergoing significant changes in morphology and, apparently also, in cell physiology. The messenger substances of the neurons have also been inherited up to the highest stages of evolutionary development with just the addition of several varieties of peptides. In this regard, the neuron in the vertebrate gut well retains the features of the neuron prototype in the most primitive metazoans and deserves to be termed the *archineuron*.

In the guinea pig myenteric plexus, there are solitary neurons in association with the nerve strands and autonomic ground plexus. Such solitary neurons are frequently demonstrated in the submucous plexus and seem to preserve the archineuronal situation better than the neurons gathered in compact ganglia.

Vertebrate gut neurons are ensheathed by enteroglial cells. It is still unknown at what stage of phylogeny this glial cell sheath appears. The neurons of *Hydra* possess no glial sheath. It must be pointed out that the enteroglial cell sheath in the mammals examined so far is provided with numerous openings through which the neuronal processes and surrounding tissues seem to interact (Endo and Kobayashi 1987). Further studies are needed to demonstrate the possible significance of glial elements in enteric nerve plexuses.

30.8 Summary

The enteric nervous system comprises ganglia, nerve strands, and an autonomic ground plexus. The latter consists of nerve bundles ensheathed by enteroglial cells. Unlike the sympathetic and parasympathetic nervous systems, the enteric nervous system is highly independent of the central nervous system; hence, Langley (1921) classified it as the "third division of the autonomic nervous system." Furthermore, it has a close relationship with the enteric paraneurons which directly receive information about the food.

Enteric neurons are multipolar and unmyelinated. They have *no definable axon*. In electrophysiological features also, enteric neurons are different from ordinary neurons (neoneurons). Sodium spikes are obscure and there are many *nonspiking* neurons.

The secretory granules, rough endoplasmic reticulum, and Golgi complexes are conspicuously developed in enteric neurons. These cell organelles, as the "secretory machinery," produce neuropeptides, including VIP/PHI, and somatostatin in addition to the classic transmitter, acetylcholine.

The *origin* of enteric neurons has been much disputed. The occurrence and morphology of *neuroendocrine complexes* in the mammalian and avian gut may imply a common origin of enteric neurons and paraneurons; however, recent transplantation studies have supported a *neuroectodermal origin* of the neurons and an endodermal origin of the paraneurons.

Phylogenetically, enteric neurons can be regarded as the direct descendants of the most primitive neurons in *Hydra*, preserving their cell-biological features as well as messenger substances. In this sense they may be termed *archineurons*.

References

Acher R (1984) Evolution of neurohormonal peptides: from genetic machinery to functional tailoring. In: Falkmer S, Håkanson R, Sundler F (eds) Evolution and tumour pathology of the neuroendocrine system. Elsevier, Amsterdam, pp 181–201

Acker H, Pietruschka F (1977) Meaning of the type I cell for the chemoreceptive process. An electrophysiological study on cultured type I cells of the carotid body. In: Acker H, Fidone S, Pallot D, Eyzaguirre C, Lübbers DW, Torrance RW (eds) Chemoreception in the Carotid Body. Springer-Verlag, Berlin, pp 92–98

Ackerman SH (1982) Ontogeny of gastric acid secretion in the rat: evidence for multiple response systems. Science 217:75–77

Adachi T, Hisano S, Daikoku S (1985) Intragranular colocalization of immunoreactive methionine-enkephalin and oxytocin within the nerve terminals in the posterior pituitary. J Histochem Cytochem 33:891–899

Adams DR (1986) The bovine vomeronasal organ. Arch Histol Jpn 49:211–225

Adams WE (1958) The comparative morphology of the carotid body and carotid sinus. Thomas, Springfield

Ádám-Vizi V, Vizi ES (1978) Direct evidence of acetylcholine releasing effect of serotonin in the Auerbach plexus. J Neural Transmission 42:127–138

Ades HW (1974) Anatomy of the inner ear. In: Keidel WD, Neff WD (eds) Auditory system. Anatomy physiology (ear) (Handbook of sensory physiology V/1). Springer, Berlin, pp 125–158

Akaike N, Noma A, Sato M (1976) Electrical responses of frog taste cells to chemical stimuli. J Physiol (Lond) 254:87–107

Akert K, Sandri C (1968) An electron-microscopic study of zinc iodide-osmium impregnation of neurons I. Staining of synaptic vesicles at cholinergic junctions. Brain Res 7:286–295

Alberts JR (1974) Producing and interpreting experimental olfactory deficits (theoretical review). Physiol Behav 12:657–670

Albinus M, Sewing KF (1974) The effect of histamine-H_2-receptor-antagonists on gastric acid secretion, gastric mucosal blood flow and acid base balance in cats. Arch Pharmacol 285:337–354

Alfejew S (1924) Die embryonale Histogenese der Zellformen des lockeren Bindegewebes der Säugetiere. Folia Haematol 30:111–172

Allison AC, Warwick RT (1949) Quantitative observations on the olfactory system of the rabbit. Brain 72:186–197

Alm PE, Bloom GD (1982) Cyclic nucleotide involvement in histamine release from mast cells. A re-evaluation. Life Sci 30:213–218

Alm PE, Bloom GD (1983) Catecholamine inhibition of rat mast cell histamine secretion. A process independent of cyclic AMP levels and counteracted by glucose. Life Sci 32:307–314

Aloe L, Levi-Montalcini R (1977) Mast cells increase in tissues of neonatal rats injected with the nerve growth factor. Brain Res 133:358–366

Alroy J, Leav I, DeLellis RA, Weinstein RS (1975) Distinctive intestinal mast cell neoplasms of domestic cats. Lab Invest 33:159–167

Al-Saffar A, Rosell S (1981) Effects of neurotensin and neurotensin analogues on the migrating myoelectrical complexes in the small intestine of rats. Acta Physiol Scand 112:203–208

Altenähr E (1971) Electron microscopical evidence for innervation of chief cells in human parathyroid gland. Experientia 27:1077

Altman J (1985) New visions in photoreception. Nature 313:264–265

Altschuler RA, Reeks KA, Marangos PJ, Fex J (1985) Neuron-specific enolase-like immunoreactivity in inner hair cells but not outer hair cells in the guinea pig organ of Corti. Brain Res 327:379–384

Alumets J, Håkanson R, Sundler F, Chang KJ (1978) Leu-enkephalin-like material in nerves and enterochromaffin cells in the gut. An immunohistochemical study. Histochemistry 56:187–196

Alumets J, Håkanson R, Sundler F, Thorell J (1979) Neuronal localisation of immunoreactive enkephalin and β-endorphin in the earthworm. Nature 279:805–806

Alzheimer A (1911) Über eigenartige Krankheitsfälle des späteren Alters. Z Ges Neurol Psychiat 4:356–385

Amoore JE, Johnston JW, Rubin M (1964) The stereochemical theory of odor. Sci Am 210:42–49

Andersen AE, Nafstad PHJ (1968) An electron microscopic investigation of the sensory organs in the hard palate region of the hen (*Gallus domesticus*). Z Zellforsch 91:391–401

Anderson PAV (1980) Epithelial conduction: its properties and functions. Prog Neurol 15:161–203

Andres KH (1965) Differenzierung und Regeneration von Sinneszellen in der Regio olfactoria. Naturwissenschaften 52:500

Andres KH (1966) Über die Feinstruktur der Rezep-

toren an Sinushaaren. Z Zellforsch 75:339–365
Andres KH (1970) Anatomy and ultrastructure of the olfactory bulb in fish, amphibia, reptiles, birds and mammals. In: Wolstenholme GEW, Knight J (eds) Taste and smell in vertebrates. J & A Churchil, London, pp 177–186
Andres KH, Düring M (1973) Morphology of cutaneous receptors. In: Iggo A (ed) Somatosensory system. Springer, Berlin Heidelberg New York, pp 3–28
Andrew A (1963) A study of the developmental relationship between enterochromaffin cells and the neural crest. J Embryol Exp Morphol 11:307–324
Andrew A (1964) The origin of intramural ganglia I. The early arrival of precursor cells in the presumptive gut of chick embryos. J Anat 98:421–428
Andrew A (1971) The origin of intramural ganglia IV. The origin of enteric ganglia: a critical review and discussion of the present state of the problem. J Anat 108:169–184
Andrew A (1974) Further evidence that enterochromaffin cells are not derived from the neural crest. J Embryol Exp Morphol 31:589–598
Andrew A (1975) APUD cells in the endocrine pancreas and intestine of chick embryos. Gen Comp Endocrinol 26:485–495
Andrew A (1976a) Intestinal endocrine cells of chicks around the time of hatching. Cell Tiss Res 172:541–551
Andrew A (1976b) Endocrine cells of the stomach of chicks around the time of hatching. Cell Tiss Res 172:553–561
Andrew A (1976c) Pancreatic APUD cells in older chick embryos, with special reference to their identity. Histochemistry 49:213–219
Andrew A (1981) APUD cells and paraneurons: embryonic origin. Adv Cell Neurobiol 2:3–32
Andrews NJ, Rinno-barmada S, Burdett K, Elder JB (1981) Effects of porcine gastric fundic factor, somatostatin, substance P, glucagon, neurotensin, bombesin, VIP, motilin, and pentagastrin on jejunal glucose absorption in the rat. Gut 24:326–332
Anghileri LJ, Tuffet-Anghileri AM (eds) (1982) The role of calcium in biological systems, vol I. CRC Press, Florida
Anholt RRH (1987) Primary events in olfactory reception. Trends Biochem Sci 12:58–62
Anniko M (1983) Embryonic development of vestibular sense organs and their innervation. In: Romand R (ed) Development of auditory and vestibular systems. Academic Press, New York London pp 375–423
Anuras S, Cooke AR (1978) Effects of some gastrointestinal hormones on two muscle layers of duodenum. Am J Physiol 234:E60–E63
Apps DK (1982) Proton-translocating ATPase of chromaffin granule membranes. Fed Proc 41: 2775–2780
Ariëns Kappers CU, Huber GC, Crosby EC (1967) The comparative anatomy of the nervous system of vertebrates, including man, vol 1. Hafner, New York, pp 343–412
Arimura A, Sato H, Dupond A, Nishi N, Schally AV (1975) Somatostatin: abundance of immunoreactive hormone in rat stomach and pancreas. Science 189:1007–1009
Arnold J (1865) Über die Structur des Ganglion intercaroticum. Virchows Arch Path Anat 33:190–209
Asada N, Kanno T (1985) Influence of pH change on redox state and CCk-induced secretion in isolated perfused rat pancreas. Am J Physiol 248:C235–C240
Asada N, Saito T, Kanno T, Nakao K, Yoshimasa T, Imura H (1983) Parallel time courses of acetylcholine-induced relases of catecholamine and enkephalin-like immunoreactive peptides from perfuse dog adrenal glands. Biomed Res 4:413–416
Asanuma A (1983) Effects of calcium ions in the inner side of skin on the lateral line organ of xenopus laevis (in Japanese). Tsurumi Shigaku (Tokyo) 9: 439–448
Asplund K (1976) Glucoreceptor deficiency in diabetes. Lancet 1:418
Atack CV, Ericson LE, Melander A (1972) Intracellular distribution of amines and calcitonin in the sheep thyroid gland. J Ultrastruct Res 41:484–498
Atlas SA, Kleinert H D, Camargo MJ, Januszewicz A, Sealey JE, Laragh JH, Schilling JW, Lewicki JA, Johnson LK, Maack T (1984) Purification, sequencing and synthesis of natriuretic and vasoactive rat atrial peptide. Nature 309:717–719
Atwater I (1980) Effects of the calcium ionophore, A23187, on glucose-induced electrical activity in mouse pancreatic B-cells. J Physiol 303:80P (Abstract)
Atwater I, Dawson CM, Scott A, Eddleston G, Rojas E (1980) The nature of the oscillatory behavior in electrical activity from the pancreatic B-cell. In: Malaisse WJ, Taljedal IB (eds) Biochemistry and biophysics of the pancreatic B-cell. Thieme-Stratton, New York, pp 100–107
Aunis D, Hesketh JE, Devilliers (1979) Freeze-fracture study of the chromaffin cell during exocytosis: evidence for connections between the plasma membrane and secretory granules and for movements of plasma membrane-associated particles. Cell Tiss Res 197:433–441
Aunis D, Hesketh JE, Devilliers G (1980) Immunohistochemical and immunocytochemical localization of myosin, chromogranin A and dopamine β-hydroxylase in nerve cells in culture and in adrenal glands. J Neurocytol 9:255–274
Aurbach GD (1959) Isolation of parathyroid hormone after extraction with phenol. J Biol Chem 234:3179–3181
Aurbach GD, Kentmann HT, Niall HD, Tregear JL, O'Riordan JLH, Marcus R, Marx SJ, Potts JT (1972) Structure, synthesis and mechanism of action of parathyroid hormone. Recent Prog Hormone Res 28:353–398
Aures D, Håkanson R, Owman Ch (1970) Formation and storage of amines in endocrine cells of gastric mucosa. J Neuro-Visceral Relations 31:337–349
Avenet P, Lindemann B (1987) Patch-clamp study of isolated taste receptor cells of the frog. J Membrane Biol 97:223–240
Avery J (1981) Repair potential of the pulp. J Endodont 7:205–212

Axelrod J (1972) Dopamine-β-hydoxylase: regulation of its synthesis and release from nerve terminals. Pharmacol Rev 24:233–243

Axelrod J (1974) The pineal gland: a neurochemical transducer. Science 184:1341–1348

Axelrod J, Quay WB, Baker PC (1965) Enzymatic synthesis of the skin lightening agent, melatonin in amphibians. Nature 208:386

Ayer-Le Lièvre C, Fontaine-Perus J (1982) The neural crest: its relations with APUD and paraneuron concepts. Arch Histol Jpn 45:409–427

Bader CR, Bertrand D, Dupin E, Kato AC (1983) Development of electrical membrane properties in cultured avian neural crest. Nature 305:808–810

Bader MF, Ciesielski-Treska J, Thierse D, Hesketh JE, Aunis D (1981) Immunocytochemical study of microtubules in chromaffin cells in culture and evidence that tubulin is not an integral protein of the chromaffin granule membrane. J Neurochem 37:917–933

Bader MF, Georges E, Mushynski WE, Trifaró JM (1984a) Neurofilament proteins in cultured chromaffin cells. J Neurochem 43:1180–1193

Bader MF, Berniner-Valentin F, Rousset B, Aunis D (1984b) The adrenal paraneurone: tubulin organization. Can J Physiol Pharmacol 62:502–511

Bagnara JT, Taylor JD, Hadley ME (1968) The dermal chromatophore unit. J Cell Biol 38:67–79

Bailey CH, Chen MC, Weiss KR, Kupfermann I (1982) Ultrastructure of a histaminergic synapse in *Aplysia*. Brain Res 238:205–210

Baird IL (1974) Anatomical features of the inner ear in submammalian vertebrates. In: Keidel WD, Neff WD (eds) Auditory system. Anatomy physiology (ear) (Handbook of Sensory Physiology V/1). Springer, Berlin, pp 159–212

Baker BL (1974) Functional cytology of the hypophysial pars distalis and pars intermedia. In: Knobil E, Sayrer WH (eds) The pituitary gland and its neuroendocrine control (Handbook of physiology 7/IV/1). American Physiological Society, Washington DC, pp 45–80

Baker PF, Knight DE (1978) Calcium-dependent exocytosis in bovine adrenal medullary cells with leaky plasma membranes. Nature 276:620–622

Baker PH, Knight DE (1981) Calcium control of exocytosis and endocytosis in bovine adrenal medullary cells. Phil Trans Roy Soc Lond B 296:83–103

Balabaud C, Noel M, Dangoumau J (1977) Influence de la secretine sur la cholerese chez le rat. J Pharmacol (Paris) 8:191–196

Balfour FM (1877) The development of elasmobranch fishes. J Anat 11:406–490

Bancroft FC, Levine L, Tashijian AH Jr (1969) Control of growth hormone production by a clonal strain of rat pituitary cells: stimulation by hydrocortisone. J Cell Biol 43:432–441

Bänder A (1950) Über zwei verschiedene chromaffine Zelltypen im Nebennierenmark und ihre Beziehung zum Adrenalin- und Arterenolgehalt. Verh Anat Ges 48:172–176

Banks P, Helle K (1965) The release of protein from the stimulated adrenal medulla. Biochem J 97: 40C–41C

Baraspiri L (1976) Synthesis of gastrointestinal peptide hormones. Acta Physiol Acad Sci Hung 47:299–309

Barberini F, Correr S (1984) Ultrastructure of Rathke's pituitary cleft. In: Motta PM (ed) Ultrastructure of endocrine cells and tissues. Martinus Nijhoff Boston Hague Dordrecht Lancaster, pp 57–63

Barbezat GO (1973) Stimulation of intestinal secretion by polypeptide hormones. Scand J Gastroenterol Suppl 22:1–21

Barbezat GO, Daniel M, Bank S, Grant B, Vinik A (1974) Effect of histamine H-receptor blockade on fasting serum gastrin and fasting gastric acid secretion in the dog. Nature 249:666–668

Barchas JD, Akil H, Elliott GR, Holman RB, Watson SJ (1978) Behavioral neurochemistry: neuroregulators and behavioral states. Science 200:964–973

Barer R, Heller H, Lederis K (1963) The isolation, identification and properties of the hormonal granules of the neurohypophysis. Proc Roy Soc B 158:388–416

Bargmann W (1943) Die Epiphysis Cerebri. In: Möllendorffs Handbuch der mikroskopischen Anatomie des Menschen VI/4. Springer, Berlin, pp 309–502

Bargmann W (1949) Über die neurosekretorische Verknüpfung von Hypothalamus and Neurohypophyse. Z Zellforsch 34:610–634

Bargmann W (1966) Neurosecretion. Int Rev Cytol 19:183–201

Bargmann W (1972) On the innervation of vertebrate endocrine organs. In: Proc 4th Int Congr Endocrinol (Int Congr Ser No 273). Excerpta Medica, Amsterdam, pp 220–273

Bargmann W, Knoop A (1957) Elektronenmikroskopische Beobachtungen an der Neurohypophyse. Z Zellforsch 46:242–251

Bargmann W, Lindner E, Andres KH (1967) Über Synapsen an endokrinen Epithelzellen und die Definition secretorischer Neurone. Untersuchungen am Zwishenlappen der Katzenhypophyse. Z Zellforsch 77:282–298

Barka T (1980) Biologically active polypeptides in submandibular glands. J Histochem Cytochem 28: 836–859

Barker DS, Welker WI (1969) Receptive fields of first-order somatic sensory neurons innervating rhinarium in coati and racoon. Brain Res 14:367–386

Barmada SR, Burdett K, Elder JB (1981) Studies of a gastric fundic mucosal factor which affects absorption from the antrum and jejunum in the rat. J Physiol 315:1–8

Baron M, Eyzaguirre C (1977) Effects of temperature on some membrane characteristics of carotid body cells. Am J Physiol 233 (Cell Physiol 2): C35–C46

Barrantes FJ (1979) Endogenous chemical receptors: some physical aspects. Ann Rev Biophys Bioeng 8:287–321

Bartholomew C, Murphy JJ, McGeeney KF, Fitzgerald O (1977) Exocrine pancreatic response to the venom of the scorpion, tityus trinitatis. Gut 18:623–625

Bartter FC, Schwartz WB (1967) The syndrome of

inappropriate secretion of antidiuretic hormone. Am J Med 42:790–806
Baserga R (1981) The cell cycle. New Engl J Med 304:453–459
Baumgarten HG, Holstein AF, Owman C (1970) Auerbach's plexus of mammals and man: electron microscopic identification of three different types of neuronal processes in myenteric ganglia of the large intestine from rhesus monkeys, guinea-pigs and man. Z Zellforsch 106:376–397
Baumgartner H, Gibb JW, Hörtnagl H, Snider SR, Winkler H (1974) Labelling of chromaffin granules in adrenal medulla with ^{35}S-sulfate. Mol Pharmacol 10:678–685
Beauville M, Raynaud P, Vernay M (1974) Secretion gastrique du lapin et theorie des deux composants. Biol Gastroenterol (Paris) 7:215–219
Becker AE, Drukker J, Meijer AEFH (1967) Histochemical characteristics of chemoreceptor organs (glomera). Histochemie 11:195–204
Becker KL, Monaghan KG, Silva OL (1980) Immunocytochemical localization of calcitonin in Kultchitsky cells of human lung. Arch Pathol Lab Med 104:196–198
Beidler LM (1970) Physiological properties of mammalian taste receptors. In: Wolsternholme GEW, Knight J (eds) Taste and smell in vertebrates. J & A Churchill, London, pp 51–67
Beidler LM, Smallman RL (1965) Renewal of cells within taste buds. J Cell Biol 27:263–272
Belpaire F, Laduron P (1968) Tissue fractionation and catecholamines. I. Latency and activation properties of dopamine β-hydroxylase in adrenal medulla. Biochem Pharmacol 17:411–421
Benchimol S, Cantin M (1978) Ultrastructural radioautography of the incorporation of tritiated leucine by the rat adrenal medulla *in vivo*. Cell Tiss Res 193:179–199
Benchimol S, Cantin M (1982) Ultrastructural radioautography of synthesis and migration of proteins and catecholamines in the rat adrenal medulla. Cell Tiss Res 225:293–314
Benedeczky I, Smith AD (1972) Ultrastructural studies on the adrenal medulla of golden hamster: origin and fate of secretory granules. Z Zellforsch 124:367–386
Bennett A, Stamford IF, Unger WG (1972) Prostaglandin E_2 and gastric acid secretion in man. Proc Physiol Soc J Physiol 226:96–98
Bennett A, Stamford IF, Unger WG (1973) Prostaglandin E_2 and gastric acid secretion in man. J Physiol 229:349–360
Bennett HS (1941) Cytological manifestations of secretion in the adrenal medulla of the cat. Am J Anat 69:333–381
Bennett HS (1956) The concepts of membrane flow and membrane vesiculation as mechanisms for active transport and ion pumping. J Biophys Biochem Cytol 2 Suppl: 99–103
Bennett HS (1977) A review of "chromaffin, enterochromaffin and related cells", with some comments on the APUD and paraneuron concepts. Arch Histol Jpn 40 Suppl:317–325
Bennett T (1971) The neuronal and extra-neuronal localizations of biogenic amines in the cervical region of the domestic fowl (*Gallus gallus domesticus* L.) Z Zellforsch 112:443–464
Bensch KG, Gordon GB, Miller LR (1965a) Studies on the bronchial counterpart of the Kultschitzky (argentaffin) cell and innervation of bronchial glands. J Ultrastr Res 12:668–686
Bensch KG, Gordon GB, Miller LR (1965b) Electron microscopic and biochemical studies on the bronchial carcinoid tumor. Cancer 18:592–602
Bensch KG, Corrin B, Pariente R, Spencer H (1968) Oat-cell carcinoma of the lung: its origin and relationship to bronchial carcinoid. Cancer 22:1163–1172
Bensley RR (1911) Studies on the pancreas of the guinea-pig. Am J Anat 12:297–388
Berelowitz M, Le Roith D, Von Schenk H, Newgard C, Szabo M, Frohman LA, Shiloach J, Roth J (1982) Somatostatin-like immunoreactivity and biological activity is present in *Tetrahymena pyriformis*, a ciliated protozoan. Endocrinology 110: 1939–1944
Berglindh T (1977) Absolute dependence on chloride for acid secretion in isolated gastric glands. Gastroenterology 73:874–880
Bernardi P, Azzone GF (1983) Electroneutral $H^+ - K^+$ exchange in liver mitochondria regulation by membrane potential. Biochim Biophys Acta 724: 212–223
Bernd P, Gershon MD, Nunez EA, Tamir H (1979) Localization of a highly specific neuronal protein, serotonin binding protein, in thyroid parafollicular cells. Anat Rec 193:257–268
Bernd P, Gershon MD, Nunes EA, Tamir H (1981) Separation of dissociated thyroid follicular and parafollicular cells: association of serotonin binding protein with parafollicular cells. J Cell Biol 88:499–508
Berneis KH, Pletscher A, Da Prada M (1969) Metal-dependent aggregation of biogenic amines: a hypothesis for their storage and release. Nature 224: 281–283
Berridge M (1986) Second messenger dualism in neuromodulation and memory. Nature 323:294–295
Berridge MJ, Irvine RF (1984) Inositol trisphosphate, a novel second messenger in cellular signal transduction. Nature 312:315–321
Bessman SP, Mohan C, Zaidise I (1986) Intracellular site of insulin action: mitochondrial Krebs cycle. Proc Natl Acad Sci USA 83:5067–5070
Bevis PJR, Thorndyke MC (1978) Endocrine cells in the oesophagus of the Ascidian *Styela clava*, a cytochemical and immunofluorescence study. Cell Tiss Res 187:153–158
Bevis PJR, Thorndyke MC (1979) A cytochemical and immunofluorescence study of endocrine cells in the gut of the Ascidian *Styela clava*. Cell Tiss Res 199: 139–144
Bevis PJR, Thorndyke MC (1981) Stimulation of gastric enzyme secretion by porcine cholecystokinin in the Ascidian *Styela clava*. Gen Com Endocrinol 45:458–464
Biales B, Dichter M, Tischler A (1976) Electrical

excitability of cultured adrenal chromaffin cells. J Physiol 262:743–753
Biales B, Dichter MA, Tischler A (1977) Sodium and calcium action potential in pituitary cells. Nature 267:172–174
Bienkowski RS (1983) Intracellular degradation of newly synthesized secretory proteins. Biochem J 214:1–10
Biggio G, Piccardi MP, Porceddu ML, Gessa GL (1976) Changes in gastro-intestinal serotonin content associated with fasting and satiation. Experientia 33:745–746
Bijlsma WA, Schotman P, Jennekens FGI, Gispen WH (1983) Regeneration and protein synthesis in the spinal cord: the influence of ACTH-like peptides on cell-free protein synthesis. Neurosci Lett 38: 297–302
Birbeck MSC, Barnicot NA (1959) Electron microscopic studies on pigment formation in human hair follicles. In: Gordon M (ed) Pigment cell biology. Academic Press, New York, pp 549–561
Biscoe TJ (1971) Carotid body: structure and function. Physiol Rev 51:437–495
Bishop JW, Osamura RY, Tsutsumi Y (1985) Multiple hormone production in an oat cell carcinoma of the larynx. Acta Pathol Jpn 35:915–923
Bissette G, Manberg P, Nemeroff CB, Prange AJ (1978) Minireview: neurotensin, a biologically active peptide. Life Sci 23:2173–2182
Bixby JL, Spitzer NC (1983) Enkephalin reduces calcium action potentials in Rohon-Beard neurons in vivo. J Neurosci 3:1014–1018
Blair EL, Grund ER, Lund PK, Piercy A, Reed JD, Sanders DJ, Shale D, Shaw B, Wilkinson J (1977) Comparison of vagal and meat stimulation on gastric acid secretion and serum gastrin. J Physiol 266:157–172
Blaschko H, Welsch AD (1953) Localization of adrenaline in cytoplasmic particles of the bovine adrenal medulla. Naunyn-Schmiedeberg's Arch Exp Pathol Pharmacol 219:17–22
Blaschko H, Comline RS, Schneider FH, Silver M, Smith AD (1967) Secretion of a chromaffin granule protein, chromogranin, from the adrenal gland after splanchnic stimulation. Nature 215:58–59
Bloch B (1917) Das problem der Pigmentbildung in der Haut. Arch Dermatol Syph. 124:129–208
Blois MS, Kallman RF (1964) Incorporation of C^{14} from 3,4-dihydroxyphenylanine-2′-C^{14} into the melanin of mouse melanomas. Cancer Res 24: 863–868
Bloom FE (1983) The endorphins: a growing family of pharmacolongically pertinent peptides. Annu Rev Pharmacol Toxicol 23:151–170
Bloom FE (1984a) The functional significance of neurotransmitter diversity. Am J Physiol 246: C184–C194
Bloom FE (1984b) Neurotransmitter molecular diversity: some views of its functional significance. Biomed Res 5 Suppl: 49–56
Bloom G (1954) Studies on the olfactory epithelium of the frog and the toad with the aid of light and electron microscopy. Z Zellforsch 41:89–100
Bloom G, Engström H (1952) The structure of the epithelial surface in the olfactory region. Exp Cell Res 3:699–701
Bloom SR, Polak JM, Pearse AGE (1973) Vasoactive intestinal peptide and watery diarrhoea syndrome. Lancet 2:14–16
Bloom W (1931) A new type of granular cell in the islets of Langerhans of man. Anat Rec 49:363–371
Bock E (1978) Nervous system specific proteins. J Neurochem 30:7–14
Böck P (1973) Das Glomus caroticum der Maus. Ergeb Anat 48–1. Springer, Berlin, pp 1–82
Böck P (1980a) Histochemical demonstration of adenine nucleotides in carotid body type I cells. In: Eränkö O, Soinila S, Päivärinta H (eds) Histochemistry and cell biology of autonomic neurons, SIF cells, and paraneurons (Adv Biochem Psychopharmacol 25). Raven Press, New York, pp 235–239
Böck P (1980b) Adenine nucleotides in the carotid body. Cell Tiss Res 206:279–290
Böck P (1980c) Identification of paraneurons by labelling with quinacrine (Atebrin). Arch Histol Jpn 43:35–44
Böck P (1982) The paraganglia (Handb mikrosk Anat Mensch VI/8). Springer, Berlin
Böck P (1986) Fine structure of the neuro-insular complex type II in the cat. Arch Histol Jpn 49:189–197
Böck P, Gorgas K (1976) Enterochromaffin cells and enterochromaffin-like cells in the cat pancreas. In: Fujita T(ed) Endocrine gut and pancreas. Elsevier, Amsterdam, pp 13–24
Bodenmüller H, Schaller HC (1981) Conserved amino acid sequence of a neuropeptide, the head activator, from coelenterates to humans. Nature 293:579–580
Boeckh J (1976) Aspects of nervous coding of sensory quality in the olfactory pathway of insects. In: Proceedings of XV International Congress of Entomology. Entomological Society of America, Washington DC, pp 308–322
Boeke J (1949) The sympathetic endformation, its synaptology, the interstitial cells, the periterminal network, and its bearing on neurone theory. Discussion and critique. Acta Anat 8:18–61
Boer HH, Schot LPC, Roubos EW, ter Maat A, Lodder JC, Reichelt D, Swaab DF (1979) ACTH-like immunoreactivity in two electrotonically coupled giant neurons in the pond snail *Lymnaea stagnalis*. Cell Tiss Res 202:231–240
Boer HH, Schot LPC, Veenstra JA, Reichelt D (1980) Immunocytochemical identification of neural elements in the central nervous systems of a snail, some insects, a fish, and a mammal with an antiserum to the molluscan cardio-excitatory tetrapeptide FMRFamide. Cell Tiss Res 213:21–27
Boer HH, Schot LPC, Steinbusch HWM, Montagne C, Reichelt D (1984) Co-existence of immunoreactivity to anti-dopamine, antiserotonin and antivasotocin in the cerebral giant neuron of the pond snail *Lymnaea stagnalis*. Cell Tiss Res 238:411–412
Bons N, Pérézi N (1981) Immunocytochemical characterization of the neurosecretoy systems producing mesotocin and vasotocin in the brain of several Lacertidae. C R Acad Sci Ser III 293:645–648
Boxer MI (1983) The possible role of opiates in habitua-

tion of responses recorded in the reticular formation of the frog, to peripheral stimulation. Neurosci Lett 40:61–65

Boyd JD (1956) Observations on the human pharyngeal hypophysis. J Endocrinol 14:66–77

Braganza JM, Gibbs ACC, Howat HT (1975) The influence of secretin on the secretion of pepsin in response to acid stimulants in the anaesthetized cat. J Physiol 252:791–801

Braganza JM, Gibbs ACC, Howat HT (1976) The interaction of secretin and stimulants of gastric acid secretion in anaesthetized cats. J Physiol 258:73–81

Braganza JM, Howat HT, Kay GH (1976) A comparison of the pepsin stimulating effects of secretin preparations. J Physiol 258:63–72

Brand SJ, Morgan RGH (1981) The release of rat intestinal cholecystokinin after oral trypsin inhibitor measured by bioassay. J Physiol 319:325–343

Brandt BL, Hagiwara S, Kidokoro Y, Miyazaki S (1976) Action potentials in the rat chromaffin cell and effects of acetylcholine. J Physiol 263:417–439

Branton WD, Mayeri E, Brownwell P, Simon SB (1978) Evidence for local hormonal communication between neurones in *Aplysia*. Nature (Lond) 274: 70–72

Brauser B, Bucher T, Dolivo M (1970) Redox transition of cytochromes and pyridine nucleotides upon stimulation of an isolated rat ganglion. FEBS Lett 8:297–300

Breathnach AS (1971) Embryology of human skin: A review of ultrastructural studies. J Invest Dermatol 57:133–143

Breathnach AS, Robins J (1970) Ultrastructural observations on Merkel cells in human foetal skin (abstr). J Anat 106:411

Breckenridge BM, Crawford EJ (1961) The quantitative histochemistry of the brain. Enzymes of glycogen metabolism. J Neurochem 7:234–240

Breckenridge BM, Norman JH (1965) The conversion of phosphorylase b to phosphorylase a in brain. J Neurochem 12:51–57

Brehm H Von (1963) Morphologische Untersuchungen an Epithelkörperchen (Glandulae parathyreoideae) von Anuren. I. Z Zellforsch 61:376–400

Brewer HB, Ronan R (1970) Bovine parathyroid hormone: amino acid sequence. Proc Natl Acad Sci USA 67:1862–1869

Brimijoin S, Lundberg JM, Brodin E, Hökfelt T, Nilsson G (1980) Axonal transport of substance P in the vagus and sciatic nerves of the guinea-pig. Brain Res 191:443–457

Brisson P, Collin JP (1980) Systèmes aminergiques des mollusques gastropodes pulmonés IV– Paraneurones et innervation catécholaminergiques de la région du carrefour des voies génitales; étude radioautographique. Biol Cell 38:211–220

Brodal A (1981) Neurological anatomy; in relation to clinical medicine, 3rd ed. Oxford University Press, New York

Brooks JC, Treml S (1983) Catecholamine secretion by chemically skinned cultured chromaffin cells. J Neurochem 40:468–473

Broomfield CE, Hardy PM (1977) The synthesis of locust adipokinetic hormone. Tetrahed Lett 25: 2201–2204

Brown AG, Iggo A (1963) The structure and function of cutaneous "touch corpuscles" after denervation of type I cutaneous receptors. J Physiol 165:28P–29P

Brown EM, Hurwitz S, Aurbach GD (1976) Preparation of viable isolated bovine parathyroid cells. Endocrinology 99:1582–1588

Brown JR, Hill RG (1983) Multiple receptors for substance P. Trends Pharmacol Sci 4:512–514

Brownfield MS, Poff BC, Holzwarth MA (1985) Ultrastructural immunocytochemical co-localization of serotonin and PNMT in adrenal medullary vesicles. Histochemistry 83:41–46

Brownstein MJ, Robinson AG, Gainer H (1977) Immunological identification of rat neurophysin precursors. Nature 269:259–261

Bryan J, Wilson L (1971) Are cytoplasmic microtubules heteropolymers? Proc Natl Acad Sci USA 68:1762–1766

Bubenik GA, Brown GM, Uhlir I, Grota LG (1974) Immunohistological localization of N-acetylindolealkylamines in pineal gland, retina and cerebellum. Brain Res 81:233–242

Bubenik GA, Brown GM, Grota LG (1976) Differential localization of N-acetylated indolealkylamines in CNS and the harderian gland using immunohistology. Brain Res 118:417–427

Bucher NLR, Weir GC (1976) Insulin, glucagon, liver regeneration and DNA synthesis. Metabolism 25 (Suppl): 1423–1425

Buijs RM (1985) Extrahypothalamic pathways of a neurosecretory system: their role in neurotransmission. In: Kobayashi H, Bern HA, Urano A (eds) Neurosecretion and biology of neuropeptides. Jpn Sci Soc Press, Tokyo/Springer, Berlin, pp 279–286

Buijs RM, Van Heerikhuize JJ (1982) Vasopressin and oxytocin release in the brain—a synaptic event. Brain Res 252:71–76

Bullock TH (1974) An essay on the discovery of sensory receptors and the assignment of their functions together with an introduction to electroreceptors. In: Fessard A (ed) Electroreceptors and other specialized receptors in lower vertebrates (Handbook of sensory physiology III/3). Springer, Berlin Heidelberg New York, pp 1–12

Burgess PR, Perl ER (1973) Cutaneous mechanoreceptors and nociceptors. In: Iggo A (ed) Somatosensory system (Handbook of Sensory Physiology II). Springer, Berlin Heidelberg New York, pp 29–78

Burgess PR, English KB, Horch KW, Stensaas LJ (1974) Patterning in the regeneration of type I cutaneous receptors. J Physiol (Lond) 236:57–82

Burgoyne RD, Norman KM (1985) Presence of tropomyosin in adrenal chromaffin cells and its association with chromaffin granule membranes. FEBS Lett 179:25–28

Burke R D (1983) Development of the larval nervous system of the sand dollar, *Dendraster excentricus*. Cell Tissue Res 229:145–154

Burnstock G (1972) Purinergic nerves. Pharmacol Rev 24:509–581

Burnstock G, Costa M (1975) Adrenergic neurons. Their organization, function and development in the peripheral nervous system. Chapman and Hall,

London
Burnstock G, Campbell G, Satchell D, Smythe A (1970) Evidence that adenosine triphosphate or a related nucleotide is the transmitter substance released by non-adrenergic inhibitory nerves in the gut. Br J Pharmacol 40:668–688
Burt DS, Stanworth DR (1983) Changes in cellular levels of cyclic AMP in rat mast cells during secretion of histamine induced by immunoglobulin E decapeptide and ACTH (1–24) peptide. Biochim Biophys Acta 762:458–465
Bussolati G, Pearse AGE (1967) Immunofluorescent localization of calcitonin in the "C" cells of pig and dog thyroid. J Endocrinol 37:205–209
Bussolati G, Pearse AGE (1970) Immunofluorescent localization of the gastrin-secreting G cells in the pyloric antrum of the pig. Histochemie 21:1–4
Bussolati G, Rost WD, Pearse AGE (1969) Fluorescence metachromasia in polypeptide hormone-producing cells of the APUD series, and its significance in relation to the structure of the precursor protein. Histochem J 1:517–530
Bussolati G, Capella C, Solcia E, Vassallo G, Vezzadini P (1971) Ultrastructural and immunofluorescent investigations of the secretin cell in the dog intestinal mucosa. Histochemie 26:218–227
Cajal SR (1893) Sur les ganglions et plexus nerveux de l'intestin. C R Soc Biol (Paris) 45:217–223
Cajal SR (1911) Histologie du système nerveux de l'homme et des vertébres 2. Maloine, Paris
Caldara R, Ferrari C, Romussi M, Bierti L, Gandini S, Curtarelli G (1978) Effect of dopamine infusion on gastric and pancreatic secretion and on gastrin release in man. Gut 19:724–728
Calvert R (1972a) Electron microscopic observations on the contribution of the ultimobranchial bodies to thyroid histogenesis in the rat. Am J Anat 133: 269–290
Calvert R (1972b) Transitional cells in the postnatal thyroid gland of the rat. Anat Rec 174:341–360
Cammermeyer J, Fenton IM (1981) Improved preservation of neuronal glycogen by fixation with iodoacetic acid-containing solutions. Exp Neurol 72:429–445
Campbell AK (1983) Intracellular calcium. Its universal role as regulator. John Wiley & Sons, Chichester New York Brisbane Toronto Singapore
Cannon WB, Rosenblueth A (1937) Autonomic neuroeffector system. Macmillan, New York
Cantin M (ed) (1984) Cell biology of the secretory process. Karger, Basel München Paris London New York Tokyo Sydney
Cantin M, Genest J (1985) The heart and the atrial natriuretic factor. Endocrin Rev 6:107–127
Cantin M, Gutkowska J, Thibault G, Milne RW, Ledoux St, MinLi S, Chapeau C, Garcia R, Hamet P, Genest J (1984) Immunocytochemical localization of atrial natriuretic factor in the heart and salivary glands. Histochemistry 80:113–127
Cantino D (1970) An histochemical study of the nerve supply to the developing alimentary tract. Experientia 26:766–767
Capella C, Solcia E (1971) Optical and electron microscopical study of cytoplasmic granules in human carotid body, carotid body tumours and glomus jugulare tumours. Virchows Arch B Zellpathol 7:37–53
Capella C, Hage E, Solcia E, Usellini L (1978) Ultrastructural similarity of endocrine-like cells of the human lung and some related cells of the gut. Cell Tiss Res 186:25–37
Capen CC (1971) Fine structural alterations of parathyroid glands in response to experimental and spontaneous changes of calcium in extracellular fluids. Am J Med 50:598–611
Capen CC, Roth SI (1973) Ultrastructural and functional relationships of normal and pathologic parathyroid cells. Pathobiol Ann 3:129–175
Carabba VH, Sorokin SP, Hoyt RF (1985) Development of neuro-epithelial bodies in intact and cultured lungs of fetal rats. Am J Anat 173:1–27
Carlei F, Polak JM (1984) Antibodies to neuron-specific enolase for the delineation of the entire diffuse neuroendocrine system in health and disease. Semin Diagn Pathol 1:59–70
Caro LG, Palade GE (1964) Protein synthesis, storage, and discharge in the pancreatic exocrine cell. an antoradiographic study. J Cell Biol 20: 473–495
Carvalheira AF, Pearse AGE (1967) The cytology and cytochemistry of the "C" cells in the thyroid gland of the pig. J Roy Microsc Soc 86:203–209
Carvalheira AF, Welsch U, Pearse AGE (1968) Cytochemical and ultrastructural observations on the argentaffin and argyrophil cells of the gastrointestinal tract in mammals and their place in the APUD series of polypeptide-secreting cells. Histochemie 14:33–46
Casanova S, Corrado F, Vignoli G (1974) Endocrine-like cells in the epithelium of the human male urethra. J Submicrol Cytol 6:435–438
Case RM (1978) Synthesis, intracellular transport and discharge of exportable proteins in the pancreatic acinar cell and other cells. Biol Rev 53:211–354
Case RM, Goebell H (eds) (1976) Stimulus-secretion coupling in the gastrointestinal tract. William Clowes and Sons, Colchester Beccles, London
Caspary WF, Winckler K, Lankisch PG, Creutzfeldt W (1975) Influence of exocrine and endocrine pancreatic function on intestinal brush border enzymatic activities. Gut 16:89–92
Castel M, Gainer H, Dellmann HD (1984) Neuronal secretory systems. Int Rev Cytol 88:303–459
Castel M, Morris J, Ben-Barak Y, Timberg R, Sivan N, Gainer H (1985) Ultrastructural localization of immunoreactive neurophysins using monoclonal antibodies and protein A-gold. J Histochem Cytochem 33:1015–1025
Cegrell L (1968) The occurrence of biogenic monoamines in the mammalian endocrine pancreas. Acta Physiol Scand Suppl 314. Grahns Boktryckeri Lund, pp 1–60
Cervetto L, MacNichol EF (1972) Inactivation of horizontal cells in turtle retina by glutamate and aspartate. Science 178:767–768
Chan-Palay V, Palay SL (1972) The stellate cells of the rat's cerebellar cortex. Z Anat Entw-Gesch 136: 224–248

Chang TM, Chey WY (1980) Radioimmunoassay of secretin. A critical review and current status. Digest Dis Sci 25:529–552

Chang WWL, Bencosme SA (1968) Selective staining of secretory granules of adrenomedullary cells by silver methanamine: a light and electron microscopic study. Canad J Physiol Pharmacol 46:745–747

Chang WWL, Leblond CP (1971) Renewal of the epithelium in the descending colon of the mouse I. Presence of three cell populations: vacuolated-columnar, mucous and argentaffin. Am J Anat 131: 73–119

Chang TL, Gainer H, Russell JT, Peng Loh Y (1982) Proopiocortin-converting enzyme activity in bovine neurosecretory granules. Endocrinology 111:1607–1614

Cheng H, Leblond CP (1974a) Origin, differentiation and renewal of the four main epithelial cell types in the mouse small intestine I. Columnar cell. Am J Anat 141:461–480

Cheng H, Leblond CP (1974b) Origin, differentiation and renewal of the four main epithelial cell types in the mouse small intestine III. Entero-endocrine cells. Am J Anat 141:503–520

Cheng KW, Friesen HG (1970) Physiological factors regulating secretion of neurophysin. Metabolism 19:876–890

Chessman DW (1970) Structural elucidation of a gonadotrophin-inhibiting substance from the bovine pineal gland. Biochem Biophys Acta 207:247–253

Chey WY, Escoffery R (1976) Secretin cells in the gastrointestinal tract. Endocrinology 98:1390–1395

Chi EY, Lagunoff D, Koehler JK (1976) Freeze-fracture study of mast cell secretion. Proc Natl Acad Sci USA 73:2823–2827

Chiba T, Black AC, Williams TH (1977) Evidence for dopamine storing interneurons and paraneurons in rhesus monkey sympathetic ganglia. J Neurocytol 6:441–453

Childs GV (1986) Functional ultrastructure of gonadotropes: a review. In: Ganten D, Pfaff D (eds) Morphology of hypothalamus and its connections. Current topics in neuroendocrinology, vol 7. Springer-Verlag, Berlin, pp 49–97

Chiu AY, Strumwasser F (1981) An immunohistochemical study of the neuropeptidergic bag cells of *Aplysia.* J Neurosci 1:812–826

Choi HSH, Anderson PJ (1985) Immunohistochemical diagnosis of olfactory neuroblastoma. J Neuropathol Exp Neurol 44:18–31

Chrétien M, Seidah NG (1984) Precursor polyproteins in endocrine and neuroendocrine systems. Int J Peptide Protein Res 23:335–341

Chrétien M, Boileau G, Lazure C, Seidah NG (1984) Processing of peptide hormone and neuropeptide precursors. In: Cantin M (ed) Cell Biology of the Secretory Process. Karger, Basel, pp 214–246

Christiansen J, Agaard P (1972) Gastric secretion of calcium in patients with primary hyperparathyroidism before and after removal of parathyroid adenomas. Scand J Gastroenterol 7:639–641

Chubb IW (1977) The release of non-transmitter substances. In: Cottrell GA, Usherwood PNR (eds) Synapses. Blackie, Glasgow, pp 264–290

Ciaccio CM (1906) Sur une nouvelle espèce cellulaire dans les glandes de Lieberkühn. C R Seances Soc Biol Ses Fil 60:76–77

Ciaccio CM (1907) Sopra speciali cellule granulose delle mucosa intestinale. Arch Ital Anat 6:482–498

Cimarusti DL, Saito K, Vaughn JE, Barber R, Roberts E, Thomas PE (1979) Immunocytochemical localization of dopamine-β-hydroxylase in rat locus coeruleus and hypothalamus. Brain Res 162:55–67

Clabough JW (1973) Cytological aspects of pineal development in rats and hamsters. Am J Anat 137:215–230

Clarke NB, Khairallah LH (1972) Ultrastructure of the parathyroid gland of freshwater turtles. J Morphol 138:131–140

Clayton CJ, McNeill TH, Sladek JR (1981) A comparison of neuropeptide immunocytochemistry in fluid-fixed and freeze-dried brains. Cell Tiss Res 220:223–230

Cobbs WH, Pugh EN (1985) Cyclic GMP can increase rod outer-segment light-sensitive current 10-fold without delay of excitation. Nature 313:585–587

Cocchia D (1981) Immunocytochemical localization of S-100 protein in the brain of adult rat. Cell Tiss Res 214:529–540

Cocchia D, Miani N (1980) Immunocytochemical localization of the brain-specific S-100 protein in the pituitary gland of adult rat. J Neurocytol 9:771–782

Cocchia D, Michetti F (1981) S-100 antigen in satellite cells of the adrenal medulla and the superior cervical ganglion of the rat: an immunochemical and immunocytochemical study. Cell Tiss Res 215:103–112

Cocchia D, Michetti F, Donato R (1981) Immunochemical and immunocytochemical localization of S-100 antigen in normal human skin. Nature 294:85–87

Cochrane DE, Douglas WW (1974) Calcium-induced extrusion of secretory granules (exocytosis) in mast cells exposed to 48/80 or the ionophores A-23187 and X-537A. Proc Natl Acad Sci USA 71:408–412

Cochrane DE, Distel DL, Lansman JB, Paterson BM (1982) Stimulus-secretion coupling in rat mast cells: inactivation of extracellular calcium dependent. J Physiol 323:423–435

Coggeshall RE (1972) Autoradiographic and chemical localization of 5-hydroxytryptamine in identified neurons in the leech. Anat Rec 172:489–498

Cohn DV, Elting J (1983) Biosynthesis, processing, and secretion of parathormone and secretory protein-I. Recent Prog Horm Res 39:181–203

Collin JP (1969a) Contribution à l'étude de l'organe pinéal. De l'épiphyse sensorielle à la glande pinéale: Modalités de transformation et implications fonctionnelles. Ann Stat Biol Besse-en-Chandesse Suppl 1:1–359

Collin JP (1969b) La cupule sensorielle de l'organe pinéal de la lamproie de planer. L'ultrastructure des cellules sensorielles et ses implications fonctionelles. Arch Anat Microsc Morphol Exp 58:145–182

Collin JP (1976) La rudimentation des photorecepteurs dans l'organe pinéal des vertébrés. Colloq

Internat CNRS 266: 393–408
Collin JP, Juillard MT, Falcon J (1977) Localization of 5-hydroxytryptamine and protein(s) in the secretion granules of the rudimentary photoreceptor cells in the pineal of *Lacerta*. J Neurocytol 6: 541–554
Collip JB (1925) The extraction of a parathyroid hormone which will prevent or control parathyroid tetany and which regulates the level of blood calcium. J Biol Chem 63: 395–438
Combs JW (1966) Maturation of rat mast cells. An electron microscope study. J Cell Biol 31: 563–575
Combs JW (1971) An electron microscope study of mouse mast cells arising *in vivo* and *in vitro*. J Cell Biol 48: 676–684
Combs JW, Lagunoff D, Benditt EP (1965) Differentiation and proliferation of embryonic mast cells of the rat. J Cell Biol 25: 577–592
Comroe JH (1984) The peripheral chemoreceptors. In: Fenn WO, Rahn H (eds) Respiration (Handbook of Physiology 3/I). American Physiological Society, Washington DC, pp 557–583
Cook DL (1984) Electrical pacemaker mechanisms of pancreatic islet cells. Fed Proc 43: 2368–2372
Cook RD, King AS (1969) A neurite-receptor complex in the avian lung: electron microscopical observation. Experientia 25: 1162–1164
Cooke HJ (1986) Neurobiology of the intestinal mucosa. Gastroenterology 90: 1057–1081
Cooke IM, Stuenkel EL (1985) Electrophysiology of invertebrate neurosecretory cells. In: Poisner AM, Trifaro JM (eds) The Electrophysiology of the Secretory Cell. Elsevier, Amsterdam, pp 115–164
Copp DH, Cameron EG, Cheney BA, Davidson AGE, Henze KG (1962) Evidence for calcitonin–a new hormone from the parathyroid that lowers serum calcium. Endocrinology 70: 638–649.
Costa JL (1968) Histological, histochemical, and cytological observations on the adrenal gland of the wild white-footed mouse, *Peromyscus leucopus*. Anat Rec 162: 275–288
Costa M, Furness JB (1973) The simultaneous demonstration of adrenergic fibres and enteric ganglion cells. Histochem J 5: 343–349
Costa M, Furness JB (1983) The origins, pathways and terminations of neurons with VIP-like immunoreactivity in the guinea-pig small intestine. Neuroscience 8: 665–676
Costa M, Furness JB, Llewellyn-Smith IJ, Davies B, Oliver J (1980) An immunohistochemical study of the projections of somatostatin-containing neurones in the guinea-pig intestine. Neuroscience 5: 841–852
Costa M, Furness JB, Llewellyn-Smith IJ, Cuello AC (1981) Projections of substance P-containing neurons within the guinea-pig small intestine. Neuroscience 6: 411–424
Costa M, Furness JB, Cuello C (1985) Separate populations of opioid containing neurons in the guinea-pig intestine. Neuropeptides 5: 445–448
Costoff A, McShan WH (1969) Isolation and biological properties of secretory granules from rat anterior pituitary glands. J Cell Biol 43: 564–574
Cottrell GA, Osborne NN (1970) Subcellular localization of serotonin in an identified serotonin-containing neurone. Nature (Lond) 225: 470–472
Coupland RE (1962) Nerve endings on chromaffin cells in the rat adrenal medulla. Nature 194: 310–312
Coupland RE (1965a) The natural history of the chromaffin cell. Longmans, London
Coupland RE (1965b) Electron microscopic observations on the structure of the rat adrenal medulla. I. Ultrastructure and organization of chromaffin cells in the normal adrenal medulla. J Anat 99: 231–254
Coupland RE (1965c) Electron microscopic observations on the structure of the rat adrenal medulla. II. Normal innervation. J Anat 99: 255–272
Coupland RE (1971) Observations on the form and size distribution of chromaffin granules and on the identity of adrenaline- and noradrenaline-storing chromaffin cells in vertebrates and man. Mem Soc Endocrinol 19: 611–635
Coupland RE (1972) The chromaffin system. In: Blaschko H, Muschol E (eds) Catecholamines. Springer, Berlin, pp 16–45
Coupland RE (1975) Blood supply of adrenal gland. In: Blaschko H, Smith AD, Sayer G (eds) Adrenal gland (Handbook of physiology 7/VI). American Physiological Society, Washington DC, pp 283–294
Coupland RE, Health ID (1961) Chromaffin cells, mast cells and melanin. II. The chromaffin cells of the liver capsule and gut in ungulates. J Endocrinol 22: 71–76
Coupland RE, Hopwood D (1966) The mechanisms of the differential staining reaction for adrenaline- and noradrenaline-storing granules in tissues fixed in glutaraldehyde. J Anat 100: 227–243
Coupland RE, Kobayashi S (1976) Recent studies on the fixation of adrenaline and noradrenaline, and on amine synthesis and storage in chromaffin cells. In: Coupland RE, Fujita T (eds) Chromaffin, enterochromaffin and related cells. Elsevier, Amsterdam, pp 59–81
Coupland RE, Selby JE (1976) The blood supply of the mammalian adrenal medulla: a comparative study. J Anat 122: 539–551
Coupland RE, Pyper AS, Hopwood D (1964) A method for differentiating between noradrenaline- and adenaline-storing cells in the light and electron microscope. Nature 201: 1240–1242
Coupland RE, Kobayashi S, Crowe J (1976a) On the fixation of catecholamine including adrenaline in tissue sections. J Anat 122: 403–413
Coupland RE, Kobayashi S, Kent C (1976b) Observations on the localization of recently synthesized catecholamines in chromaffin cells after the injection of L-2,5 6-^{3}H-dopa. J Endocrinol 69: 136–148
Cox BM (1982) Endogenous opioid peptides: a guide to structures and terminology. Life Sci 31: 1645–1658
Crepel F, Dupont JL, Gardette R (1983) Voltage clamp analysis of the effect of excitatory amino acids and derivatives on Purkinje cell dendrites in rat cerebellar slices maintained *in vitro*. Brain Res 279: 311–315
Crowe R, Burnstock G (1981) Comparative studies of quinacrine-positive neurones in the myenteric plexus of stomach and intestine of guinea-pig, rabbit and rat. Cell Tiss Res 221: 93–107

Crowe R, Whitear M (1978) Quinacrine fluorescence of Merkel cells in *Xenopus laevis*. Cell Tiss Res 190:273–283

Csaba G, Baráth P (1974) Localization of ^{3}H-5-HTP in cells of the adrenal medulla. Z Mikrosk-Anat Forsch 88:849–856

Cuello AC, Del Fiacco M, Paxinos G, Somogyi P, Priestly JV (1981) Neuropeptides in striato-nigral pathways. J Neural Transmission 51:83–96

Cummins JT, Bull R (1971) Spectrophotometric measurements of metabolic responses in isolated rat brain cortex. Biochim Biophys Acta 253:29–38

Currie MG, Geller DM, Cole BR, Boylan JG, YuSheng W, Holmberg SW, Needleman P (1983) Bioactive cardiac substances: potent vasorelaxant activity in mammalian atria. Science 221:71–73

Currie MG, Geller DM, Cole BR, Siegel NR, Fok KF, Adams SP, Eubanks SR, Galluppi GR, Needleman P (1984) Purification and sequence analysis of bioactive atrial peptides (atriopeptins). Science 223: 67–69

Curtarelli G (1978) Effect of dopamine infusion on gastric and pancreatic secretion and on gastric release in man. Gut 19:724–728

Cutz E, Conen PE (1972) Endocrine-like cells in human fetal lungs: an electron microscopic study. Anat Rec 173:115–122

Cutz E, Chan W, Wong V, Conen PE (1974) Endocrine cells in rat fetal lungs: ultrastructural and histochemical study. Lab Invest 30:458–464

Cutz E, Chan W, Wong V, Conen PE (1975) Ultrastructure and fluorescence histochemistry of endocrine (APUD-type) cells in tracheal mucosa of human and various animal species. Cell Tiss Res 158:425–437

Cutz E, Chan W, Track NS, Goth A, Said SI (1978) Release of vasoactive intestinal polypeptide in mast cells by histamine liberators. Nature 275:661–662

Cutz E, Chan W, Track NS (1981) Bombesin, calcitonin and leu-enkephalin immunoreactivity in endocrine cells of human lung. Experimentia 37:765–767

Cutz E, Gillan JE, Track NS (1984) Pulmonary endocrine cells in the developing human lung and during neonatal adaption. In: Becker KL, Gazdar AF (eds) The endocrine lung in health and disease. WB Saunders, Philadelphia, pp 210–231

Dahl JL, Weibel VJ (1979) Changes in tubulin heterogeneity during postnatal development of rat brain. Biochem Biophys Res Commun 86:822–828

Dahlström A, Fuxe K (1966) Monoamines and the pituitary gland. Acta Endocrinol 51:301–314

Dahlström A, Lundberg J, Bylock A, Hansson HA, Goldstein M (1977) Immunohistochemical studies on the subcellular distribution of dopamine β-hydroxylase in central and peripheral tissues. Proc Int Soc Neurochem 6:249

Daikoku S, Ikeuchi C, Miki S (1975) Electron-microscopic studies on developing intramural ganglia of the small intestine in human and rabbit fetuses. Acta anat 91:429–454

Daikoku S, Hisano S, Kawano H (1983a) Ontogenetic studies on the topotgraphical heterogeneity of somatostatin-congaining neurons in rat hypothalamus. Cell Tiss Res 233:347–354

Daikoku S, Chikamori M, Adachi T, Okamura Y, Nishiyama T, Tsuruo Y (1983b) Ontogenesis of hypothalamic immunoreactic ACTH cells in vivo and in vitro: Role of Rathke's pouch. Dev Biol 97:81–88

Daikoku S, Chikamrri M, Adachi T, Maki Y (1983c) Effect of basal diencephalon on the development of Rathke's pouch in rat: a study in combined organ cultures. Dev Biol 90:198–202

Daikoku S, Ikeuchi C, Miki S (1986) Electron microscopic studies on developing intramural ganglia of the small intestine in human and rabbit fetuses. Acta Anat 91:429–454

Dainaka J, Ichikawa A, Koibuchi Y, Nakagawa M, Tomita K (1986) Effect of the tridecamer of compound 48/80, a Ca^{2+}-dependent histamine releaser, on phospholipid metabolism during the early stage of histamine release from rat mast cells. Biochem Pharmacol 35:3739–3744

Dallos P (1984) Peripheral mechanisms of hearing. In: Dalian-Smith I (ed) Sensory process, part 2 (Handbook of physiology I/III). American Physiological Society, Bethesda Maryland, pp 595–637

Daly IB, Hebb C (1966) Pulmonary and bronchial vascular systems: their reactions under controlled conditions of ventilation and circulation. E Arnold, London

Daly MJ, Humphray JM, Stables R (1982) Effects of H_2-receptor antagonists and anticholinoceptor drugs on gastric and salivary secretion induced by bethanechol in the anaesthetized dog. Br J Pharmacol 76:361–365

D'Amico-Martel A, Noden DM (1983) Contributions of placodal and neural crest cells to avian cranial peripheral ganglia. Am J Anat 166:445–468

Dammann HG, Besterman HS, Bloom SR, Schreiber HW (1981) Gut-hormone profile in totally pancreatectomised patients. Gut 22:103–107

Daniels AJ, Williams RJP, Wright PE (1978) The character of the stored molecules in chromaffin granules of the adrenal medulla: a nuclear magnetic resonace study. Neuroscience 3:573–585

Daniels AJ, Johnson LN, Williams RJP (1979) Uptake of manganese by chromaffin granules *in vitro*. J Neurochem 33:923–929

Danisch F (1924) Zur Histogenese der sogenannten Appendixkarzinoide. Beitr Path Anat Allgem Path 72:687–709

Darian-Smith I (1984) The sense of touch: performance and peripheral neural processes. In: Darian-Smith I (ed) Sensory processes, part 2 (Handbook of physiologyl/III). American Physiological Society, Bethesda Maryland, pp 739–788

Dastoli FR, Price S (1966) Sweet-sensitive protein from bovine taste buds: isolation and assay. Science 154:905–907

Datyner NB, Gage PW (1982) Secretion of acetylcholine in response to graded depolarization of motor nerve terminals. J Physiol (Paris) 78:412–416

Davis MD, Hadley ME (1976) Spontaneous electrical potentials and pituitary hormone (MSH) secretion. Nature 261:422–423

Davis MD, Hadley ME (1978) Pars intermedia electrical potentials: changes in spike frequency induced by regulatory factors of melanocyte hormone (MSH) secretion. Neuroendocrinology 26:277–282
Dawes GS, Comroe JH (1954) Chemoreflexes from the heart and lungs. Physiol Rev 34:167–201
Dayer AM, Kapanci Y, Rademakers A, Marangos PJ, De Mey J, Will JA (1983) Neuron-specific enolase and serotonin distribution in the fetal Rhesus monkey lung by immunocytochemistry. Fed Proc 42:798
Dean PM, Matthews EK (1968) Electrical activity in pancreatic islet cells. Nature 219:389–390
Dean PM, Matthews EK (1970a) Glucose-induced electrical activity in pancreatic islet cells. J Physiol 210:255–264
Dean PM, Matthews EK (1970b) Electrical activity in pancreatic islet cells: effect of ions. J Physiol 210:265–275
Debas HT (1981) The colon as an endocrine organ. Digest Dis Sci 26:193–194
De Castro F (1929) Über die Struktur und Innervation des Glomus caroticum beim Menschen und bei den Säugetieren. Anatomisch-experimentelle Untersuchungen. Z Anat Entw-Gesch 89:250–265
De Castro F (1951) Sur la structure de la synapse dans les chemocepteurs: leur mécanisme d'excitation et rôle dans la circulation sanguine locale. Acta Physiol Scand 22:14–43
De Duve C, Wattiaux R (1966) Functions of lysosomes. Ann Rev Physiol 28:435–492
Defeudis FV (1984) GABA and endocrine regulation-relation to neurologic-psychiatric disorders. Neurochem Int 6:1–16
Deguchi T (1979) A circadian oscillator in cultured cells of chicken pineal gland. Nature 282:94–96
DeHan R, Graziadei PPC (1971) Functional anatomy of frog's taste organs. Experientia 27:823–826
DeHan RS, Graziadei P (1973) The innervation of frog's taste organ "a histochemical study". Life Sci 13:1435–1449
Dejours P (1962) Chemoreflexes in breathing. Physiol Rev 42:335–358
De Kock LL (1954) The intra-glomerular tissues of the carotid body. Acta Anat 21:101–116
DeLellis RA (ed) (1984) Advances in immunohistochemistry. Masson, New York Chicago, pp 141–158
De Lorenzo AJD (1970) The olfactory neuron and the blood-brain barrier. In: Wolstenholme GEW, Knight J (eds) Taste and Smell in Vertebrates. J & A Churchill, London, pp 151–176
Demling VL (1972) Gastrointestinal hormones. Georg Thieme Verlag, Stuttgart
De Potter WP, Smith AD, De Shaepdryver AF (1970) Subcellular fractionation of splenic nerve: ATP, chromogranin A and dopamine β-hydroxylase in noradrenergic vesicles. Tissue Cell 2:529–546
De Robertis E (1964) Histophysiology of synapses and neurosecretion. Pergamon Press, Oxford London Edinburgh New York Paris Frankfurt
De Robertis E (1967) Ultrastructure and cytochemistry of the synaptic region. Science 156:907–914
De Robertis E, Franchi CM (1958) Submicroscopic morphology and function of the synapse. Exp Cell Res 5 Suppl: 347–369
De Robertis E, Vaz Ferreira A (1957) Electron microscopic study of the excretion of catechol-containing droplets in the adrenal medulla. Exp Cell Res 12:568–574
Deschner EE, Lipkin M (1966) An autoradiographic study of the renewal of argentaffin cells in human rectal mucosa. Exp Cell Res 43:661–665
Dey RD, Hoffpauir JM (1986) Ultrastructural colocalization of the bioactive mediators 5-hydroxytryptamine and bombesin in endocrine cells of human fetal airways. Cell Tiss Res 246:119–124
Dias-Amado L (1925a) Sur l'existence de cellules argentaffines dans le tissu conjonctif des villosités intestinales. C R Soc Biol (Paris) 93:1548–1549
Dias-Amado L (1925b) Sur la signification des cellules de Nicolas. C R Soc Biol (Paris) 93:1550–1551
Dichter MA, Tischler AS, Greene LA (1977) Nerve growth factor-induced increase in electrical excitability and a cetylcholine sensitivity of a rat pheochromocytoma cell line. Nature 268:501–504
Diepen R (1962) Der Hypothalamus. In: Bargmann W (ed) Möllendorffs Handbuch der mikroskopischen Anatomie des Menshen. Bd IV/7. Springer-Verlag, Berlin Heidelberg
Dierickx K (1962) the dendrites of the preoptic neurosecretory nucleus of *Rana temporaria* and the osmoreceptor. Arch Int Pharmacodyn 140:708–725
Dijkgraaf S, Kalmijn AJ (1966) Versuche zur biologischen Bedeutung der Lorenzinischen Ampullen bei den Elasmobranchiern. Z Vergl Physiol 53:187–194
Diner O (1967) L'expulsion des granules de la médullosurrénalle chez le Hamster. C R Acad Sci D 265:616–619
Dipolo R, Rojas H, Vergara J, Lopez R, Caputo C (1983) Measurements of intracellular ionized calcium in squid giant axons using calcium-selective electrodes. Biochim Biophys Acta 728:311–318
Dixon JS, Gosling JA, Ramsdale DR (1973) Urethral chromaffin cells: a light and electron microscopic study. Z Zellforsch 138:397–406
Docherty K, Steiner F (1982) Post-translational proteolysis in polypeptide hormone biosynthesis. Annu Rev Physiol 44:624–638
Dodt E, Nauheim B (1973) The parietal eye (pineal and parapineal organs) of lower vertebrates. In: Jung R (ed) (Handbook of sensory physiology VII/3B). Springer, Berlin, pp 113–140
Doerr-Schott J, Jean-Claude G (1981) Ultrastructural identification of gastrin-like immunoreactive nerve fibres in the brain of *Xenopus laevis* by means of colloidal gold or ferritin immunocytochemical methods. Cell Tiss Res 216:581–589
Dogiel AS (1894) Die Nervenendigungen in den Nebennieren der Säugethiere. Arch Anat Entw-Gesch 90–101
Dogiel AS (1896) Zwei Arten sympatischer Nervenzellen. Anat Anz 11:679–687
Dogiel AS (1899) Über den Bau der Ganglien in den Geflechten des Darmes und der Gallenblase des Menschen und der Säugetiere. Arch Anat Physiol (Leipzig) Anat Abt 130–158

Domschke S, Domschke W (1980) New histamine H_2-receptor antagonists. Hepato-Gastroenterology 27:163–168
Domschke VW (1976) Motilin—Wirkungsspektrum und Wirkungsanalyse. Z Gastroenterol 14:112–115
Domschke W (1977) Motilin: Spectrum and mode of gastrointestinal actions. Am J Digest Dis 22:454–461
Domschke W, Classen M, Demling L (1972a) Circadian rhythmicity of gastric secretion and cyclic 3′, 5′-adenosine monophosphate contents of gastric mucosa in rats. Scand J Gastroenterol 7:39–41
Domschke W, Domschke S, Classen M, Demling L (1972b) Cyclic 3′, 5′-AMP and gastric secretion in the rat: effect of adrenalectomy and glucocorticoids. Horm Metab Res 4:226–227
Donev SR (1984) Ultrastructural evidence for the presence of a glial sheath investing the islets of Langerhans in the pancreas of mammals. Cell Tiss Res 237:343–348
Dóra E, Gyulai L, Kovách AGB (1984) Determinants of brain activation-induced cortical NAD/NADH responses in vivo. Brain Res 229:61–72
Douglas WW (1968) Stimulus-secretion coupling: the concept and clues from chromaffin and other cells. Br J Pharmacol 34:451–474
Douglas WW (1974a) Exocytosis and the exocytosis-vesiculation sequence: with special reference to neurohypophysis, chromaffin and mast cells, calcius and calcium ionophores. In: Thorn NA, Peterson OH (eds) Secretory Mechanisms of Exocrine Glands. Munksgaard, Copenhagen, pp 116–133
Douglas WW (1974b) Mechanism of release of neurohypophysial hormones: stimulus-secretion coupling. In: Knobil E, Sawyer WH (eds) The pituitary gland and its neuroendocrine control, part 1. (Handbook of physiology 7/IV). American Physiological Society, Washington DC, pp 191–224
Douglas WW (1975a) Secretomotor control of adrenal medullary secretion: synaptic, membrane, and ionic events in stimulus-secretion coupling. In: Blaschko H, Sayers G, Smith AD (eds) Adrenal gland (Handbook of physiology 7/VI). American Physiological Society, Washington DC, pp 367–388
Douglas WW (1975b) Stimulus-secretion coupling in mast cells: regulation of exocytosis by cellular and extracellular calcium. In: Carafoll E et al. (eds) Calcium transport in contraction and secretion. North-Holland, Amsterdam, pp 167–174
Douglas WW, Kanno T (1967) The effect of amethocaine on acetylcholine-induced depolarization and catecholamine secretion in the adrenal chromaffin cell. Br J Pharmac Chemother 30:612–619
Douglas WW, Rubin RP (1961) The role of calcium in the secretory response of the adrenal medulla to acetylcholine. J Physiol 159:40–57
Douglas WW, Taraskevich PS (1978) Action potentials in gland cells of rat pituitary pars intermedia: inhibition by dopamine, an inhibitor of MSH secretion. J Physiol 285:171–184
Douglas WW, Taraskevich PS (1980) Calcium component to action potentials in rat pars intermedia cells. J Physiol 309:623–630
Douglas WW, Taraskevich PS (1982) Slowing effects of dopamine and calcium-channel blockers on frequency of sodium spikes in rat pars intermedia cells. J Physiol 326:201–211
Douglas WW, Taraskevich PS (1985) The electrophysiology of adenohypophyseal cells. In: Poisner AM, Trifaró JM (eds) The electrophysiology of the secretory cell. Elsevier, Amsterdam New York Oxford, pp 63–92
Douglas WW, Ueda Y (1973) Mast cell secretion (histamine release) induced by 48/80: calcium-dependent exocytosis inhibited strongly by cytochalasin only when glycolysis is rate-limiting. J Physiol 234: 97–98
Douglas WW, Kanno T, Sampson SR (1967a) Effects of acetylcholine and other medullary secretagogues and antagonists on the membrane potential of adrenal chromaffin cells: an analysis employing techniques of tissue culture. J Physiol 188:107–120
Douglas WW, Kanno T, Sampson SR (1967b) Influence of the ionic environment on the membrane potential of adrenal chromaffin cells and on the depolarizing effect of acetylcholine. J Physiol 191: 107–121
Douglas WW, Nagasawa J, Schulz R (1971) Electron microscopic studies on the mechanism of secretion of posterior pituitary hormones and significance of microvesicles ('synaptic vesicles'): evidence of secretion by exocytosis and formation of microvesicles as a by-product of this process. Mem Soc Endocrinol 19:353–378
Dowdall MJ, Boyne AF, Whittaker VP (1974) Adenosine triphosphate: a constituent of cholinergic synaptic vesicles. Biochem J 140:1–12
Dowling JE (1970) Organization of vertebrate retinas. Invest Ophthalmol 9:655–680
Dowling JE, Dubin MW (1984) The vertebrate retina. In: Darian-Smith I (ed) Sensory processes, part 1. (Handbook of physiology 1/III) American Physiological Society, Bethesda Maryland, pp 317–340
Dowling JE, Ripps H (1972) Adaptation in skate photoreceptors. J Gen Physiol 60:698–719
Dowling JE, Ripps H (1973) Neurotransmission in the distal retina: the effect of magnesium on horizontal cell activity. Nature 242:101–103
Dreiling DA, Greenstein A (1972) Pancreatic function in patients with Zollinger-Ellison syndrome. Am J Gastroenterol 58:68–70
Dreiling DA, Messer J (1978) The secretin story. Am J Gastroenterol 70:455–479
Dreyer MV (1975) Cytological changes in the chemoreceptor cells of the avian (*Gallus domesticus*) carotid body after exposure to a high percentage of carbon dioxide gas mixture. University of Pretoria, Pretoria.
Drummond GI, Bellward G (1970) Studies on phosphorylase b kinase from neural tissues. J Neurochem 17:475–482
Dubois PM, Bégeot M, Dubois MP, Herbert DC (1978) Immunocytological localization of LH, FSH and TSH and their subunits in the pituitary of normal and anencephalic human foetuses. Cell Tiss Res 9: 249–265
Dubrasquet M, Bataille D, Gespach C (1982) Oxynto-

modulin (glucagon-37 or bioactive enteroglucagon): A potent inhibitor of pentagastrin-stimulated acid secretion in rats. Bioscience Reports 2: 391–395

Ducret F (1978) Particularités structurales du système optique chez deux Chaetognathes (*Sagitta tasmanica* et *Eukrohnia hamata*) et incidences phylogenetiques. Zoomorphologie 91:201–215

Dudel J (1983) Graded or all-or-nothing release of transmitter quanta by local depolarizations of nerve terminals on crayfish muscle? Pflügers Arch 398: 155–164

Dudel J, Parnas I, Parnas H (1983) Neurotransmitter release and its facilitation in crayfish muscle. VI. Release determined by both, intracellular calcium concentration and depolarization of the nerve terminal. Pflügers Arch 399:1–10

Dufy B, Vincent JD, Fleury H, Pasquier PD, Gourdji D, Tixier-Vidal A (1979) Membrane effects of thyrotropin-releasing hormone and estrogen shown by intracellular recording from pituitary cells. Science 204:509–511

Duke HN (1954) The site of action of anoxia on the pulmonary blood vessels of the cat. J Physiol 125: 373–382

Dunn LA, Holz RW (1983) Catecholamine secretion from digitonin-treated adrenal medullary chromaffin cells. J Biol Chem 258:4989–4993

Dunn RF (1973) The ultrastructure of the vertebrate retina. In: Friedmann I (ed) The ultrastructure of sensory organs. North-Holland/American Elsevier, Amsterdam London New York, pp 153–265

Düring MV, Andres KH (1976) The ultrastructure of taste and touch receptors of the frog's taste organ. Cell Tiss Res 165:185–198

Düring MV, Karduck A, Richter HG (1974) The fine structure of the inner ear in caiman crocodilus. Z Anat Entw 145:41–65

Du Shane GP (1935) An experimental study of the origin of pigment cells in Amphibia. J Exp Zool 72:1–31

Du Shane GP (1943) The embryology of vertebrate pigment cells. I. Amphibia. Quart Rev Biol 18:109–127

Dustin P (1978) Microtubules. Springer-Verlag, Berlin Heidelberg New York

Dutt B, Magee DF (1972) Pepsin secretion by Heidenhain pouches in dogs. Am J Physiol 223:480–483

Duve H, Thorpe A (1982) The distribution of pancreatic polypeptide in the nervous system and gut of the blowfly, *Calliphora vomitoria* (Diptera). Cell Tiss Res 227:67–77

Duve H, Thorpe A (1984) Immunocytochemical mapping of gastrin/CCK-like peptides in the neuroendocrine system of the blowfly *Calliphora vomitoria* (Diptera). Cell Tiss Res 237:309–320

Dyball REJ, Koizumi K (1969) Electrical activity in the supraoptic and paraventricular nuclei associated with neurohypophysial hormone release. J Physiol 201:711–722

Dyer RF (1978) Endoneurial mast cells in peripheral nerves of the armadillo dermis. Cell Tiss Res 192: 97–105

Eakin RM (1972) Structure of invertebrate photoreceptors. In: Dartnall HJA (ed) Photochemistry of vision (Handbook of sensory physiology VII/1). Springer, Berlin Heidelberg New York, pp 625–684

Eakin RM (1979) Evolutionary significance of photoreceptors: in retrospect. Am Zool 19:647–653

Eakin RM (1982a) Continuity and diversity in photoreceptors. In: Westfall JA (ed) Visual cells in evolution. Raven Press, New York, pp 91–105

Eakin RM (1982b) Morphology of invertebrate photoreceptors. Meth Enzymol 81:17–25

Eakin RM, Bush FE (1957) Development of the amphibian pituitary with special reference to the neural lobe. Anat Rec 129:279–295

Ebels I, Balemans MGM (1986) Physiological aspects of pineal functions in mammals. Physiol Rev 66: 581–605

Eberts TJ, Ransburg RC (1979) Primary intracranial endodermal sinus tumor. J Neurosurg 50:246–252

Eccles JC (1986) Chemical transmission and Dale's principle. Proc Brain Res 68:3–13

Eddlestone GT, Goncalves A, Bangham JA, Rojas E (1984) Electrical coupling between cells in islets of Langerhans from mouse. J Membrane Biol 77: 1–14

Edinger L (1906) Einiges vom "Gehirn" des Amphioxus. Anat Anz 28:417–428

Edvinsson L, Emson P, McCulloch J, Tatemoto K, Uddman R (1983) Neuropeptide Y: cerebrovascular innervation and vasomotor effects in the cat. Neurosci Lett 43:79–84

Edwards DL, Grafstein B (1983) Intraocular tetrodotoxin in goldfish hinders optic nerve regeneration. Brain Res 269:1–14

Ehinger B (1982) Neurotransmitter systems in the retina. Retina 2:305–321

Ehrhart M, Grube D, Bader MF, Aunis D, Gratzl M (1986) Chromogranin A in the pancreatic islet: cellular and subcellular distribution. J Histochem Cytochem 34:1673–1682

Ehrlich P (1877) Beiträge zur Kenntnis der Anilinfärbungen und ihrer Verwendung in der mikroskopischen Technik. Arch Mikrosk Anat 13:263–277

Ekengren B (1975) Aminergic nuclei in the hypothalamus of the roach *Leuciscus rutilus*. Cell Tiss Res 159:493–502

Ekholm R, Ericson LE (1968) The ultrastructure of the parafollicular cells of the thyroid gland in the rat. J Ultrastruct Res 23:378–402

Elfvin LG (1967) The development of the secretory granules in the rat adrenal medulla. J Ultrastruct Res 17:45–62

Elfvin LG (1968) A new granule-containing nerve cell in the inferior mesenteric ganglion of the rabbit. J Ultrastruct Res 22:37–44

Elfvin LG, Appelgren LE, Ullberg S (1966) High resolution autoradiography of the adrenal medulla after injection of tritiated dihydroxyphenylalanine (dopa). J Ultrastruct Res 14:277–293

Elliott WM, Youson JH (1987) Immunohistochemical observations of the endocrine pancreas during metamorphosis of the sea lamprey, *Petromyzon marinus* L. Cell Tiss Res 247:351–357

El-Salhy (1981) On the phylogeny of the gastro-entero-pancreatic (GEP) neuroendocrine system.

Acta Universitatis Upsaliensis, Uppsala, pp 1–39
El-Salhy M, Abou-El-Ela R, Falkmer S, Grimelius L, Wilander E (1980) Immunohistochemical evidence of gastro-entero-pancreatic neurohormonal peptides of vertebrate type in the nervous system of the larva of a dipteran insect, the hoverfly, *Eristalis aeneus*. Regul Pept 1:187–204
El-Salhy M, Falkmer S, Kramer KJ, Speirs RD (1983) Immunohistochemical investigations of neuropeptides in the brain, corpora cardiaca, and corpora allata of an adult lepidopteran insect, *Manduca sexta* (L). Cell Tiss Res 232:295–317
Enami M (1956) Studies in neurosecretion. VIII Changes in the caudal neurosecretory system of the loach (*Misgurnus anguillicaudatus*) in response to osmotic stimuli. Proc Jpn Acad 32:759–764
Enami M, Imai K (1958) Studies in neurosecretion, XII. Electron microscopy of the secrete granules in the caudal neurosecretory system of the eel. Proc Jpn Acad 34:164–168
Endo Y (1984) Ontogeny of endocrine cells in the gut of the insect *Periplaneta americana*. Cell Tiss Res 238:421–423
Endo Y, Kobayashi S (1987) A scanning electron microscopic study on the autonomic groundplexus in the lamina propria mucosae of the guinea-pig small intestine. Arch Histol Jpn 50:243–250
Endo Y, Nishiitsutsuji-Uwo J (1981) Gut endocrine cells in insects: The ultrastructure of the gut endocrine cells of the lepidopterous species. Biomed Res 2:270–280
Endo Y, Nishiitsutsuji-Uwo J (1982) Exocytotic release of secretory granules from endocrine cells in the midgut of insects. Cell Tiss Res 222:515–522
Endo K, Koizumi M, Kanazawa Y, Yoshida S, Miura M, Kanno T, Koizumi F, Goto Y (1980) Plasma secretin level and exocrine pancreatic secretion in response to intraduodenal infusion of 1-Phenyl-1-hydroxy-n-pentane and HCl by using a new four lumen tube. Tohoku J Exp Med 131:363–371
Endo Y, Nishiitsutsuji-Uwo J, Iwanaga T, Fujita T (1982a) Ultrastructural and immunohistochemical identification of pancreatic polypeptide-immunoreactive endocrine cells in the cockroach midgut. Biomed Res 3:454–456
Endo Y, Iwanaga T, Fujita T, Nishiitsutsuji-Uwo J (1982b) Localization of pancreatic polypeptide (PP)-like immunoreactivity in the central and visceral nervous systems of the cockroach *Periplaneta*. Cell Tiss Res 227:1–9
Endo Y, Uchida T, Kobayashi S (1986) Somatostatin neurons in the small intestine of the guinea pig: a light and electron microscopic immunocytochemical study combined with nerve lesion experiments by laser irradiation. J Neurocytol 15:725–731
English KB (1974) Cell types in cutaneous type I mechanoreceptors (Haarscheiben) and their alterations with injury. Am J Anat 141:105–126
English KB (1977a) The ultrastructure of cutaneous type I mechanoreceptors (Haarscheiben) in cats following denervation. J Comp Neurol 172:137–163
English KB (1977b) Morphogenesis of Haarscheiben in rats. J Invest Dermatol 69:58–67
Epple A (1964) Zur vergleichenden Zytologie des Inselorgans. Verh Deut Zool Ges 1963 Ergänz Zool Anz 27:461–470
Epple A (1982) Functional principles of vertebrate endocrine systems. Verh Deutsch Zool Ges 1982: 117–126
Epple A, Brinn JE (1986) Pancreatic islets. In: Pang PKT, Schreibman MP (eds) Vertebrate endocrinology: fundamentals and biomedical implications, vol 1. Academic Press, New York, pp 279–317
Eränkö L, Eränkö O (1972) Effect of hydrocortisone on histochemically demonstrable catecholamines in the sympathetic ganglia and extra adrenal chromaffin tissue of the rat. Acta Physiol Scand 84:125–133
Eränkö O (1952) On the histochemistry of the adrenal medulla of the rat, with special reference to acid phosphatase. Acta Anat 16 Suppl 17:1–60
Eränkö O (1955) Distribution of adrenalin and noradrenalin in the adrenal medulla. Nature 175:88
Eränkö O (1976) SIF cells, chromaffin cells, granule-containing cells, and interneurons. In: Eränkö O (ed) SIF cells. Structure and function of the small intensely fluorescent sympathetic cells. DHEW Publ., Washington DC, pp 1–7
Eränkö O, Härkönen M (1963) Histochemical demonstration of fluorogenic amines in the cytoplasm of sympathetic ganglion cells of the rat. Acta Physiol Scand 58:285–286
Eränkö O, Härkönen M (1965) Monoamine-containing small cells in the superior cervical ganglion of the rat and an organ composed of them. Acta Physiol Scand 63:511–512
Ericson LE (1968) Degranulation of the parafollicular cells of the rat thyroid by vitamine D_2-induced hypercalcemia. J Ultrastruct Res 24:145–149
Ericson LE (1970) Subcellular localization of 5-hydroxytryptamine in the parafollicular cells of the mouse thyroid gland: an autoradiographic study. J Ultrastruct Res 31:162–177
Ericson LE (1972) Formation and storage of 5-hydroxytriptamine in thyroid parafollicular cells. J Ultrastr Res 41:467–483
Ericson LE, Sundler F (1984) Thyroid parafollicular cells ultrastructural and functional correlations. In: Motta PM (ed) Ultrastructure of endocrine cells and tissues. Martinus Nijhoff publishers, Boston The Hague Dordrecht Lancaster, pp 276–285
Ericson LE, Håkanson R, Larson B, Owman CH, Sundler F (1972) Fluorescence and electron microscopy of amine-storing enterochromaffin-like cells in tracheal epithelium of mouse. Z Zellforsch 124:532–545
Erlandsen SL, Hegre OD, Parsons JA, McEvoy RC, Elde RP (1976) Pancreatic islet hormones: distribution of cell types in the islet and evidence for the presence of somatostatin and gastrin within the D-cell. J Histochem Cytochem 24:883–897
Erspamer V (1983) Amphibian skin peptides in mammals—looking ahead. Trends Neurosci 6:200–201
Erspamer V, Asero B (1952) Identification of enteramine, the specific hormone of the enterochromaffin cell system as 5-hydroxy-tryptamine. Nature 169: 800–801
Erspamer V, Melchiorri P, Erspamer CF, Negri L (1977) Polypeptides of the amphibian skin active on

the gut and their mammalian counterparts. Adv Exp Med Biol 106:51–64

Esterhuizen AC, Spriggs TLB, Lever JD (1968) Nature of islet-cell innervation in the cat pancreas. Diabetes 17:33–36

Euler USv, Hamberg U (1949) L-Noradrenaline in the suprarenal medulla. Nature 163:642–643

Euler USv, Lishajko F, Stjärne L (1963) Catecholamines and adenosine triphosphate in isolated adrenergic nerve granules. Acta Physiol Scand 59:495–496

Evans CJ, Erdelyi E, Weber E, Barchas JD (1983) Identification of pro-opiomelanocortin-derived peptides in the human adrenal medulla. Science 221: 957–960

Everts TJ, Ransburg RC (1979) Primary intracranial sinus tumor. J Neurosurg 50:246–252

Ewen SWB, Bussolati G, Pearse AGE (1972) Uptake of L-dopa and L-5 hydroxytryptophan by endocrine-like cells in the rat larynx. Histochem J 4: 103–110

Eyzaguirre C, Koyano H, Taylor JR (1965) Presence of acetylcholine and transmitter release from carotid body chemoreceptors. J Physiol 178:463–476

Eyzaguirre C, Fitzgerald RS, Lahiri S, Zapata P (1983) Arterial chemoreceptors. In: Shepherd JT, Abboud FM (eds) Peripheral circulation and organ blood flow, part 2 (Handbook of physiology 2/III). American Physiological Society, Bethesda, pp 557–621

Fabri PJ, Gower WR, Weber D, Tuttle S (1986) Hyperparathyoid glands contain G-17 and G-34 gastrin. J Surg Res 41:333–337

Fahrenholz C (1936) Tastzellen und Tastorgane in der Neunaugenhaut. Z Mikrosk-Anat Forsch 39:116–134

Fahrenkrug J, Galbo H, Holst JJ, Schaffalitzky de Muckadell OB (1978) Influences of the autonomic nervous system on the release of vasoactive intestinal polypeptide from the porcine gastrointestinal tract. J Physiol 280:405–422

Falck B, Owman C (1968) 5-hydroxytryptamine and related amines in endocrine cell systems. In Adv Pharmacol 6A:211–231

Falck B, Hillarp NA, Thieme G, Torp A (1962) Fluorescence of catecholamines and related compounds condensed with formaldehyde. J Histochem Cytochem 10:348–354

Falck B, Jacobson S, Olivecrona H (1966) Fluorescent dopa reaction of nevi and melanomas. Arch Dermatol 94:363

Falck B, Müntzing J and Rosengren A-M (1969) Adrenergic nerves to the dermal melanophores of the rainbow trout, *Salmo gairdneri*. Z Zellforsch 99: 430–434

Falck BB, Larson CV, Merkelenburg E, Rosengren, K Svenaeus (1964) On the presence of a second specific cell system in mammalian thyroid gland. Acta Physiol Scand 62:491–492

Falkmer S (1980) Some recent aspects on the phylogeny of the GEP paraneurons. Biomed Res Suppl 1:17–23

Falkmer S, Matty AJ (1966) Blood sugar regulation in the haglish, *Myxine glutinosa*. Gen Comp Endocrinol 6:334–346

Falkmer S, Pihl E (1968) Structural lability of zinc-containing secretion granules of pancreatic beta-cells after exposure to hydrogen sulphide. Diabetologia 4:239–243

Falkmer S, Östberg, Van Noorden S (1978) Entero-insular endocrine systems of cycostomes: A clue to hormone evolution. In: Bloom SR (ed) Gut hormones. Churchil Livingstone, Edinburgh London New York, pp 57–63

Falkmer S, Carraway RE, El-Salhy M, Emdin SO, Grimelius L, Rehfeld JF, Reinecke M, Schwartz TW (1981) Phylogeny of the gastro-entero-pancreatic neuroendocrine system: a review. In: Grossman MI, Brazier MAB, Lechago J (eds) Cellular basis of chemical messengers in the digestive system. Academic Press, New York, pp 21–42

Falkmer S, Gustafsson MKS, Sundler F (1985) Phylogenetic aspects on the neuroendocrine system: a minireview with particular reference to cells storing neurohormonal peptides in some primitive protostomian invertebrates (flatworms, annelids). Nord Psykiatr Tidsskr 39 Suppl 11:21–30

Farber IC, Grinvald A (1983) Identification of presynaptic neurons by laser photostimulation. Science 222:1025–1027

Farbman AI (1965a) Fine structure of the taste bud. J Ultrastruct Res 12:328–350

Farbman AI (1965b) Electron microscope study of the developing taste bud in fungiform papilla. Devel Biol 11:110–135

Farbman AI (1971) Development of the taste bud. In: Beidler LM (ed) Taste (Handbook of sensory physiology IV/2). Springer, Berlin, pp 51–62

Farbman AI, Hellekant G, Nelson A (1985) Structure of taste buds in foliate papillae of the rhesus monkey, *Macaca mulatta*. Am J Anat 172:41–56

Farquhar MG (1978) Recovery of surface membrane in anterior pituitary cells. Variations in traffic detected with anionic and cationic ferritin. J Cell Biol 77:R35–R42

Farquhar MG (1982) Membrane recycling in secretory cells: pathway to the Golgi complex. In: Evered D, Collins GM (eds) Membrane recycling. Pitman, London, pp 157–183

Farquhar MG (1983) Complexity of intracellular biosynthetic and recycling membrane traffic. In: Crisolía S, Guerri C, Samson F, Norton S, Reinoso-Suárez F (eds) Ramón y Cajal's contribution to the neurosciences. Elsevier, Amsterdam, pp 67–89

Farquhar MG, Rinehart JF (1954) Cytologic alterations in the anterior pituitary gland following thyroidectomy: An electron microscope study. Endocrinology 55:857–875

Fasolo A, Gaudino G (1981) Somatostatin immunoreactive neurons and fibers in the hypothalamus of the newt. Gen Comp Endocrinol 43:256–263

Fenwick EM, Marty A, Neher E (1982a) A patch-clamp study of bovine chromaffin cells and of their sensitivity to acetylcholine. J Physiol 331:577–597

Fenwick EM, Marty A, Neher E (1982b) Sodium and calcium channels in bovine chromaffin cells. J Physiol 331:599–635

Ferreira MN, Leblond CP (1971) Argentaffin and

other "endocrine" cells of the small intestine in the adult mouse. II. Renewal. Am J Anat 131:331–351
Fesenko EF, Kolesnikov SS, Lyubarsky AL (1985) Induction by cyclic GMP of cationic conductance in plasma membrane of retinal rod outer segment. Nature 313:310–313
Fetissof F, Dubois MP, Arbeille-Brassart B, Lanson Y, Boivin F, Jobard P (1983) Endocrine cells in the prostate gland, urothelium and Brenner tumors. Virchows Arch (Cell Pathol) 42:53–64
Fetissof F, Berger G, Dubois MP, Arbeille-Brassart B, Lansac J, Sam-Giao M, Jobard P (1985) Endocrine cells in the female genital tract. Histopathology 9:133–145
Fewtrell CMS, Foreman JC, Jordan CC, Oehme P, Renner H, Stewart JM (1982) The effects of substance P on histamine and 5-hydroxytryptamine release in the rat. J Physiol 330:393–411
Feyrter F (1938) Über diffuse endokrine epitheliable Organe. Zbl Inn Med 545:561
Feyrter F (1946) Über die These von den peripheren endokrinen Drüsen. Wien Z Inn Med 1:10–38
Feyrter F (1951) Über das urogenitale Helle-Zellen-System des Menschen. Z Mikrosk-Anat Forsch 57: 324–344
Feyrter F (1952) Zum Begriff der Helle-Zellen-Systeme. Frankf Z Pathol 63:259–266
Feyrter F (1953) Über die peripheren endokrinen (parakrinen) Drüsen des Menschen. 2nd ed. Wilhelm Maudrich, Wien, Düsseldorf
Feyrter F (1962) Über den derzeitigen Stand der Lehre von den peripheren endokrinen (parakrinen) Drüsen. Acta Neuroveg 25:63–89
Fidone SJ, Gonzalez C (1986) Initiation and control of chemoreceptor activity in the carotid body. In: Cherniack NS, Widdicombe JG (eds) Control of breathing, part 1 (Handbook of physiology 3/II). American Physiological Society, Bethesda Maryland, pp 247–297
Fidone S, Gonzalez C, Yoshizaki K (1982) Effects of low oxygen on the release of dopamine from the rabbit carotid body *in vitro*. J Physiol (Lond) 333: 93–110
Fillion G, Nosál R, Uvnäs B (1971) The presence of a sulphomucopolysaccharide-protein complex in adrenal medullary cell granules. Acta Physiol Scand 83:286–288
Fischer JA, Oldham SB, Sizemore GW, Arnaud CD (1972) Calcium-regulated parathytroid hormone peptidase. Proc Natl Acad Sci USA 69:2341–2345
Fisher AWF, Wong K, Gill V, Lederis K (1984) Immunocytochemical localization of urotensin I neurons in the caudal neurosecretory system of the white sucker (*Catostomus commersoni*). Cell Tiss Res 235:19–23
Fischer-Colbrie R, Lassmann H, Winkler H (1985) Immunological studies on the distribution of chromogranin A and B in endocrine and nervous tissues. Neuroscience 16:547–555
Fitzpatrick TB, Quevedo WC, Levene AL, McGovern VJ, Mishima Y, Oettle AG (1966) Terminology of vertebrate melanin-containing cells: 1965. Science 152:88–89
Flock Å (1971) Sensory transduction in hair cells. In: Loewenstein WR (ed) Principles of receptor physiology (Handbook of sensory physiology I/1). Springer, Berlin Heidelberg New York, pp 396–441
Foesel S, Sewing KFr (1978) Enhancement of electrically stimulated guinea-pig gallbladder contration by subthreshold concentrations of gastrointestinal hormones in vitro. Experientia 34:205–206
Fontaine J (1979) Multistep migration of calcitonin cell precursors during ontogeny of the mouse pharynx. Gen Comp Endocrinol 37:81–92
Foreman JC (1980) Receptor-secretion coupling in mast cells. Trends Pharmacol Sci 2:460–462
Foreman JC, Garland LG (1974) Desensitization in the process of histamine secretion induced by antigen and dextran. J Physiol 239:381–391
Foreman JC, Mongar JL, Gomperts BD (1973) Calcium ionophores and movement of calcium ions following the physiological stimulus to a secretory process. Nature 245:249–251
Foreman JC, Garland LG, Mongar JL (1976) The role of calcium in secretory processes: model studies in mast cells. In: Duncan CJ (ed) Calcium in biological systems (30th Symp Soc for Experimental Biology). Cambridge University Press, Cambridge, pp 193–218
Foreman JC, Hallett MB, Monger JO (1977) The relationship between histamine secretion and ^{45}calcium-uptake by mast cells. J Physiol 271: 193–214
Forsberg JE, Rosengren E, Sjöberg NO (1964) On the occurrence of 5-hydroxytryptamine containing cells in the vaginal and vestibular epithelium of the rabbit. Z Zellforsch 63:302–308
Forsling ML, Martin MJ, Burton AM (1971) The effect of hydration on vasopressin and neurophysin release in the rat. J Endocrinol 51:413–414
Forssmann WG (1970) Ultrastructure of hormone-producing cells of the upper gastrointestinal tract. In: Creutzfeldt C (ed) Origin, chemistry, physiology and pathophysiology of the gastrointestinal hormones. FK Schattauer, Stuttgart New York, pp 31–70
Forssmann WG, Mutt V (1985) Cardiodilatin-immunoreactive neurons in the hypothalamus of Tupaia. Anat Embryol 172:1–5
Forssmann WG, Orci L (1969) Ultrastructure and secretory cycle of the gastrin-producing cell. Z Zellforsch 101:419–432
Forssmann WG, Orci L, Pictet R, Renold AE, Rouiller C (1969) The endocrine cells in the epithelium of the gastrointestinal mucosa of the rat. An electron microscope study. J Cell Biol 40:692–715
Forssmann WG, Wacker P, Daldrup J (1971) Ultrastrukturelle autoradiographie zum stoffwechsel der biogenen Amine in Magen-Darm-Trakt und in anderen Organen. Verh Anat Ges Anat Anz 128:19–29
Forssmann WG, Yanaihara N, Helmstaedter V, Grube D (1976) Differential demonstration of the motilin-cell and the enterochromaffin cell. Scand J Gastroenterol 11 Suppl 39:43–45
Forssmann WG, Hock D, Lottspeich F, Henschen A, Kreye V, Christmann M, Reinecke M, Metz J, Carlquist M, Mutt V (1983) The right auricle of the heart is an endocrine organ: cardiodilatin as a peptide

hormone candidate. Anat Embryol 168:307–313
Forssmann WG, Hock D, Kirchheim F, Metz J, Mutt V, Reinecke M (1984) Cardiac hormones: Morphological and functional aspects. Clin Exp Hypertension A6:1873–1878
Forte JG, Forte TM, Machen TE (1975) histamine-stimulated hydrogen ion secretion by *in vitro* piglet gastric mucosa. J Physiol 244:15–31
Foster CL (1971) Relationship between ultrastructure and function in the adenohypophysis of the rabbit. In: Heller H, Lederis K (eds) Subcellular organization and function in endocrine tissues. Cambridge Univ Press, Cambridge, pp 125–146
Fox CA, Andrade AN, Schwyn RC, Rafols JA (1971/1972) The aspiny neurons and the glia in the primate striatum: a Golgi and electron microscopic study. J Hirnforsch 13:341–362
Fox H, Whitear M (1978) Observations on Merkel cells in amphibians. Biol Cell 32:223–232
Fox H, Kazzaz B, Langley FA (1964) Argyrophil and argentaffin cells in the female genital tract and in ovarian mucinous cysts. J Path Bact 88:479–488
Fox JET, Daniel EE, Jury J, Track NS, Chiu S (1983) Cholinergic control mechanisms for immunoreactive motilin release and motility in the canine duodenum. Can J Physiol Pharmacol 61:1042–1049
Frazier WT, Kandel ER, Kuphermann I, Waziri R, Coggeshall RE (1967) Morphological and functional properties of identified neurons in the abdominal ganglion of *Aplysia Californica*. J Neurophysiol 30:1288–1351
Frexinos J, Vignoli F, Ribet A (1972) Les cellules endocrines de la muqueuse gastrique chez l'homme. Etude ultrastructurale. Biol Gastro-enterol (Paris) 5:21–36
Fried G, Terenius L, Brodin E, Efendic S, Dockray G, Fahrenkrug J, Goldstein M, Hökfelt T (1986) Neuropeptide Y, enkephalin and noradrenaline coexist in sympathetic neurons innervating the bovine spleen. Biochemical and immunohistochemical evidence. Cell Tiss Res 243:495–508
Friend BJ (1976) Morphology and location of dense-core vesicles in the stomatogastric ganlion of the lobster, *Panulirus interruptus*. Cell Tiss Res 175:369–390
Fritsch HAR, Van Noorden S, Pearse AGE (1980) Calcitonin-like immunochemical staining in the alimentary tract of *Ciona intestinalis* L. Cell Tiss Res 205:439–444
Fritsch HAR, Van Noorden S, Pearse AGE (1982) Gastro-intestinal and neurohormonal peptides in the alimentary tract and cerebral complex of *Ciona intestinalis* (Ascidiaceae). Their relevance to the evolution of the diffuse neuroendocrine system. Cell Tiss Res 223:369–402
Fröhlich F (1949) Die "helle Zelle" der Bronchialschleimhaut und ihre Beziehungen zum Problem der Chemoreceptoren. Frankf Z Pathol 60:517–558
Fujie E (1968) Ultrastructure of the pineal body of the domestic chicken, with special reference to the changes induced by altered photoperiods. Arch Histol Jpn 29:271–303
Fujii H, Isono H (1972) Ultrastructural observations on the parathyroid glands of the hen (*Gallus domesticus*). Arch Histol Jpn 34:155–165
Fujii R (1966) Correlation between fine structure and activity in fish melanophore. In: Porta GD, Mühlbock O (eds) Structure and control of the melanocyte. Springer-Verlag, Berlin Heidelberg New York, pp 114–123
Fujii R (1969) Chromatophores and pigments. In: Hoar WS, Randall DJ (eds) Fish physiology, vol III. Academic Press, New York, pp 307–353
Fujii R, Oshima N (1986) Control of chromatophore movements in teleost fishes. Zool Sci 3:13–47
Fujii R, Taguchi S (1969) The responses of fish melanophores to some melanin-aggregating and dispersing agents in potassium-rich medium. Annot Zool Jpn 42:176–182
Fujii R, Taguchi S (1970) Ultrastructure of nerve melanophore relationships in the guppy, *Lebistes reticulatus*. Annot Zool Jpn 43:123–131
Fujii S (1979) Development of pancreatic endocrine cells in the rat fetus. Arch Histol Jpn 42:467–479
Fujimoto S (1982) Effects of *p*-chlorophenylalanine on rabbit taste receptor cells. Arch Histol Jpn. 45:365–373
Fujimoto S, Murray RG (1970) Fine structure of degeneration and regeneration in denervated rabbit vallate taste buds. Anat Rec 168:393–414
Fujimoto S, Hattori T, Kimoto K, Yamashita S, Fujita S, Kawai K (1980) Tritiated thymidine autoradiographic study on origin and renewal of gastrin cells in antral area of hamsters. Gastroenterology 79:785–791
Fujita H (1975) Fine structure of the thyroid gland. Int Rev Cytol 40:197–280
Fujita H (1976) Ontogenesis of granule formation in adrenal chromaffin cells of the chick. In: Coupland RE, Fujita T (eds) Chromaffin, enterochromaffin and related cells. Elsevier, Amsterdam, pp 1–12
Fujita H, Honma Y (1968) Some observations on the fine structure of the endostyle of larval lampreys, ammocoetes of *Lampetra japonica*. Gen Comp Endocrinol 11:111–131
Fujita H, Kurihara H, Miyagawa J (1983) Ultrastructural aspects of the effects of calcium ionophore A23187 on incubated anterior pituitary cells of rats. Cell Tiss Res 229:129–136
Fujita S (1966) Application of light and electron microscopic autoradiography to the study of cytogenesis of the forebrain. In: Hassler R, Stephan H (eds) Evolution of the forebrain. Thieme, Stuttgart, pp 180–196
Fujita T (1959) Histological studies on the neuro-insular complex in the pancreas of some mammals. Z Zellforsch 50:94–109
Fujita T (1962) Über das Inselsystem des Pankreas von *Chimaera monstrosa*. Z Zellforsch 57:487–494
Fujita T (1964a) Cytologic studies on the pancreatic islets in *Chimaera monstrosa*. In: Brolin SE, Hellman B, Knutson H (eds) The structure and metabolism of the pancreatic islets. Proc 3rd Int Symp. Stockholm Aug 1963. Pergamon Press, London, pp 11–15
Fujita T (1964b) The identification of the argyrophil cells of pancreatic islets with D-cells. Arch Histol Jpn 25:189–197
Fujita T (1965) Über die Granula der Gewebsmast-

zellen der Ratte. Z Zellforsch 66: 66–82
Fujita T (1966) D-Zellen der Pankreasinseln beim Diabetes mellitus mit besonderer Berücksichtigung ihrer Argyrophilie. Z Zellforsch 69: 363–370
Fujita T (1968) D cell, the third endocrine element of the pancreatic islet. Arch Histol Jpn 29: 1–40
Fujita T (1973) Insulo-acinar portal system in the horse pancreas. Arch Histol Jpn 35: 161–171
Fujita T (1976) The gastro-enteric endocrine cell and its paraneuronic nature. In: Fujita T, Coupland RE (eds) Chromaffin, enterochromaffin and related cells. Elsevier, Amsterdam Oxford New York, pp 191–208
Fujita T (1977) Concept of paraneurons. Arch Histol Jpn 40 Suppl: 1–12
Fujita T (1980) Paraneuron, its current implications. Biomed Res 1 Suppl: 3–9
Fujita T (1983a) New aspects of cells secreting neuropeptides. In: Shizume K, Imura H, Shimizu N (eds) Endocrinology. Excerpta Medica, Amsterdam Oxford Princeton, pp 35–43
Fujita T (1983b) Messenger substances of neurons and paraneurons: Their chemical nature and the routes and ranges of their transport to targets. Biomed Res 4: 239–256
Fujita T (1985) Neurosecretion and new aspects of neuroendocrinology. In: Kobayashi H, Bern HA, Urano A (eds) Neurosecretion and the biology of neuropeptides (Proc 9th Int Symp.). Japan Scientific Societies Press, Tokyo/Springer, Berlin Heidelberg New York Tokyo, pp 521–528
Fujita T, Kobayashi S (1971) Experimentally induced granule release in the endocrine cells of dog pyloric antrum. Z Zellforsch 116: 52–60
Fujita T, Kobayashi S (1973) The cells and hormones of the GEP endocrine system—the current of studies. In: Fujita T (ed) Gastro-entero-pancreatic endocrine system: a cell-biological approach. Igaku-Shoin, Tokyo, pp 1–16
Fujita T, Kobayashi S (1975) Paraneurons: new sisters of neurons (in Japanese). Igaku no Ayumi 94: 638–640
Fujita T, Kobayashi S (1977) Structure and function of gut endocrine cells. Int Rev Cytol Suppl 6: 187–233
Fujita T, Kobayashi S (1978) Neuroendocrine correlations in the GEP endocrine system. In: Coupland RE, Forssman WG (eds) Peripheral neuroendocrine interaction. Springer, Berlin Heidelberg New York, pp 97–105
Fujita T, Kobayashi S (1979a) Proposal of a neurosecretory system in the pancreas. An electron microscope study in the dog. Arch Histol Jpn 42: 277–295
Fujita T, Kobayashi S (1979b) Current views on the paraneurone concept. TINS 2: 27–30
Fujita T, Murakami T (1973) Microcirculation of monkey pancreas with special reference to the insulo-acinar portal system. A scanning electron microscope study of vascular casts. Arch Histol Jpn 35: 255–263
Fujita T, Takaya K (1968) Metachromatic reaction of pancreatic B cells to toluidine blue; influence of pH on staining. Stain Technol 43: 329–332
Fujita T, Watanabe Y (1973) The effects of islet hormones upon the exocrine pancreas. In: Fujita T (ed) Gastro-entero-pancreatic endocrine system—A cell-biological approach. Igaku-Shoin, Tokyo, pp 164–173
Fujita T, Yanatori Y (1976) Intestinal endocrine cell reception of luminal stimuli. In: Fujita T (ed) Endocrine gut and pancreas. Elsevier, Amsterdam, pp 237–243
Fujita T, Hasegawa N, Koga Y, Kameda Y, Takaya K (1968) Is it insulin that is stained by aldehyde fuchsin and pseudoisocyanin in the sections of the pancreas? Arch Histol Jpn 29: 313–325
Fujita T, Kobayashi S, Sasagawa T (1971) Gastroenteric hormones: their source cells and secretory mechanism (in Japanese). Igaku no Ayumi 76: 647–651
Fujita T, Osaka M, Yanatori Y (1974) Granule release of entero-chromaffin (EC) cells by cholera enterotoxin in the rabbit. Arch Histol Jpn 36: 367–378
Fujita T, Yanatori Y, Murakami T (1976) Insuloacinar axis, its vascular basis and its functional and morphological changes caused by CCK-PZ and caerulein. In: Fujita T (ed) Endocrine gut and pancreas. Elsevier, Amsterdam, pp 347–357
Fujita T, Matsunari Y, Muraki S, Sato K, Shimoji K (1978) Effect of intraluminally administered lidocaine upon the pancreozymin-producing endocrine cells of the canine duodenum. In: Ito M (ed) Integrative control functions of the brain, vol 1. Kodansha Scientific, Tokyo, pp 296–297
Fujita T, Kobayashi S, Muraki S, Sato K, Shimoji K (1979) Gut endocrine cells as chemoreceptors. In: Miyoshi A (ed) Gut peptides. Secretion, function and clinical aspects. Kodansha, Tokyo/Elsevier, Tokyo Amsterdam New York Oxford, pp 47–52
Fujita T, Kobayashi S, Yui R, Iwanaga T (1980) Evolution of neurons and paraneurons. In: Ishii S, Hirano T, Wada M (eds) Hormones, adaptation and evolution. Jpn Sci Soc Press, Tokyo/Springer, Berlin, pp 35–43
Fujita T, Kobayashi S, Fujii S, Iwanaga T, Serizawa Y (1981a) Langerhans islets as the neuro-paraneuronal control center of the exocrine pancreas. In: Grossman MI, Brazier MAB, Lechago J (eds) Cellular basis of chemical messengers in the digestive system. Academic Press, New York London Toronto Sydney San Francisco, pp 231–242
Fujita T, Kobayashi S, Serizawa Y (1981b) Intercellular canalicule system in pancreatic islet. Biomed Res Suppl 2: 115–118
Fujita T, Tanaka K, Tokunaga J (1981c) SEM atlas of cells and tissues. Igaku-Shoin, Tokyo New York
Fujita T, Iwanaga T, Kusumoto Y, Yoshie S (1982) Paraneurons and neruosecretion. In: Farner DS, Lederis K (eds) Neurosecretion: molecules, cells, systems. Plenum, New York London, pp 3–13
Fujita T, Iwanaga T, Nakajima T (1983a) Immunohistochemical detection of nervous system-specific proteins in normal and neoplastic paraneurons in the gut and pancreas. In: Miyoshi A (ed) Gut peptides and ulcer (Proc Hiroshima Symp 1982). Biomedical Research Foundation, Tokyo, pp 81–88
Fujita T, Kanno T, Yanaihara N (eds) (1983b) Brain-gut axis (Biomed Res 4 Suppl)

Fujita T, Kobayashi S, Uchida T (1984) Secretory aspect of neurons and paraneurons. Biomed Res Suppl 5:1–8
Fujita T, Kuramoto H, Yui R, Iwanaga T (in press) Immunocytochemical and phylogenetical aspects. In: Heym C (ed) Histochemistry and cell biology of autonomic neurons and paraganglia. Exp Brain Res
Furness J, Costa M (1979) Projections of intestinal neurons showing immunoreactivity for vasoactive intestinal polypeptide are consistent with these neurons being the enteric inhibitory neurons. Neurosci Lett 15:199–204
Furness JB, Costa M (1980) Types of nerves in the enteric nervous system. Neuroscience 5:1–20
Furness JB, Costa M (1982) Neurons with 5-hydroxytriptamine-like immunoreactivity in the enteric nervous system: their projections in the guinea-pig small intestine. Neuroscience 7:341–349
Furness JB, Costa M, Emson PC, Håkanson R, Moghimzadeh E, Sundler F, Taylor IL, Chance RE (1983a) Distribution, pathways and reactions to drug treatment of nerves with neuropeptide Y and pancreatic polypeptide-like immunoreactivity in the guinea-pig digestive tract. Cell Tiss Res 234:71–92
Furness JB, Costa M, Miller RJ (1983b) Distribution and projections of nerves with enkephalin-like immunoreactivity in the guinea pig small intestine. Neuroscience 8:653–664
Furshpan EJ, MacLeish PR, O'Lague PH, Potter DD (1976) Chemical transmission between rat sympathetic neurons and cardiac myocytes developing in microcultures: evidence for cholinergic, adrenergic and dual-function neurons. Proc Natl Acad Sci USA 73:4225–4229
Furshpan EJ, Potter DD, Landis SC (1982) On the transmitter repertoire of sympathetic neurons in culture (Harvey lectures series 76). Academic Press, New York London Toronto Sydney San Francisco, pp 149–191
Furutani Y, Morimoto Y, Shibahara S, Noda M, Takahashi H, Hirose T, Asai M, Inayama S, Hayashida H, Miyata T, Numa S (1983) Cloning and sequence analysis of cDNA for ovine corticotropin-releasing factor precursor. Nature 301: 537–540
Fuxe K, Hökfelt T, Said SI, Mutt V (1977) Vasoactive intestinal polypeptide and the nervous system. Immunohistochemical evidence for localization in central and peripheral neurons, particularly intracortical neurons of the cerebral cortex. Neurosci Lett 5:241–246
Gabella G (1969) Taste buds and adrenergic fibers. J Neurol Sci 9:237–242
Gabella G (1979) Innervation of the gastrointestinal tract. Int Rev Cytol 59:129–192
Gainer H, Sarne Y, Brownstein MJ (1977) Biosynthesis and axonal transport of rat neurohypophysial proteins and peptides. J Cell Biol 73:366–381
Gallacher DV (1983) Substance P is functional neurotransmitter in the rat parotid gland. J Physiol 342:483–498
Gammil S, Weichert R (1973) Common origin for all neuroendocrine tumors. Acta Radiol Ther Phys Biol 12:321–326
Ganong WF (1985) Review of medical physiology. Lange Medical, Los Altos, California
Ganten D, Speck G, Schelling P, Unger T (1981) The brain renin-angiotensin system. In: Martin JB, Reichlin S, Bick KL (eds) Implications for brain functions and neurological disease (Advances in biomedical pharmacology, vol 28). Raven Press, New York, pp 359–372
Gati T (1976) Biological determination of gastrointestinal peptide hormones. Acta Physiol Acad Sci Hung 47:345–354
Geerdink HG, Drukker J (1973) Uptake of L-DOPA by cells in the taste buds of the vallate papilla of the mouse. Histochemie 36:219–223
Geissler D, Martinek A, Margolis RU, Margolis RK, Skrivanek JA, Ledeen R, König P, Winkler H (1977) Composition and biogenesis of complex carbohydrates of ox adrenal chromaffin granules. Neuroscience 2:685–693
Georges D, Dubois MP (1979) Immunological evidence for an ACTH-like antigen in the nervous ganglion of a protochordate *Ciona intestinalis* (Ascidiacea). Arch Anat Microsc Morphol Exp 68:121–125
Georges D, Dubois MP (1980) Mise en évidence par les techniques d'immunofluorescence d'un antigène de type LH-RH dans le système nerveux de *Ciona intestinalis* (Tunicier ascidiace). C R Acad Sci [D] (Paris) 290:29–31
Gerschenfeld HM, Piccolino M (1979) Pharmacology of the connectins of cones and L-horizontal cells in the vertebrate retina. In: Schmitt FO, Worden FG (eds) The neurosciences. Fourth study program. MIT Press, Cambridge and London, pp 213–226
Gershon MD (1981) The enteric nervous system. Ann Rev Neurosci 4:227–272
Gershon MD, Erde SM (1981) The nervous system of the gut. Gastroenterology 80:1571–1594
Gershon MD, Nunez EA (1970) Histochemical and radioautographic studies of serotonin and parafollicular cells in the thyroid gland of prehibernating bats. Endocrinology 86:160–166
Gershon MD, Ross LL (1966a) Radioisotopic studies of the binding, exchange, and distribution of 5-hydroxytryptamine synthesized from its radioactive precursor. J Physiol 186:451–476
Gershon MD, Ross LL (1966b) Location of sites of 5-hydroxytriptamine storage and metabolism by radioautography. J Physiol 186:477–492
Gershon MD, Drakontides AB, Ross LL (1965) Serotonin: synthesis and release from the myenteric plexus of the mouse intestine. Science 149:197–199
Gershon MD, Belshaw BE, Nunez EA (1971) Biochemical, histochemical and ultrastructural studies of thyroid serotonin, parafollicular and follicular cells during development in the dog. Am J Anat 132:5–20
Gershon MD, Payette RF, Rothman TP (1983) Development of the enteric nervous system. Fed Proc 42:1620–1625
Gesser BP, Larsson LI (1985) Changes from enkephalin-like to gastrin/cholecystokinin-like immunoreactivity in snail neurons. J Neurosci 5:1412–1417
Getchell TV (1974) Unitary responses in frog olfac-

tory epithelium to sterically related molecules at low concentrations. J Gen Physiol 64:241–261

Getchell, TV (1986) Functional properties of vertebrate olfactory receptor neurons. Physiol Rev 66: 772–818

Ghatak NR, Hirano A, Zimmerman HM (1971) Ultrastructure of craniopharyngioma. Cancer 27:1465–1475

Ghosh MN, Schild HO (1958) Continuous recording of acid gastric secretion in the rat. Br J Pharmacol 13:54–61

Giannattasio G, Zanini A, Meldolesi J (1979) Complex carbohydrate of secretory organelle. In: Margolis RM, Margolis RK (eds) Complex carbohydrates of nervous tissue. Plenum Press, New York, pp 327–345

Gibson AW, Burke RD (1985) The origin of pigment cells in embryos of the sea urchin *Strongylocentrotus purpuratus*. Devel Biol 107:414–419

Gilman AG (1984) G-proteins and dual control of adenylate cyclase. Cell 36:577–579

Ginzberg RD, Gilula NB (1980) Synaptogenesis in the vestibular sensory epithelium of the chick embryo. J Neurocytol 9:405–424

Girod C (1977) Apport de L'immunohistochimie a l'etude cytologique de l'adenohypophyse. Bull Assoc Anat 62

Girod C (1984) Fine structure of the pituitary pars distalis. In: Motta PM (ed) Ultrastructure of endocrine cells and tissues. Martinus Nijhoff Publishers, Boston Hague Dordrecht Lancaster, pp 12–28

Girod C, Trouillas J, Dubois MP (1985) Immunocytochemical localization of S-100 protein in stellate cells (folliculo-stellate cells) of the anterior lobe of the normal human pituitary. Cell Tiss Res 241:505–511

Girod C, Trouillas J, Raccurt M, Dubois MP (1986) Immunocytochemical localization of S-100 protein in stellate cells (folliculo-stellate cells) of the adenohypophysis in the monkeys *Macaca irus* and *Cercopithecus aethiops*. Cell Tiss Res 246:237–242

Girod C, Durand N, Raccurt M (1987) Immunostaining of a cell type in the islets of Langerhans of the monkey *Macaca irus* by antibodies against S-100 protein. Cell Tiss Res 247:11–16

Glenner GG, Cohen LA (1960) Histochemical demonstration of a trypsin-like esterase activity in mast cells. Nature 185:846

Glenner GG, Grimley PM (1974) Tumors of the extra-adrenal paraganglion system (including chemoreceptors). Fasc 9. In: Atlas of tumor pathology, 2nd ser. Armed Forces Institute of Pathology.

Gley E (1891) Sur les fonction de la glande thyroide chez le lapin et chez le chien. C R Soc Biol Paris 48:843–847

Gold G, Grodsky GM (1984) The secretory process in B cells of the pancreas. In: Cantin M (ed) Cell biology of the secretory process. Karger, Basel, pp 359–388

Goldberg LI, Volkman PH, Kohli JD (1978) A comparison of the vascular dopamine receptor with other dopamine receptors. Ann Rev Pharmacol Toxicol 18:57–79

Goldfine ID, Williams JA (1984) Actions of cholecystokinin and insulin on the acinar pancreas. In: Cantin M(ed) Cell biology of the secretory process. Karger, Basel, pp 389–422

Golding DW, Whittle AC (1977) Neurosecretion and related phenomena in annelids. Int Rev Cytol Suppl 5:189–302

Gomori G (1941) Observations with differential stains on human islets of Langerhans. Am J Pathol 17:395–406

Gomori G (1950) Aldehyde-fuchsin: a new stain for elastic tissue. Am J Clin Pathol 20:665–666

Gomperts BD (1976) The plasma membrane: models for its structure and function. Academic Press, New York

Goniakowska-Witalinska L (1981) Neuroepithelial bodies in the lung of the tree frog, *Hyla arborea* L. A scanning and transmission electron microscopic study. Cell Tiss Res 217:435–441

Goodall McC (1951) Studies of adrenaline and noradrenaline in mammalian heart and suprarenals. Acta Physiol Scand Suppl 85:7–51

Goormaghtigh N (1936) On the existence of abdominal vagal paraganglia in the adult mouse. J Anat 71:77–90

Goormaghtigh N (1940) Histological changes in the ischemic kidney with special reference to the juxtaglomerular apparatus. Am J Pathol 16:409–416

Goossens N, Vandenberghe MP (1974) The basal cells in the papilla fungiformes of the tongue of the common frog, *Rana temporaria* L. Arch Histol Jpn 36: 173–179

Goossens N, Dierickx K, Vandesande F (1980) Immunocytochemical localization of somatostatin in the brain of the lizard, *Ctenosauria pectinata*. Cell Tiss Res 208:499–505

Gorbman A (1959) Problems in the comparative morphology and physiology of the vertebrate thyroid gland. In: Gorbman A (ed) Comparative endocrinology. Wiley, New York, pp 266–282

Gordon-Weeks PR (1981) Properties of nerve endings with small granular vesicles in the distal colon and rectum of the guinea-pig. Neuroscience 6:1793–1811

Gorgas K (in press) The morphology of dogfish chromaffin bodies (*Scyliorhinus catalus* L.). In: Heym C (ed) Histochemistry and cell biology of autonomic neurons and paraganglia. Exp Brain Res Suppl

Gorgas K, Böck P (1976) Morphology and histochemistry of the adrenal medulla. I. Various types of primary catecholamine-storing cells in the mouse adrenal medulla. Histochemistry 50:17–31

Gotoh H, Schwartz JH (1982) Specificity of axonal transport in C2, a histaminergic neuron of *Aplysia californica*. Brain Res 242:87–98

Gottschaldt KM, Vahle-Hinz C (1981) Merkel cell receptors; Structure and transducer function. Science 214:183–186

Gould RP, Hodges RD (1971) Studies on the fine structure of the avian parathyroid glands and ultimobranchial bodies. Mem Soc Endocrinol 19:567–604

Gould VE, Moll R, Moll I, Lee I, Franke WW (1985) Biology of disease. Neuroendocrine (Merkel) cells of the skin: hyperplasias, dysplasias, and neoplasms. Lab Invest 52:334–353

Gozes I, Littauer UZ (1978) Tubulin microheterogeneity increases with rat brain maturation. Nature 276:411–413

Graf L (1976) Isolation and evolution of the gastrointestinal hormones. Acta Physiol Acad Sci Hung 47:285–298

Grafe P, Mayer CJ, Wood JD (1980) Synaptic modulation of calcitonin-dependent potassium conductance in myenteric neurons in the guinea-pig. J Physiol 305:235–248

Gratzl M, Torp-Pedersen C, Dartt DA, Treiman N, Thorn NA (1980) Isolation and characterization of secretory granules from bovine neurohypophyses. Hoppe Seyler's Z Physiol Chem 361:1615–1628

Gray P, Attwell D (1985) Kinetics of light-sensitive channels in vertebrate photoreceptors. Proc Roy Soc Lond (Biol) 223:379–388

Graziadei PPC (1971) The olfactory mucosa of vertebrates. In: Beidler M (ed) Olfaction (Handbook of sensory physiology IV/I). Springer, Berlin Heidelberg New York, pp 27–58

Graziadei PPC (1973a) Cell dynamics in the olfactory mucosa. Tiss Cell 5:113–131

Graziadei PPC (1973b) The ultrastructure of vertebrates olfactory mucosa. In: Friedmann I (ed) The ultrastructure of sensory organs. North-Holland, Amsterdam London/American Elsevier, New York, pp 267–305

Graziadei PPC, Graziadei GAM (1976) Olfactory epithelium of *Necturus maculosus* and *Ambystoma tigrinum*. J Neurocytol 5:11–32

Graziadei PPC, Monti-Graziadei GA (1978) Continuous nerve cell renewal in the olfactory system. In: Jacobson M (ed) Development of sensory system (Handbook of sensory physiology IX). Springer, Berlin Heidelberg New York, pp 55–83

Graziadei PPC, Metcalf JF (1971) Autoradiographic and ultrastructural observations on the frog's olfactory mucosa. Z Zellforsch 116:305–318

Graziadei PPC, Tucker D (1970) Vomeronasal receptors in turtles. Z Zellforsch 105:498–514

Green JD, Breeman VLv (1955) Electron microscopy of the pituitary and observations on neurosecretion. Am J Anat 97:177–227

Green JP, Day M (1960) Heparin, 5-hydroxytryptamine, and histamine in neoplastic mast cell. Biochem Pharmacol 3:190

Green JP, Day M, Robinson JD (1962) Some acidic substances in neoplastic mast cells and in the pineal body. Biochem Pharmacol 11:957–960

Greenberg MJ, Price D A (1983) Invertebrate neuropeptides: Native and naturalized. Ann Rev Physiol 45:271–288

Greenberg ME, Ziff EB, Greene LA (1986) Stimulation of neuronal acetylcholine receptors induces rapid gene transcription. Science 234:80–83

Greene LA, Tischler AS (1976) Establishment of a noradrenergic clonal line of rat adrenal pheochromocytoma cells which respond to nerve growth factor. Proc Natl Acad Sci USA 73:2424–2428

Greene LA, Rein G (1977) Synthesis, storage and release of acetylcholine by a noradrenergic pheochromocytoma cell line. Nature 268:349–351

Grell KG (1968) Protozoologie, 2nd ed. Springer, Berlin Heidelberg New York

Grenachev H (1879) Untersuchungen über das Sehorgan der Arthropoden, insbesondere der Spinner, Insekten und Crustaceen. Vanden-hoeck & Ruprecht, Göttingen. (Cited from Eakin 1982b)

Griffiths EC, McDermott JR (1984) Biotransformation of neuropeptides. Neuroendocrinology 39:573–581

Grillner S, McClellan A, Sigvardt K (1982) Mechanosensitive neurons in the spinal cord of the lamprey. Brain Res 235:169–173

Grillo MA (1966) Electron microscopy of sympathetic tissues. Pharmacol Rev 18:387–399

Grillo MA, Jacobs L, Comroe JH (1974) A combined fluorescence histochemical and electron microscopic method for studying special monoamine-containing cells (SIF cells). J Comp Neurol 153:1–14

Grimelius L (1968a) A silver nitrate stain for A-cells in human pancreatic islets. Acta Soc Med Upsala 73:243–270

Grimelius L (1968b) The argyrophil reaction in islet cells of adult human pancreas studied with a new silver nitrate procedure. Acta Soc Med Upsala 73:271–294

Grimmelikhuijzen CJP (1982) Oxytocin/vasopressin-like immunoreactivity is present in the nervous system of hydra. Neuroscience 7:3191–3199

Grimmelikhuijzen CJP (1983) Coexistence of neuropeptides in hydra. Neuroscience 9:837–845

Grimmelikhuijzen CJP (1985) Antisera to the sequence Arg-Phe-amide visualize neuronal centralization in hydroid polyps. Cell Tiss Res 241:171–182

Grimmelikhuijzen CJP, Spencer AN (1984) FNRF-amide immunoreactivity in the nervous system of the medusa *Polyorchis penicillatus*. J Comp Neurol 230:361–371

Grimmelikhuijzen CJP, Sundler F, Rehfeld JF (1980) Gastrin/CCK-like immunoreactivity in the nervous system of coelenterates. Histochemistry 69:61–68

Grimmelikhuijzen CJP, Balfe A, Emson PC, Powell D, Sundler F (1981a) Substance P-like immunoreactivity in the nervous system of hydra. Histochemistry 71:325–333

Grimmelikhuijzen CJP, Carraway RE, Rökaeus Å, Sundler F (1981b) Neurotensin-like immunoreactivity in the nervous system of hydra. Histochemistry 72:199–209

Grimmelikhuijzen CJP, Dockray GJ, Yanaihara N (1981c) Bombesin-like immunoreactivity in the nervous system of hydra. Histochemistry 73:171–180

Grimmelikhuijzen CJP, Dockray GJ, Schot LPC (1982) FMRFamide-like immunoreactivity in the nervous system of hydra. Histochemistry 73:499–508

Grodsky GM, Fanska R, West L, Manning M (1974) Anomeric specificity of glucose-stimulated insulin release: Evidence for a glucoreceptor? Science 186:536–538

Grodsky GM, Fanska R, Lundquist I (1975) Interrelationships between α and β anomers of glucose affecting both insulin and glucagon secretion in the perfused rat pancreas II. Endocrinology 97:

573–580
Grönblad M, Åkerman KE, Eränkö O (1979) Induction of exocytosis from glomus cells by incubation of the carotid body of the rat with calcium and ionophore A23187. Anat Rec 195:387–395
Grönblad M, Åkerman KE, Eränkö O (1980) Exocytosis of amine-storing granules from glomus cells of the rat carotid body induced by incubation in potassium-rich media or media containing calcium and ionophore A23187. In: Eränkö O, Soinila S, Päivärinta H (eds) Histochemistry and cell biology of autonomic neurons, SIF cells, and paraneurons (Adv Biochem Psychopharmacol 25) pp 227–233
Grönblad M, Liesi P, Rechardt L (1983) Serotonin-like immunoreactivity in rat carotid body. Brain Res 276:348–350
Gross DS, Baker BL (1979) Developmental correlation between hypothalamic gonadotropin-releasing hormone and hypophysial luteinizing hormone. Am J Anat 154:1–9
Grossman MI (1963) Integration of neural and hormonal control of gastric secretion. Physiologist 6: 349–357
Grossman MI (1974) Gastrointestinal hormones: spectrum of actions and structure-activity relations. In: Chey WY, Brooks FP (eds) Endocrinology of the gut. Charles B Slack, Thorofare, New Jersey, pp 65–75
Grossman MI (1979) Vagal stimulation and inhibition of acid secretion and gastrin release: which aspects are cholinergic? In: Rehfeld J, Amtrup E (eds) Gastrin and the Vagus. Academic Press, London, pp 105–113
Grossman MI, Brown JC, Said S, Adelson J, Rothman SS, Lim TM, Chance RE, Gerring EL, Gregory H, Dreutzfeldt W, Kokas E, Nasset ES, Sasaki, H, Falona GR, Unger RH, Andersson S, Thompson JC, Glass GBJ (1974) Candidate hormones of the gut. Gastroenterology 67:730–755
Grube D (1976a) Die endokrinen Zellen des Magendarmepithels und der Stoffwechsel der biogenen Amine im Magendarmtrakt. Progr Histochem Cytochem 8/3:1–128
Grube D (1976b) Biogenic monoamines in the GEP endocrine system of various mammals. In: Fujita T (ed) Endocrine gut and pancreas. Elsevier, Amsterdam, pp 119–132
Grube D (1980) Immunoreactivities of gastrin (G-) cells: II. Non-specific binding of immunoglobulins to G-cells by ionic interactions. Histochemistry 66:149–167
Grube D (1986) The endocrine cells of the digestive system: amines, peptides, and modes of action. Anat Embryol 175:151–162
Grube D, Bohn R (1983) The microanatomy of human islets of Langerhans, with special reference to somatostatin (D-) cells. Arch Histol Jpn 46: 327–353
Grube D, Forssmann WG (1979) Morphology and function of the enteroendocrine cells. Horm Metab Res 11:589–606
Grube D, Maier V, Raptis S, Schlegel W (1978a) Immunoreactivity of the endocrine pancreas: evidence for the presence of cholecystokinin-pancreozymin within the A-cell. Histochemistry 56:13–35
Grube D, Voigt KH, Weber E (1978b) Pancreatic glucagon cells contain endorphin-like immunoreactivity. Histochemistry 59:75–79
Grube D, Aunis D, Bader F, Cetin Y, Jörns A, Yoshie S (1986) Chromogranin A (CGA) in the gastro-entero-pancreatic (GEP) endocrine system. I. CGA in the mammalian endocrine pancreas. Histochemistry 85:441–452
Grundfest H (1957) Electrical inexcitability of synapses and some consequences in the central nervous system. Physiol Rev 37:337–361
Grynszpan-Winograd O (1969) Persistance des synapses au niveau de cellules chromaffines en mitose, pendent le development. C R Acad Sci [D] (Paris) 269:348–350
Grynszpan-Winograd O (1974) Adrenaline and noradrenaline cells in the adrenal medulla of the hamster: a morphological study of their innervation. J Neurocytol 3:341–361
Grynszpan-Winograd O (1975) Ultrastructure of the chromaffin cell. In: Blaschko H, Smith AD, Sayer G (eds) Adrenal gland (Handbook of Physiology 7/VI). American Physiological Society, Washington DC, pp 295–308
Gu J, Polak JM, Tapia FJ, Marangos PJ, Pearse AGE (1981) Neuron-specific enolase in the Merkel cells of mammalian skin: the use of specific antibody as a simple and reliable histologic marker. Am J Pathol 104:63–68
Gunji T (1982) Morphological research on the sensitivity of dentin. Arch Histol Jpn 45:45–67
Gustafsson MKS (1984) Synapses in *Diphyllobothrium dendriticum* (Cestoda). An electron microscopical study. Ann Zool Fennici 21:167–175
Gustafsson MKS, Wikgren MC, Karhi TJ, Schot LPC (1985) Immunocytochemical demonstration of neuropeptides and serotonin in the tapeworm *Diphyllobothrium dendriticum*. Cell Tiss Res 240: 255–260
Gustafsson MKS, Lehtonen MAI, Sundler F (1986) Immunocytochemical evidence for the presence of "mammalian" neurohormonal peptides in neurones of the tapeworm *Diphyllobothrium dendriticum*. Cell Tiss Res 243:41–49
Guth L (1971) Degeneration and regeneration of taste buds. In: Beidler LM (ed) Taste (Handbook of sensory physiology IV/2). Springer, Berlin Heidelberg New York, pp 63–74
Habener JF, Potts JT (1976) Chemistry, biosynthesis, secretion, and metabolism of parathyroid hormone. In: Aurbach GD (ed) Parathyroid gland (Handbook of physiology 7/VII). American Physiological Society, Washington DC, pp 313–342
Habener JF, Kemper B, Potts JT, Rich A (1975) Preproparathyroid hormone identified by cell-free translation of messenger RNA from hyperplastic human parathyroid tissue. J Clin Invest 56:1328–1333
Habener JF, Potts JT, Rich A (1976) Pre-proparathyroid hormone; evidence for an early biosynthetic precursor of proparathyroid hormone. J Biol Chem 251:3893–3899
Habener JF, Amherdt M, Ravazzola M, Orci L (1979)

Parathyroid hormone biosynthesis. Correlation of conversion of biosynthetic precursors with intracellular protein migration as determined by electron microscope autoradiography. J Cell Biol 80:715–731

Habener JF, Rosenblatt M, Potts JT (1984) Parathyroid hormone: Biochemical aspects of biosynthesis, secretion, action and metabolism. Physiol Rev 64: 985–1053

Hacki VWH (1978) Sekretin-Radioimmunoassay, Physiologie und Pathophysiologie beim Menschen. Z Gastroenterol 16(5):294–303

Haddad A, Guaraldo SP, Pelletier G, Brasiliero ILG, Marchi F (1980) Glycoprotein secretion in the hypothalamo-neurohypophysial system of the rat. Cell Tiss Res 209:399–422

Hafeez MA, Quay WB (1969) Histochemical and experimental studies of 5-hydroxytryptamine in pineal organs of teleosts (*Salmo gairdneri* and *Atherinopsis californiensis*). Gen Comp Edocrinol 13:211–217

Hagadorn I (1979) Book review. Gen Comp Endocrinol 37:271–272

Hage E (1972) Electron microscopic identification of endocrine cells in the bronchial epithelium of human foetuses. Acta Path Microbiol Scand, Section A 80:143–144

Hage E, Hage J, Juel G (1977) Endocrine-like cells of the pulmonary epithelium of the human adult lung. Cell Tiss Res 178:39–48

Hager H, Tafuri WL (1959) Elektronenoptischer Nachweis sog. neurosekretorischer Elementargranula in marklosen Nervenfasern des Plexus myentericus (Auerbach) des Meerschweinchens. Naturwissenschaften 45:332–333

Hagiwara S, Ohmori H (1982) Studies of calcium channels in rat clonal pituitary cells with patch electrode voltage clamp. J Physiol (Lond) 331:231–252

Håkanson R, Owman C (1966) Pineal DOPA decarboxylase and monoamine oxidase activities as related to the monoamine stores. J Neurochem 13: 597–605

Håkanson R, Owman Ch (1967) Concomitant histochemical demonstration of histamine and catecholamines in enterochromaffin-like cells of gastric mucosa. Life Sci 6:759–766

Håkanson R, Lindstrand K, Nordgren L, Owman CH (1969) Histamine-containing epithelial cells in rat stomach: a possible storage site for the intrinsic factor. Eur J Pharmacol 8:315–325

Håkanson R, Owman Ch, Sjöberg NO, Sporrong B (1970) Amine mechanisms in enterochromaffin and enterochromaffin-like cells of gastric mucosa in various mammals. Histochemie 21:189–220

Håkanson R, Larsson LI, Sjöberg NO, Sundler F (1974) Amine producing endocrine-like cells in the epithelium of urethra and prostate of the guinea-pig: A chemical, fluorescence histochemical, and electron microscopic study. Histochemistry 38: 259–270

Håkanson R, Sundler F, Ekman R (1980) New peptide hormone candidates possess a unique amino terminus. Trends Pharmacol Sci 1:337–341

Håkanson R, Böttcher G, Ekblad E, Panula P, Simonsson M, Dohlsten M, Hallberg T, Sundler F (1986) Histamine in endocrine cells in the stomach. A survey of several species using a panel of histamine antibodies. Histochemistry 86:5–17

Halter F, Galeazzi R, Leuenberger J, Smith GM (1977) Intragastric titration acid in the perfused rat stomach preparation following parenteral and intraluminal stimulation. Eur J Pharmacol 43:247–252

Halter J, Kulkosky P, Woods S, Makous W, Chen M, Porte D (1975) Afferent receptors, taste perception, and pancreatic endocrine function in man. Diabetes 24 Suppl 2:414

Hamburger V (1961) Experimental analysis of the dual origin of the trigeminal ganglion in the chick embryo. J Exp Zool 148:91–117

Hamperl H (1925) Über die "gelben (chromaffinen)" Zellen im Epithel des Verdauungstraktes. Z Mikrosk Anat Forsch 2:506–535

Hanbauer I, Hellstrom S (1978) The regulation of dopamine and noradrenaline in the rat carotid body and its modification by denervation and by hypoxia. J Physiol (Lond) 282:21–34

Hannover A (1840) Über die Netzhaut und ihre Gehirnsubstanz bei Wirbelthieren. Arch Anat Physiol Wiss Med: 320–345

Hansen JT, Brokaw J, Christie D, Karasek M (1982) Localization of enkephalin-like immunoreactivity in the cat carotid and aortic body chemoreceptors. Anat Res 203:405–410

Hansen K (1978) Insect chemoreception. In: Hazelbauer GL (ed) Taxis and Behavior. Chapman and Hall, London, pp 231–292

Hanson AM (1925) The hormone of the parathyroid gland. Proc Soc exp Biol Med 22:560–561

Hanssen LE (1980) The effect of atropine on secretin release and aspirated bicarbonate secretion after duodenal acidification in man. Scand J Gastroenterol 15:465–469

Hanyu S, Iwanaga T, Kano K, Fujita T (in press) The distribution of serotonin-immunoreactive paraneurons in the lower urinary tract of dogs. Am J Anat

Häppölä O, Soinila S, Päivärinta H, Joh TH, Panula P (1985) Histamine-immunoreactive endocrine cells in the adrenal medulla of the rat. Brain Res 339: 393–396

Harper AA (1972) The control of pancreatic secretion. Gut 13:308–317

Harris GG, Flock A (1967) Spontaneous and evoked activity from the *Xenops laevis* lateral line. In: Cahn PH (ed) Lateral line detectors. Indiana University Press, Bloomington, pp 135–161

Harris GG, Frischkopf LS, Flock Å (1970) Receptor potentials from hair cells of the lateral line. Science 167:76–79

Hartschuh W, Grube D (1979) The Merkel cell—A member of the APUD cell system? Fluorescence and electron microscopic contribution to the neurotransmitter function of the Merkel cell granules. Arch Dermatol Res 265:115–122

Hartschuh W, Weihe E (1977) The effect of denervation on Merkel cell in cats. Neurosci Lett 5:227–232

Hartschuh W, Weihe E (1980) Fine structural analysis of the synaptic junction of Merkel cell: Axon-complexes. J Invest Dermatol 75:159–165

Hartschuh W, Weihe E, Buchler M, Helmstaedter V, Feurle GE, Forssmann WG (1979) Met-enkephalin-like immunoreactivity in Merkel cells. Cell Tiss Res 201:343–348

Hartschuh W, Weihe E, Yanaihara N, Reinecke M (1983) Immunohistochemical localization of vasoactive intestinal polypeptide (VIP) in Merkel cells of various mammals: evidence for a neuromodulator function of the Merkel cell. J Invest Dermatol 81: 361–364

Hartschuh W, Reinecke M, Weihe E, Yanaihara N (1984) VIP-immunoreactivity in the skin of various mammals: immunohistochemical, radioimmunological and experimental evidence for a dual localization in cutaneous nerves and Merkel cells. Peptides 5:239–245

Hartwig HG, Baumann C (1974) Evidence for photosensitive pigments in the pineal complex of the frog. Vision Res 14:597–598

Hartwig HG, Oksche A (1982) Neurobiological aspects of extraretinal photoreceptive systems: structure and function. Experientia 38:991–996

Hartwig HG, Brisson P, Lyncker I, Collin JP (1980) Aminergic systems in pulmonate gastropod molluscs III. Microspectrofluorometric characterization of the monoamines in the reproductive system. Cell Tiss Res 210:223–234

Harty RF, McGuigan JE (1980) Effects of carbachol and atropine on gastrin secretion and synthesis in rat antral organ culture. Gastroenterology 78: 925–930

Hasegawa K, Koga M (1977) Induction of liver cell proliferation in intact rats by amines and glucagon. Life Sci 21:1723–1728

Hashimoto K (1972) The ultrastructure of the skin of human embryos, X. Merkel tactile cells in the finger and nail. J Anat 111:99–120

Hashimoto K, Satoh S, Takeuchi O (1971) Effect of dopamine on pancreatic secretion in the dog. Br J Pharmacol 43:739–746

Haskins JT, Price CH, Blankenship JE (1981) A light and electron microscopic investigation of the neurosecretory bag cells of *Aplysia*. J Neurocytol 10: 729–747

Hattori T, Niki H, Fujita S (1977) Tritiated thymidine autoradiographic study on the origin and renewal of argentaffin cells in the pyloric gland of hamsters. Cell Tiss Res 181:15–25

Hauge A, Staub NC (1969) Prevention of hypoxic vasoconstriction in cat lung by histamine-releasing agent 48/80. J Appl Physiol 26:693–699

Hawke SD, Farley RD (1971) Antennal chemoreceptors of the desert burrowing cockroach, *Arenivaga* SP. Tiss Cell 3:649–664

Hawkins JE (1976) Drug ototoxicity. In: Keidel WD, Neff WD (eds) Auditory system, clinical and topics (Handbook of sensory physiology V/3). Springer, Berlin Heidelberg New York, pp 707–748

Haydon DA, Taylor J (1963) The stability and properties of bimolecular lipid leaflets in aqueous solutions. J Theor Biol 4:281–296

Haynes LW (1980) Peptide neuroregulators in invertebrates. Prog Neurobiol 15:205–245

Hedeskov CJ (1980) Mechanism of glucose-induced insulin secretion. Physiol Rev 60:442–509

Hehn G von (1970) Über den Feinbau des Hyponeuralen Nervensystems des Seesternes (*Asterias rubens* L). Z Zellforsch 105:137–154

Heidenhain M (1911) Plasma und Zelle. Fischer, Jena

Heidenhain R (1870) Untersuchungen über den Bau der Labdrüsen. Arch Mikrosk Anat 6:368–406

Heiligenberg W (1973) Electrolocation of objects in the electric fish *Eigenmannia* (Rhamphichthyidae, Gymnotoidei). J Comp Physiol 87:137–164

Heine H, Förster FJ (1975) Relationships between mast cells and preterminal nerve fibers. Z Mikrosk-Anat Forsch 89:934–937

Heitz PH, Polak JM, Kasper M, Timson CM, Pearse AGE (1977) Immunoelectron cytochemical localization of motilin and substance P in rabbit bile duct enterochromaffin (EC) cells. Histochemistry 50: 319–325

Hellman B, Hellerström C (1960) The islets of Langerhans in ducks and chickens with special reference to the argyrophil reaction. Z Zellforsch 52:278–290.

Helmpel FG, Kariman K, Saltzman HA (1980) Redox transitions in mitochondria of cat cerebral cortex with seizures and hemorrhagic hypotension. Am J Physiol 238:H249–H256

Henderson JR (1969) Why are the islets of Langerhans? Lancet III:469–470

Henderson JR, Daniel PM (1978) Portal circulations and their relation to countercurrent system. Quart J Exp Physiol 63:355–369

Hendry SHC, Jones EG, Beinfeld MC (1983) Cholecystokinin-immunoreactive neurons in rat and monkey cerebral cortex make symmetric synapses and have intimate associations with blood vessels. Proc Natl Acad Sci USA 80:2400–2404

Henle J (1865) Über das Gewebe der Nebenniere und der Hypophyse. Z Vet Med 24:143–152

Henquin JC, Meissner HP (1983) Dibutyryl cyclic AMP triggers Ca^{2+} influx and Ca^{2+}-dependent electrical activity in pancreatic B cells. Biochem Biophys Res Commun 112:614–620

Henquin JC, Meissner HP (1984a) Significance of ionic fluxes and changes in membrane potential for stimulus-secretion coupling in pancreatic B-cells. Experientia 40:1043–1052

Henquin JC, Meissner HP (1984b) The ionic, electrical, and secretory effects of endogenous cyclic adenosine monophosphate in mouse pancreatic B cells: studies with forskolin. Endocrinology 115: 1125–1134

Henquin JC, Schmeer W, Meissner HP (1983) Forskolin, an activator of adenylate cyclase, increases Ca^{2+}-dependent electrical activity induced by glucose in mouse pancreatic B cells. Endocrinology 112:2218–2220

Herndon RM (1963) The fine structure of the Purkinje cell. J Cell Biol 18:167–180

Herrick K, Rubinstein LJ (1979) The cytological differentiating potential of pineal parenchymal neoplasms. (True pinealomas). A clinicopathological study of 28 tumours. Brain 102:289–320

Hervonen A, Pelto-Huikko M, Linnoila I (1980) Ultrastructural localization of enkephalin-like immunoreactivity in the rat adrenal medulla. Am J Anat

157:445–448
Hess A (1968) electron microscopic observations of normal and experimental cat carotid bodies. In: Torrance RW (ed) Arterial chemoreceptors. Blackwell, Oxford, pp 51–56
Heuser JE, Rees TS (1973) Evidence for recycling of synaptic vesicle membrane during tramsmitter release at the frog neuromuscular junction. J Cell Biol 57:315–344
Heward CB, Hadley ME (1975) Structure-activity relationships of melatonin and related indoleamines. Life Sci 17:1167–1178
Heym CH (1978) Functional morphology of monoamine-storing cells in the rat superior cervical ganglia. In: Coupland RE, Forssmann WG (eds) Peripheral neuroendocrine interaction. Springer, Berlin, pp 70–85
Hiatt RB, Goodman I, Overweg NIA (1970) Serotonin and intestinal motility. Am J Surg 119:527–529
Hiji Y (1975) Selective elimination of taste responses to sugars by proteolytic enzymes. Nature (Lond) 256:427–429
Hijiya K, Okada Y, Tankawa H (1977) Ultrastructural study of the alveolar brush cell. J Electron Microsc 26:321–329
Hillarp NÅ (1959) The construction and functional organization of the autonomic innervation apparatus. Acta Physiol Scand 46 Suppl 157:1–38
Hillarp NÅ (1960) Different pools of catecholamines stored in the adrenal medulla. Acta Physiol Scand 50:8–22
Hillarp NÅ, Hökfelt B (1953) Evidence of adrenaline and noradrenaline in separate adrenal medullary cells. Acta Physiol Scand 30:55–68
Hillarp NÅ, Thieme G (1959) Nucleotides in the catecholamine granules of the adrenal medulla. Acta Physiol Scand 45:328–338
Hillarp NÅ, Lagerstedt S, Nilson B (1953) The isolation of a granular fraction from the suprarenal medulla, containing the sympathomimetic catechol amines. Acta Physiol Scand 29:251–263
Hillman DE (1969) New ultrastructural findings regarding a vestibular ciliary apparatus and its possible functional significance. Brain Res 13:407–412
Hillman DE (1972) Observations on morphological features and mechanical properties of the peripheral vestibular receptor system in the frog. Prog Brain Res 37:69–75
Hillman DE, Lewis ER (1971) Morphological basis for a mechanical linkage in otolithic receptor transduction in the frog. Science 174:416–419
Hirano S, Ohashi T, Fujita T (1974) Embryological study on a possible neural crest origin of mast cells (Japanese abstract). Acta Anat Nippon 49:32
Hirano T (1982) Neural regulation of adrenal chromaffin cell function in the mouse.-Stress effect on the distribution of [^{3}H]-dopamine in denervated adrenal medulla. Brain Res 238:45–54
Hirano T, Kobayashi S (1978) ACTH controls [^{3}H]-dopamine uptake in the adrenal chromaffin cell. Neurosci Lett 9:337–340
Hirano T, Kobayshi S, Kent C, Coupland RE (1977) Autoradiographic demonstration of a zonal distribution of ^{3}H-dopamine-derived radioactivity in the mouse adrenal medulla perfusion-fixed with glutaraldehyde. Arch Histol Jpn 40:107–119
Hirata K, Nada O (1975) A monoamine in the gustatory cell of the frog's taste organ. A fluorescence histochemical and electron microscopic study. Cell Tiss Res 159:101–108
Hirata Y (1966) Fine structure of the terminal buds on the barbels of some fishes. Arch Histol Jpn 26:507–523
Hirsch PF, Gauthier GF, Munson PL (1963) Thyroid hypocalcemic principle and recurrent laryngeal nerve injury as factors affecting the response to parathyroidectomy in rats. Endocrinology 73:244–252
Hirschowitz BI, Hutchison G, Sachs G (1972) Kinetics of atropine inhibition of hitamine-stimulated gastric secretion in the dog. Am J Physiol 222:1316–1321
Hirst GDS, McKirdy HC (1975) Synaptic potentials recorded from neurones of the submucous plexus of guinea-pig small intestine. J Physiol Lond 249: 369–385
Hisada M (1984) Neuronal mechanism of coordinative simultaneity of two behavioral acts in crayfish-uropod steering and abdominal posture movement. In: Aoki K, Ishii S, Morita H (eds) Animal bahavior: neurophysiological and ethological approaches. Japan Sci Soc Press, Tokyo/Springer, Berlin, pp 95–107
Hisada M, Takahata M, Nagayama T (1984a) Structure and output connection of local non-spiking interneurons in crayfish. Zool Sci 1:41–49
Hisada M, Takahata M, Nagayama T (1984b) Local non-spiking interneurons in the arthropod motor control systems: characterization and their functional significance. Zool Sci 1:681–700
Hodgkiss JP, Lees GM (1983) Morphological studies of electrophysiologically-identified myenteric plexus neurons of the guinea-pig ileum. Neurosci 8:593–608
Hoffman HH, Schnitzlein HN (1969) The number of nerve fibers in the vagus nerves of man. Anat Rec 139:429–435
Hoffman PN, Lasek RJ (1975) The slow component of axonal transport: identification of major structural polypeptides of the axon and their generality among mammalian neurons. J Cell Biol 66:351–366
Hofstein R, Hesse G, Shashoua VE (1983) Proteins of the extracellular fluid of mouse brain: extraction and partial characterization. J Neurochem 40: 1448–1455
Hökfelt T, Efendić S, Hellerström C, Johanson O, Luft R, Arimura A (1975) Cellular localization of somatostatin in endocrine-like cells and neurons of the rat with special references to the A_1-cells of the pancreatic islets and to the hypothalamus. Acta Endocrinol 80 Suppl 200:1–41
Hökfelt T, Elfvin LG, Elde R, Schultzberg M, Goldstein M, Luft R (1977) Occurrence of somatostatin-like immunoreactivity in some peripheral sympathetic noradrenergic neurons. Proc Natl Acad Sci USA 74:3587–3591
Hökfelt T, Lunberg JM, Schultzberg M, Johansson O,

Skirboll L, Änggård A, Fredholm B, Hamberger B, Pernow B, Rehfeld J, Goldstein M (1980a) Cellular localization of peptides in neural structures. Proc Roy Soc Lond B 210:63–77
Hökfelt T, Johansson O, Ljungdahl Å, Lundberg JM, Schultzberg M (1980b) Peptidergic neurones. Nature (Lond) 284:515–521
Hökfelt T, Lundberg JM, Schultzberg M, Johansson O, Ljungdahl A, Rehfeld J (1980c) Coexistence of peptides and putative transmitters in neurons. In: Costa E, Trabucchi M (eds) Neural peptides and neuronal communication. Raven Press, New York, pp 1–23
Hökfelt T, Johansson O, Goldstein M (1984) Chemical anatomy of the brain. Science 225:1326–1334
Hökfelt T, Holets VR, Staines W, Meister B, Melander T, Schalling M, Schultzberg M, Freedman J, Björklund H, Olson L, Lindh B, Elfvin LG, Lundberg JM, Lindgren JÅ, Samuelsson B, Pernow B, Terenius L, Post C, Everitt B, Goldstein M (1986) Coexistence of neuronal messengers–an overview. In: Hökfelt T, Fuxe K, Pernow B (eds) Progress in brain research, vol 68. Elsevier, Amsterdam, pp 33–70
Hokin LE (1985) Receptors and phosphoinositide-generated second messengers. Ann Rev Biochem 54:205–235
Holets V, Elde R (1982) The differential distribution and relationship of serotoninergic and peptidergic fibers to sympathoadrenal neurons in the intermediolateral cell column of the rat: a combined retrograde axonal transport and immunofluorescence study. Neuroscience 7:1155–1174
Hollenberg MD (1981) Membrane receptors and hormone action I. New trends related to receptor structure and receptor regulation. Trends Pharmacol Sci 2:320–323
Hollenberg MD (1982) Membrane receptors and hormone action II. New perspectives for receptor-modulated cell function. Trends Pharmacol Sci 3: 25–28
Holman ME, Hirst GDS (1977) Junctional transmission in smooth muscle and the autonomic nervous system. In: Kandel ER (ed) Cellular biology of neurons, part 1. (Handbook of Physiology 1/I). Amer Physiol Soc, Bethesda, MD, pp 417–461
Holmgren S, Vaillant C, Dimaline R (1982) VIP-, substance P-,gastrin/CCK-, bombesin-, somatostatin- and glucagon-like immunoreactivities in the gut of the rainbow trout, *Salmo gairdneri*. Cell Tiss Res 223:141–153
Holzwarth MA (1984) The distribution of vasoactive intestinal peptide in the rat adrenal cortex and medulla. J Autonom Nerv Syst 11:269–283
Hong SS, Magee DF (1970) Pharmacological studies on the regulation of pancreatic secretion in pigs. Ann Surg 172:41–48
Hopfield JJ (1984) Neurons with graded response have collective computational properties like those of two-state neurons. Proc Natl Acad Sci USA 81: 3088–3092
Hopkins CR, Farquhar MG (1973) Hormone secretion by cells dissociated from rat anterior pituitaries. J Cell Biol 59:276–303
Hopwood D (1971) The histochemistry and electron histochemistry of chromaffin tissue. Prog Histochem Cytochem 3:1–66
Horai T, Nishihara H, Hattori S, Tateishi R (1979) Malignant melanoma producing serotonin. Cancer 43:294–298
Horch KW, Whitehorn D, Burgess PR (1974) Impulse generation in type I cutaneous mechanoreceptors. J Neurophysiol 37:267–281
Horvath E, Kovacs K (1976) Ultrastructural classification of pituitary adenomas. Canad J Neurol Sci 3:9–21
Hosoya Y, Matsushita M (1979) Identification and distribution of the spinal and hypophyseal projection neurons in the paraventricular nucleus of the rat: a light and electron microscopic study with the horseradish peroxidase method. Exp Brain Res 35:315–331
Hotz J, Goebell H, Ziegler R (1975) The role of calcium in gastric acid and pepsin secretion. In: Case RM, Goebell H (eds) Stimulus-secretion coupling in the gastrointestinal tract (Proceedings of an International Conference held at Titisee, Germany). University Park Press, Baltimore, pp 159–175
Howard A, Pelc SR (1953) Synthesis of desoxyribonucleic acid in normal and irradiated cells and its relation to chromosome breakage. Heredity Suppl 6:261–273
Howell SL, Tyhurst M (1982) The insulin storage granule. In: Poisner AM, Trifaró JM (eds) The secretory granule. Elsevier/Biomedical Press, Amsterdam New York Oxford, pp 155–172
Howell SL, Kostianovsky M, Lacy PE (1969) Beta granule formation in isolated islets of Langerhans. J Cell Biol 42:695–705
Hoyle G (1983) Forms of modulatable tension in skeletal muscles. Comp Biochem Physiol 76A: 203–210
Hoyle G, Field LH (1983a) Defense posture and lefposition learning in a primitive insect utilize catch-like tension. J Neurobiol 14:285–298
Hoyle G, Field LH (1983b) Elicitation and abrupt termination of behaviorally significant catchlike tension in a primitive insect. J Neurobiol 14:299–312
Hozumi I, Iwanaga T, Fujita T, Yamakuni T, Takahashi Y (1986) A Purkinje cell-specific protein (spot 35 protein) showing wide distribution in the endocrine system of some mammals. An immunohistochemical study. Acta Histochem Cytochem 19: 545–553
Huber GC (1900) A contribution on the minute anatomy of the sympathetic ganglia of the different classes of vertebrates. J Morphol 16:27–90
Hudspeth AJ (1983a) Transduction and tuning by vertebrate hair cells. Trends Neurosci 6:366–369
Hudspeth AJ (1983b) The hair cells of the inner ear. Sci Am 248:42–52
Hudspeth AJ (1983c) Mechanoelectrical transduction by hair cells in the acousticolateralis sensory system. Ann Rev Neurosci 6:187–215
Hudspeth AJ, Jacobs R (1979) Stereocilia mediate transduction in vertebrate hair cells. Proc Natl Acad Sci USA 76:1506–1509
Hughes RL, Woollacott RM (1978) Ultrastructure of

potential photoreceptor organs in the larva of *Scrupocellaria bertholetti* (Bryozoa). Zoomorphologie 91:225–234

Hung KS (1980) Innervation of rabbit fetal lungs. Am J Anat 159:73–83

Hung KS, Chapman AL, Mestemacher MA (1979) Scanning electron microscopy of bronchiolar neuroepithelial bodies in neonatal mouse lungs. Anat Rc 193:913–926

Hung KS, Loosli CG (1975) Bronchiolar neuroepithelial bodies in the neonatal mouse lungs. Am J Anat 140:191–200

Ichikawa A (1959) Electron microscope study on secretion of the rat adenohypophysis. (Japanese text with English abstract). Acta Anat Nippon 34:460–482

Ichikawa M (1933) Remarks on the investigation of the origin of vesceral cartilages by means of staining. Proc Imp Acad Tokyo 9:117–119

Ide C, Munger BL (1978) A cytologic study of grandry corpuscle development in chicken toe skin. J Comp Neurol 179:301–324

Ide C, Munger BL (1980) The cytologic composition of primate laryngeal chemosensory corpuscles. Am J Anat 158:193–209

Iggo A (1963a) New specific sensory structures in hairy skin. Acta Neuroveget 24:175–180

Iggo A (1963b) An electrophysiological analysis of afferent fibers in primate skin. Acta Neuroveg 24:225–240

Iggo, A, Muir AR (1969) The structure and function of a slowly adapting touch corpuscle in hairy skin. J Physiol 200:763–796

Iijima T (1981) Occurrence of uranaffin-positive synaptic vesicles in both adrenergic and non-adrenergic nerves of the rat anococcygenus muscle. Cell Tiss Res 220:427–433

Iijima T, Hasegawa K (1985) Quinacrine-induced dilation of the rat cecum and degeneration of large granular vesicle-containing neurons in the myenteric plexus. Cell Tiss Res 241:383–389

Illes P, North RA (1982) Effects of divalent cations and normorphine on spontaneous excitatory junction potentials in the mouse vas deferens. Br J Pharmacol 75:599–604

Imura H, Kato Y, Nakai Y, Nakao K, Tanaka I, Jingami H, Koh T, Yoshimasa T, Tsukada T, Suda M, Sakamoto M, Morii N, Takahashi H, Tojo K, Sugawara A (1985) Endogeneous opioids and related peptides: from molecular biology to clinical medicine. J Endocrinol 107:147–157

Inagaki S, Kubota Y, Kito S, Kangawa K, Matsuo H (1986) Immunoreactive atrial natriuretic polypeptides in the adrenal medulla and sympathetic ganglia. Regul Peptides 15:249–260

Inagami T, Yokosawa H, Hirose S (1978) Definitive evidence for renin in rat brain by affinity chromatographic separation from protease. Clin Sci Molec med 55:121S-123S

Inage T (1973) Fluorescence histochemical and electron microscopic observations on the enterochromaffin cells in the dog pancreas. In: Fujita T (ed) Gastro-entero-pancreatic endocrine system. A cell-biological approach. Igaku-Shoin, Tokyo, pp 96–100

Inokuchi H, Fujimoto S, Kawai K (1983) Cellular kinetics of gastrointestinal mucosa, with special reference to gut endocrine cells. Arch Histol Jpn 46: 137–157

Inoue K, Tanaka S, Kurosumi K (1985) Mitotic activity of gonadotropes in the anterior pituitary of the castrated male rat. Cell Tiss Res 240:271–276

Inoue Y, Setoguti T (1986) Immunocytochemical localization of parathormone in the mammalian parathyroid gland using the protein A-gold technique. Cell Tiss Res 243:3–7

Inoue Y, Akita M, Shimai K (1968) The mast cells in the brain of the *Triturus pyrrhogaster* (Boie). Keio J Med 17:235–249

Isaacson P (1976) Mast cells in benign nerve sheath tumors. J Pathol 119:193–196

Ishida H (1977) Fine structural study on the postnatal development of the rat tracheal mucosa with special reference to the brush cells. Yokohama Med Bull 28:123–147

Ishikawa H, Shiino M, Arimura A Rennels EG (1977) Functional clones of pituitary cells derived from Rathke's pouch epithelium of fetal rats. Endocrinology 100:1227–1230

Ishimori A, Sakurada H, Yamagata S (1972) Gastrectomy and secretory function of the stomach. Tohoku J Exp Med 107:1–13

Ishimura K, Lederis K (1983) Immunohistochemistry of the caudal neurosecretory system in the sucker *Catostomus commersoni*. Fol Endocrinol Jpn 59: 354

Ishizaka T, Foreman JC, Sterk AR, Ishizaka K (1979) Induction of calcium flux across the rat mast cell membrane by bridging IgE receptors. Proc Natl Acad Sci USA 76:5858–5862

Ishizaka T, Hirata F, Ishizaka K, Axelrod J (1980) Stimulation of phospholipid methylation, Ca^{2+} influx, and histamine release by bridging of IgE receptors on rat mast cells. Proc Natl Acad Sci USA 77:1903–1906

Ishizaki H, Suzuki A (1980) Prothoracicotropic hormone. In: Miller TA (ed) Neurohormonal techniques in insects. Springer, New York, pp 244–276

Ishizuka Y, Nozawa Y (1983) Concerted stimulation of PI-turnover, Ca^{2+}-influx and histamine release in antigen-activated rat mast cells. Biochem Biophys Res Commun 117:710–717

Isobe T, Okuyama T (1978) The amino-acid sequence of S-100 protein (PAP I-b protein) and its relation to the calcium-binding proteins. Eur J Biochem 89: 379–388

Isobe T, Okuyama T (1981) The amino-acid sequence of the α subunit in bovine brain S-100 protein. Eur J Biochem 116:79–86

Isobe T, Tsugita A, Okuyama T (1978) The amino acid sequence and the subunit structure of bovine brain S-100 protein (PAP I-b). J Neurochem 30:921–923

Isobe T, Ishioka N, Okuyama T (1981) Structural relation of two S-100 proteins in bovine brain; subunit composition of S-100a protein. Eur J Biochem 115:469–474

Isobe T, Ishioka N, Kocha T, Okuyama T (1983) Chemical structure and molecular evolution of S-100 proteins. In: Peeters M (ed) Protides of the

biological fluids, vol 30. Pergamon Press, London, pp 21–24
Isono H, Shoumura S (1973) Fine structure of the new parathyroid gland after EDTA administration; mechanism of formation and extrusion of secretory granules. J Electron Micros 22:191–204
Isono H, Sakurai S, Fujii H, Aoki S (1971) Ultrastructural changes in the parathyroid gland of the phosphate-treated newt, *Triturus pyrrhogaster* (Bioé) Arch Histol Jpn 33:357–370
Israel JM, Jaquet P, Vincent JD (1985) The electrical properties of isolated human prolactin-secreting adenoma cells and their modification by dopamine. Endocrinology 117:1448–1455
Ito M, Kameda Y, Tagawa T (1986) An ultrastructural study of the cysts in chicken ultimobranchial glands, with special reference to C-cells. Cell Tiss Res 246:39–44
Ito S, Winchester RJ (1963) The fine structure of the gastric mucosa in the bat. J Cell Biol 16:541–577
Itoh N, Obata K, Yanaihara N, Okamoto H (1983) Human preprovasoactive intestinal polypeptide contains a novel PHI-27-like peptide, PHM-27. Nature 304:547–549
Iversen LL (1977) Inactivation of neurotransmitters. In: Cottrell GA, Usherwood PNR (eds) Synapses. Blackie, Glasgow, pp 137–153
Iversen LL (1983) Neuropeptides–what next? Trends Neurosci 6:293–316
Iversen LL (1984) Amino acids and peptides: fast and slow chemical signals in the nervous system? Proc R Soc Lond [Biol] 221:245–260
Iwanaga T, Fujita T (1984) Sustentacular cells in the fetal human adrenal medulla are immunoreactive with antibodies to brain S-100 protein. Cell Tiss Res 236:733–735
Iwanaga T, Kobayashi S, Fujita T, Yanaihara N (1980) Immunocytochemistry and ultrastructure of Segi's cap. Biomed Res 1:117–129
Iwanaga T, Fujita T, Nishiitsutsuji-Uwo J, Endo Y (1981) Immunohistochemical demonstration of PP-, somatostatin-, enteroglucagon- and VIP-like immunoreactivities in the cockroach midgut. Biomed Res 2:202–207
Iwanaga T, Fujita T, Ito S (1982a) Immunohistochemical staining of enteroendocrine paraneurons with anti-brain tubulin antiserum. Biomed Res 3:99–101
Iwanaga T, Fujita T, Masuda T, Takahashi Y (1982b) S-100 protein-immunoreactive cells in the lymph node and spleen of the rat. Arch Histol Jpn 45:393–397
Iwanaga T, Ito S, Yamada Y, Fujita T (1983) Cytochemical demonstration of somatostatin 28-(1–14)-like immunoreactivity in the rat hypothalamus and gastro-entero-pancreatic endocrine system. Endocrinology 113:1839–1844
Iwanaga T, Takahashi-Iwanaga H, Fujita T, Yamakuni T, Takahashi Y (1985a) Immunohistochemical demonstration of a cerebellar protein (spot 35 protein) in some sensory cells of guinea pig. Biomed Res 6:329–334
Iwanaga T, Takahashi Y, Fujita T (1985b) Immunohistochemical localization of S-100 protein in the retina, ciliary body and iris of human fetuses. Cell Tiss Res 239:505–510
Iwanaga T, Fujita T, Takeda N, Endo Y, Lederis K (1986) Urotensin I-like immunoreactivity in the midgut endocrine cells of the insects *Gryllus bimaculatus* and *Periplaneta americana*. Cell Tiss Res 244:565–568
Iwanaga T, Hanyu S, Fujita T (1987) Serotonin-immunoreactive cells of peciliar shape in the urethral epithelium of the human penis. Cell Tiss Res 249:51–56
Iwasaki Y, Ito S, Shibata A (1980) Radioimmunoassay of neurotensin and the distribution and concentration of gut neurotensin in rat and dog. Tohoku J Exp Med 130:129–137
Iwayama T (1970) Changes in the cell population of taste buds during degeneration and regeneration of their sensory innervation. Z Zellforsch 110:487–495
Iwayama T, Nada O (1967) Histochemical observations on the phosphatases of the tongue, with special reference to taste buds. Arch Histol Jpn 28:151–163
Iwayama T, Nada O, Wasano T (1967) Phosphatase activities of the degenerated and regenerated taste buds, caused by the transection of the glossopharyngeal nerve. Arch Histol Jpn 28:542–543
Jacobowitz DM, Laties AM (1968) Direct adrenergic innervation of a teleost melanophore. Anat Rec 162:501–504
Jacobs S, Cuatrecasas P (1981) Insulin receptor: structure and function. Endocrinol Rev 2:251–263
Jahnke V, Lundquist PG, Wersäll J (1969) Some morphological aspects of sound perception in birds. Acta Otolaryngol 67:583–601
Jalowy B (1934) Über die Regeneration der Nervenendigungen in den Tasthaaren des Meerschweinchens (*Cavia cobaya*). Z Zellforsch 21:149–168
Jamieson JD, Palade GE (1964) Specific granules in atrial muscle cells. J Cell Biol 23:151–172
Jamieson JD, Palade GE (1967a) Intracellular transport of secretory proteins in the pancreatic exocrine cell. I. Role of the peripheral elements of the Golgi complex. An autoradiographic study. J Cell Biol 34:577–596
Jamieson JD, Palade GE (1967b) Intracellular transport of secretory proteins in the pancreatic exocrine cell. II. Transport to condensing vacuoles and zymogen granules. J Cell Biol 34:597–615
Jan LY, Jan YN (1982) Peptidergic transmission in sympathetic ganglia of the frog. J Physiol 327:219–246
Jan LY, Jan YN (1983) A LHRH-like peptidergic neurotransmitter capable of 'action at a distance' in autonomic ganglia. Trends Neurosci 6:320–325
Janakidevi K, Dewey VC, Kidder GW (1966) The biosynthesis of catecholamines in two genera of protozoa. J Biol Chem 241:2576–2578
Janscó N (1968) Desensitization with capsaicin and related acylamides as a tool for studying the function of pain receptors. In: Phamacology of pain (Proc. 3rd Int. Phamacol. Meet. 1966, vol 9). Pergamon Press, Oxford New York, pp 33–35
Janscó N, Janscó-Gábor A, Szolcsányi J (1967) Direct evidence for neurogenic inflammation and its pre-

vention by denervation and by pretreatment with capsaicin. Br J Pharmac Chemother 31:138–151

Janscó G, Kiraly E, Janscó-Gábor A (1977) Pharmacologically induced selective degeneration of chemosensitive primary sensory neurones. Nature 270:741–743

Javor T (1976) Physiological and pathophysiological aspects of gastrointestinal peptide hormones. Acta Physiol Acad Sci Hung 47:317–322

Jobsis FF, Rosenthal M (1978) Behavior of the mitochondrial respiratory chain *in vivo*. In: Elliot K, O'Conner M (eds) Cerebral vascular smooth muscle and its control. Elsevier/Excerpta Medica/North-Holland and Elsevier/North-Holland, Amsterdam, pp 149–169

Jobsis FF, O'Conner M, Vitale A, Vreman H (1983) Intracellular redox changes in functioning cerebral cortex. I. Metabolic effects of epileptiform activity. J Neurophysiol 34:735–749

Johansen T (1979) Utilization of adenosine triphosphate in rat mast cells during histamine release induced by the ionophore A23187. Br J Pharmacol 65:103–109

Johansen T (1980) Calcium requirement in compound 48/80 induced histamine release from rat mast cell *in vitro*. Life Sci 27:369–375

Johansson I, Lundell L, Svensson SE (1972) Inhibitory and stimulatory effects of secretin on gastric secretion in conscious rats. Br J Pharmacol 46:94–100

Johnson AR, Erdös EG (1973) Release of histamine from mast cells by vasoactive peptides. Proc Soc Exp Biol Med 142:1252–1256

Johnson RG, Carty S, Scarpa A (1982) A model of biogenic amine accumulation into chromaffin granules and ghosts based on coupling to the electrochemical proton gradient. Fed Proc 41: 2746–2754

Johnston MC (1966) A radioautographic study of the migration and fate of cranial neural crest cells in the chick embryo. Anat Rec 156:143–155

Jones DG (1978) Some current concepts of synaptic organization. Adv Anat Embryol Cell Biol 55:3–69

Jones DP, Kennedy FG (1982) Intracellular oxygen supply during hypoxia. Am J Physiol 243:C247–C253

Jørgensen AO, Subrahmanyan L, Turnbull C, Kalnins VI (1976) Localization of the neurofilament protein in neuroblastoma cells by immunofluorescent staining. Proc Natl Acad Sci USA 73:3192–3196

Jørgensen JM, Flock Å, Wersäll J (1972) The Lorenzinian ampullae of *Polyodon spathula*. Z Zellforsch 130:362–377

Jørgensen KD, Diamant B, Jørgensen KH, Thim L (1982a) Pancreatic spasmolytic polypeptide PSP: III. Pharmacology of a new porcine pancreatic polypeptide with spasmolytic and gastric acid secretion inhibitory effects. Regul Peptide 3:231–243

Jørgensen KH, Thim L, Jacobsen HE (1982b) Pancreatic spasmolytic polypeptide (PSP): I. Preparation and initial chemical characterization of a new polypeptide from porcine pancreas. Regul Peptide 3:207–219

Joselsson L (1983) Invertebrate neuropeptide hormones. Int J Pept Protein Res 21:459–470

Joss JMP (1977) Hydroxyindole-O-methyltransferase (HIOMT) activity and the uptake of ^{3}H-melatonin in the lamprey, *Geotria australis* Gray. Gen Comp Endocrinol 31:270–275

Judge DM, McGavran MH, Trapukdi S (1976) Fume induced fluorescence in diagnosis of nasal neuroblastoma. Arch Otolaryngol 102:97–98

Juillard MT, Collin JP (1976) L'organe pinéal aviaire: étude ultracytochimique et pharmacologique d'un "pool" granulaire de 5-hydroxytryptamine chez la Perruche (*Melopsittacus undulatus*, Shaw). J Microsc Biol Cell 26:133–138

Julesz J (1976) Radiommunoassay of the gastrointestinal hormones. Acta Physiol Acad Sci Hung 47: 355–367

Jurecka W, Lassmann H, Hörandner H (1978) The proliferation of adrenal medullary cells in newborn and adult mice. A light and electron microscopic autoradiographic study. Cell Tiss Res 189:305–312

Kadowaki T, Kasuga M, Akanuma K (1985) Hormone receptors in adipocytes (in Japanese). Seitai no Kagaku (Tokyo) 36:580–593

Kagayama K (1965) The follicular cell in the pars distalis of the dog pituitary gland: an electron microscopic study. Endocrinology 77:1053–1060

Kagayama M, Douglas WW (1974) Electron microscope evidence of calcium-induced exocytosis in mast cells treated with 48/80 or the ionophores A-23187 and X-537A. J Cell Biol 62:519–526

Kahn CR (1985) Insulin receptors. In: Alberti KGMM, Krall LP (eds) The Diabetes Annual 1. Elsevier, Amsterdam New York Oxford, pp 446–462

Kajihara H, Akimoto T, Iijima S (1978) On the chromaffin cells in dog adrenal medulla, with special reference to the small granule chromaffin cells (SGC cells). Cell Tiss Res 191:1–14

Kakidani H, Furutani Y, Takahashi H, Noda M, Morimoto Y, Hirose T, Asai M, Inayama S, Nakanishi S, Numa S (1982) Cloning and sequence analysis of cDNA for porcine B-neo-endorphin/dynorphin precursor. Nature 298:245–249

Kakudo K, Uematsu K, Matsuno Y, Mitsunobu M, Toyosaka A, Okamoto E, Fukuchi M (1984) Malignant pheochromocytoma with ACTH production. Acta Pathol Jpn 34:1403–1410

Kameda Y (1968) Parafollicular cells of the thyroid as studied with Davenport's silver impregnation. Arch Histol Jpn 30:83–94

Kameda Y (1970) Increased mitotic activity of the parafollicular cells of the dog thyroid in experimentally induced hypercalcemia. Arch Histol Jpn 32: 179–192

Kameda Y (1971) The occurrence and distribution of the parafollicular cells in the thyroid, parathyroid IV and thymus IV in some mammals. Arch Histol Jpn 33:283–299

Kameda Y (1973) Electron microscopic studies on the parafollicular cells and parafollicular cell complexes in the dog. Arch Histol Jpn 36:89–105

Kameda Y (1977) Electron microscopical and immunohistochemical study of parafollicular cell complex with reference to parafollicular cell as a paraneuron. Arch Histol Jpn 40 Suppl: 133–145

Kameda Y (1982a) Immunohistochemical study of C cell follicles in dog thyroid glands. Anat Rec 204: 855–60

Kameda Y (1982b) Degranulation and appearance of vesicular inclusions in canine C cells after administration of antithyroid drug. Cell Tiss Res 225: 315–318

Kameda Y, Ikeda A (1978) Immunohistochemical study of a large molecular fragment of tyroglobulin in parafollicular cells. Cell Tiss Res 187:429–438

Kameda Y, Ikeda A (1979) C cell (parafollicular cell)–immunoreactive thyroglobulin: Purification, identification and immunological characterization. Histochemistry 60:155–168

Kameda Y, Ikeda A (1980) Immunohistochemical study of the C-cell complex of dog thyroid gland with reference to the reactions of calcitonin, C-thyroglobulin and 19S thyroglobulin. Cell Tiss Res 208:405–415

Kameda Y, Oyama H, Endoh M, Horino M (1982) Somatostatin immunoreactive C cells in thyroid glands from various mammalian species. Anat Rec 204:161–170

Kameya T, Shimosato Y, Hayashi H, Tsumuraya M (1977) Growth and differentiation of hormone-producing human tumors in nude mice. In: Nomura T, Ohsawa N, Tamaoki N, Fujiwara K (eds) Proceedings of the second international workshop on nude mice. University of Tokyo Press, Tokyo/Gustav-Fischer, Stuttgart, pp 405–416

Kameya T, Bessho T, Tasumuraya M, Yamaguchi K, Abe K, Shimosato Y, Yanaihara N (1983a) Production of gastrin releaseing peptide by medullary carcinoma of the thyroid. An immunohistochemical study. Virchows Arch (Pathol Anat) 401:99–108

Kameya T, Yamaguchi K, Abe K, Nakajima T, Shimase J, Shimosato Y, Yanaihara N (1983b) Immunoreactive gastrin-releasing peptide in human neuroendocrine tumors studies by immunohistochemistry. Biomed Res 4 Suppl: 301–306

Kanaseki T, Kadota K (1969) The "vesicle in a basket". A morphological study of the coated vesicle isolated from the nerve endings of the guinea pig brain, with special reference to the mechanism of membrane movements. J Cell Biol 42:202–220

Kanazawa Y, Kawazu S, Kosaka K (1976) Mechanism of insulin release: Studies in living B cell in monolayer culture. In: Fujita T (ed) Endocrine gut and pancreas. Elsevier, Amsterdam, pp 301–312

Kaneko A, Tsukahara N, Uchizono K (eds) (1982) Neurotransmitters in the retina and the visual centers. Biomed Res Found, Tokyo

Kaneko T, Kaku K (1983) Regulation of pineal function by vasoactive intestinal polypeptide (VIP) and peptide with N-terminal histidine and C-terminal isoleucine (PHI). Biomed Res 4 Suppl: 251–254

Kaneko T, Cheng P, Oka H, Oda T, Yanaihara N, Yanaihara C (1980) Vasoactive intestinal polypeptide stimulates adenylate cyclase and serotonin *N*-acetyltransferase activities in rat pineal *in vitro*. Biomed Res 1:84–87

Kangawa K, Matsuo H (1984) Purification and complete amino acid sequence of alpha-human atrial natriuretic polypeptide (α-hANP). Biochem Biophys Res Comm 118:131–139.

Kangawa K, Fukuda A, Minamino N, Matsuo H (1984) Purification and complete amino acid sequence of beta-rat atrial natriuretic polypeptide (β-rANP) of 5,000 daltons. Biochem Biophys Res Commun 119:933–940

Kannan H, Yamashita H (1983) Electrophysiological study of paraventricular nucleus neurons projecting to the dorsomedial medulla and their response to baroreceptor stimulation in rats. Brain Res 279: 31–40

Kanno T (1976) The relationship between changes in electrophysiological properties of adrenal chromaffin cells and catecholamine release. In: Coupland RE, Fujita T (eds) Chromaffin, enterochromaffin and related cells. Elsevier, Amsterdam New York Oxford, pp 47–57

Kanno T (1977) Physiology of paraneurons. Arch Histol Jpn 40 Suppl: 13–29

Kanno T (1980) Process of reception, conduction and extrusion in paraneurons. Biomed Res 1 Suppl: 10–16

Kanno T (1981) Stimulus-extrusion coupling in mast cells (in Japanese). Saibou 13:240–245

Kanno T (1984) Physiology of brain-gut peptides. In: Yanaihara N, Nakagawa H (eds) Recent advances in gut hormone research 5, Brain-gut peptide (in Japanese), Chugai Igaku, Tokyo, pp 33–43

Kanno T (1985) Electrophysiology and stimulus-secretion coupling of the pancreatic acinar cell. In: Poisner AM, Trifaro JM (eds) The Electrophysiology of the secretory cell. Elsevier, Amsterdam, pp 3–24

Kanno T, Douglas WW (1967) Effect of rapid application of acetylcholine or depolarizing current on transmembrane potentials of adrenal chromaffin cells. Can Fed Biol Soc 10:39

Kanno T, Saito A (1976) The potentiating influences of insulin on pancreozymin-induced hyperpolarization and amylase release in the pancreatic acinar cell. J Physiol 261:505–521

Kanno T, Saito A (1978) Influence of external potassium concentration on secretory responses to cholecystokinin-pancreozymin and ionophore A23187 in the pancreatic acinar cell. J Physiol 278: 251–263

Kanno T, Saito A (1979) Cellular mechanism of insulin enhancement of cholecystokinin-induced responses of the pancreatic acinar cell. In: Baba S, Kaneko T, Yanaihara N (eds) Proinsulin, insulin, C-peptide. (Proceedings of the Symposium on Proinsulin, Insulin and C-peptide. Tokushima, 1978) Excerpta Medica, Amsterdam Oxford, pp 148–154

Kanno T, Saito A (1981) Marked reduction of the cytochromes induced in bullfrog brain by perfusing with a calcium-deficient or a high potassium solution. Biomed Res 2:699–704

Kanno T, Saito A (1982) Direct effect of CCK-8 stimulating respiratory chain in bullfrog brain. Biomed Res 3:313–318

Kanno T, Saito T (1985) Parrallel reduction response of cytochromes to cholecystokinin-octapeptide and neuropeptide Y, and parallel oxidation response to vasoactive intestinal polypeptide, both coincide with electrical activity in perfused bullfrog brain. In: Kobayashi H, Bern HA, Urano A (eds) Neurosecretion and the biology of neuropeptides. Jpn Sci Soc Press, Tokyo/Springer, Berlin, pp 350–358

Kanno T, Saito T (1986) Vasoactive intestinal polypeptide nullifies cytochrome reduction response to afferent activation of the perfused bullfrog brain. Brain Res 369:275–284

Kanno T, Sampson SR, Douglas WW (1966) Membrane potentials recorded from chromaffin cells of the adrenal medulla. Fed Proc 25:290

Kanno T, Cochrane DE, Douglas WW (1973) Exocytosis (secretory granule extrusion) induced by injection of calcium into mast cells. Can J Physiol Pharmacol 51:1001–1004

Kanno T, Ueda N, Saito A (1976) Insulo-acinar axis: a possible role of insulin potentiating the effects of pancreozymin in the pancreatic acinar cell. In: Fujita T (ed) Endocrine gut and pancreas. Elsevier, Amsterdam, pp 335–345

Kanno T, Saito A, Yonezawa H (1980) Unidirectional cellular processes of stimulus-secretion coupling in a cell secereting cholecystokinin-pancreozymin. In: Eränkö O, Soinila S, Päivärinta H (eds) Histochemistry and cell biology of autonomic neurons, SIF cells, and paraneurons. Raven Press, New York, pp 327–332

Kanno T, Ikei N, Nakase Y (1981) Simultaneous measurements of cholecystokinin-induced changes in absorption spectrum of cytochromes, NAD (P)H-fluorescence, and enzyme output in isolated perfused rat pancreas. Biomed Res 2:381–389

Kanno T, Saito A, Ikei N (1983a) Dose-dependent effects of acetylcholine stimulating respiratory chain and secretion of isolated perfused rat pancreas. Biomed Res 4:175–186

Kanno T, Shibuya I, Asada N (1983b) The role of extracellular calcium and other ions in the secretory responses of the exocrine pancreas to cholecystokinin. Biomed Res 4:295–320

Kanno T, Matsumoto T, Kawahara J, Hattori C, Saito T, Kato Y, Eguchi J (1984) Extracellular calcium responsible for the secretory response of the exocrine pancreas to cholecystokinin. In: Case RM, Lingard JM, Young JA (eds) Secretion: mechanisms and control. Manchester University Press, Manchester, pp 181–186

Kanno T, Saito T, Saito A (1985) Cytochrome reduction coincides with electrical activity in perfused bullfrog brain. Brain Res 338:237–242

Kanno Y (1985) Modulation of cell communication and carcinogenesis. Jpn J Physiol 35:693–707

Kappers JA (1960) The development topographical relations and innervation of the epiphysis cerebri in the albino rat. Z Zellforsch 52:163–215

Karasek M (1983) Ultrastructure of the mammalian pineal gland: its comparative and functional aspects. In: Reiter RJ (ed) Pineal Reasearch Reviews, vol 1. Alan R Liss, New York, pp 1–48

Kare MD (1971) Comparative study of taste. In: Beidler LM (ed) Taste (Handbook of sensory physiology IV/2). Springer, Berlin Heidelberg New york, pp 278–292

Karlson P (1982) Was sind Hormone? Der Hormonbegriff in Geschichte und Gegenwart. Naturwissenschaften 69:3–14

Kashiwayanagi M, Miyake M, Kurihara K (1983) Voltage-dependent Ca^+ channel and Na^{2+} channel in frog taste cells. Am J Physiol 244:C82–C88

Kataoka K (1973) An electron microscope study of the gastro-enteric endocrine cells of the frog, *Rana nigromaculata nigromaculata*. In: Fujita T (ed) Gastro-entero-pancreatic endocrine system. Igaku-Shoin, Tokyo, pp 39–48

Kataoka K (1974) An electron microscope study on a neuro-endocrine complex in the proventricular mucosa of the finch, *Uroloncha striata* var *domestica*. Arch Histol Jpn 36:391–400

Kataoka K, Fujita H (1974) The occurrence of endocrine cells in the intestine of the lancelet, *Branchiostoma japonicum*. An electron microscope study. Arch Histol Jpn 36:401–406

Kataoka K, Yamamoto M, Yamamoto T, Ochi J (1982) Intercellular canalicular system and intercellular junctions in the pancreatic islet of the Mongolian garbil. Biomed Res 3:235–238

Kataoka S, Yamamoto TY (1981) Diurnal changes in the fine structure of photoreceptors in and abalone, *Nordotis discus*. Cell Tiss Res 218:181–189

Kataoka S, Yamamoto TY (1983) Fine structure and formation of the photoreceptor in *Octopus ocellatus*. Biol Cell 49:45–54

Kater SB, Murphy AD, Rued JR (1978) Control of the salivary glands of *Helisoma* by identified neurones. J Exp Biol 72:91–106

Katsuki Y (1982) Receptive mechanisms of sound in the ear. Cambridge Univ Press, Cambridge

Katsuki Y (1985) Recept progress in research on auditory mechanism. Biomed Res 6:1–12

Katsuki Y, Yanagisawa K (1982) Chemoreception in the lateralline organ. In: Hara Th (ed) Chemoreception in fishes. Elsevier, Amsterdam, pp 227–242

Katz FN, Rothman JE, Lingappa JE, Blobel VR, Lodish HF (1977) Membrane assembly in vitro: Synthesis, glycosylation, and asymmetric insertion of a transmembrane protein. Proc Natl Acad Sci USA 74:3278–3282

Kaul S, Vollrath L (1974) The goldfish pituitary. II. Innervation. Cell Tiss Res 154:231–249

Kawamoto M (1967) Zur Morphologie der Hypophysis cerebri von Teleostiern. Arch Histol Jpn 28:123–150

Kawamura T, Yamashita S (1981) Chemical and thermal responses from buccal and maxillary nerves in the minnow, *Pseudorasbora parva*. Comp Biochem Physiol 69A:187–195

Kawamura T, Ishibashi Y, Mori S (1971) Haarscheibe, with special reference to its Merkel cells and perineurale Pigmenthulle (periaxonale Pigmenthulle). Jpn J Dermatol 81:363–376

Kazimierczak W, Diamant B (1978) Mechanisms of histamine release in anaphylactic and anaphylactoid reactions. Prog Allergy 24:295–365

Kazzaz BA (1974) Argentaffin and argyrophil cells in the prostate. J Path 112:189–193

Kelly RB, Hooper JE (1982) Cholinergic vesicles. In: Poisner AM, Trifaró JM (eds) The secretory

granule. Elsevier, Amsterdam New York Oxford, pp 81–118
Kemper B, Habener JF, Mulligan RC, Potts JT, Rich A (1974) Pre-proparathyroid Hormone: A direct translation product of parathyroid messenger RNA. Proc Nat Acad Sci USA 71:3731–3735
Kenigsberg RL, Trifaró JM (1980) Presence of a high affinity uptake system for catecholamines in cultured bovine adrenal chromaffin cells. Neuroscience 5:1547–1556
Kent C, Coupland RE (1981) On the uptake of exogenous catecholamines by adrenal chromaffin cells and nerve endings. Cell Tiss Res 221:371–383
Kent C, Coupland RE (1984) On the uptake and storage of 5-hydroxytryptamine, 5-hydroxytryptophan and catecholamines by adrenal chromaffin cells and nerve endings. Cell Tiss Res 236:189–195
Kent GC (1973) Comparative anatomy of the vertebrates. C V Mosby Co/Toppan Co, Saint Louis/ Tokyo
Kern HF, Hofmann HV, Kern D (1971) Licht- und elektronen-mikroskopische Untersuchung der Langerhansschen Inseln von Nutria (*Myocastor coypus*), mit besonderer Berücksichtigung der neuroinsulären Komplexe. Z Zellforsch 113:216–229
Kessler JA (1984) Environmental co-regulation of substance P, somatostatin and neurotransmitter synthesizing enzymes in cultured sympathetic neurons. Brain Res 321:155–159
Keuning FJ (1944) The development of the intramural nerve elements of the digestive tract in tissue culture. Acta neerl Morphol 5:237–247
Kidokoro Y (1975) Spontaneous Calcium action potentials in a clonal pituitary cell line and their relationship to prolactin secretion. Nature 258:741–742
Kidokoro Y (1984) Acetylcholine-induced potential fluctuations and underlying currents in rat adrenal chromaffin cells. Fed Proc 43:2364–2367
Kidokoro Y (1985) Electrophysiology of adrenal chromaffin cells. In: Poisner AM, Trifaro JM (eds) The electrophysiology of the secretory cell. Elsevier, Amsterdam, pp 195–218
Kikuchi K, Hilding D (1965) The development of the organ of Corti in the mouse. Acta Otolaryngol 60:207–222
Kikuta A, Murakami T (1982) Microcirculation of the rat adrenal gland: A scanning electron microscope study of vascular casts. Am J Anat 164:19–28
Kikuta A, Murakami T (1984) Relationship between chromaffin cells and blood vessels in the rat adrenal medulla: A transmission electron microscopic study combined with blood vessel reconstructions. Am J Anat 170:73–81
Kilpatrick DL, Jones BN, Kojima K, Udenfriend S (1981) Identification of the octapeptide Met-enkephalin-Arg^6-Gly^7-Leu^8 in extracts of bovine adrenal medulla. Biochem Biophys Res Commun 103:698–705
Kimmel JR, Pollock HG, Hazelwood RL (1968) Isolation and characterization of chicken insulin. Endocrinology 83:1323–1330
Kimmel JR, Hayden LJ, Pollock HG (1975) Isolation and characterization of a new pancreatic polypeptide hormone. J Biol Chem 250:9369–9376
Kimura A, Beider LM (1961) Microelectrode study of taste receptors of rat and hamster. J Cell Comp Physiol 58:131–140
Kimura RS (1966) Hairs of the cochlear sensory cells and their attachment to the tectorial membrane. Acta Otolaryngol 61:55–72
Kimura S, Lewis RV, Stern AS, Rossier J, Stein S, Udenfriend S (1980) Probable precursor of [Leu] enkephalin and [Met] enkephalin in adrenal medulla. peptides of 3–5 kilodaltons. Proc Natl Acad Sci USA 77:1681–1685
Kinnamon SC, Roper SD (1986) Evidence for a role of voltage-sensitive potassium channels in taste transduction. Biophy J 49:21*a*
King AS, McLelland J, Cook RD, King DZ, Walsh C (1974) The ultrastructure of afferent nerve endings in the avian lung. Resp Physiol 22:21–40
Kirkeby S, Rømert P (1977) Argyrophilic cells in the larynx of the guinea-pig demonstrated by the method of Grimelius. J Anat 123:87–92
Kirpekar SM, Nobiletti J, Trifaró JM (1983) Effects fo 6-hydroxydopamine on bovine adrenal chromaffin cells in culture. Br J Pharmacol 79:947–952
Kirschner N (1975) Biosynthesis of the catecholamines. In: Greep RO, Astwood EB (eds) Adrenal gland (Handbook of Physiology 7/VI). Am Physiol Soc, Washington DC, pp 341–355
Kirschner N, Sage HJ, Smith WJ, Kirschner AG (1966) Release of catecholamines and specific protein from adrenal glands. Science 154:529–531
Kirschner N, Sage HJ, Smith WJ, Kirschner AG (1967) Mechanism of secretion from the adrenal medulla. 2. Release of catecholamines and storage vesicle protein in response to chemical stimulation. Mol Pharmacol 3:254–265
Kisch B (1956) Electron microscopy of the atrium of the heart. I. Guinea pig. Exp Med Surg 14:99–112
Kiss J, Léránth Cs, Halász B (1984) Serotoninergic endings on VIP-neurons in the suprachiasmatic nucleus and on ACTH-neurons in the arcuate nucleus of the rat hypothalamus. A combination of high resolution autoradiography and electon microscopic immunocytochemistry. Neurosci Lett 44: 119–124
Kitamura Y, Shimada M, Hatanaka K, Miyano Y (1977) Development of mast cells from grafted bone marrow cells in irradiated mice. Nature 268:442–443
Kitamura Y, Shimada M, Go S, Matsuda H, Hatanaka K, Seki M (1979a) Distribution of mast-cell precursors in hematopoietic and lymphopoietic tissues of mice. J Exp Med 150:482–490
Kitamura Y, Hatanaka K, Shimada M, Go S, Matsuda H (1979b) Local control of mast cell differentiation in mice. In: Baum SJ, Ledney GD (eds) Experimental hematology today 1979, Springer, New York Heidelberg Berlin, pp 45–50
Kjærgaad J (1973) Anatomy of the carotid glomus and carotid glomus-like bodies (non-chromaffin paraganglia). FADL's Forlag, Copenhagen.
Klaeveman HL, Conlon TP, Levy AG, Gardner JD (1975) Effects of gastrointestinal hormones on adenylate cyclase activity in human jejunal mucosa.

Gastroenterology 68:667–675
Klein C, Van Noorden S (1980) Pancreatic polypeptide (PP)- and glucagon cells in the pancreatic islet of *Xiphophorus helleri* H. (Teleostei). Cell Tiss Res 205:187–198
Klöppel G, Schäfer HJ (1976) Effects of sulphonylureas on histochemical and ultrastructural calcium distribution in B cells of mice. Diabetologia 12: 227–235
Knight DE, Kesteven NT (1983) Evoked transient intracellular free Ca^{2+} changes and secretion in isolated bovine adrenal medullary cells. Proc Roy Soc Lond B 218:177–199
Knoth J, Zallakian M, Njus D (1982) Mechanism of proton-linked monoamine transport in chromaffin granule ghosts. Fed Proc 41:2742–2745
Kobayashi M (1971) Fine stucture of the basal granular cell in the mid-gut epithelium of the silkworm, *Bombyx mori* (Japanese text with English summary). J Sericult Sci Jap 40:101–106
Kobayashi S (1968) Fine structure of the carotid body of the dog. Arch Histol Jpn 30:95–120
Kobayashi S (1969a) Catecholamines in the avian carotid body. Experientia 25:1075–1076
Kobayashi S (1969b) On the fine structure of the carotid body of the bird, *Uroloncha domestica*. Arch Histol Jpn 31:9–19
Kobayashi S (1971a) Comparative cytological studies of the carotid body 1. Demonstration of monoamine-storing cells by correlated chromaffin reaction and fluorescence histochemistry. Arch Histol Jpn 33: 319–339
Kobayashi S (1971b) Comparative cytological studies of the carotid body. 2. Ultrastructure of the synapses on the chief cell. Arch Histol Jpn 33:397–420
Kobayashi S (1975) Uptake and intracellular localization of exogenous L-DOPA, L-Leucine and their metabolites in the gastro-enteric endocrine cells of the mouse studied by electron microscope autoradiography. Arch Histol Jpn 37:313–333
Kobayashi S (1976) An autoradiographic study of the mouse carotid body using tritiated leucine, dopa, dopamine and ATP with special reference to the chief cell as a paraneuron. Arch Histol Jpn 39: 295–317
Kobayashi S (1977) Adrenal medulla: chromaffin cells as paraneurons. Arch Histol Jpn Suppl 40:61–79
Kobayashi S (1979) Cellular background in gut hormone secretion. In: Miyoshi A (ed) Gut peptides. Secretion, function and clinical aspects. Kodansha, Tokyo, pp 53–58
Kobayashi S (1980) Synaptic vesicles and secretory granules. A morphologist's view on their identity and difference. Biomed Res Suppl 1:30–33
Kobayashi S, Coupland RE (1977) Two populations of microvesicles in the SGC (small granule chromaffin) cells of the mouse adrenal medulla. Arch Histol Jpn 40:251–259
Kobayashi S, Fujita T (1969) Fine structure of mammalian and avian pancreatic islets with special reference to D cells and nervous elements. Z Zellforsch 100:340–363
Kobayashi S, Fujita T (1973) Emiocytotic granule release in the basal-granulated cells of the dog induced by intraluminal application of adequate stimuli. In: Fujita T (ed) Gastro-entero-pancreatic endocrine system: a cell-biological approach. Igaku-Shoin, Tokyo, pp 49–58
Kobayashi S, Nishisaka T (1985) Myenteric enkephalin neurons around the laser-photocoagulation necrosis: an immunocytochemical investigation in the guinea pig jejunum and proximal colon. Arch Histol Jpn 48:239–254
Kobayashi S, Sasagawa T (1976) Morphological aspects of the secretion of gastro-enteric hormones. In: Fujita T (ed) Endocrine gut and pancreas. Elsevier, Amsterdam, pp 255–271
Kobayashi S, Segi M (1984) Gut paraneurons and Segi's cap. In: Motta PM (ed) Ultrastructure of endocrine cells and tissues. Martinus Nijhoff, Boston/The Hague Dordrecht, Lancaster, pp 127–135
Kobayashi S, Serizawa Y (1979) Stress-induced degranulation accompanied by vesicle formation in the adrenal chromaffin cells of the mouse. Arch Histol Jpn 42:375–388
Kobayashi S, Uchida T (1982) Tubulo-vesicular structures in the basal cytoplasm of gut basal-granulated cells revealed by zinc iodide-osmium tetroxide staining. Okajimas Folia Anat Jpn 58:1009–1018
Kobayashi S, Uchida T (1983) Adrenal chromaffin cells. Immunocytochemical demonstration of catecholamines and opioid peptides. Gunma Symp Endocrinol 20:39–49
Kobayashi S, Uehara M (1970) Occurrence of afferent synaptic complexes in the carotid body of the mouse. Arch Histol Jpn 32:193–201
Kobayashi S, Watanabe Y (1971) Selective staining of A cells in the freeze-dried and formaldehyde-fixed adrenal medulla with toluidine blue and pseudoisocyanin. Arch Histol Jpn 33:59–66
Kobayashi S, Fujita T, Sasagawa T (1970) The endocrine cells of human duodenal mucosa. An electron microscope study. Arch Histol Jpn 31:477–494
Kobayashi S, Fujita T, Sasagawa T (1971) Electron microscope studies on the endocrine cells of the human gastric fundus. Arch Histol Jpn 32:429–444
Kobayashi S, Kent C, Coupland RE (1978a) Observations on the localization of labelled amino acid in mouse adrenal chromaffin cells after the injection of L-[4, 5-^{3}H]-leucine. J Endocrinol 78:21–29
Kobayashi S, Serizawa Y, Fujita T, Coupland RE (1978b) SGC (small granule chromaffin) cells in the mouse adrenal medulla: light and electron microscopic identification using semi-thin and ultra-thin sections. Endocrinol Jpn 25:467–476
Kobayashi S, Fujita T, Ito S (1979) Secretory granules in the insulinoma cell. an interpretation of their fine structural changes. In: Baba S, Kaneko T, Yanaihara N (eds) Proinsulin, insulin, C-peptide. Excerpta Medica, Amsterdam, pp 394–401
Kobayashi S, Iwanaga T, Fujita T (1980a) Segi's cap: huge aggregation of basal-granulated cells discovered by Segi (1935) on the intestinal villi of the human fetus. Arch Histol Jpn 43:79–83
Kobayashi S, Iwanaga T, Fujita T, Yanaihara N (1980b) Do enterochromaffin (EC) cells contain motilin? Arch Histol Jpn 43:85–98

Kobayashi S, Yoshimura M, Shibahara T, Nakase Y, Yaono S (1981) Scanning spectrophotometry for dynamic study of the organ oxidative metabolism. Biomed Res 2:390–397
Kobayashi S, Uchida T, Ohashi T, Fujita T, Imura H, Mochizuki T, Yanaihara C, Yanaihara N (1983a) Met-enkephalin-Arg-Gly-Leu-like immunoreactivity in adrenal chromaffin cells and carotid body chief cells of the dog and monkey. Biomed Res 4:201–210
Kobayashi S, Uchida T, Ohashi T, Fujita T, Nakao K, Yoshimasa T, Imura H, Mochizuki T, Yanaihara C, Yanaihara N, Verhofstad AAJ (1983b) Immunocytochemical demonstration of the co-storage of noradrenalin with Met-enkephalin Arg^6-Phe^7 and Met-enkephalin-Arg^6-Gly^7-Leu^8 in the carotid body chief cells of the dog. Arch Histol Jpn 46:713–722
Kobayashi S, Ohashi T, Fujita T, Nakao K, Yoshimasa T, Imura H, Mochizuki T, Yanaihara C, Yanaihara N, Verhofstad AAJ (1983c) An immunohistochemical study on the co-storage of Met-enkephalin-Arg^6-Gly^7-leu^8 and Met-enkephalin-Arg^6-Phe^7 with adrenaline and/or noradrenaline in the adrenal chromaffin cells of the rat, dog and cat. Biomed Res 4:433–442
Kobayashi S, Ohashi T, Uchida T, Nakao K, Imura H, Yanaihara N, Verhofstad AAJ (1984a) Co-storage of adrenaline and noradrenaline with Met-enkephalin-Arg^6-Gly^7-Leu^8 and Met-enkephalin-Arg^6-Phe^7 in chromaffin cells of hamster adrenal medulla. Arch Histol Jpn 47:319–336
Kobayashi S, Suzuki M, Uchida T, Yanaihara N (1984b) Enkephalin neurons in the guinea pig duodenum: a light and electron microscopic immunocytochemical study using an antiserum to methionine-enkepalin-Arg^6-Gly^7-Leu^8. Biomed Res 5:489–506
Kobayashi S, Suzuki M, Yanaihara N (1985) Enkephalin neurons in the guinea pig proximal colon: an immunocytochemical study using an antiserum to methionine-enkephalin-Arg^6-Gly^7-Leu^8. Arch Histol Jpn 48:27–44
Kobayashi S, Suzuki M, Endo T, Tsuji S, Daniel EE (1986) Framework of the enteric nerve plexuses: an immunocytochemical study in the guinea pig jejunum using an antiserum to S-100 protein. Arch Histol Jpn 49:159–188
Kohn A (1895) Studien über die Schilddrüse. Arch mikrosk Anat 44:366–422
Kohn A (1898) Die chromaffin Zellen des Sympathicus. Anat Anz 15:393–400
Kohn A (1900) Die Epithelkörperchen. Ergebn Anat 9:194–252
Kohn A (1903) Die Paraganglien. Arch Mikrosk Anat 62:263–365
Koketsu K (1984) Modulation of receptor sensitivity and action potentials by transmitters in vertebrate neurones. Jpn J Physiol 34:945–960
Kokue E, Hayama T (1976) The endogenous secretin in chicken: minor physiological role in exocrine pancreatic secretion. Jpn J Physiol 26:1–8
Kolmer W (1921) Das "Sagittalorgan" der Wirbeltiere. Z Anat Entw-Gesch 60:652–717
Kondo H (1971) An electron microscopic study on innervation of the carotid body of guinea pig. J Ultrasturct Res 37:544–562
Kondo H (1974) On the granule-containing cells in the aortic wall of the young chick. Anat Rec 178:253–266
Kondo H (1977) Innervation of the chief cells of the carotid body. An ultrastructural review. Arch Histol Jpn Suppl 40:221–230
Kondo H (1985) Immunohistochemical analysis of the localization of neuropeptides in the adrenal galnd. Arch Histol Jpn 48:453–481
Kondo H, Yui R (1981) Enkephalin-like immunoreactivity in the SIF cells of sympathetic ganglia of frogs. Biomed Res 2:338–340
Kondo H, Yui R (1984) Co-existence of enkephalin and adrenalin in the frog adrenal gland. Histochemistry 80:243–246
Kondo H, Iwanaga T, Nakajima T (1982) Immunocytochemical study on the localization of neuron-specific enolase and S-100 protein in the carotid body of rats. Cell Tiss Res 227:291–295
Kondo H, Iwanaga T, Nakajima T (1983) An immunocytochemical study on the localization of S-100 protein in the retina of rats. Cell Tiss Res 231:527–532
Kondo H, Kuramoto H, Iwanaga T, Fujita T (1985) Cerebellar purkinje cell-specific protein-like immunoreactivity in noradrenalin-chromaffin cells and ganglion cells but not in adrenaline-chromaffin cells in the rat adrenal medulla. Arch Histol Jpn 48:421–426
Kondo H, Kuramoto H, Fujita T (1986) Neuropeptide tyrosine-like immunoreactive nerve fibers in the carotid body chemoreceptor of rats. Brain Res 372:353–356
Konishi S, Tsunoo A, Yanaihara N, Otsuka M (1980) Peptidergic excitatory and inhibitory synapses in mammalian sympathetic ganglia: roles of substance P and enkephalin. Biomed Res 1:528–536
Konturek SJ, Llanos OL, Rayford PL, Thompson JC (1977) Vagal influence on gastrin and gastric acid responses to gastric and intestinal meansl. AM J Physiol 232:E542–E546
Korc M, Owerbach D, Quinto C, Rutter WJ (1981) Pancreatic islet acinar cell interaction: amylase messenger RNA levels are determined by insulin. Science 213:351–353
Korkala O, Hervonen A (1973) Origin and development of the catecholamine-storing cells of the human fetal carotid body. Histochemie 37:287–297
Korte GK (1980) Ultrastructure of the tastebuds of the redeared turtle, *Chrysemys scripta elegans*. J Morphol 163:231–252
Kose W (1907) Die Paraganglien bei den Vögeln. Zweiter Teil. Arch Mikrosk Anat 69:665–790
Kostron H, Winkler H, Geissler D, König P (1977) Uptake of calcium by chromaffin granules in vitro. J Neurochem 28:487–497
Kotani M, Seiki K, Higashida M, Imanishi Y, Yamashita A, Miyamoto M, Horii I (1968) Demonstration of thyroglobulin in dog lymph. Endocrinol 82:1047–1049
Kramer KJ, Tager HS, Childs CN (1980) Insulin-like, glucagon-like, peptides in insect hemolymph. Insect Biochem 10:179–182

Kratzing JE (1971) Olfactory and vomeronasal organs in rats, sheep, and a lizard. Thesis, University of Queensland.

Kratzing JE (1978) The olfactory apparatus of the bandicoot (*Isoodon macrourus*): fine structure and presence of a septal olfactory organ. J Anat 125: 601–613

Kreidl A (1893) Weitere Beiträge zur Physiologie des ohrlabyrinthes. II. Versuche an Krebsen. S B Akad Wiss Wien (III) 102:149–174

Kreisman NR, Sick TJ, Lamanna JC, Rosenthal M (1981a) Local tissue oxygen tension—cytochrome a, a_3 redox relationships in rat cerebral cortex in vivo. Brain Res 218:161–174

Kreisman NR, Lamanna JC, Rosenthal M, Sick TJ (1981b) Oxidative metabolic responses with recurrent seizures in rat cerebral cortex: role of systemic factors. Brain Res 218:175–188

Kreulen DL (1984) Integration in autonomic ganglia. Physiologist 27:49–55

Krieger DT (1983) Brain peptides: what, where, and why? Science 222:975–985

Krisch B (1986) Ultrastructure of regulatory neuroendocrine neurons and functionally related structures. Current Topics Neuroendocrinol 7:251–290

Krstić R (1975) Scanning electron microscope observations of the canaliculi in the rat pineal gland. Experientia 31:1072–1074

Kruggel W, O'Connoe DT, Lewis RV (1985) The amid terminal sequences of bovine and human chromogranin A and secretory protein I are identical. Biochem Biophys Res Commun 127:380–383

Krylov SS, Anichkov SV (1968) The effect of metabolic inhibitors on carotid chemoreceptors. In: Torrance RW (ed) Arterial chemoreceptors. Blackwell, Oxford, pp 103–113

Kubben FJGM, Assche CLMBJ van, Bosman FT (1986) FMRF-amide immunoreactivity in the mammalian gastroenteropancreatic neuroendocrine system. Histochemistry 84:439–444

Kubozoe T, Daikoku S, Takita S (1969) Electron microscopic observations on Auerbach's plexus in a 12mm human embryo. J NeuroVisc Rel 31:291–307

Kuffler SW (1980) Slow synaptic responses in autonomic ganglia and the pursuit of a peptidergic transmitter. J Exp Biol 89:257–286

Kuffler SW, Sejnowski TJ (1983) Peptidergic and muscarinic excitation at amphibian sympathetic synapses. J Physiol 341:257–278

Kukreja SC, Johnson PA, Ayala G, Banerjee P, Bowser EN, Hargis GK, Williams, GA (1976) Role of calcium and beta-adrenergic system in control of parathyroid hormone secretion. Proc Soc Exp Biol Med 151:325–328

Kull H (1912) Die "basalgekörnten Zellen" des Dünndarmepithels. Arch Mikrosk Anat 81:185–195

Kull H (1925) Die chromaffinen Zellen des Verdauungstraktus. Z Mikr-Anat Forsch 2:163–200

Kultschitzky N (1896) Über acidophile Zellen im Epithel des Darmkanals. Med Umschau (Moskau) 46:90–91

Kultschitzky N (1897) Zur Frage über den Bau des Darmkanals. Arch Mikrosk Anat 49:7–35

Kummer W, Addieks K (1982) The paraganglion supracardiale ragi: an introvagal paraganglion in the rat. Cell Tiss Res 224:455–458

Kuramoto H (1987) An immunohistochemical study of chromaffin cells and nerve fibers in the adrenal gland of the bullfrog, *Rana catesbeiana*. Arch Histol Jpn 50:15–38

Kuramoto H, Yamada J, Kitamura N, Yamashita T, Yanaihara N (1983) An immunohistochemical and ultrastructural study of Segi's cap, a large aggregation of gut endocrine cells, in bovine fetuses. Arch Histol Jpn 46:701–711

Kuramoto H, Kondo H, Fujita T (1985) Substance P-like immunoreactivity in adrenal chromaffin cells and intra-adrenal nerve fibers of rats. Histochemistry 82:507–512

Kuramoto H, Kondo H, Fujita T (1986) Neuropeptide tyrosine (NPY)-like immunoreactivity in adrenal chromaffin cells and intraadrenal nerve fibers of rats. Anat Rec 214:321–328

Kuramoto H, Kondo H, Fujita T (1987) Calcitonin gene-related peptide (CGRP)-like immunoreactivity in scattered chromaffin cells and nerve fibers in the adrenal gland of rats. Cell Tiss Res 247:309–315

Kurihara K (1984) Molecular mechanism of chemoreception. Protein, Nucleic Acid and Enzyme 26: 1175–1187 (in Japanese)

Kurihara K, Yoshii K, Miyake M (1981) Molecular mechanisms of transduction in chemoreception. In: Cagen RH, Kare MR (eds) Biochemistry of taste and smell. Academic Press, New York, pp 249–285

Kurose M, Saeki K (1981) Histamine release induced by neurotensin from rat peritoneal mast cells. Eur J Pharmacol 76:129–136

Kurosumi K (1961) Electron microscopic analysis of the secretion mechanism. Int Rev Cytol 11:1–124

Kurosumi K (1968) Functional classification of cell types of the anterior pituitary gland accomplished by electron microscopy. Arch Histol Jpn 29:329–362

Kurosumi K (1971) Mitosis of the rat anterior pituitary cells: an electron microscope study. Arch Histol Jpn 33:145–160

Kurosumi K (1974) Adenohypophysis. In: Kurosumi K, Fujita H (eds) An atlas of electron micrographs. Functional morphology of endocrine glands. Igaku-Shoin, Tokyo, pp 1–100

Kurosumi K (1976) Paraneuronic and related cells in the mammalian skin. In: Coupland RE, Fujita T (eds) Chromaffin, enterochromaffin and related cells. Elsevier, Amsterdam Oxford New York, pp 171–189

Kurosumi K (1986) Cell classification of the rat anterior pituitary by means of immunoelectron microscopy. J Clin Electron Microsc 19:299–319

Kurosumi K, Inoue (1986) Ultrastructure of anterior pituitary cells. In: Ganten D, Pfaff D (eds) Morphology of hypothalamus and its connections. (Current topics in neuroendocrinology, vol 7) Springer, Berlin, pp 99–134

Kurosumi K, Oota Y (1966) Corticotrophs in the anterior pituitary glands of gonadectomized and thyroidectomized rats as revealed by electron mi-

croscopy. Endocrinology 79:808–814
Kurosumi K, Oota Y (1968) Electron microscopy of two types of gonadotraphs in the anterior pituitary glands of persistent estrous and diestrous rats. Z Zellforsch 85:34–46
Kurosumi K, Kurosumi U, Suzuki H (1969) Fine structure of Merkel cells and associate nerve fibers in the epidermis of certain mammalina species. Arch Histol Jpn 30:295–313
Kurosumi K, Kurosumi U, Inoue K (1979) Morphological and morphometric studies with the electron microscope on the Merkel cells and associated nerve terminals of normal and denervated skin. Arch Histol Jpn 42:243–261
Kurosumi K, Taniguchi Y, Inoue K (1984) Electron microscopic immunocytochemistry by the protein A-gold method of two types of rat anterior pituitary ACTH cells. Arch Histol Jpn 47:279–291
Kusumoto Y (1980) Calcitonin and somatostatin are localized in different cells in the canine thyroid gland. Biomed Res 1:237–241
Kusumoto Y (1981) The effects of local anesthetic on the degranulation of rat peritoneal mast cells. Biomed Res 2:572–582
Kusumoto Y, Iwanaga T, Ito S, Fujita T (1979) Juxtaposition of somatostatin cell and parietal cell in the dog stomach. Arch Histol Jpn 42:459–465
Kuwano R, Iwanaga T, Nakajima T, Masuda T, Takahashi Y (1983) Immunocytochemical demonstration of hydroxyindole O-methyltransferase (HIOMT), neuron-specific enolase (NSE) and S-100 protein in the bovine pineal gland. Brain Res 274:171–175
Lacy PE (1970) Functional morphology of the islet cells. Nobel Symp 13:109–120
Lacy PE, Greider MH (1972) Ultrastructural organization of mammalian pancreatic islets. In: Steiner DF, Freinkel N (eds) Endocrine pancreas (Handbook of physiology 7/I). American Physiological Society, Washington DC, pp 77–89
Lacy PE, Howell SL, Young DA, Fink CJ (1968) New hypothesis of insulin secretion. Nature 219:1177–1179
Lacy PE, Finke EH, Codilla RC (1975) Cinematographic studies on β granule movement in monolayer culture of islet cells. Lab Invest 33:570–576
Lagercrantz H (1976) On the composition and function of large dense cored vesicles in sympathetic nerves. Neuroscience 1:81–92
Lagercrantz H, Stjärne L (1974) Evidence that most noradrenaline is stored without ATP in sympathetic large dense core nerve vesicles. Nature 249:843–844
Lagunoff D, Benditt EP (1961) Histochemical examinations of chymotrypsin-like esterases. Nature 192:1198–1199
Lagunoff D, Phillips MT, Iseri OA, Benditt EP (1964) Isolation and preliminary characterization of rat mast cell granules. Lab Invest 13:1331–1344
Lam DMK (1972) Biosynthesis of acetylcholine in turtle photoreceptors. Proc Natl Acad Sci USA 69: 1987–1991
Lam DMK, Frederick JM, Hollyfield JG, Sarthy PV, Marc RE (1982) Identification of neurotransmitter candidates in invertebrate and vertebrate photoreceptors. In: Westfall JA (ed) Visual Cells in Evolution. Raven Press, New York, pp 65–80
Landis SC (1980) Developmental changes in the neurotransmitter properties of dissociated sympathetic neurons: a cytochemical study of the effects of medium. Dev Biol 77:349–361
Landis SC, Keefe D (1983) Evidence for neurotransmitter plasticity *in vivo*: Developemtal changes in properties of cholinergic sympathetic neurons. Dev Biol 77:349–372
Lane EB, Whitear M (1977) On the occurrence of Merkel cells in the epidermis of teleost fishes. Cell Tiss Res 182:235–246
Lane MA (1907) The cytological characters of the areas of Langerhans. Am J Anat 7:409–422
Lang RE, Thölken H, Ganten D, Luft FC, Ruskoaho H, Unger TH (1985) Atrial natriuretic factor—a circulating hormones stimulated by volume loading. Nature 314:264–266
Lange RH (1976) Crystallography of islet secretory granules. A contribution to the problem of chemical composition of secretion granules. In: Fujita T (ed) Endocrine gut and pancreas. Elsevier, Amsterdam, pp 167–178
Langley JN (1921) The autonomic nervous system. Part 1. Heffner, Cambridge
Langslow DR, Kimmel JR, Pollock HG (1973) Studies of the distribution of a new avian pancreatic polypeptide and insulin among birds, reptiles, amphibians and mammals. Endocrinology 93:558–565
Lapham LW (1968) Tetraploid DNA content of Purkinje neurons of human cerebellar cortex. Science 159:310–312
Laros CD (1971) Local chemical regulation of the flow resistance in the bronchial tree and pulmonary circulation. Respiration 28:120–136
Larsson JO (1981) Regional differences in the kinetics of the canine parietal cell mass in response to pentagastrin. Gynecol Obstet 153:169–176
Larsson JO (1982) Difference in kinetics between the distal and proximal parts of the canine parietal cell mass in response to pentagastrin. J Physiol 327: 507–517
Larrsson L-I, Sundler F, Håkanson R (1975) Immunohistochemical Localization of human pancreatic polypeptide (HPP) to population of islet cells. Cell Tiss Res 156:167–171
Larsson L-I (1977a) Ontogeny of peptide-producing nerves and endocrine cells of the gastro-duodeno-pancreatic region. Histochemistry 54:133–142
Larsson L-I (1977b) Ultrastructural localization of a new neuronal peptide (VIP). Histochemistry 54: 173–176
Larsson L-I (1978) Endocrine pancreatic tumors. Human Pathol 9:401–416
Larsson L-I (1979) Structure and function of putative paracrine cells. In: Miyoshi A (ed) Gut peptides. Secretion, function and clinical aspects. Kodansha, Tokyo/Elsevier, Amsterdam New York Oxford, pp 112–117
Larsson L-I (1981) Adrenocorticotropin-like and α-melanotropin-like peptides in a subpopulation of human gastrin cell granules: bioassay, immuno-

assay and immunocytochemical evidence. Proc Natl Acad Sci USA 78:2990–2994
Larsson L-I (1985) Differential changes in calcitonin, somatostatin and gastrin/cholecystokinin-like immunoreactivities in rat thyroid parafollicular cells during ontogeny. Histochemistry 82:121–130
Larsson L-I, Sundler F, Håkanson R (1976) Pancreatic polypeptide a postulated new hormone: identification of its cellular storage site by light and electron microscopic immunocytochemistry. Diabetologia 12:211–216
Larsson L-I, Goltermann N, Magistris Ld, Rehfeld JF, Schwartz TW (1979) Somatostatin cell processes as pathways for paracrine secretion. Science 205:1393–1395
Laurent P (1974) Arterial chemoreceptive structures in fish. In: Paintal AS (ed) Morphology and mechanisms of chemoreceptors. University of Delhi, Delhi, pp 275–281
Lauweryns JM, Bock V (1981) Origin and development of neuroepithelial bodies in fetal rabbit lungs. Experientia 37:1329–1330
Lauweryns JM, Cokelaere M (1973a) Intrapulmonary neuro-epithelial bodies: hypoxia sensitive neuro (chemo-)receptors. Experientia 29:1384–1386
Lauweryns JM, Cokelaere M (1973b) Hypoxia-sensitive neuro-epithelial bodies: Intrapulmonary secretory neuroreceptors, modulated by the CNS. Z Zellforsch 145:521–540
Lauweryns JM, Goddeeris P (1975) Neuroepithelial bodies in the human child and adult lung. Amer Rev Respir Disease 111:469–476
Lauweryns JM, Lommel AV (1982) Morphometric analysis of hypoxia-induced synaptic activity in intrapulmonary neuroepithelial bodies. Cell Tiss Res 226:201–214
Lauweryns JM, Peuskens JC (1972) Neuro-epithelial bodies (neuroreceptor or secretory organs?) in human infánt bronchial and bronchiolar epithelium. Anat Rec 172:471–481
Lauweryns JM, Cokelaere M, Theunynck P (1972) Neuro-epithelial bodies in the respiratory mucosa of various mammals. A light optical, histochemical and ultrastructural investigation. Z Zellforsch 135:569–592
Lauweryns JM, Cokelaere M, Theunynck P (1973) Serotonin producing neuroepithelial bodies in rabbit respiratory mucosa. Science 180:410–413
Lauweryns JM, Cokelaere M, Deleersnyer M, Liebens M (1977) Intrapulmonary neuro-epithelial bodies in newborn rabbits. Influence of hypoxia, hyperoxia, hypercapnia, nicotine, reserpine, L-DOPA and 5-HTP. Cell Tiss Res 182:425–440
Lauweryns JM, Cokelaere M, Lerut T, Theunynck P (1978) Cross-circulation studies on the influence of hypoxia and hypoxaemia on neuro-epithelial bodies in young rabbits. Cell Tiss Res 193:373–386
Lauweryns JM, Ranst LV, Verhofstad AAJ (1986) Ultrastructural localization of serotonin in the intrapulmonary neuroepithelial bodies of neonatal rabbits by use of immunoelectron microscopy. Cell Tiss Res 243:455–459
Lauweryns JM, Ranst LV, Lloyd RV, O'Connor DT (1987) Chromogranin in bronchopulmonary neuroendocrine cells. Immunocytochemical detection in human, monkey, and pig respiratory mucosa. J Histochem Cytochem 35:113–118
Lawson WB. Zeidler A, Rubenstein A (1979) Taste detection and preferences in diabetics and their relatives. Psychosomat Med 41:219–227
Leake LD, Griffith SG, Burnstock G (1985) 5-hydroxytryptamine-like immunoreactivity in the peripheral and central nervous systems of the leech *Hirudo medicinalis*. Cell Tiss Res 239:123–130
Leatherland JF, Renfree MB (1983) Structure of the pars distalis in the adult tammar wallaby (*Macropus eugenii*). Cell Tiss Res 229:155–174
Leblond CP (1981) The life history of cells in renewing systems. Am J Anat 160:114–158
Leboulenger F, Lleroux P, Tonon MC, Coy DH, Vaudry H, Pelletier G (1983) Coexistence of vasoactive intestinal peptide and enkephalins in the adrenal chromaffin granules of the frog. Neurosci Lett 37:221–225
Leboulenger F, Charnay Y, Dubois PM, Rossier J, Coy DH, Pelletier G, Vaudry H (1984) Coexistence de neuropeptides et de catecholamines dans la surrenale: Recherche d'effets paracrines sur les cellules coticosurrenaliennes. Ann Endocrinol 45:217–227
Lechago J, Barajas L (1981) Innervation of rat antral gastrin-producing cells. Anat Rec 200:309–313
Lechago J, Bencosme SA (1973) The endocrine cells of the upper gut mucosa in dogs with transplantation of the pyloric antrum to the colon. Z Zellforsch 146:237–242
Lederis K, Letter A, McMager D, Moore G, Schlesinger D (1982) Complete amino acid sequence of urotensin I, a hypotensive and corticotropin-releasing neuropeptide from *Catostomus*. Science 218:162–164
Le Douarin NM (1975) The neural crest in the neck and other parts of the body. In: Bergsma D (ed) Morphogenesis and malformation of face and brain. Birth Defects, Orig Article Ser 11:19–50
Le Douarin NM (1980) The ontogeny of the neural crest in avian embryo chimaeras. Nature 286:663–669
Le Douarin NM (1982) The neural crest. Cambridge University Press, Cambridge London New York
Le Douarin NM, Le Lièvre C (1970) Demonstration de l'origine neural des cellules à calcitonine du corps ultimobranchial chez l'embryon du poulet. C R Acad Sci Ser (Paris) D 270:2857–2860
Le Douarin NM, Smith J (1983) Differentiation of avian autonomic ganglia. In: Elfvin L-G (ed) Autonomic ganglia. Wiley, Chichester, pp 427–452
Le Douarin NM, Teillet MA (1973) The migration of neural crest cells to the wall of the digestive tract in avian embryo. J Embryol Exp Morphol 30:31–48
Le Douarin NM, Le Lièvre C, Fontaine J (1972) Recherches expérimentales sur l'origine embryologique du corps carotidien chez les Oiseaux. C R Acad Sci D (Paris) 275:583–586
Le Douarin NM, Fountaine J, Le Lièvre C (1974) New studies on the neural crest origin of the avian ultimobranchial glandular cells—interspecific combinations and cytochemical characterization of C cells based on the uptake of biogenic amine precur-

sors. Histochemistry 38:297–305
Leduque P, Paulin C, Chayvialle JA, Dubois PM (1981) Immunocytological evidence of motilin- and secretin-containg cells in the human fetal gastro-entero-pancreatic system. Cell Tiss Res 218:519–527
Leedale GF, Meeuse BJ, Pringsheim EG (1965) Euglena structure and physiology of spirogyra I and II. Arch Mikrobiol 50:68–102
Lee KY, Chang TM, Chey WY (1981) Effect of electrical stimulation of the vagus on plasma motilin concentration in dog. Life Sci 29:1093–1097
Lee TJF, Saito A, Berezin I (1984) Vasoactive intestinal polypeptide-like substance: the potential transmitter for cerebral vasodilation. Science 224: 898–901
Leeman SE, Carraway RE (1982) Neurotensin: discovery, isolation, characterization, synthesis and possible physiological roles. In: Nemeroff CB, Prange AJ (eds) Neurotensin, a brain and gastrointestinal peptide. Ann New York Acad Sci 400:1–15
Lefranc G, Pradal G, L'Hermite A (1977) Etude cytochimique des cellules endocrines intestinales foetales au cours de leur dégénérescence et de leur differenciation in vitro. Histochemistry 53:157–163
Lehy T, Willems G (1975) Combind use of immunocytochemical and autoradiographic techniques for studying the renewal of gastrin cells. Biomed Express 23:433–446
Lehy T, Willems G (1976) Population kinetics of antral gastrin cells in the mouse. Gastroenterology 71:614–619
Leibson L (1979) Endocrinology evolution and evolutionary endocrinology. Perspect Biol Med 23:25–43
Leitner JW, Sussman KE, Vatter AE, Schneider FH (1975) Adenine nucleotides in the secretory granule fraction of rat islets. Endocrinology 96:662–667
Le lièvre C (1976) Contribution des crêtes neurales à la genèse des structures céphaliques et cervicales chez les oiseaux. Thèse, Nantes
Le Lièvre C, Le Douarin NM (1975) Mesenchymal derivatives of the neural crest: analysis of chimaeric quail and chick embryo. J Embryol Exp Morphol 34:125–154
Lemaire S, Day R, Dumont M, Chouinard L, Calvert R (1984) Dynorphin and enkephalins in adrenal paraneurones. Opiates in the adrenal medulla. Can J Physiol Pharmacol 62:484–492
Lembeck F, Folkers K, Donnerer J (1981) Analgesic effect of antagonists of substance P. Biochem Biophys Res Commun 103:1318–1321
Lendon RG, Dixon JS, Gosling JA (1976) The distribution of endocrine-like cells in the human male and female urethral epithelium. Experientia 32: 377–378
Lentz RD, Lapham LW (1969) A quantitative cytochemical study of the DNA content of neurons of the rat cerebellar cortex. J Neurochem 16:379–384
Lentz TL (1966) The cell biology of *Hydra*. North-Holland, Amsterdam; Interschience, New York
Lentz TL (1968) Primitive nervous system. Yale University Press, New Haven London
Leonhardt H (1980) Ependym und circumventriculäre organe. In: Oksche A (ed) Neuroglia I. (Handbook der mikroskopischen Anatomie des Menschen IV/10) Springer, Berlin Heidelberg New York, pp 177–666
Lerner AB, Case JD, Takahashi Y, Lee TH, Mori W (1958) Isolation of melatonin, the pineal gland factor that lightens melanocytes. J Am Chem Soc 80:2587
Le Roith D, Shiloach J, Roth J, Lesniak MA (1980) Evolutionary origins of vertebrate hormones: substances similar to mammalian insulins are native to unicellular eukaryotes. Proc Natl Acad Sci USA 77:6184–6188
Le Roith D, Shiloach J, Roth J, Lesniak MA (1981) Insulin or a closely related molecule is native to *Escherichia coli*. J Biol Chem 256:6533–6536
Le Roith D, Liotta AS, Roth J, Shiloach J, Lewis ME, Pert CB, Krieger DT (1982a) Corticotropin and β-endorphin-like materials are native to unicellular organisms. Proc Natl Acad Sci USA 79:2086–2090
Le Roith D, Shiloach J, Roth AJ (1982b) Is there an earlier phylogenetic precursor that is common to both the nervous and endocrine systems? Peptides 3:211–215
Le Roith D, Shiloach J, Roth J (1984) Coexistence of multiple neuroactive materials in individual neurons: do recent evolutionary studies provide a rational interpretation? In: Chan-Palay V, Palay SL (eds) Coexistence of neuroactive substances in neurons. John Wiley and Sons, New York, pp 411–421
Lever JD (1955) Electron microscopic observations on the normal and denervated adrenal medulla of the rat. Endocrinology 57:621–635
Lever JD, Graham JDP, Spriggs TLB (1967) Electron microscopy of nerves in relation to the arteriolar wall. In: Symp Electr Activ Innerv Blood Vessels (Cambridge, 1966) Bibl Anat 8. Karger, Basel New York, pp 51–55
Levi-Montalcini R, Aloe L (1980) Tropic, trophic and transforming effects of nerve growth factor. In: Eränkö O, Soinila S, Päivärinta H (eds) Histochemsitry and cell biology of Autonomic neurons, SIF cells and paraneurons (Adv Biochem Psychopharmacol 25). Raven Press, New York, pp 3–15
Levi-Montalcini R, Aloe L (1983) The effect of nerve growth factor on autonomic ganglion cells. In: Elfvin L-G (ed) Autonomic ganglia. Wiley, Chichester, pp 401–426
Levi-Montalcini R, Cohen S (1960) Effects of the extract of the mouse submaxillary salivary glands on the sympathetic system of mammals. Ann New York Acad Sci 85:324–341
Levine RA, Hall RC (1976) Cyclic AMP in secretin choleresis. Evidence for a regulatory role in man and baboons but not in dogs. Gastroenterology 70:537–544
Levine RB, Truman JW (1983) Peptide activation of a simple neural circuit. Brain Res 279:335–338
Levitan IB (1985) Phosphorylation of ion channels. J Membrane Biol 87:177–190
L'Heureux MV, Melius P (1956) Differential centrifugation of bovine parathyroid tissue. Biochim

Biophys Acta 20:447–448

Lichtenberger LM, Lechago J, Dockray GJ, Passaro E (1975) Culture of Zollinger-Ellison tumor cells. Gastroenterology 68:1119–1126

Liem R (1982) Simultaneous separation and purification of neurofilament and glial filament proteins from brain. J Neurochem 38:142–150

Liem RKH, Yen SH, Salomon GD, Shelanski ML (1978) Intermediate filaments in nervous tissues. J Cell Biol 79:637–645

Lignon J, Le Douarin G (1978) Small intensely fluorescent (SIF) cells and myocardic cells in the ammocoete heart: a correlative histofluorescence, light and electron microscopic study with special reference to the action of reserpine. Biol Cell 31:169–176

Like AA, Chick WL (1969) Mitotic division in pancreatic beta cells. Science 163:941–943

Limacher W, Wild P, Manser E, Lutz H (1979) Identification of immunoreactive sites in bovine parathyroid cells to antibodies raised against the NH_2-terminal sequence of parathyroid hormone. J Histochem Cytochem 27:1209–1214

Lin DC, Tobin KD, Grumet M, Lin S (1980) Cytochalasins inhibit nuclei-induced actin polymerization by blocking filament elongation. J Cell Biol 84: 455–460

Lindeman HH, Ades HW, Bredberg G, Engström H (1971) The sensory hairs and the tectorial membrane in the development of the cat's organ of Corti. Acta Otolaryngol 72:229–242

Lingappa VR, Shields D, Woo SLC, Blobel G (1978) Nascent chicken ovalbumin contains the functional equivalent of a signal sequence. J Cell Biol 79:567–572

Linnoila RI, Diaugustine RP, Hervonen A, Miller RJ (1980) Distribution of [Met^5]- and [Leu^5]-enkephalin-, vasoactive intestinal polypeptide-, and substance P-like immunoreactivities in human adrenal glands. Neuroscience 5:2247–2259

Lippmann W (1974) Oral gastric acid secretion-inhibitory activity and anti-ulcer activity of synthetic prostagland in analogues: C-15 epimers of 15-hydroxy-15-methyl-9-oxoprostanoic acid. Prostaglandins 7:223–229

Lippmann W (1974) Inhibition of gastric acid secretion and ulcer formation in the rat by orally-administered 11-deoxyprostagland in analogues: 15-hydroxy-16,16-dimethyl-9-oxoprost-5,13-dienoic acids. Prostaglandins 7:231–246

Lipton JS, Petterborg LJ, Reiter RJ (1981) Influence of propranolol, phenoxybenzamine or phentolamine on the in vivo noctural rise of pineal melatonin levels in the Syrian hamster. Life Sci 28:2377–2382

Lipton P, Robacker K (1983) Glycolysis and brain function: $[K^+]_o$ stimulation of protein synthesis and K^+ uptake require glycolysis. Fed Proc 42:2875–2880

Lissmann HW, Machin KE (1958) The mechanism of object location in *Gymnarchus niloticus* and similar fish. J Exp Biol 35:451–486

Lissman HW, Mullinger AM (1968) Organization of ampullary electric receptors in Gymnotidae (Pisces). Proc Roy Soc Lond B 169:345–378

Litzén M, Hammerström L, Ullberg S (1965) Autoradiographic distribution of 5-hydroxy-triptamine and 5-hydroxytryptophan in the mouse. Biochem Pharmacol 14:313–321

Livett BG (1984) The secretory process in adrenal medullary cells. In: Cantin M (ed) Cell biology of the secretory process. Karger, Basel, pp 309–358

Livett BG, Boksa P (1984) Receptors and receptor modulation in cultured chromaffin cells. Can J Physiol Pharmacol 62:467–476

Livett BG, Dean DM, Whelan LG, Udenfriend S, Rossier J (1981) Co-release of enkephalin and catecholamines from cultured adrenal chromaffin cells. Nature 289:317–319

Ljungdahl Å, Hökfelt T, Nilsson G (1978) Distribution of substance P-like immunoreactivity in the central nervous system of the rat. I. Cell bodies and nerve terminals. II. Light microscopic localization in relation to catecholamine-containing neruons. Neuroscience 3:861–943, 945–976

Llinás R (1982) Calcium in synaptic transmission. Sci Am 247(4):38–47

Llinás R (1983) The role of calcium in neuronal function. In: The Fourth Study Program of NRP (Neuroscineces Research Program). pp 555–571

Llinás R, Sugimori M (1980) Electrophysiological properties of *in vitro* Purkinje cell somata in mammalian cerebellar slices. J Physiol 305:171–195

Llinás R, Sugimori M (1980) Electrophysiological properties of *in vitro* Purkinje cell dendrites in mammalian cerebellar slices. J Physiol 305:197–213

Lloyd PE, Kupfermann I, Weiss KR (1984) Evidence for parallel actions of a molluscan neuropeptide and serotonin in mediating arousal in *Aplysia*. Proc Natl Acad Sci USA 81:2934–2937

Lloyd RV, Wilson BS (1983) Specific endocrine tissue marker defined by a monoclonal antibody. Science 222:628–630

Lloyd RV, Warner TFCS, Mervak T et al. (1984) Immunohistochemical detection of chromogranin and neuronspecific enolase in pancreatic endocrine neoplasmas. Am J Surg Pathol 8:607–614

Lodhi S, Weiner ND, Schacht J (1979) Interactions of neomycin with monomolecular films of polyphosphoinositides and other lipids. Biochem Biophys Acta 557:1–8

Loewenstein WR (1981) Junctional intercellular communication: the cell-to-cell membrane channel. Physiol Rev 61:829–913

Loh YP, Gainer H (1982) Processing of normal and non-glycosylated forms of toad pro-opiocortin by rat intermediate (pituitary) lobe pro-opiocortin converting enzyme activity. Life Sci 31:3043–3050

Lombardi T, Montesano R, Wohlwend A, Amherdt M, Vassalli J-D, Orci L (1985) Evidence for polarization of plasma membrane domains in pancreatic endocrine cells. Nature 313:694–696

Lopez LC, Frazier ML, Su CJ, Kumar A, Saunders GF (1983) Mammalian pancreatic preproglucagon contains three glucagonrelated peptides. Proc Natl Acad Sci USA 80:5485–5489

Lorković H, Hörr R (1985) Electrogenic Na-K pump in denervated mouse soleus muscles: Prolonged activation by briefly applied acetylcholine. Neurosci Lett 56:175–181

Lösecke W, Naumann W, Sterba G (1982) Immunocytochemische Darstellung von Neuropeptiden (Neurophysin und Oxtocin) mit der Protein-A-Gold Technik. Acta Histochem 71:201–208

Luciano L, Reale E (1969) A new cell type ("brush cell") in the gall bladder epithelium of the mouse. J Submicr Cytol 1:43–52

Luciano L, Reale E, Ruska H (1968) Über eine "chemorezeptive" Sinneszelle in der Trachea. Z Zellforsch 85:350–375

Luciano L, Castellucci M, Reale E (1981) The brush cells of the common bile duct of the rat. Thin section, freeze-fracture and scanning electron microscopy. Cell Tiss Res 218:403–420

Lucy JA (1969) Lysosomal membrane. In: Dingle JT, Fell H (eds) Lysosomes in biology and pathology 2. North Holland, Amsterdam, pp 313–341

Ludwin SK, Kosek JC, Eng LF (1976) The topographical distribution of S-100 and GFA proteins in the adult rat brain: an immunohistochemical study using horseradish peroxidase-labeeled antibodies. J Comp Neurol 165:197–208

Lukaszyk A, Reiter RJ (1975) Histophysiological evidence for the secretion of polypeptides by the pineal gland. Am J Anat 143:451–464

Lundberg JM, Hamberger B, Schultzberg M, Hökfelt T, Granberg PO, Efendić S, Terenius L, Goldstein M, Luft R (1979a) Enkephalin-and somatostatin-like immunoreactivities in human adrenal medulla and pheochromocytoma. Proc Natl Acad Sci USA 76:4079–4083

Lundberg JM, Hökfelt T, Schultzberg M, Uvnäs-Wallensten K, Köhler C, Said SI (1979b) Occurrence of vasoactive intestinal polypeptide (VIP)-like immunoreactivity in certain cholinergic neurons of the cat: evidence from combined immunohistochemistry and acetylcholinesterase staining. Neuroscience 4: 1539–1559

Lundberg JM, Hökfelt T, Änggård A, Pernow B, Emson P (1979c) Immunohistochemical evidence for substance P immunoreactive nerve fibers in the taste buds of the cat. Acta Physiol Scand 107:389–391

Lundberg JM, Řokaeus Å, Hökfelt T, Rosell S, Brown M, Goldstein M (1982) Neurotensin-like immunoreactivity in the preganglionic sympathetic nerves and in the adrenal medulla of the cat. Acta Physiol Scand 114:153–155

Lundberg JM, Terenius L, Hökfelt T, Martling CR, Tatemoto K, Mutt V, Polak J, Bloom S, Goldstein M (1982) Neuropeptide Y (NPY)-like immunoreactivity in peripheral noradrenergic neurons and effects of NPY on sympathetic function. Acta Physiol Scand 116:477–480

Lundberg JM, Hökfelt T, Hemsén A, Theodorsson-Norheim E, Pernow J, Hamberger B, Goldstein M (1986) Neuropeptide Y-like immunoreactivity in adrenaline cells of adrenal medulla and in tumors and plasma of pheochromocytoma patients. Regul Peptides 13:169–182

Luschka H (1862) Uber die drusenartige natur des sogenannten ganglion intercaroticum. Arch Anat Physiol Leipzig 405–414 (cited from Adams, 1958)

Lyne AG, Hollis DE (1971) Merkel cells in sheep epidermis during fetal development. J Ultrastuct Res 34:464–472

MacCallum WG, Voegtlin C (1909) On the relation of tetany to the parathyroid glands and to calcium metabolism. J Exp Med 11:118–151

Maclean DB, Lewis SF, Engler D (1984) Studies of the axoplasmic transport of somatostatin in the vagus nerve of the rat. Endocrinology 114:1145–1154

MacMahon HW, Werch J, Sorger K (1967) Tumorlet of bronchus. With a 12 year follow-up. Arch Pathol 83:359–363

Magistretti PJ (1985) VIP-containing neurons in the cerebral cortex: Celluar actions and interactions with the noradrenergic system. In: Ritchie JM (ed) Ionic channels in neural membranes. Alan R Liss, New York, pp 323–331

Magistretti PJ, Schorderet M (1984) VIP and noradrenaline act synergistically to increase cyclic AMP in cerebral cortex. Nature 308:280–282

Magistretti PJ, Morrison JH, Shoemaker WJ, Sapin V, Bloom FE (1981) Vasoactive intestinal polypeptide induces glycogenolysis in mouse cortical slices: A possible regulatory mechanism for the local control of energy metabolism. Proc Natl Acad Sci USA 78:6535–6539

Magistretti PJ, Morrison JH, Shoemaker WJ, Bloom FE (1984) Morphological and functional correlated of VIP neurons in cerebral cortex. Peptides 5:213–218

Malaisse WJ (1973) Insulin secretion: multifactorial regulation for a single process of release. Diabetologia 9:167–173

Malaisse WJ, Malaisse-Lagae F, Lacy PE (1970) Participation of microtubular-microfilamentous system in insulin secretion. Ann Meeting Eur Assoc Study Daibetes, Warsaw

Malaisse WJ, Sener A, Koser M, Herchuelz A (1976) Stimulus-secretion coupling of glucose-induced insulin release. Metabolism of α- and β-D-glucose in isolated islets. J Biol Chem 251:5936–5943

Malaisse WJ, Sener A, Herchuelz A, Hutton JC (1979) Insulin relase: The fuel hypothesis. Metabolism 28:373–386

Malvaldi G, Viola-Magni MP (1972) DNA turnover in adrenal medullary cells of different strains of rats and its enhancement after intermittent exposure to cold. Cell Tiss Kinet 5:103–113

Manger WM, Gifford RW (1977) Pheochromocytoma. Springer, New York

Mann SJ, Straile WE (1963) Tylotrich (hair) follicle: association with a slowly adapting tactile receptor in the cat. Science 147:1043–1045

Manocchio I (1960) Metachromatische färbung der a-zellen in pancreas-inseln von "*Canis familiaris*". Zbl Allg Path (Anat) 101:1–4

Marangos PJ, Polak JM, Pearse AGE (1982) Neuron-specific enolase: a probe for neurons and neuroendocrine cells. Trends Neurosci 5:193–196

Marchevsky AM, Kleinerman J (1983) Quantitative and immunocytochemical studies of APUD cells in airways and gut: the effects of priming with L-dopa and 5-HTP. Lung 161:67–76

Margolis RK, Jaanus SD, Margolis RU (1973) Stimulation by acetylcholine of sulfated mucopolysaccharide release from the perfused cat adrenal gland. Mol Pharmacol 9:590–594

Margolis RU, Margolis RK (1973) Isolation of chondroitin sulfate and glycopeptides from chromaffin granules of adrenal medulla. Biochem Pharmacol 22:2195–2197

Marik F, Code CF (1975) Control of the interdigestive myoelectric activity in dogs by the vagus nerves and pentagastrin. Gastroenterology 69:387–395

Markl H (1974) The perception of gravity and of angular acceleration in invertebrates. In: Kornhuber HH (ed) Vestibular system, part 1: basic mechanisms (Handbook of sensory physiology VI/1). Springer, Berlin, pp 17–74

Martin R, Voigt KH (1981) Enkephalins co-exist with oxytocin and vasopressin in nerve terminals of rat neurohypophysis. Nature 289:502–504

Martin R, Voigt KH (1985) Opioid peptides as cofactors in vasopressin and oxytocin nerve endings of the neruohypophysis. In: Kobayashi H, Bern HA, Urano A (eds) Neurosecretion and the biology of neuropeptides. Jpn Sci Soc Press, Tokyo/Springer, Berlin, pp 26–33

Martin TW, Lagunoff D (1984) Mast cell secretion. In: Cantin M (ed) Cell biology of the secretory process. Karger, Basel, pp 481–516

Martinez J, Lee A, Moossy J, Maroon JC (1980) Pituitary adenomas: clinicopathological and immunohistochemical study. Ann Neurol 7:24–36

Marty A (1981) Ca-dependent K channels with large unitary conductance in chromaffin cell membrane. Nature 291:497–500

Mascorro JA, Yates RD (1970) Microscopic observations on abdominal sympathetic paraganglia. Texas Rep Biol Med 28:59–68

Masson P (1924) Appendicite neurogène et carcinoide. Ann Anat Pathol 1:3–59

Masson P (1928) Carcinoids (argentaffin-cell tumors) and nerve hyperplasia of the appendicular mucosa. Am J Pathol 4:181–211

Masson P (1948) Pigment cells in man. In: Biology of melanomas. Spec Publ, New York Acad Sci 6:15–51

Masson P, Berger L (1923) Sur un nouveau mode de sécrétion interne: la neurocrinie. C R Hebd Séanc Acad Sci Paris 176:1748–1750

Matschinsky FM, Pagliara As, Hover BA, Haymond MW, Stillings SN (1975) Differential effects of alpha- and beta-D-glucose on insulin and glucagon secretion from the isolated perfused rat pancreas. Diabetes 24:369–372

Matsuba I, Tanese T, Abe M (1982) Human pancreatic islet cell clones secreting insulin, glucagon and somatostatin: immunocytochemical and functional studies. Arch Histol Jpn 45:111–119

Matsuo Y, Seki A (1978) The coordination of gastrointestinal hormones and the autonomic nerves. Am J Gastroenterol 69:21–50

Matsushima S, Reiter RJ (1975) Ultrastructural observations of pineal gland capillaries in four rodent species. Am J Anat 143:265–282

Matsushima S, Reiter RJ (1977) Fine structural features of adrenergic nerve fibers and endings in the pineal gland of the rat, ground squirrel and chinchilla. Am J Anat 148:463–477

Matsuyama T, Tanaka R, Shima K, Tarui S (1979) Anomeric specificity in the response of gut glucagon-like immunoreactive materials to glucose. Horm Metab Res 11:214–216

Matsuzawa T (1966) Experimental morphological studies on the parafollicular cells of the rat thyroid ground, with special reference to the source of thyrocalcitonin. Arch Histol Jpn 27:521–544

Matthews EK (1985) Electrophysiology of pancreatic islet β-cells. In: Poisner AM, Trifaro JM (eds) The electrophysiology of the secretory cell. Elsevier, Amsterdam New York Oxford, pp 93–112

Matthews EK, Dean PM (1970) Electrical activity in islet cells. In: Falkmer S, Hellman B, Taljedal I-B (eds) the structure and metabolism of the pancreatic islets. Pergamon Press, Oxford, pp 305–313

Matthews EK, Sakamoto Y (1975) Electrical characteristics of pancreatic islet cells. J Physiol 246:421–437

Matthews MR, Raisman G (1969) The ultrastructure and somatic efferent synapses of small granule-containing cells in the superior cervical ganglion. J Anat 105:255–282

Maue RA, Dionne VE (1987) Patch-clamp studies of isolated mouse olfactory receptor neurona. J Gen Physiol 90:95–125

Maul GG, Brumbaugh JA (1971) On the possible function of coated vesicles in melanogenesis of the regenerating fowl feather. J Cell Biol 48:41–48

Mayer EA, Elashoff J, Walsh JH (1982) Characterization of bombesin effects on canine gastric muscle. Am J Physiol 243:G141–147

Mayevsky A (1984) Brain NADH redox state monitored in vivo by fiber optic surface fluorometry. Brain Res Rev 7:49–68

McArdle JJ (1983) Molecular aspects of the trophic influence of nerve on muscle. Porg Neurobiol 21:135–198

McCann FV, Cole JJ, Pettengill OS, Russell JAG, Sorenson GD (1985) Electrical properties of human pulmonary endocrine-like tumor cells: small cell carcinoma of the lung. In: Poisner AM, Trifaro JM (eds) The electrophysiology of the secretory cell. Elsevier, Amsterdam New York Oxford, pp 241–266

McCulloch J, Kelly PAT (1983) A functional role for vasoactive intestinal polypeptide in anterior cingulate cortex. Nature 304:438–440

McCumbee WD, Hazelwood RL (1977) Biological evaluation of the third pancreatic hormone (APP): hepatocyte and adipocyte effects. Gen Comp Endocrinol 33:518–525

McDonald DM (1976) Structure and function of reciprocal synapses interconnecting glomus cells and sensory nerve terminals in the rat carotid body. In: Coupland RE, Fujita T (eds) Chromaffin, enterochromaffin and related cells. Elsevier, Amsterdam Oxford New York, pp 375–394

McDonald DM (1977) Structure-function relation-

ships of chemoreceptive nerves in the carotid body. Am Rev Respir Dis 115(2):193–207
McDonald DM (1981) Peripheral chemoreceptors. Structure-function relationships of the carotid body. In: Hornbein TF (ed) Regulation of breathing. Part 1. Marcel Dekker, New York, pp 105–319
McDonald DM, Blewett RW (1981) Location and size of carotid body-like organs (paraganglia) revealed in rats by the permeability of blood vessels to Evans blue dye. J Neurocytol 10:607–643
McDonald DM, Mitchell RA (1975) The innervation of glomus cells, ganglion cells and blood vessels in the rat carotid body: a quantitative ultrastructural analysis. J Neurocytol 4:177–230
McGuigan JE (1968) Gastric mucosal intracellular localization of gastrin by immunofluorescence. Gastroenterol 55:315–327
McKenzie JC, Tanaka I, Misono KS, Inagami T (1985) Immunocytochemical localization of atrial natriuretic factor in the kidney, adrenal medulla, pituitary, and atrium of rat. J Histochem Cytochem 33:828–832.
McLaughlin S, Eisenberg M (1975) Antibiotics and membrane biology. Ann Rev Biophys Bioeng 4: 335–366
McLeay LM, Titchen DA (1977) Acid and pepsin secretion of separated gastric pouches during perfusion of antral pouches with cholinergic stimulating and blocking agents and lignocaine. J Physiol 264:215–227
McMaster D, Lederis K (1983) Isolation and amino acid sequence of two urotension II peptides from *Catostomus commersoni* urophyses. Peptides 4: 367–373
McMillan PJ, Hooker WM, Deftos LJ (1974) Distribution of calcitonin-containing cells in the human thyroid. Am J Anat 140:73–80
McNeilly AS, Legros JJ, Forsling ML (1972) Release of oxytocin, vasopressin and neurophysin in the goat. J Endocr 52:209–210
McQueen DS, Ribeiro JA (1981) Effect of adenosine on carotid chemoreceptor activity in the cat. Br J Pharmacol 74:129–136
McQueen DS, Ribeiro JA (1983) On the specificity and type of receptor invoved inearotid body chemoreceptor activation by adenosine in the cat. Br J Pharmacol 80:347–354
McShan WH, Hartley MW (1965) Production, storage and release of anterior pituitary hormones. Ergeb Physiol Biol Chem Exp Pharmakol 56:264–296
Mei M (1978) Vagal glucoreceptors in the small intestine of the cat. J Physiol 282:485–506
Meisenberg G, Simmons WH (1983) Peptides and the blood-brain barrier. Life Sci 32:2611–2623
Meissner HP, Atwater I (1976) The kinetics of electrical activity of B-cells in response to a 'square wave' stimulation with glucose or glibenclamide. Hormone Metab Res 8:11–16
Meissner HP, Henquin JC (1984) The sodium pump of mouse pancreatic β-cells: electrogenic properties and activation by intracellular sodium. In: Blaustein MP, Lieberman M (eds) Electrogenic transport: fundamental principles and physiological implications. Raven Press, New York, pp 295–306
Meissner HP, Schmelz H (1974) Membrane potentials of beta-cells in pancreatic islets. Pflügers Arch 351:195–206
Melander A, Owman C, Sundler F (1971) Concomitant depletion of dopamine and secretory granules from cells in the ultimobranchial gland of vitamin d_2-treated chicken. Histochemie 25:21–31
Melander A, Ericson LE, Sundler F, Westgren U (1975) Intrathyroidal amines in the regulation of thyroid activity. Rev Physiol Biochem Pharmacol 73:39–71
Melander T, Hökfelt T, Rökaeus Å, Fahrenkrug J, Tatemoto K, Mutt V (1985) Distribution of galanin-like immunoreactivity in the gastrointestinal tract of several mammalian species. Cell Tiss Res 239:253–270
Meldolesi J, Huttner WB, Tsien RY, Pozzan T (1984) Free. cytoplasmic Ca^{2+} and neurotransmitter release: Studies on PC12 cells and synaptosomes exposed to α-latrotoxin. Proc Natl Acad Sci USA 81:620–624
Melkonian M, Robenek H (1980) Eyespot membranes of *Chlamydomonas reinhardii*: a freeze-fracture study. J Ultrastruct Res 72:90–102
Menco BPM (1977) A qualitative and quantitative investigation of olfactory and nasal respiratory mucosl surfaces of cow and sheep based on various ultrastructural and biochemical methods. H Veenman & Zonen BV, Wageningen
Menco BPM, Dodd GH, Davey M, Bannister LH (1976) Presence of membrane particles in freeze-etched bovine olfactory cilia. Nature 263:597–599
Menco BPM, Leunissen JLM, Bannister LH, Dodd GH (1978) Bovine olfactory and nasal respiratory epithelium surfaces. Cell Tiss Res 193:503–524
Merkel F (1875) Tastzellen und Tastkörperchen bei den Hausthieren und beim Menschen. Arch Mikrosk Anat 11:636–652
Merkel F (1880) Über die Endigungen der sensiblen Nerven in der Haut der Wirbelthiere. H. Schmidt, Rostock
Metzger H (1977) The cellular receptor for IgE. In: Cuatrecasas P, Greaves MF (eds) Receptor and recognition series A, vol 4. Champman Hall, London, pp 75–102
Meyer JH, Kelly GA (1976) Canine pancreatic responses to intestinally perfused proteins and protein digests. Am J Physiol 231:682–691
Meyer JH, Kelly GA, Jones RS (1976) Canine pancreatic response to intestinally perfused oligopeptides. Am J Physiol 231:678–681
Meyer JH, Kelly GA, Spingola LJ, Jones RS (1976) Canine gut receptors mediating pancreatic responses to luminal L-amino acids. Am J Physiol 231:669–677
Meyrick B, Reid L (1968) The alveolar brush cells in rat lung–a third pneumocyte. J Ultrastr Res 23:71–80
Michell B (1986) Inositol phosphates: profusion and confusion. Nature 319:176–177
Michetti F, Dell'Anna E, Tiberio G, Cocchia D (1983) Immunochemical and immunocytochemical study of S-100 protein in rat adipocytes. Brain Res

262:352–356

Mihara M, Hashimoto K, Ueda K, Kumakiri M (1979) The specialised junctions between Merkel cell and neurite: an electron microscope study. J Invest Dermatol 73:325–334

Milcu SM, Pavel S, Neacsu C (1963) Biological and chromatographic characterization of a polypeptide with pressor and oxytocic activities: solated from bovine pineal gland. Endocrinology 72:563–566

Miledi R (1973) Transmitter release induced by injection of calcium ions into nerve terminals. Proc R Soc B 183:421–425

Milutinović S, Schulz I, Rosselin G (1976) The interaction of secretin with pancreatic membranes. Biochim Biophys Acta 436:113–127

Mirolli M (1983) Inward and outward currents in isolated dendrites of crustacea coxal receptors. Cell Mol Neurobiol 3:355–370

Mishima Y (1960) New technic for comprehensive demonstration of melanin, premelanin and tyrosinase site. J Invest Dermatol 34:355–360

Mishima Y (1964) Electron microscopic cytochemistry of melanosomes and mitochondria. J Histochem Cytochem 12:784–790

Mitschke H (1973) Immunohistologische, cytochemische und elektronenoptische Untersuchungen beim Zollinger-Ellison-Syndrom. Virchows Arch Path Anat 361:335–348

Mitznegg P, Bloom SR, Christofides N, Besterman H, Domschke W, Domschke S, Wunsch E, Demling L (1976) Release of motilin in man. Scand J Gastroenterol 11:53–56

Mitznegg P, Bloom SR, Domschke W, Haecki WH, Domschke S, Belohlavek D, Wunsch E, Demling L (1977) Effect of secretin on plasma motilin in man. Gut 18:468–471

Mitznegg VP (1978) Die radioimmunologische Bestimmung von Motilin. Z Gastroenterol 16:325–329

Miyagami H, Watanabe Y, Sawada Y, Kato K, Shiono K, Kondo K, Kidokoro T (1977) Ultrastructures of G cells and the mechanism of gastrin release before and after selective vagotomy with pyloroplasty. Arch Histol Jpn 40:51–62

Mizobe F, Livett BG (1983) Nicotine stimulates secretion of both catecholamines and acetylcholinesterase from cultured adrenal chromaffin cells. J Neurosci 3:871–876

Moffett DF, Jöbsis FF (1976) Response of toad brain respiratory chain enzymes of ouabain, elevated potassium, and electrical stimulus. Brain Res 117: 239–255

Møller M, Ingild A, Böck E (1978) Immunohistochemical demonstration of S-100 protein and GFA protein in interstitial cells of rat pineal gland. Brain Res 140:1–13

Molnar M, Skinner JE (1983) Evoked and spontaneous extracellular potassium shifts in the cerebral cortex of unanaesthetized cats. Acta Physiol Hungarica 61:265–279

Monaghan P, Roberts JDB (1985) Immunocytochemical evidence for neuroendocrine differentiation in human breast carcinomas. J Pathol 147:281–289

Monti-Bloch L, Stensaas LJ, Eyzaguirre C (1983) Carotid body grafts induce chemosensitivity in muscle nerve fibers of the cat. Brain Res 270:77–92

Moodley I, Mongar JL, Foreman JC (1982) Histamine release induced by dextran: the nature of the dextran receptor. Eur J Pharmacol 83:69–81

Moody MF, Robertson JD (1960) The fine structure of some retinal photoreceptors. J Biophys Biochem Cytol 7:87–92

Moore BW (1965) A soluble protein characteristic of the nervous system. Biophys Res Commun 19:739–744

Moore BW (1982) Chemistry and biology of the S-100 protein. Scand J Immunol 15 Suppl 9:53–74

Moore GL, Thornhill JA, Gill V, Lederis K, Lukowiak K (1981) An arginine vasotocin-like neuropeptide in present in the nervous system of the marine mollusc *Aplysia californica*. Brain Res 206:213–218

Moran DT, Rowley III JC, Jafek BW, Lovell MA (1982) The fine structure of the olfactory mucosa in man. J Neurocytol 11:721–746

Morgan SE, Pallot DJ, Willshaw P (1981) The effect of ventilation with different concentrations of oxygen upon the synaptic vesicle density in nerve endings of the rat carotid body. Neuroscience 6:1461–1467

Morgan WW, Reiter RJ (1977) Pineal noradrenaline levels in the Monogolian gerbil and in different strains of laboratory rats over a lighting regimen. Life Sci 21:555–558

Morgan WW, Reiter RJ, Pfeil KA (1976) Hamster pineal noradrenaline: Levels over a regulated lighting period and the influence of superior cervical ganglionectomy. Life Sci 19:437–440

Mori H, Kurokawa M (1980) Morphological and biochemical characterisation of neurofilaments isolated from the rat peripheral nerve. Biomed Res 1:24–31

Morii H, Fujita T, Okinaka S (1963) Effect of vagotomy and atropine on recovery from induced hypocalcemia. Endocrinol 72:173–179

Morimoto K, Sato M (1975) Noradrenaline as a chemical transmitter from taste cells to sensory nerve terminals in frog. Proc Jpn Acad 51:347–352

Morimoto K, Sato M (1982) Role of monoamines in afferent synaptic transmission in frog taste organ. Jpn J physiol 32:855–871

Morinaga S, Tsumuraya M, Nakajima T, Shimosato Y, Yoshinaga K, Yamaguchi K, Abe K, Ohno H, Kanai A (1985) Immunohistochemical and immunocytochemical localization of atrial natriuretic polypeptide in human adult and fetal hearts. Acta Histochem Cytochem 18:605–613

Morita Y, Bergmann G (1971) Physiologische Untersuchungen und weitere Bemerkungen zur Struktur des lichtempfindlichen Pineal-organs von *Pterophyllum scalare* Cuv et Val (Cichlidae, Teleostei). Z Zellforsch Mikrosk Anat 119:289–294

Morley JE, Levine AS, Silvis SE (1981) Intraventricular calcitonin inhibits gastric acid secretion. Science 214:671–673

Morley JS (1977) Information about peptide hormone receptors from structure-activity studies. In: Bonfils S, Fromageot P, Rosselin G (eds) First Int Symp on Hormonal Receptors in digestive tract Physiology, INSERM Sym No. 3. pp 3–11

Morley JS (1981) Nomenclature. Neuropeptides 1: 231–235
Morris JF, Nordmann JJ, Dyball REJ (1978) Structure-function correlation in mammalian neurosecretion. Int Rev Exp Pathol 18:1–45
Morrison JD (1983) Morphogenesis of photoreceptor outer segments in the developing kitten retina. J Anat 136:521–533
Morrison JH, Magistretti PJ (1983) Monoamines and peptides in cerebral cortex contrastin principles of cortical organization. Trends Neurosci 6:146–151
Morrissey JJ, Cohn DV (1979) Regulation of secretion of parathormone and secretory protein-I from separate intracellular pools by calcium, dibutyryl cyclic AMP, and (1)-isoproterenol. J Cell Biol 82:93–102
Mortensen NJMcC, Morris JF (1977) The effect of fixation conditions on the ultrastructural appearance of gastrin cell granules in the rat gastric pyloric antrum. Cell Tiss Res 176:251–263
Moura CS, Miraglia T (1978) Histochemical observations on the taste buds of the marmosets (*Callithrix jacchus* and *Callithrix penicillata*). Z Mikrosk Anat Forsch 92:587–597
Moxey PC, Trier JS (1977) Endocrine cells in the human fetal small intestine. Cell Tiss Res 183:33–50
Mozsik G, Nagy L, Tarnok F, Vizi F, Kutas J (1974) H^+ secretion and Na^+-K^+-dependent ATPase system in the human gastric mucosa. Experientia 30:1024
Mukhopadhyay AK, Johnson LR, Copeland EM, Weisbrodt NW (1975) Effect of secretin on electrical activity of small intestine. Am J Physiol 229:484–488
Mulligan RM (1950) Chemodectoma in the dog. Am J Pathol 26:680–681
Mullinger AM (1964) The fine structure of ampullary electic receptors in *Amiurus*. Proc Roy Soc B 160: 345–359
Munger BL (1965) The intraepidermal innervation of the snout skin of the opossum: a light and electron microscopic study, with observations on the nature of Merkel's *Tastzellen*. J Cell Biol 26:79–96
Munger BL (1971) Patterns of organization of peripheral sensory receptors. In: Loewenstein WR (ed) Principles of receptor physiology (Handbook of sensory physiology 1). Springer, Berlin Heidelberg New York, pp 523–556
Munger BL (1977) Neural-epithelial interactions in sensory receptors. J Invest Dermatol 69:27–40
Munger BL (1981) Morphological characterization of islet cell diversity. In: Cooperstein SJ, Watkins D (eds) The islets of Langerhans: Biochemistry, physiology, and pathology. Academic Press, New York London Toronto Sydney San Francisco, pp 3–34
Munger BL, Roth SI (1963) The cytology of the normal parathyroid glands of man and Virginia deer. A light and electron microscopic study with morphologic evidence of secretory activity. J Cell Biol 16:379–400
Munger BL, Caramia F, Lacy PE (1965) The ultrastructural basis II-5 for the identification of cell types in the pancreatic islets. II. Rabbit, dog and opossum. Z Zellforsch 67:776–798
Munson PL (1976) Physiology and pharmacology of thyrocalcitonin. In: Aurbach GD (ed) Parathyroid gland. (Handbook of Physiology 7/VII) American Physiological Society, Washington DC, pp 443–464
Murakami M, Ohtsuka T, Schimazaki H (1975) Effects of aspartate and glutamate on the bipolar cells in the carp retina. Vision Res 15:456–458
Murakami T, Itoshima T, Shimada Y (1974) Peribiliary portal system in the monkey liver as evidenced by the injection replica scanning electron microscope method. Arch Histol Jpn 37:245–260
Murphy RA, Watson AY, Metz J, Forssmann WG (1980) The mouse submandibular gland: an exocrine organ for growth factors. J Histochem Cytochem 28:890–902
Murray RG (1971) Ultrastructure of taste receptors. In: Beidler LM (ed) Taste. (Handbook of sensory physiology. IV/3) Springer, Berlin Heidelberg New York, pp 31–50
Murray RG (1973) The ultrastructure of taste buds. In: Friedmann I (ed) The ultrastructure of sensory organs. North-Holland, Amsterdam, pp 1–81
Murray RG, Murray A, Fujimoto S (1969) Fine structure of gustatory cells in rabbit taste buds. J Ultrastruct Res 27:444–461
Murray RW (1974) The ampullae of Lorenzini. In: Fessard A (ed) Electroreceptors and other specialized receptors in lower vertebrates (Handbook of sensory physiology III/3). Springer, Berlin Heidelberg New York, pp 125–146
Muscholl E, Rahn KH, Watzka M (1960) Nachweis von Noradrenalin im Glomus caroticum. Naturwiss 47:325
Nabeyama A, Murata F, Matsuda H, Ogata T (1971) Ultrastructural classification of endocrine-like cells in the mucosa of human stomach. Tohoku J Exp Med 103:17–47
Nada O, Hirata K (1975a) The occurrence of the cell type containing a specific monoamine in the taste bud of the rabbit's foliate papilla. Histochemistry 43:237–240
Nada O, Hirata K (1975b) Specific effect of 5, 6-dihydroxytryptamine on the monoamine fluorophore of the frog's gustatory cells. Histochemistry 45:121–125
Nada O, Hirata K (1975c) Ultrastructural evidence for the uptake of 5, 6-dihydroxytryptamine in the frog's gustatory cell. Am J Anat 144:393–398
Nada O, Hirata K (1976) Pharmaco-histochemical studies on a specific monoamine in the gustatory epithelia of the rabbit. Histochemistry 50:111–117
Nada O, Hirata K (1977) The monoamine-containing cell in the gustatory epithelium of some vertebrates. Arch Histol Jpn Suppl 40:197–206
Nada O, Toyohara T (1987) An immunohistochemical study of serotonin-containing nerves in the colon of rats. Histochem 86:229–232
Nafstad PHJ (1971a) Comparative ultrastructural study on Merkel cells and dermal basal cells in poultry (*Gallus domesticus*). Z Zellforsch Mikrosk Anat 116:342–348
Nafstad PHJ (1971b) On the ultrastructure of neuroepithelial interactions: the dermal innervation in the snout of the pig. Z Zellforsch Mikrosk Anat 122:

528–537
Nafstad PHJ (1987) Evidence of turnover of mammalian Merkel cell. J Anat 151:57–64
Nafstad PHJ, Baker RE (1973) Comparative ultrastructural study of normal and grafted skin in the frog, *Rana pipiens*, with special reference to neuroepithelial connections. Z Zellforsch 139:451–462
Nagahama S, Kurihara K (1985) Norepinephrine as a possible transmitter involved in synaptic transmission in frog taste organs and Ca dependence of its release. J Gen Physiol 85:431–442
Nagahama S, Kobatake Y, Kurihara K (1982) Effect of Ca^{2+}, cyclic GMP, and cyclic AMP added to artificial solution perfusing lingual artery on frog gustatory nerve responses. J Gen Physiol 80:785–800
Nagahara Y (1940) Experimentelle Studien über die histologischen Veränderungen des Geruchsorgans nach der Olfactoriusdurchschneidung. Japan J Med Sci Pt V Pathol 5:165
Nagasawa J, Douglas WW (1972) Thorium dioxide uptake into adrenal medullary cells and the problem of recapture of granule membrane following exocytosis. Brain Res 37:141–145
Nagasawa H, Kataoka H, Hori Y, Isogai A, Tamura S, Suzuki A, Guo F, Zhong X, Mizoguchi A, Fujishita M, Takahashi SY, Ohnishi E, Ishizaki H (1984a) Isolation and some characterization of the prothoracicotropic hormone from *Bombyx mori.* Gen Comp Endocrinol 53:143–152
Nagasawa H, Kataoka H, Isogai A, Tamura S, Suzuki A, Ishizaki H, Mizoguchi A, Fujiwara Y, Suzuki A (1984b) Amino-terminal amino acid sequence of the silkworm prothoracicotropic hormone: homology with insulin. Science 226:1344–1345
Nagasawa J (1977) Exocytosis: the common release mechanism of secretory granules in glandular cells, neurosecretory cells, neurons and paraneurons. Arch Histol Jpn Suppl 40:31–47
Nagasawa J, Douglas WW, Schultz RA (1971) Micropinocytotic origin of coated and smooth microvesicles ("synaptic vesicles") in neurosecretory terminals of posterior pituitary glands demonstrated by incorporation of horseradish peroxidase. Nature 232:341–342
Nagata Y (1984) Memory and learning. In: Oomura Y, Shimazu H, Ito M (eds) Structure and function of the brain, part 2 (in Japanese). Igaku-Shoin, Tokyo, pp 702–713
Nagatsu I, Kondo Y (1974) Immunoelectron-microscopic localization of phenylethanolamine-N-methyltransferase in the bovine adrenal medulla. Histochemistry 42:351–358
Nagatsu I, Karasawa Y, Kondo Y, Inagaki S (1979) Immunocytochemical localization of tyrosine hydroxylase, dopamine β-hydroxylase and phenylethanolamine-N-methyltransferase in the adrenal glands of the frog and rat by a peroxidase-antiperoxidase method. Histochemistry 64:131–144
Nagayama T, Takahata M, Hisada M (1983) Local spikeless interaction of motoneuron dentrites in the crayfish *procambarus clarkii* Girard. J Comp Physiol 152:335–345
Nagayama T, Takahata M, Hisada M (1984) Functional characteristics of local non-spiking interneurons as the premotor elements in crayfish. J Comp Physiol A 154:499–510
Nahorski SR, Rogers KJ (1972) An enzymatic fluorometric micromethod for determination of glycogen. Anal Biochem 49:492–497
Nakae PK (1970) Classifications of anterior pituitary cell types with immunoenzyme histochemistry. J Histochem Cytochem 18:9–20
Nakagami K, Yamazaki Y, Tsunoda Y (1968) An electron microscopic study of the human fetal parathyroid gland. Z Zellforsch 85:89–95
Nakagami K, Warshawsky H, Leblond CP (1971) The elaboration of protein and carbohydrate by rat parathyroid cells as revealed by electron microscope autoradiography. J Cell Biol 51:596–610
Nakai Y, Uchida M, Iwashita T (1976) Fine-structural localization of adenosinetriphosphatase in the mouse adrenal medulla under normal conditions and following treatment with reserpine. In: Coupland RE, Fujita T (eds) Chromaffin, enterochromaffin and related cells. Elsevier, Amsterdam Oxford New York, pp 35–46
Nakai Y, Naito N, Shioda S, Ochiai H (1977a) Fine-structural and functional aspects of the cerebrospinal fluid-contacting subependymal cells in the anuran hypothalamus. Arch Histol Jpn 40:245–259
Nakai Y, Ochiai S, Shioda S, Ochi J (1977b) Cytological evidence for different types of cerebrospinal fluid-contacting subependymal cells in the preoptic and infundibular recesses of the frog. Cell Tiss Res 176:317–334
Nakajima T (1981) Active peptides in amphibian skin. TIPS Reviews 2:202–205
Nakajima T, Yamaguchi H, Takahashi K (1980) S-100 protein in folliculostellate cells of the rat pituitary anterior lobe. Brain Res 191:523–531
Nakajima T, Kameya T, Watanabe S, Hirota T, Sato Y, Shimosato Y (1982a) An immunoperoxidase study of S-100 protein distribution in normal and neoplastic tissue. Am J Surg Pathol 6:715–727
Nakajima T, Watanabe S, Sato Y, Shimosato Y, Motoi M, Lennert K (1982b) S-100 protein in Langerhans cells, interdigitating reticulum cells and histiocytosis X cells. Gann 73:429–432
Nakajima T, Sato Y, Watanabe S, Shimosato Y, Ishihara K, Isobe T (1982c) Immunoelectron microscopical demonstration of S-100 protein in epidermal Langerhans cells. Biomed Res 3:226–231
Nakajima T, Kameya T, Watanabe S, Hirota T, Shimosato Y, Isobe T (1984) S-100 protein distribution in normal and neoplastic tissues. In: DeLellis RA (ed) Advances in immunohistochemistry. Masson, Chicago, pp 141–158
Nakamura T, Gold GH (1987) A cyclic nucleotide-gated conductance in olfactory receptor cilia. Nature 325:442–444
Nakane PK (1968) Simultaneous localization of multiple tissue andtigens using the peroxidase-labeled antibody method: a study on pituitary glands of the rat. J Histochem Cytochem 16:557–560
Nakane PK (1970) Classifications of anterior pituitary

cell types with immunoenzyme histochemistry. J Histochem Cytochem 18:9–20
Nakanishi S, Inoue A, Kita T, Nakamura M, Chang ACY, Cohen SN, Numa S (1979) Nucleotide sequence of cloned cDNA for bovine corticotropin-β-lipotropin precursor. Nature 278:423–427
Nakashima Y, Appert HE, Howard JM (1977) The effects of calcitonin on pancreatic exocrine secretion in dogs. Surg Gynecol Obstet 144:71–76
Narayanan CH, Narayanan Y (1980) Neural crest and placodal contributions in the development of the glossopharyngeal-vagal complex in the chick. Anat Rec 196:71–82
Neher E, Marty A (1982) Discrete changes of cell membrane capacitance observed under conditions of enhanced secretion in bovine adrenal chromaffin cells. Proc Natl Acad Sci USA 76:6712–6716
Nelson SR, Schulz DW, Passonneaun JV, Lowry OH (1968) Control of glycogen levels in brain. J Neurochem 15:1271–1279
Nemeroff CB, Youngblood WW, Manberg PJ, Prange AJ Jr., Kizer JS (1983) Regional brain concentrations of neuropeptides in huntington's chorea and schizophrenia. Science 221:972–975
Neutra M, Leblond CP (1966) Radioautographic comparison of the uptake of galactose-^{3}H and glucose-^{3}H in the Golgi region of various cells secreting glycoproteins or mucopolysaccharides. J Cell Biol 30:137–150
Newson B, Ahlman H, Dahlström A, Gupta TKD, Nyhus LM (1979) Are there sensory neurons in the mucosa of the mammalian gut? Acta Physiol Scand 105:521–523
Nexø E, Olsen PS, Poulsen K (1984) Exocrine and endocrine secretion of renin and epidermal growth factor from the mouse submandibular glands. Regul Peptides 8:327–334
Niall HD, Keutmann HT, Sauer RT, Hogan M, Dawson BF, Aurbach GD, Potts JT (1970) The amino acid sequence of bovine parathyroid hormone I. Hoppe-Seylers Z Physiol Chem 351:1586–1588
Niall HD, Sauer RT, Jacobs J, Keutmann HT, Segre GV, O'Riordan JLH, Aurbach GD, Potts JT (1974) The amino acid sequence of the amino terminal 37 residueds of human parathyroid hormone. Proc Natl Acad Sci USA 71:384–388
Nichols DH (1986) Mesenchyme formation from the trigeminal placodes of the mouse embryo. Am J Anat 176:19–31
Nicolas A (1890) Sur les cellules à grains du fond des glandès de Lieberkühn chez quelques mammifères et chez le lézard. Bull Soc Sci Nancy 2:45–49
Nicoll RA (1982) Neurotransmitters can say more than just 'yes' or 'no'. Trends NeuroSci 5:369–374
Nihei K, Iwanaga T, Yanaihara N, Mochizuki T, Fujita T (1983) Preproenkephalin A occurs in the enterochromaffin (EC) cells of the porcine intestine: an immunocytochemical study using antisera to Met-Enkephalin-Arg^6-Gly^7-Leu^8 and to serotonin. Biomed Res 4:393–398
Nihei K, Iwanaga T, Fujita T, Mochizuki T, Yanaihara N (1984) Distribution of Met-enkephalin-Arg^6-Gly^7-Leu^8-like immunoreactivity in the gut of rats and pigs. Endorphin Enkephalin 3:434–439
Niijima A (1980) Effect of serotonin (5-HT) on the firing rate of afferent discharges recorded from the mesenteric nerves in the rat. Biomed Res Suppl 1:95–97
Niijima A, Fukuda A, Taguchi T, Okuda J (1983) Suppression of afferent activity of the hepatic vagus nerve by anomers of D-glucose. Am J Physiol 244: R611–R614
Niki A, Niki H (1976) The pancreatic endocrine cell recognition of D-glucose anomers. In: Fujita T (ed) Endocrine gut and pancreas. Elsevier, Amsterdam, pp 225–236
Niki A, Niki H (1980) Hexose anomers, insulin release and diabetes mellitus. Biomed Res 1:189–206
Niki A, Niki H, Miwa I, Okuda J (1974) Insulin secretion by anomers of D-glucose. Science 186:150–151
Niki A, Niki H, Miwa I, Lin BJ (1976) Interaction of alloxan and anomers of D-glucose on glucose-induced insulin secretion and biosynthesis in vitro. Diabetes 25:574–579
Niki A, Niki H, Miwa I (1979) Effect of anomers of D-mannose on insulin release from perfused rat pancreas. Endocrinology 105:1051–1054
Nilsson G (1979) Some mechanisms with an inhibitory effect on gastric acid secretion. Scand J Gastroenterol 55:29–35
Nilsson G, Larsson L-I, Håkanson R, Brodin E, Pernow B, Sundler F (1975) Localization of substance P-like immunoreactivity in mouse gut. Histochemistry 43:97–99
Nilsson O (1977) Studies on the ultrastructure of the human parathyroid glands in various pathological conditions. Acta Pathol Microbiol Scand A Suppl 263:1–88
Nilsson O, Ahlman H, Geffard M, Dahlström A, Ericson LE (1987) Bipolarity of duodenal enterochromaffin cells in the rat. Cell Tiss Res 248:49–54
Nilsson SEG (1965) The ultrastructure of the receptor outer segments in the retina of the leopard frog (*Rana pipiens*). J Ultrastruct Res 12:207–231
Nisell O (1951) The influence of blood gases on the pulmonary vessels of the cat. Acta physiol scand 23:85–90
Nishi S, North RA (1973) Intracellular recording from the myenteric plexus of the guinea-pig ileum. J Physiol 231:471–491
Nishi S, Soeda H, Koketsu K (1967) Release of acetylcholine from sympathetic preganglionic nerve terminals. J Neurophysiol 30:114–134
Nishiitsutsuji-Uwo J, Endo Y (1981) Gut endocrine cells in insects: The ultrastructure of the endocrine cells in the cockroach midgut. Biomed Res 2:30–44
Nishizuka Y (1984) The role of protein kinase C in cell surface signal transduction and tumour promotion. Nature 308:693–698
Noda M, Furutani Y, Takahashi H, Toyosato M, Hirose T, Inayama S, Nakanishi S, Numa S (1982) Cloning and sequence analysis of cDNA for bovine adrenal preproenkepahlin. Nature 295:202–206
Noda M, Takahashi H, Tanabe T, Toyosato M, Kikyotani S, Furutani Y, Hirose T, Takashima H, Inayama S, Miyata T, Numa S (1983a) Structural homology of *Torpedo californica* acetylcholine receptor subunits. Nature 302:528–532

Noda T, Yamamoto T, Miyata M, Nishimura Y (1983b) The mode of cerebellar activation of neurons in the cat motor cortex: an intracellular HRP study. Brain Res 268:153–157

Noden DM (1975) An analysis of the migratory behavior of avian cephalic neural crest cells. Devel Biol 42:106–130

Noe BD, Baste CA, Bauer GE (1977) Studies on proinsulin and proglucagon biosynthesis and conversion at the subcellular level. I. Fractionation procedure and characterization of the subcellular fraction. J Cell Biol 74:578–588

Noguchi R, Kobayashi S (1977) On the vascular architecture of the carotid labyrinth in *Cynops pyrrhogaster* and *Onychodactylus japonicus*. Arch Histol Jpn 40:347–360

Nolan JA, Trojanowski JQ, Hogue-Angeletti R (1985) Neurons and neuroendocrine cells contain chromogranin: Detection of the molecule in normal bovine tissues by immunochemical and immunohistochemical methods. J Histochem Cytochem 33: 791–798

Nonidez JF (1932) The origin of the "parafollicular" cells, a second epithelial component of the thyroid gland of the dog. Am J Anat 49:479–505

Nonidez JF (1935) The presence of depressor nerves in the aorta and carotid of birds. Anat Rec 62:47–74

Norberg KA, Ritzén M, Ungerstedt U (1966) Histochemical studies on a special catecholamine-containing cell type in sympathetic ganglia. Acta Physiol Scand 67:260–270

Norris EH (1937) The parathyroid glands and the lateral thyroid in man: their morphogenesis, histogenesis, topographic anatomy and prenatal growth. Contr Embryol Carneg Inst 26:247–294

Norris EH (1946) Anatomical evidence of prenatal function of the human parathyroid glands. Anat Rec 96:129–141

North RA (1982) Electrophysiology of the enteric nervous system. Neuroscience 7:315–325

North RA, Williams JT (1983) How do opiates inhibit neurotransmitter release? Trends NeuroSci 6:337–339

Novikoff AB (1964) GERL, its form and function in neurons of rat spinal ganglia. Biol Bull 127:358A

Nozue T, Kayano T (1977) Multiple APUD system (neural crest) tumors caused by endotoxin in suckling mice. Experientia 33:516–517

Nunez EA, Gershon MD (1972) Synthesis and storage of serotonin by parafollicular (C) cells of the thyroid gland of active, prehibernating and hibernating bats. Endocrinology 90:1008–1024

Nunez EA, Gershon MD (1976) Secretion by parafollicular cells beginning at birth: ultrastructural evidence from developing canine thyroid. Am J Anat 147:375–392

Nunez EA, Gershon MD (1978a) Secretion by parafollicular cells beginning at birth: Ultrastructural evidence from developing canine thyroid. Am J Anat 147:375–392

Nunez EA, Gershon MD (1978b) Cytophysiology of thyroid parafollicular cells. Int Rev Cytol 52:1–80

Nurse CA, Mearow KM, Holmes M, Visheau B, Diamond J (1983) Merkel cell distribution in the epidermis as determined by quinacrine fluorescence. Cell Tiss Res 228:511–524

Obata K (1977) Biochemistry and physiology of amino acid transmitters. In: Kandel ER (ed) Cellular biology of neurons, part 1. (Handbook of physiology 1/I) American Physiological Society, Bethesda Maryland, pp 625–650

Oberndorfer S (1907) Karzinoide Tumoren des Ditnndarms. Frankfurt Z Pathol 1:426–432

Ochiai T, Suzuki M (1981) Ultrastructural studies of the Merkel cells during fetal and postnatal development. J Invest Dermatol 77:437–443

O'Connor DT (1983) Chromogranin: wide spread immunoreactivity in polypeptide hormone producing tissues and in serum. Regul Peptide 6:263–280

O'Connor DT, Burton D, Deftos LJ (1983) Chromogranin A: immunohistology reveals its universal occurrence in normal polypeptide hormone producing endocrine glands. Life Sci 33:1657–1663

O'Connor DT, Frigon RP, Sokoloff RL (1984) Human chromogranin A: purification and characterization from catecholamine storage vesicles of human pheochromocytoma. Hypertension 6:2–12

Odartchenko N, Hedinger C, Ruzicka J, Weber E (1970) Cytokinetics of argentaffin cells in mouse intestinal mucosa. Virchows Arch Abt B Zellpathol 6:132–136

Ogata T, Ogata A (1923) Über die Henlesche Chromreaktion der sog, chromaffinen Zellen und den mikroskopischen Nachweis des Adrenalins. Ziegler Beitr 71:376–387

Ohhashi T, Jacobowitz DM (1983) The effects of pancreatic polypeptides and neuropeptide Y on the rat vas deferens. Peptides 4:381–386

Ohkawa H, Watanabe M (1977) Effects of gastrointestinal hormones on the electrical and mechanical activities of the cat small intestine. Jap J Physiol 27:71–79

Ohkawa H, Watanabe M (1977) Effects of gastrointestinal hormones on the electrical and mechanical activity of the cat stomach. Tokoku J Exp Med 122:287–298

Ohmori H (1984a) Studies of inonic currents in the isolated vestibular hair cell of the chick. J Physiol (Lond) 350:561–581

Ohmori H (1984b) Mechanoelectrical transducer has discrete conductances in the chick vestibular hair cell. Proc Natl Acad Sci USA 81:1888–1891

Ohtani O, Ohtsuka A (1985) Three-dimensional organization of lymphatics and their relationship to blood vessels in rabbit small intestine. A scanning electron microscopic study of corrosion casts. Arch Histol Jpn 48:255–268

Ohtani O, Kikuta A, Ohtsuka A, Taguchi T, Murakami T (1983) Microvasculature as studied by the microvascular corrosion casting/scanning electron microscope method. I. Endocrine and digestive system. Arch Histol Jpn 46:1–42

Ohtani O, Ushiki T, Kanazawa H, Fujita T (1986) Microcirculation of the pancreas in the rat and rabbit with special reference to the insulo-acinar portal system and emissary vein of the islet. Arch Histol Jpn 49:45–60

Okamoto E, Ueda T (1967) Embryogenesis of

intramural ganglia of the gut and its relation to Hirschsprung's disease. J Pediatr Surg 2:437–443
Okamoto H, Nose K, Itoh N, Sei T, Yamamoto H (1979) The translational control of proinsulin synthesis in pancreatic islets. In: Baba S, Kaneko T, Yanaihara N (eds) Proinsulin, insulin, C-peptide. (Int Congr Ser 468.) Excerpta Medica, Amsterdam, pp 27–35
Okamoto T, Subimura M, Kubo N (1976) Distribution of endocrine cells in duck digestive tracts. J Fac Fish Anim Hubs, Hiroshima University 15: 127–134
Okamoto T, Yamada J (1981) Light and electron microscopic studies on the endocrine cells in the duck proventriculus. Jpn J Vet Sci 43:863–870
Okeda R, Song SJ, Nakajima T, Matsutani M (1984) Pinealoma. Observation of an autopsy case by electron microscopy and cell markers. Acta Pathol Jpn 34:911–918
Oksche A (1982) Neuroanatomical pattern of endocrine and oscillatory systems of the brain: retrospect and prospect. In: Aaschoff J, Daan S, Ggroos G (eds) Vertebrate circadian systems. Springer, Berlin Heidelberg New York, pp 31–41
Oksche A, Hartwig HG (1979) Pineal sense organs–components of photoneuroendocrine systems. Prog Brain Res 52:113–130
Oksche A, Vaupel-von Harnack M (1966) Elektronenmikroskopische Untersuchungen zur Frage der Sinneszellen im Pinealorgan der Vögel. Z Zellforsch 69:41–60
Okuma Y, Osumi Y (1982) Neurotensin-induced release of endogenous noradrenaline from rat hypothalamic slices. Life Sci 30:77–84
Olschowka JA, Jacobowitz DM (1983) The coexistence and release of bovine pancreatic polypeptide-like immunoreactivity from noradrenergic superior cervical ganglia neurons. Peptides 4:231–238
Olson L, Marmfors T (1970) Growth characteristics of adrenergic nerves in the adult rat: fluorescence histochemical and ^{3}H-noradrenaline uptake studies using tissue transplanations to the anterior chamber of the eye. Acta physiol scand Suppl 348:1–112
Olson L, Ålund M, Norberg KA (1976) Fluorescence–microscopical demonstration of a population of gastrointestinal nerve fibers with a selective affinity for quinacrine. Cell Tiss Res 171:407–423
Olsson R (1973) Fine structure of the eta cells of the milk fish, *Chanos chanos*. In: Szentágothai J, Hajós F (eds) 7th Conf Europ Comp Endocrinol. Akadémiai Kiadó, Budapest, p 158
Olsson R, Fujita H (1976) The follicular prolactin cells of the primitive teleost *Engraulis japonica*. Cell Tiss Res 172:185–194
Ookawara S, Suzuki K, Yoshida Y, Ooneda G (1974) Monoamine-storing cells in the media of the thoracic aorta of *Gallus domesticus*. Cell Tiss Res 151: 309–316
Oomura Y, Ohta M, Kita H, Ishibashi S, Okajima T (1978) Hypothalamic neuron response to glucose, phlorizin, and cholecystokinin. In: Ryall RW, Kelly JS (eds) Iontophoresis and Transmitter Mechanisms in the Mammalian Central Nervous System. Elsevier, Amsterdam, pp 120–123
Orci L, Pictet R, Forssmann WG, Renold AE, Rouiller C (1968) Structural evidence for glucagon producing cells in the intestinal mucosa of the rat. Diabetologia 4:56–67
Orci L, Baetens D, Dubois MP, Rufener C (1975) Evidence for the D-cell of the pancreas secreting somatostatin. Horm Metab Res 7:400–405
Orci L, Perrelet LA, Friend DS (1977) Freeze-fractrure of membrane fusions during exocytosis in pancreatic B cells. J Cell Biol 75:23–30
Orth DN (1973) Establishment of human malignant melanoma-clonal cell lines that secrete ectopic adrenocorticotropin. Nature Biol 242:26–28
Osaka M, Kobayashi S (1976) Duodenal basal-granulated cells in the human fetus with special reference to their relationship to nervous elements. In: Fujita T (ed) Endocrine gut and pancreas. Elsevier Scientific Publishing Company, Amsterdam New York, pp 145–158
Osaka M, Sasagawa T, Fujita T (1973) Emiocytotic granule release in the human antral endocrine cells. In: Fujita T (ed) Gastro-entero-pancreatic endocrine system: a cell-biological approach. Igaku-Shoin, Tokyo, pp 59–63
Osaka M, Sasagawa T, Fujita T (1974) Granule release from endocrine cells in acidified human duodenal bulb: an electron microscope study of biopsy materials. Arch Histol Jpn 37:73–94
Osaka M, Fujita T, Yanatori Y (1975) On the possible role of intestinal hormones as the diarrhogenic messenger in cholera. Virchows Arch B (Cell Pathol) 18:287–296
Osborne MP (1977) Role of vesicles with some observations on vertebrate sensory cells. In: Cottrell GA, Usherwood PNR (eds) Synapses. Blackie, Glasgow, pp 40–63
Osborne MP, Butler PJ (1975) New theory for receptor mechanism of carotid body chemoreceptors. Nature 254:701–703
Osborne NN (1983) Dale's principle and communication between neurones, Pergamon Press, Oxford
Osborne NN, Patel S (1984) Localization of histamine-immunoreactivity in specific neurones of the gatropod CNS including a cell which has been identified as histaminergic. Histochemistry 81:207–208
Oshima N, Fujii R (1984) A precision photoelectric method for recording chromatophore responses *in vitro*. Zool Sci 1:545–552
Oshima K, Gorbman A (1969) Pars intermedia: unitary electrical activity regulated by light. Science 163:195–197
Östberg Y (1976) The entero-insular endocrine organ in a cyclostome, *Myxine glutinosa*. Umeå University Medical Dissertations. 15
Östberg Y, Van Noorden S, Pearse AGE (1975) Cytochemical, immunofluorescence, and ultrastructural investigations on polypeptide hormone localization in the islet parenchyma and bile duct mucosa of a cyclostome, *Myxine glutinosa*. Gen Comp Endocrinol 25:274–291
Otsuka M, Konishi S (1983) Substance P—the first peptide neurotransmitter? Trends Neurosci 6:317–

320
Otsuka M, Konishi S, Yanagisawa M, Tsunoo A, Akagi H (1982) Role of substance P as a sensory transmitter in spinal cord and sympathetic ganglia. Substance P in the nervous system (Ciba Foundation symposium 91). Pitman, London, pp 13–34
Otuski M, Williams JA (1983) Direct modulation of pancreatic CCK receptors and enzyme secretion by insulin in isolated pancreatic acini from diabetic rats. J Am Diabet Assoc 32:241–246
Owada K, Kawata M, Akaji K, Takagi A, Moriga M, Kobayashi H (1985) Urotensin II-immunoreactive neurons in the caudal neurosecretory system of freshwater and seawater fish. Cell Tiss Res 239: 349–354
Owman C, Rüdeberg C, Ueck M (1970) Fluoreszenz-mikroskopischer Nachweis biogener Monoamine in der Epiphysis cerebri von *Rana esculenta* and *Rana pipiens*. Z Zellforsch 111:550–558
Owman C, Owman T, Sjöberg NO (1971) Short adrenergic neurons innervating the female urethra of the cat. Experientia 27:313–315
Ozawa S (1981) Action potential and hormone secretion in the clonal anterior pituitary cell. In: Kanno T (ed) Paraneurons, their features and functions. Excerpta Medica, Amsterdam, pp 140–143
Ozawa S (1985) Electrophysiology of clonal anterior pituitary cells. In: Poisner AM, Trifaró JM (eds) The electrophysiology of the secretory cells. Elsevier, Amsterdam New York Oxford, pp 221–239
Ozawa S, Miyazaki S (1979) Electrical excitability in the rat clonal pituitary cell and its relation to hormone secretion. Jpn J Physiol 29:411–426
Ozawa S, Sand O (1978) Electrical activity of rat anterior pituitary cells in vitro. Acta Physiol Scand 102:330–341
Padykula HA, Gauthier GT (1970) The ultrastructure of the neuromuscular junctions of mammalian red, white and intermediate skeletal muscle fibers. J Cell Biol 46:27–41
Palade GE (1959) Functional changes in the structure of cell components. In: Hayashi T (ed). Subcellular Particles. Ronald Press, New York, pp 64–80
Palade GE (1975) Intracellular aspects of the process of protein secretion. Science 189:347–358
Palade GE (1982) Problems in intracellular membrane traffic. In: Evered D, Collins GM (eds) Membrane recycling. Pitman, London, pp 1–14
Palay SL (1957) The fine structure of the neurohypophysis. In: Waelsch H (ed) Ultrastructure and cellular chemistry of neural tissue. Hoeber, New York, pp 31–49
Palme F (1934) Die Paraganglien über dem Herzen und im Endigungsgebiet des Nervus depressor. Z Mikrosk Anat Forsch 36:391–420
Palmer P (1965) Ultrastructural alterations of Merkel cells following denervation (abst). Anat Rec 151: 396–397
Pangborn RM, Gee SC (1961) Relative sweetness of α- and β-forms of selected sugars. Nature 191:810–811
Panula P, Kaartinen M, Macklin M, Costa E (1985) Histamine containing peripheral neuronal and endocrine systems. J Histochem Cytochem 33: 933–941
Paran N, Mattern CFT (1975) The distribution of acetylcholinesterase in buds of the rat vallate papilla as determined by electron microscope histochemistry. J Comp Neurol 159:29–44
Paran N, Mattern CFT, Henkin RI (1975) Ultrastructure of the taste bud of the human fungiform papilla. Cell Tiss Res 161:1–10
Parat M (1924) Contribution a l'histo-physiologie des organes digestifs de l'embryon. L'apparition corrélative de la cellule de kultschitzky et de la sécrétine chex l'embryon. C R Seances Soc Biol Ses Fil 90: 1023–1024
Pardee AB, Dubrow R, Hamlin JL, Kletzien RF (1978) Animal cell cycle. Annu Rev Biochem 47:715–750
Parent A, Poitras D (1974) Morphological organization of monoamine-containing neurons in the hypothalamus of the painted turtle (*Chrysemys picta*). J comp Neurol 154:379–394
Parratt JR, West GB (1957) 5-hydroxytryptamine and tissue mast cells. J Physiol 137:169–178
Partanen M, Hervonen A (1979) The formaldehyde-induced fluorescence of the developing hypogastric (main pelvic) ganglion of the rat. Short adrenergic neurons and the effect of testosterone. Histochemistry 62:249–258
Partanen S, Rechardt L, Bäck N (1975) Histochemical observations on uptake of L-dopa into endocrine cells of the rat pituitary gland during the postnatal development. Cell Tiss Res 156:451–461
Pascal JP, Vaysse N (1976) Action du systeme adrenergique sur le pancreas exocrine. Biol Gastroenterol (Paris) 9:243–254
Pavel S (1971) Evidence for the ependymal origin of arginine vasotocin in the bovine gland. Endocrinology 89:613–614
Pavel S (1978) Arginine vasotocin as a pineal hormone. J Neural Transm Suppl 13:135–155
Pavel S, Goldstein R, Ghinea E, Calb M (1977) Chromatographic evidence for vasotocin biosynthesis by cultured pineal ependymal cells from rat fetuses. Endocrinology 100:205–208
Payne CM, Cromey DW (in press) Ultrastructural similarities between *Chlorohydra viridssima* and human neurosecretory granules: a cytochemical study using the uranaffin reaction. Cytobios 50: 191–203
Pazoles CJ (1982) Anion and proton transport in chromaffin granules. Fed Proc 41:2769–2774
Pearse AGE (1966a) Common cytochemical properties of cells producing polypeptide hormones, with particular reference to calcitonin and the thyroid C cells. Vet Rec 79:587–590
Pearse AGE (1966b) The Cytochemistry of the thyroid C cells and their relationship to calcitonin. Proc Roy Soc B 164:478–487
Pearse AGE (1968) Common cytochemical and ultrastructural characteristics of cells producing polypeptide hormones (APUD series) and their relevance to thyroid and ultimobranchial C cells and calcitonin. Proc Roy Soc [Biol] 170:71–80
Pearse AGE (1969) The cytochemistry and ultrastructure of polypeptide hormone-producing cells of the

APUD series and the embryologic, physiologic and pathologic implications of the concept. J Histochem Cytochem 17:303–313
Pearse AGE (1975a) The endocrine division of the nervous system: A logical extension of the APUD concept. Fol Anat Iugoslav 4:5–20
Pearse AGE (1975b) Neurocristopathy, neuroendocrine pathology and the APUD concept. Z Krebsforsch 84:1–18
Pearse AGE (1976) Neurotransmission and the APUD concept. In: Coupland RE, Fujita T (eds) Chromaffin, enterochromaffin and related cells. Elsevier, Amsterdam, pp 147–154
Pearse AGE (1977) The diffuse neuroendocrine system and the APUD concept: related "endocrine" peptides in brain, intestina, pituitary, placenta, and anuran cutaneous glands. Med Biol 55:15–125
Pearse AGE (1981) The diffuse neuroendocrine system: falcification and verification of a concept. In: Grossman M, Brazier MAB, Lechago J (eds) Cellular basis of chemical messengers in the digestive system. Academic Press, New York London Toronto Sydney San Francisco, pp 13–19
Pearse AGE, McGregor MM (1964) Functional cytology of the pituitary gland. Ann Rep Brit Emp Cancer Camp 2:665–666
Pearse AGE, Polak JM (1971a) Neural crest origin of the endocrine polypeptide (APUD) cells of the gastrointestinal tract and pancreas. Gut 12:783–788
Pearse AGE, Polak JM (1971b) Cytochemical evidence for the neural crest origin of mammalian ultimobranchial C cells. Histochemie 27:96–102
Pearse AGE, Polak JM (1971c) Cytochemical evidence for the neural crest origin of mammalian ultimobranchial C cells. Histochemie 27:96–102
Pearse AGE, Polak JM (1974) Endocrine tumors of neural crest origin: Neurolophomas, apudomas and the APUD concept. Med Biol 52:3–18
Pearse AGE, Takor Takor T (1976) Neuroendocrine embryology and the APUD concept. Clin Endocrinol Suppl 5:229s–244s
Pearse AGE, Takor Takor T (1979) Embryology of the diffuse neuroendocrine system and its relationship to the common peptides. Fed Proc 38:2288–2294
Pearse AGE, Welbourn RB (1973) The apudomas. Br J Hosp Med 1973:617–624
Pearse AGE, Polak JM, Heath CM (1974a) Polypeptide hormone production by "carcinoid" apudomas and their relevant cytochemistry. Virchows Archiv B Cell Pathol 16:95–109
Pearse AGE, Polak JM, Bloom SR, Adams C, Dryburgh JR, Brown JC (1974b) Enterochromaffin cells of the mammalian small intestine as the source of motilin. Virchows Arch B Cell Path 16:111–120
Pearse AGE, Polak JM, Bloom SR (1977) The newer gut hormones. Cellular sources, physiology, pathology, and clinical aspects. Gastroenterology 72: 746–761
Pearse BMF (1976) Clathrin: a unique protein associated with intracellular transfer of membrane by coated vesicles. Proc Natl Acad Sci USA 73:1225–1259
Pearse BMF, Bretscher MS (1981) Membrane recycling by coated vesicles. Annu Rev Biochem 50:85–101
Peng TC, Cooper CW, Petrusz P, Volpert EM (1975) Identification of C-cells in normal and goitrous rat thyroid tissues using antiserum to rat thyrocalcitonin and the immunoperoxydase bridge technique. Endocrinology 97:1537–1544
Pestarino M (1982) Occurrence of different secretin-like cells in the digestive tract of the ascidian *Styela plicata* (Urochordata, Ascidiacea) Cell Tiss Res 226:231–235
Pestarino M (1983) Prolactinergic neurons in a protochordate. Cell Tiss Res 233:471–474
Pestarino M (1984a) Immunocytochemical demonstration of prolactin-like activity in the neural gland of the ascidian *Styela plicata*. Gen Comp Endocrinol 54:444–449
Pestarino M (1984b) Secretin-like immunoreactivity in the neural complex of a protochordate. Acta Zoologica (Stockh) 65:113–118
Pestarino M (1985a) Localization of α-MSH-like immunoreactive cells in the neural gland of the ascidian *Styela plicata*. Cell Tiss Res 240:497–500
Pestarino M (1985b) Occurrence of β-endorphin-like immunoreactive cells in the neural complex of a protochordate. Cellular Mol Biol 31:27–31
Pestarino M, Tagliafierro G (1983) Immunocytochemical localization of secretin-like cells in the digestive tract of the ascidian *Halocynthia papillosa*. Basic Appl Histochem 27:141–148
Peters A, Kaiserman-Abramof IR (1970) The small pyramidal neuron of the rat cerebral cortex. The perikaryon, dendrites and spines. Am J Anat 127:321–356
Peters A, Kimerer LM (1981) Bipolar neurons in rat visual cortex: a combined Golgi-electron microscope study. J Neurocytol 10:921–946
Peute J, Meij JCA (1973) Ultrastructural and functional aspects of the nucleus infundibularis ventralis in the green frog, *Rana esculenta*. Z Zellforsch 144:191–217
Pévet P (1977) On the presence of different populations of pinealocytes in the mammalian pineal gland. J neural Transm 40:289–304
Pévet P (1983) Anatomy of the pineal gland of mammals. In: Relkin R (ed) The pineal gland. Elsevier, New York, pp 1–75
Pévet P, Dogterom J, Buijs RM, Reinharz A (1979) Is it the vasotocin or vasotocin-like peptide which is present in the mammalian pineal and subcommissural organ? J Endocrinol 80:49P
Pfeiffer EFv, Raptis S, Ziegler R (1976) Wechselbeziehungen zwischen gastrointestinalen Hormone and endkriner Regulation. Z Gastroenterol 14:70–94
Pfenninger KH (1973) Synaptic morphology and cytochemistry. Prog Histochem Cytochem 5:1–86
Phillips JH, Apps DK (1980) Stoichiometry of catecholamine/proton exchange across the chromaffin granule membrane. Biochem J 192:273–278
Phillis JW, Kostopoulos GK (1975) Adenosine as a putative transmitter in the cerebral cortex. Studies with potentiators and antagonists. Life Sci 17: 1085–1094
Pick J (1959) The discovery of the carotid body. J

History Med 14:61–73

Pickel VM, Joh TH, Field PM, Becker CG, Reis DJ (1975a) Cellular localization of tyrosine hydroxylase by immunohistochemistry. J Histochem Cytochem 23:1–12

Pickel VM, Reis DJ, Marangos PJ, Zomzely-Neurath C (1975b) Immunocytochemical localization of nervous system specific protein (NSP-R) in rat brain. Brain Res 105:184–187

Pickel VM, Joh TH, Reis DJ (1976) Monoamine-synthesizing enzymes in central dopaminergic, noradrenergic and serotonergic neurons: Immunocytochemical localization by light and electron microscopy. J Histochem Cytochem 24:792–806

Pickel VM, Joh TH, Reis DJ (1977a) A serotonergic innervation of noradrenergic neurons in nucleus locus coeruleus: demonstration by immunocytochemical localization of the transmitter specific enzymes tyrosine and tryptophan hydroxylase. Brain Res 131:197–214

Pickel VM, Rein DJ, Leeman SE (1977b) Ultrastructural localization of substance P in neurons of rat spinal cord. Brain Res 122:534–540

Pickering BT, Swann RW (1984) Secretion in the posterior pituitary gland. In: Cantin M (ed) Cell Biology of the Secretory Process. Karger, Basel, pp 247–275

Pictet R, Rutter WJ (1972) Development of the embryonic endocrine pancreas. In: Steiner DF, Freinkel N (eds) Endocrine pancreas. (Handbook of physiology 7/I) American Physiological Society, Washington DC, pp 25–66

Pinkus F (1902) Über bisher unbekannten Nervenapparat am Haarsystem des Menschen: Haarsscheiben. Dermatol Z 9:465–469

Pinkus F (1905) Über Hautsinnesorgane neben dem menschlichen Haar (Haarscheiben) und ihr vergleichend-anatomische Bedeutung. Archi Mikrosk Anat Entw-Gesch 65:121–180

Poisner AM, Douglas WW (1968) Adenosine triphosphate and adenosine triphosphatase in hormone-containing granules of posterior pituitary gland. Science 160:203–204

Poisner AM, Trifaro JM (eds) (1985) The electrophysiology of the secretory cell. Elsevier, Amsterdam

Polak JM, Rost FWD, Pearse AGE (1971) Fluorogenic amine tracing of neural crest derivatives forming the adrenal medulla. Gen Comp Endocrinol 16:132–136

Polak JM, Bloom SR, Kuzio M, Brown JC, Pearse AGE (1973) Cellular localization of gastric inhibitory polypeptide in the duodenum and jejunum. Gut 14:284–288

Polak JM, Pearse AGE, Adams C, Caraud JC (1974a) Immunohistochemical and ultrastructural studies on the endocrine polypeptide (APUD) cells of the avian gastrointestinal tract. Experientia 30:564–567

Polak JM, Pearse AGE, Le Lièvre C, Fontaine J, Le Douarin NM (1974b) Immunocytochemical confirmation of the neural crest origin of avian calcitonin-producing cells. Histochemistry 40:209–214

Polak JM, Pearse AGE, Joffe S, Bloom SR (1975) Quantification of secretin release by acid, using immunocytochemistry and radioimmunoassay. Experientia 31:462–464

Polak JM, Pearse AGE, Gibson S (1976) Cellular endocrinology of the gut and pancreas. In: Fujita T (ed) Endocrine gut and pancreas. Elsevier, Amsterdam New York Oxford, pp 89–102

Polak JM, Sullivan SN, Bloom SR, Facer P, Pearse AGE (1977a) Enkephalin-like immunoreactivity in the human gastrointestinal tract. Lancet I:972–975

Polak JM, Sullivan SN, Bloom SR, Buchan AMJ, Facer P, Brown MR, Pearse AGE (1977b) Specific localization of neurotensin to the N-cell in human intestine by radioimmunoassay and immunocytochemistry. Nature 270:183–184

Pollard HB, Creutz CE, Fowler V, Scott, J, Pazoles CJ (1982) Calcium-dependent regulation of chromaffin granule movement, membrane contact, and fusion during exocytosis. Cold Spring Habor Symp Quant Biol 46:819–834

Polyak SL (1941) The retina. University of Chicago Press, Chicago

Polyak SL (1957) The vertebrate visual system. In: Klüver H (ed) University of Chicago Press, Chicago

Polyzonis MB (1970) Ultrastructural study of human parathyroid adenoma and the occurrence of abnormal cilia in the adenoma cells. Pathol Europ 5:454–469

Poritsky R, Singer M (1977) Intraperitoneal transplants of taste buds in the newt. Anat Rec 188:219–228

Post RM, Gold P, Rubinow DR, Ballenger JC, Bunney WE, Goodwin FK (1982) Peptides in the cerebrospinal fluid of neuropsychiatric patients: an approach to central nervous system peptide function. Life Sci 31:1–15

Potter DD, Landis SC, Furshpan EJ (1980) Dual function during development of rat sympathetic neurones in culture. J Exp Biol 89:57–71

Potter DD, Furshpan EJ, Landis SC (1981) Multiple-transmitter status and "Dale's principle". Neurosci Commentaries 1:1–9

Potter DD, Furshpan EJ, Landis SC (1983) Transmitter status in cultured rat sympathetic neurons: plasticity and multiple function. Fed Proc 42:1626–1683

Prenant A (1896) Sur le développement des glandes accessoires de la glande thyroïde et celui de la glande carotidienne. Anat Anz 12:242–244

Prentice FD, Wood JG (1975) Adrenergic innervation of cat adrenal medulla. Anat Rec 181:689–704

Pretl K (1944) Zur Frage der Endokrinie der menschlichen Vorsteherdrüse. Virchows Arch Pathol Anat 312:392–404

Price DA, Greenberg MJ (1977) Structure of a molluscan cardioexcitatory neuropeptide. Science 197:670–671

Price JL, Powell TPS (1970) The mitral and short axon cells of the olfactory bulb. J Cell Sci 7:631–652

Purves HD (1966) Cytology of the adenohypophysis. In: Harris GW, Donovan BT (eds) The pituitary gland, vol 1. Univ of California Press, Berkeley, pp 147–232

Quay WB (1974) Pineal canaliculi: demonstration,

twenty-four-hour rhythmicity and experimental modification. Am J Anat 139:81–93
Quay WB (1975) Perfusion of intercellular spaces in the pars distalis and their variation with time of day. J Endocr 66:385–390
Quirion R, Rioux F, Regoli D (1978) Chronotropic and inotropic effects of neurotensin on spontaneously beating auricles. Can J Physiol Pharmacol 56:671–673
Quirion R, Rioux F, Regoli D, St-Pierre S (1979) Neurotensin-induced coronary vessels constriction in perfused rat hearts. Eur J Pharmacol 55:221–223
Rabié A, Thomasset M, Legrand Ch (1983) Immunohistochemical detection of calcium-binding protein in the cochlear and vestibular hair cells of the rat. Cell Tiss Res 232:691–696
Racodot J, Olivier L, Porcile E, Droz B (1965) Appareil de Golgi et origine des grains de sécrétion dans les cellules adenohypophysaires chez le rat. Étude radioautoradiographique en microscopie électronique après injection de leucine tritiée. C R Acad Sci (Paris) D261:2972–2974
Radke R, Stach W (1986) Are the islets of Langerhans neuro-paraneuronal control centers of the exocrine pancreas? Arch Histol Jpn 49:411–420
Raikhlin NT, Kvetnoy IM, Tolkachev VN (1975) Melatonin may be synthesised in enterochromaffin cells. Nature 255:344–345
Ralph CL (1978) Introductory Animal Physiology. McGraw Hill, New York St Louis
Ramirez J, Hubel KA, Clifton JA (1966) Intestinal factors affecting pancreatic exocrine secretion in the rat. Am J Physiol 211:R260–R263
Ramsdale DR (1974) Further observations on urethral chromaffin cells. An electron microscopic study. Cell Tiss Res 148:499–504
Ranadive NS, Dhanani N (1980) Movement of calcium ions and release of histamine from rat mast cells. Int Arch Allergy Appl Immunol 61:9–18
Rapoport SI, Fredericks WR, Ohno K, Pettigrew KD (1980) Quantitative aspects of reversible osmotic opening of the blood-brain barrier. Am J Physiol 238:R421–R431
Rasmussen H (1981) Calcium and cAMP as synarchic messengers. John Wiley & Sons, New York Chichester Brisbane Toronto Singapore
Rasmussen H, Barrett Q (1984) Calcium messenger system: an integrated view. Physiol Rev 64:938–984
Rasmussen H, Craig LC (1961) Isolation of a parathyroid polypeptide from acetic acid extracts of bovine parathyroid glands. J Biol Chem 236:1083–1086
Rasmussen H, Goodman DBP (1977) Relationship between calcium and cyclic nucleotides in cell activation. Physiol Rev 57:421–509
Rasmussen H, Waisman DM (1982) Modulation of cell function in the calcium messenger system. Rev Physiol Biochem Pharmacol 95:111–148
Ratzenhofer M, Leb D (1965) Über die Feinstruktur der argentaffinen und der anderen Erscheinungsformen der "hellen Zellen" Feyrter's im Kaninchen-Magen. Z Zellforsch 67:113–150
Ravasio RA, Delouvee A, Yamada KM, Timpl R, Thiery JP (1983) Neural crest cell migration: requirements for exogeneous fibronectin and high cell density. J Cell Biol 96:462–473
Ravazzola M (1976) Intracellular localization of calcium in the chromaffin cells of the rat adrenal medull. Endocrinology 98:950–953
Ravazzola M, Orci L, Habener JF, Potts JT (1978) Parathyroid secretory protein: immunocytochemical localization within cells that contain parathyroid hormone. Lancet Aug:371–372
Ravazzola M, Siperstein A, Moody AJ, Sundby F, Jacobsen H, Orci L (1979) Glicentin immunoreactive cells: their relationship to glucagon-producing cells. Endocrinology 105:499–508
Ravazzola M, Benoit R, Ling N, Guillmin R, Orci L (1983) Immunocytochemical localization of prosomatostatin fragments in maturing and mature secretory granules of pancreatic and gastrointestinal D cells. Proc Natl Acad Sci USA 80:215–218
Rawles ME (1947) Origin of pigment cells from neural crest in mouse embryo. Physiol Zoöl 20:248–266
Rawles ME (1948) Origin of melanophores and their role in development of color patterns in vertebrates. Physiol Rev 28:383–408
Rawles ME (1955) Pigmentation in autoplastic and homoplastic grafts of skin from fetal and newborn hooded rats. Am J Anat 97:79–127
Rayford PL, Miller TA, Thompson JC (1976) Secretin, cholecystokinin and newer gastrointestinal hormones. (second of two ports) N Engl J Med 294: 1157–1163
Redick JA, Thomas JA, Van Orden LS, Van Orden DE, Kopin IJ (1974) Immunohistochemical localization of dopamine β-hydroxylase in adrenal chromaffin granules. Neuropharmacology 13:1005–1014
Rehfeld JF (1972) Gastrointestinal hormones and insulin secretion. Scand J Gastroenterol 7:289
Rehfeld JF, Stadil F (1973) The effect of gastrin on basal- and glucose-stimulated insulin in man. J Clin Invest 52:1415–1426
Reinecke M (1981) Immunohistochemical localization of polypeptide hormones in endocrine cells of the digestive tract of *Branchiostoma lanceolatum*. Cell Tiss Res 219:445–456
Reinecke M, Carraway RE, Falkmer S, Feurle GE, Forssmann WG (1980) Occurrence of neurotensin-immunoreactive cells in the digestive tract of lower vertebrates and deuterostomian invertebrates. A correlated immunohistochemical and radioimmunochemical study. Cell Tiss Res 212:173–183
Reinharz AC, Czernichow P, Vallotton MB (1975) Neurophysins I and II from the bovine posterior pituitary lobe and neurophysin-like proteins from bovine pineal gland. Ann NY Acad Sci 248:172–183
Reinharz AC, Noteborn HPJM, Klein M, Vallotton MB, Ebels I, Pévet P (1985) Presence of immunoreactive neurophysins of a higher molecular weight in addition to the 10,000 form in the ovine pineal gland. Peptides NY 6:199–203
Reiter RJ (1981) The mammalian pineal gland: Structure and function. Am J Anat 162:287–313
Rémy CH, Dubois MP (1979) Localisation par im-

munofluorescence de peptides analogues à l'α-endorphine dans les ganglions infra-oesophagiens du lombricide *Dendrobaena subrubicunda* Eisen. Experientia 35:137–138

Rémy C, Dubois MP (1981) Immunohistological evidence of methionine enkephalin-like material in the brain of the migratory locust. Cell Tiss Res 218:271–278

Rémy C, Girardie J (1980) Anatomical organization of two vasopressin-neurophysin-like neurosecretory cells throughout the central nervous system of the migratory locust: Gen Comp Endocrinol 40: 27–35

Rémy C, Girardie J, Dubois MP (1979) Vertebrate neuropeptide-like substances in the suboesophageal ganglion of two insects: *Locusta migratoria* R. and F. (Orthoptera) and *Bombyx mori* L.(Lepidoptera). Immunocytological investigation. Gen Comp Endocrinol 37:93–100

Reuter H (1983) Calcium channel modulation by neurotransmitters, enzymes and drugs. Nature 301: 569–574

Reuter M, Lindroos P (1979) The ultrastructure of the nervous system of *Gyratrix hermaphroditus* (Turbellaria, rhabdocoela) I. The brain. Acta Zool (Stockh) 60:139–152

Reuter M, Karhi T, Schot LPC (1984) Immunocytochemical demonstration of peptidergic neurons in the central and peripheral nervous systems of the flatworm *Microstomum lineare* with antiserum to FNRF-amide. Cell Tiss Res 238:431–436

Reuter M, Wikgren M, Lehtonen M (1986) Immunocytochemical demonstration of 5-HT-like and FMRF-amide-like substances in whole mounts of *Microstomum lineare* (Turbellaria). Cell Tiss Res 246:7–12

Reutter K (1971) Die Geschmacksknospen des Zwergwelses, *Amiurus nebulosus* (Lesueur). Morphologische und histochemische Untersuchungen. Z Zellforsch 120:280–308

Rhodin J (1966) Ultrastructure and function of the human tracheal mucosa. Amer Rev Resp Dis 93:1–15

Rhodin J, Dalhamn T (1956) Electron microscopy of the tracheal ciliated mucosa in rat. Z Zellforsch 44:345–412

Ribalet E, Beigelman PM (1980) Calcium action potentials and potassium permeability activation in pancreatic B-cells. Am J Physiol 239:C124–C133

Ribalet E, Beigelman PM (1981) Effects of divalent cations on B-cell electrical activity. Am J Physiol 241:C59–C67

Ribalet E, Beigelman PM (1982) Effects of sodium on B-cell electrical activity. Am J Physiol 242:C296–C303

Ribeiro JA (1978) ATP; related nucleotides and adenosine on neurotransmission. Life Sci 22:1373–1380

Ricci A, Sanders KM, Portmore J, Van der Kloot WG (1975) Effects of the ionophores, X-537A and A-23187 on catecholamine release from the in vitro frog adrenal. Life Sci 16:177–184

Richards JG, Da Prada M (1977) Uranaffin reaction: a new cytochemical technique for the localization of adenine nucleotides in organelles storing biogenic amines. J Histochem Cytochem 25:1322–1336

Richards JG, Tranzer J (1970) The ultrastructural localisation of amine storage sites in the central nervous system with the aid of a specific marker, 5 hydroxydopamine. Brain Res 17:463–469

Richardson KC (1958) Electronmicroscope observations on Auerbach's plexus in the rabbit, with special reference to the problem of smooth muscle innervation. Am J Anat 103:99–136

Richardson KC (1960) Studies on the structure of autonomic nerves in the small intestine, correlating the silver-impregnated image in light microscopy with the permanganate-fixed ultrastructure in electronmicroscopy. Am J Anat 94:457–472

Richardson KC (1964) The fine structure of the albino rabbit iris with special reference to the identification of adrenergic and cholinergic nerves and nerve endings in its intrinsic muscles. Am J Anat 114:173–205

Riemer J. Kolling K, Mayer CJ (1977) The effect of motilin on the electrical activity of rabbit circular duodenal muscle. Pflugers Arch 372:243–250

Rinderknecht H, Renner IG, Douglas AP, Adham NF (1978) Profiles of pure pancreatic secretions obtained by direct pancreatic duct cannulation in normal healthy human subjects. Gastroenterology 75:1083–1089

Rindi G, Buffa R, Sessa F, Tortora O, Solcia E (1986) Chromogranin A, B and C immunoreactivities of mammalian endocrine cells. Distribution, distinction from costored hormones/prohormones and relationship with the argyrophil component of secretory granules. Histochem 85:19–28

Ritzén M, Hammerström L, Ullberg S (1965) Autoradiographic distribution of 5-hydroxytriptamine and 5-hydroxytryptophan in the mouse. Biochem Pharmacol 14:313–321

Robberecht P, Deschodt-Lanckman M (1981) Les recepteurs du tube digstif aux peptides hormonaux. Gastroenterol Clin Biol 5:290–299

Robbins N (1967) The role of the nerve in maintenance of frog taste buds. Exp Neurol 17:364–380

Roberts GW, Ferrier IN, Lee Y, Crow TJ, Johnstone EC, Owens DGC, Bacarese-Hamilton AJ, McGregor G, O'Shaughnessey D, Polak JM, Bloom SR (1983) Peptides, the limbic lobe and schizophrenia. Brain Res 288:199–211

Robertson DR (1967) The ultimobranchial body in *Rana pipiens*. III. Sympathetic innervation of the secretory parenchyma. Z Zellforsch 78:328–340

Robertson DR (1968) The ultimobranchial body in *Rana pipiens*. VII. Cellular responses in denervated glands in autoplastic transplants. Z Zellforsch 90:273–288

Robinson AG, Zimmermann EA, Frantz AG (1971) Physiologic investigation of posterior pituitary binding proteins neurophysin I and neurophysin II. Metabolism 20:1148–1155

Robinson AG, Michelis MF, Warms PC, Davis BB (1974) Natriuretic effect of posterior pituitary neurophysin. J Clin Endocrinol 39:913–918

Rodrigo J, Polak JM, Terenghi G, Cervantes C, Ghatei MA, Mulderry PK, Bloom SR (1985) Calcitonin gene-related peptide (CGRP)-immunoreactive sensory and motor nerves of the mammalian palate. Histochem 82:67–74

Rogers DC, Haller CJ (1978) Innervation and cytochemistry of the neuroepithelial bodies in the ciliated epithelium of the toad lung (*Bufo marinus*). Cell Tiss Res 195:395–410

Röhlich P, Anderson P, Uvnäs B (1971) Electron microscope observations on compound 48/80—induced degranulation in rat mast cells. Evidence for sequential exocytosis of storage granules. J Cell Biol 51:465–483

Role LW, Perlman RL (1983) Catecholamine uptake into isolated adrenal chromaffin cells: inhibition of uptake by acetylcholine. Neuroscience 10:987–996

Romagnano MA, Hamill RW (1984) Spinal sympathetic pathway: an enkephalin ladder. Science 225:737–739

Romand R (1983) Development of the cochlea. In: Romand R (ed) Development of Auditory and Vestibular Systems. Academic Press, New York London, pp 47–48

Romeis B (1940) Hypophyse. (Möllendorff's Handbuch mikrosk Anat Menschen VI/3) Springer, Berlin

Romijn HJ (1975) Structure and innervation of the pineal gland of the rabbit, *Oryctolagus cuniculus* (L). III. An electron microscopic investigation of the innervation. Cell Tiss Res 157:25–51

Rosa P, Hille A, Lee RWH, Zanini A, De Camilli P, Huttner WB (1985) Secretogranins I and II: two tyrosine-sufated secretory proteins common to a variety of cells secreting proteins by the regulated pathway. J Cell Biol 101:1999–2011

Rosenfeld MG, Mermod JJ, Amara SG, Swanson LW, Sawchenko PE, Rivier J, Vale WW, Evans RM (1983) Production of a novel neuropeptide encoded by the calcitonin gene via tissue-specific RNA processing. Nature 304:129–135

Rosenthal M, Somjen G (1973) Spreading depression, sustained potential shifts and metabolic activity of cerebral cortex of cats. J Neurophysiol 36:739–749

Rosenzweig MR, Bennett EL (eds) (1976) Neural mechanism of learning and memory. MIT Press, Cambridge

Ross G (1970) Cardiovascular effects of secretin. Am J Physiol 218:1166–1170

Rost FWD, Polak JM (1969) Fluorescence microscopy and microspectrofluorometry of malignant melanomas, naevi and normal melanocytes. Virchows Arch Abt A Pathol Anat 347:321–326

Rost FWD, Polak JM, Pearse AGE (1969) The melanocyte: its cytochemical and immunological relationship to cells of the endocrine polypeptide (APUD) series. Virchows Arch Abt B Zellpath 4: 93–101

Rost FWD, Polak JM, Pearse AGE (1973) The cytochemistry of normal and malignant melanocytes and their relationship to cells of the endocrine polypeptide (APUD) series. Pigment Cell 1:55–65

Rostene WH (1984) Neurobiological and neuroendocrine functions of the vasoactive intestinal peptide (VIP). Prog Neurobiol 22:103–129

Roth A (1973) Electroreceptors in brachiopterygii and dipnoi. Naturwissenschaften 60:106

Roth A, Tscharntke H (1976) Ultrastructure of the ampullary electroreceptors in lungfish and brachiopterygii. Cell Tiss Res 173:95–108

Roth J, Bendayan M, Orci L (1978) Ultrastructural localization of intracellular antigens by the use of protein A-gold complex. J Histochem Cytochem 26:1074–1081

Roth J, LeRoith D, Shiloach J, Rosenzweig JL, Lesniak MA, Havrankova J (1982) The evolutionary origins of hormones, neurotransmitters, and other extracellular chemical messengers. Implications for mammalian biology. N Engl J Med 306:523–527

Roth SI, Capen CC (1974) Ultrastructural and functional correlations of the parathyroid gland. Int Rev exp Pathol 13:161–221

Roth SI, Schiller AL (1976) Comparative anatomy of the parathyroid glands. In: Aurbach GD (ed) Parathyroid gland. (Handbook of physiology 7/VII). pp 281–311

Roth TF, Porter KR (1964) Yolk protein uptake in the oocyte of the mosquito *Aedes aegypti* L. J Cell Biol 20:313–332

Rothman JE, Lodish HF (1977) Synchronised transmembrane insertion and glycosylation of a nascent membrane protein. Nature 269:775–780

Rovasio RA, Delouvee A, Yamada KM, Timpl R and Thiery JP (1983) Neural crest cell migration: requirements for exogenous fibronectin and high cell density. J Cell Biol 96:462–473

Rubel EW (1978) Ontogeny of structure and function in the vertebrate auditory system. In: Jacovson M (ed) Development of sensory systems (Handbook of sensory physiology IX). Springer, Berlin Heidelberg New York, pp 135–237

Ruben RJ (1967) Development of the inner ear of the mouse: a radioautographic study of terminal mitoses. Acta Otolaryngol Suppl 220:1–44

Rubin RP (1970) The role of calcium in the release of neurotransmitter substances and hormones. Pharmacol Rev 22:389–428

Rubin RP (1974) Calcium and the secretory process. Plenum Press, New York London

Rubin RP (1982) Calcium and cellular secretion. Plenum Press, New York London

Rubin W (1972) An unusual intimate relationship between endocrine cells and other types of epithelial cells in the human stomach. J Cell Biol 52:219–227

Rubin W, Schwarz B (1978) An electron microscopic radioautographic identification of the "Enterochromaffin-Like" APUD cells in murine oxyntic glands. Demonstration of a metabolic difference between rat and mouse gastric A-like cells. Gastroenterology 76:437–449

Rubinstein LJ (1981) Cytogenesis and differentiation of pineal neoplasms. Human Pathol 12:441–448

Rundle S, Somogyi P, Fischer-Colbrie R, Hagn C, Winkler H Chubb IW (1986) Chromogranin A, B and C: immunohistochemical localization in ovine pituitary and the relationship with hormone-containing cells. Regul Peptides 16:217–233

Russell IJ, Sellick PM (1976) Measurement of potassium and chloride ion concentrations in the cupulae of the lateral lines of *Xenopus lavis*. J Physiol (Lond) 257:245–255

Russel JT, Brownstein MJ, Gainer H (1980) Biosynthesis of vasopressin, oxytocin and neurophysins: isolation and characterization of two common precursors (propressophysin and pro-oxyphysin). Endocrinology 107: 1880–1891

Sachs H, Fawcett CP, Takabatake Y, Portanova R (1969) Biosynthesis and release of vasopressin and neurophysin. Recent Progr Hormone Res 25: 447–491

Saibil HR (1986) From photon to receptor potential: the biochemistry of vision. News Physiol Sci 1: 122–125

Saito H, Saito S, Ohuchi T, Oka M, Sano T, Hosoi E (1984) Co-storage and co-secretion of somatostatin and catecholamine in bovine adrenal medulla. Neurosci Lett 52: 43–47

Saito T, Kanno T, Tatemoto K, Mutt V (1983) Dose-related effect of neuropeptide Y stimulating respiratory chain in the bullfrog brain. Biomed Res Suppl 4: 173–178

Sakaue M, Saito N, Tanaka C (1987) Immunohistochemical localization of gamma-aminobutyric acid (GABA) in the rat pancreas. Histochem 86: 365–369

Sakmann B, Neher E (1984) Patch clamp techniques for studying ionic channels in excitable membranes. Annu Rev Physiol 46: 455–472

Salpeter MM, Farquhar MG (1981) High resolution analysis of the secretory pathway in mammotrophs of the rat anterior pituitary. J Cell Biol 91: 240–246

Samson WK, Lumpkin MD, McCann SM (1982) Motilin stimulates growth hormone release in vitro. Brain Res Bull 8: 117–121

Sand O, Ozawa S, Hagiwara S (1975) Electrical and mechanical stimulation of hair cells in the mudpuppy. J Comp Physiol A102: 13–26

Sandström IV (1880) Om en ny körtel hos menniskan och ästkilliga däggdjur. Uppsala Läk-Förem 15: 441–471

Sano T, Saito H, Inaba H, Hizawa K, Saito S, Yamanoi A, Mizunuma Y, Matsumura M, Yuasa M, Hiraishi K (1983) Immunoreactive somatostatin and vasoactive intestinal polypeptide in adrenal pheochromocytoma. Cancer 52: 282–289

Sano T, Hirose T, Hizawa K (1984) S-100 protein-immunoreactive cells in the lamina propria and carcinoid tumors of the appendix. Biomed Res 5: 157–164

Sano Y, Knoop A (1959) Elektronenmikroskopische Untersuchungen am kaudalen neurosekretorischen System von *Tinca vulgaris*. Z Zellforsch 49: 464–492

Sano Y, Ueda S, Yamada H, Takeuchi Y, Goto M, Kawata M (1983) Immunohistochemical demonstration of serotonin-containing CSF-contacting neurons in the submammalian paraventricular organ. Histochem 77: 423–430

Sant'Agnese PA di, Mesy Jensen KL de (1984a) Somatostatin and/or somatostatinlike immunoreactive endocrine-paracrine cells in the human prostate gland. Arch Pathol Lab Med 108: 693–696

Sant'Agnese PA di, Mesy Jensen KL de (1984b) Endocrine paracrine cells of the prostate and prostatic urethra. An ultrastructural study. Human Pathol 15: 1034–1041.

Saper CB, Loewy AD, Swanson LW, Cowan WM (1976) Direct hypothalamo-autonomic connections. Brain Res 117: 305–312

Sar M, Stumpf WE, Miller RJ, Chang KJ, Cuatrecasas P (1978) Immunohistochemical localization of enkephalin in rat brain and spinal cord. J Comp Neurol 182: 13–38

Sasagawa T, Kobayashi S, Fujita T (1973) Electron microscope studies on the endocrine cells of the human gut and pancreas. In: Fujita T (ed) Gastro-entero-pancreatic endocrine system. Igaku-Shoin, Tokyo, pp 17–38

Sato A (1952) Innervation of corpus suprarenale in human adult. Tohoku J Exp Med 55: 259–271

Sato S (1966) An electron microscopic study on the innervation of the intracranial artery of the rat. Am J Anat 118: 873–890

Sauer RT, Niall HD, Hogan ML, Keutman HT, O'Riordan JLH, Pott JT (1974) The aminoacid sequence of porcine parathyroid hormone. Biochemistry 13: 1994–1999

Sawada Y, Audhya TK, Walther R (1979) Bovine neurophysin lipid complex. Their isolation, characterization and reaggregation. Int J Pept Protein Res 14: 317–325

Saxod R (1978) Ultrastructure of Merkel corpuscles and so-called "transitional" cells in the white leghorn chicken. Am J Anat 151: 453–474

Schacht J (1976) Biochemistry of neomycin ototoxicity. J Acoust Soc Amer 59: 940–944

Schack L (1932) Über die gelben Zellen in menschlichen Wurmfortsatz. Beitr Pathol Anat 90: 441–478

Schafer R (1971) Antennal sense organs of the cockroach, *Leucophaea maderae*. J Morph 134: 91–104

Schaffalitzky de muckadell OB, Fahrenkrug J (1978) Secretion pattern of secretin in man: regulation by gastic acid. Gut 19: 812–818

Schaller HC (1979) Neuropeptides in *Hydra*. Trends in Neurosci 2: 120–122

Schaller HC, Bodenmüller H (1981) Isolation and amino acid sequence of a morphogenetic peptide from hydra. Proc Natl Acad Sci USA 78: 7000–7004

Schaper A (1895) Über die sogenannten Epithelkörper (Glandulae parathyreoideae) in der seitlichen Nachbarschaft der Schilddrüse und der Umgebung der Arteria carotis der Säuger und des Menschen. Arch Mikr Anat 46: 239–279

Schapiro H, Woodward ER (1965) Inhibition of the secretin mechanism by local anesthetics. Am Surg 31: 139–141

Scharrer B (1969) Neurohumors and neurohormones: definitions and terminology. J Neuro-Viscer. Relations Suppl 9: 1–20

Scharrer B (1978) Peptidergic neurons: facts and trends. Gen Comp Endocrinol 34: 50–62

Scharrer E (1928) Die Lichtempfindlichkeit blinder Elritzen I. Untersuchungen über das Zwischenhirn der Fische. Z Vergl Physiol 7: 1–38

Scharrer E (1933) Ein inkretorisches Organ im Hypothalamus der Erdkröte, *Bufo vulgaris* Laur. Z wiss Zool Abt A 144: 1–11

Scharrer E, Scharrer B (1937) Über Drüsen-nervenzellen und neurosekretorische Organe bei Wirbellosen und

Wirbeltieren. Biol Rev 12:185–216
Scharrer E, Scharrer B (1940) Secretory cells within the hypothalamus. Res Publ Assoc Nerv Ment Dis 20:170–194
Scharrer E, Scharrer B (1945) Neurosecretion. Physiol Rev 25:171–181
Scheller RH, Rothman BS, Mayeri E (1983) A single gene encodes multiple peptide-transmitter candidates involved in a stereotyped behavior. Trends NeuroSci 6:340–345
Schepp W, Ruoff HT, Miederer SE (1983) Cellular origin and release of intrinsic factor from isolated rat gastric mucosal cells. Biochem Biophys Acta 763:426–433
Scheuermann DW (1987) Morphology and cytochemistry of the endocrine system in the lung. In: Bourne GH, Jeon KW, Friedlander M (eds) International review of cytology. A survey of cell biology. Academic Press, Orlando, pp 35–88
Schimke RN (1976) Multiple endocrine adenomatosis syndromes. Adv Int Med 21:249–265
Schlaepfer WW, Lynch RG (1977) Immunofluorescence studies of neurofilaments in the rat and human peripheral and central nervous system. J Cell Biol 74:241–250
Schmechel D, Marangos PJ (1978) Neurone-specific enolase is a molecular marker for peripheral and central neuroendocrine cells. Nature 276:834–836
Schmechel D, Marangos PJ, Brightman M (1978) Neurone-specific enolase is a molecular marker for peripheral and central neuroendocrine cells. Nature 276:834–836
Schmidt JE (1905) Beiträge zur normalen und pathologischen Histologie einiger Zellarten der Schleimhaut des menschlichen Darmkanales. Arch Mikrosk Anat 66:12–40
Schmidt R, Zimmermann H, Whittaker VP (1980) Metal ion content of cholinergic synaptic vesicles isolated from the electric organ of *Torpedo*: effect of stimulation-induced transmitter release. Neuroscience 5:625–638
Schmidt-Nielsen K (1975a) Animal physiology, adaptation and environment. Cambridge University Press, London
Schmidt-Nielsen K (1975b) Physiologische Funktionen bei Tieren. Gustav Fischer, Stuttgart
Schmidt-Wilcke HA, Haake U, Riecken EO (1974) Investigations on the relationship between maximal acid output and parietal cells in gastric mucosal biopsies with special reference to duodenal ulcer. Acta Hepato-Gastroenterology 21:297–302
Schneider FH (1971) Acetaldehyde-induced catecholamine secretion from the cow adrenal medulla. J Pharmacol Exp Ther 177:109–118
Schneider FH, Smith AD, Winkler H (1967) Secretion from the adrenal medulla: biochemical evidence for exocytosis. Br J Pharmacol 31:94–104
Schneider KC (1890) Histologie von *Hydra fusca* mit besonderer Berücksichtigung des Nervensystems der Hydropolypen Arch Mikr Anat 35:321–379
Schofield GC (1968) Anatomy of muscular and neural tissues in the alimentary canal. In: Code CF (ed) Alimentary canal, Motility (Handbook of physiology 6/IV) American Physiological Society, Washington DC, pp 1579–1627
Schöne H, Steinbrecht RA (1968) Fine structure of statocyst receptor of astacus fluviatilis. Nature 220: 184–186
Schot LPC, Boer HH (1982) Immunocytochemical demonstration of peptidergic cells in the pond snail *Lymnaea stagnalis* with an antiserum to the molluscan cardioactive tetrapeptide FMRF-amide. Cell Tiss Res 225:347–354
Schot LPC, Boer HH, Swaab DF, Van Noorden S (1981) Immunocytochemical demonstration of peptidergic neurons in the central nervous system of the pond snail *Lymnaea stagnalis* with antisera raised to biologically active peptides of vertebrates. Cell Tiss Res 216:273–291
Schuetze SM (1983) The discovery of the action potential. Trends Neuro Sci 6:164–168
Schultz EW (1960) Repair of olfactory mucosa, with special reference to regeneration of olfactory cells (sensory neurons). Am J Pathol 37:1–19
Schultzberg M, Hökfelt T, Lundberg JM, Terenius L, Elfvin LG, Elde R (1978a) Enkephalin-like immunoreactivity in nerve terminals in sympathetic ganglia and adrenal medulla and in adrenal medullary gland cells. Acta Physiol Scand 103:475–477
Schultzberg M, Lundberg JM, Hökfelt T, Terenius L, Brandt J, Elde RP, Goldstein M (1978b) Enkephalin-like immunoreactivity in gland cells and nerve terminals of the adrenal medulla. Neuroscience 3:1169–1186
Schultzberg M, Hökfelt T, Lundberg JM (1979) Peripheral peptide-containing neurons. In: Miyoshi A (ed) Gut peptides, secretion, function and clinical aspects. Kodansha, Tokyo Elsevier, Amsterdam New York Oxford, pp 281–292
Schultzberg M, Hökfelt T, Lundberg JM, Dalsgaad CJ, Elfvin L-G (1983) Transmitter histochemistry of autonomic ganglia. In: Elfvin L-G (ed) Autonomic ganglia, John Wiley & Sons, Chichester New York Brisvane Toronto Singapore, pp 205–233
Schultzberg M, Foster GA, Gage FH, Bjorklund A, Hokfelt T (1986) Coexistence during ontogeny and transplantation. Proc Brain Res 68:129–145
Schumacher S (1938) Über die Bedeutung der arteriovenösen Anastomosen und der epitheloiden Muskelzellen (Quellzellen). Z Mikrosk Anat Forsch 43:107–130
Schwalbe G (1872) Beiträge zue Kenntnis der Drüsen in dem Darmwandungen, in's Besondere der Brunner'schen Drüsen. Arch Mikrosk Anat 8:92–140.
Schwartz JH, Elste A, Shapiro E, Gotoh H (1986) Biochemical and morphological correlates of transmitter type in C2, an identified histaminergic neuron in *Aplysia*. J Comp Neurol 245:401–421
Scott AM, Atwater I, Rojas E (1981) A method for the simultaneous measurement of insulin release and β-cell membrane potential in single mouse islets. Diabetologia 21:470–475
Scott DE, Kurobisch-Dudley G, Paull WK, Kozlowski GP (1977) The ventricular system in neuroendocrine mechanisms III. Supraependymal neuronal networks in the primate brain. Cell Tiss Res 179: 235–254

Scott SA, Cooper E, Diamond J (1981) Merkel cells as targets of the mechanosensory nerves in salamander skin. Proc R Soc Lond (Biol) 211:455–470

Segal MM (1983) Neurotransmission: Specification of synaptic action. Trends NeuroSci 6:118–121

Segi M (1935) Über eine aus Chromaffine-Darmzellen bestehende Struktur auf den Zotten von Menschenembryo (In Japanese with German summary). Acta Anat Nippon 8:276–280

Segi M (1936) Über die Entwicklung der verschieden Granulazellen im Darmepithel des Menschenembryo (In Japanese with German summary). Acta Anat Nippon 9:850–937

Seiji M, Fitzpatrick TB, Birbeck MS (1961) The melanosome: a distinctive subcellular particle of mammalian melanocytes and the site of melanogenesis. J Invest Dermatol 36:243–252

Seizinger BR, Liebisch DC, Gramsch C, Herz A, Weber E, Evans CJ, Esch FS, Böhlen P (1985) Isolation and structure of a novel C-terminally amidated opioid peptide, amidorphin, from bovine adrenal medulla. Nature 313:57–59

Sejnowski TJ (1982) Peptidergic synaptic transmission in sympathetic ganglia. Fed Proc 41:2923–2928

Sekine Y, Yui R (1981) Immunohistochemical study of the pancreatic endocrine cells of the ray, *Dasyatis akajei*. Arch Histol Jpn 44:95–101

Semm P, Vollrath L (1979) Electrophysiology of the guinea-pig pineal organ: Sympathetically influenced cells responding differently to light and darkness. Neurosci Lett 12:93–96

Serizawa Y, Kobayashi S, Fujita T (1979) Neuro-insular complex type I in the mouse. Re-evaluation of the pancreatic islet as a modified ganglion. Arch Histol Jpn 42:389–394

Setoguti T, Yonemoto Y (1965) An observation of the mast cells, especially the dendrite-type tissue mast cell in Triturus *pyrrhogaster* (Boie). Okajimas Folia Anat Jpn 40:540–550

Setoguti T, Isono H, Sakurai S (1970) Electron microscopic study on the parathyroid gland of the newt *Triturus pyrrhogaster* (Boié) in natural hibernation. J Ultrastruct Res 31:46–60

Setoguti T, Inoue Y, Kato K (1981) Electron-microscopic studies on the relationship between the frequency of parathyroid storage granules and serum calcium levels in the rat. Cell Tiss Res 219:459–467

Setoguti T, Inoue Y, Shin M, Matsumura H (1984) Effects of epinephrine treatment on the rat parathyroid gland, with special reference to the frequency of storage granules. Acta Anat 118:54–59

Setoguti T, Inoue Y, Shin M (1985) Effects of short-term treatment with calcium on the parathyroid gland of the rat, under particular consideration of the alteration of storage granules. Cell Tiss Res 240:9–17

Shannon WA, Roth SI (1974) An ultrastructural study of acid phosphatase activity in normal, adenomatous and hyperplastic (chief cell type) human parathyroid glands. Am J Path 77:493–501

Shaw G, Weber K (1981) The distribution of the neurofilament triplet proteins within individual neurones. Exp Cell Res 136:119–125

Shaw HM, Heath T (1973) The phases of pancreatic secretion in rats. Q J Exp Physiol 58:229–237

Shaw JE, Urquhart J (1972) Parameters of the control of acid secretion in the isolated blood-perfused stomach. J Physiol 226:107–108

Shelanski ML, Liem RKH (1979) Neurofilaments. J Neurochem 33:5–13

Shibasaki S, Ito T (1969) Electron microscopic study on the human pancreatic islets. Arch Histol Jpn 31:119–154

Shibuya I, Kanno T (1985) Calcium-dependent secretory and redox response to CCK-8 in isolated perfused rat pancreas. Am J Physiol 248:C228–C234

Shimada T (1981) Lymph and blood capillaries as studied by a new SEM technique. Biomed Res Suppl 2:243–248

Shimizu K, Kinoshita H, Yamamoto T, Kimura H, Ochi J (1982) Silver impregnation and fluorescence histochemistry of the subependymal cells in the preoptic recess of the bullfrog (*Rana catesbeiana*). Acta Histochem Cytochem 15:520–526

Shimizu K, Kimura H, Yamamoto T, Ochi J (1983) Immunohistochemical demonstration of the serotonin-containing subependymal cells in the frog hypothalamus. Histochemistry 79:23–29

Shimosegawa T, Koizumi M, Toyota T, Goto Y, Yanaihara C, Yanaihara N (1986) An immunohistochemical study of methionine-enkephalin-Arg^6-Gly^7-Leu^8-like immunoreactivity-containing liquor-contacting neurons (LCNs) in the rat spinal cord. Brain Res 379:1–9

Shiotani Y (1980) An electron microscopic study on stellate cells in the rabbit adenohypophysis under various endocrine conditions. Cell Tiss Res 213:237–246

Shirasawa N, Kihara H, Yamaguchi S, Yoshimura F (1983) Pituitary folliculo-stellate cells immunostained with S-100 protein antiserum in postnatal, castrated and thyroidectomized rats. Cell Tiss Res 231:235–249

Shirasawa N, Yamaguchi S, Yoshimura F (1984) Granulated folliculo-stellate cells and growth hormone cells immunostained with anti-S-100 protein serum in the pituitary glands of the goat. Cell Tiss Res 237:7–14

Shkolnick LJ, Schwartz JH (1980) Genesis and maturation of serotonergic vesicles in identified giant cerebral neuron of *aplysia*. J Neurophysiol 43:945–967

Shusdziarra V, Harris V, Conlon JM, Arimura A, Unger R (1978) Pancreatic and gastric somatostatin release in response to intragastric and intraduodenal nutrients and HCl in the dog. J Clin Invest 62:509–518

Sidhu GS (1979) The endodermal origin of digestive and respiratory tract APUD cells. Am J Pathol 96:5–17

Siegelbaum SA, Tsien RW (1983) Modulation of gated ion channels as a mode of transmitter action. Trends Neurosci 6:307–313

Siegrist G, Dolivo M, Dunant Y, Foroglou-Kerameus C, Ribaupierre F de, Rouiller C (1968) Ultrastructure and function of the chromaffin cells in the superior cervical ganglion of the rat. J Ultrastruct

Res 25:381–407

Siggins GR, Gruol DL (1986) Mechanism of transmitter action in the vertebrate central nervous system. In: Bloom SR (ed) Intrinsic regulatory system of the brain. (Handbook of Physiology 1/IV) Am Physiol Society, Bethesda MD, pp 1–114

Sikri KL, Varndell IM, Hamid QA, Wilson BS, Kameya T, Ponder BAJ, Lloyd RV, Bloom SR, Polak JM (1985) Medullary carcinoma of the thyroid. An immunocytochemical and histochemical study of 25 cases using eight separate markers. Cancer 56:2481–2491

Simard LC (1937) Les complexes neuro-insulaires du pancreas humain. Arch Anat Micr 33:49–64

Simon B, Kather H (1977) Histamine-sensitive adenylate cyclase of human gastric mucosa. Gastroenterology 73:429–431

Singleton JW (1969) Humoral effects of the pancreas upon the gastrointestinal tract. Gastroenterology 56:342–362

Sippl JH (1961) The association of pheochromocytoma with carcinoma of the thyroid gland. Am J Med 31:163–166

Sjodin L, Andersson S (1972) Inhibition of gastric secretion by acidification of the duodenal bulb before and after transection of the pylorus. Scand J Gastroenterol 7:151–156

Sjodin L, Miura S (1974) Secretin-induced inhibition of acid secretion before and after vagal denervation of canine gastric pouches. Scand J Gastroenterol 9:185–190

Sjoerdsma A (1959) Serotonin. New Engl J Med 261:181–188

Sjöstrand FS (1953a) The ultrastructure of the outer segments of rods and cones of the eye as revealed by the electron microscope. J Cell Comp Physiol 42: 15–44

Sjöstrand FS (1953b) The ultrastructure of the inner segments of the retinal rods of the guinea pig retina as revealed by electron microscopy. J Cell Comp Physiol 42:45–70

Sjöstrand FS (1958) Ultrastructure or retinal rod synapses of the guinea pig eye as revealed by three-dimensional reconstruction serial sections. J Ultrastruct Res 2:122–170

Sjöstrand FS (1961) Electron microscopy of the retina. In: Smelser GK (ed) The Structure of the Eye. Academic Press, New York, pp 1–28

Skirboll L, Hökfelt T, Rehfeld J, Cuello AC, Dockray G (1982) Coexistence of substance P- and cholecystokinin-like immunoreactivity in neurons of the mesencephalic periaqueductal central gray. Neurosci Lett 28:35–39

Skotland T, Ljones T (1979) Dopamine β-monooxygenase: Structure, mechanism and properties of the enzyme-bound copper. Inorgan Perspect Biol Med 2:151–180

Skotland T, Ljones T, Flatmark T, Sletten K (1977) NH_2-terminal sequence of dopamine β-hydroxylase from bovine adrenal medulla. Biochem Biophys Res Commun 74:1483–1489

Slayback JB, Swena EM, Thomas JE, Smith LL (1967) The pancreatic secretory respon se to topical anesthetic block of the small bowl. Surgery 61:591–595

Sloboda RD, Dentler WL, Bloodgood RA, Teizer BR, Granett S, Rosenbaum JL (1976) Microtubule-associated proteins (MAPS) and the assembly of microtubules *in vitro*. In: Goldman R, Pollard T, Rosenbaum J (eds) Cell Motility. Cold Spring Harbor Lab, New York, pp 1171–1212

Smith AD, De Potter WP, Moerman EJ, De Schaepdryver AF (1970) Release of dopamine β-hydroxylase and chromogranin A upon stimulation of splenic nerve. Tissue Cell 2:547–568

Smith CG (1951) Regeneration of sensory olfactory epithelium and nerves in adult frogs. Anat Rec 109:661–671

Smith DE (1958) Nature of the secretory activity of the mast cell. Am J Physiol 193:0573–0575

Smith DE (1963) The tissue mast cell. Int Rev Cytol 14:327–386

Smith KR (1967) The structure and function of *Haarscheibe*. J Comp Neurol 131:459–474

Smith KR (1970) The ultrastructure of the human Haarscheibe and Merkel cell. J Invest Dermatol 54:150–159

Smith KR (1977) The Haarscheibe. J Invest Dermatol 69:68–74

Smith PH (1975) Structural modification of Schwann cells in the pancreatic islets of the dog. Am J Anat 144:513–517

Smith PH, Madson KL (1981) Interactions between autonomic nerves and endocrine cells of the gastroenteropancreatic system. Diabetologia 20:314–324

Smith PH, Merchant FW, Johnson DG, Fujimoto WY, Williams RH (1977) Immunocytochemical localization of a gastric inhibitory polypeptide-like material within A-cells of the endocrine pancreas (1). Am J Anat 149:585–590

Smith RE, Farquhar MG (1966) Lysosome function in the regulation of the secretory process in cells of the anterior pituitary gland. J Cell Biol 31:319–347

Smoller CG (1965) Neurosecretory processes extending into third ventricle: secretory or sensory? Science 147:882–884

Snape WJ, Matarazzo SA, Cohen S (1978) Effect of eating and gastrointestinal hormones on human colonic myoelectrical and motor activity. Gastroenterology 75:373–378

Sneddon P, Burnstock (1984) Inhibition of excitatory junction potentials in guinea-pig vas deferens by α, β-methylene-ATP: further evidence for ATP and noradrealine as cotransmitters. Eur J Pharmacol 100:85–90

Sneddon P, Westfall DP (1984) Pharmacological evidence that adenosine triphosphate and noradrenaline are co-transmitters in the guinea-pig vas deferens. J Physiol 347:561–580

Sofroniew MV (1980) Projections from vasopressin, oxytocin and neurophysin neurons to neural targets in the rat and human. J Histochem Cytochem 28: 475–478

Sofroniew MV, Weindl A (1978a) Extrahypothalamic neurophysin-containing perikarya, fiber pathways and fiber clusters in the rat brain. Endocrinology 102:334–337

Sofroniew MV, Weindl A (1978b) Projections from

the parvocellular vasopressin- and neurophysin-containing neurons of the suprachiasmatic nucleus. Am J Anat 153:391–429

Sofroniew MV, Weindl A, Schinko I, Wetzstein R (1979) The distribution of vasopressin-, oxytocin-, and neurophysin-producing neurons in the guinea pig brain. I. The classical hypothalamo-neuro-hypophyseal system. Cell Tiss Res 196:367–384

Soga J (1973) Carcinoids: Their changing concepts and a new histologic classification. In: Fujita T (ed) Gastro-entero-pancreatic endocrine system. A cell-biological approach. Igaku-Shoin, Tokyo, pp 101–119

Soga J (1976) Neoplasms of GEP endocrine cells: The present-day concept of carcinoids. In: Fujita T (ed) Endocrine gut and pancreas. Elsevier Scientific Publishing Company, Amsterdam, pp 387–394

Soga J, Tazawa K (1971) Pathological analysis of carcinoids. Histologic reevaluation of 62 cases. Cancer 28:990–998

Solcia E, Sampietro R (1965a) On the nature of the metachromatic cells of pancreatic islets. Z Zellforsch 65:131–138

Solcia E, Sampietro R (1965b) Cytologic observations on the pancreatic islets with reference to some endocrine-like cells of the gastrointestinal mucosa. Z Zellforsch 68:689–698

Solcia E, Sampietro R (1968) New method for staining secretory granules and 5-hydroxytryptamine in the thyroid C cells. In: Taylor S (ed) Calcitonin: proceedings of the symposium on thyrocalcitonin and the C cells. William Heinemann Medical Books, London, pp 127–132

Solcia E, Vassallo G, Sampietro R (1967) Endocrine cells in the antro-pyloric mucosa of the stomach. Z Zellforsch 81:474–486

Solcia E, Vassallo G, Capella C (1968) Selective staining of endocrine cells by basic dyes after acid hydrolysis. Stain Technol 43:257–263

Solcia E, Capella C, Vassallo G (1969) Lead-haematoxylin as a stain for endocrine cells. Significance of staining and comparison with other selective methods. Histochemie 20:116–126

Solcia E, Polak JM, Pearse AGE, Forssmann WG, Larsson LI, Sundler F, Lechago J, Grimelius L, Fujita T, Creutzfeldt W, Gepts W, Falkmer S, Lefranc G, Heitz Ph, Hage E, Buchan AMJ, Bloom SR, Grossman MI (1978) Lausanne 1977 classification of gastroentero-pancreatic endocrine cells. In: Bloom SR (ed) Gut hormones. Chrchill/Livingstone, Edinburgh London New York, pp 40–48

Solcia E, Creutzfeldt W, Falkmer S, Fujita T, Greider MH, Grossman MI, Grube D, Håkanson R, Larsson LI, Lechago J, Lewin K, Polak JM, Rubin W (1981) Human gastroenteropancreatic endocrine-paracrine cells. Santa Monica 1980 classification. In: Grossman MI, Brazier MAB, Lechago J (eds) Cellular basis of chemical messengers in the digestive system. Academic Press, New York, pp 159–165

Solcia E, Capella C, Buffa R, Tenti P, Guidorindi, Cornaggia M (1986) Antigenic markers of neuroendocrine tumors. Their diagnostic and prognostic value. In: Fenoglio-Preiser CM, Weinstein RS (eds) New concepts in neoplasia as applied to diagnostic pathology. International academy of pathology monograph. William & Wilkins, Baltimore, pp 142–261

Solcia E, Capella C, Buffa R, Usellini L, Frigerio B, Fontana P (1979) Endocrine cells of the gastrointestinal tract and related tumors. In: Ioachim HL (ed) Pathobiology annual 9. Raven Press, New York, pp 163–203

Soll AH, Walsh JH (1979) Regulation of gastric acid secretion. Annu Rev Physiol 41:35–53

Solomon TE, Petersen H, Elashoff J, Grossman MI (1979) Effects of chemical messenger peptides on pancreatic growth in rats. In: Gut peptides: secretion, function and clinical aspects. Proceedings of the Hiroshima Symposium on Gut Hormones. Kodansha, Tokyo/Elsevier North-Holland Biomedical Press, Amsterdam New York Oxford, pp 213–219

Somers G, Blondel E, Orci L, Malaisse WJ (1979) Motile events in pancreatic endocrine cells. Endocrinology 104:255–264

Sonnenberg H, Veress AT (1984) Cellular mechanism of release of atrial natriuretic factor. Biochem Biophys Res Commun 124:443–449

Sotelo C, Palay SL (1968) The fine structure of the lateral vestibular nucleus in the rat I. Neurons and neuroglial cells. J Cell Biol 36:151–179

Spataro AC, Bosmann HB (1976) Mechanism of action of disodium cromogycate-mast cell calcium ion influx after a histamine releasing stimulus. Biochem Pharmacol 25:505–510

Stach W (1986) Demonstration of eight structurally defined neurone types in the pig enteral system. In: Heym CH (ed) Experimental Brain Res Suppl. Springer, Berlin

Stadler H, Whittaker VP (1978) Identification of vesiculin as a glycosaminoglycan. Brain Res 153:408–413

Stanka P, Kinzel V, Mohr U (1969) Elektronenmikroskopisch Untersuchung über die Prämelanosomentstehung an Melanomazellen *in vitro*. Virchows Arch B Zellpathol 2:91–102

Stanulla H, Wöckel W (1969) Nichtchromaffin Paragangliome (Chemodektome). Gustav Fischer, Jena

Starck D (1982) Vergleichende Anatomie der Wirbeltiere; auf evolutionsbiologischer Grundlage, Band 3: Organe des aktiven Bewegungsapparates, der Koordination, der Umweltbeziehung, des Stoffwechsels und der Fortpflanzung. Springer, Berlin

Starzl TE, Porter KA, Putnam CW (1976) Insulin, glucagon, and the control of hepatic structure, function, and capacity for regeneration. Metabolism Suppl 25:1429–1434

Stasiewicz J, Adler M, Delcout A (1980) Pancreatic and gastrointestinal hormones in chronic pancreatitis. Hepato-Gastroenterology 27:152–160

Stefansson K, Wollmann RL, Moore BW, Arnason BGW (1982) S-100 protein in human chondrocytes. Nature 295:63–64

Steinbusch HWM, Verhofstad AAJ, Joosten HWJ (1978) Localization of serotonin in the central nervous system by immunohistochemistry: description of a specific and sensitive technique and some applications. Neuroscience 3:811–819

Steiner AL, Goodman AD, Powers SR (1968) Study of a kindred with pheochromocytoma, medullary thyroid carcinoma, hyperparathyroidism and Cushing's disease: multiple endocrine neoplasia, type 2. Medicine 47:371–409

Steiner DF (1977) Insulin today. Diabetes 26:322–340

Steiner DF, Oyer PE (1967) The biosynthesis of insulin by a human islet cell adenoma. Proc Natl Acad Sci USA 57:473–480

Stening GF, Grossman MI (1969) Hormonal control of Brunner's glands. Gastroenterology 56:1047–1052

Sterba G, Naumann W, Hoheisel G (1980) Exohypothalamic axons of the classic neurosecretory system and their synapses. In: McConnell, Ronijn B, Poll, Corner (eds) Adaptive capabilities of the nervous system (Progress in Brain Research) Elsevier North-Holland Biomedical Press, pp 141–158

Stern AS, Lewis RV, Kimura S, Rossier J, Gerber LD, Brink L, Stein S, Udenfriend S (1979) Isolation of the opioid heptapeptide Met-enkephalin [Arg^6-Phe^7] from bovine adrenal medullary granules and striatum. Proc Natl Acad Sci USA 76:6680–6683

Stieda L (1881) Untersuchungen über die Entwickelung der Glandula Thymus, Glandula Thyreoidea und Glandula carotica. Leipzig, Englemann, p 38 (cited from Adams, 1958)

Stilling H (1892) Du ganglion intercarotidien. Inaug Diss Trav Fac Univ, Lausanne, pp 321–331

Stjärne L, Lishajko F (1966) Comparison of spontaneous loss of catecholamines and ATP *in vitro* from isolated bovine adrenomedullary, vesicular gland, vas deferens and splenic nerve granules. J Neurochem 13:1213–1216

Stoeckel ME, Porte A (1970) Origine embryonnaire et différentitation sécrétoire des cellules à calcitionine (Cellules C) dans la thyroïde foetale du rat. Z Zellforsch 106:251–268

Stoeckel ME, Porte A (1984) Fine structure and development of the pars tuberalis in mammals. In: Motta PM (ed) Ultrastructure of endocrine cells and tissues. Martinus Nijhoff, Boston Hague Dordrecht Lancaster, pp 29–38

Stöhr P (1935) Zur Innervation der menschlichen Nebenniere. Z Anat Entw-Gesch 104:475–490

Stone JV, Mordue W, Batley KE, Morris HR (1976) Structure of locust adipokinetic hormone, a neurohormone that regulates lipid utilisation during flight. Nature 263:207–211

Stone TW, Taylor DA (1977) Microiontophoretic studies of effects of cyclic nucleotides on excitability of neurons in rat cerebral-cortex. J Physiol 266: 523–543

Storch VN, Welsch UN (1970) Electron microscopic observations on the taste-buds of some bony fishes. Arch Histol Jpn 32:145–153

Straus E (1978) Radioimmunoassay of gastrointestinal hormones. Gastroenterology 74:141–152

Strocchi P, Marotta CA, Bonventre J, Gilbert JM (1981) The subunit composition of cerebellar tubulin: evidence for multiple beta tubulin messenger RNAs. Brain Res 211:206–210

Sugawara K, Negishi K (1973) Effects of some amino acids on light-induced responses in the isolated carp retina. Vision Res 13:2479–2489

Sugawara M (1987) Ultrastructural evidence for exocytotic release of atrial natriuretic polypeptide in the rat atrial muscle with the use of the tannic acid method. Biomed Res 8:9–17

Sundler F, Håkanson R, Alumets J, Walles B (1977) Neuronal localization of pancreatic polypeptide (PP) and vasoactive intestinal peptide (VIP) immunoreactivity in the earthworm (*Lumbricus terrestris*). Brain Res Bull 2:61–65

Sundler F, Håkanson R, Ekblad E, Uddman R, Wahlstedt C (1986) Neuropeptide Y in the peripheral adrenergic and enteric nervous systems. Int Rev Cytol 102:243–269

Suzuki F, Nakajima T, Kato K (1982) Peripheral distribution of nervous system-specific S-100 protein in rat. J Biochem 92:835–838

Suzuki N (1977) Intracellular responses of lamprey olfactory receptors to current and chemical stimulation. In: Katsuki Y, Sato M, Takagi SF, Oomura Y (eds) Food intake and chemical senses. Japan Scientific Societies Press, Tokyo, pp 13–22

Suzuki T, Ito S, Yamada Y, Matsuzuka F, Matsubayashi S, Miyauchi A (1985) Ultrastructural demonstration of calcitonin in osmium-fixed human medullary carcinoma of thyroid by the protein A-colloidal gold technique. Virchows Arch Pathol Anat 407:407–417

Suzuki Y, Takeda M (1984) Ultrastructure of taste buds in birds. Jap J Oral Biol 26:669–678

Swadlow HA (1982) Impulse conduction in the mammalian brain: physiological properties of individual axons monitored for several months. Science 218: 911–913

Swann AC (1983) Stimulation of brain Na^+, K^+-ATPase by norepinephrine in vivo: prevention by receptor antagonists and enhancement by repeated stimulation. Brain Res 260:338–341

Swanson LW (1977) Immunohistochemical evidence for a neurophysin-containing autonomic pathway arising in the paraventricular nucleus of the hypothalamus. Brain Res 128:346–353

Sydbom A, Fredholm B, Uvnäs B (1981) Evidence against a role of cyclic nucleotides in the regulation of anaphylactic histamine release in isolated rat mast cells. Acta Physiol Scand 112:47–56

Szabo T (1965) Sense organs of the lateral line system in some electric fish of the Gymnotidae, Mormyridae and Gymnarchidae. J Morph 117:229–249

Szabo T (1972) Ultrastructural evidence for a mechanoreceptor function of the ampullae of Lorenzini. J Microscopie 14:343–350

Szabo T (1974) Anatomy of the specialized lateral line organs of electroreception. In: Fessard A (ed) Electroreceptors and other specialized receptors in lower vertebrates (Handbook of sensory physiology III/3). Springer, Berlin Heidelberg New York, pp 13–58

Szabo T, Fessard A (1974) Physiology of electroreceptors. In: Fessard A (ed) Electroreceptors and other specialized receptors in lower vertebrates (Handbook of sensory physiology III/3). Springer, Berlin

Heidelberg New York, pp 59–124

Sziji I, Csapó Z, László FA, Kovács K (1969) Medullary cancer of the thyroid gland associated with hypercorticism. Cancer 24:167–173

Tabata M, Tamura T, Niwa H (1975) Origin of the slow potential in the pineal organ of the rainbow trout. Vision Res 15:737–740

Tachibana T (1978) The Merkel cell in the labial ridge epidermis of anuran tadpole. 1. Fine structure, distribution and cytochemical studies. Anat Rec 191: 487–502

Tachibana T (1979) The Merkel cell in the labial ridge epidermis of the anuran tadpole II. Electron microscope observation on the appearance and differentiation of the Merkel cell. Arch Histol Jpn 42:129–140

Tachibana T, Ishizeki K (1981) Merkel cell development in the wound healing in the labial mucosa of adult rabbits. Arch Histol Jpn 44:151–165

Tachibana T, Nawa T (1980) Merkel cell differentiation in the labial mucous epithelium of the rabbit. J Anat 131:145–155

Tachibana T, Sakakura Y, Ishizeki K, Iida S, Nawa T (1982) Migration of Merkel cells in the labial mucous epithelium of adult rabbits following mental nerve resection. Cell Tiss Res 223:659–664

Tachibana T, Sakakura Y, Ishizeki K, Iida S, Nawa T (1983) An experimental study of the influence of sensory nerve fibers on Merkel cell differentiation in the labial mucosa of the rabbits. Arch Histol Jpn 46:469–477

Tachibana T, Ishizeki K, Sakakura Y, Nawa T (1984) Ultrastructural evidence for a possible secretory function of Merkel cells in the barbels of a teleost fish, *Cyprinus carpio*. Cell Tiss Res 235:695–697

Tager HS, Markese J (1976) Glucagon-like and insulin-like hormones of the insect neurosecretory system. Biochem J 156:515–520

Tager HS, Steiner DF (1974) Peptide Hormones. Annu Rev Biochem 43:509–538

Taha AAM, King AS (1986) Aortico-pulmonary bodies in the domestic fowl: ultrastructure, innervation and secretion. J Anat 149:41–53

Taira K (1985) Endocrine-like cells in the laryngeal mucosa of adult rabbit demonstrated by electron microscopy and by the Grimelius silver-impregnation method. Biomed Res 6:377–385

Taira K, Shibasaki S (1978) A fine structure study of the nonciliated cells in the mouse tracheal epithelium with special reference to the relation of "brush cells" and "endocrine cells". Archi Histol Jpn 41: 351–366

Takagi I, Yamada K, Karasawa N, Nagatsu I (1984) Immunohistochemical localization of tyrosine hydroxylase and calcitonin in the ultimobranchial body of the grass parakeet. Acta Histochem Cytochem 17:359–370

Takagi SF (1971) Degeneration and regeneration of the olfactory epithelium. In: Beidler LM (ed) Olfaction. (Handbook of sensory physiology IV/1). Springer, Berlin Heidelberg New York, pp 75–94

Takagi SF, Yajima T (1964) Electical responses to odours of degenerating olfactory epithelium. Nature 202:1220

Takagi SF, Yajima T (1965) Electrical activity and histological change in the degenerating olfactory epithelium. J Gen Physiol 48:559–569

Takahashi I, Honda R, Dodds WJ, Sarna S, Toouli J, Itoh Z, Chey WY, Hogan WJ, Greiff D, Baker K (1983) Effect of motilin on the opossum upper gastrointestinal tract and sphincter of Oddi. Am J Physiol 245:G476–481

Takahashi K, Yamaguchi H, Ishizeki J, Nakajima T, Nakazato Y (1981) Immunohistochemical and immunoelectron microscopic localization of S-100 protein in the interdigitating reticulum cells of the human lymph node. Virchows Archiv (Cell Pathol) 37:125–135

Takahashi S, Iwanaga T, Takahashi Y, Nakano Y, Fujita T (1984) Neuron-specific enolase, neurofilament protein and S-100 protein in the olfactory mucosa of human fetuses: an immunohistochemical study. Cell Tiss Res 238:231–234

Takahashi S, Yui R (1983) Gastrin-releasing peptide, (GRP) and serotonin in the human fetal lung: an immunohistochemical study. Biomed Res 4:315–320

Takahata M, Hisada M (1982) Statocyst interneurons in the crayfish *Procambarus clarkii* Girard. J Comp Physiol 149:287–300, 301–306

Takai Y, Kikkawa U, Kaibuchi K, Nishizuka Y (1984) Membrane phospholipid metabolism and signal transduction for protein phosphorylation. Adv Cyclic Nucl Prot Phosph 18:119–158

Takami H, Yamaguchi K, Abe K, Adachi I, Arai S, Kameya T, Yamachika K, Tashiro M, Nakagawa Y, Abe O (1978) A case of Zollinger-Ellison syndrome whose pancreatic tumor produced multiple hormones and review of 48 cases reported in the Japanese literature. Jap J Clin Oncol 8:75–89

Takamine J (1901) Adrenalin, the active principle of the suprarenal glands and its mode of preparation. Am J Pharmacy 73:523–531

Takaya K (1969) The relationship between mast cells and histamine in phylogeny with special reference to reptiles and birds. Arch Histol Jpn 30:401–420

Takaya K (1970) Two types of chromaffin cells in the adrenals of the bullfrog, *Rana catesbiana*. A comparative study. Arch Histol Jpn 32:69–80

Takaya K, Fujita T (1967) Mast cells free of histamine in *Rana catesbiana*. Nature 215:776–777

Takeda M (1972) Fine structure of developing taste buds in human fetal circumvallate papilae (Japanese text with English abstract). Acta Anat Nippon 47: 325–337

Takeda M (1976) An electron microscopic study on the innervation in the taste buds of the mouse circumvallate papillae. Arch Histol Jpn 39:257–269

Takeda M (1977) Uptake of 5-hydroxytryptophan by gustatory cells in the mouse taste bud. Arch Histol Jpn 40:243–250

Takeda M, Hoshino T (1975) Fine structure of taste buds in the rat. Arch Histol Jpn 37:395–413

Takeda M, Kitao K (1980) Effect of monoamines on the taste buds in the mouse. Cell Tiss Res 210:71–78

Takeda M, Suzuki Y (1983) An autoradiographic study of the taste bud following injections of labeled

biogenic amine precursors. J Electron Microsc 32: 357–360
Takeda M, Shishido Y, Kitao K, Suzuki M (1982) Monoamines of taste buds in the fungiform and foliate papillae of the mouse. Arch Histol Jpn 45:239–246
Takeichi M (1967) The fine structure of ependymal cells. Part II. An electron microscopic study of the soft-shelled turtle paraventricular organ with special reference to the fine structure of ependymal cells and so-called albuminous substance. Z Zellforsch 76:471–485
Takeuchi Y, Kimura H, Sano Y (1982) Immunohistochemical demonstration of the distribution of serotonin neurons in the brainstem of the rat and cat. Cell Tiss Res 224:247–267
Takor T, Pearse AGE (1975) Neuroectodermal origin of avian hypothalamo-hypophyseal complex: the role of the ventral neural ridge. J Embryol Exp Morphol 34:311–325
Tamir H, Huang YL (1974) Binding of serotonin to soluble protein from synaptosomes. Life Sci 14: 83–93
Tamir H, Kuhar MJ (1975) Association of serotonin-binding protein with projections of the midbrain raphe nuclei. Brain Res 83:169–172
Tanaka K, Sakai N, Terayama Y (1979) Organ of Corti in the human fetus: scanning and transmission electronmicroscope studies. Ann Otol Rhinol Laryngol 88:749–758
Tani S, Mutoh N (1979) Effects of adrenergic agonists on gastric secretion in the rat. Biochem Pharmacol 28:678–683
Taniguchi H, Okada Y, Seguchi H, Shimada C, Seki M, Tsutou A, Baba S (1979) High concentration of gamma-aminobutyric acid in pancreatic beta cells. Diabetes 28:629–633
Tapper DN (1965) Stimulus-response relationships in the cutaneus slowly-adapting mechanoreceptor in hairy skin of the cat. Exp Neurol 13:364–385
Taraskevich PS, Douglas WW (1977) Action potentials occur in cells of anterior pituitary gland and are stimulated by the hypopysiotropic peptide thyrotropin-releasing hormone. Proc Natl Acad Sci USA 74:4064–4067
Taraskevich PS, Douglas WW (1978) Catecholamines of supposed inhibitory hypophysiotrophic function suppress action potentials in prolaction cells. Nature 276:832–834
Taraskevich PS, Douglas WW (1980) Electrical behaviour in a line of anterior pituitary cell (GH cells) and the influence of the hypothalamic peptide, thyrotropin-releasing factor. Neuroscience 5:421–431
Taraskevich PS, Douglas WW (1984) Electrical activity in adenohypophyseal cells and effects of hypophyseotropic substances. Fed Proc 43:2373–2378
Tarvin JT, Pace CS (1981) Glucose-induced electrical activity in the pancreatic B-cell: effect of veratridine. Am J Physiol 240:C127–C134
Tashjian AH Jr., Yasumura Y, Levine L, Sato GH, Parker ML (1968) Establishment of clonal strains of rat pituitary tumor cells that secrete growth hormone. Endocrinology 82:342–352
Tashjian AH Jr., Bancroft FC, Levine L (1970) Production of both prolactin and growth hormone by clonal strains of rat pituitary tumor cells. Differential effects of hydrocortisone and tissue extracts. J Cell Biol 47:61–71
Tasso F, Picard D, Dreifuss JJ (1976) Ultrastructural identification of granules containing oxytocin and vasopressin. Nature 260:621–622
Tatemoto K (1982) New Peptides, their structures and functions. In: Bloom SR, Polak JM, Lindenlaub E (eds) Systemic role of regulatory peptides. Symposium Oosterbeek, F.K. Schattauer Verlag, Stuttgart, pp 507–531
Taugner R, Forssmann WG, Billich H, Boll U, Ganten D, Seller H (1978) Innervation of the juxtaglomerular apparatus and the effect of renal nerve stimulation. In: Coupland RE, Forssman WG (eds) Peripheral neuroendocrine interaction. Springer, Berlin Heidelberg New York, pp 153–163
Taugner R, Whalley A, Angermüller S, Bührle CP, Hackenthal E (1985) Are the renin-containing granules of juxtaglomerular epithelioid cells modified lysosomes? Cell Tiss Res 239:575–587
Taxi J (1959) Sur la structure des travées du plexus d'Auerbach: confrontation des données fournies par le microscope ordinaire et par le microscope électronique. Ann Sci Nat Zool 1:571–594
Taxi J (1965) Contribution a l'étude des connexions des neurones moteurs du système nerveux autonome. Ann Sci Nat Zool Paris 7:413–674
Tehver J (1930) Über die enteruchromaffinen Zellen der Haussäugetiere. Z Mikrosk-Anat Forsch 21: 462–496
Teicher MH, Shaywitz BA, Lumia AR (1984) Olfactory and vomeronasal system mediation of maternal recognition in the developing rat. Dev Brain Res 12:97–110
Terzakis JA, Sommers SC, Andersson B (1972) Neurosecretory appearing cells of human segmental bronchi. Lab Invest 26:127–132
Theiler K, Müntener M (1974) Die Entwicklung der Nebennieren der Maus. I. Pränatale Entwicklung. Z Anat Entw-gesch 144:195–203
Theodosis DT, Dreifuss JJ, Orci L (1978) A freeze-fracture study of membrane events during neurohypophysial secretion. J Cell Biol 78:542–553
Theodosis DT, Legendre P, Vincent JD (1983) Immunocytochemically identified vasopressin neurons in culture show slow, calcium-dependent electrical responses. Science 221:1052–1054
Thibault G, Garcia R, Seidah NG, Lazure C, Cantin M, Chrétien M, Genest J (1983) Purification of three rat atrial natriuretic factors and their amino acid composition. FEBS Lett 164:286–290
Thibault G, Garcia R, Cantin M, Genest J, Lazure C, Seidah NG, Chrétien M (1984) Primary structure of a high *Mr* form of rat atrial natriuretic factor. FEBS Let 167:352–357
Thiele JC (1977) Human parathyroid glands: a freeze fracture and thin section study. Curr Top Pathol 65:31–80
Thiele J (1980) Changes in the plasma membrane associated with endocrine activity. A thin section and freeze-fracture study on the human parathyroid

chief cell. Virchows Arch B Cell Pathol 34:219–237
Thim L, Jorgensen KH, Jorgensen KD (1982) Pancreatic spasmolytic polypeptide (PSP): II. Radioimmunological determination of PSP in porcine tissues, plasma and pancreatic juice. Regul Pept 3: 221–230
Thomas JA, Van Orden LS, Redick JA, Kopin IJ (1973) Immunological localization of dopamine β-hydroxylase on the chromaffin granule membrane. In: Usdin E, Snyder S (eds) Frontiers in catecholamine research. Pergamon Press, Oxford, pp 79–81
Thomas TB (1940) Islet tissue in the pancreas of the elasmobranchii. Anat Rec 76:1–17
Thompson JH (1971) Serotonin and the alimentary tract. Res Commun Chem Pathol Pharmacol 2: 687–781
Thorn NA, Russell JT, Vilhardt H (1975) Hexosamine, calcium and neurophysin in secretory granules and the role of calcium in hormone release. Ann NY Acad Sci 248:202–217
Thorn NA, Russel JT, Treiman M (1982) The neurosecretory granule. In: Poisner AM, Trifaró JM (eds) The secretory granule. Elsevier Biomedical, Amsterdam New York Oxford, pp 119–151
Thorndyke MC, Probert L (1979) Calcitonin-like cells in the pharynx of the ascidian *Styela clava*. Cell Tiss Res 203:301–309
Thornhill JA, Lukowiak K, Cooper KE, Veale WL, Edstrom JP (1981) Arginine vasotocin, an endogenous neuropeptide of *Aplysia*, suppresses the gill withdrawal reflex and reduces the evoked synaptic input to central gill motor neurons. J Neurobiol 12:533–544
Thuneberg L (1982) Interstitial cells of Cajal. Intestinal pacemaker cells? Adv Anat Embryol Cell Biol 71:1–130
Till JE, McCulloch EA (1980) Hemopoietic stem cell differentiation. Biochim Biophys Acta 605:431–459
Tischler AS, Dichter MA, Biales B (1977a) Electrical excitability of oat cell carcinoma. J Pathol 122: 153–156
Tischler AS, Dichter MA, Biales B, Greene LA (1977b) Neuroendocrine neoplasms and their cells of origin. N Engl J Med 296:919–926
Tixier-Vidal A, Picart R (1967) Étude quantitative par radioautographie au microscope électronique de l'utilisation de la DL-leucine-^{3}H par les cellules de l'hypophyse du canard en culture organotypique. J Cell Biol 35:501–514
Tobe T, Izumikawa F, Sano M, Tanaka C (1976) Release mechanisms of 5-HT in mammalian gastrointestinal tract-especially, vagal release of 5-HT. In: Fujita T (ed) Endocrine gut and pancreas. Elsevier, Amsterdam, pp 371–380
Toker C (1972) Trabecular carcinoma of the skin. Arch Dermatol 105:107–110
Tokuyasu K, Yamada E (1959) The fine structure of the retina studied with the electron microscope. IV. Morphogenesis of outer segments of retinal rods. J Biophys Biochem Cytol 6:225–230
Tokuyasu K, Yamada E (1960) The fine structure of the retina studied with the electron microscope V. Abnormal retinal rods and their morphologenesis. J Biophys Biochem Cytol 7:187–190
Tomben AS (1977) Type 2 neurosecretory axons (secretory end-feet) at the base of the cerebral ganglion of *Perinereis cultrifera*, Grube (Annelida: polychaeta). Gen Comp Endocrinol 32:407–440
Tomita T, Pollock HG (1981) Four pancreatic endocrine cells in the bullfrog (Rana atesbeiana). Gen Comp Endocrinol 45:355–363
Toneguzz F, Ghosh HP (1983) Synthesis and glycosylation invitro of glycoprotein of vesicular stomatitis-virus. Proc Natl Acad Sci USA 74:1516–1520
Toner PG (1964) Fine structure of argyrophil and argentaffin cells in the gastro-intestinal tract of the fowl. Z Zellforsch 63:830–839
Torrance RW (1977) Convergence of stimuli in arterial chemoreceptors. Adv Exp Med Biol 78:203–207
Toyoshima K, Tandler B (1986) Dividing type II cell in rabbit taste bud. Anat Rec 214:161–164
Toyoshima K, Honda E, Nakahara S, Shimamura A (1984) Ultrastructural and histochemical changes in the frog taste organ following denervation. Arch Histol Jpn 47:31–42
Tramezzani JH, Chiocchio S, Wassermann GF (1964) A technique for light and electron microscopic identification of adrenalin and noradrenalin-storing cells. J Histochem Cytochem 12:890–899
Tranzer JP, Richards JG (1976) Ultrastructural cytochemistry of biogenic amines in nervous tissue: methodologic improvements. J Histochem Cytochem 24: 1178–1193
Triepel J, Grimmelikhuijzen CJP (1984) Mapping of neurons in the central nervous system of the guinea pig by use of antisera specific to the molluscan neuropeptide FMRFamide. Cell Tiss Res 237:575–586
Trifaró JM (1982) The cultured chromaffin cell: A model for the study of biology and pharmacology of paraneurones. Trends Pharmacol Sci 3:389–392
Trifaró JM (1984) Symposium the adrenal paraneurone: Its biology and pharmacology. Can J Physiol Pharmacol 62:465–466
Trifaró JM, Lee RWH (1980) Morphological characteristics and stimulus-secretion coupling in bovine adrenal chromaffin cell cultures. Neuroscience 5: 1533–1546
Trifaró JM, Poisner AM (1982) Common properties in the mechanisms of synthesis, processing and storage of secretory products In: Poisner AM, Trifaró JM (eds) The secretory granule. Elsevier Biomedical Press, Amsterdam, pp 392–403
Trifaró JM, Poisner AM, Douglas WW (1967) The fate of chromaffin granule during catecholamine release from the adrenal medulla. I. Unchanged efflux of phospholipid and cholesterol. Biochem Pharmacol 16:2095–2100
Trifaró JM, Keningsberg RL, Cote A, Lee RWH, Hikita T (1984) Adrenal paraneurone contractile proteins and stimulus-secretion coupling. Can J Physiol Pharmacol 62:493–501
Tritt SH, Lowe IP, Byrne JH (1983) A modification of the glyoxylic acid induced histofluorescence technique for demonstration of catecholamines and se-

rotonin in tissues of *Aplysia californica*. Brain Res 259:159–162
Trotier D, MacLeod P (1983) Intracellular recordings from salamander olfactory receptor cells. Brain Res 268:225–237
Tsacopoulos M, Orkand RK, Coles JA, Levy S, Poitry S, (1983) Oxygen uptake occurs faster than sodium pumping in bee retina after a light flash. Nature 301:604–606
Tsai TLA (1958) A histological study of sensory nerves in the liver. Acta Neuroveg 17:354–385
Tsien RY (1980) New calcium indicators and buffers with high selectivity against magnesium and protons: design, synthesis, and properties of prototype. Biochemistry 19:2396–2404
Tsien RY (1981) A non-disruptive technique for loading calcium buffers and indicators into cells. Nature 290:527–528
Tsubouchi S (1981) Epithelial cell migration of the intestine and "Segi's cap" (in Japanese). Cell (Saibo) 13:452–462
Tsubouchi S, Leblond CP (1979) Migration and turnover of entero-endocrine and caveolated cells in the epithelium of the descending colon, as shown by radioautography after continuous infusion of ^{3}H-thymidine into mice. Am J Anat 156:431–451
Tsuchiya M, Okazaki I, Kashiwazaki K, Nagata H, Nakamura M, Ohya Y, Asakura H (1980) Gastric mucosal collagenase in restraint-induced gastric lesions and the effects of proglumid. Excerpta Medica 87–94
Tsuruo Y, Kawano H, Nishiyama T, Hisano S, Daikoku S (1983) Substance P-like immunoreactive neurons in the tuberoinfundibular area of rat hypothalamus. Light and electron microscopy. Brain Res 289:1–9
Tsutsumi Y, Osamura Y, Watanabe K, Yanaihara N (1982) Immunohistochemical studies on gastrin releasing peptide (GRP) and ACTH in the lung: normal and pathological conditions. J Histochem Cytochem 30:601 (Abstract)
Tsutsumi Y, Osamura Y, Watanabe K, Yanaihara N (1983) Simultaneous immunohistochemical localization of gastrin releasing peptide (GRP) and calcitonin (CT) in human bronchial endocrine-type cells. Virchows Arch Pathol Anat 400:163–171
Tsuzuki Y (1947) Relation of sweetness to chemical structure of carbohydrate (in Japanese). Kagaku 17:342–346
Turner JD, Cottrell GA (1977) Properties of an identified histamine-containing neurone. Nature 267: 447–448
Tweedle CD (1978) Ultrastructure of Merkel cell development in aneurogenic and control amphibian larvae (*Ambystoma*) Neuroscience 3:481–486
Uchida T (1980) Ultrastructural and histochemical studies on the taste buds in some reptiles. Arch Histol Jpn 43:459–478
Uchida T (1985) Serotonin-like immunoreactivity in the taste bud of the mouse circumvallate papilla. Jpn J Oral Biol 27:132–139
Uchida T, Kobayashi S, Yanaihara N (1985) Occurrence and projections of three subclasses of Met-enkephalin-Arg^{6}-Gly^{7}-Leu^{8} neurons in the guinea-pig duodenum: immunoelectron microscopic study of the co-storage of Met-enkephalin-Arg^{6}-Gly^{7}-Leu^{8} with substance P or PHI (1–15). Biomed Res 6:415–422
Uchizono K (1964) On different types of synaptic vesicles in the sympathetic ganglia of amphibia. Jpn J Physiol 14:210–219
Uchizono K (1968) Axon identification in the cerebellar cortex of the cat. Arch Histol Jpn 29:399–424
Uchizono K (1975) Excitation and inhibition. Synaptic morphology. Igaku-Shoin, Tokyo
Uddman R, Edvinsson L, Ekblad E, Håkanson R, Sundler F (1986) Calcitonin gene-related peptide (CGRP): perivascular distribution and vasodilatory effects. Regul Peptide 15:1–23
Ueck M (1981) Variation in structure and function of the pineal systems. In: Oksche A, Pévet P (eds) The pineal organ: photobiology biochronometry endocrinology. Elsevier North-Holland Biomedical Press, Amsterdam, pp 151–168
Ueck M (1979) Innervation of the vertebrate pineal. Prog Brain Res 52:45–88
Ueck M, Wake K (1977) The pinealocyte—a paraneuron? A review. Arch Histol Jpn 40 Supple: 261–278
Ueck M, Wake K (1979) The pinealocyte—a paraneuron. Prog Brain Res 52:141–147
Uga S (1969) A study on the cytoarchitecture of taste buds of rat circumvallate papillae. Arch Histol Jpn 31:59–72
Unsicker K (1971) On the innervation of mammalian endocrine glands (anterior pituitary and parathyroids). Z Zellforsch 121:283–291
Unsicker K (1973a) Fine structure and innervation of the avian adrenal gland. I. Fine structure of adrenal chromaffin cells and ganglion cell. Z Zellforsch 145:389–416
Unsicker K (1973b) Fine structure and innervation of the avian adrenal gland. II. Cholinergic innervation of adrenal chromaffin cells. Z Zellforsch 145:417–442
Unsicker K (1973c) Fine structure and innervation of the avian adrenal gland. III. Non-cholinergic nerve fibers. Z Zellforsch 145:557–575
Unsicker K (1976) Chromaffin, small granule-containing and ganglion cells in the adrenal gland of reptiles. A comparative ultrastructural study. Cell Tiss Res 165:477–508
Unsicker K (1984) Innervation of endocrine tissues. In: Motta PM (ed) Ultrastructure of endocrine cells and tissues. Martinus Nijhoff, Boston Hague Dordrecht Lancaster, pp 321–332
Unsicker K, Chamley JH (1977) Growth characteristics of postnatal rat adrenal medulla in culture. A study correlating phase contrast, microcinematographic, histochemical, and electron microscopical observations. Cell Tiss Res 177:247–268
Unsicker K, Habura-Flüh O, Zwarg U (1978) Different types of small granule-containing cells and neurons in guinea-pig adrenal medulla. Cell Tiss Res 189:109–130
Usellini L, Capella C, Frigerio B, Rindi G, Solcia E (1984) Ultrastructural localization of secretin in endocrine cells of the dog duodenum by the im-

munogold technique. Comparison with ultrastructurally characterized S cells of various mammals. Histochem 80:435–441

Usellini L, Capella C, Malesci A, Rindi G, Solcia E (1985) Ultrastructural localization of cholecystokinin in endocrine cells of the dog duodenum by the immunogold technique. Histochemistry 83:331–336

Ushiki T, Iwanaga T, Masuda T, Takahashi Y, Fujita T (1984) Distribution and ultrastructure of S-100-immunoreactive cells in the human thymus. Cell Tiss Res 235:509–514

Usukura J, Yamada E (1978) Observations on the cytolemma of the olfactory receptor cell in the newt. 1. Freeze replica analysis. Cell Tiss Res 188: 83–98

Uvnäs B, Åborg CH (1980) In vitro studies on a two-pool storate of adrenaline and noradrenaline in granule material from bovine adrenal medulla. Acta Physiol Scand 109:345–354

Uvnäs B, Åborg CH, Goiny M (1985) The kinetics of adrenal catecholamine secretion elicited by splanchnic nerve stimulation or by ach is consistent with non-exocytotic, multivesicular release on cation exchange basis. Acta Physiol Scand 123:249–259

Uvnäs-Wallensten K, Efendić S, Luft R (1978) The occurrence of somatostatin-like immunoreactivity in the vagal nerves. Acta Physiol Scand 102:248–250

Vale W, Rivier C, Brown M (1977) Regulatory peptides of the hypothalamus. Ann Rev Physiol 39: 473–527

Valentine KA, Hollenberg MD (1984) Membrane receptors hormone action. In: Cantin M (ed) Cell Biology of the Secretory Process. Karger, Basel, pp 1–51

Valenzuela JE (1976) Dopamine as a possible neurotransmitter in gastric relaxation. Gastroenterology 71:1019–1022

Van Campenhout E (1932) Further experiments on the origin of the enteric nervous system in the chick. Physiol Zool 5:333–353

Vandell IM, Sikri K, Scheimberg I, Minith CD, Dixon JE, Bloom SR, Polak JM (1984) Localisation of peptide biosynthetic sites using hybridisation histochemistry. Dig Dis Sci Suppl 29(8):A92

Vanderhaeghen JJ, Lotstra F, Vierendeel G, Gilles C, Deschepper C, Verbanck P (1981) Cholecystokinin in the central nervous system and neurohypoplysis. Peptides Suppl 2:81–88

Vanderhaeghen JJ, Lotstra F, Liston DR, Rossier J (1983) Proenkephalin, [Met]enkephalin, and oxytocin immunoreactivities are colocalized in bovine hypothalamic magnocellular neurons. Proc Nat Acad Sci USA 80:5139–5143

Vandesande F, Dierickx K, De Mey J (1975) Identification of the vasopressin-neurophysin II and the oxytocin-neurophysin I producing neurons in the bovine hypothalamus. Cell Tiss Res 156:189–200

Vanfleteren JR (1982) A monophyletic line of evolution? Ciliary induced photoreceptor membranes. In: Westfall JA (ed) Visual cells in evolution. Raven Press, New York, pp 107–136

Van Noorden S, Falkmer S (1980) Gut-islet endocrinology-some evolutionary aspects. Invest Cell Pathol 3:21–35

Van Noorden S, Pearse AGE (1976) The localisation of immunoreactivity to insulin, glucagon and gastrin in the gut of *Amphioxus* (*Branchiostoma*) *lanceolatus*. In: Grillo TAI, Leibson L, Epple A (eds) The evolution of the pancreatic islets. Pergamon Press, Oxford, pp 163–178

Van Noorden S, Greenberg J, Pearse AGE (1972) Cytochemical and immunofluorescence investigations on polypeptide hormone localization in the pancreas and gut of the larval lamprey. Gen Com Endocrinol 19:192–199

Van Noorden S, Polak JM, Pearse AGE (1977a) Single cellular origin of somatostatin and calcitonin in the rat thyroid gland. Histochemistry 53:243–247

Van Noorden S, Östberg Y, Pearse AGE (1977b) Localization of somatostatin-like immunoreactivity in the pancreatic islets of the hagfish, *Myxine glutinosa* and the lamprey *Lampetra fluviatilis*. Cell Tiss Res 177:281–285

Van Orden LS III, Burke JP, Redick JA, Rybarczyk KE, Van Orden DE, Baker HA, Hartman BK (1977) Immunocytochemical evidence for particulate localization of phenylethanolamine-N-methyltransferase in adrenal medulla. Neuropharmacology 16:129–133

Varga L (1976) Metabolism of the gastrointestinal peptide hormones. Acta Physiol Acad Sci Hung 47:335–344

Varndell IM, Tapia FJ, De Mey J, Rush RA, Bloom SR, Polak JM (1982) Electron immunocytochemical localization of enkephalin-like material in catecholamine-containing cells of the carotid body, the adrenal medulla, and in pheochromocytomas of man and other mammals. J Histochem Cytochem 30:682–690

Varndell IM, Polak JM, Allen JM, Terenghi G, Bloom SR (1984) Neuropeptide tyrosine (NPY) immunoreactivity in norepinephrine-containing cells and nerves of the mammalian adrenal gland. Endocrinology 114:1460–1462

Varró A, Lonovics J, Hajnal F, Somogyi J (1983) Subcellular localization of bioactive cholecystokinin-like peptides in the rat cerebral cortex. J Neurochem 41:890–892

Varró V (1976a) Introduction. Acta Physiol Acad Sci Hung 47:277–278

Varró V (1976b) Clinical significance and perspectives of gastrointestinal peptide hormones. Acta Physiol Acad Sci Hung 47:369–378

Vaughan MK, Vaughan GM, Blask DE, Barnett MP, Reiter RJ (1976) Arginine vasotocin: Structure-activity relationships and influence on gonadal growth and function. Am Zoologist 16:25–34

Veenstra JA, Yanaihara N (1984) Immunocytochemical localization of gastrin-releasing peptide/bombesin-like immunoreactive neurons in insects. Histochemistry 81:133–138

Veenstra JA, Romberg-Privee HM, Schooneveld H (1984) Immunocytochemical localization of peptidergic cells in the neuro-endocrine system of the Colorado potato beetle, *Leptinotarsa decemlineata*, with antisera against vasopressin, vasotocin and oxytocin. Histochemistry 81:29–34

Verhofstad AAJ, Jonsson G (1983) Immunohistochemical and neurochemical evidence for the presence of serotonin in the adrenal medulla of the rat. Neuroscience 10:1443–1453

Verhofstad AAJ, Steinbusch HWM, Penke B, Varga J, Joosten HWJ (1980) Use of antibodies to norepinephrine and epinephrine in immunohistochemistry. In: Eränkö O, Soinila S, Päivärinta H (eds) Histochemistry and cell biology of autonomic neurons, SIF cells and paraneurons (Adv Biochem Psychopharmacol 25). Raven Press, New York, pp 185–193

Verhofstad AAJ, Steinbusch HWM, Penke B, Varga J, Joosten HWJ (1981) Serotonin-immunoreactive cells in the superior cervical ganglion of the rat. Evidence for the existence of separate serotonin- and catecholamine-containing small ganglionic cells. Brain Res 212:39–49

Verhofstad AAJ, Steinbusch HWM, Joosten HWJ, Penke B, Verga J, Goldstein M (1983) Immunocytochemical localization of noradrenaline, adrenaline and serotonin. In: Polak JM, Van Noorden S (eds) Immunocytochemistry. Practical applications in pathology and biology. Wright, Bristol, pp 143–168

Vern BA, Schuette WH, Whitehouse WC (1981) Effects of brain stem stimulation on cortical NAPH fluorescence, blood flow and O_2 consumption in the cat. Exp Neurol 71:581–600

Verna A (1971) Infrastructure des divers types de terminasions nerveuses dans le glomus carotidien du lapin. J Microsc 10:59–66

Verna A (1979) Ultrastructure of the carotid body in the mammals. Int Rev Cytol 60:271–330

Verner JV, Morrison AB (1974) Endocrine pancreatic islet disease with dianhoea. Arch Int Med 133:492–500

Vessale G, Generali F (1900) Fonction parathyroidienne et fonction thyroidienne. Arch Ital Biol 33:154–156

Vidon N, Hecketsweiler P, Butel J, Bernier JJ (1978) Effect of continuous jejunal perfusion of elemental and complex nutritional solutions on pancreatic enzyme secretion in human subjects. Gut 19:194–198

Vigh B, Vigh-Teichmann I (1973) Comparative ultrastructure of the cerebrospinal fluid-contacting neurons. Int Rev Cytol 35:189–251

Vigh B, Vigh-Teichmann I (1974) Vergleich der Ultrastruktur der Liquorkontaktneurone und Pinealozyten. Verh Anat Ges 68:433–443

Vigh B, Vigh-Teichmann T(1981) Light- and electron-microscopic demonstration of immunoreactive opsin in the pinealocytes of various vertebrates. Cell Tiss Res 221:451–463

Vigh B, Vigh-Teichmann I (1982) Comparison between CSF contacting neurons and cells of the radial nerve of some echinoderms. Verh Anat Ges 76:461–463

Vigh B, Vigh-Teichmann I, Röhlich P, Oksche A (1983) Cerebrospinal fluid-contacting neurons. sensory pinealocytes and Landolt's clubs of the retina as revealed by electron-microscopic immunoreaction against opsin. Cell Tiss Res 233:539–548

Vigh-Teichmann I (1969) Hydrencephalocriny of neurosecretory material in amphibia. In: Sterba G (ed) Zirkumventrikuläre Organe und Liquor. Fischer, Jena, pp 269–272

Vigh-Teichmann I, Vigh B (1974) The infundibular cerebrospinal-fluid contacting neurons. Adv Anat Cell Biol 50:1–91

Vigh-Teichmann I, Vigh B (1983) The system of cerebrospinal fluid-contacting neurons. Arch Histol Jpn 46:427–468

Vigh-Teichmann I, Röhlich P, Vigh B (1969) Licht- und elektronenmi kroskopische Untersuchungen am Recessus praeopticus-Organ von Amphibien. Z Zellforsch 98:217–232

Vigh-Teichmann I, Vigh B, Aros B (1970) Enzymhistochemische Studien am Nervensystem. IV. Acetylcholinesteraseaktivität im Liquorkon takt-Neuronesystem vershiedener Vertebraten. Histochemie 21:322–337

Vijayan E, Samson WK, McCann SM (1978) Effects of intraventricular injection of gastrin on release of LH, prolactin, TSH and GH in conscious ovariectomized rats. Life Sci 23:2225–2232

Vila-Porcile E (1972) Le réseau des cellules folliculo-stellaires et les follicules de l'adénohypophse du rat (Pars distalis). Z Zellforsch 129:328–369

Viveros OH, Arqueros L, Kirschner N (1968) Release of catecholamines and dopamine β-oxidase from the adrenal medulla. Life Sci 7:609–618

Viveros OH, Arqueros L, Connett J, Kirshner N (1969) Mechanism of secretion from the adrenal medulla. 4. The fate of the storage vesicles following insulin and reserpine administration. Mol Pharmacol 5:69–82

Vizi ES, Török T, Seregi A, Serfözö P, Ádám-Vivi V (1982) Na-K-activated ATPase and the release of acetylcholine and noradrenaline. J Physiol (Paris) 78:399–406

Vizi SE, Bertaccini G, Impicciatore M, Knoll J (1973) Evidence that acetylcholine released by gastrin and related polypeptides contributes to their effect on gastrointestinal motility. Gastroenterology 64:268–277

Vollrath L (1969) Zur innervation endokriner Drüsen. Beitrage der moderen Morphologie und verglichenden Anatomie. Müncher Med Wochenschr 111:1464–1468

Vollrath L (1973) Synaptic ribbons of a mammalian pineal gland. Circadian changes. Z Zellforsch 145:171–183

Vollrath L (1981) The pineal organ (Möllendorff's Handbuch der mikroskopischen Anatomie des Menschen. VI/7). Springer, Berlin Heidelberg New York

Wagner JA, Carlson SS, Kelly RB (1978) Chemical and physical characterization of cholinergic synaptic vesicles. Biochemistry 17:1199–1206

Wakade AR, Wakade TD (1982) Secretion of catecholamines from adrenal gland by a single electrical shock: Electrotonic depolarization of medullary cell membrane. Proc Natl Acad Sci USA 79:3071–3074

Wake K (1973) Acetylcholinesterase-containing nerve cells and their distribution in the pineal organ of the goldfish, *Carassius auratus.* Z Zellforsch 145:287–298

Wake K, Ueck M, Oksche A (1974) Acetylcholinesterase-containing nerve cells in the pineal complex and subcommissural area of the frogs, *Rana ridibunda* and *Rana esculenta.* Cell Tiss Res 154:423–442

Wallace EF, Krantz MJ, Lovenberg W (1973) Dopamine β-hydroxylase: a tetrameric glycoprotein. Proc Natl Acad Sci USA 70:2253–2255

Walls GL (1942) The vertebrate eye and its adaptive radiation. Cranbrook Institute of Science, Bloomfield Hills Michigan

Waltman B (1966) Electrical properties and fine structure of the ampullary canals of Lorenzini. Acta Physiol Scand 66 Suppl 264:1–60

Ward AS, Bloom SR (1974) The role of secretin in the inhibition of gastric secretion by intraduodenal acid. Gut 15:889–897

Ward PH, Keumann VK, Chiang L (1978) Partial characterization of pepsins and gastricsins and their zymogens from human and toad gastric mucosae. Comp Biochem Physiol 61B:491–498

Warner TFCS, Uno H, Hafez R, Burgess J, Bolles C, Lloyd RV, Oka M (1983) Merkel cells and Merkel cell tumors. Ultrastructure, immunocytochemistry and review of the literature. Cancer 52:238–245

Wartenberg H (1968) The mammalian pineal organ: Electron microscopic studies on the fine structure of pinealocytes, glial cells and on the perivascular component. Z Zellforsch 86:74–97

Wasano K, Yamamoto T (1978) Monoamine-containing granulated cells in the frog lung. Cell Tiss Res 193:201–209

Wasserman RH, Corradino RH, Taylor AN (1968) Vitamin D-dependent calcium-binding protein: purification and some properties. J Biol Chem 243:3978–3986

Watanabe H, Passonneau JV (1973) factors affecting the turnover of cerebral glycogen and limit dextrin *in vivo.* In: Schmitt FO, Worden FG, Adelman G, Dennis SG (eds) The organization of the cerebral cortex. MIT Press, Cambridge, pp 153–161

Watanabe YG, Daikoku S (1979) An immunohistochemical study on the cytogenesis of adenohypophysial cells in fetal rats. Devel Biol 68:557–567

Watari N (1973) Innervation of pancreatic islet cells. In: Yagi K, Yoshida S (eds) Neuroendocrine Control. University of Tokyo Press, Tokyo, pp 165–188

Watari N, Hotta Y, Mabuchi Y (1977) Comparative light and electron microscopic study of the adrenal gland of the snake. Arch Histol Jpn Suppl 40:81–85

Watson JH, Brinkman GL (1964) Electron microscopy of the epithelial cells of normal and bronchitic human bronchus. Amer Rev Resp Dis 90:851–866

Watzka M (1943) Die Paraganglien. In: Möllendorff's Handbuch der mikroskopischen Anatomie des Menschen VI/4. Springer, Berlin, pp 262–308

Weber E, Evans CJ, Barchas JD (1983) Multiple endogenous ligands for opioid receptors. Trends Neuro Sci 6:333–336

Weichert RF (1970) The neural ectodermal origin of the peptide-secreting endocrine glands. A unifying concept for the etiology of multiple endocrine adenomatosis and the inappropriate secretion of peptide hormone by non endocrine tumors. Am J Med 49:232–241

Weindl A, Sofroniew M (1976) Demonstration of extrahypothalamic peptide secreting neurons. A morphological contribution to the investigation of psychotropic effects of neurohormones. Pharmakopsych 9:226–234

Weinreich D, Weiner C, McCaman R (1975) Endogenous levels of histamine in single neuron isolated from CNS of *Aplysia californica.* Brain Res 84:341–345

Weinstein H, Malamed A, Sachs H (1961) Isolation of vasopressin-containing granules from the neurohypophysis of the dog. Biochim Biophys Acta 50:386–389

Wellmann KF, Volk BW (1977a) Historical review. In: Volk BW, Wellmann KF (eds) The diabetic pancreas. Plenum Press, New York, pp 1–14

Wellmann KF, Volk BW (1977b) Histology, cell types, and functional correlation of islets of Langerhans. In: Volk BW, Wellmann KF (eds) The Diabetic Pancreas. Plenum Press, New York, pp 99–119

Welsch U, Storch V (1973) Einführung in cytologie und histologie der tiere. Gastav Fishcher, Stuttgart

Werblin FS, Dowling JE (1969) Organization of the retina of the mudpuppy, *Necturus maculosus* II. Intracellular recording. J Neurophysiol 32:339–355

Wermel EM, Kacharova EA (1948) The role of the basal granular cells of the mucosa of the small intestine in the production of secretin. Anat Rec 101:595–604

Wermer P (1954) Genetic aspects of adenomatosis of endocrine glands. Am J Med 16:363–371

Werner G, Mountcastle VB (1965) Neural activity in mechanoreceptive cutaneous afferents: stimulus-response relationships, Weber functions, and information transmission. J Neurophysiol 28:359–397

Wersäll J, Bagger-Sjöbäck D (1974) Morphology of the vestibular sense organ. In: Kornhuber HH (ed) Vestibular system, part 1: basic mechanisms (Handbook of sensory physiology VI/1). Springer, Berlin, pp 123–170

Wersäll J, Flock Å (1964) Suppression and restoration of the microphonic output from the lateral line organ after local application of streptomycin. Life Sci 3:1151–1155

Wersäll J, Björkroth B, Flock Å, Lundquist PG (1973) Experiments on ototoxic effects of antibiotics. Adv Oto-Rhino-Laryngol 20:14–41

Westfall JA (1973) Ultrastructural evidence for a granule-containing sensory-motor-interneuron in *Hydra littoralis.* J Ultrastruct Res 42:268–282

Westfall JA, Kinnamon JC (1978) A second sensory-motor-interneuron with neurosecretory granules in *Hydra.* J Neurocytol 7:365–379

Westhead EW, Livett BG (1983) Storage and secretion by adrenal chromaffin cells. Trends Neurosci 6:254–255

Weston JA (1970) The migration and differentiation of neural crest cells. Adv Morph 8:41–114

Weymouth RJ, Sheridan MN (1966) Fine structure of

human parathyroid glands: normal and pathological. Acta Endocrinol 53:529–546
Wharton J, Polak JM, Bloom SR, Ghatei MA, Solcia E, Brown MR, Pearse AGE (1978) Bombesin-like immunoreactivity in the lung. Nature (Lond) 273: 769–770
Wharton J, Polak JM, Pearse AGE, McGregor GP, Bryant MG, Bloom SR, Emson PC, Bisgard GE, Will JA (1980) Enkephalin-, VIP- and substance P-like immunoreactivity in the carotid body. Nature 284:269–271
Wharton J, Polak JM, Cole GA, Marangos PJ, Pearse AGE (1981) Neuron-specific enolase as an immunocytochemical marker for the diffuse neuroendocrine system in human foetal lung. J Histochem Cytochem 29:1359–1364
Wheatley DN (1982) The centriole: A central enigma of cell biology. Elsevier Biomed Press, Amsterdam New York Oxford
Whitear M, Lane EB (1981) Fine structure of Merkel cells in lampreys. Cell Tiss Res 220:139–151
Whittaker VP (1965) The application of subcellular fractionation techniques to the study of brain function. Progr Biophys 15:39–96
Whittaker VP (1972) The storage and release of acetylcholine. Proc Biochem Soc 128:73P–74P
Whittle AC, Golding DW (1975) The infracerebral gland and cerebral neurosecretory system—a probable neuroendocrine complex in *Phyllodocid polychaetes*. Gen Comp Endocrinol 24:87–98
Whur P, Herscovics A, Leblond CP (1969) Radioautographic visualization of the incorporation of galactose-^{3}H and mannose-^{3}H by rat thyroids in vitro in relation to the stages of thyroglobulin synthesis. J Cell Biol 43:289–311
Wiedenmann B, Franke W (1985) Identification and localization of synaptophysin, an integral membrane clycoprotein of Mr 38,000 characteristic of presynaptic vesicles. Cell 47:1017–1028
Wiedenmann B, Franke WW, Kuhn C, Moll R, Gould VE (1986) Synaptophysin: A marker protein for neuroendocrine cells and neoplasms. Proc Natl Acad Sci USA 83:3500–3504.
Wikgren MC, Reuter M (1985) Neuropeptides in a microturbellarian-whole mount immunocytochemistry. Peptides 6 (Suppl 3):471–476
Wilbur DL, Yee JA, Raigue SE (1978) Hypophysial portal vascular infusion of TRH in the rat: an ultrastructural and radioimmunoassay study. Am J Anat 151:277–294
Wilcox BJ, Seybold VS (1982) Localization of neuronal histamine in rat brain. Neurosci Lett 29:105–110
Willard DM (1936) The innervation of the adrenal glands of mammals: a contribution to the study of nerve endings. Quart J Microsc Sci 78:475–485
Willems G, Galand P, Vansteenkiste Y, Zeitoun P (1972) Cell population kinetics of zymogen and parietal cells in the stomach of mice. Z Zellforsch 134:505–518
Williams TH (1967) Electron microscopic evidence for an autonomic interneuron. Nature 214:309–310
Williams TH, Palay SL (1969) Ultrastructure of the small neurons in the superior cervical ganglion. Brain Res 15:17–34
Williams TH, Black AC, Chiba T, Jew J (1976) Interneurons/SIF cells in sympathetic ganglia of various mammals. In: Coupland RE, Fujita T (eds) Chromaffin, enterochromaffin and related cells. pp 95–116
Williams V, Morriss F (1970) Formaldehyde-induced fluorescence as a mean for differentiating epinephrine cells from norepinephrine cells in adrenal medulla. Stain Technol 45:205–213
Williamson JR (1986) Inositol lipid metabolism and intracellular signaling mechanisms. News Physiol Sci 1:72–76
Williamson JR, Lacy PE, Grisham JW (1961) Ultrastructural changes in the islets of the rat produced by tolbutamide. Diabetes 10:460–469
Williamson JR, Cooper RH, Joseph SK, Thomas AP (1985) Inositol trisphosphate and diacylglycerol as intracellular second messengers in liver. Am J Physiol 248:C203–C216
Wilson AJ, Furness JB, Costa M (1979) A unique population of uranaffin-positive intrinsic nerve endings in the small intestine. Neurosci Lett 14:303–308
Wilson BS, Lloyd RV (1984) Detection of chromogranin in neuroendocrine cells with a monoclonal antibody. Am J Pathol 115:458–468
Wilson DE (1972) Prostaglandins and the gastrointestinal tract. Prostaglandins 1:281–293
Wilson L (1975) Action of drugs on microtubules. Life Sci 17:303–310
Wilson SP, Kirshner N (1983) Calcium-evoked secretion from digitonin-permeabilized adrenal medullary chromaffin cells. J Biol Chem 258:4994–5000
Wilson SP, Klein RL, Chang KJ, Gasparis MS; Viveros OH, Yang WH (1980) Are opioid peptides co-transmitters in noradrenergic vesicles of sympathetic nerves? Nature 288:707–709
Winbladh L (1966) Light microscopical and ultrastructural studies of the pancreatic islet tissue of the lamprey (*Lampetra fluviatilis*). Gen Comp Endocrinol 6:534–543
Winder CV (1937) On the mechanism of stimulation of carotid gland chemoreceptors. Am J Physiol 118: 389–398
Winkelmann RK (1977) The Merkel cell system and a comparison between it and the neurosecretory. J Invest Dermatol 69:41–46
Winkelmann RK, Breathnach AS (1973) The Merkel cell. J Invest Dermatol 60:2–15
Winkler H (1969) Isolierung und Charakterisierung von chromaffinen Noradrenalin-Granula aus Schweine-Nevennierenmark. Naunyn-Schmiedebergs Arch Pharmakol Exp Pathol 263:340–357
Winkler H (1976) The composition of adrenal chromaffin granules: an assessment of controversial results. Neuroscience 1:65–80
Winkler H, Carmichael SW (1982) The chromaffin granule. In: Poisner AM, Trifaro JM (eds) The secretory granules. Elsevier Biomedical Press, Amsterdam New York Oxford, pp 3–79
Winkler H, Hörtnagl H (1973) Composition and molecular organization of chromaffin granules. In: Usdin E, Snyder S (eds) Frontiers in catecholamine

research. Pergamon, Oxford, pp 415–421
Winkler H, Smith AD (1975) The chromaffin granule and the storage of catecholamines. In: Blaschko H, Sayers G, Smith AD (eds) Adrenal gland (Handbook of physiology 7/VI). American Physiological Society, Bethesda Maryland, pp 321–339
Winkler H, Westhead EW (1980) The molecular organization of adrenal chromaffin granules. Neuroscience 5: 1803–1823
Winkler H, Hörtnagl H, Hörtnagl H, Smith AD (1970) Membranes of the adrenal medulla. Behavior of insoluble proteins of chromaffin granules on gel electrophoresis. Biochem J 118: 303–310
Winkler H, Schöpf JAL, Hörtnagl H, Hörtnagl H (1972) Bovine adrenal medulla. Subcellular distribution of newly synthesised catecholamines, nucleotides and chromogranins. Naunyn-Schmiedeberg's Arch Pharmacol 273: 43–61
Woldemussie E, Moran NC (1979) Calcium dependence of 48/80-induced histamine from rat mast cells. Fed Proc 27: 605
Wolfe DE (1965) The pineal cell: an electron microscopic study of its intracellular relationships and intracellular morphology in the pineal body of the albino rat. Prog Brain Res 10: 332–386
Wolfe DE, Potter LT, Richardson KC, Axelrod J (1962) Localizing tritiated norepinephrine in sympathetic axons by electron microscopic autoradiography. Science 138: 440–442
Wollheim CB, Sharp GWG (1981) Regulation of insulin release by calcium. Physiol Rev 61: 914–973
Wood JD (1981) Physiology of the enteric nervous system: In: Johnson LR, Christensen J, Grossman MI, Jacobson ED, Schultz SG (eds) Physiology of the gastrointestinal tract. Raven, New York, pp 1–38
Wood JD (1984) Enteric neurophysiology. Am J Physiol 247: G585–G598
Wood JG (1963) Identification of and observations on epinephrine and norepinephrine containing cells in the adrenal medulla. Am J Anat 112: 285–304
Wood JG, Barnett RJ (1964) Histochemical demonstration of norepinephrine at a fine structure level. J Histochem Cytochem 12: 197–209
Wood JG, Lentz TL (1964) Histochemical localization of amines in *Hydra* and in the sea anemone. Nature 201: 88–90
Wood JD, Mayer CJ (1978) Intracellular study of electrical activity of Auerbach's plexus in guinea-pig small intestine. Pflügers Arch 374: 265–275
Woods SC, Porte JR D (1974) Neural control of the endocrine pancreas. Physiol Reviews 54: 596–619
Wu SM, Dowling JE (1978) L-aspartate: evidence for a rolw in cone photoreceptor synaptic transmission in the carp retina. Proc Natl Acad Sci USA 75: 5205–5209
Wurtman RJ, Axelrod J (1966) A 24-hour rhythm in the content of norepinephrine in the pineal and salivary gland of the rat. Life Sci 5: 665–669
Wurtman RJ, Axelrod J, Fischer JE (1964) Melatonin synthesis in the pineal gland: effect of light mediated by the sympathetic nervous system. Science 143: 1328–1330
Wysocki CJ (1979) Neurobehavioral evidence for the involvement of the vomeronasal system in mammalian reproduction. Neurosci Behav Rev 3: 301–341
Wysocki CJ, Katz Y, Bernhard R (1983) Male vomeronasal organ mediates female-induced testosterone surges in mice. Biol Reprod 28: 917–922
Yagi K, Azuma T, Matsuda K (1966) Neurosecretory cell: capable of conducting impulse in rats. Science 154: 778–779
Yamada J, Kuramoto H, Iwanaga T, Yamashita T, Misu M, Yanaihara N (1981) Segi's cap, a large aggregation of endocrine cells on the intestinal villi, occurs also in cattle and pig fetuses. Arch Histol Jpn 44: 193–197
Yamada J, Ito H, Kuramoto H, Kitamura N, Yamashita T (1983) Segi's cap, a large aggregation of peptide and amine-containing cells on the fetal intestinal villi; the history and present status of its study. Biomed Res Suppl 4: 159–165
Yamada Y, Ito S, Matsubara Y, Kobayashi S (1977) Immunohistochemical demonstration of somatostatin-containing cells in the human, dog and rat thyroids. Tohoku J Exp Med 122: 87–92
Yamaguchi K, Abe K, Aadachi I, Tanaka N, Tanaka M, Miyakawa S, Kameya T, Kimura T, (1977) A case of small cell carcinoma of the lung producing ADH, ACTH, MSH and calcitonin: Successful treatment of severe hyponatremia with furosemide and hypertonic saline. Jpn J Clin Oncol 7: 111–118
Yamakuni T, Usui H, Iwanaga T, Kondo H, Odani S, Takahashi Y (1984) Isolation and immunohistochemical localization of a cerebellar protein. Neurosci Lett 45: 235–240
Yamakuni T, Araki K, Takahashi Y (1985) The developmental changes of mRNA levels for a cerebellar protein (spot 35 protein) in rat brains. FEBS Lett 188: 127–130
Yamamoto M (1976) An electron microscopic study of the olfactory mucosa in the bat and rabbit. Arch Histol Jpn 38: 359–412
Yamamoto M, Kataoka K (1986) Electron microscopic observation of the primary cilium in the pancreatic islets. Arch Histol Jpn 49: 449–457
Yamamoto M, Uchida K, Ito T (1976) Regeneration of olfactory bipolar neurons in the normal olfactory epithelium of the adult bat. Okajimas Folia Anat Jpn 53: 291–304
Yamamoto T, Ozawa H (1977) Adenylate cyclase in paraneurons. A histochemical study. Arch Histol Jpn Suppl 40: 49–60
Yamamoto T, Tasaki K, Sugawara Y, Tonosaki A (1965) Fine structure of the octopus retina. J Cell Biol 25: 345–359
Yamamoto T, Kataoka K, Ochi J, Honma Y (1982) The fine structure of the follicle-forming cells in the adenohypophysis of the ironfish, a natural hybrid between the funa and the goldfish (Teleostei, Cyprinidae). Arch Histol Jpn 45: 355–364
Yamamoto T, Kataoka K, Shimizu K, Ochi J (1983) Fluorohistochemistry and electron microscopy of the chromaffin cells in the adrenal gland of the clawed toad, *Xenopus laevis*. Acta Histochem Cytochem 16: 270–276
Yamasaki H, Fujita T, Ohara Y, Komoto S (1970)

Electron microscope studies on the release of histamine from rat peritoneal mast cells. Arch Histol Jpn 31: 393–408
Yamasaki H, Kubota Y, Takagi H, Tohyama M (1984) Immunoelectron-microscopic study on the fine structure of substance-P-containing fibers in the taste buds of the rat. J Com Neurol 227: 380–392
Yamashita H, Inenaga K, Koizumi K (1984) Possible projections I-5 from regions of paraventricular and supraoptic nuclei to the spinal cord: electrophysiological studies. Brain Res 296: 373–378
Yamauchi A (1976) Autonomic nervous system and paraneurons (in Japanese). Igaku-no-Ayumi (Tokyo) 97: 33–35
Yamauchi A (1977) On the recepto-endocrine property of granule-containing (GC) cells in the autonomic nervous system. Arch Histol Jpn Suppl 40: 147–161
Yanagisawa K, Shiozawa K (1980) Stimulus reception in hair cells of the lateral-line organ. Biomed Res Suppl 1: 128–129
Yanaihara C, Kadowaki M, Hoshino M, Kishida S, Suzuki T, Yanaihara N (1987) Production of region-specific antisera to human prorenin and renin using synthetic peptides. Biomed Res 8: 95–101
Yanaihara N, Nokihara K, Yanihara C, Iwanaga T, Fujita T (1983) Immunocytochemical demonstration of PHI and its co-existence with VIP in intestinal nerves of the rat and pig. Arch Histol Jpn 46: 575–581
Yanatori Y, Fujita T (1976) Hypertrophy and hyperplasia in the endocrine and exocrine pancreas of rats fed soybean trypsin inhibitor or repeatedly injected with pancreozymin. Arch Histol Jpn 39: 67–78
Yates RD, Wood JG, Duncan D (1962) Phase and electron microscopic observations on two cell types in the adrenal medulla of the Syrian hamster. Texas Rep Biol Med 20: 494–502
Yau KW, Nakatani K (1985) Light-induced reduction of cytoplasmic free calcium in retinal rod outer segment. Nature 31: 579–582
Yen SH, Fields KL (1981) Antibodies to neurofilament, glial filament, and fibroblast intermediate filament proteins bind to different cell types of the nervous system. J Cell Biol 88: 115–126
Yntema CL (1937) An experimental study of the origin of the cells which constitute the VIIth and VIII ganglia and nerves in the embryo of *Amblystoma punctatum*. J Exp Zool 75: 75–101
Yntema CL (1944) Experiments on the origin of the sensory ganglia of the facial nerve of the chick. J Comp Neurol 81: 147–167
Yokoh Y (1974) On sensory cells in the human otocyst. Acta Anat 87: 72–76
Yokoh S, Aoji O, Matsumo Z, Yoshida H (1969) Electron microscopic histochemistry of heavy metals in islets of Langerhans of rabbits. Diabetologia 5: 137–142
Yokota R, Burnstock G (1978) A dividing granule-containing cell in the pelvic ganglion of the guinea pig. Cell Tiss Res 192: 187–192
Yoshida K, Iwanaga T, Fujita T (1983) Gastro-entero-pancreatic (GEP) endocrine system of the flatfish, *Paralichtys olivaceus*: an immunocytochemical study. Arch Histol Jpn 46: 259–266
Yoshida Y, Takahashi Y (1980) Compositional changes in soluble proteins of cerebral mantle, cerebellum, and brain stem of rat brain during development: a two-dimensional gel electrophoretic analysis. Neurochem Res 5: 81–96
Yoshida Y, Takei T, Hattori A, Kaku T, Yokokawa K, Mori M (1984) Merkel cell tumor of the skin. Ultrastructural and immunohistochemical study. Acta Pathol Jpn 34: 1433–1440
Yoshie S, Iwanaga T, Fujita T (1985) Coexistence of bombesin and 5-hydroxytryptamine in the cutaneous gland of the frog, *Bombina orientalis*. Cell Tiss Res 239: 25–29
Yoshimura F, Yachi H, Ishikawa H, Ohtsuka Y, Kiguchi Y (1972) Morphological and functional differentiation of isolated parafollicular and follicular epithelial cells of the rat thyroids during culture. Endocrinol Jpn 19: 151–161
Yoshioka T, Asanuma A, Yanagisawa K, Katsuki Y (1978) The chemical receptive mechanism in the lateral-line organ. Jpn J Physiol 28: 557–567
Young BA, Dickson GR (1979) A little known cell in the gastrointestinal tract. Med Hypotheses 5: 695–698
Young RW (1967) The renewal of photoreceptor cell outer segments. J Cell Biol 33: 61–72
Young RW (1971) Shedding of discs from rod outer segments in the rhesus monkey. J Ultrastruct Res 34: 190–203
Young RW (1976) Visual cells and the concept of renewal. Am J Med Sci 272: 700–725
Young SH, Poo M (1983) Spontaneous release of transmitter from growth cones of embryonic neurones. Nature 305: 634–647
Yui R (1983) Immunohistochemical studies on peptide neurons in the hypothalamus of the bullfrog *Rana catesbeiana*. Gen Comp Endocrinol 49: 195–209
Yui R, Fujita T (1986) Immunocytochemical studies on the pancreatic islets of the ratfish *Chimaera monstrosa*. Arch Histol Jpn 49: 369–377
Yui R, Fujita T, Ito S (1980) Insulin-, gastrin-, pancreatic polypeptide-like immunoreactive neurons in the brain of the silkworm, *Bombys mori*. Biomed Res 1: 42–46
Yui R, Yamada Y, Kayamori R, Fujita T (1981) Calcitonin-immunoreactive neurons in the brain of the bullfrog, *Rana catesbeiana* with special reference to their liquor-contacting and neurosecretory nature. An Immunochemical and immunohistochemical study. Biomed Res 2: 208–216
Yui R, Iwanaga T, Kuramoto H, Fujita T (1985) Neuropeptide immunocytochemistry in protostomian invertebrates, with special reference to insects and molluscs. Peptides 6 (Suppl 3): 411–415
Yunge L, Ballak M, Beuzeron J, Lacasse J, Cantin M (1980) Ultrastructural cytochemistry of atrial and ventricular cardiocytes of the bullfrog (*Rana catesbeiana*). Relationship of specific granules with reninlike activity of the myocardium. Can J Physiol Pharmacol 58: 1463–1476
Zabel M, Schäfer H (1986) Ultrastructural localization of calcitonin and somatostatin in C cells of

rabbit thyroid. Cell Tiss Res 245:667–672
Zaccone G (1984) Immunohistochemical demonstration of neuron-specific enolase in the nerve endings and skin receptors of marine eels. Histochem J 16:1231–1236
Zaccone G (1986) Neuron-specific enolase and serotonin in the Merkel cells of conger-eel (*Conger conger*) epidermis. An immunohistochemical study. Histochemistry 85:29–34
Zaccone G, Fasulo S, Lo Cascio P, Licata A (1986) 5-hydroxytryptamine immunoreactivity in the epidermal sacciform gland cells of the clingfish *Lepadogaster candollei* Risso. Cell Tiss Res 246:679–682
Zachrisson BU, Schulz-Haudt SD (1967) Biologically active substances of the mast cell. J Periodont Res 2:21
Zahid ZR, Golding DW (1975) Structure and ultrastructure of the central nervous system of the polychaete *Nephtys* with special reference to neurosecretion. Acta Zool (Stockh) 56:11–28
Zalawski AA (1970) Regeneration of taste buds in the eingual epithelium after excision of the vallate papilla. Exp Neurol 26:621–629
Zalin R (1979) The cell cycle, myoblast differentiations, and prostaglandin as a developmental signal. Dev Biol 71:274–288
Zatz M (1978) Sensitivity and cyclic nucleotides in the rat pineal gland. J Neural Transm Suppl 13:97–114
Zerihun L, Harris M (1981) Electrophysiological identification of neurones of paraventricular nucleus sending axons to both the neurohypophysis and the medulla in the rat. Neurosci Lett 23:157–160
Zimmermann H, Whittaker VP (1974) Effect of electrical stimulation on the yield and composition of synaptic vesicles from the cholinergic synapses of the electric organ of torpedo: a combined biochemical, electrophysiological and morphological study. J Neurochem 22:435–450
Zipser B (1980) Identification of specific leech neurones immunoreactive to enkephalin. Nature 283:857–858
Zollinger RM, Ellison EH (1955) Primary peptic ulcerations in the jejunum associated with islet cell tumors of the pancreas. Ann Surg 142:709–728
Zomzely-Neurath CE (1982) Nervous-system-specific proteins: 14-3-2 protein, antigen alpha and neuron-specific enolase. In: Bock E (ed) Nervous-system-specific proteins. Blackwell, Oxford, pp 1–40

List of Abbreviations

ACTH	adrenocorticotrophic hormone
ADP, AMP, ATP	adenosine triphosphate, etc.
ANP (ANF)	atrial natriuretic peptide (factor)
APP, BPP	avian, bovine pancreatic polypeptide
APUD	amine precursor uptake and decarboxylation
CCK	cholecystokinin
CGRP	calcitonin gene related peptide
CLIP	corticotropin-like intermediate lobe peptide
CRF	corticotropin releasing factor
CSF	cerebrospinal fluid
DBH	dopamine-β-hydroxylase
DNA	deoxyribonucleic acid
EC	enterochromaffin
EPSP	excitatory postsynaptic potential
FIF	formaldehyde-induced fluorescence
FMRF	FMRF-amide
FSH	follicle stimulating hormone
GABA	γ-aminobutylic acid
GEP	gastroenteropancreatic
GH	growth hormone
GIP	gastric inhibitory peptide
GRP	gastrin releasing peptide
GTP	guanosine triphosphate
HCG	human chorionic gonadotropin
HIOMT	hydroxyindol-*o*-methyltransferase
5-HT	5-hydroxytryptamine
5-HTP	5-hydroxytryptophan
IPSP	inhibitory postsynaptic potential
LH	luteinizing hormone
LHRH	luteinizing hormone releasing hormone
MEA	multiple endocrine adenomatosis
MEN	multiple endocrine neoplasm
MSH	melanocyte-stimulating hormone
NFP	neurofilament protein
NGF	nerve growth factor
NPY	neuropeptide Y
NSE	neuron-specific enolase
PHI	peptide HI
PI	phosphatidylinositol
PIP	phosphatidylinositol-4-phosphate
PNMT	phenylethanolamine-*N*-methyltransferase
PP	pancreatic peptide
PTH	parathyroid hormone
RNA	ribonucleic acid
PTTH	prothoracicotrophic hormone
RH (RF)	releasing hormone (factor)
SGC	small granule chromaffin

SIF	small intensely fluorescent
SP	secretory protein
TH	tyrosine hydroxylase
TPI	triphosphoinositide
TRH	thyrotropin releasing hormone
TSH	thyroid stimulating hormone
VIP	vasoactive intestinal polypeptide

Subject Index

Page references in bold type indicate main descriptions